Industrial Heat Pump-Assisted Wood Drying

Advances in Drying Technology

Series Editor:
Arun S. Mujumdar
McGill University, Quebec, Canada

Handbook of Industrial Drying, Fourth Edition
Arun S. Mujumdar

Advances in Heat Pump-Assisted Drying Technology
Vasile Minea

Computational Fluid Dynamics Simulation of Spray Dryers: An Engineer's Guide
Meng Wai Woo

Handbook of Drying of Vegetables and Vegetable Products
Min Zhang, Bhesh Bhandari, and Zhongxiang Fang

Intermittent and Nonstationary Drying Technologies: Principles and Applications
Azharul Karim and Chung-Lim Law

Thermal and Nonthermal Encapsulation Methods
Magdalini Krokida

Intelligent Control in Drying
Alex Martynenko and Andreas Bück

For more information about this series, please visit: www.crcpress.com

Industrial Heat Pump-Assisted Wood Drying

Vasile Minea

CRC Press is an imprint of the
Taylor & Francis Group, an **informa** business

CRC Press
Taylor & Francis Group
6000 Broken Sound Parkway NW, Suite 300
Boca Raton, FL 33487-2742

First issued in paperback 2020

ISBN-13: 978-1-138-04125-7 (hbk)
ISBN-13: 978-0-367-78098-2 (pbk)

Visit the Taylor & Francis Web site at
http://www.taylorandfrancis.com

and the CRC Press Web site at
http://www.crcpress.com

Contents

Preface

Drying of wood is one of the most common, but complex and energy-intensive processes existing in many countries. It involves a mix of fundamental sciences and technologies, being based on extensive experimental observations and operating experience.

The book *Industrial Heat Pump-Assisted Wood Drying* contains 17 chapters intending at updating some of conventional and unconventional wood drying technologies. It focuses on various aspects of heat pump-assisted wood drying ranging from technological advancements in heat pumping in general to optimal dryer–heat pump coupling and control strategies, system modeling and simulation, and laboratory and in-field experiences.

The book emphasizes several coupling and system design concepts, and operating and control strategies that can be applied to improve the energy efficiency and environmental sustainability of heat pump-assisted wood drying processes.

Fundamental thermophysical and energetic aspects of conventional wood drying processes are first explained and summarized.

Several modern aspects of industrial heat pump-assisted wood drying technology (e.g., advanced components, kiln energy requirements, modeling, available software) are then presented with a special emphasis on dryer and heat pump optimum coupling, control, and energy efficiency.

A number of solved exercises are included in most of the chapters, such as practical and numerical examples for process and/or system/components calculation/design. Most promising advancements in industrial heat pump-assisted wood drying technology and some R&D challenges and future requirements are also presented.

This book is intended to serve both the practicing engineers involved in the selection and/or design of sustainable drying systems and the researchers as a reference work that covers the wide field of wood drying principles, and the various commonly used wood drying methods and equipment, including heat pumps.

The main aim of this book is to provide, on the one hand, general information about the wood physical drying processes to heat pump technicians and, on the other hand, basic information about the air-to-air mechanical vapor compression heat pump technology to industrial and academia drying R&D communities. That is because only a good understanding of both wood dryers' and heat pumps' technologies, followed by a strong collaboration between wood drying and heat pump specialists, may ensure the development of new, advanced integrated concepts, and successful future industrial implementations.

The final scope of this book is to increase the number of feasible, successful industrial implementation of heat pump-assisted wood dryers.

Vasile Minea
Hydro-Québec Research Institute
Québec, Canada

Acknowledgments

I gratefully acknowledge Professor Arun Mujumdar for his gentle acceptation to include this book in the new series of works entitled "Advances in Drying Science and Technology." As one of the most energy-efficient heat recovery technologies, air-to-air mechanical vapor compression heat pumps may reduce the global energy consumption and greenhouse gas emissions in wood industry throughout the world.

I express my thanks to Professor Arun Mujumdar and to the publisher, CRC Press; this book will disseminate, on the one hand, basic information about the wood physical drying processes to heat pump technicians and, on the other hand, some general information about the air-to-air mechanical vapor compression heat pump technology to academia and industrial drying R&D communities.

Finally, I thank the publisher, CRC Press, for its sustained involvement, assistance, patience, and strong support throughout the entire process. Specifically, I would like to mention Allison Shatkin, senior editor of the book, *Materials Science and Chemical Engineering*, and Laurie Oknowski, project coordinator of Editorial Project Development at Taylor & Francis Group for their courteous, professional, and helpful support.

Vasile Minea
Hydro-Québec Research Institute
Quebec, Canada

Author

Vasile Minea is a PhD graduate of civil, industrial, and agricultural installation engineering faculty from the Bucharest Technical Construction University, Romania. He worked as a professor at that university for more than 15 years, teaching courses such as heating, ventilation and air-conditioning systems for civil, agricultural, and industrial buildings, as well as thermodynamics, heat transfer, and refrigeration. During this period, his R&D works focused on heat exchangers, heat pump and heat recovery systems, development and experimentation of advanced compression–absorption/resorption heat pump concepts, as well as on the usage of solar energy for comfort cooling processes and industrial cold and ice production. Since 1987, Dr Minea has been working as a scientist researcher at the Hydro-Québec Research Institute, Canada. His research activity mainly focuses on commercial and industrial refrigeration, heat recovery and geothermal heat pump systems, low-enthalpy power generation cycles, and heat pump drying. During the past 15 years, he collaborated with the Canadian and American heat pump drying industry and R&D drying community in developing laboratory- and industrial-scale experimental prototypes. Drying of various products such as vegetables, agricultural and biological products, and wood has been theoretically and experimentally studied, and its results have been published in several drying conference proceedings and in prestigious journals such as *Drying Technology*, *International Journal of Refrigeration*, *Applied Thermal Engineering*, and *IEA Heat Pump Centre Newsletter*.

1 Introduction

Drying is one of the methods employed to preserve and/or improve the quality of many products (such as woods and agro foods) after harvesting. It is also an energy-intensive industrial operation. In developed countries, the energy consumption for drying may range, for example, from 10% to 15% in the United States, Canada, France, and the United Kingdom, and from 20% to 25% in Denmark and Germany. For particular industrial sectors, the energy consumption for drying may vary from as high as 60% in farm corn production, 50% in the manufacturing of finished textile fabrics, and 35% in pulp and paper industry, to as low as 3%–5% in chemical processes (Mujumdar and Huang 2007). This means that the major costs for industrial dryers may come from the energetic operation rather than from their initial investment costs.

Drying is not only an energy-intensive operation but also a complex, nonlinear coupled heat and mass transient transfer process phenomenon, including that related to drying of hygroscopic and porous material such as woods and agro foods (Chua et al. 2002; Perré and Keey 2014). Consequently, improving our knowledge of interconnected physical, chemical, and thermodynamic processes involved in drying as well as the energy efficiency of drying equipment are the most relevant objectives of today and future research and development (R&D) activities throughout the world.

In particular, industrial wood drying is the most time- and energy-demanding process at a sawmill and an essential step in the manufacturing of most commercial wood products, having a great impact on the parameters that affect the selling price of the sawn timber. The main objective of industrial wood drying is to add value to the dried lumber by minimizing the development of defects and achieve low and uniform moisture content suitable for the intended end use and/or before machining, gluing, and finishing are carried out (McCurdy 2005; Vikberg 2015).

During the past few decades, industrial wood drying has made great progress through the increased knowledge and awareness of both sawmill managers and kiln operators as to how to operate the kilns to achieve the best drying results. On the other hand, the R&D performed by sawmills, kiln manufacturers, and universities has also played a major role in that progress. As a result, the kiln design, process control, and drying schedules have been considerably improved.

Today, the main objective of the development of drying technology is to further improve the drying process by the application of new technologies and advanced measurement, modeling, and simulation techniques.

Among other methods aiming at reducing global energy consumption (electrical and fossil) in the wood drying industry, there are heat pump-assisted dryers. These systems generally integrate air-to-air mechanical vapor compression heat pumps (acting as cooling, dehumidifying, and heating devices) with slightly modified conventional drying enclosures (also called kilns, chambers, or compartments). Heat pumps recover sensible and latent heat by condensing moisture from a part of the moist drying air and supplying it back to the wood dryers by heating the same air streams. Such a process may accelerate the drying cycles, preserve the quality of dried lumber boards, and reduce the overall energy consumption.

Heat pump-assisted industrial dryers are mainly used to dry hardwoods at relatively low temperatures (up to 55°C) and softwoods at medium and high temperatures (between 80°C and 110°C), generally in combination with fossil (oil, natural gas) or renewable (biomass) fuels, or electricity as supplementary (and backup) energy sources.

This book focuses on heat pump-assisted batch drying of the most common wood species (such as softwoods and hardwoods), a technology aiming at reducing the global energy consumption in wood industry and improving the final quality of dried products. Some of the general principles, such as dryer heat pump coupling, operation, and control, may be applied to dry many other materials, such as agro food products.

The main scope of this book is to provide the basic information about the wood physical drying processes to heat pump technicians and some general information about the air-to-air mechanical vapor compression heat pump technology to industrial and academia drying R&D communities. This is because only a good understanding of both wood dryer and heat pump technologies, followed by a strong collaboration between wood drying and heat pump specialists, may ensure the development of new, advanced integrated concepts and successful future industrial implementations.

REFERENCES

Chua, K.J., S.K. Chou, J.C. Ho, N.A. Hawlader. 2002. Heat pump drying: Recent developments and future trends. *Drying Technology* 20(8):1579–1610.

McCurdy, M.C. 2005. Efficient kiln drying of quality softwood timber. PhD Thesis, Chemical and Process Engineering, University of Canterbury, Christchurch, New Zealand.

Mujumdar, A.S., L.X. Huang. 2007. Global R&D needs in drying. *Drying Technology* 25:647–658.

Perré, P., R.B. Keey. 2014. Drying of wood: Principles and practices. In *Handbook of Industrial Drying*, 4th edition, edited by A. S. Mujumdar, Advances in Drying Science and Technology, CRC Press, Boca Raton, FL, pp. 822–877.

Vikberg, T. 2015. Industrial wood drying—Airflow distribution, internal heat exchange and moisture content as input and feedback to the process. PhD Thesis, Division of Wood Science and Engineering, Department of Engineering Sciences and Mathematics, Luleå University of Technology, Luleå, Sweden.

2 Wood

2.1 WOOD STRUCTURE AND PROPERTIES

Green wood of living and freshly felled trees is a natural and renewable resource prevalent in many human activities, such as wood frame houses and furniture. It is a porous, hygroscopic substance consisting of dead and living cells, and is filled with either water or air (Onn 2007). Its physical and chemical properties, as well as drying characteristics, depend on the species, tree structure, and amount of moisture content.

The properties of living trees are related to directions such as (i) axial (parallel to the stem), (ii) radial (horizontal rays from the center of the tree that connect layers from pith to bark for storage and transfer of food), and (iii) tangential to the bark.

The cells are built of cellulose (as a structural component), hemicellulose (that surrounds cellulose microfibers which together form the wood fiber framework), and lignin (that binds the cellulose structure together and provides the rigidity and plastic nature of the wood). Wood cells consist of two thin walls surrounding cavities. The primary wall is called pit and serves as a means of communication between cells by allowing liquids (sap) to pass through in the living tree and moisture to move during the drying of lumber (Bian 2001). The secondary wall consists of layers added after cell formation. The cells have three main purposes: (i) holding the wood structure, (ii) transporting the water, and (iii) transporting the nutrients. Generally, there are three types of cells: (i) vessels, which are pipelike structures of indeterminate lengths formed by individual cells connected endwise, with their average lengths ranging from 0.2 to 0.5 mm and their diameter from 20 to 400 μm; (ii) fibers (present only in hardwoods), which are long narrow cells (on average, with lengths of 1–2 mm and diameters ranging from 0.01 to 0.05 mm) with closed and mostly pointed ends; and (iii) parenchyma cells, which perform the function of storage tissues in wood.

The amount of cell wall substance per unit volume of wood determines its specific gravity, and, therefore, woods with thicker cell walls are heavy and those with thinner cell walls are light (Bian 2001). Cell wall substance, with a relative density of about 1.5, is heavier than water (relative density of 1.0), but since air is normally present in many cell cavities, a piece of wood usually weighs less than an equal volume of water and the wood floats. However, if most cell cavities are filled with water, the wood will weigh more than an equal volume of water and, consequently, the wood will sink. The complex system of cells of different types and sizes that transport vast amounts of water from the roots to the leaves is composed of long and narrow tubes nearly parallel to the long axis of the tree trunk. They make up more than 90% of the volume of a live tree and can hold more water than wood.

In wood, there are mainly two types of pits: simple and inter-tracheid bordered pit pairs (specialized valves designed to seal and isolate individual tracheids if they become damaged by which water is transported from one fiber to another). The bordered pit pairs are circular openings (called pit apertures) in adjacent cell walls located on the radial surface of the fibers. The spaces between apertures and membranes (called pit cavities) are narrow and abrupt toward the cell lumen. For softwood, of which the composition is dominated by tracheids, the center of pit membrane (called torus) is thickened, a feature that does not exist in hardwood, where pits are often bordered.

From the outermost to the center, a cross section of mature trees consists of three distinguishable parts: (i) bark, (ii) wood, and (iii) pith (Figure 2.1). The bark is divided into two layers: an outer dead part and an inner thin living portion which carries food from the leaves to the growing parts of the tree. The cambium layer, lying immediately under the bark (a thin living layer of cell division between the bark and the wood), consists of a single layer of cells that can be seen only with a microscope. The cambial cells divide each growing season to form a new concentric layer of cells (called growth ring) both for the bark and for the woody stem (Bian 2001). The woody part of mature trees is composed of three zones (Figure 2.1): (i) the sapwood (the peripheral portion with continuous channels for water, nutrients, and resin transport); (ii) heartwood (the inner portion of the cross section of wood, usually darker in color than the outer layer); and (iii) pith (normally located at the center of the stem; it varies in size and color, and its structure may be solid, porous chambered, or hollow) (Cech and Pfaff 2000). Above the fiber saturation point, the transport of water is very different in sapwood and heartwood. However, below the fiber saturation point, the water transport is similar.

2.1.1 Sapwood

Sapwood, the youngest and growing part of trees containing both living and dead tissues, functions as a sap and liquid nutrient conductor from the roots to the leaves and wood storage of tree. The sapwood annulus ranges in width from a few millimeters to a maximum of about 50 mm, depending on the species, tree age, and height of the stem, with a general tendency for the width to be smaller in mature trees (Keey et al. 2000). The sapwood is perishable, but it is protected in the living tree by the sap which saturates the tissue and excludes the oxygen needed by degrading organisms. Generally, the sapwood contains much more moisture than the heartwood.

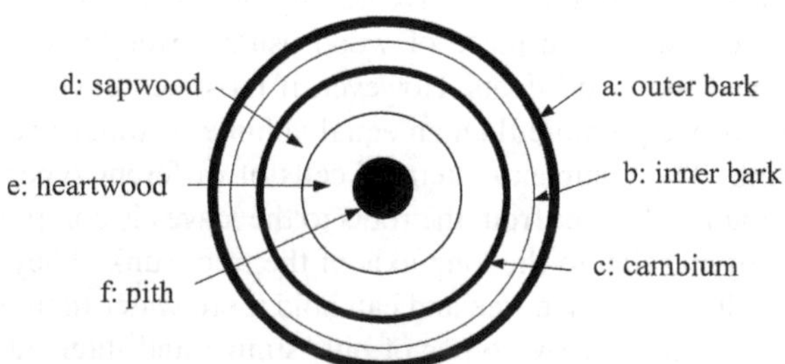

FIGURE 2.1 Schematic cross section of tree trunk.

Since the major function of the sapwood is to transport water from the roots to the leaves, its cells normally contain large quantities of water even after the tree has been felled and converted to lumber. Usually, only a portion of this water is retained in the cells after transition to the heartwood, and, consequently, there is a large difference in moisture content between the heartwood and the sapwood of most species (Cech and Pfaff 2000). The sapwood of freshly cut logs characteristically has a higher moisture content, and, normally, it is lighter in color and thus, normally, is readily distinguishable from the darker heartwood.

2.1.2 Heartwood

In young stems, all cells are active in conducting sap from the roots to the leaves. As the tree grows, the wood nearest the pith becomes aged and undergoes changes. It stops conducting sap, and a gradual transition to heartwood takes place, accompanied by the formation of materials such as tannins and gums. From a growth point of view, one can say that heartwood is dead. However, in terms of utilization as raw material, heartwood is extremely important, even more than sapwood.

In older trees, a harder and, sometimes, darker zone forms, by a gradual change of sapwood at the center of the stem. This is the heartwood of which all cells are dead and physiologically inactive. It no longer takes part in the transport and storage of food, but instead provides mechanical rigidity to the stem.

The heartwood region expands as the tree increases in girth, being surrounded by an annulus of sapwood, which is typically 10–50 mm wide. The often dark color of heartwood reflects the enrichment of the wood by various resins and carbohydrates (known as extractives) that include sugars, organic acids, and tannins. The heartwood contains lower moisture than the sapwood, and, in some cases, it is more resistant to fungal attacks. However, it is more difficult to withdraw water from the heartwood because of its high resistance to fluids' movement.

2.1.3 Earlywood and Latewood

The trees' growth rings contain two kinds of fibers (tracheids): earlywood and latewood. They are visible because the size and wall thickness of tracheid cells vary systematically across each ring.

Through the growing season, there are changes in the structure of wood. Cells produced at the beginning of the growing season (spring and start of summer) are commonly larger in diameter and thin-walled cells as these are principally involved in sap conduction. Thus, this wood (called earlywood) appears less dense than the latewood produced toward the end of the season.

The cells present in the later part of the ring are formed when growth is slower. Their radial diameters are smaller and the cell walls are thicker. This wood is called latewood where, compared to earlywood, there are smaller and fewer pits, and the permeability to moisture movement is different.

The distinction between earlywood and latewood is generally associated with differences in the distribution and size of large diameter, thin-walled cells (vessels) and small-diameter, thick-walled cells (fibers).

Earlywood cells, formed during summer, are darker in color and are important for the strength of the tree, whereas latewood cells, formed in the late summer and autumn, are thicker-walled and rectangular cells, with a much reduced radial width and, thus, contribute more to structural support. The transition from earlywood to latewood is abrupt in species such as European larch and Douglas-fir, but quite imperceptible in the spruce.

2.2 TYPES OF WOODS

According to their ease of drying and proneness to drying degrade (particularly cracks and splits), timbers can be classified as follows (Oliveira et al. 2012):

a. Highly refractory woods, such as oak, beech, and subalpine pine, that are slow and difficult to dry;
b. Moderately refractory woods, such as Sydney blue gum (Bootle 1994) and other timbers of medium density, potentially suitable for furniture, that show a moderate tendency to crack and split during seasoning; they can be seasoned free from defects with moderately rapid drying conditions (e.g., at maximum dry-bulb temperatures of 85°C);
c. Non-refractory woods, such as low-density *Pinus radiata*, that can be rapidly dried to be free from defects even by applying high dry-bulb temperatures of more than 100°C in industrial kilns; if not dried rapidly, such wood timbers may develop discoloration (blue stain) and mold on the surface.

According to their botanical origin, trees are classified into two subdivisions: (i) gymnosperms, the needle-leaved, coniferous trees, as spruce and pine, known as softwoods, and (ii) angiosperms, the broad-leaved, deciduous trees known in the lumber trade as hardwoods.

In the lumber commercial terminology, the adjective *hard* is not directly associated with the hardness nor *soft* with the softness of the wood. In other words, these terms do not apply to the hardness or density of the woods. In practice, hardwoods, such as aspen or basswood, have softer wood than softwoods such as red pine or black spruce (Cech and Pfaff 2000). Some softwoods, such as Southern yellow pine, are harder than some hardwoods, such as basswood or cottonwood.

The terms softwood and hardwood originated in the medieval timber trade, which was unaware of soft hardwoods such as balsa and despite the relative hardness of familiar softwoods, such as yew. This book follows well-established conventions in using the terms softwoods and hardwoods for gymnosperms and angiosperms, respectively (Keey et al. 2000).

2.2.1 SOFTWOODS

Softwoods, such as Norway spruce and Scots pine (two common softwood species in Europe), and fir, spruce, yellow pine, white spruce (more common in North America), are lighter and, generally, have a simple and uniform structure with few cell types. Softwoods, such as spruce and pine, are the main wooden building materials in countries such as the United States, Canada, and Norway, used as exterior cladding, structural parts, and interior linings and floors.

Softwoods have a simple, uniform structure with few fiber cell types vertically aligned, with both vertical and horizontal enlarged cavities surrounded by cells. Pointed cells (called tracheids; arranged in regular rows, generally ranging from 1 to 7 mm in length, depending on the species and location within the stem, being shorter near the pith, and longer near the cambium) are dominant, constituting 90%–95% of the stem volume. These cells provide both mechanical support and hollow conduits for sap flow (Cech and Pfaff 2000; Keey et al. 2000; Nijdam and Keey 2002).

In the drying of softwoods, it may be important to segregate heartwood from sapwood because of the large differences in moisture saturation and drying behavior. In spite of the high difference in the green moisture content, the drying time can sometimes be comparable for sapwood and heartwood (Salin 1989; Perré and Martin 1994). However, with species such as *P. radiata*, even though the heartwood is much less permeable than the sapwood, the latter takes twice the time to dry (Haslett 1998). The presence of heartwood influences the drying of lumber in other ways. Sometimes, for some tropical hardwood species, there is reduced shrinkage, a desirable feature, due to the bulking of cell walls by extractives.

Most softwoods, such as conifers, contain some resin ducts which are continuous tubes mainly extending in the longitudinal direction that are ineffective for the movement of liquids or gases (Cech and Pfaff 2000). The density of softwoods ranges from 350 to 700 kg/m^3 (Keey et al. 2000). Highly permeable softwoods, such as *P. radiata*, are typically much lighter and easier to process than hardwoods. Generally, the green moisture content in softwoods lies in the range of 150%–200% (dry basis) in sapwood and in the range of 30%–80% (dry basis) in heartwood. In Norwegian spruce and Scotch pine, for example, which are the common species in Nordic European countries such as Sweden, the moisture content is 70–90 wt%, depending on the season. Softwood lumber intended for framing in construction is usually dried to an average moisture content of 15%, not to exceed 19% (dry basis). Softwood lumber is dried to a moisture content of 10%–12% (dry basis) for many other uses, such as for appearance grades, and as low as 7%–9% (dry basis) for furniture, cabinets, and millwork.

Although there are about a hundred times more species of hardwood trees than of softwood trees, the ability to be dried and processed faster and more easily than hardwoods makes softwoods the main supply of commercial wood today.

2.2.2 Hardwoods

Hardwoods, species derived from the botanic group of angiosperms, are porous materials with a greater variety of more complex cell structures than softwoods, mainly because large diameter cells (called vessels, like pipes with diameters varying between 0.05 and 0.25 mm through which liquids can pass) form long and narrow vertical tubes (generally shorter in length compared to softwood cells) with more or less open ends (Cech and Pfaff 2000). Hardwood fibers vary in length from about 0.8 to 2.0 mm; their primary function in trees is to provide mechanical support. The vessels can be solitary or grouped together in various patterns and interconnected in order to create a continuous conducting system up the stem; they are irregular and of varying lengths.

Because more cell types are present in hardwoods, they are more complex and harder than softwoods (Bian 2001) and have a much greater diversity in structure among species. Therefore, the drying behavior of hardwoods is more variable than that of softwoods. For example, in oak, the rays constitute about 30% of the volume of wood, whereas in basswood, they form only about 5% of the volume of wood.

In North American hardwoods, for example, the vessel volume ranges from 6% to 55%, the fiber volume from 29% to 76%, the ray volume from 6% to 31%, and the axial parenchyma volume from 0% to 23%.

Hardwoods are classified as ring-porous or diffuse-porous trees depending upon the arrangement of the vessels within the annual ring (Cech and Pfaff 2000).

Ring-porous hardwoods (such as oak) are characterized by layers of large vessels formed during the early spring, followed by the formation of denser fibrous wood containing smaller scattered vessels later in the growing season. In diffuse-porous woods, the vessels are distributed more uniformly throughout the annual rings.

The density of hardwoods is largely determined by the proportion of fibers to other cell types. In low-density hardwoods, the thin-walled fibers with large lumens occupy a minor proportion of the volume, whereas in high-density hardwoods, the fibers, which are smaller and more numerous, and have thicker walls and lumens contracted, occupy a large proportion of the volume. The density of hardwoods varies between 450 and 1,250 kg/m^3 (Keey et al. 2000); thus, the permeability of hardwoods is much less than that of softwoods, making them more difficult to dry.

Unlike softwoods, the green moisture contents of sapwood and heartwood of hardwoods are roughly comparable, and there is little variation with age and position in the tree's stem. Depending on the species and its density, the green moisture content of hardwoods ranges from 50% to about 100% (dry basis). For furniture, cabinets, and millwork, hardwood lumber is usually dried to 6% or 8% moisture content (dry basis) (Cech and Pfaff 2000).

Hardwoods are usually dried at low temperatures (maximum 55°C), a process that consumes up to 70% of the total energy required for primary wood transformation. Usual energy sources are fossil fuels (oil, natural gas, and propane) and bark. Electrically driven low-temperature heat pumps are also used in combination with fossil fuels or electricity as backup energy sources.

2.3 WOOD PROCESSING

Timber is the wood in any of its stages from felling to readiness for use as structural material for construction or wood pulp for production.

In the United States and Canada, timber often refers to the wood contents of standing, live trees that can be used for lumber (the product of timber cut into boards, supplied either rough or finished for furniture and other items requiring additional cutting and shaping) or fiber production, although it can be used to describe sawn lumber whose smallest dimension is not less than 127 mm. In the United Kingdom, Australia, and New Zealand, timber is a term used for sawn wood products, such as those for floor boards. In the United Kingdom, lumber has several other meanings, including unused or unwanted items.

Sawmills produce lumber from forested green timber through the following processes (Anderson 2014): (i) when the timber arrives at a sawmill, it is roughly handled, that is, sorted and stored in the lumber yard; (ii) the bark is separated from the timber debarking before the sawing process; (iii) timber logs are usually square edged, quarter or radially sawn, that is, to different types of lumber boards, before further processing takes place; (iv) the sawn lumber boards are sorted depending on the quality and length; (v) the pre-drying is done by natural (yard) (see Section 6.3) or accelerated (see Section 6.4) methods then with artificial techniques in facilities called kilns (see Section 6.5); and (vi) after drying, the lumber is sorted once more, in some cases grinded, and finally packaged for transportation.

Between the forests and sawmills, the wood passes through the following processes:

a. Logging, consisting of cutting, skidding, on-site processing, and loading of trees or logs onto trucks. There are several logging methods: tree-length logging (when trees are felled and delimbed, then transported), full-tree logging (when trees are felled and transported to the roadside with top and limbs intact, then delimbed, topped, and bucked at the landing), and cut-to-length logging (a process consisting of felling, delimbing, bucking, and sorting at the stump area for trees up to 900 mm in diameter);
b. Transportation of felled logs by truck, rail, or river to sawmills to be cut into timber, then scaled and decked (a process for sorting the logs by species, size, and end use, such as lumber, plywood, and chips); in Sweden, for example, 45% of the volume of the harvested forest is brought to sawmills as sawlogs; the purchase of sawlogs is the largest expense for sawmills, which, probably, accounts for more than 60% of the total costs (Vikberg 2015).

The main purpose of sawmills is to produce lumber boards of different dimensions and qualities based on a simple operation principle: wood logs enter on one end and dimensional lumber exits on the other end. The aim is to use the wood as effectively as possible and to satisfy the customer needs for particular material characteristics. Sawmill technology and operation have significantly changed in recent years, emphasizing increasing profits through waste minimization and increased energy efficiency as well as improving operator safety. Today, most sawmills are large facilities in which most aspects of the work are computerized.

Most of modern sawmills have the following subdivisions: (i) timber intake, (ii) debarking, (iii) sawing, (iv) sorting, (v) drying, (vi) alignment and sorting, and (vii) packing and storage. When the timber arrives at sawmills, it is debarked (separated from the bark), roughly sorted, and stored in the lumber yard. Then the timber logs are sawn into different types of lumber boards, depending on the dimension, length, quality, and species. Bark, sawdust, and wood chips can be energy sources, a way to lower greenhouse gas emissions by reducing the consumption of oil and gas. Due to several physical differences between the earlywood and the latewood, the way in which the boards are cut from the logs can have an important bearing on subsequent drying behavior. The sawing pattern governs the orientation of the growth rings relative to the drying surfaces. In the so-called flat-sawn lumber, the annual rings lie parallel to the exposed surfaces (McCurdy et al. 2002), whereas in the quarter-sawn lumber, the growth rings run at right angles to them (Nijdam and Keey 2002).

Softwood and hardwood lumber that is finished/planed and cut to standardized width and depth specified in millimeters (or inches) is termed as dimensional lumber (Figure 2.2). Lumber's nominal dimensions are given in terms of green (not dried), rough (unfinished) dimensions. The finished size is smaller, as a result of drying (which shrinks the wood), and planning to smooth the wood. However, the difference between "nominal" and "finished" lumber sizes can vary. Therefore, various standards have specified the difference between the nominal size and the finished size of lumber.

Common lumber sizes (width × depth, expressed in inches; 1 inch = 25.4 mm) such as 2 × 4, 2 × 6, 2 × 8, 2 × 10, 2 × 12, and 4 × 4 are used in the United States and Canada for modem constructions. The sizes for the dimensional lumber made from hardwoods may vary from the sizes for softwoods. The length of solid dimensional lumber boards is usually specified separately from the width and depth, typically available up to lengths of 3,658 mm (or 24 feet; 1 feet = 12 inches = 12 × 25.4 mm^2) (Cech and Pfaff 2000).

The sawn lumbers are sorted into different lumber packages, then naturally dried outdoor in ambient air and/or artificially dried in kilns to remove the moisture in excess.

Generally, for drying, several quasi-identical species are grouped. The amount of individual species in the mix depends on the region under consideration. In Canada, for example, the spruce–pine–fir group comprises eight different species: white spruce, Engelmann spruce, black spruce, red spruce, Lodgepole pine, Jack pine,

FIGURE 2.2 Example of a 2 × 4 inch (1 inch = 25.4 mm) dimensional lumber board.

subalpine fir, and balsam fir. Although processing the spruce–pine–fir mix together has several advantages from the production and marketing point of view, kiln operators are well aware of the challenges associated with the drying characteristics of each species. In many cases, mills have established that if the percentage of (subalpine) fir in the spruce–pine–fir mix is higher than 10%, then the species must be sorted and dried separately. This is because the (subalpine) fir may have a high incidence of wet wood, which are localized areas within a board with higher moisture content and lower dry-ability. Some white spruce may also exhibit very high initial moisture content and collapse during drying (Oliveira et al. 2012).

Drying is one of the essential processes in the sawmills where the wood quality and suitability to different end user products for construction (mostly from coniferous) and cabinets, furniture, and high-grade flooring (with hardwoods) are considerably affected by significantly upgrading and adding value to sawn timber materials (Vikberg 2015).

After drying, the finished lumber boards are sorted again, grinded, packaged for transportation in standard sizes, and finally shipped to market. Individual lumber boards exhibit a wide range in quality and appearance with respect to knots, slope of grain, shakes, and other natural characteristics. Therefore, they vary considerably in strength, utility, and value. Boards are usually supplied in random widths and lengths with a specified thickness, and sold by the board foot (equivalent to 0.002359 m^3). In Australia, many boards are sold to timber yards in packs with a common profile (dimensions), but not necessarily consisting of the same length boards.

2.4 OUTLOOK ON WOOD RESOURCES AND MARKETS

The majority of the world's forests are either tropical (47%) or boreal (33%) with the remainder in temperate (11%) and subtropical (9%) regions. These forests provide environmental services and protection such as watersheds, soil erosion control, microclimate, and biodiversity, and act as terrestrial carbon sinks. Primary regions comprise North America, Europe (including Russia), and China, and the plantation forestry in Southern countries such as Brazil, Chile, Uruguay, Argentina, New Zealand, Australia, and South Africa.

The total world's forest cover was 3.9 billion ha (2000), equivalent to 0.6 ha per capita which included natural (95%) and plantation (5%) forests (FAO 2016) with an estimated wood volume of 386 billion m^3. The harvest of wood from these areas today exceeds 3.4 billion m^3 annually, approximately half being used as fuel material.

Historically, about 78% of the world's accessible forests supplied wood for the construction (generally, light and strong compared to other materials available); wood is also a convenient and abundant source of fuel for heating and cooking (McCurdy 2005). In addition, green wood supplies salient raw materials for products such as timber, lumber, and fiber for newspapers, books, and magazines, many of them being recoverable and/or recyclable (Bian 2001; Anderson 2014).

Commodities, such as solid wood, veneer, and plywood, are produced from forestry resources through primary processing. Secondary processing, which usually requires a high-quality resource, includes, among others, sawmilling and drying aiming at producing higher value products such as furniture and building structures (Walker 1993).

The majority of the international wood trade is geared toward the United States, Canada, Japan, and Europe, and also toward some developing countries, particularly China, India, and Russia. In 2002, the value of global imports of wood-based forest products, including fuel wood and charcoal, was US$141.4 billion (FAO 2004).

2.4.1 The United States

In 2015, the gross domestic product (GDP) of the United States was of US$17,600 billion (2009 dollars), and the statistics for wood and wood products (production, imports, exports, and consumption) is shown in Table 2.1 (Howard and McKeever. 2016) (www.fpl.fs.fed.us/documnts/fplrn/fpl_rn343.pdf. Accessed December 4, 2017).

The United States softwood industry is highly competitive, being rated among the most efficient lumber industries in the world. The current annual revenue is of US$7 billion with an annual growth of 15%–20% (www.ibisworld.com/industry/default.aspx?indid=396. Accessed March 2015). Wood industry, including small to medium-size sawmills with less than 25 workers, employs more than 75,000 people.

End-use markets of interest in the United States include new single family, multifamily, and mobile home construction, repair and remodeling of existing residential structures, low-rise nonresidential building and other types of nonresidential construction, furniture and other manufactured wood products, and packaging and shipping (Table 2.2). These end-use markets account for 80%–90% of all solid wood products consumption.

TABLE 2.1
Statistics for Wood and Wood Products in the United States (2016) (in 1,000 m³)

Wood and Wood Products	Production	Imports	Exports	Consumption
Sawn softwood	54,000	21,802	2,906	72,896
Coniferous logs	130,026	700	7,643	123,083
Sawn hardwood	18,804	950	2,900	16,854
Hardwood logs	37,450	1,575	3,800	35,225

TABLE 2.2
Wood Products Market Share in the United States by End Use (2016) (in million m³)

	Residential Construction	Nonresidential Construction	Manufacturing (Furniture, etc.)	Packaging and Shipping	Total Reported End Uses	Others
Sawn softwood	63	9	8	9	89	11
Sawn hardwood	10	14	18	43	85	15

Approximately 85% (68.8 million m^3) of the overall lumber production in the United States is kiln dried (Rice 1994), and only 13 major species account for 82% of all kiln-dried lumber. Hardwoods account for 11.7 million m^3 (of which about 73% of the kiln-dried species comprise red and white oak, yellow poplar, maple, red alder, and cherry), and softwoods account for 57.1 million m^3 (of which about 84% of the kiln-dried species are southern yellow and ponderosa pine, Douglas-fir, western fir, and western hemlock).

In the United States, almost 70% of kilns are operated with dry-bulb temperatures between 70°C and 82°C, with a further 20% operating at higher temperatures. A number of hardwoods, including red alder, aspen, black gum, and yellow poplar, are being dried under high-temperature schedules, usually with final equilibration and conditioning treatments. Between 2010 and 2015, the annual market growth was of 4.3%. In 2014, the top four companies in the industry accounted for about 14.2% of industry revenue.

2.4.2 Canada

About 45% of the Canadian territory is wooded, representing 10% of the planet forests. In 2005, Canada was the second largest producer of lumber in the world, with 16% of worldwide production, and the top world exporter with 30% of the international lumber trade. Canada sells its forestry products to over 100 countries, namely, to the United States, the European Union, and Japan (FPInnovations 2010).

The Canadian wood industry provides 3% of the country's GDP, 10% of the total foreign exchanges, and 57% of the annual commercial surplus (Canada Statistics 2002). The Canadian sawmills and wood preservation industry contributed nearly CAN$7.6 billion to the national GDP in 2006. This is the equivalent of 4.4% of the manufacturing sector's GDP and 0.7% of the entire Canadian economy. This industry is also a key player in the export market, accounting for more than 4% of total Canadian merchandise exports.

The construction lumber is the main Canadian wood product export (Gaston 2006). The double-digit growth in production and exports of sawn wood and panels is attributable to increased sales to the United States due to a recovering economy and housing market. Recent trends related to industrial kiln drying with an especial emphasis on the drying of softwood dimension lumber (Oliveira et al. 2012).

Increased lumber supply from Russia and other countries around the world has significantly changed the flow of forest products to certain traditional markets. For example, between 1995 and 2007 the number of softwood sawmills in Canada (as well as in the United States) was reduced (from 1,311 to 900) by 24.5% (Spelter 2007).

In 2002, Canadian sawmills supplied 72 million m^3 of resinous timber (including white and black spruce, jack pine, and balsam fir), of which approximately 24% was produced in the province of Québec (eastern Canada) (Canada Statistics 2002).

About 85% of this production is presently dried with bark boilers, 10% with fossil fuels (natural gas, propane, or oil), and only 3% with low-temperature heat pumps.

The potential reduction in energy use by dry kilns in Canada would be 5.5 PJ/year (1 petajoule = 10^{15} J) or 335 kT/year (1 kiloton = 10^6 kg) in carbon dioxide emissions.

Furthermore, it was estimated that CO_2 emissions could be reduced by an additional 90 kT/year through a decrease in the amount of lumber that is downgraded.

2.4.3 The European Union

In many European countries, wood is a resource of great importance for the gross national products. In Sweden, for example, 57% of the land area is classified as productive forest land where the annual harvest is of about 80 million m^3. Scots pine and Norway spruce, two species of greatest commercial interest, account for 39.2% and 41.5%, respectively; Birch for 12.0%; and other species for the rest (Vikberg 2015).

The European Union's wood-based industries cover a range of downstream activities, including woodworking industries, large parts of the furniture industry, pulp and paper manufacturing and converting industries, and printing industry. Together, some 415,000 enterprises are active in wood-based industries across the European Union; they represent nearly 20% of manufacturing enterprises across the European Union, many being small or medium-sized enterprises.

The overall round wood production of the European Union was estimated to be 447 million m^3 in 2015 (http://ec.europa.eu/eurostat/statistics-explained/index. Accessed January 7, 2018), of which about 22% is used as fuelwood and the remainder for sawn wood, veneers, and pulp and paper production. Among the European Union countries, Sweden produces the most round wood (74 million m^3) (2015), followed by Finland, Germany, and France, each producing between 51 and 59 million m^3).

From 2010 to 2015, the average total output of sawn wood across the European Union was about 100 million m^3/year, Germany (21%), and Sweden (17%), which are the leading sawn wood producers.

The economic weight of the wood-based industries in the European Union, measured by the gross value added, is about 6.2% of the total manufacturing production (2014). Within the wood-based industries, the highest share was recorded for pulp, paper, and paper products manufacturing (40.7%). Sectors such as printing and furniture manufacture each amounted to 27% of the gross value added, whereas the manufacturing of wood and wood products provided 29%. In 2014, across the European Union, the wood-based industries employed 11% of the total manufacturing workforce.

2.4.4 Russian Federation

The total forest area of Russian Federation represents 46.6% of the total area of the country and 25% of the world's reserves of wood (2015) (https://en.wikipedia.org/wiki/Forestry_in_Russia. Accessed February 24, 2018). Despite the fact that Russia contains the largest area of natural forests in the world, its current share in the world trade of forest products is below 4%. Although the forests occupy over half of the land of the country, the share of the forest sector in the 2010 GDP was only 1.3% in industrial production, 3.7% in employment, and 2.4% in export revenue (UN FAO 2012).

Following the collapse of the Soviet Union and the subsequent decline in economic activity, domestic production, as well as domestic consumption, fell as much as 40% (Backman 1998). The fall in wood exports was due to large increases in transportation costs. With a low capacity to process timber domestically, Russia increasingly became an exporter of round wood.

Although the total export value of wood products increased from US$2.4 billion in 2000 to US$8.8 billion in 2007, the percentage of this total export value that is attributable to forest products with at least basic processing showed slow growth. The export value of non-round wood forest products as a percentage of Russia's total export value of forest products rose from 37.0% in 2001 to 50.5% in 2007. These low percentages of value-added exports contributed to Russia's first-place rank in the world for the highest quantity of round wood exports between 2003 and 2005 (WTO 2010).

The Russian Federation has recently overtaken Canada and Germany to become the world's third largest producer and consumer of wood-based panels, and Thailand to become the fifth largest exporter in 2015.

2.4.5 South America

South America has continued expanding wood pulp production with an increasing number of new pulp mills being built in Brazil, Chile, and Uruguay. These three countries currently account for 14% of global wood pulp production and 74% of exports. In 2014, Brazil overtook Canada for the first time as the world's fourth largest producer of fiber furnish.

2.4.6 New Zealand

Milling of New Zealand's extensive native forests was one of the earliest industries in the European settlement of the country. The long, straight hardwood from the kauri was ideal for ship masts and spars. As the new colony was established, timber was the most common building material, and vast areas of native forest were cleared. *P. radiata* was introduced to New Zealand in the 1850s (https://en.wikipedia.org/wiki/Forestry_in_New_Zealand. Accessed January 8, 2018). This species reaches maturity in 28 years, much faster than that in its native California. It was found to grow well in the infertile acidic soil of the volcanic plateau, where attempts for agriculture had failed.

Plantation forests of various sizes can now be found in almost all regions of New Zealand. In 2006, their total area was 1.8 million ha, with 89% of *P. radiata* and 5% of Douglas-fir. Log harvesting in 2006 was 18.8 million m^3 (New Zealand Forestry and Wood Products Report 2016; https://gain.fas.usda.gov/.pdf. Accessed October 11, 2017).

Within the New Zealand economy, forestry accounts for approximately 4% of national GDP. On the global stage, the New Zealand forestry industry is however a relatively small contributor in terms of production, accounting for 1% of global wood supply for industrial purposes.

The forest industry is one of the larger energy users in New Zealand and the largest user of wood fuels. The total annual energy use of the forestry, logging, wood, and pulp and paper industries is around 68.8 PJ. This is made up of around 21 PJ coming from a range of energy sources (electricity, coal, and natural gas), the balance coming from wood-derived sources.

Woody biomass currently provides around 7.5% of New Zealand's primary energy supply. The balance is derived from the burning of black liquor arising from the Kraft pulp and paper process. Almost all of this energy is derived from the burning of wood residues to provide heat, only a small proportion is used for electricity generation via combined heat and power.

The log harvest for the year ending September 2015 was of 29.33 million m^3 of round wood. The annual harvest volume is expected to reach 32.4 million m^3 by 2020.

The value of all forestry exports (logs, chips, sawn timber, panels, and paper products) for the year ended March 31, 2006, was NZ$3.62 billion.

Raw log exports for the year ending September 2015 totaled 15.9 million m^3, that is, 83% of the total volume of wood exports. Over 25% of export values are to Australia, mostly paper products, followed by Japan, South Korea, China, and the United States.

2.4.7 Australia

Australia has many forests of importance due to significant features, despite being one of the driest continents.

Australia has approximately 123 million ha of native forest (80% of the total)—the majority of trees being hardwoods, typically eucalypts (75%) and acacias (8%)—and 2.0 million ha of industrial plantations, which represents together about 16% of Australia's land area (https://en.wikipedia.org/wiki/Forests_of_Australia. Accessed November 15, 2017). Australia has about 3% of the world's forest area and, globally, is the seventh country with the largest forest area.

In addition to native forests, there are two other categories of forest: (i) industrial plantations (grown on a commercial scale for wood production) and (ii) other forests that include small areas of mostly nonindustrial plantations and planted forests of various types including sandalwood plantations, small farm forestry and agroforestry plantations, plantations within the reserve system, and plantations regarded as noncommercial.

Australia's industrial plantations are a major source of commercial wood products, of which about 50% are exotic softwood (predominantly *P. radiata*), whereas the remaining 50% are hardwood (predominantly *Eucalyptus* species).

Solid wood products consumption in Australia has been relatively stable at 5–6 million m^3/year over the last decade. Australia's production of sawn timber in 2003–2004 was more than 4×10^6 m^3, of which 25% was hardwood.

The strengths of the Australian wood industry are (i) a relatively small domestic market and a current sector setup that is not well positioned to take advantage of value-added opportunities, (ii) location in the Asia-Pacific region which is a major growth area globally, (iii) good fiber and land availability, (iv) competitive wood costs (especially for softwoods), (v) relatively strong Australian economy, and (vi) established industry players, with business models and concepts proven elsewhere already.

Among the opportunities for the Australian wood industry can be mentioned (i) export value-added products; (ii) operate with fully integrated processing that is competitive and efficient, using proven business models, and implement new operating models to improve market reach; (iii) update processing facilities to improve asset efficiency; (iv) provide customized export solutions which consider product, marketing, and supply chain together (www.agriculture.gov.au/abares/forestsaustralia/australias-forests. Accessed December 12, 2017).

2.4.8 China

China has over 1.4 million ha of forests harboring over 2,800 tree species and multitudes of other plants (https://en.wwfchina.org/en/what_we_do/forest/. Accessed November 20, 2017). Forested uplands protect China's lowland river valleys by storing rainfall and gradually releasing it so as to reduce the severity of droughts and floods, preventing soils from eroding, and making possible the country's intensive irrigated agriculture system. Forests also provide 40% of the fuel for rural households. For these reasons, China has been called one of the most forest-dependent civilizations in the world.

China's timber market system has undergone reform as a result of the increasing liberalization of its markets (Sun et al. 2004). Its primary wood-processing industry and wood-consuming sectors have experienced rapid growth. The limited amount of domestic resources and the strong domestic demand for timber products in industries such as construction, furniture, and panel have caused the volume of China's imports of forest products to increase.

In China, the construction, furniture, and paper-making are the top three timber-consuming industries (Table 2.3) (Zhu et al. 2004).

China has grown in importance as both a producer and a consumer of forest products, and has recently overtaken a number of other big players in different product groups (e.g., overtaking Canada in sawn wood production and the United States in sawn wood consumption). The country is by far the largest producer and consumer of wood-based panels and paper. It is also highly significant in international trade of forest products being the world's largest importer of industrial round wood, sawn wood, and fiber furnish (pulp and paper); and the largest exporter of wood-based panels.

TABLE 2.3
Percentages of Timber Consumption by End Users in China (2002)

Industry	Timber Consumption (%) (2002)
Total timber consumption by domestic sector	100.0
Construction and housing	71.4
Furniture	11.3
Paper	7.8
Coal and mining	5.9
Vehicle, ship, and boat manufacturing	2.2
Other	1.4

The wood-processing industry is the primary conversion industry, including sawn wood and wood chips, plywood, fiberboard, particleboard, and other wood-based panel sectors. In China, there are over 10,000 sawmills, of which about 30% have annual capacities of 30,000 m^3 or more. In 2002, sawn wood production, strictly limited to the construction sector, was of $52 \times 10^6 m^3$.

In 2001, consumption of industrial round wood (logs), sawn wood, wood-based panels, pulp, and paper and paperboards ranked third, fifth, second, fourth, and second in the world, respectively.

At the same time, the Chinese forest industry has experienced rapid growth to meet the increasing demand for wood products, both at home and abroad. For example, China's plywood production has exceeded U.S. production, making China the largest plywood-producing country in the world. China has also emerged as one of the most important players in the pulp and paper market; it contributed more than 50% of the global production growth in paper and board last decade, becoming the second largest producer in the world.

In 2011, the timber demand in China, including products for industrial construction and exported logs, was 431 million m^3. The timber consumed was supplied from domestic production (57.5%) and from import (42.5%).

China is today a major net importer of timber and ranked first in log imports in the world (Zhu et al. 2004) because nearly 50% of China's total commercial timber logs is supplied by imports. The main imported products (mainly from Russia—31%, Canada—17%, New Zealand—12%, and the United States—12%) were logs, lumber, pulp and paper, and paperboards, whereas exported products were wooden furniture and wood-based panels (Nengwen 2012). On the other hand, a substantial share of the wood grown or imported into China is exported in the form of processed timber, paper, and finished or semifinished manufactured products. In 2015, China's imports of industrial round wood declined by 14%, whereas sawn wood and paper production and consumption continued to grow.

The exports (furniture, paper products, and plywood) account for nearly 10% of China's total wood consumption. The proportion of export furniture, paper, and plywood in 2003 reached 32%, 25%, and 19%, respectively, of total exports of wood products.

2.4.9 India

Because years of mismanagement has led to a serious depletion of forest resources with a total land area of $3{,}287 \times 10^6 km^2$, India has a remaining forest area of only 640,000 km^2 (www.em.gov.bc.ca. Accessed October 11, 2017).

The economic reforms which were begun in 1992 have accelerated the pace of development and industrialization leading to, among other things, greater demands for pulp, paper, and wood products.

In 2014–2015, India became the world's fifth largest paper and paperboard producer by overtaking the Republic of Korea and the fourth largest fiber furnish (pulp and recovered paper) importer by overtaking the Republic of Korea and Italy. Wood-based panel and sawn wood production (combined) grew in all five regions around the world in 2015. Global production of both panels and sawn wood increased by

3%, whereas trade grew at a lower rate (2% for both products). Global production of panels has been record high (399 million m³) and sawn wood registered the highest production (452 million m³) since 1990.

Most wood used in India is hardwood, particularly teak, which accounts for approximately 50% of domestic consumption. There are limited supplies of other species such as sal, eucalyptus, poplar, and deodar. Some private plantations, particularly rubber wood, also contribute to the domestic supply of lumber. India's coniferous stands are found in Himachal Pradesh, Kashmir, and other northern states.

Imports are mainly of teak logs from Burma, Malaysia, Thailand, and Africa. In the last 3 years, imports of radiata pine from New Zealand in log form have also become common. Radiata pine is primarily used for pallets, crates, and low-end doors and window products. Most wood in India is used for the manufacture of doors, window frames, wall panels, moldings, and furniture. The Indian construction industry accounts for approximately US$12 billion and is growing at 8% per annum. Of this, wood and wood products use accounts for approximately US$2.75 billion, of which current imports are estimated to be US$250 million. India produces 157 million m^2 of plywood and decorative veneers and 220,000 tons of particleboard, fiberboard, and medium-density fiberboard (MDF) annually.

Some Indian companies are now importing particleboard panels for use in kitchen cabinets from Italy, Australia, and New Zealand.

Because India's supplies of domestic wood are very limited, the import market for raw and semiprocessed wood is expected to increase with demand and construction activity over the next few years.

REFERENCES

Anderson, J.O. 2014. Energy and resource efficiency in convective drying systems in the process industry. PhD Thesis, Luleå University of Technology, Luleå, Sweden.

Backman, C.A. 1998. *The Forest Industrial Sector of Russia: Opportunity Awaiting.* Parthenon Publishing Group, New York.

Bian, Z. 2001. Airflow and wood drying models for wood kilns. MASc Thesis. Faculty of Graduate Studies, Department of Mechanical Engineering, The University of British Columbia, Vancouver, BC, May.

Bootle, K.R. 1994. *Wood in Australia: Types, Properties and Uses.* McGraw-Hill Book Company, Sydney, 443pp.

Canada Statistics. 2002. (www.statcan.gc.ca/eng/start, accessed March 12, 2016).

Cech, M.J., F. Pfaff. 2000. *Operator Wood Drier Handbook for East of Canada*, edited by Forintek Corp., Canada's Eastern Forester Products Laboratory, Québec city (Québec, Canada).

FAO. 2004. *FAO Forest Products Yearbook in 2002.* Food and Agriculture Organization of the United Nations, Rome, Italy.

FAO. 2016. State of the World's Forests (SOFO). 2016. (www.fao.org/publications/sofo/2016/en/, accessed October 15, 2017).

FPInnovations. 2010. Wood market statistics including pulp and paper, 2010 Edition, compiled by James Poon (https://fpinnovations.ca/products-and-services/market-and-economics/Documents/2010-wood-market-statistics-in-canada.pdf, accessed March 31, 2017).

Gaston, C. 2006. *The Competitiveness of Canadian Softwood Lumber: A Trade Flow Analysis*, FPInnovations, Vancouver, Canada.

Haslett, A.N. 1998. *Drying Radiate Pine in New Zealand.* FRI Bulletin 206 NZFRI, Rotonda, New Zealand.

Howard, J.L., D.B. McKeever. 2016. *U.S. Forest Products Annual Market Review and Prospects, 2012–2016*. United States Department of Agriculture, Washington, DC.

Keey, R.B., T.A.G. Langrish, J.C.F. Walker. 2000. *Kiln-Drying of Lumber*, edited by T.E. Timell. Springer Series in Wood Science, ISBN-13: 978-3-642-64071-1.

McCurdy, M.C. 2005. Efficient kiln drying of quality softwood timber. PhD Thesis. Chemical and Process Engineering, University of Canterbury, Christchurch, New Zealand.

McCurdy, M.C., J.I. Nijdam, R.B. Keey. 2002. Biological control of kiln brown stain in Radiata pine. *Maderas: Ciencia y Technologia* 4(2):140–147.

Nengwen, L. 2012. Overview of Chinese timber and wood products market (www.ettf.info/sites/default/files/presentation_liu_nengwen.pdf, accessed October 11, 2017).

Nijdam, J.J., R.B. Keey. 2002. New timber kiln designs for promoting uniform airflows within the wood stack. *Chemical Engineering Research Design* 80(A7):739–744.

Oliveira, L., D. Elustondo, A. Mujundar, R. Ananias. 2012. Canadian developments in kiln drying. *Drying Technology* 30(15):1792–1799.

Onn, L.K. 2007. *Studies of Convective Drying Using numerical Analysis on Local Hardwood.* Universiti Sains, Penang, Malaysia.

Rice, R.W. 1994. *Kiln Drying Lumber in the United States: A Survey of Volumes, Species, Kiln Capacity, Equipment, and Procedures, 1992–1993*, Forest Products Laboratory, Madison, WI.

Salin, J.G. 1989. Prediction of checking surface discoloration and final moisture content by numerical methods. In *Proceedings of the 2nd IUFRO International Wood Drying Conference*, Seattle, Washington, DC, August 9–13, pp. 71–78.

Spelter, H. 2007. Profile 2007: Softwood Sawmills in the United States and Canada. *Research paper FPL-RP-644*, Forest Products Laboratory, Madison, WI.

Sun, X., L. Wang, Z. Gu. 2004. A brief review of China's timber market system. *International Forestry Review* 6(3–4): 221–226.

UN FAO. 2012. *The Russian Federation Forest Sector: Outlook Study to 2030.* Food and Agricultural Organization of the United Nations Rome, Italy.

Vikberg, T. 2015. Industrial wood drying—Airflow distribution, internal heat exchange and moisture content as input and feedback to the process. PhD Thesis. Division of Wood Science and Engineering Department of Engineering Sciences and Mathematics, Lulea University of Technology, Lulea, Sweden.

Walker, J.C.F. 1993. *Primary Wood Processing: Principles and practice*. Chapman & Hall, London, pp. 197–246.

WTO. 2010. World Trade Organization Report: Trade in natural resources (https://www.wto.org/english/res_e/reser_e/wtr10_brochure_e.pdf, accessed April 4, 2017).

Zhu, C., R. Taylor, F. Guoqiang. 2004. *China's Wood Market, Trade, and Environment*. Science Press US Inc., Monmouth Junction, NJ.

3 Moisture in Wood

3.1 INTRODUCTION

The natural living trees, freshly felled timber logs, lumber boards, and wood in service are porous, hygroscopic materials. These materials contain cell wall substance; internal voids (e.g., 62% of the volume of oven-dry sugar maple); large amounts of moisture, that is, water in forms of liquid or vapor that often constitutes over 50% (sometimes up to 75%) of the wood's mass; and some extractives. In some cases, moisture may be even in greater proportion by mass than the solid wood materials.

Moisture has a significant influence on green (harvested, fresh cut) timbers and sawn logs, as well as on lumber wood boards because they continually exchange moisture according to the ambient (e.g., temperature and relative humidity) conditions (Eekelman 1968). In addition, the moisture profoundly influences the wood's strength properties, weight, stiffness, shrinkage, hardness, abrasion, specific heat, thermal conductivity, resistance against insect, fungi and decay attacks, and machinability (Kollmann et al. 1968; Perré and Keey 2015).

Since almost all the physical properties of wood are significantly influenced by its moisture content, most of it must be removed by drying processing in order to improve the lumber mechanical and structural properties. Effectively, when wood is well dried, subsequent shrinking and swelling (and other defects) are minimal, making further cutting and machining easier and more efficient. Drying is also an efficient way to prevent degradation due to biological activity, since wood with a <22% moisture content is immune to fungal and other attacks (Bian 2001).

3.2 PHYSICAL STATES OF PURE WATER

Knowing the fundamental properties of water is a starting point in understanding the moisture behavior in wood (Berit Time 1988).

A pure substance, such as water, has a homogeneous and constant chemical composition that may exist in more than one phase. Pure water can exist in the solid (ice), liquid, or vapor phase depending on the temperature and pressure conditions. Liquid water, a mixture of liquid water and water vapor, and a mixture of solid water (ice) and liquid water are all pure substances (Sauer and Howell 1983).

Molecules of pure water (H_2O) comprise one larger, negatively charged oxygen atom and two smaller, positively charged hydrogen atoms. Due to the sticky nature, water molecules hold large amounts of thermal energy. Many of the unusual properties of water, including its large heat of vaporization and its expansion upon freezing, are attributable to the hydrogen bonding in which a hydrogen atom is in a line between the oxygen atom on its own molecule and the oxygen on another molecule.

Each of the physical states of pure water is defined by at least two independent properties: temperature and pressure.

In the water pressure–temperature phase diagram, there are three lines of equilibrium (or phase boundaries) between the three state phases: sublimation (S-TP), vaporization (TP-CP), and fusion (M-TP) lines (Figure 3.1). These lines mark the conditions under which two phases may coexist at equilibrium during phase transitions.

As can be seen in Figure 3.1, along with an equilibrium phase line, a slight change in temperature or pressure may cause the phase to abruptly change from one physical state to the other. For example, the phase boundary line M-TP (sublimation line) marks the conditions under which solid (ice) and liquid water can exist at equilibrium. Along this line, a phase transition from solid (ice) to water liquid occurs. If the initial state of the solid water is below the triple point TP (state b), by supplying heat at a constant pressure lower than that of the triple point's pressure, the ice melts (sublimes) and passes directly to the vapor state (process *b-b′*) without passing through the liquid state. The triple point TP, occurring at 0.0098°C and 0.64 kPa, is the point where all lines of equilibrium intersect marking the single temperature and pressure at which the three water phases (solid, liquid, and gaseous) coexist having identical Gibbs free energies. These phases may change into each other given a slight change in temperature and pressure.

Also, water liquid and water vapor coexist in equilibrium only under the conditions along the temperature–pressure line TP-CP (vaporization line). These two states have the same pressure and the same temperature but are definitely different. In such a saturation state, therefore, pressure and temperature are not independent properties.

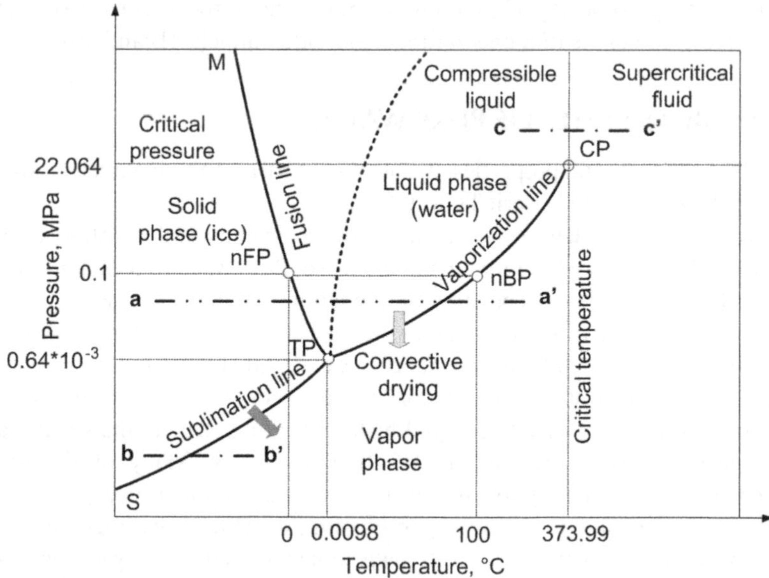

FIGURE 3.1 Typical water pressure–temperature phase diagram. TP, triple point; CP, critical point; nFP, normal fusion point; nBP, normal boiling point.

Two independent properties, such as pressure and specific volume, or pressure and quality, are required to specify the saturation state along the vaporization line (Sauer and Howell 1983).

The vaporization line TP-CP between water liquid and water vapor ends at the critical point CP characterized by the following parameters: critical temperature $T_{cr} = 647.096\,K$ (373.946°C), critical pressure $p_{cr} = 22.064\,MPa$, and critical density $\rho_{cr} = 356\,kg/m^3$. At such extremely high temperatures and pressures, the water liquid and water vapor become indistinguishable (i.e., there is no distinct change from the liquid phase to the vapor phase above this point), forming a supercritical fluid where the liquid and gaseous phases can blend continuously into each other. At temperatures above the critical temperature, the gas (supercritical vapor) cannot be liquefied. At a pressure above the critical pressure, such as during the process *c-c′*, there is no sharp distinction between the liquid and vapor phases.

Solid (ice) and water liquid phases coexist in equilibrium only under the conditions along the temperature–pressure line M-TP (fusion line) having a positive slope. If heat is supplied at a constant pressure to water at solid state a (ice), its temperature rises, the ice melts and passes from the solid (ice) to the liquid phase at one temperature, and the resulting water is warmed and, finally, evaporated at state *a′* (i.e., superheated vapor) at a higher temperature (Figure 3.1). As heat is transferred to the water, the temperature increases appreciably, the specific volume increases slightly, and the pressure remains constant. When the temperature reaches 100°C at normal atmospheric pressure, additional heat transfer results in a change of phase. Therefore, some of the liquid becomes vapor, and during this process, both the temperature and the pressure remain constant, but the specific volume increases considerably. When the last drop of liquid has vaporized, further transfer of heat will result in an increase in both temperature and specific volume of the vapor (Sauer and Howell 1983). In Figure 3.1, nFP and nBP represent the normal freezing and boiling points, respectively, at the normal atmospheric pressure (0.1 MPa).

The term saturation temperature designates the temperature at which fusion or vaporization takes place at a given pressure. Conversely, the pressure is called the saturation pressure for the given temperature. Thus, for water at 100°C, the saturation pressure is 0.1 MPa, and for water at 0.1 MPa, the saturation temperature is 100°C.

If the water exists as liquid at the saturation temperature and pressure, it is called saturated liquid. If the temperature of the liquid is lower than the saturation temperature for the existing pressure, water is called either a subcooled liquid (indicating that the temperature is below the saturation temperature for the given pressure) or a compressed liquid (indicating that the pressure is greater than the saturation pressure).

When a substance exists as vapor at the saturation temperature, it is called saturated vapor. When the vapor is at a temperature greater than the saturation temperature, it is said to be a superheated vapor. The pressure and temperature of superheated vapor are independent properties since the temperature may increase while the pressure remains constant. The common gases are highly superheated vapor. By definition, a saturated vapor exists in the presence of its liquid. In other words, a saturated vapor is in the two-phase condition where the vapor is constantly condensing into liquid and is replaced by liquid that is evaporating into vapor. As a result, there is equilibrium in the percentage of vapor at a given pressure. In the liquid–vapor mixture, the vapor may be superheated and/or the liquid may be subcooled.

When the water exists as part liquid and part vapor at the saturation temperature, its quality (symbol x) is defined as the ratio of the mass of vapor to the total mass. Quality has meaning only when the water is at saturation pressure and temperature (Sauer and Howell 1983).

The change of water from solid to liquid, as in the case of melting ice, or from liquid to vapor, is called phase change. In both cases, heat (called enthalpy or latent heat of fusion and vaporization, respectively) must be added to ice or water to change their respective phases. Heat required for these phase changes is exactly equal to the latent heat that must be removed when the vapor condenses back.

On the other hand, heat that changes the temperature only of a substance is called sensible heat.

3.3 TYPES OF MOISTURE

The attraction of wood for water arises from the presence of free, negatively charged hydroxyl groups in the celluloses, hemicelluloses, and lignin molecules in the cell wall that attract and hold water by hydrogen bonding (Cech and Pfaff 2000; Perré and Keey 2015).

Within the porous and hygroscopic structure of wood, moisture can coexist in three phases: (i) free liquid water in cell cavities (lumens), including water vapor that remains in cell cavities after the liquid water is removed; when the wood moisture content is above the fiber saturation point, the cell walls are saturated and excess water is found in the lumen; (ii) hygroscopic (or bound) water contained in void of cell walls (or associated with the cell wall structure); and (iii) vapor in cell cavities (Berit Time 1988; Skaar 1988; Onn 2007).

The relative and absolute amounts of wood substance, free and bound (imbibed) water, as well as water vapor and void spaces present in wood have a direct impact on the physical and mechanical properties of wood and, also, upon its drying characteristics.

Free (liquid) water (also called bulk, unattached, chemically unbound, or capillary moisture), held only by capillary forces, includes both bound and unbound water as well as dissolved chemicals substances that can alter the drying characteristics of wood. If the cells of wood are wetter than fiber saturation point (at about 28%–30% moisture content), there will be free water in the cell cavities. At a given temperature, the free water is removable.

The properties of free water (such as density, viscosity, and saturated vapor pressure) are very close to those of pure liquid water (Perré and Keey 2015). However, the free water is not in the same thermodynamic state as pure liquid water because it is subjected to capillary force and energy is required to overcome these forces. Free water just adds mass to the wood, and, with very few exceptions, as in collapse, its removal does not cause shrinkage or a change in strength properties of the wood.

Unbound moisture is the moisture in excess of the equilibrium moisture content corresponding to saturated moisture. Water may become bound (hygroscopic, linked, or imbibed) in a solid wood by retention as a liquid solution in voids of cell walls or by chemical adsorption on the surface of the solid material. Bound water exerts a vapor pressure less than that of pure liquid and is responsible for the shrinkage of the wood when it is removed. Water vapor remains in the air within cell lumens (cavities) after the liquid water is removed. Normally, this vapor constitutes only a small amount of total weight of wood, and thus, it is normally negligible at normal temperature and humidity.

3.4 DENSITY OF WOOD

The wood's density is a variable parameter influenced by factors such as (i) the amount of lignin in its cells; (ii) how densely the cells grow together—tightly packed cells have fewer air spaces and a higher cell count, and result in a denser wood; (iii) cells with more water capacity or more vessels for water transfer between them have a looser structure resulting in a softer wood; (iv) number of air spaces within the cells; (v) amount of moisture content; and (vi) wood chemical composition.

3.4.1 Mass Density

Wood mass (or weight) density is defined as the mass per unit volume of green, oven-dry or partially dried wood (Eekelman 1968):

$$\rho_{wood} = \frac{m_{wood}}{V_{wood}} \tag{3.1}$$

where

m_{wood} is the green (kg_{green}), oven-dry ($kg_{oven\text{-}dry}$), or partially dried ($kg_{MC\%}$) mass of wood

V_{wood} is the volume of green $\left(m^3_{green}\right)$, oven-dry $\left(m^3_{oven\text{-}dry}\right)$, or partially dried $\left(m^3_{MC\ \%}\right)$wood.

3.4.2 Basic Density

The basic density of wood is defined as the oven-dry mass of a wood sample divided by its green, swollen (saturated) volume (V_{green}) (i.e., the green volume of the wood sample when it is in equilibrium with the surrounding medium, before any shrinkage due to drying):

$$\rho_{basic} = \frac{m_{oven\text{-}dry}}{V_{green}} = \frac{Oven\text{-}dry\ mass}{Green\ (saturated)\, volume} \tag{3.2}$$

where

ρ_{basic} is the basic density of wood $\left(\frac{kg_{oven\text{-}dry}}{m^3_{green}}\right)$

In practice, V_{green} $\left(m^3_{green}\right)$ can be determined by tempering a lumber sample in water and then measuring its volume $\left(m^3_{green}\right)$. Thus, the oven-dry mass (kg) is:

TABLE 3.1
Average Basic Density for Some Important Species of Wood Utilized in Eastern Canada

Species	Basic Density $\left(\frac{kg_{oven\text{-}dry}}{m^3_{green}}\right)$
Softwoods	
Balsam fir	0.34
White spruce	0.35
Red pine	0.39
Hardwoods	
White birch	0.51
Sugar maple (hard)	0.60
White oak	0.65

$$m_{oven\text{-}dry} = \rho_{basic} * V_{green} \tag{3.3}$$

Table 3.1 shows some average basic densities for important species utilized in eastern Canada (Cech and Pfaff 2000).

3.4.3 Specific Density

When working with different wood species, specific (or relative) density is an important parameter to understand because it is a measure of the amount of structural material a wood (or tree) species allocates to support and strength. The differences in specific density between species are due to differences in the size and the thickness of the cell walls.

Specific density of a wood is defined as the ratio of the mass density of the green, oven-dry, or partially dried wood specimen to the mass density of equal volume of pure water (Sears and Zemansky 1953; Eekelman 1968):

$$\rho_{specific} = \frac{\textit{Mass density of oven-dry wood specimen}}{\textit{Mass density of equal volume of water}} = \frac{\rho_{oven\text{-}dry}}{\rho_{water}} \tag{3.4}$$

where

$\rho_{specific}$ is the specific density of a wood (–)

$\rho_{oven\text{-}dry}$ is the mass density of the wood specimen $\left(\frac{kg_{oven\text{-}dry}}{m^3_{oven\text{-}dry}} \text{ or } \frac{kg_{oven\text{-}dry}}{m^3_{green}} \text{ or } \frac{kg_{oven\text{-}dry}}{m^3_{MC\ \%}}\right)$

ρ_{water} is the density of pure water (usually, 1,000 kg/m^3 taken at 4°C) $\left(\frac{kg_{water}}{m^3_{water}}\right)$

It can be seen that the specific density of a wood specimen is always based on its mass when oven-dry and the volume of the wood when oven-dry, green or partially dried at any arbitrary value of moisture content (*MC*). In other words, when calculating the wood's specific density, the state of wood specimen (oven-dry, green or partially dried) has to be specified. The specific density of a piece (specimen) of wood thus depends upon the circumstances under which it was determined. It is highest when based on oven-dry volume measurements and least when based on green volume measurements (Eekelman 1968). The green volume of wood specimen can be determined at the moisture content desired, for example, by submerging a piece of wood and measuring the displaced volume of fluid. However, it is convenient to base specific density calculations on the oven-dry mass of the wood since it is a reproducible, constant condition that can be determined at any time. The oven-dry mass is always determined after oven-drying at 103°C until the weight of specimen stabilizes (24–72 hours) (see Section 3.6.1.1).

The average specific density, based on oven-dry mass and green volume of commercial woods, varies, for example, from 0.31 for western red cedar and black cottonwood to 0.64 for the hickories, and can be as high as 1.1 for some tropical species.

Exercise E3.1

The oven-dry mass of a wood specimen is 200 g. Its green volume, volume at 15% moisture content, and oven-dry volume are 356, 322, and 294 cm³, respectively. Calculate the specific density values based on the volumes measured when the wood sample was green, partially dried at 15% moisture content, and oven-dry, respectively.

Solution

$$\rho_{green}^{specific} = \frac{200\ g_{oven\text{-}dry}}{356\ cm_{green}^3} = 0.561$$

$$\rho_{MC=15\%}^{specific} = \frac{200\ g_{oven\text{-}dry}}{322\ cm_{MC=15\%}^3} = 0.621$$

$$\rho_{oven\text{-}dry}^{specific} = \frac{200\ g_{oven\text{-}dry}}{294\ cm_{oven\text{-}dry}^3} = 0.68$$

3.5 SPECIFIC HEAT

The specific heat of wood can be determined as a linear function of temperature (Dunlap 1912):

$$c_p = 0.266 + 0.00116 * T \tag{3.5}$$

where

T is the temperature of a wood specimen over the interval from 0°C to $T_1 = 100$°C.

For a wood sample undergoing a temperature change of $T_1 \neq 0$ to $T_2 \geq 0$, the mean specific heat (kcal/kg · K) (1 kcal = 4.1868 kJ) in this range is given by (Beall 1968):

TABLE 3.2
Average Specific Heat for Timber Specimens

Substance	Specific Heat (J/kg·K)
Birch	1,900
Maple	1,600
Oak	2,400
Red pine	1,500

$$\bar{c}_p = 0.266 + 0.00058(T_1 + T_2) \tag{3.6}$$

Table 3.2 shows the average specific heats for some current timber specimens.

Exercise E3.2

A piece of softwood timber sample undergoes a temperature change of T_1 = 20°C to T_2 = 100°C. Calculate its average specific heat.

Solution

For a timber sample undergoing a temperature change from T_1 = 20°C to T_2 = 100°C, the mean specific heat in this range is given as follows (Beall 1968) (see equation 3.6):

$$\bar{c}_p = 0.266 + 0.00058(T_1 + T_2) = 0.266 + 0.00058(20 + 100)$$

$$= 0.3356 \text{ kcal/kg}\cdot\text{K} * 4.1868 \text{ kJ/kcal} = 1.405 \text{ kJ/kg}\cdot\text{K}$$

Exercise E3.3

The mass of a piece of softwood is m = 4 kg. Calculate the mean specific heat of this piece of timber if it absorbs 540 kJ of heat when its temperature changes from T_1 = 20°C to T_2 = 100°C.

Solution

The total heat transferred to the piece of wood can be expressed as follows (Eekelman 1968):

$$Q = m * \bar{c}_p * (T_2 - T_1)$$

where
$\bar{c}_p$ is the mean specific heat (kJ/kg·K)

Thus,

$$\bar{c}_p = \frac{Q}{m * (T_1 - T_0)} = \frac{540}{4 * (100 - 20)} = 1.6875 \text{ kJ/kg}\cdot\text{K}$$

3.6 MOISTURE CONTENT

The wood contains varying amounts of moisture (i.e., water in the form of liquid or vapor) depending on different heights in the tree and growing conditions, as well as on the species, density, degree of seasoning, and many other factors.

The total amount of water existing at any time in a given living tree or piece of wood is a quality factor that is important in wood drying because its changes result in wood dimension variation. Therefore, almost all wood drying schedules are based on the rate of moisture content loss. In some low-density wood species, the moisture content may exceed 100% (dry basis). The choice of end average moisture content of lumber boards in the drying process determines the required parameters for optimum drying.

Because the green (wet) wood is usually swollen compared with its dried condition (i.e., free of moisture), and since its volume changes during the drying process because the mass of water liquid and vapor appreciably varies, it is not convenient to express the moisture content in terms of volume, but as the moisture content of wood samples in terms of the oven-dry mass that is a constant value which can be determined at any time.

The dry-basis moisture content of green wood ($MC_{d.b.}$) (expressed in decimals or percentage) is defined as the ratio of the difference between the initial green mass of a piece of wood (that appreciably varies, m_{green}) (expressed in kg_{green}) and the oven-dry mass of the same wood sample, and the unit mass of the same piece of wood completely dry ($m_{oven\text{-}dry}$) that is a constant parameter (expressed in $kg_{oven\text{-}dry}$) (Eekelman 1968). For example, 0.59 $\frac{kg_{water}}{kg_{oven\text{-}dry}}$ expresses the same moisture content as 59% (oven-dry basis).

The following definition relations are equivalent:

$$MC_{d.b.}\ (\%) = \frac{m_{green} - m_{oven\text{-}dry}}{m_{oven\text{-}dry}} * 100 \tag{3.7}$$

or

$$MC_{d.b.}\ (\%) = \frac{(Initial\ green\ mass\ of\ wood\ sample) - (Oven\text{-}dry\ mass\ of\ wood\ sample)}{Oven\text{-}dry\ mass\ of\ wood\ sample} * 100 \tag{3.8}$$

or

$$MC_{d.b.}\ (\%) = \left(\frac{Initial\ green\ mass\ of\ wood\ sample}{Oven\text{-}dry\ mass\ of\ wood\ sample} - 1 \right) * 100 \tag{3.9}$$

In relations 3.8 and 3.9, *initial green mass of wood sample* means the mass of wood sample at the moment when the moisture content is determined, that is, the mass of green lumber after yard storage and/or air-drying, boards from sawmill, during or at the end of kiln drying, and so on.

The oven-dry (bone-dry, moisture-free), quasi-constant mass of wood sample is attained after drying it in an oven at 103°C ± 2°C for 24 hours in order to extract all water (Siau 1984) (see Section 3.6.1.1).

The average amount of moisture in green wood varies with the species, density, degree of seasoning, and many other factors. Table 3.3 shows the average moisture contents of some commercial green wood species in eastern Canada (Cech and Pfaff 2000).

If the mass of moisture is higher than the mass of the wooden material, the moisture content of the wood sample will be >100%. The range of moisture content for green (i.e., un-dried) hardwood lumber can range between 45% and 150% (dry basis). For example, the average moisture content of green White and Black spruce from eastern Canada varies from 38% (heartwood) to 144% (sapwood) (dry basis), whereas the moisture content of green balsam fir varies between 88% and 173% (dry basis) (Berit Time 1988). Also, the moisture content of heartwood in Scots pine and Norway spruce ranges from 30% to 50% (dry basis), whereas the moisture content of sapwood ranges from 100% to 200% (dry basis) (Vikberg and Elustondo 2015).

In hardwood timber, both the heartwood and the sapwood have similar moisture content levels, but the sapwood of softwoods normally has significantly higher moisture content than that of heartwoods. For example, the average moisture content of the green sapwood of *Pinus ponderosa* is about 1.5 $\frac{kg_{water}}{kg_{oven\text{-}dry}}$, which is nearly four times that for its heartwood (Hoadley 1980).

Due to their different green moisture contents, the heartwood:sapwood ratio is one of the largest factors affecting the average moisture content of the sawn timber, and the drying rate of sapwood is much higher than that of heartwood (Salin 2002).

There is little information available concerning the critical moisture content (MC_{cr}) that is related to the fiber saturation moisture content, but it is not

TABLE 3.3
Average Green Moisture Contents for Some Softwoods and Hardwoods Utilized in Eastern Canada

Species	Moisture Content (%)	
	Heartwood	Sapwood
–	–	–
Softwoods		
Balsam fir	88	173
White spruce	38	144
Red pine	32	134
Hardwoods		
White birch	74	72
Sugar maple (hard)	65	72
White oak	64	78

necessarily the same value. The critical moisture content does not appear to vary significantly with relative humidity, air velocity, or slab thickness, but it does vary with wood density or porosity (Meroney 1969). Sometimes, in practice, the wet-basis moisture content ($MC_{w.b.}$), or the moisture:solid ratio based on the total mass of wet material, defined as the mass of moisture per unit mass of wet grain $\left(\frac{kg_{water}}{kg_{wet\ sample}}\right)$, is used:

$$MC_{w.b.}\ (\%) = \frac{Mass\ of\ moisture}{Unit\ mass\ of\ wet\ material} * 100 \tag{3.10}$$

The dry-basis and wet-basis-defined moisture contents are related by the expression:

$$MC_{d.b.} = \frac{MC_{w.b.}}{1 - MC_{w.b.}} \tag{3.11}$$

According to the preceding equation, 20% of the wet-basis moisture content is equivalent to 25% of the dry-basis moisture content for the same wood sample.

Exercise E3.4

The green mass of a hardwood specimen is m_{green} = 136 g and its oven-dry mass $m_{oven\text{-}dry}$ = 100 g. Determine the sample's moisture content (dry basis).

Solution

According to relation 3.7 (Eekelman 1968):

$$MC_{d.b.}\ (\%) = \frac{m_{green} - m_{oven\text{-}dry}}{m_{oven\text{-}dry}} = \frac{136\ \text{g} - 100\ \text{g}}{100\ \text{g}} * 100\% = 36\%$$

Exercise E3.5

The mass of a green wood sample is 1,000 g and its moisture content (dry basis) 75%. Determine the oven-dry mass of this wood sample.

Solution

The oven-dry mass of wood sample can be calculated from the moisture content definition (equation 3.7) (Eekelman 1968):

$$MC = \frac{1000\ \text{g} - m_{oven\text{-}dry}}{m_{oven\text{-}dry}} = 0.75$$

Result:

$$m_{oven\text{-}dry} = 571.4\ \text{g}$$

3.6.1 Measurement of Moisture Content

The wood is a biological material with great variations in properties (such as moisture content and density) between individual logs and even within the same log. However, boards with similar moisture contents, density, dimensions, and grades show a similar behavior during drying of large volumes of wood (Skog et al. 2010)

To reduce the production of off-grade products and obtain the highest possible quality of the end products, the raw logs are sorted according to their specific properties and initial moisture content in order to process the wood drying efficiently and to obtain high-quality end products (Skog et al. 2010). The logs' sort mainly consists in separating them into groups with high and low moisture content because the moisture content is one of the most important for both the processing and the end products.

Also, well-adapted drying schedules can be constructed with respect to time, energy consumption, and quality of the final products. The drying behavior of wood can thus be characterized by measuring the moisture content loss as a function of time.

Higher production rate and less time from the felling of trees in the forest to the final products, driven by economic and qualitative factors, have led to an increased demand for accurate and automated measuring devices of moisture content (Vikberg and Elustondo 2015). If the initial average moisture content of lumber stack(s) in the kiln is known, over-drying can be reduced and the finishing time can be predicted more accurately (Larsson and Moren 2003). Avoiding errors in metering the moisture content of the lumber before, during, or after the kiln drying is very important for the wood industry (Frank Controls 1986).

During drying, the moisture content estimates are traditionally achieved by periodically removing few board samples (ideally, 76 cm long, cut 350 cm from the ends of some pieces of lumber without knots, bark, pitch, and decay) and placed near the top and bottom of the lumber stack(s), on both sides of the load, and yet at intervals of 3–4.9 m along the length of the kiln.

The weighing frequently depends on the rate of moisture loss. The more rapid the loss, the more frequently the samples must be weighed, but at least once a day. The samples must be returned to their pockets immediately after weighing, that is, within minutes.

The selection, preparation, placement, and weighing of lumber board samples provide information that enables a kiln operator to reduce drying defects, improve the control of final moisture content of charge, reduce drying time while maintaining lumber quality, develop time schedules, and locate kiln performance problems.

The selection of representative lumber board samples must include the extreme conditions of the expected kiln drying behavior. The chosen board samples must represent the entire lumber charge and its variability according, for example, to sample placement at various heights and distances from the ends of lumber stacks. It is however recommended to use the average moisture content of the wetter half of the controlling wood samples (generally, from the wettest heartwood) as a factor of determining when to change drying conditions in the kiln schedule.

Each selected lumber sample is separately weighted manually or automatically, for example, by means of a precision balance (preferably electronic scale) at intervals during drying, and the moisture content at those times is calculated, for example, with electrical resistance probes inserted into sample boards that fed signal into a computerized control system allowing automatically changing the drying schedule. Another measurement system uses miniature load cells that can continuously weigh individual sample boards, and the measurements are fed into a computerized control system.

Kiln samples that dry slowly indicate the zones of low temperature, high humidity, or low air circulation. Samples that dry rapidly indicate the zones of high temperature, low humidity, or high air circulation. If the drying rates vary greatly, action should be taken to locate and eliminate the cause of such variation. Differences in drying rates between the samples on the entering-air and exiting-air sides of the stacks will determine how often to reverse air circulation—the greater the difference in the drying rate between these sides, the more frequently the direction of air circulation should be reversed.

In order to estimate the wood moisture content, several measurements methods are currently used (Frank Controls 1986; Vikberg 2012): (i) standard (oven-dry); (ii) electrical (resistive or capacitive) and advanced methods; and (iii) lumber stack(s) weighting prior, during, and after drying.

3.6.1.1 Oven-Dry Method

The standard procedure for measuring the moisture content of wood is the oven-drying method. This relatively slow method, necessitating cutting of the wood, may provide values slightly greater than true moisture content with woods containing volatile extractives. Oven-drying of lumber board samples is performed in forced convection or microwave ovens until all water is completely removed, usually less than 1 hour in microwave ovens and between 24 and 48 hours in forced convection maintained at 101°C–105°C. Oven-dried board samples are weighed by the same procedures used for freshly cut moisture sections, immediately after removing them from the oven.

Within the standard oven-dry measurement method, the following operations must be achieved (Cech and Pfaff 2000; Reeb and Milota 2016):

1. Cut a sample (specimen) of 25 mm long along the grain at least 500 mm from the end of a representative board of the batch lumber, which should be free from knots and other irregularities, such as bark and pitch pockets.
2. Weigh immediately this wood sample before drying.
3. Dry the wood sample in an oven with fan to well circulate the hot air at 103°C ∓ 2°C for approximately 24 hours.
4. Reweigh the wood sample.
5. Repeat steps 3 and 4 until the current weight equals the previous weight, that is, dry until constant weight is attained; oven-dryness is reached when the piece of wood stops losing weight (i.e., two successive weight readings taken an hour or two apart during oven-drying are identical); the wood sample is now oven dried, sometimes being referred to as bone-dry.
6. Apply one of relations 3.7, 3.8, or 3.9 for determining the moisture content $\left(\% \text{ or } \frac{kg_{water}}{kg_{oven\text{-}dry}}\right)$ of the wood sample.

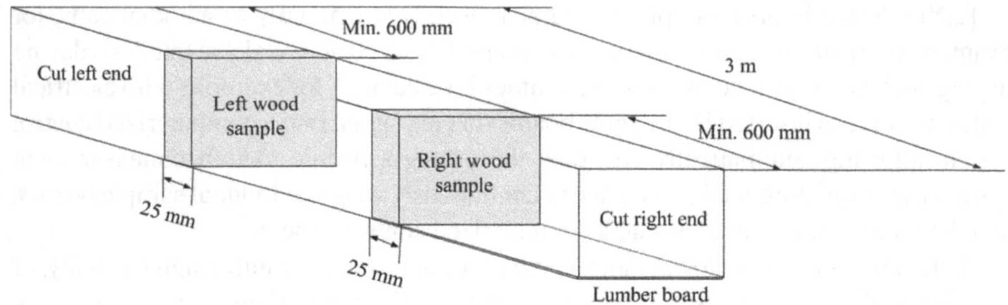

FIGURE 3.2 Schematic of wood samples.

The main steps required to determine the average moisture contents of a lumber stack may be further detailed as follows (Minea and Normand 2005):

1. Before each drying cycle, select at least six representative boards ($j = 1, 2,\ldots, 6$) of the lumber batch each of about 3 m long (see Figure 3.2).
2. Cut and discard from both ends of the six representative boards not less than 600 mm in length (Cech and Pfaff 2000).
3. From the freshly exposed ends, immediately cut the left and right wood samples to be tested, each of about 25 mm in the direction of the grain and the full width of the board; these specimens should be free from knots, rot, and other irregularities, such as bark and pitch pockets;
4. Weigh the left and right samples immediately and accurately (Cech and Pfaff 2000); note
   $m_{wet\text{-}left,j}^{initial}$ — the initial humid mass of the left sample j.
   $m_{wet\text{-}right,j}^{initial}$ — the initial humid mass of the right sample j.
5. Place the left and right samples j in an oven (with fan to well circulate the hot air) at 103°C ∓ 2°C during about 24 hours.
6. Reweigh the wood samples j after 24 hours.
7. Repeat steps 5 and 6 until the current weight of each sample j equals the previous weight, that is, dry until constant weight is attained. Oven-dryness is reached when the specimen of wood stops losing weight (i.e., two successive weight readings taken an hour or two apart during oven-drying are identical). The wood samples are now oven-dried, sometimes being referred to as bone-dry; note
   $m_{oven\text{-}dry\text{-}left,j}$ — the oven-dry mass of the left sample j.
   $m_{oven\text{-}dry\text{-}right,j}$ — the oven-dry mass of the right sample j.
8. Determine the initial moisture contents of left and right samples j:

$$MC_{left,j}^{initial} = \left(\frac{m_{wet\text{-}left,j}^{initial}}{m_{oven\text{-}dry\text{-}left,j}} - 1 \right) * 100\% \tag{3.12}$$

$$MC_{right,j}^{initial} = \left(\frac{m_{wet\text{-}right,j}^{initial}}{m_{oven\text{-}dry\text{-}right,j}} - 1 \right) * 100\% \tag{3.13}$$

The initial moisture content ($MC_{initial}$) of softwood, a value measured by using the standard oven-dry method, is generally higher than 30% (dry basis).

9. Determine the average initial moisture contents of left and right samples $j = 1, 2,\ldots, 6$:

$$MC^{initial}_{average,j} = \frac{MC^{initial}_{left,j} + MC^{initial}_{right,j}}{2} * 100\% \tag{3.14}$$

10. By definition, the initial average moisture content (dry basis) of a lumber board j is

$$MC^{initial}_{average,j} = \frac{m^{initial}_{wet,j} - m_{oven\text{-}dry,j}}{m_{oven\text{-}dry,j}} = \frac{m^{initial}_{wet,j}}{m_{oven\text{-}dry,j}} - 1 \tag{3.15}$$

where

$m^{initial}_{wet,j}$ is the initial wet (green) mass of the board j at the kiln inlet (kg)
$m_{oven\text{-}dry,j}$ is the oven-dry mass of the board (kg)

Thus,

$$m_{oven\text{-}dry,j} = \frac{m^{initial}_{wet,j}}{MC^{initial}_{average,j} + 1} \tag{3.16}$$

The oven-dry mass of the six representative lumber samples is calculated using equation 3.16.

If, for example, the initial average moisture content and the initial wet mass at the kiln inlet of a lumber board 1 are $MC^{initial}_{average,1} = 40.2\%$ and $m^{initial}_{wet,1} = 3.292$ kg, respectively, the oven-dry mass of the sample 1 will be

$$m_{oven\text{-}dry,1} = \frac{m^{initial}_{wet,1}}{MC^{initial}_{average,1} + 1} = \frac{3.292}{0.402 + 1} = 2.348 \text{ kg} \tag{3.17}$$

In the same manner, the oven-dry masses of samples $j = 2, 3,\ldots 6$ ($m_{oven\text{-}dry,2}$, $m_{oven\text{-}dry,3}, \ldots m_{oven\text{-}dry,6}$) are calculated.

11. After 8 hours of drying, the selected lumber boards are weighted again. The new wet masses corresponding to the sample weighting #2 will be $m^{weight\ \#2}_{wet,j}$.
12. Based on the wet masses $m^{weight\ \#2}_{wet,j}$, it is possible to calculate the average moisture content of the lumber boards after the second weighting:

$$MC^{weight\ \#2}_{average,j} = \left(\frac{m^{weight\ \#2}_{wet,j}}{m_{oven\text{-}dry,\ \#2}} - 1 \right) * 100 \tag{3.18}$$

13. The average moisture content after the second weighting will be

$$MC^{weight\ \#2}_{average} = \frac{MC^{weight\ \#2}_{average,1} + MC^{weight\ \#2}_{average,2} + \cdots + MC^{weight\ \#2}_{average,6}}{6} \tag{3.19}$$

14. After another 8 hour period of drying, the board samples are weighted again. The new wet masses are $m_{wet,j}^{weight\ \#3}$, and the average moisture content of the lumber board after the third weighting will be

$$MC_{average,j}^{weight\ \#3} = \left(\frac{m_{wet,j}^{weight\ \#3}}{m_{oven\text{-}dry,\#3}} - 1 \right) * 100 \tag{3.20}$$

15. The operation described at step 14 is repeated as many times ($n = 1,2,\ldots$) as needed at 8 hour intervals. The successive weighted wet masses will be $m_{wet,j}^{weight\ \#i}$, and the average moisture content of the lumber boards after each weighting operation #*i* will be

$$MC_{average,j}^{weight\ n=1,2,3\ldots} = \left(\frac{m_{wet,j}^{weight\ n=1,2,3\ldots}}{m_{oven\text{-}dry,\#n}} - 1 \right) * 100 \tag{3.21}$$

16. When the average moisture content of the *n* boards' weighting operations attains the desired value (e.g., 15%), the drying cycle can be finished.

Exercise E3.6

An industrial softwood kiln contains 96 lumber packages of 2.265 m^3 each grouped in four distinct stacks. The total wet volume of all lumber stacks is thus 217.44 m^3. If the measured initial and final wet (green) masses of each softwood stack are $m_{wet}^{initial} = 6{,}000$ kg and $m_{wet}^{final} = 5{,}490$ kg, respectively, determine

i. The total available moisture in the initial (wet, green) lumber charge
ii. The initial moisture content (dry basis)
iii. The final moisture content (dry basis)

Solution

i. The final moisture content of the softwood lumber stack is defined as follows (Eekelman 1968):

$$MC_{final} = \frac{m_{wet}^{final} - m_{oven\text{-}dry}^{final}}{m_{oven\text{-}dry}^{final}} = \frac{5{,}490 - m_{oven\text{-}dry}^{final}}{m_{oven\text{-}dry}^{final}} = 0.175$$

Thus, the final oven-dry mass of each lumber stack is:

$$m_{oven\text{-}dry}^{final} = 4{,}672 \text{ kg}$$

Consequently, the maximum available moisture (water) in each softwood stack will be

$$m_{wet}^{initial} - m_{oven\text{-}dry}^{final} = 6{,}800 - 4{,}672 = 2{,}128\ kg_{water}$$

or

$$\frac{2{,}128\ kg_{water}}{4\ stacks} = \frac{2{,}128\ kg_{water}}{4 * 2.265\ m^3} = 234.9\ \frac{kg_{water}}{m^3}$$

Within the entire charge softwood lumber, the maximum moisture (water) available will be

$$m_{water,\ total} = 217.44\ \text{m}^3 * 234.9\ \frac{kg_{water}}{m^3} = 51{,}072\ kg_{water}$$

ii. The initial moisture content (dry basis) can be calculated as follows:

$$MC_{initial} = \frac{m_{wet}^{initial} - m_{oven\text{-}dry}^{final}}{m_{oven\text{-}dry}^{final}} * 100 = \frac{6{,}800 - 4{,}672}{4{,}672} * 100 = 45.5\%$$

where

$m_{wet}^{initial}$ is the initial mass of wet wood before drying (kg)

$m_{oven\text{-}wet}^{final}$ is the final mass of wet wood after drying (kg)

iii. The final moisture content (dry basis) can be calculated as follows:

$$MC_{final} = \frac{m_{wet}^{final} - m_{oven\text{-}dry}^{final}}{m_{oven\text{-}dry}^{final}} * 100 = \frac{5{,}490 - 4{,}672}{4{,}672} * 100 = 17.5\%$$

3.6.1.2 Electrical Methods

The resistive-type meters (portable or hand), extensively used, are adaptable to rapid sorting and/or establishing the moisture content of quality lumber according to its moisture content. These devices consist in two pin-type insulated electrodes that are driven into the wood specimen being tested, which uses the relationship between moisture content and direct current electrical resistance.

This measurement method is rapid and does not require cutting the wood. However, it is reasonably accurate only for values in the range from 7% to 25% moisture content, that is, below the fiber saturation point. Below 7% moisture content (dry basis), the electrical resistance is so high that it becomes difficult to measure. It is also uncertain being affected by parameters such as species, temperature, and moisture gradients. The capacitive-type meters, operating on the relationship between the temperature and the dielectric properties of wood, including water, use surface contact-type electrodes (Cech and Pfaff 2000).

3.6.1.3 Advanced Measurement Methods

More advanced methods, based on computer software, imply large investments and increased demand of maintenance. To get reliable results, it is important to provide the software with correct input data in terms of wood species, basic density, moisture content, and kiln characteristics (volume, etc.) (Vikberg and Elustondo 2015).

Some sawmills use three-dimensional (3D) and X-ray scanners to sort logs (Skog et al. 2010). Because the one-level energy X-ray scanners lack the ability to distinguish between wood and water substance, dual-energy X-ray scanners have been recently developed (Hultnas and Fernandez-Cano 2012).

By combining 3D and X-ray scanning in a log-sorting station, it was possible to measure the green sapwood density and to estimate the dry sapwood density and, accordingly, the moisture content of *Scots pine* sawlogs (Skog et al. 2010). To determine the average moisture content of heartwood and sapwood of different Swedish Scots pine lumber boards, Vikberg et al. (2012) used 3D dimensional scanning together with discrete X-rays.

Microwaves can also be used to predict different wood parameters (Schajer and Orhan 2006).

For other hygroscopic materials (such as foods), advanced methods of moisture profile measurement have been developed (Srikiatden and Roberts 2007):

i. Destructive methods consist of immediately freezing the sample in liquid nitrogen at various times during dehydration, allowing the sample to rise to −4°C for cutting the sample along its characteristic dimension and then determining the moisture content of each section by the vacuum oven method.
ii. The radiography (nondestructive) method is based on a radiation source (X- or gamma rays, or neutrons) such as an accelerator containing a detector which records the intensity of the beam after passing through the material's sample.
iii. Electric reflectometry methods involve a sensing probe which transmits electromagnetic waves and measures the time for the electromagnetic waves to travel through the length of the sample. The velocity of the propagating wave is a function of the material's dielectric properties, which in turn is a function of the material's moisture content.
iv. Fiber optic method is a method in which a light is sent through a fiber optic into the material's sample, and then the intensity of the reflected light is measured; this technique shows reasonable results for the moisture distribution in bread samples (Thorvaldsson and Skjoldebrand 1996).
v. Nuclear magnetic resonance is the most common nondestructive method to determine the material structure, component saturation, and material properties including self-diffusion coefficient. This method is based on the measurement of resonant, radio frequency absorption by nonzero nuclear spins in the presence of an applied static magnetic field. During nuclear magnetic resonance, a radiofrequency excitation causes magnetic moments to align briefly and then returns to equilibrium, which can be recorded by a radiofrequency receiver.
vi. Magnetic resonance imaging is a spectroscopic method used to generate internal images based on the magnetic properties of nuclei. Pulsed linear magnetic field gradients are applied to produce a frequency variation across the sample, which can be converted into spatial coordinates. This method is capable of providing the mass moisture transport and accurate moisture content profiles in modeling of food, as well as fruits and vegetables, such as apple, potato, and sweet corn.

3.7 FIBER SATURATION POINT

The wood cells hold most of the free (unbound) water, a component that is relatively easy to remove. As green (wet) timber begins drying when exposed to the ambient or drying air, free water leaves fairly easily the lumens before bound water (of which content remains constant).

When in-use dry lumber is exposed to ambient moisture, water molecules bond within the cell walls. To accommodate this water, the cell structure swells, and, eventually, the cell walls become saturated with water, yet the lumens (cavities) remain dry. The moisture content at which the cell lumen (cavity) is completely devoid of liquid water (i.e., free water is completely gone), while the cell walls are saturated with bound water (Bian 2001), is called the fiber saturation point. In other words, the moisture content level which corresponds to the lumens containing no free water (only water vapor), while no bound water has been desorbed from the cell wall material, is known as the fiber saturation point, a value that decreases with the increase of wood's temperature. At this point, moisture migration throughout the wood is accelerated (Berit Time 1988).

If capillary condensation effects in pores less than 0.1 µm in equivalent cylindrical diameter are neglected, the fiber saturation point can also be defined as the equilibrium moisture content of a wood sample in an environment of 99% relative humidity (Keey et al. 2000).

The definition of fiber saturation point applies to cells or localized regions, but not to the moisture content of an entire board (Cech and Pfaff 2000). This is because during drying, the center of a lumber board may be at a relatively high moisture content (e.g., at 40%, dry basis), and the surface fibers may be much dryer (e.g., at 12%).

It is thus virtually impossible to determine the exact fiber saturation point by direct methods because of the difficulty in producing, throughout a piece of wood, a uniform moisture content at which all cell walls are saturated, but there is no free water in the lumens. For practical purposes, however, for most commercial wood species, it may be stated that the fiber saturation point for wood at room temperature is approximately 25%–35% based upon the species oven-dry mass (or dry basis) (note: in eastern Canada, it is of 25%) (Skaar 1988; Cech and Pfaff 2000; Keey et al. 2000).

At and above the fiber saturation point, there is practically no change in wood volume (shrinkage or swelling) with changes in moisture content.

Below the fiber saturation point (i.e., when the bound water begins to leave the cell wall material upon a process called desorption), many properties of wood show considerable change as it is dried. Among them, the following can be mentioned: (i) shrinkage begins since the moisture removed comes from the cell walls; in other words, only after water begins to leave the cell walls, the wood starts to shrink and its strength begins to increase; (ii) mechanical properties, such as strength, hardness, and toughness, increase (Desch and Dinwoodie 1996); (iii) there is a rapid increase in wood electric resistance with the loss of bound water; (iv) there is a decrease in the drying rate as the moisture movement of bound water is slow compared to unbound water; and (v) more energy is required to vaporize (or evaporate) the bound water because the attraction between the wood and water must be overcome.

For all species of woods, the moisture content at fiber saturation point $\left(FSP, \frac{kg_{water}}{kg_{oven\text{-}dry}}\right)$ can be expressed as a temperature function by the following empirical equation (Siau 1984):

$$MC_{FSP} = 0.30 - 0.001 * (T - 20) \tag{3.22}$$

where
T is the temperature (°C)

At temperatures below 0°C, the fiber saturation point falls, for example, from 32% at −2°C to 21% (dry basis) at −30°C (Shubin 1990). At these temperatures, the saturation vapor pressure over ice is very low, and the moisture from the frozen cell wall can sublime.

For softwoods such as Sitka spruce, the fiber saturation point varies from about 31% (dry basis) at 25°C–23% (dry basis) at 100°C (Stamm 1964), which means that the moisture becomes less strongly bound with rising temperature (Keey et al. 2000). At 120°C, the value of the fiber saturation point falls to 5%, and at 150°C, it is below 2% (dry basis). Experimental measurement of the fiber saturation point above 130°C is difficult because of the onset of thermal degradation with some species.

For species with a high extractive content, such as redwood and mahogany, the fibers remain saturated at 22%–24% moisture content, whereas speçies such as birch may have a moisture content up to 35% at fiber saturation point (Hoadley 1980; Perré and Keey 2015).

3.8 EQUILIBRIUM MOISTURE CONTENT

Living trees with fully saturated cell walls contain large quantities of water and moisture contents higher than 30% (dry basis). When a green wood from a freshly felled tree is exposed to the atmospheric condition, its moisture content becomes thermodynamically unstable and the cell structures dry until it reaches equilibrium with the ambient atmosphere.

Generally, the moisture contained in wet hygroscopic solids (such as woods and foods) exerts a vapor pressure depending upon the nature of moisture, the nature of solid, and the temperature. A wet solid exposed to a continuous supply of fresh air continues to lose moisture until the vapor pressure of the moisture in the solid is equal to the partial pressure of the vapor in the air. The solid and the ambient air of given relative humidity and temperature are then said to be in equilibrium, and the moisture content of the solid is called the equilibrium moisture content (EMC) under the prevailing conditions. It is approximately proportional to the ambient relative humidity (a function of temperature) and, to a lesser degree, the temperature of the surrounding air where lumber is stored, manufactured, shipped, or used for a sufficiently long period to reach equilibrium (Siau 1984; Mackay and Oliveira 1989; Cech and Pfaff 2000). In other words, as green wood dries, most of the water is removed, and the moisture remaining tends to come to equilibrium with the relative humidity of the surrounding air.

Under constant external conditions, wood will continue to dry until equilibrium moisture content is reached below the fiber saturation point.

Lumber drying is accomplished by vaporizing the moisture from the surface of the wood which loses moisture rapidly forming a dry outer shell surrounding the wet inner core (Pang 1994; Onn 2007). The surface fibers of most wood species reach moisture content equilibrium with the surrounding air soon after drying begins. This is the beginning of the difference (gradient) in moisture content between the inner and outer ports of lumber boards. Consequently, the surface of the wood must be drier than the interior if moisture is to be removed. Moisture will move from an area of higher moisture content to an area of lower moisture content within the wood. When the surface moisture vaporizes (or evaporates), moisture moves from the interior toward these locations. This process continues until the wood reaches its equilibrium moisture content. At this point, the moisture content is equal throughout the piece of wood.

Thicker lumber exposed to the same drying conditions will take longer to reach its equilibrium moisture content than thinner lumber.

The equilibrium moisture content gradually increases with the relative humidity until the cell wall becomes saturated when the relative humidity reaches 100%. Further exposure to the ambient air for indefinitely long periods will not bring about any additional loss of moisture. Thus, at equilibrium moisture content the wood neither gains nor loses moisture at a given temperature and relative humidity (Cech and Pfaff 2000). However, by using convective air drying processes, it is possible to dry the wood to moisture content lower than the equilibrium moisture content associated with the relative humidity and temperature of the drying air.

The amount of equilibrium moisture content of a natural green wood from a freshly felled tree or a given board (piece) of wood varies from one location to another and upon the season, and depends significantly on surrounding (ambient or drying) air relative humidity (that measures the capacity of air to absorb moisture and can be expressed in terms of a wet-bulb temperature or wet-bulb depression), and, to a lesser degree, on dry-bulb temperature.

Wood seeks its equilibrium moisture content in relation to the relative humidity and temperature of its surroundings. Thus, as wood is dried below its fiber saturation point, the amount of moisture leaving the wood will be determined by the relative humidity of the atmosphere surrounding the wood. Table 3.4 shows the equilibrium moisture content over a range of air relative humidity (Cech and Pfaff 2000).

It can be seen that the equilibrium moisture content is approximately proportional to the ambient relative humidity (Skaar 1988). At a given relative humidity and temperature, the equilibrium moisture content of many wood species is nearly the same regardless of their mass or relative density.

TABLE 3.4
Wood Equilibrium Moisture Contents at Different Relative Humidities of the Surrounding Air

Air relative humidity (%)	10	20	30	40	50	60	70	80	90
Equilibrium moisture content (%)	2–3	4	6	7–8	9–10	10–11	12–13	15–16	18–21

For example, at 21°C and a relative humidity of 55%, the approximate equilibrium moisture content is 10.5% (dry basis), whereas at 85°C and a relative humidity of 55%, the equilibrium moisture content is about 6.4% (dry basis) (Cech and Pfaff 2000). At 42.4°C and 95% relative humidity, the equilibrium moisture content ranges from 22% (dry basis) for pine to 33% (dry basis) for oak (Shubin 1990; Perré and Keey 2015).

The equilibrium moisture content is more sensitive to changes in relative humidity than it is to changes in temperature. For example, keeping the relative humidity constant at 40% and varying the temperature from −1°C to 93°C cause a small change in the equilibrium moisture content from 7.7% to 4.3% (dry basis), whereas changing the relative humidity from 20% to 80% and keeping the temperature constant at 49°C cause a change in the equilibrium moisture content from 4% to 14.1% (dry basis) (Keey et al. 2000).

Equilibrium moisture content also depends on wood species (generally, very slightly), chemical composition, wood physical structure and density, drying history of wood, proportion of cell wall constitutes and extractive content, exposure history or hysteresis, mechanical stress, presence of decay, the direction of sorption in which the moisture change takes place (i.e., adsorption or desorption), and the presence of decay (Skaar 1988). However, for practical purposes, these factors are usually ignored (Cech and Pfaff 2000).

For the majority of Australian states, the equilibrium moisture content is recommended to be 10%–12% (dry basis), although extreme cases are up to 15%–18% for some places in Queensland, Northern Territory, Western Australia, and Tasmania. However, the equilibrium moisture content can be as low as 6%–7% in dry centrally heated houses and offices, or in permanently air-conditioned buildings (AS/NZS 4787. 2001).

The hygroscopic behavior of common commercial woods is sufficiently uniform. At 65% relative humidity and 26°C, the equilibrium moisture content for 15 coniferous species (including both heart and sapwood) has been found to vary from 10.7% to 13.5% (dry basis), with a mean of 12.2% and a standard deviation of 0.6% (Harris 1961).

The following equation can be used to predict the equilibrium moisture content within 1% of moisture content (Simpson 1971; Simpson and Rosen 1981):

$$EMC = \left[\frac{K_1 * K_2 * RH}{1 + K_1 * K_2 * RH} + \frac{K_2 * RH}{1 - K_2 * RH}\right]\frac{18}{K_3} \tag{3.23}$$

where

K_1, K_2, and K_3 are experimental constants dependent on the air dry-bulb temperature

RH is the relative humidity of the airflow (decimals)

To estimate the equilibrium moisture content of woods as a function of relative humidity, the following equation can also be used (Hailwood and Horrobin 1946):

$$EMC = \frac{RH}{A + B * RH - C * RH^2} \tag{3.24}$$

where

RH is the ambient air relative humidity (decimals)

A, B, and C are the curve fitting parameters

3.9 MOISTURE SORPTION

Sorption is a wood surface phenomenon contributing at attaining the equilibrium moisture content by either adsorption or desorption. Adsorption occurs when water vapor (moisture) molecules from the ambient (surrounding) air condense to the wood free surfaces by physical and/or chemical processes. Desorption takes place on the wood free surfaces which are already covered by adsorbed water molecules. In other words, sorption is a concept based on a dynamic equilibrium resulting from an equal rate of adsorption (condensation) of the water vapor molecules and the rate of vaporization (desorption) of the water molecules.

The sorption process of wood consists of two parts: (i) an instantaneous equilibrium between the moisture content in the wood cavities and the surrounding cell walls, and (ii) a slow sorption due to the creation of sorption sites by molecular bond breaking or rearrangement and absorption into the cell wall.

Most of the water adsorbed by wood is held by different hydroxyl groups in amorphous areas (that have free sorption sites available) of the cellulose and by the hemicellulose in the micro-fibrils. The amount of water adsorbed is thus dependent upon the chemical composition of the wood fiber, and, above a certain vapor pressure, this process is governed by capillary condensation.

When the surrounding environmental air conditions (e.g., temperature and relative humidity) change, the sorption physical process is reversible.

3.9.1 Sorption Isotherms

The sorption (adsorption and desorption) isotherms are the graphs of equilibrium moisture contents measured experimentally under isothermal conditions. They express the minimum value of wood moisture content that can be reached by wood in relation to the relative humidity of the surrounding air.

Adsorption isotherms are obtained by exposing the wood specimens to the air of increasing relative humidity, whereas desorption isotherms are obtained by exposing the wood to the air of decreasing relative humidity (Figure 3.3a). It can be seen in Figure 3.3b that the equilibrium moisture content is lower during desorption than during adsorption (Bian 2001).

To experimentally obtain sorption isotherms, a dry wood specimen is first exposed to moving air under vacuum conditions that helps establishing equilibrium moisture quickly. The initial relative humidity of the air with which the wood piece is in equilibrium should be brought to extremely low values, so that the moisture content of the wood specimen is close to zero at the beginning. The wood specimen is then

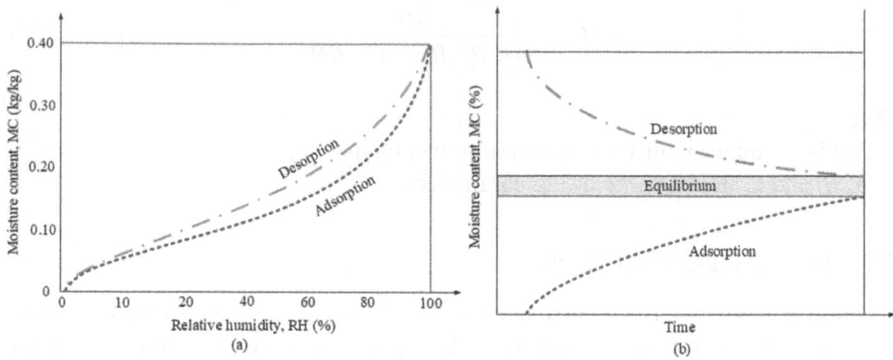

FIGURE 3.3 Typical representations of desorption and adsorption isotherms. (a) Moisture content variation as a function of ambient air temperature; (b) moisture content variation with time.

exposed to successively greater air relative humidity in a thermostatically controlled atmosphere, generally, at room temperature.

Modern experimental techniques, such as neutron imaging or a combination of impedance and acoustic emission from wood specimens (Margaritis and King 1971), can be used to measure the moisture adsorption and desorption rates (Wang and Brennan 1995).

Adsorption physical process is similar with water vapor condensation, but the enthalpy (latent heat) of adsorption is higher than the enthalpy (latent heat) of condensation. Therefore, more thermal energy is needed to release adsorbed water vapor molecules. The difference of these latent heats is defined as the enthalpy (differential heat) of sorption of liquid water by wood (Skaar 1979). However, above the fiber saturation point, the condensing enthalpy of water in wood is essentially the same as that of liquid water (Skaar 1988; Perré and Keey 2015).

The wood sorption enthalpy is slightly higher than 1,000 kJ/kg for oven-dry wood and decreases rapidly with the moisture content to reach zero at the fiber saturation point (Perré and Keey 2015). It can be indirectly determined based on Clausius–Clapeyron equation or directly by a calorimetric method (Skaar 1988; Wadsö 1994).

When water vapor molecules are adsorbed, the enthalpy (latent heat) of adsorption is released. Correspondingly, when water vapor molecules are desorbed from wood, the enthalpy (latent heat) of desorption is required. Ideally, under steady-state conditions, the rate of condensation (adsorption) of the vapor on a dry wood surface must be equal to the rate of evaporation (desorption) from the surface wet regions.

Adsorption equilibrium is obtained when the number of molecules arriving on the wood surface is equal to the number of molecules leaving the surface. For low relative humidity (up to about 10%), the water molecules are adsorbed in a single layer (Kaminski and Kudra 1996).

To develop theoretical sorption isotherms, two general approaches have been proposed: (i) sorption is assumed to be a surface phenomenon in which water molecules are condensed in one or more layers on sorption sites, and (ii) polymer–water system is considered as a solution in which some of the water molecules form hydrates with sorption sites within the cell wall, and the remaining water molecules form a solution (Schniewind 1989).

Many empirical and semiempirical equations, such as Guggenheim–Anderson–de Boer equation (Marinos-Kouris and Maroulis 2006), have been proposed and tested for describing the moisture sorption isotherms of materials such as woods and foods (Skaar 1988).

To calculate the variations in equilibrium moisture content according to the temperature for a given relative humidity, the following empirical formula has been proposed (Keey 1978):

$$\frac{\partial(MC)}{\partial T} = -A * MC \tag{3.25}$$

where

MC is the wood moisture content $\left(\frac{kg_{water}}{kg_{oven\text{-}dry}}\right)$

A is a coefficient that lies between 0.005/°C and 0.01/°C for air relative humidity varying between 10% and 90% for materials such as woods

T is the temperature (°C)

An approximate calculation for spruce at 50% relative humidity gives a change in moisture content per °C equal to 0.05 wt% (A is assumed equal to 0.0075) if the temperature decreases from 20°C to 0°C (Keey 1978).

3.9.2 Sorption Hysteresis

A plot of wood water moisture content as a function of equilibrium relative humidity at constant temperature yields the moisture sorption (adsorption and desorption) not-identical and nonreversible isotherms (see Figure 3.3a). The difference between the reproducible absorption and desorption isotherms is termed sorption hysteresis (Skaar 1979; Schniewind 1989).

The adsorption and desorption isotherms for a piece of wood diverge with increasing relative humidity, the corresponding equilibrium moisture content being higher on desorption than on adsorption (Perré and Keey 2015), with the values differing by their greatest moisture amount under near-saturation conditions. The ratio of the value for adsorption to the value for desorption normally ranges between 0.75 and 0.85 (Schniewind 1989).

For example, when various New Zealand woods are subject to environmental change cycles between 50% and 80% relative humidity, the adsorption and desorption isotherms are apart by 1%–2% moisture content at a given relative humidity (Orman 1955).

Many theoretical, semiempirical, and empirical equations have been proposed and tested to describe the sorption isotherms, but none of them can describe the phenomenon of hysteresis.

The hysteresis can be explained based on the incomplete wetting (where, at low relative humidity levels, the attraction forces of the water molecules within porous wood compete with the attraction forces of air molecules) and bottleneck (where, once the pore is filled, desorption cannot take place until the pressure falls to a specific value) capillary theories of adsorption (Cohan 1944; Keey et al. 2000).

An alternative theory to explain the hysteresis is based on the idea that the sorption sites are less available after wood has been dried from green timber (Stamm 1964; Walker 1993).

For European spruce, sorption hysteresis decreases with increasing temperature and disappears at temperatures between 75°C and 100°C (Skaar 1979).

3.9.3 Wood in Use

Since it is a hygroscopic material, wood in use (or service) (such as construction, flooring, or furniture products) is always undergoing slight changes in moisture content in response to more or less cyclical fluctuating diurnal, seasonal, or annual variations of the ambient air temperature and relative humidity.

The primary reason for drying wood to a moisture content equivalent to its average equilibrium moisture content under the use conditions is to minimize the dimensional changes in the final in-use products. In a given geographic location, these conditions vary for exterior and interior in-field uses.

The in-service moisture content of exterior wood primarily depends on the outdoor dry- and wet-bulb temperature and relative humidity, and exposure to rain or sun, whereas that of interior wood primarily depends on indoor dry- and wet-bulb temperature relative humidity, which in turn is a complex function of moisture sources, ventilation rate, dehumidification (e.g., air conditioning), and outdoor humidity conditions.

Therefore, wood must be dried to slightly lower moisture content than in-service conditions where moderate increases or decreases in moisture content can be anticipated. To minimize the changes in wood moisture content in order to avoid problems such as floor buckling or cracks in furniture (Cech and Pfaff 2000), the lumber or wood products are usually dried to moisture contents as close as possible to the average moisture equilibrium content conditions to which it will be exposed in-service.

In North America, for example, dry wood lumber that shall be used in outdoor conditions (e.g., in construction industry) must have a maximum moisture content of 19% (dry basis), but for lumber that shall be used in dry indoor climates, this must be around 8% (dry basis) (Skaar 1988; Cech and Pfaff 2000).

If lumber is dried to excessively low moisture content, it will later absorb moisture and swell. Conversely, if insufficiently dried, it will continue to lose moisture after it is placed in service, and consequently, it will shrink and often check or warp. This is because the water vapor may be attracted to condense (adsorb) on a dry surface of a piece of wood.

Correct drying, handling, and storage of wood contribute at minimizing moisture content undesirable changes that might occur after drying when the wood is in service. If moisture content is controlled within reasonable limits, major problems from dimensional changes can be avoided. Generally, no significant dimensional changes will occur if wood is fabricated or installed at moisture content corresponding to the average atmospheric conditions to which it will be exposed (Keey et al. 2000).

REFERENCES

AS/NZS 4787. 2001. Australian/New Zealand Standard. Timber Assessment of Drying Quality (www.saiglobal.com/.pdf, accessed October 28, 2017).

Beall, F.C. 1968. Specific heat of wood—Further research required to obtain meaningful data. *U.S. Forest Service Research Note FPL-0184*, February.

Berit Time. 1988. Hygroscopic moisture transport in wood. EngD Thesis. Department of Building and Construction Engineering, Norwegian University of Science and Technology, Trondheim, Norway.

Bian, Z. 2001. Airflow and wood drying models for wood kilns. MASc Thesis. Faculty of Graduate Studies, Department of Mechanical Engineering, the University of British Columbia, Vancouver, BC.

Cech, M.J., F. Pfaff. 2000. *Operator Wood Drier Handbook for East of Canada*, edited by Forintek Corp., Canada's Eastern Forester Products Laboratory, Québec city, Québec, Canada.

Cohan, L.H. 1944. Hysteresis and the capillary theory of adsorption of vapors. *Journal of the American Chemical Society* 66:98–105.

Desch, H.E., J.M. Dinwoodie. 1996. *Timber: Structure, Properties, Conversion and Use*, 7th edition. Macmillan Press Ltd., London, 306pp.

Dunlap, F. 1912. The specific heat of wood. U.S. Department of Agriculture, *Forest Service Bulletin 110*.

Eckelman, C.A. 1968. Wood moisture calculations, Purdue Forestry & Natural Resources, Department of Forestry & Natural Resources, FRN-156b, (www.agriculture.purdue.edu/fnr/faculty/eckelman/documents/FNR-156b.pdf, accessed May 15, 2016).

Frank Controls. 1986. Lumber Drying Theory (www.frankcontrols.com/downloads/ddpbook.pdf, accessed September 1, 2016).

Hailwood, A.J., S. Horrobin. 1946. Absorption of water by polymers: Analysis in terms of a simple model. *Transactions of Faraday Society* 42B:84–102.

Harris, J.M. 1961. The dimensional stability shrinkage and intersection point and related properties of New Zealand timbers. *NZFS Tech*, Paper 36, NZFRI Rotorua NZ.

Hoadley, R.B. 1980. *Understanding Wood, A Craftsman's Guide to Wood Technology*. The Taunton Press, Newtown, CT, 256pp.

Hultnas, M., V. Fernandez-Cano. 2012. Determination of the moisture content in wood chips of Scots pine and Norway spruce using Mantex Desktop Scanner based on dual energy X-ray absorptiometry. *Journal of Wood Science* 58(4):309–314.

Kaminski, W., T. Kudra. 1996. *Material-Moisture Equilibrium*. Lecture Notes, Faculty of Process and Environmental Engineering, Technical University of Lodz, Lodz, Poland.

Keey, R.B. 1978. *Introduction to Industrial Drying Operations*. Pergamon Press, Oxford.

Keey, R.B., T.A.G. Langrish, J.C.F. Walker. 2000. *Kiln Drying of Lumber*, Springer Series in Wood Science, edited by T.E. Timell, Springer, New York (ISBN-13:978-3-642-64071-1).

Kollmann, P.P., W.A. Cote. 1968. *Principles of Wood Science and Technology: Solid Wood*, Vol. 1, Springer, Berlin.

Larsson, R., T. Moren. 2003. Implementation of adaptive control system in industrial dry kilns. In *Proceedings of the 8th International IUFRO Wood Drying Conference*, Brasov, Romania, August 24–29, pp. 397–400.

Mackay J.F.G., L.C. Oliveira. 1989. *Kiln Operator's Handbook for Western Canada*. Forintek Canada, Vancouver, BC, SP-31, 53pp.

Margaritis, A., C.J. King. 1971. Measurement of rates of moisture transport within the solid matrix of hygroscopic porous materials. *Industrial & Engineering Chemistry Fundamentals* 10(3):510–515.

Marinos-Kouris, D., Z.B. Maroulis. 2006. Transport properties in the drying of solids. In *Handbook of Industrial Drying*, edited by A.S. Mujumdar, CRC Press, Boca Raton, FL, pp. 82–119.

Meroney, R.N. 1969. The state of moisture transport rate calculations in wood drying. *Wood and Fiber* 1(1):64–74.

Minea, V., D. Normand. 2005. Séchage de bois avec pompe à chaleur basse température (in French), Hydro-Quebec Research Institut Report.

Onn, L.K. 2007. Studies of convective drying using numerical analysis on local hardwood, University Sains, Malaysia (https://core.ac.uk/download/pdf/32600391.pdf, accessed June 15, 2016).

Orman, H.R. 1955. The response of New Zealand timbers to fluctuations in atmospheric moisture conditions. *NZFS Tech. Paper 8,* NZFRI Rotorua, New Zealand.

Pang, S. 1994. High-temperature drying of Pinus radiata boards in a batch kiln. PhD Thesis. University of Canterbury, Christchurch, New Zealand.

Perré, P., R.B. Keey. 2015. Drying of wood principles and practices, In *Handbook of Industrial Drying*, 4th edition, edited by A.S. Mujumdar, CRC Press, pp. 872–877. ISBN: 978-1-4665-9665-8, eBook ISBN: 978-1-4665-9666-5.

Reeb, J., M. Milota. 2016. Moisture content by oven-dry method for industrial testing (http://ir.library.oregonstate.edu/xmlui.pdf, accessed July 31, 2016).

Salin, J.G. 2002. The timber final moisture content variation as a function of the natural variation in wood properties and of the position in the kiln load: An evaluation using simulation models. In *Proceedings of the 4th COST E15 Workshop*, Santiago de Compostela, Spain, May 30–31.

Sauer, H.J. Jr., R. H. Howell. 1983. *Heat Pump Systems*. John Wiley & Sons, New York.

Schajer, G.S., F.B. Orhan. 2006. Measurement of wood grain angle, moisture content and density using microwaves. *Holz Roh Werkst* 64(6):483–490.

Schniewind, A.P. (Ed.). 1989. *Concise Encyclopedia of Wood and Wood-Based Materials.* Pergamon Press, Oxford, 354pp.

Sears, F.W., M.W. Zemansky. 1953. *College Physics*, 2nd edition, Addison-Wesley Publishing Co., Cambridge, MA, 912pp.

Shubin, G.S. 1990. *Drying and Thermal Treatment of Wood.* Moskow, Russia: Lesnaya Promyshlennost, URSS, 337pp.

Siau, J.F. 1984. Wood Drying (http://home.eng.iastate.edu.pdf, accessed October 28, 2017).

Skaar, C. 1979. Moisture sorption hysteresis in wood. In *Proceedings of Symposium on Wood Moisture Content—Temperature and Humidity Relationships.* Forest Products Laboratory, USDA Forest Service. Blacksburg, Virginia, October 29.

Skaar, C. 1988. *Wood-Water Relations.* Springer Series in Wood Science, edited by T.E. Timell, Springer-Verlag, Berlin/Heidelberg.

Skog, J., T. Vikberg, J. Oja. 2010. Sapwood moisture content measurements in *Pinus sylvestris* sawlogs combining X-ray and three-dimensional scanning. *Wood Material Science and Engineering* 5:9196.

Simpson, W.T. 1971. Equilibrium moisture content prediction for wood. *Forest Product Journal* 21(5):48–49.

Simpson, W.T., H.N. Rosen. 1981. Equilibrium moisture content of wood at high temperature. *Wood and Fiber* 13(3):150–158.

Srikiatden, J., J. Roberts. 2007. Moisture transfer in solid food materials: A review of mechanisms, model, and measurements. *International Journal of Food Properties* 10:739–777.

Stamm, A.J. 1964. *Wood and Cellulose Science.* Ronald Press, New York, 317pp.

Thorvaldsson, K., C. Skjoldebrand. 1996. Method and instrument for measuring local water content inside food. *Journal of Food Engineering* 29:1–11.

Vikberg, T., J. Oja, L. Antti. 2012. Moisture content measurement in *Scots Pine* by microwave and X-rays. *Wood and Fiber Science* 44(3):280–285.

Vikberg, T. 2012. Moisture content measurement in the wood industry. Licentiate Thesis. Luleå University of Technology, Luleå, Sweden.

Vikberg, T., D. Elustondo. 2015. Basic density determination for Swedish softwood and its influence on average moisture content packages estimated by measuring their mass. *Wood Material Science and Engineering* 11(4):248–253 (doi:10.1080/17480272.2015.1090481).

Wadsö, L. 1994. Micro-calorimetric investigations of building materials. Report TVBM-3063. Division of Building Materials, Lund Institute of Technology, University of Lund, Lund, Sweden.

Walker, J.C.F. 1993. *Primary Wood Processing Principles and Practice.* Chapman & Hall, London, 596pp.

Wang, N., J.G. Brennan. 1995. A mathematical model of simultaneous heat and moisture transfer during drying of potato. *Journal of Food Engineering* 24:47–60.

4 Moist Air

4.1 COMPOSITION OF ATMOSPHERIC AIR

Atmospheric (ambient) dry air at the standard sea level temperature, pressure, and density of 15°C, 101.325 kPa, and 1.225 kg/m^3, respectively, and constant gravity acceleration (9.807 m/s^2) (U.S. Standard Atmosphere) is a mixture of nitrogen (78.084%), oxygen (20.9476%), argon (0.934%), and carbon dioxide (0.0314%), substances that can be approximated as perfect gases (Table 4.1) (ASHRAE 2009).

In addition to the previous gases, approximate percentages of other gaseous components of dry air by volume are neon (0.001818%), helium (0.000524%), methane (0.00015%), sulfur dioxide (0%–0.0001%), hydrogen (0.00005%), and minor components such as krypton, xenon, and ozone (0.0002%) (ASHRAE 2009).

4.2 PROPERTIES OF MOIST AIR

Moist air is a two-phase (binary) mixture of dry air (i.e., atmospheric air with all water vapor and contaminants removed) and water vapor that needs to be treated separately from dry air since its physical properties are different. The homogeneous mixture of dry air and water vapor that do not react with each other and do not change in chemical composition is considered a pure substance. Moreover, because none of the components of dry air is highly soluble in liquid water, dry air composition can be considered invariable, even if small variations in the amounts of individual components occur with time, geographic location, and altitude (McQuiston and Parker 1988).

Since in commercial wood dryers the working pressures of the moist air are low, both the dry air and the water vapor can be assumed to behave like ideal gases, and the moist air as an ideal gas mixture (Sauer and Howell 1983; Tsilingiris 2008).

If parameters such as absolute humidity (humidity ratio), enthalpy, and specific volume at standard atmospheric pressure and at temperatures ranging from −50 to 50°C are used in the ideal gas equations, errors less than 0.7% are generally achieved (ASHRAE 2009).

The amount of water vapor varies from zero (dry air) to a maximum, depending on the mixture's temperature and pressure (ASHRAE 2009). The maximum amount of water vapor corresponds to the saturated air, a state of equilibrium between the moist air and the water liquid. Consequently, all mass flow calculations are based on dry air because this mass flow is constant, whereas moist air mass flow varies through the process when moisture is added (or removed).

The thermodynamic properties of moist air depend on the thermodynamic temperature related to the Celsius scale as follows (ASHRAE 2009):

$$T = t + 273.16 \tag{4.1}$$

TABLE 4.1
Composition of Dry Air

Substance	Molar Mass (kg/kmol)	Composition Mole Fraction (Goff 1949) (%)	U.S. Standard Atmosphere (%)
Nitrogen	28.016	78.09	78.084
Oxygen	32	20.95	20.9476
Argon	39.944	0.93	0.934
Carbon dioxide	4.401	0.03	0.0314

where
T is the absolute temperature T(K)
t is the Celsius temperature (°C)

In other words, Kelvin and Celsius scales are related as follows:

$$1\text{K} = °\text{C} + 273.16 \tag{4.2}$$

The properties of the air flowing inside the wood dryers are major factors in determining the rate of removal of moisture. The capacity of air to remove moisture is principally dependent upon its initial dry- and wet-bulb temperatures and relative humidity. The greater the dry-bulb temperature and lower the relative humidity, the greater the moisture removal capacity of the air.

The knowledge of air thermophysical and transport at intermediate temperatures levels up to 150°C is fundamental to understanding the behavior of air during the whole drying process of wood as well as when it passes through moisture condensers (i.e., heat pump evaporator coils). This allows accurate prediction of heat and mass transfer phenomena during the complex drying processes involved (Tsilingiris 2008), including during heat pump-assisted drying.

4.2.1 Equation of State

An ideal (or perfect) gas is defined as a gas at sufficiently low density where intermolecular forces and the associated energy are negligibly small.

The ideal gas law (or equation of state) expresses the relation between pressure, temperature, volume, and the individual gas constant (Sauer and Howell 1983):

$$pV = mRT \tag{4.3a}$$

or

$$p = \frac{m}{V}RT = \rho RT \tag{4.3b}$$

where
p is the absolute pressure (Pa = N/m^2)
V is the volume of gas (m^3)

m is the mass of gas (kg)
ρ is the gas density (kg/m^3)
R is the individual gas constant (J/kg · K)
T is the absolute temperature (K)

The individual gas constant R (Table 4.2) is independent of temperature, but depends on the particular gas and is related to the molecular mass of the gas (Table 4.1).

Even no real gas exactly satisfies equation of state 4.3a (or 4.3b) over any finite range of temperature and pressure, almost all real gases approach ideal behavior at low pressures. Thus, equation of state 4.3a (or 4.3b) provides a good approximation to real gas behavior at low pressures.

The moist air is considered to be a mixture of two independent perfect gases, dry air and water vapor. Consequently, each of them is assumed to obey the perfect gas equation of state as follows (ASHRAE 2009):

For dry air:

$$p_{dry\ air}V = n_{dry\ air}\mathcal{R}T \tag{4.4}$$

For water vapor:

$$p_{water\ vapor}V = n_{water\ vapor}\mathcal{R}T \tag{4.5}$$

where

$p_{dry\ air}$ is the partial pressure of dry air (Pa)
$p_{water\ vapor}$ is the partial pressure of water vapor (Pa)
V is the total mixture volume (m^3)
$n_{dry\ air}$ is the number of moles of dry air (–)
$\mathcal{R}$ is the gas universal gas constant (8,314.459 J/mol * K)
$n_{water\ vapor}$ is the number of moles of water vapor (–)
T is the absolute temperature (K)

The mixture also obeys the perfect gas equation according to the following equation:

$$pV = n\mathcal{R}T \tag{4.6}$$

or

$$\left(p_{dry\ air} + p_{water\ vapor}\right)V = \left(n_{dry\ air} + n_{water\ vapor}\right)\mathcal{R}T \tag{4.7}$$

TABLE 4.2
Individual Gas Constant and Molecular Mass of Dry Air and Water Vapor

Component	Individual Gas Constant, R (J/kg · K)	Molecular Mass, M_{gas} (kg/kmol)
Dry air	287.055	28.9645
Water vapor	461.520	18.01528

where

$p = p_{dry\,air} + p_{water\,vapor}$ is the total mixture pressure (Pa)
$n = n_{dry\,air} + n_{water\,vapor}$ is the total number of moles in the mixture (–)

From the previous equations, the mole fractions of dry air and water vapor are, respectively, as follows:

$$x_{dry\,air} = \frac{p_{dry\,air}}{\left(p_{dry\,air} + p_{water\,vapor}\right)} = \frac{p_{dry\,air}}{p} \tag{4.8a}$$

and

$$x_{water\,vapor} = \frac{p_{water\,vapor}}{\left(p_{dry\,air} + p_{water\,vapor}\right)} = \frac{p_{water\,vapor}}{p} \tag{4.8b}$$

The relation between individual (R_{gas}) and the universal ($\mathcal{R}$) gas constants is the same for all ideal gases:

$$R_{gas} = \frac{\mathcal{R}}{M_{gas}} \tag{4.9}$$

where

$\mathcal{R} = 8{,}314.47$ is the universal gas constant (J/kmol·K)
M_{gas} is the molecular mass of the gas (kg/kmol) (see Table 4.1)

Table 4.2 shows the values of the molecular mass and the individual gas constants for dry air and water vapor, respectively, all based on the carbon-12 scale (Harrison 1965; ASHRAE 2009).

4.2.2 Density

The density of a substance (ρ) is defined as the mass per unit volume (kg/m^3) (Sauer and Howell 1983):

$$\rho = \frac{m}{V} \tag{4.10}$$

where

m is the mass of substance (kg)
V is the volume of the substance (m^3)

The density of dry air, water vapor, and a mixture of dry air and water vapor (moist or humid air) can be calculated based on the ideal gas law (equation 4.3a or 4.3b).

The density of dry air can be also expressed as

$$\rho_{dry\,air} = \frac{p_{dry\,air}}{R_{dry\,air} * T} \approx 0.0035 * \frac{p_{dry\,air}}{T} \tag{4.11}$$

where

$\rho_{dry\ air}$ is the density of dry air (kg/m^3)
$R_{dry\ air} = 287.055$ J/kg·K is the dry air gas constant (Table 4.2)
$p_{dry\ air}$ is the partial pressure of dry air (Pa)
T is the absolute dry-bulb temperature (K)

Similarly, the density of water vapor can be expressed as

$$\rho_{water\ vapor} = \frac{p_{water\ vapor}}{R_{water\ vapor} * T} \approx 0.0022 * \frac{p_{water\ vapor}}{T} \tag{4.12}$$

where

$\rho_{water\ vapor}$ is the density of water vapor (kg/m^3)
$p_{water\ vapor}$ is the partial pressure water vapor (Pa)
$R_{water\ vapor} = 461.52$ J/kg·K is the water vapor gas constant (Table 4.2)
T is the absolute dry-bulb temperature (K)

In the case of moist air, the amount of water vapor (a relatively light gas compared to diatomic oxygen (O_2) and nitrogen (N_2), the dominant components in air) influences density. When the water vapor content increases in the moist air, the amount of O_2 and N_2 decreases per unit volume and, thus, the density decreases because the mass is decreasing.

Humid air is less dense than dry air because a molecule of water ($M \approx 18$ units) is less massive than either a molecule of nitrogen ($M \approx 29$ units) (about 78% of the molecules in dry air) or a molecule of oxygen ($M \approx 32$ units) (about 21% of the molecules in dry air).

4.2.3 Specific Volume

The specific volume (v) of a substance is defined as the volume per unit mass (Sauer and Howell 1983).

The specific volume (v) of a moist air mixture is expressed in terms of a unit mass of dry air:

$$v = \frac{V}{m_{dry\ air}} = \frac{V}{M_{dry\ air} * n_{dry\ air}} = \frac{V}{28.9645 * n_{dry\ air}} \tag{4.13}$$

where

V is the total volume of the moist air (m^3)
$m_{dry\ air}$ is the total mass of dry air (kg)
$M_{dry\ air} = 28.9645$ kg/kmol is the molar mass of dry air (Table 4.2)
$n_{dry\ air}$ is the number of moles of dry air (–)

Based on the ideal gas equation of state (equations 4.3a or 4.3b) and the dry air unit mass, the specific volume (v) of a moist air mixture at the total $p = p_{dry\ air} + p_{water\ vapor}$ can be also expressed as

$$v = \frac{R_{air} * T}{p - p_{water\ vapor}} \tag{4.14}$$

where

$$R_{air} = \frac{\mathcal{R}}{M_{air}} = \frac{8{,}314.47}{28.9645} = 287.055 \text{ J/kg}\cdot\text{K}$$

T is the temperature (°C)
p is the total pressure (Pa)
$\mathcal{R} = 8{,}314.47$ is the universal gas constant (J/kmol · K)
$M_{air} = 28.9645$ kg/kmol is the molar mass of air (Table 4.2)

4.2.4 Pressure

The pressure is defined as the normal component of force (N) per unit area (N/m^2 = Pa). In stationary fluids, the pressure is the same in all directions.

In wood drying applications, the absolute pressure of air can be calculated as follows (Figure 4.1) (Sauer and Howell 1983):

$$p_{abs} = p_{gauge} + p_{atmospheric} \tag{4.15}$$

where
p_{gauge} is the pressure measured (read) by means of a pressure gauge (Pa)
$p_{atmospheric}$ is the atmospheric pressure at a given geographic location (Pa)

Standard atmospheric pressures for altitudes up to 10,000 m are calculated from

$$p = 101.325\left(1 - 2.25577 * 10^{-5} z\right)^{5.2559} \tag{4.16}$$

where
z is the altitude from the sea level (m)

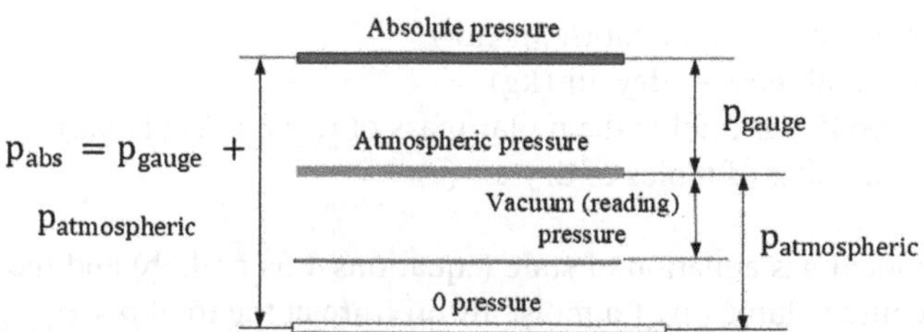

FIGURE 4.1 Schematic representation of current pressure designations used in moist air thermodynamic processes.

According to Gibbs–Dalton law, the pressure for a mixture of perfect gases as the atmospheric moist air is equal to the sum of the partial pressures of the constituents (McQuiston and Parker 1988):

$$p_{tot} = p_{N_2} + p_{O_2} + p_{CO_2} + \cdots + p_{water\ vapor} \tag{4.17}$$

Because the various constituents of the air may be considered to be dry air, it follows that the total pressure of moist air is the sum of the partial pressures of the dry air and the water vapor:

$$p_{tot} = p_{dry\ air} + p_{water\ vapor} \tag{4.18a}$$

where

p_{tot} is the total pressure (Pa)
$p_{dry\ air}$ is the partial pressure of the dry air (Pa)
p_{water} is the partial pressure of the water vapor (Pa)

The dry air partial pressure can be expressed as

$$p_{dry\ air} = \rho_{dry\ air} * R_{dry\ air} * T \tag{4.18b}$$

where

$R_{dry\ air} = 286.9\,\text{J/kg}\cdot\text{K}$ is the dry air constant (Table 4.2)

Similarly, the water vapor partial pressure can be expressed as

$$p_{water\ vapor} = \rho_{water\ vapor} * R_{water\ vapor} * T \tag{4.19}$$

where

$R_v = 461.5\,\text{J/kg}\cdot\text{K}$ is the water vapor constant

Unlike other gases in air, the water vapor may condense under common conditions. Since the boiling point of water at normal atmospheric pressure (101.3 kPa) is 100°C, the vapor partial pressure of water is low compared to dry air partial pressure in moist air. Common values for vapor pressure in moist air are in the range from 0.5 to 3.0 kPa.

Maximum vapor pressure possible before water vapor start condensing at an actual dry-bulb temperature is called saturation pressure (p_s).

4.2.5 Dry-Bulb Temperature

Dry-bulb temperature (*DBT*) of moist air is the temperature as registered by the immersion of an ordinarily thermometer in the mixture (Harriman et al. 2001; Harriman and Judge 2002; Mujumdar 2007; Tsilingiris 2008).

The dry-bulb temperature of atmospheric moist air can be expressed as a function of altitude (from −5,000 to +11,000 m) as follows:

$$T = 15 - 0.00652 * z \tag{4.20}$$

where
z is the altitude from the sea level (m)

4.2.6 Enthalpy

The enthalpy of a substance undergoing state changes conveniently combines three thermodynamic properties, i.e., the internal energy ($U = m * u$), pressure (p), and volume ($V = m * v$).

This property combination is an extensive property (Sauer and Howell 1983; Mujumdar 2007) defined in terms of unit mass as specific enthalpy

$$h = u + pv \tag{4.21}$$

or in terms of the substance total mass, as total enthalpy:

$$H = m * h = U + pV \tag{4.22}$$

where
m is the total mass (kg)
h is the specific enthalpy (J/kg)
H is the total enthalpy (J)
u is the specific internal energy (J/kg)
v is the specific volume (m^3/kg)
U is the total internal energy (J)
V is the total volume (m^3)

The specific (u) or total (U) internal energy of all substances is due to the energies of molecular motion and relative position of the constituent atoms and molecules. Their absolute values are unknown, but numerical values relative to an arbitrarily defined baseline at a particular temperature can be determined. For convenience, the internal energy is lumped with flow energies in forcing a substance into a system against pressure and forcing it out, getting the new property called enthalpy.

The terms pv and pV are the specific (per unit mass) or total work, respectively, associated to any steady flows involving forcing streams against the system pressure.

From the definition of enthalpy and the equation of state 4.3a (or 4.3b) of ideal gases, it follows that

$$h = u + pv = u + RT \tag{4.23}$$

where
R is the gas constant (J/kg · K)

Since R is a constant and u is a function of temperature only, it follows that the specific enthalpy (h) of an ideal gas is a function of temperature only and independent of the pressure, regardless of what kind of process is considered. From the definition of specific heat at constant pressure $\left(c_p = \frac{dh}{dT}\right)$ results,

$$dh = c_p dT \tag{4.24}$$

Absolute values of enthalpy of a substance, like the internal energy, are not known. Since absolute values of the enthalpy cannot be fixed, relative values of enthalpy at other conditions may be calculated by arbitrarily setting the enthalpy to zero at a convenient reference state. One convenient and commonly used reference state for zero enthalpy is liquid water under its own vapor pressure of 611.2 Pa at the triple-point temperature of 273.16 K (0.01°C).

With the assumption of perfect gas behavior, the specific enthalpy is a function of temperature only. If 0°C is selected as the reference state where the enthalpy of dry air is zero, and if the specific heat $c_{p,dry\ air}$ and $c_{p,water\ vapor}$ are assumed to be constant, the following relations result:

$$h_{dry\ air} = c_{p,dry\ air} * T \tag{4.25}$$

$$h_{water\ vapor} = h_{water\ vapor,\ sat} + c_{p,water\ vapor} * T \tag{4.26}$$

where

$h_{water\ vapor,\ sat} = 2{,}501.3$ kJ/kg is the enthalpy of saturated water vapor at 0°C
$\overline{c}_{p,\ dry\ air} = 1.006\ \text{kJ/kg} \cdot \text{K}$ is the mean specific heat of the dry air at constant pressure (kJ/kg·K)
T is the air temperature (°C)
$\overline{c}_{p,\ water\ vapor} = 1.86\ \text{kJ/kg} \cdot \text{K}$ is the mean specific heat of the water vapor at constant pressure (J/kg·K)

On the other hand, since the enthalpy is an extensive property, the specific enthalpy of the humid air (mixture of two perfect gases) is the sum of the partial enthalpy of the constituents (dry air and water vapor). Therefore, if the influences of residual enthalpy of mixing and of other effects are neglected (Mujumdar 2007), the specific enthalpy of moist air can be written as follows (McQuiston and Parker 1988):

$$h_{moist\ air} = h_{dry\ air} + \omega * h_{water\ vapor,\ sat} \tag{4.27}$$

where

$h_{dry\ air} \approx 1.006 * T$ is the specific enthalpy of dry air $\left(\frac{kJ}{kg_{dry\ air}}\right)$

ω is the absolute humidity of the moist air $\left(\frac{kg_{water}}{kg_{dry\ air}}\right)$

$h_{water\ vapor,\ sat} \approx 2{,}501.3 + 1.86 * T$ is the specific enthalpy of saturated water vapor $\left(\frac{kJ}{kg_{dry\ air}}\right)$ at the mixture dry-bulb temperature (T)

The moist air-specific enthalpy then becomes $\left(\frac{kJ}{kg_{dry\ air}}\right)$:

$$h_{moist\ air} = 1.006 * T + \omega * (2{,}501.3 + 1.86 * T) \tag{4.28}$$

4.2.7 Relative Humidity

Relative humidity (*RH*) is the ratio of the mole fraction of water vapor ($x_{water\ vapor}$) in a given moist air sample to the mole fraction ($x_{water\ vapor,sat}$) in a saturated air sample at the same temperature and total pressure (McQuiston and Parker 1988):

$$RH = \frac{x_{water\ vapor}}{x_{water\ vapor,\ sat}} \tag{4.29}$$

where

$x_{water\ vapor}$ is the mole fraction of water vapor in a given moist air sample (–)
$x_{water\ vapor,sat}$ is the molar fraction of the saturated water vapor at the same temperature (–)

The molar fraction of water vapor ($x_{water\ vapor}$) is defined as the ratio of the number of water vapor moles ($n_{water\ vapor}$) to the total number of moles of the moist air mixture ($n_{dry\ air} + n_{water\ vapor}$):

$$x_{water\ vapor} = \frac{n_{water\ vapor}}{n_{dry\ air} + n_{water\ vapor}} \tag{4.30}$$

The relative humidity of moist air is also defined as the ratio of the partial pressure of water vapor at a specific temperature in the moist air to the equilibrium saturated vapor pressure of water at the same temperature and total pressure:

$$RH = \frac{p_{water\ vapor}}{p_{water\ vapor,\ sat}} * 100\ (\%) \tag{4.31}$$

where

$p_{water\ vapor}$ is the partial pressure of water vapor (Pa)
$p_{water\ vapor,sat}$ is the equilibrium partial pressure of water vapor at saturation (Pa)

The equilibrium vapor pressure increases as temperature increases. If the partial pressure of the water falls below or equal to the equilibrium pressure of water, then the *RH* will be 100% and condensation would occur.

The relative humidity indicates how close the air is to saturated conditions. For example, an air sample at 50% relative humidity contains 50% of the maximum amount of water that it could contain at its current dry-bulb temperature (Harriman et al. 2001).

Relative humidity is normally expressed as a percentage and depends on temperature and the pressure of the system of interest. Increasing the temperature of the air increases its capacity for holding moisture, and, therefore, more kilograms of moisture (water) per kilogram of dry air are required to saturate the air. Increasing the temperature of the air without adding more moisture will cause the relative humidity to decrease. For example, if air with a relative humidity of 100% at 60°C is heated to 70°C, its relative humidity drops to 64% because of the increased moisture-holding capacity of the hotter air.

Relative humidity does not directly indicate the drying capacity of the kiln drying air. Therefore, drying schedules are expressed in terms of dry-bulb temperature, and either equilibrium moisture content or wet-bulb depression, or both.

4.2.8 Absolute Humidity

The amount of moisture (water vapor) in the surrounding air has a crucial importance on the rate at which the wet wood will dry out. However, the relative humidity does not give an indication of how much water vapor there actually is in a moist air sample.

Consequently, for drying, and other thermodynamic processes with moist air, another essential variable related to relative humidity is the absolute humidity which is influenced by the amount of water in the humid air.

The absolute humidity (also called humidity ratio, mass mixing ratio, or specific humidity) is defined as the fraction of the mass of water in the moist air ($m_{water\ vapor}$) to the mass of dry air ($m_{dry\ air}$):

$$\omega = \frac{m_{water\ vapor}}{m_{dry\ air}} = \frac{m_{dry\ air} - m_{water\ vapor}}{m_{dry\ air}} \tag{4.32}$$

where

ω is the air absolute humidity $\left(\frac{kg_{water}}{kg_{dry\ air}}\right)$

The absolute humidity is thus the moisture content of the humid air (expressed as the mass of water per unit mass of dry air) $\left(\frac{kg_{water}}{kg_{dry\ air}}\right)$, whereas the relative humidity is the ratio (expressed as a percentage) of the moisture content of the moist air at a specified temperature to the moisture content of moist air if it is saturated at the same temperature and total pressure.

The absolute humidity is independent of air temperature, but the drying capacity of air depends on both the absolute humidity and the ambient temperature. The absolute humidity is essential for humidity control of drying systems because it clearly defines the total mass of water in the humid air, and therefore it allows definition of work that must be done to change the moisture condition of the air (Harriman et al. 2001).

The reason for not using the total mass of moist air but the mass of dry air is that for any range of temperatures and pressures, the dry air will be permanently in the gaseous phase, while water vapor can be easily condensed. In other words, it is easier to follow changes in absolute humidity throughout a wood kiln on a dry-air basis, since the mass of circulating moisture-free air is not altered by the drying process.

By considering both dry air and water vapor as perfect gases at temperature T and volume V, their respective masses can be expressed, respectively, as follows:

$$m_{dry\ air} = \frac{p_{dry\ air} * V}{R_{dry\ air} * T} = \frac{p_{dry\ air} * V * M_{water\ vapor}}{\mathcal{R}T} \tag{4.33}$$

$$m_{water\ vapor} = \frac{p_{water\ vapor} * V}{R_{water\ vapor} * T} = \frac{p_{water\ vapor} * V * M_{water\ vapor}}{\mathcal{R}T} \tag{4.34}$$

where

$p_{dry\ air}$ is the partial pressure of dry air (Pa)
$p_{water\ vapor}$ is the partial pressure of water vapor (Pa)
$R_{dry\ air}$ is the individual constant of dry air (287.055 J/kg · K)
$R_{dry\ air}$ is the individual constant of water vapor (461.520 J/kg · K)
$M_{water\ vapor}$ is the molecular mass of dry air (29.9645 kg/kmol)
$M_{water\ vapor}$ is the molecular mass of water vapor (18.01528 kg/kmol)
V is the total volume (m^3)
T is the temperature (°C)
$\mathcal{R}$ is the gas universal gas constant (8,314.459 J/mol * K)

Consequently, the absolute humidity will be as follows:

$$\omega = \frac{M_{water\ vapor}}{M_{dry\ air}} * \frac{p_{water\ vapor}}{p_{dry\ air}} = 0.6219 \frac{p_{water\ vapor}}{p - p_{water\ vapor}} \tag{4.35}$$

where

p is the total pressure of moist air (Pa)

By combining equations 4.34 and 4.35, the relation between the relative and absolute humidity can be expressed as follows:

$$RH = \frac{\omega * p_{dry\ air}}{0.6219 * p_{water\ vapor,\ sat}} \tag{4.36}$$

where

$p_{water\ vapor,\ sat}$ is the saturation vapor pressure, i.e., the maximum value that $p_{water\ vapor}$ can take (Pa)

The total mass of dry air/water vapor mixture can be finally written in terms of absolute humidity ω and $m_{dry\ air}$ as follows:

$$m_{dry\ air} + m_{water\ vapor} = m_{dry\ air}(1 + \omega) \tag{4.37}$$

4.2.9 Wet-Bulb Temperature

Theoretically, for a given pressure and temperature of an air–water vapor mixture, any of the parameters previously defined (dry air and water vapor partial pressures, relative and absolute humidity, and enthalpy) would be acceptable to completely specify the state of the moist air, except at saturation. However, in drying practice, there is no way to measure any of these parameters. Because the measurement of pressure and temperature is relatively easy, the concept of adiabatic saturation provides a convenient practical solution to this problem (McQuiston and Parker 1988).

A current, practical method to determine the air relative and absolute humidity is to measure its wet- and dry-bulb temperatures.

The wet-bulb temperature is a thermodynamic property of a given humid air sample independent of measurement techniques and one of the most important parameter in drying. For any state of moist air, the wet-bulb temperature (*WBT*) is the temperature at which a small wet object surrounded by a small mech soaked in water would reach, by adiabatic evaporation cooling when exposed to an airflow at dry-bulb temperature (*DBT*) and absolute humidity (ω), until saturation at the same temperature, while total pressure (p) is constant (ASHRAE 2009).

Because the measurement of pressure and dry-bulb temperature is relatively easy, the adiabatic saturation process provides a convenient method of determining the state of moist air.

The wet-bulb temperature (*WBT*) is approximately the adiabatic saturated temperature a small wet-object would reach, by evaporative cooling, when exposed to an airflow. In other words, the wet-bulb temperature is the steady temperature reached by a small amount of liquid evaporating into a large amount of rapidly moving unsaturated moist air mixture. It is measured by blowing the air rapidly past a thermometer bulb surrounded by a small mech soaked (saturated wick) in water and shielded from the effects of radiation. If the air is unsaturated, some liquid is evaporated from the wick into the air stream, carrying with it the associated latent heat from a small reservoir of water. Evaporation from the wet-bulb wick cools the wet-bulb thermometer, which therefore shows a lower temperature than the dry-bulb thermometer. This latent heat is taken from within the liquid in the wick, and the wick is cooled. As the temperature of the wick is lowered, sensible heat is transferred by convection from the air stream and by radiation from the surroundings. At steady state, the net heat flow to the wick is zero and the temperature is constant (Mujumdar 2007). To prevent a zone of stagnant damp air formation around the sleeve, a minimum air velocity of at least 3 m/s is needed over the surface which needs to be shielded from excessive radiation and receive no conduction.

This is the principle of the adiabatic process achieved at constant pressure schematically represented in Figure 4.2a and c. In such a thermodynamic process (ASHRAE 2009), (i) the adiabatic saturation state is reached when adding liquid water to humid air until saturation at steady state and without letting it exchange heat (neither work) with the environment; (ii) the absolute humidity of air increases from the value ω_1 of unsaturated inlet air to $\omega_{2,sat}$ corresponding to the leaving moist air at saturated state 2 having the temperature $DBT_2 = WBT_2$; in other words, the temperature WBT_2 is the adiabatic saturation temperature where the value of relative humidity RH_2 is 100%; this means that when the moist air

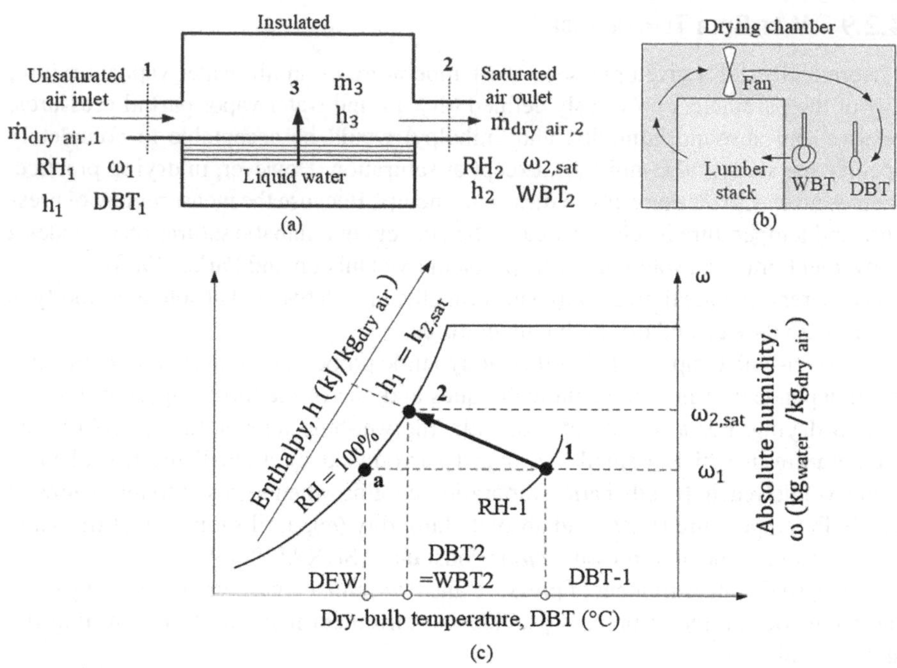

FIGURE 4.2 (a) Principle of the adiabatic process in a psychrometer; (b) dry- (DBT) and wet-bulb (WBT) thermometers of a psychrometer installed inside a lumber dryer (note: schematics not to scale); (c) adiabatic saturation process shown in Mollier diagram.

is saturated, both wet-bulb and dry-bulb temperatures are equal; (iii) the mass of water added per unit mass of dry air is $\omega_{2,\,sat} - \omega_1$; and (iv) the energy added to the moist air is $(\omega_{2,\,sat} - \omega_1) * h_{water}$, where h_{water} is the specific enthalpy (kJ/kg) of water added at temperature WBT_2.

To determine the moist air's relative humidity, dry- and wet-bulb temperatures are currently measured with psychrometers consisting of two similar thermometers exposed to the circulating drying air: one conventional and one of which the bulb is covered by a clean cotton wick covering the bulb thoroughly and continuously wetted with water from a small reservoir of water.

When such a psychrometer is placed in a lumber kiln (Figure 4.2b), water evaporates from the wick drawing heat for evaporation from the bulb as well as from the surrounding air. Evaporation process cools the bulb and the surrounding air because a portion of their sensible heat is lost to the evaporative process. Within a few minutes, the evaporation rate becomes constant (assuming the wick stays wet and air continues to flow across the wick). At that point, the thermometer reading will have fallen to equilibrium temperature (saturation) called wet-bulb temperature. When air is fully saturated, it cannot absorb more water vapor, so no evaporation can take place. It is an adiabatic saturation process of simultaneous heat and mass transfer from the wet bulb, where saturated air is expelled at a temperature equal to that of the injected water.

The difference between the two thermometer readings is called the wet-bulb depression, which is directly related to the relative humidity of the air. The greater the rate of evaporation from the wet-bulb wick, the greater is the cooling effect, and this varies directly with the amount of moisture in the air, such that a greater wet-bulb depression is equivalent to a lower relative humidity.

The mass and energy balances for adiabatic saturation process are (Figure 4.2a, b) as follows:

Mass balance for dry air:

$$\dot{m}_{dry\,air,1} = \dot{m}_{dry\,air,2} \tag{4.38}$$

Mass balance for water vapor:

$$\dot{m}_{dry\,air,1} * \omega_1 + \dot{m}_3 = \dot{m}_{dry\,air,2} * \omega_{2,sat} \tag{4.39}$$

Energy balance:

$$\dot{m}_{dry\,air,1} * h_1 + \dot{m}_{dry\,air,3} * h_3 = \dot{m}_{dry\,air,2} * h_2 \tag{4.40}$$

Because the enthalpy of the supply water (h_3) is very small (small flow rate at liquid state), the energy balance may be approximated to $h_1 = h_{2sat} = h_{process}$, i.e., an isenthalpic process for the humid air. With the perfect gas model, and neglecting the sensible enthalpy of water vapor with respect to its latent enthalpy, yields (in J/kg)

$$\bar{c}_{p,\,dry\,air}\left(DBT_1\right) + \omega_1 * h_{process} = \bar{c}_{p,\,dry\,air}\left(DBT_2\right) + \omega_2 * h_{process} \tag{4.41}$$

where

$\bar{c}_{p,\,dry\,air}$ is the air average isobaric specific heat (J/kg·K).

Therefore, if the process is strictly adiabatic, conservation of enthalpy at constant total pressure requires that

$$h_1 + (\omega_{sat} - \omega_1) * h_{water} = h_{2,\,sat} \tag{4.42}$$

where

ω_{sat}, h_{water}, and $h_{2,\,sat}$ are functions only of wet-bulb temperature WBT_2 for a fixed value of air pressure.

In practice, the difference between the dry- and wet-bulb temperatures (*DBT–WBT*), denoted as the wet-bulb depression (sometimes used as a control parameter in wood drying), is used to determine the relative humidity from a Mollier diagram of moist air. Although such a psychrometric measurement is convenient, in industrial practice it is difficult to determine the air relative humidity to accuracy greater than ±3%.

A higher difference between the dry-bulb and wet-bulb temperatures indicates lower relative humidity. For example, if the dry-bulb temperature is 100°C and wet-bulb temperature is 60°C, the relative humidity is read as 17% from a Mollier diagram.

4.2.10 Dew Point

Dew point (*DEW*) is the temperature of saturated moist air corresponding to thermodynamic state 1 at the same total pressure (p) and same absolute humidity (ω) (see Figure 4.2c).

The dew point temperature can be easily attained if the moist air is cooled from the temperature of humid air at state 1 until the saturated state of air (a) is reached (i.e., the state of neutral equilibrium between moist air and the condensed water liquid (Hyland and Wexler 1983) with no change in absolute humidity ($\omega_1 = \omega_a$)) (McQuiston and Parker 1988; Harriman et al. 2001).

The most important significance of the dew point temperature in drying technology is that if air at state 1 is in contact with a surface having a temperature of *DEW*, or lower, condensation of moisture will occur on that surface.

In a psychrometric diagram (see Section 4.9.10 and Figure 4.3), *DEW* point is obtained by moving horizontally to the left to intersect the saturation curve corresponding to 100% relative humidity, then proceeding straight down to the bottom of the chart and read *DEW* temperature. Unlike relative humidity, which changes widely and constantly with temperature, *DEW* point is not affected by changes in dry-bulb temperature as the air moves through cooling and dehumidification equipment. In cooling and dehumidification processes occurring in heat pump-assisted wood dryers, *DEW* point changes only by removing (water vapor).

Exercise E4.1

The dry-bulb temperature and relative humidity of air inside a lumber drying chamber are 50°C and 45%, respectively. If the surface average temperature of the heat pump finned-tube evaporator (moisture condenser) is 15°C, determine if the water vapor may condensate.

Solution

By using the Mollier psychrometric diagram (see Figure 4.3) or an air properties calculator (www.psychrometric-calculator.com/HumidAirWeb.aspx. Accessed December 10, 2017), the wet-bulb temperature (*WBT*) and the *DEW* point temperature corresponding to the given moist air's state ($DBT = 50°C$ and $RH = 45\%$) are determined:

$$WBT = 36.3°C$$

$$DEW = 34.78°C$$

Since the surface average temperature of the heat pump finned-tube evaporator (moisture condenser) is 15°C, i.e., lower than *DEW* point temperature, some moisture (water vapor) from the moist air will condensate.

4.2.11 Psychrometric Diagram

For moist air, both thermodynamic tables with properties and psychrometric diagrams are available. For quick visualization, as well as for determining changes

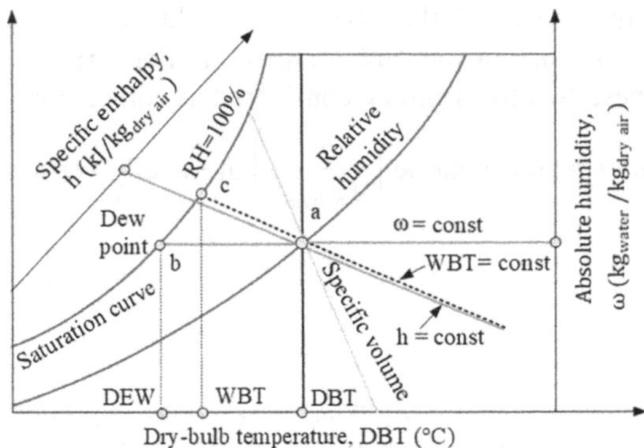

FIGURE 4.3 Basic thermodynamic lines in Mollier psychrometric diagram.

in significant properties such as temperature, absolute humidity, and enthalpy for any drying process, processes performed with the drying moist air can be plotted on psychrometric diagrams (Sauer and Howell 1983; McQuiston and Parker 1988; ASHRAE 2009).

The psychrometric (Mollier or Carrier) diagrams (Figure 4.3) represent the relationship between the temperature, relative and absolute humidity, enthalpy, and other thermodynamic properties of humid air in order to facilitate engineering calculations of dry air/water vapor mixtures (Solukhin and Kuts 1983; McQuiston and Parker 1988; Mollier 1923; ASHRAE 2009).

Based on the fact that a unit mass of dry air normally remains constant in drying processes, even when the amount of water increases or decreases, in the North American diagram, the horizontal abscissa is the dry-bulb temperature (*DBT*, in °C) and the vertical axis is the absolute humidity (ω) $\left(\frac{kg_{water}}{kg_{dry\text{-}air}}\right)$ (see Section 4.2.8).

In Europe, absolute humidity was chosen for abscissas and enthalpy for ordinates, although not in rectangular but in oblique direction.

In practice, thermodynamic properties of moist air at the beginning and the end of a specific process, as well as the path of the process, are of primary interest.

In Figure 4.3, the curved line on the upper-left boundary of the diagram is the saturation curve that corresponds to 100% relative humidity, i.e., to the maximum water vapor content of moist air at a given dry-bulb temperature.

Lower relative humidity values (%) are represented as a series of curves in the region below and to the right of the saturation line, where unsaturated air is able of holding additional water vapor, as proportions of the vertical distance from the base (horizontal axis) of the diagram, where the water vapor content is zero, to the saturation curve.

The enthalpy scale, located to the left side of the saturation line, allows drawing the lines of constant specific enthalpy $\left(\frac{kJ}{kg_{dry\ air}}\right)$. It is usual to assign zero enthalpy for dry air and liquid water at 0°C.

It can be seen in Figure 4.3 that there is a slight deviation of slope between constant enthalpy and constant wet-bulb temperature lines. However, for most engineering analysis, these two lines can be considered as coincident.

Lines of constant specific volume $\left(\frac{m^3}{kg_{dry\,air}}\right)$ are nearly vertical on the psychrometric chart.

Exercise E4.2

Table E4.2.1 shows three variants of drying air with known properties.

Determine the other properties of the drying air by using Mollier psychometric diagram.

Solution

Table E4.2.2 groups together calculated thermodynamic properties of the three given states of the drying air.

TABLE E4.2.1
Known Properties of a Drying Air

	Dry-Bulb Temperature (°C)	Absolute Humidity $\left(\frac{kg_{water}}{kg_{dry\,air}}\right)$	Wet-Bulb Temperature (°C)
V1	80	0.12	–
V2	45	–	26.7
V3-saturated	21.1	–	–

TABLE E4.2.2
The Other Properties of the Given Variants of Drying Air

	V1	V2	V3
Dry-bulb temperature (°C)	80	45	21.1
Wet-bulb temperature (°C)	57.2	26.7	21.1
DEW point (°C)	55.8	21	21.1
Absolute humidity $\left(\frac{kg_{water}}{kg_{dry\,air}}\right)$	0.12	0.0145	0.0158
Relative humidity (%)	34.5	24	100
Specific enthalpy $\left(\frac{J}{kg_{dry\,air}}\right)$	398.8	85.8	61.2
Specific volume $\left(\frac{m^3}{kg_{dry\,air}}\right)$	1.013	0.9124	0.8494

4.2.12 Basic Thermodynamic Processes

The psychrometric diagram can be used to represent and understand numerous processes with moist air currently occurring in lumber kilns, generally at constant atmospheric pressure of 101.325 kPa.

4.2.12.1 Air Humidification

If inside a lumber drying chamber, moisture (water vapor) is added to the drying air at state 1 without heat supply, the state of the air changes adiabatically along a constant enthalpy line (process 1-2) up to the state 2, while the dry-bulb temperature of the air decreases from *DBT*-1 to *DBT* -2, and the absolute humidity increases from ω_1 to ω_2 (Figure 4.4).

If the air mass flow rate is $\dot{m}_{dry\,air}\left(\frac{kg_{dry\,air}}{s}\right)$, the mass flow rate of moisture (water) added to moist air $\left(\frac{kg_{water}}{s}\right)$ can be calculated as follows:

$$\dot{m}_{dry\,air} = \dot{m}_{dry\,air}\left(\omega_2 - \omega_1\right) \tag{4.43}$$

4.2.12.2 Sensible Cooling and Dehumidifying

Sensible cooling and latent dehumidification thermodynamic processes of the moist air are accomplished in heat pump-assisted wood dryers by refrigerant evaporators that act as moisture condensers. The latent dehumidification takes place when the moist air is cooled to temperatures below its dew point. In this case, some of the water vapor condenses and leave the air stream.

Figure 4.5a schematically shows a sensible cooling and dehumidifying device where, even if the heat exchanger is dry when starting operation, moisture from the air condenses on the cold surface of the coil, such that thereafter the surface is wetted. The air flowing across the surface changes state from 2 to *a* then to 3 (Figure 4.5b) giving up heat to the coil and the refrigerant carries it away.

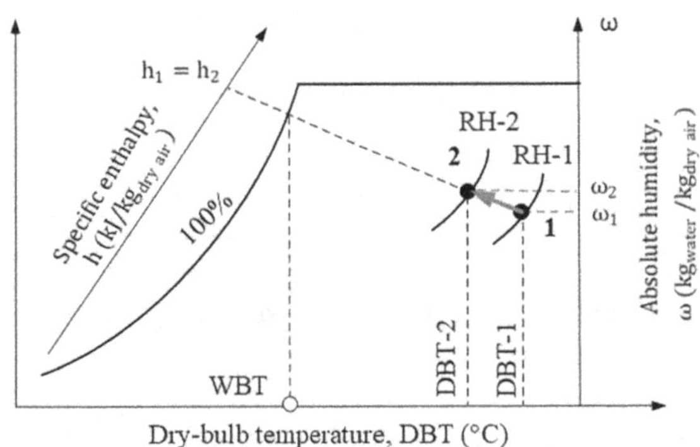

FIGURE 4.4 Air humidification process as in a wood drying chamber. DBT, dry-bulb temperature; WBT, wet-bulb temperature.

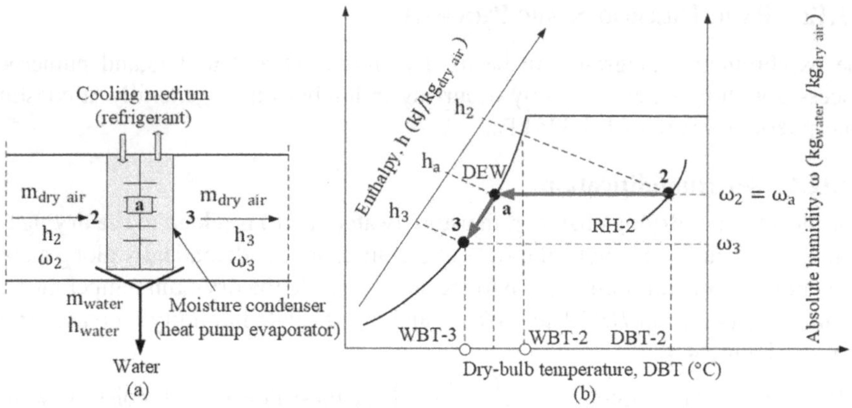

FIGURE 4.5 Moist air cooling and dehumidifying: (a) schematic of moisture condenser; (b) processes represented in a Mollier diagram; DBT, dry-bulb temperature. DEW, dew point; h, specific enthalpy; $\dot{m}_{dry\ air}$, dry air mass flow rate; WBT, wet-bulb temperature; ω, absolute humidity.

Cooling and dehumidifying processes involve both sensible and latent heat transfer, where sensible heat transfer is associated with the decrease in dry-bulb temperature from *DBT*-2 to *DBT*-3 = *WBT*-3, and the latent heat transfer is associated with the decrease in absolute humidity from $\omega_2 = \omega_a$ to ω_3. Both processes thus result in a reduction of both the dry-bulb temperature and the absolute humidity.

Although the actual process path will vary considerably depending on the type of surface, surface temperature, and flow conditions, the heat and mass transfer can be expressed in terms of the initial (2) and final (3) states. Since both sensible and latent heats are removed from the moist air, to calculate the energy required to accomplish the change it is most convenient to use the enthalpy difference $(h_2 - h_3)$ between the entering and leaving air conditions.

The water mass rate $\left(\frac{kg_{water}}{s}\right)$ and thermal power balance equations for the entire process 2-*a*-3 are (McQuiston and Parker 1988) as follows:

$$\dot{m}_{dry\ air} * \omega_2 = \dot{m}_{dry\ air} * \omega_3 + \dot{m}_{water} \tag{4.44}$$

$$\dot{m}_{dry\ air} * h_2 = \dot{m}_{dry\ air} * h_3 + \dot{Q}_{23} + \dot{m}_{water} * h_3 \tag{4.45}$$

Thus,

$$\dot{m}_{water} = \dot{m}_{dry\ air} * (\omega_2 - \omega_3) \tag{4.46}$$

$$\dot{Q}_{23} = \dot{m}_{dry\ air}\left[(h_2 - h_3) - (\omega_2 - \omega_3)h_3\right] \tag{4.47}$$

where

$\dot{m}_{water}$ is the water flow rate extracted from the moist air $\left(\frac{kg_{water}}{s}\right)$

$\dot{Q}_{23}$ is the total heat transfer rate (or thermal power) extracted from the moist air (W)

The last term on the right-hand side of equation 4.45 is usually small compared the others and is often neglected.

By referring again to Figure 4.5b, the sensible heat transferred can be expressed as follows:

$$\dot{Q}_{2-a} = \dot{m}_{dry\,air}\left(h_2 - h_a\right) = \dot{m}_{dry\,air} * c_{p,\,dry\,air} * \left(DBT_2 - DBT_a\right) \tag{4.48}$$

Similarly, by neglecting the energy of the condensate, the latent heat transfer is given by

$$\dot{Q}_{a-3} = \dot{m}_{dry\,air}\left(h_a - h_3\right) = \dot{m}_{dry\,air} * \left(\omega_2 - \omega_3\right) * h_{fg} \tag{4.49}$$

where

h_{fg} is the water-specific enthalpy (latent heat) of vaporization $\left(\frac{J}{kg_{dry\,air}}\right)$

The total heat transfer rate extracted from the moist air will be

$$\dot{Q}_{23} = \dot{Q}_{sens} + \dot{Q}_{latent} = \dot{Q}_{2-a} + \dot{Q}_{a-3} \tag{4.50}$$

4.2.12.3 Adiabatic Mixing

A common process in wood drying systems is the adiabatic mixing of two moist airstreams of two different thermodynamic states 3 and 2 (McQuiston and Parker 1988) (Figure 4.6a, b). The final state of adiabatic mixing process (*M*) is governed by the three following balance equations.

Mass balance on the dry air:

$$\dot{m}_{dry\,air,\,2} + \dot{m}_{dry\,air,\,3} = \dot{m}_{dry\,air,\,M} \tag{4.51a}$$

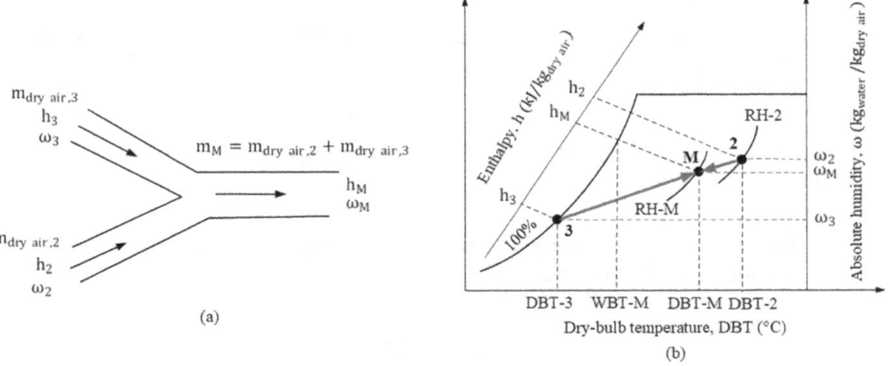

FIGURE 4.6 Adiabatic mixing process of two air streams: (a) schematic of mixing process; (b) process represented in a Mollier diagram. DBT, dry-bulb temperature; h, specific enthalpy; $\dot{m}_{dry\,air}$, dry air mass flow rate; WBT, wet-bulb temperature; ω, absolute humidity.

Energy balance:

$$\dot{m}_{dry\,air,2} * h_2 + \dot{m}_{dry\,air,3} * h_3 = \dot{m}_{dry\,air,M} * h_M \quad (4.51b)$$

Mass balance on water vapor:

$$\dot{m}_{dry\,air,2} * \omega_2 + \dot{m}_{dry\,air,3} * \omega_3 = \dot{m}_{dry\,air,M} * \omega_M \quad (4.51c)$$

From equation 4.51b, results:

$$\frac{\dot{m}_{dry\,air,\,2}}{\dot{m}_{dry\,air,\,3}} = \frac{h_M - h_3}{h_2 - h_M} = \frac{\overline{M3}}{\overline{M2}} \quad (4.51d)$$

From equation 4.51c, results:

$$\frac{\dot{m}_{dry\,air,\,2}}{\dot{m}_{dry\,air,3}} = \frac{\omega_M - \omega_3}{\omega_2 - \omega_M} = \frac{\overline{M3}}{\overline{M2}} \quad (4.51e)$$

According to expressions 4.51d and 4.51e, the state point of the resulting mixture (M) lies on the straight line connecting the state points of the two streams being mixed (2 and 3) and divides the line into two segments inversely proportional to the ratio of the masses of dry air in the two streams.

Exercise E4.3

Inside a wood heat pump-assisted dryer, the drying air leaves the heat pump evaporator at 13°C dry-bulb temperature, 100% relative humidity, and V_3 = 24 m³/s volumetric flow rate (state 3 in Figure 4.6a, b). This air stream is mixed with indoor drying air with dry- and wet-bulb temperatures of 40°C and 32°C, respectively, and volumetric flow rate $\dot{V}_2 = 72\ \text{m}^3/\text{s}$ (state 2), resulting the total airflow rate at state M (Figure 4.6a, b).

Determine analytically and graphically the specific enthalpy and absolute humidity of mixed air (state M).

Solution

Given data allow placing both states 2 and 3 on the psychometric chart, as shown in Figure 4.6b.

1. From a Moist Air Properties Software Calculator, the air specific volumes at states 3 and 2 are, respectively:

$$\rho_3 = 0.824 \frac{m^3}{kg_{dry\,air}}$$

$$\rho_2 = 0.934 \frac{m^3}{kg_{dry\,air}}$$

Consequently, the mass flow rates of the air streams are as follows:

$$\dot{m}_3 = \frac{\dot{V}_3}{\rho_3} = \frac{24}{0.824} = 29.12 \frac{kg_{dry\ air}}{s}$$

$$\dot{m}_2 = \frac{\dot{V}_2}{\rho_2} = \frac{72}{0.934} = 77.1 \frac{kg_{dry\text{-}air}}{s}$$

$$\dot{m}_M = \dot{m}_3 + \dot{m}_2 = 29.12 + 77.1 = 106.2 \frac{kg_{dry\ air}}{s}$$

From the psychometric chart or from a Moist Air Software Calculator, we also can determine

$$h_3 = 115.1 \frac{kJ}{kg_{dry\ air}}$$

$$\omega_3 = 0.029 \frac{kg_{water}}{kg_{dry\ air}}$$

$$h_2 = 36.6 \frac{kJ}{kg_{dry\text{-}air}}$$

$$\omega_3 = 0.00933 \frac{kg_{water}}{kg_{dry\ air}}.$$

The specific enthalpy of mixed air can be calculated as follows:

$$h_M = \frac{\dot{m}_3 * h_3 + \dot{m}_2 * h_2}{\dot{m}_M} = \frac{29.12*115.1 + 77.1*36.6}{29.12 + 77.1} = 58.11 \frac{kJ}{kg_{dry\ air}}$$

$$\omega_M = \frac{\dot{m}_3 * \omega_3 + \dot{m}_2 * \omega_2}{\dot{m}_M} = \frac{29.12*0.029 + 77.1*0.00933}{29.12 + 77.1} = 0.0147 \frac{kg_{water}}{kg_{dry\ air}}$$

In Figure 4.6b, on the straight line 3–2 lying states 3 and 2, the location of state *M* after air mixing is proportional to

$$\frac{\dot{m}_{dry\ air,2}}{\dot{m}_{dry\ air,3}} = \frac{25.7}{87.4} = 0.377 = \frac{\overline{M3}}{\overline{M2}}$$

From the psychometric chart, we obtain $DBT_M = 31.5°C$

$$WBT_M = 26.6°C$$

$$RH_M = 72\%$$

$$h_M = 85.8 \frac{kJ}{kg_{dry\ air}}$$

$$\omega_M = 0.02114 \frac{kg_{water}}{kg_{dry\ air}}$$

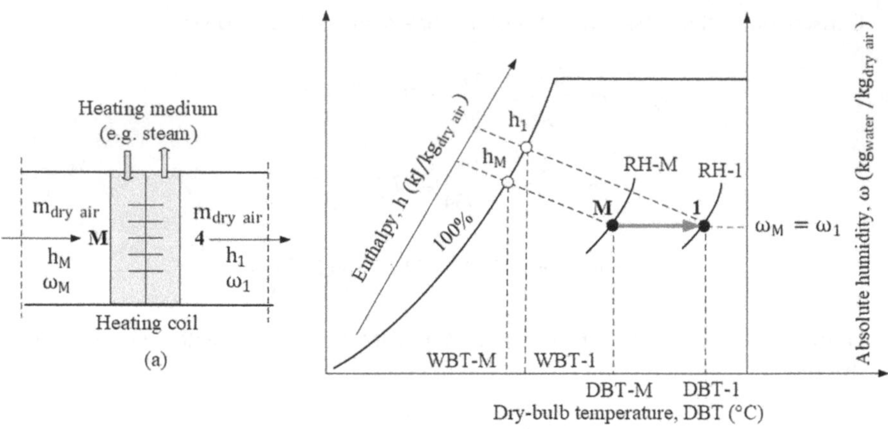

FIGURE 4.7 Sensible air heating process: (a) schematic of heating coil; (b) thermodynamic process in the psychometric diagram (for legend, see Figure 4.6).

4.2.12.4 Sensible Heating

When moist air is heated from state M to state 1 by flowing through a heat exchanger (Figure 4.7a), the process yields a straight horizontal line ($\overline{M1}$) on the psychrometric chart since the absolute humidity is constant ($\omega_M = \omega_1$) (ASHRAE 2009).

Under steady-state flow conditions, the energy balance becomes

$$\dot{m}_{dry\,air} h_M + \dot{Q}_{M1} = \dot{m}_{dry\,air} h_1 \tag{4.52}$$

and

$$\omega_M = \omega_1 \tag{4.53}$$

For steady-state flow conditions, the required thermal power addition is a positive number expressed as follows (McQuiston and Parker 1988):

$$\dot{Q}_{M1} = \dot{m}_{dry\text{-}air}\left(h_1 - h_M\right) = \dot{m}_{dry\text{-}air} * \overline{c}_{p,\,dry\text{-}air}\left(DBT_1 - DBT_M\right) \tag{4.54}$$

Exercise E4.4

In a heat pump-assisted wood dryer, drying air with a mass flow rate of 11.3 kg/s (state M in Figure 4.7a, b), dry-bulb temperature $DBT_M = 42°C$, and wet-bulb temperature $WBT_M = 27°C$ passes through the heat pump's condenser where a sensible heating process takes place. The dry temperature of drying air leaving the condenser (at state 1) is $DBT_M = 49°C$.

Determine the thermal power required to achieve the sensible heating process $\overline{M1}$.

Solution

From a Moist Air Properties Software, can be determined:
$h_M = 87.4$ kJ/kg

$$h_1 = 115\,\text{kJ/kg}$$

$$\dot{Q}_{M1} = \dot{m}_{dry-air}(h_1 - h_M) = 11.3(115 - 87.4) = 311.88\,\text{kW}$$

REFERENCES

ASHRAE. 2009. *ASHRAE Handbook – Fundamentals*, SI Edition, Supported by ASHRAE Research, American Society of Heating, Refrigerating and Air-Conditioning Engineers, Inc., Atlanta, GA.

Goff, J.A. 1949. Standardization of the thermodynamic properties of moist air. *ASHVE Transactions* 55:459–484.

Harrison, L.P. 1965. Fundamental concepts and definitions relating to humidity. In *Humidity and Moisture Measurement and Control in Science and Industry*, vol. 3, pp. 3–69, edited by A. Wexler and W.A. Wildhack, Reinhold, New York.

Harriman, L.G., G.W. Brundrett, R. Kittler. 2001. ASHRAE humidity control design guide for commercial and institutional buildings.

Harriman, L.G., J. Judge. 2002. Dehumidification equipment advances. *ASHRAE Journal* 44(8): 22–29.

Hyland, R.W., A. Wexler. 1983. Formulations for the thermodynamic properties of saturated phases of H_2O from 173.15 K to 473.15 K. *ASHRAE Transactions* 89(2A):500–519.

McQuiston, F.C., J.D. Parker. 1988. *Heating, Ventilating, and Air Conditioning - Analysis and Design*, 3rd edition. John Wiley & Sons, Hoboken, NJ.

Mollier, R. 1923. A new diagram for moist air mixtures. *Z VDI* 67:869–872.

Mujumdar, A.S. 2007. Principles, classification and selection of dryers. In *Handbook of Industrial Drying*, 3rd edition, edited by A.S. Mujumdar, CRC Press - Taylor & Francis Group, Boca Raton, FL, pp. 4–32.

Sauer, H.J. Jr., R.H. Howell. 1983. *Heat Pump Systems.* John Wiley & Sons, Hoboken, NJ.

Solukhin, R.I., P.C. Kuts. 1983. Introduction to the technology of industrial drying. *Nauk I Tekhnika*, Minsk.

Tsilingiris, P.T. 2008. Thermo-physical and transport properties of humid air at temperature range between 0 and 100°C. *Energy Conversion and Management* 49:1098–1110.

5 Importance of Wood Drying

5.1 INTRODUCTION

The fibrous and hygroscopic mass of freshly felled trees and recently cut green timbers contain up to 75% of water which affects the wood's physical properties as density, strength, and hardness. Most of this water must be removed to a desirable level of moisture content, usually below 20% (dry basis) before the wood is processed into end-use products. Drying the wood reduces the risk of dimensional changes when the wood is in place and significantly upgrades, sometimes up to 700%, the commercial value of raw timber.

In use, the wood moisture content is still constantly changing with changes in temperature and relative humidity of its surroundings, which results in changes in dimensions. Wooden materials used outdoor (as those for construction) or indoor (as those for furniture) absorb or desorb moisture until they are in equilibrium with their surroundings. Such equilibration processes cause unequal shrinkage in the wood and can cause damage to the wood if equilibration occurs too rapidly (Pang 1994; McCurdy 2005). Therefore, wood products should be dried to final moisture content about midrange (within about 2%) of the expected equilibrium moisture content of its in-use surroundings that can vary considerably by product, geographic location, and relative humidity of the inside or outside ambient air (Table 5.1).

Wood products (e.g., softwood construction lumber) intended for use outside buildings, but protected from direct precipitation, will stabilize with the surrounding environment at about 12%–19% moisture content (dry basis) in the humid southern regions, but may stabilize to as low as 6% moisture content (dry basis) in the arid zones.

Wood products (e.g., hardwood furniture, cabinetry, and paneling) used inside heated buildings (e.g., homes and offices) that are only occasionally heated should be dried to about 6%–8% moisture content (Bian 2001).

5.2 ADVANTAGES OF WOOD DRYING

Deformations and volumetric changes in wood may be found during the natural growth, timber drying, loading and unloading, and in-service due to changes in ambient air relative humidity and temperature.

Proper drying under controlled conditions prior to use is of great importance in timber use, especially in countries where climatic conditions vary considerably at different times of the year. For this, the drying operations require proper equipment and control procedures of drying degrade and final lumber moisture content and quality.

TABLE 5.1
Approximate Relationship between Relative Humidity and Equilibrium Moisture Content at Room Temperature

Relative Humidity (%)	Equilibrium Moisture Content (%)
0	0
15	3.5
30	6
50	9
65	12
80	16
90	20
95	24
100	28

Improper drying of wood may lead to irreversible damages to quality and, hence, to nonsalable products. In addition, drying defects caused by an interaction of wood properties with processing factors could further increase the overall cost of drying.

Among the advantages of dried lumber over the undried green timber, the following can be mentioned (Erickson 1954; Walker 1993; Walker et al. 1993; Desch and Dinwoodie 1996; Onn 2007; Mujumdar and Huang 2007; Oliveira et al. 2012; Anderson 2014):

a. It prevents the wood degradation after felling and sterilizes it for export.
b. Dried timber is lighter, and the handling, storage, and transportation costs are reduced.
c. Dried lumber works, machines, finishes, and glues are better than green timber.
d. It increases the wood's strength making it more compatible with the normal service conditions as a building material, and, at less than 20% moisture content, wood is immune to microorganisms (as insects) and fungal attacks, and also to subsequent shrinking, swelling, warping and cracking.
e. It improves mechanical and structural support of building and objects such as furniture and joinery because dried timber is stronger than green timber.
f. It improves lumber boards' dimensional stability for various finishing processes, such as polishing and painting, while paints last longer on dry timber.
g. It improves the electrical and thermal insulation properties of wood.

5.3 WOOD DRYING DEFECTS

Industrial wood drying consists of not only removing moisture from green wood but also ensuring that its quality is adequate for end use. In other words, the lumber must be dried with minimal development of degrade such as splits, checks, warp, discoloration, and any other change in wood that lowers the wood's usefulness and, therefore, its value.

In practice, conventional drying processes often result in significant quality problems from wood shrinks and cracks, both externally and internally leading to unusable dried products (Perré and Keey 2015). Often, degrade losses exceed all other operating costs in drying. Even drying quality and costs are limited by the kiln type; with the proper equipment and tools, drying costs can be minimized, and drying degrade could be under 2% (Onn 2007).

In commercial kiln drying of lumber, there are two major drying defects that cause significant losses for wood processors due to rejection or downgrade of the wood products: (i) excessive moisture content variation in the dried lumber boards and (ii) residual drying stresses and boards' distortions after drying.

If wood is not correctly dried, it can shrink leading to surface, end and internal checks (lengthwise separation of wood fibers that extends across the annual growth rings), splits (separation of wood fibers along the grain forming a crack or fissure), warp (distortion in lumber and other wood products causing departure from their original planes), case-hardening (a condition of varying degrees of stress set in wood such that the outer wood fibers are under compressive stress and the inner fibers under tensile stress), honeycomb (checks, often not visible on the surface, which occur most often in the interior of the wood, usually along the wood rays), and collapse resulting from the physical flattening of fibers above the fiber saturation point.

Drying too fast (i.e., air dry-bulb temperature too high, relative humidity too low, and/or velocity too fast) will result in failure of the wood (checking, splitting, honeycomb, etc.). In almost all cases, these failures will occur or become inevitable during the first drying period.

Drying too slowly (i.e., air relative humidity too high, velocity too low, and, rarely, dry-bulb temperature too low) will result in staining and discoloration, as well as warp and aggravation of surface checks and end splits. The risk is greatest above 40% (dry basis) of moisture content, but below 30% (dry basis), the risk is probably nil.

5.3.1 Shrinkage and Swelling

As many other hygroscopic materials, any piece of wood exchanges moisture with the surrounding air according to the level of atmospheric relative humidity. Too rapid loss or gain of moisture in wood products may alter the mechanism by which moisture migrates through the material and cause shrinking (when moisture is withdrawn from the wood by the dryer surrounding air and, thus, the wood's moisture content decreases) or swelling (when the dried wood absorbs water from moist air and its moisture content increases), usually dependent on the internal vapor pressure and linearly proportional to the change in moisture content below the hygroscopic limit (Stamm 1964; Zogzas et al. 1994; McCurdy 2005; Mujumdar and Huang 2007). Swelling typically occurs laterally and significantly diminishes the wood's structural performance (Berit Time 1988).

Shrinkage can be defined by the total variation of dimension between the green state and the oven-dry condition or in terms of variation of dimension divided by the variation in moisture content (known as shrinkage coefficient).

The amount of shrinkage varies strongly between and within species and tends to increase with basic density. Dense woods contain more cell wall material per cubic foot and, therefore, would be expected to show greater shrinkage on drying. The denser the wood, the greater the shrinkage that can be expected for a given change in moisture content. Substantial variation in shrinkage occurs within a given species, and therefore, it is impossible to predict accurately the shrinkage of an individual piece of wood.

Depending on the lumber being dried, the physical properties of the material, such as density and porosity, are affected by shrinkage during drying at various levels. As wood dries, shrinkage continues until the equilibrium moisture content is reached. Wood shrinkage is mainly responsible for wood ruptures and distortion of shape.

When a log is sawn into lumber, the drying process starts and soon after stacking in the air-drying yard, shrinkage of the board begins. When wood is at or above fiber saturation point, the cell wall is in a non-shrunken condition and wood is capable of holding large quantities of water. Below fiber saturation point, as water is removed from the cell wall by diffusion, the microfibrils are drawn closer together and the cell shrinks. As an approximation, volumetric change in wood below fiber saturation point is directly proportional to the volume of water gained or lost from the cell wall. In most cases, shrinkage takes place only if moisture is lost below fiber saturation point. The stresses set up in the surface zones of the lumber by shrinkage may cause deformation or failure.

Because the amount of shrinkage varies with the species of wood and the grain patterns of the lumber, a change in shape usually results. If the drying stresses exceed the strength of the wood, failures can develop, such as various types of splits and cracks.

Shrinkage is the driving force for drying stress, that is, without shrinkage, no drying stresses would develop. Drying stresses originate from shrinkage is responsible for mechanical degradation of the material. The cause of drying stresses in the differential shrinkage between the outer part of a board (the shell) and the interior part (the core) can also cause drying defects. Early in drying, the fibers in the shell dry first and begin to shrink. However, the core has not yet begun to dry and shrink. Consequently, the core prevents the shell from shrinking; thus, the shell goes into tension and the core into compression. This means that one of the principal difficulty experienced in the kiln drying of timber is the tendency of its outer layers to dry out more rapidly than the interior ones. If these layers are allowed to dry much below fiber saturation point while the interior is still saturated, drying stresses are set up because the shrinkage of the outer layers is restricted by the wet interior. The shrinkage strain is proportional to the difference between the local moisture content and the local value of the moisture content at fiber saturation point at the same temperature (Perré and Keey 2015).

The shrinkage results mainly in a reduction in wood's cell diameter, and, in turn, this results in a much greater shrinkage across the wood grain than along the grain.

Only moisture contained within the cell wall structure has an effect on shrinkage, and, consequently, all water in the cell cavity (lumen) must be removed before any shrinkage of the cell wall can occur. This is the basis for the concept of fiber saturation point (Cech and Pfaff 2000).

Shrinkage (volume change) of wood is not uniform in all dimensions (because of orientation of the micro-fibrils) and is generally expressed as a percentage of the green dimension. The reduction in size parallel to the growth ring, or circumferentially, is called tangential shrinkage, while the reduction in size parallel to the wood rays, or radially, is called radial shrinkage. Consequently, flat-sawn boards shrink more across their width than edge-grain or quarter-sawn boards. Because of the cell's tissues arrangement, the difference between earlywood and latewood structure, and the presence of ray cells, the greatest dimensional change occurs in a direction tangential to the growth rings (2%–10%), while longitudinal (along the grain) shrinkage is so slight (0.1%–0.3%, i.e., up to about 1/40 that of tangential shrinkage) as to be usually neglected (Walker et al. 1993; Keey et al. 2000; Perré and Keey 2015). The most commonly used explanation for the shrinkage and swelling anisotropy is the occurrence of wood rays in the radial direction. The effect is found to be more pronounced for hardwoods than for softwoods (Perré and Keey 2015).

In practice, it is practically impossible to completely eliminate dimensional change in wood, but elimination of change in size may be approximated by chemical modification. For example, wood can be treated with chemicals to replace the hydroxyl groups with other hydrophobic functional groups of modifying agents (Stamm 1964). Wood modification with acetic anhydride is among the high anti-shrink or anti-swell efficient existing processes attainable without damage to wood, but slow to be commercialized due to the cost, corrosion, and the entrapment of the acetic acid in wood.

As an approximation, volumetric change (shrinkage or swelling) of wood below fiber saturation point is directly proportional to the volume of water gained or lost from the cell wall and can be expressed as follows (Cech and Pfaff 2000):

$$S_v = SG(FSP - MC) \tag{5.1}$$

where

S_v is the total specific volumetric shrinkage $\left(\frac{m^3}{kg_{oven\text{-}dry}}\right)$

SG is the basic specific density of wood $\left(\frac{m^3}{kg_{green\ wood}}\right)$

FSP is the fiber saturation point $\left(\frac{kg_{water}}{kg_{oven\text{-}dry}}\right)$

MC is the moisture content of wood below fiber saturation point $\left(\frac{kg_{water}}{kg_{oven\text{-}dry}}\right)$

Assuming a fiber saturation point of 30%, each percentage point decreases in fiber saturation; the point below 30% would result in 1/30 of the total volumetric shrinkage.

The average shrinkage of a large number of pieces is more predictable, and to estimate this shrinkage from the green condition to any moisture content, the following equation can be used (Cech and Pfaff 2000):

$$MC_{green} = S_0 \frac{\left(30 - MC_{fin}\right)}{(30)} \tag{5.2}$$

where

MC_{green} is the percent shrinkage expected from the green condition (%).
S_0 is the total shrinkage (tangential or radial, %) (generally, given in tables).
MC_{fin} is the final overall average moisture content (below 30%, dry basis).

5.3.2 Other Drying Defects of Wood

Differential drying between the lumber board surface and the core combined with heterogeneous shrinkage can result in stress fracture of the wood grain, and distortion that can manifest as surface and end checks and splits, collapses, honeycomb, and warp.

Checks can develop into serious degrade if the lumber is cycled between humid and dry conditions, such as caused by long fan reversals cycles at air relative high humidity.

Surface checks, a result of using too low air relative humidity and/or too high velocity, occur especially during the first drying stage, most often on the face of flat-sawn boards. Surface checks, invisible from the surface, but visible when the lumber is planned, are more likely to occur with bacterial infected wood or lumber.

End checks, which appear on the ends of boards, frequently beginning in the log yard, are similar to surface checks and occur because the rapid longitudinal movement of moisture causes the board end to dry very quickly, developing high stresses, therefore fracturing. Nearly all end checks can be prevented with end coating on the logs and on freshly sawn lumber ends. End coating the freshly sawn lumber immediately after trimming to length may retard the end drying that causes these failures. To minimize end checking, several end coatings methods are available for use on lumber being air dried such as (i) applying emulsified waxes by either brushing or spraying at both lumber boards' ends and (ii) sprinkling the dry log deck with water.

When a surface or end check goes through from face to face (a defect called split), the drying rate must be slowed or else the checks will develop into internal checks.

Collapse, common in low-density woods with high green moisture contents, is distortion, flattening, or crushing of wood cells. The water is pulled out of the cells so fast that the cells collapse inward, reducing the lumen size. Very slow drying is the only potential preventative measure of collapse. However, almost all collapse can be recovered at the end of drying by steaming the lumber at 100% relative humidity.

It can be noted that removing the surface water faster than moisture is able to move from the board core, may dry the board's surface layer, and seal the pores, a sealing process called case-hardening (Frank Controls 1986). In other words, the effects of case-hardening are due to the drying stresses from shrinkage in one zone (board surface) acting on another (board core) at different moisture contents or due to a poor conditioning treatment. Once lumber is case-hardened, it takes more energy and time to dry, lumber must then be dried at low relative humidity, requiring more air venting. In addition, the case-hardened wood may warp considerably when the stress is released by sawing.

Reverse case-hardening occurs during rapid heating-up in high-humidity conditions, for example, by injecting steam at the end of a drying schedule. The rapid heating-up makes the surface of the wood hot and soggy, allowing the surface fibers to stretch and elongate the cells. As the air relative humidity is lowered and the wood dries, the surface fibers and cells do not return to their original shape. Therefore, the interior of the wood shrinks more than the surface causing honeycombing. Honeycomb is thus an internal crack that occurs in the later stages of kiln drying when the core of a board is in tension. It is caused when the core is still at relatively high moisture content and drying temperatures are too high for too long during this critical drying period (Cech and Pfaff 2000).

Wood degrade during air drying may also be caused by fungal infection (causing blue or sap stain, mold, mildew, and decay, which means that the lumber was dried too slowly at high moisture contents), insect infestation (resulting in pith flecks, pinholes, and grub holes left in the wood), chemical action (causing brown stain, gray stain, and sticker marking), or ultraviolet radiation (that degrades the surface of lumber and causes it to turn gray, a discoloration limited to a thin layer).

Differential shrinkage caused by differences in radial, tangential, and longitudinal shrinkage is a major cause of several types of warp (e.g., bow, crook, twist, oval, diamond, and cup). Warp, defined as deviation of the face or edge of a board from flatness, or any edge that is not at right angles to the adjacent face or edge, is generally a result of drying at an excessively high relative humidity or of poor stacking. Poor stacking (i.e., having stickers out of alignment, having bolsters out of alignment or not under every sticker or not level, or having too few stickers) can accentuate warp, especially waviness or bow along the lumber's length. Warp can also be caused by the difference in longitudinal shrinkage when juvenile wood is present on one edge or face of a board and normal wood is present on the opposite side.

Crook, twist, and cup are not too sensitive to stacking, except when stacking is very poor. Humidity and natural factors (tension wood, compression wood, and juvenile wood) dominate the causes of warp in most operations.

The wood color changes (darkening) during drying are a major issue for wood processors for production of high-value, appearance-grade timber. The development of surface discoloration such as kiln brown stain can cause significant economic loss and is particularly problematic in high-temperature drying (McCurdy and Pang 2007). Chemical discoloration is the result of oxidative and enzymatic reactions with chemical constituents in wood. Brown stain in pines and darkening in many hardwoods is a common problem when drying temperatures are too high. A deep grayish-brown chemical discoloration can occur in many hardwood species if initial drying is too slow. Methods of sap displacement (Kreber et al. 2001) and biological treatment (McCurdy et al. 2002) have been tested as a way to reduce the concentration of reactants and, therefore, discoloration. These methods have been successfully used for laboratory scale experiments, but are not viable for a commercial operation (McCurdy and Pang 2007). Other methods such as compression rolling (Kreber and Haslett 1997) and chemical inhibitors (Kreber et al. 1999) also show benefits but are not commercially viable (McCurdy and Pang 2007). Some studies have shown that vacuum drying (Wastney et al.1997) and modified drying atmospheres (Pang and Li 2005s) can reduce the formation of kiln brown stain. These methods require special drying equipment that many operators may be reluctant to adopt or the owners are unable to invest in (McCurdy and Pang 2007).

The simplest solution to kiln brown stain is to reduce the temperature of the wood during the drying process using low-temperature drying schedules. The reduction in temperature, however, comes with lower production levels and higher energy use for drying the same volume of timber. Some studies have suggested that slower straight drying schedules are not commercially viable; but, with better understanding of the kiln brown stain formation and quantification of the affecting factors (Kreber and Haslett 1998; McCurdy et al. 2005), optimized drying schedules are possible for producing bright colored wood at acceptable drying costs.

Fungal stains (often called blue or sap stain) are results of too slow drying processes above 40% moisture content (dry basis). Blue stain fungi do not cause decay of the sapwood, and fungi generally do not grow in heartwood. Immediate stacking after drying prevents fungal stains, and then, the stacked lumber is placed in a location where the relative humidity throughout the lumber pile is below 92%. Another common type of stain develops under stickers as a result of the contact of the sticker with the board. Sticker stains are imprints of the sticker that are darker or lighter than the wood between the stickers and can be caused by either chemical or fungal action, or both (McCurdy and Pang 2007).

5.4 CONTROL OF WOOD DRYING DEFECTS AND QUALITY

The new measurement instruments and computer devices offer today efficient means for improved control of lumber final quality. In addition, almost all defects of dried wood can be prevented by proper operation of the dryer, assuming that the equipment is correctly designed and in acceptable operating parameters. Also, good drying schedules may identify the cause of wood degrade and determine the cost of fixing of wrong practices.

In controlling the wood drying defects and quality, some of the following factors and/or parameters must be carefully considered:

a. Control the drying rate from the moment the lumber is sawn until the moisture content decreases below 40% (dry basis), for example, by measuring the average moisture content loss per hour or per day and comparing the drying rate to the accepted standard; it is also important to measure the average moisture content at the end of drying cycles; often, incorrect final average moisture content leads to catastrophic losses in secondary manufacturing processes.

b. Properly choose the drying schedule's air dry- and wet-bulb temperatures, relative and absolute humidity, and velocity passing through the lumber stack(s) (that are the most critical parameters in wood drying) because they can be responsible for wood drying defects (Perré and Keey 2015).
c. After the initial free water has been removed, excessive surface vaporization from lumber boards sets up high moisture gradients from the interior to the surface; this can cause overdrying and excessive shrinkage and, consequently, high tension within the material, resulting in cracking and warping; in these cases, surface vaporization should be retarded through the employment of high air relative humidity while maintaining the highest safe rate of internal moisture movement by supplying more heat; in other words, the successful control of drying defects in a wood drying process consists in maintaining a balance between the rate of vaporization of moisture from the lumber board surfaces and the rate of outward movement of moisture from the interior of the wood boards.
d. Drying too quickly, drying at too high dry-bulb temperature or at too low relative humidity, or lumber boards' poor stacking, may accentuate the wood defects resulting in uneven final moisture contents (too wet or too dry lumber) and not at correct levels (that generally must be within 2% of the equilibrium moisture content corresponding to lumber storage, manufacturing, or infield use).
e. The maximum drying temperature to which a given wood species can be dried (depending on the material thermal sensitivity) must be carefully selected; high values of dry-bulb drying temperatures may not only accelerate the internal moisture movement and activate the viscoelastic creep, but they can also increase the risk of collapse, discoloration, or even thermal degradation of the wood constituents.
f. Provide low moisture content gradients by reducing the wet-bulb depression as much as possible, condition which imposes a relatively high value of equilibrium moisture content; wood drying schedules generally must never exceed dry-bulb temperatures and temperature depression of 115°C and 45°C, respectively.
g. Provide high air velocity for good uniformity of drying throughout the lumber stack(s), even it may increase the electricity consumption and produce excessively high external heat transfer rates; in this case, the internal moisture movement has to be increased by exceeding the boiling point of water (i.e., by drying at high temperature), whereas external moisture vaporization should be reduced.
h. When stresses are severe enough that checks occur on the wood surface, rewetting the surface with wet steam (with hardwood) or water (with softwoods) may help the lumber surface to swell slightly, relieving the stress.

REFERENCES

Anderson, J.O. 2014. Energy and resource efficiency in convective drying systems in the process industry. Doctoral Thesis, Luleå University of Technology, Luleå, Sweden.

Berit Time. 1988. Hygroscopic moisture transport in wood. Doctoral Thesis, Department of Building and Construction Engineering, Norwegian University of Science and Technology, Trondheim, Norway.

Bian, Z. 2001. Airflow and wood drying models for wood kilns. MASc Thesis. Faculty of Graduate Studies, Department of Mechanical Engineering, the University of British Columbia, Vancouver BC, Canada.

Cech, M.J., F. Pfaff. 2000. *Operator Wood Drier Handbook for East of Canada*, edited by Forintek Corp., Canada's Eastern Forester Products Laboratory, Québec city, Québec, Canada.

Desch. H.E, J.M. Dinwoodie. 1996. *Structure, Properties, Conversion and Use. Timber.* 7th edition. Macmillan Education, London, p. 320.

Erickson, H.D. 1954. Mechanisms of moisture movement in woods. In *Proceedings of the 6th Annual Meeting of the Western Dry Kiln Clubs at Eureka*, California, May 14.

Frank Controls. 1986. Lumber drying theory (www.frankcontrols.com, accessed June 15, 2015).

Haslett, A.N. 1998. *Drying radiata pine in New Zealand*, New Zealand Forest Research Institute Limited.

Keey, R.B., T.A.G. Langrish, J.C.F. Walker. 2000. *Kiln Drying of Lumber*, Springer Series in Wood Science, edited by T.E. Timell, ISBN-13:978-3-642-64071-1.

Kreber, B., A.N. Haslett. 1997. A study of some factors promoting kiln brown stain formation in radiata pine. *Holz als Roh und Werkstoff* 55(4):215–220.

Kreber, B., A.N. Haslett. 1998. The current story on kiln brown stain, *FRI Bulletin* Issue No.23.

Kreber, B., A.N. Haslett, A.G. McDonald. 1999. Kiln brown stain in radiata pine: A short review on cause and methods for prevention. *Forest Products Journal* 49(4):66–70.

Kreber, B., M.R. Stahl, A.N. Haslett. 2001. Application of a novel de-watering process to control kiln brown stain in radiata pine. *Holz als Roh und Werkstoff* 59:29–34.

McCurdy, M.C. 2005. Efficient kiln drying of quality softwood timber, PhD Thesis. Chemical and Process Engineering, University of Canterbury, Christchurch, New Zealand.

McCurdy, M.C., R.B. Keey. 2002. The effect of growth-ring orientation on moisture movement in the high temperature drying of softwood boards. *Holz als Roh- und Werkstoff* 60(5):363–368.

McCurdy, M.C., S. Pang. 2007. Optimization of kiln drying for softwood, through simulation of wood stack drying, energy use, and wood color change. *Drying Technology* 25:1733–1740.

Mujumdar, A.S., L.X. Huang. 2007. Global R&D needs in drying. *Drying Technology* 25:647–658.

Oliveira, L., D. Elustondo, A.S. Mujumdar, R. Ananias, 2012. Canadian development in kiln drying. *Drying Technology* 30(15):1792–1799.

Onn, L.K. 2007. Studies of convective drying using numerical analysis on local hardwood, University Sains, Malaysia (https://core.ac.uk/download/pdf/32600391.pdf, accessed June 15, 2016).

Pang, S. 1994. High-temperature drying of pinus radiata boards in a batch kiln. PhD Thesis. University of Canterbury, Christchurch, New Zealand.

Pang, S., J. Li. 2005. Drying of Pinus radiata sapwood in oxygen free medium for bright colour wood. In *Proceedings of the 9th IUPRO Wood Drying Conference*, Nanjing, China.

Perré, P., R. Keey. 2015. Drying of wood: Principles and practice. In *Handbook of Industrial Drying*, 3rd edition, edited by A.J. Mujumdar, Dekker, New York, pp. 821–877.

Stamm, A.J. 1964. *Wood and Cellulose Science*. Ronald Press Company, New York.

Walker, J.C.F. 1993. *Primary Wood Processing Principles and Practice*, Chapman & Hall, London.

Walker, J.C.F., B.G. Butterfield, J.M. Harris, T.A.G. Langrish, J.M. Uprichard. 1993. *Primary Wood Processing – Principles and Practice*, edited by J.C.F. Walker, Chapman and Hall, London, 595pp. ISBN 978-1-4020-4393-2.

Wastney, S., R. Bates, B. Kreber, A. Haslett. 1997. The potential of vacuum drying to control kiln brown stain in radiata pine. *Holzforschung und Holzverwertung* 49(3):56–58.

Zogzas, N.P., Z.B. Maroulis, D. Marinos-Kouris. 1994. Densities, shrinkage and porosity of some vegetables during air drying. *Drying Technology* 12(7):1653–1666.

6 Wood Drying Methods

6.1 INTRODUCTION

Green wood (also called natural, unseasoned or timber) is an organic, hygroscopic, capillary porous, anisotropic, nonuniform, and heterogeneous mixture of solids, liquids, and gases. It contains considerable amounts of moisture (i.e., liquid water and vapor) depending on the relative humidity of the surrounding air.

For industrial (e.g., laminated and fiberboard products, furniture) and building (e.g., flooring, paneling) applications, many national legislations require the moisture be removed from lumber for minimal spoilage, quality preservation, and cost transportation reduction (Mujumdar 2015).

Wood drying is thus a key operation in sawmill production, affecting the product quality and value, and wood material suitability to different end-user products. The lumber drying process aims at a compromise among high quality, low lead time and low energy usage.

Historically, quality and drying time have always been prioritized before the efficiency of energy usage, even if wood drying is a process that consumes a lot of energy (as much as 70% of the total energy required in wood processing) and requires a lot of time (Mujumdar and Huang 2007).

6.2 WOOD DRYER SELECTION

Optimum selection of energy efficient dryers is dependent on (Law and Mujumdar 2007) (i) the properties and quality of the product to dry; (ii) relative cost of fossil fuel/electricity; (iii) ability to control the dryer; (iv) possibility of using new drying techniques, for example, heat pump-assisted, dielectric, pulse combustion, or superheated steam.

In drying industry, about 100 distinct dryers in different shapes and sizes are available throughout the world. About 85% of them are of forced, convective type with hot and dry air or direct combustion gases as the drying mediums. Dryers may consume up to 25% of the national industrial energy in most of the developed countries. Our knowledge base with regard to dryers has continued to expand over the years, but scale-up of most of the dryer types continues to be complex and empirical. Dryers are often equipment and product dependent because of the highly nonlinear nature of the thermodynamic transport processes (Mujumdar 2015).

Most of the economic studies focusing on different wood drying technologies are based on the total energy, capital, insurance/risk, environmental impacts, labor, maintenance, and product degrade costs for the task of removing water from the wood fiber. Since thousands of different types of wood products manufacturing plants exist around the globe, which may be integrated (lumber, plywood, paper, etc.) or stand-alone (lumber only), the true costs of the drying processes can be determined by comparing the total operating costs and risks of sawmills with and without drying.

Generally, industrial dryers are classified according to criteria as follows: (i) type of drying unit (i.e., batch or continuous, tray, rotating drum, fluidized bed, pneumatic, and spray); (ii) physical form of the dried material (fibrous, powder, granules, slurry, etc.); and (iii) methods of heat transfer (basically, convection, conduction, radiation, or dielectric).

The selection of wood drying process and equipment represents a compromise between the initial capital and operating costs. The choice is also dependent on production throughput, flexibility, and energy cost (Mujumdar 2015).

Typically, industrial timber drying kilns were selected and designed based only on experience, and trial and error. Therefore, it is necessary to develop methods for designing kilns to meet the increasing need to dry lumber quickly while ensuring minimal degradation and energy consumption.

Among the conventional methods used to dry lumber, the most common are natural or accelerated yard air drying, and artificial (kiln) drying.

In practice, natural and accelerated yard air drying and kiln drying are usually combined for drying hardwood and softwood species, at low, medium, and high temperatures (Cech and Pfaff 2000). In the United States and Canada, for example, the majority (more than 80%) of lumber production is kiln dried based on indirect steam heating, 7% on heat pump-assisted (dehumidifying drying) and 5% on direct-fired technologies. Less than 1% uses vacuum, radio-frequency heating or other unconventional drying methods. Dehumidifying drying, which has been used for several years in Europe for drying hardwoods, is becoming more and more popular in North America (Rosen 1995).

In both natural air and kiln forced-air convective drying, the water vapor in the atmospheric air has a lower partial pressure compared to the air close to the lumber boards, and the equilibrium principle provides the moisture movement from the lumber boards into the ambient air. As a result, the bound and free water inside and between the wooden cells will move toward the surface. A large difference in partial pressure will cause a faster moisture movement. However, if moisture movement is done too fast, an ununiformed distribution of the water can result in cracks and lumber deformation (Anderson 2014).

6.3 NATURAL AIR DRYING

The natural convection air drying (i.e., by exposition to atmospheric conditions) of fresh cut trees, green timbers, and lumber stack(s) is traditionally one of the most popular and cost-effective drying method. This slow, low initial cost, low energy consumption, and weather-dependent drying process uses natural, free energy as those supplied by sun and wind. If the weather is rainy and cold, drying takes place slowly; and if it is too hot, it can result in cracks and other quality problems. In regions with no suitable weather conditions, air drying can take months, or even years.

In the natural (yard) drying process, the ambient (atmospheric) air is used as moisture transport medium of evaporated moisture (water), but there is no control of the drying time and the product final quality, which may lead to fluctuation in the industrial production. However, pre-drying of wood in the sawmill yards at relatively low temperatures ahead of forced convective air, high-temperature kilns is advantageous for energy savings whenever woods have to be dried to high temperatures (Mujumdar 2015).

The natural (yard) drying process operates until wood moisture content drops, depending on species, thickness, location, and the time of year the lumber is stacked, at between 20% and 30%, prior transferring it to a wood kiln if lower final moisture content is required. By air drying wood for a year or more, moisture levels under 20% moisture content could be achieved, depending on the climate, species, and lumber thickness.

The major objective of natural air drying is to reduce the chances of the wood to develop mold, stain, or decay during transit and storage, and even use. Because many uses did not require additional drying, air drying may provide the correct moisture content for many wood products.

The natural air drying process consists of stacking of sawn timber (with the layers of boards as close to edge-to-edge as possible) separated by uniform stickers allowing free circulation of air forming packages) on raised foundations (i.e., concrete blocks, treated timbers or used railroad ties), in a clean, open (e.g., sawmill yard), dry, and shady place, with no artificial heating or control of ambient air velocity and relative humidity (Cech and Pfaff 2000) (Figure 6.1). The stickers must be uniform in size, aligned one on top of the other, and the end of each board must rest on a sticker. On the topmost layer of lumber, thicker and longer pieces (not shown in Figure 6.1) are used to support a small, sloped roof to protect the lumber piles from precipitations and sun rays.

The lumber stacks are exposed to the ambient (outdoor) air at low dry-bulb temperatures (typically, from 24°C to 38°C) and relative humidity (from 60% to 90%).

FIGURE 6.1 Example of one lumber board stack prepared for natural (yard) air drying (note: lumber stack only shown; reproduced with permission from Group Crête).

According to the wood species and size, the site selection and preparation, yard layout (that must be dry and clean), as well as proper stacking and orientation of lumber stack(s) are essential to make air drying efficient. An efficient yard air drying should be situated on high, well-drained ground with no obstruction to prevailing winds, placed away from buildings, trees, or other objects that can block the wind to penetrate between the lumber packages (Onn 2007). Lumber piles have to never be stacked over vegetative-covered ground since the bottom layer will always be exposed to air with a higher humidity.

The lumber piles must be oriented parallel with the prevalent direction of wind for better circulation and therefore more uniform drying. Adequate alleys and spaces for the transport of lumber to the piles must be provided between rows and piles, sufficiently large, straight and continuous across the yard, and perpendicular to the prevailing wind direction.

During natural air drying, the outdoor air enters the lumber piles and removes moisture from the wood boards. As the cool, damp air leaves the piles, fresh air enters and the drying process continues. Through the lumber packages, the air circulation is developed by natural convection because warm, dry air enters the sides of the lumber stacks, moves thorough them, and evaporates the moisture from the boards' surfaces. Through the process of evaporation, the air becomes cooler, moister, and heavier, and, thus, moves toward the bottom of the stacks. If the prevailing wind moves freely, the cool, moist air is blown away, generally laterally across the packages, and replaced with warmer, drier air. Downward movement of air can also be allowed by building vertical spacing boards within the pile packages. Because the top boards are more exposed to wind than the lower boards, they will dry quicker. Moreover, the air near the surface of the ground and the bottom of the pile is generally cooler and, consequently, has a higher relative humidity than the air near the top of the pile. This cold moist air that enters the sides of the lumber pile in the lower portions will not induce rapid drying as the hotter, drier air entering the upper parts of the lumber pile (Cech and Pfaff 2000). Increasing the height of the open pile foundations to allow more space under the pile will increase the drying rate providing a more rapid and economical drying process (Onn 2007).

The rate of natural air drying and the minimum moisture content attainable depend on factors such as (i) the weather (e.g., lumber piled during spring and summer will reach 20% moisture content faster that that piled during the fall or winter), (ii) wood species and specific gravity, (iii) climatic conditions (lumber in the pile dries most rapidly when the air dry-bulb temperature is high), (iv) board thickness, (v) yard layout and piling arrangement, (vi) season of stacking, and (vii) configuration of air movement.

For high-value lumber, the drying rate can be controlled by coating the ends of logs with substances relatively impermeable to moisture (as special mineral oils), generally not soak in more than 1–2 mm below the surface (Cech and Pfaff 2000).

Table 6.1 shows some approximate air drying periods for freshly cut (green) 25 mm thick softwood lumber boards to attain 20% (dry basis) moisture content. Low-density lumber species (such as Pine, Spruce, and Soft maple), stacked in favorable locations and times of the year, may require 15–30 days for air drying, while

TABLE 6.1
Approximate Air Drying Periods for 25-mm-Thick Lumber

Species	Days Required to Dry Lumber up to 20% (d.b.)
Red oak	100–300
White oak	150–300
Southern pine	40–150
Yellow poplar	60–150

slow drying species (such as red and white oaks), in northern locations stacked at unfavorable times of the year, could require up to 300 days. However, southern pine and yellow poplar will dry much faster in the yard (Walker 1993; Cech and Pfaff 2000). On the other hand, heavier hardwoods require longer drying periods to reach the desired average moisture content.

The benefits of natural (yard) air drying are as follows (Cech and Pfaff 2000):

a. Relatively low initial capital and operating costs, mainly associated with storing the wood and the slower process of getting the wood to market;
b. Reduction of fossil energy consumptions and costs, and of associated atmospheric emissions;
c. Less drying defaults resulting in higher quality (particularly for hardwoods), brighter lumber with more uniform moisture content throughout the boards, and more easily workable wood;
d. Further reduction of drying defects and the chance that mold and decay may develop in transit, storage, or subsequent use;
e. Reduction of the subsequent kiln capacity and drying time by 30% or more;
f. Reduction of lumber weight and shipping costs;
g. In regions where annual mean temperatures are not very low and space is not particularly limited, air drying of easy-to-dry lumber is an economical and energy conserving method to withdraw most of the water from lumber prior to kiln drying.

The natural air drying's drawbacks can be mentioned as follows: (i) the output is dependent on the local climate (ambient dry- and wet-bulb temperatures, relative humidity, wind velocity, rainfall, sunshine, etc.); (ii) drying rate is uncontrolled and, depending on the weather conditions, the natural drying process may take several months or years; (iii) during the cold winter months (as in Nordic countries), drying rates are very slow and, thus, it is generally difficult to dry the timber to required target moisture contents; moreover, the drying time (t_{drying}) varies with the thickness of lumber boards (δ) approximately at the power of 1.5 ($t_{drying} \approx \delta^{1.5}$); this means that, for example, a 50 mm thick lumber board will require about three times longer time to lose a given amount of moisture than 25 mm thick lumber board; (iv) under improper drying procedures, significant wood defects, degrade or losses exceeding 10%, such as fungal infection, insect infestation, discoloration, and shrinkage (Onn 2007) can occur, especially if drying is prolonged beyond the time needed to lower the moisture content to 25% or less; (v) the lack of absolute control of drying conditions, for example, hot and dry low velocity winds, strong winds in combination with low relative humidity, high humidity, heavy rain, and fog may increase degradation and volume losses as a result of severe surface checking and end splitting; such conditions may also facilitate the growth of fungal stains, aggravate chemical stains, brown discoloration, and gray stain; (vi) higher costs of carrying a large inventory of high-value lumber for extended periods.

6.4 ACCELERATED NATURAL AIR DRYING

In order to increase the wood drying rate and reduce the drying time, stickered packages of lumber boards are sometimes placed in unheated, permanent roofed structures (sheds) having open sides (Figure 6.2). The roof constructed over the wood stacks helps protecting the lumber from rain and direct solar insolation.

Because the atmospheric air is forced through the wood piles by multiblade fans, this drying method is known as accelerated air drying, similar to the natural (yard) air drying (Section 6.3). The multiblade fans are installed on one side to pull outside air through the spaces between lumber boards and piles to provide uniform airflow at velocities of 2.5–3.6 m/s and, thus, accelerating the drying process (Onn 2007).

The accelerated air drying is most advantageous, particularly for fast pre-drying of large quantities of low-degrade-risk wood lumber (up to 700 m^3) at low temperatures (20°C–45°C) and up to moisture contents of 25% (dry basis), prior entering the conventional kiln dryers.

Such a process reduces the total drying time of lumber, while the over-cost of capital investment in building materials (sheds and roofs) is minimal compared to conventional natural (yard) air drying.

FIGURE 6.2 View of a shed industrial wood natural air drying facility (www.google.fr/search?q=images+shed+industrial+wood+yard+natural+air+drying. Accessed October 29, 2017).

However, the accelerated air drying method increases the cost of pre-drying because of the cost of additional equipment (e.g., fans and motors) and associated electrical electricity consumption. To minimize electrical energy consumption, controllers of relative humidity (humidistats) are installed to stop the fans when higher air velocities are undesirable or ineffective (Onn 2007).

The efficiency of accelerated air drying is greatest for lumber with high initial moisture contents and thin wood boards, but depends on the local climate. The air relative humidity is influenced solely by the evaporation of moisture from the wood and by the ambient atmospheric conditions. During periods of high humidity and/or too low ambient temperature, the removal of moisture from the wood board surface and the circulating air could be retarded, and fans may have little effect on drying rate. By venting, some control over extremes of relative humidity can also be provided (Cech and Pfaff 2000).

6.5 CONVENTIONAL KILN WOOD DRYING

Today, open yard natural air drying is no longer widely accepted as a only one quality drying method for wood lumber mainly because (i) the quality and availability of the wood resource are lower than in the past; (ii) users' demands for quality have risen; (iii) the value of many lumber species has increased, thus losses in quality are more expensive; and (iv) as a result of low to moderate interest rates, the air drying for a too long time may be expensive and increase inventory costs.

Therefore, after yard natural drying, drying the wood under controlled conditions in conventional kilns (where hot air as drying medium at controlled velocity, dry- and wet-bulb temperatures and relative humidity to remove moisture from the lumber board stacks) is generally used at higher temperatures compared to those of the ambient conditions became the current technology in almost all modern sawmills, even such a drying technique is responsible for large heat and electricity consumptions, and power demands.

The kiln drying method is well suited to dry timbers provided to construction and furniture industries of varied species (as green Beech, Birch, and Maple) and thickness, including refractory hardwoods that are more liable than other species to check and split.

The goal is to lower the initial moisture content of the wood (60%–120%, dry basis) up to optimal moisture content between 6% and 12% (dry basis) so that it can be stored and/or transported without damage. In North America, for example, more than 88% of the overall lumber production is kiln-dried (Bian 2001). Moreover, the kilns improve the overall thermal efficiency of the drying process by offering opportunities of recovering a part of heat losses (Bian 2001; Anderson and Westerlund 2014).

Compared to the natural (yard) air drying, kiln drying presents the following advantages: (i) for most large-scale drying operations, the conventional kiln drying is more efficient than air (yard) drying; (ii) better control of the final moisture content regardless of weather conditions; lumber can be dried to any desired low moisture content, but moisture contents of less than 18% (dry basis) are difficult to attain; (iii) reduction in drying time obtained by using higher drying temperatures

(Cech and Pfaff 2000); (iv) with dry-bulb drying temperatures above 60°C, all the fungi and insects are killed; (v) higher throughput; and (vi) controlled humidification is possible by including equalization periods to reduce the variation in final moisture content, and conditioning treatment to relieve residual stresses.

Among the disadvantages of conventional kilns, the following can be mentioned: (i) much higher capital, operation, and maintenance costs as compared to natural (yard) air drying; and (ii) heat losses and inefficiencies in heat transfer usually limit the drying dry-bulb temperatures between 82°C and 93°C.

6.5.1 Kiln Construction and Operation

A conventional dry kiln is a large industrial building (also called room, enclosure, or chamber) that provides an appropriate artificial environment allowing optimizing the wood drying process. The majority of the conventional wood dryers (kilns) employ batch loaded, stationary air forced convection operation with indirect heating using steam from oil-, natural gas- or biomass-fired boilers.

The industrial wood lumber batch kilns may be either package-type (side loader, as most of hardwood lumber kilns in which fork trucks can be used to load lumber packages into the kiln) or track-type (or tram, as for most softwood lumber kilns) constructions.

Process controllers, which include air dry- and wet-bulb temperature, relative humidity, and velocity sensors, as well as heating coils and spray devices, allow controlling the process parameters until the lumber reaches the required moisture content in the least amount of time, while avoiding the creation of wood defects. The control modes can be manual, semiautomatic or fully automatic, depending on the dryer size and wood species to dry.

The kiln geometry and construction contribute at reducing (fan and heating) energy requirements, shorten the drying times, and providing more uniform moisture content throughout the lumber stack(s) (Keey and Nijdam 2002), while the quality of kiln construction directly affect the drying efficiency.

Traditionally, industrial prefabricated lumber kilns were usually available in a variety of sizes and shapes (e.g., capacities between 10 and 200 m^3) of which form and construction were designed to supply and control the air dry- and wet-bulb temperatures, relative and absolute humidity, and circulation (mass flow rate and velocity).

Today, the modern kilns are compartment-type, closed chambers in sizes able to hold as much as 350–400 m^3 of stickered lumber. They are constructed of sheet steel metal airtight and thermally insulated panels in order to reduce the energy consumption, prevent condensation on the walls, and for better control of leaks out of the structure. Steel and aluminum are also used for support of fan decks, motors, and heating coils (Cech and Pfaff 2000).

Industrial batch lumber kilns are generally built on minimum 25 cm thick concrete foundations (depending on the type of soil and loads to be imposed), extending from 1.22 m below grade and to 15 cm or more above grade. Most kiln floors are made of poured concrete, usually 5 cm thick.

The insulation is composed of multiple layers of either aluminum or stainless steel in the inner and outer surfaces sandwiching a central core of insulation material, usually of mineral glass, wool, or expanded polyurethane foams. The kiln insulation protects the wood dryers against loss of heat and moisture. It also provides protection against the effects of high dry-bulb temperatures and relative/absolute humidity, the expansion and contraction caused by frequent and wide changes in temperature, and the frequently corrosive action of acidic vapor from the wood being dried. Kiln doors are insulated panel-type construction elements with rubber gaskets around the openings to keep infiltration and leakage to a minimum (Cech and Pfaff 2000).

Generally, the conventional kilns are charged with lumber having similar drying characteristics. However, differences between the drying characteristics of boards always exist because of stack(s) composition (heartwood and/or sapwood), different species (density, permeability, and anatomical properties), thickness, grain (flat sawn or quarter sawn), initial moisture content, etc. Density of wood allows determining which species can be dried together in a mixed load. Species such as soft maple and yellow poplar with low density are relatively easy to dry, while others, such as the oaks, black walnut, and beech, are more likely to check or honeycomb during kiln drying.

The lumber is normally stacked externally (outdoor) in a uniform length rectangular piles on a low, flatbed trolley, with the timber placed in flat layers of up to 2.4 m width. Each layer is separated by 12.7 mm (½″) to 31.75 mm (1¼″) thick stickers (strips, spacers) vertically aligned at 0.6 m intervals along the length of the timber stack to provide spaces of uniform hydraulic resistance for the drying air to flow between the board layers (Bian 2001).

Stickers are small wooden boards of uniform thickness that separate the lumber boards. Stickers should be more wide than thick (e.g., 44.45 mm or 1 3/4″). However, for example, 45 × 19 mm stickers may be acceptable for fast-drying softwoods, but would be too wide for slow-drying hardwoods, for which 30 × 30 mm might be more appropriate (Keey et al. 2000). Stickers must also be properly placed, depending on thickness of the lumber, as far apart as possible (generally, between 600 and 900 mm for almost any size lumber) to ensure good air circulation at equal distances across each layer of lumber and aligned on top of one another from the bottom of the stack to the top.

The stacks' boards are butted up with their long faces incident to the airflow. Conventional wood kilns can be loaded with separate stacks (packages, piles) of lumber introduced on single or twin trays (Rasmussen 1988). With single-track kilns (generally, with 2.4 m wide stack), more uniform drying can be achieved, but the twin-track kilns (i.e., double-width stack of 4.8 m) may increase kiln capacity. In Australia and New Zealand, for example, a typical high-temperature drying kiln has a timber charge volume of 50–70 m^3 held on a single or double track.

After stacking, the lumber packages are railed and placed inside the kiln enclosure (Figure 6.3). The batch kilns are generally charged with lumber stacks in one operation, and the lumber remains stationary during the entire drying cycle, while temperature and relative humidity are kept as uniform as possible throughout the

FIGURE 6.3 Picture showing a single lumber package railed inside a conventional wood kiln. (Source: https://www.pinterest.ca/dwayneowens146/wood-kiln/.)

kiln and controlled over wide ranges based on specific schedules taking into account the moisture content and/or the drying rate of the lumber. Inside the kilns, lumber boards should be also properly stacked in the same way as for natural (yard) air drying, that is, in packages (stacks) of boards of uniform lengths, shorter pieces placed above the longer ones, and uniform-sized stickers at equal distance across each layer of lumber allowing a more uniform drying rate for each piece of lumber (Keey et al. 2000).

In northern sawmills (e.g., those located in Canada and Sweden), the major part of the energy in lumber production is used for the drying process, and electricity is used for electrically driven transportation, sawing, grinding, fans for the drying kilns, room lighting, etc. (Anderson 2014). The ideal option for sawmills is to use their local wood waste to fire the boilers for kiln operations, thus reducing global energy costs.

Heat required for kiln warming-up, wood drying, and supplementary (back-up) heating is generally transferred by forced convection to the exposed lumber boards' surfaces via the heated air, and the vaporized moisture carried away by the same drying medium.

Heat can be supplied inside the wood kilns either directly (by using combustion gases or, occasionally, electrical heating elements) or indirectly (generally, by using steam, hot water or oil flowing through plain or fined heat exchangers controlled by on/off or proportional pneumatic valves, or electrical actuators) at temperatures of drying medium ranging from 50°C to 150°C, depending on the wood species.

Heat is typically introduced by steam (regulated by air- or electrically operated valves) flowing through heat exchangers of copper (the best conductor, but expensive and subject to damage) or iron tubes (the most rigid and serviceable but subject to corrosion), and aluminum (an excellent heat conductor, but much more subject to damage) fins. The need for steam for drying operations provides an ideal opportunity to salvage the energy potential of wood residues by means of waste-fired boilers.

Steam coils are of various types as (Cech and Pfaff 2000) follows: (i) plain-return header (either horizontal or vertical, most commonly used); (ii) single-return header; and (iii) multiple-return header. In multiple track kilns where the air circuit from the fans passes through more than one lumber stack, it is customary to install booster coils vertically between the tracks in order to prevent an excessive temperature drop across the lumber packages.

In electrically heated kilns, the heating elements are located in the same position as the steam heated coils, that is, above the false ceiling adjacent to the circulation fans.

Auxiliary energy can be supplied via radio-frequency, microwave and infrared electrical devices, generally under vacuum conditions, or heat pumps, and from superheated steam (Mujumdar 2015), and solar energy.

For kiln structure and wood charge warming-up, the boilers of high-temperature wood kilns located in cold climates should be able of producing kiln temperatures of up to 116°C within 6–8 hours in a fully loaded kiln operating under normal winter drying conditions. Steam at 204.28 kPa has a saturated temperature of 121°C, but, due to heat losses and pressure drop in the system, this pressure is not sufficient to achieve drying temperature of 116°C. Even with steam at 302.28 kPa and 134°C saturated temperature, the heat-up period would be more than 8 hours, defeating the advantages of short drying times and increased production capacity. The heat-up period could be reduced by increasing the number of heating coils and the volume of steam, but this would be prohibitive in terms of the increased cost of the kiln and the requirements for larger and more expensive piping, valves, steam traps, etc. For an efficient high-temperature system, a steam pressure of at least 698 kPa is recommended. Some sawmill plants use high-pressure steam up to 1,034 kPa and, by means of reducing valves, are able to operate several kilns at lower pressure, depending on the capacity of the boiler. High-pressure boilers also offer more flexibility with respect to maximum drying temperature (Cech and Pfaff 2000).

The supply steam lines from the boiler to the kiln must be properly insulated to reduce steam consumption and maintained on a slight slope. The large volumes of condensate must be allowed to drain freely via steam traps, usually located outside the kiln, which remove the condensed water from the heating system without wasting steam or reducing its pressure. Commonly, the hot condensate is returned to the boiler.

6.5.1.1 Fans

Any optimized kiln should remove water inside the wood in the least amount of time to a desired moisture content and with a low moisture variation, while utilizing the least amount of energy, and avoiding wood drying defects that can occur during convection drying (Mackay and Oliveira 1989). Because forced air circulation is required to carry heat from the heating coils to the lumber being dried, and to carry away the moisture vaporized from the wood, the kilns are fitted with overhead fans powered by electrical motors which may be installed outside the kiln chamber or inside it.

The overhead fans circulating the air have to overcome the pressure drops accompanying the changes in air direction, the losses at the inlet and outlet faces of the stack(s), and the pressure loss in forcing the air through the passageways in the lumber stack(s) (Keey et al. 2000). The dryer's fans are situated above the kiln stack(s) in a false ceiling (fan deck) in such a way as to ensure even distribution of the air and heat needed for drying (see Figure 6.4). The drying air moves perpendicularly to the length of the lumber packages and parallel to the stickers that separate each layer of lumber boards (Pang and Haslett 1995; Cech and Pfaff 2000).

The overhead fans create a negative pressure in the plenum of the drying chamber on the downstream side and a positive pressure on the upstream side. This pressure difference forces the air through the stack of lumber. To obtain as uniform airflow distribution as possible to the air-inlet face of the stack, the plenum spaces at each side of the stack have to be sufficiently wide (Nijdam et al. 2000).

The fans are alternately right hand and left hand, so that two adjacent fans force air toward each other. Air circulation is periodically reversed by a timer to provide effective airflow and obtain more uniform heating and drying.

From the false ceiling, the fans direct the drying air through the lumber stacks via baffles or curtains installed at either side of the lumber packages. The baffles and kiln walls deflect the air stream downward along one side of the lumber stack(s) and, then, through each lumber stack from one side to the other. Inward-swinging side baffles or curtains and contoured, right-angled bends to the plenum space from the ceiling space improve the uniformity of the airflow and, in addition, minimize the bypass of drying air around the lumber stack(s).

Modern kilns are equipped with variable speed fans or with two- or three-speed settings that can provide considerable savings in electrical energy consumption

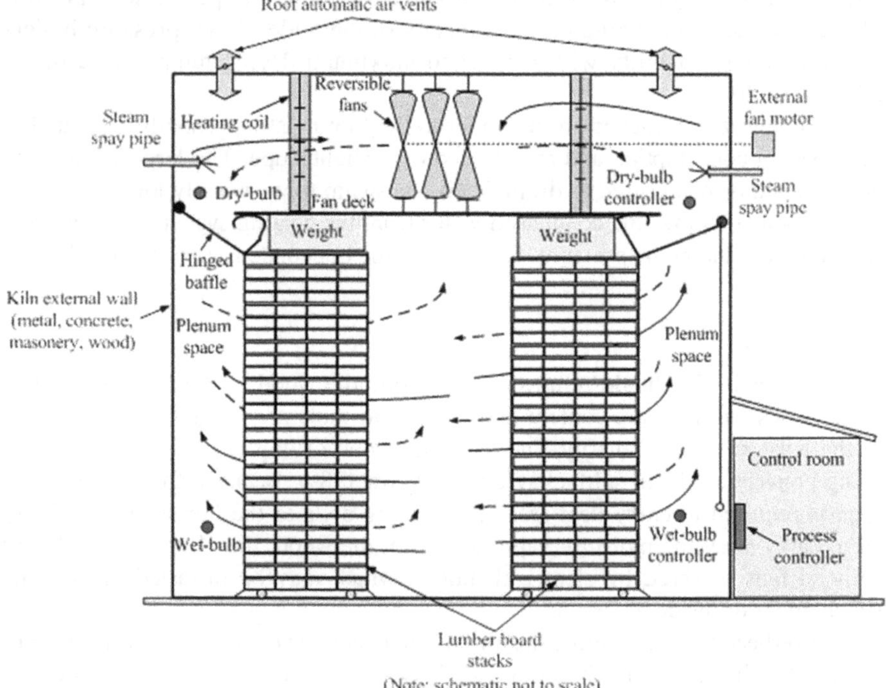

FIGURE 6.4 Schematic front view of a conventional batch convective-type wood kiln dryer (note: schematic not to scale).

by enabling the use of high air velocities when the moisture content is above the fiber saturation point, and of substantially lower air velocities after the moisture content has been reduced (Cech and Pfaff 2000). A typical conventional wood kiln (see Figure 6.4) provides control over the air dry- and wet-bulb temperatures, relative and absolute humidity, and air velocity (mass flow rate). In such conventional wood kilns, the air state changes over time corresponding to predetermined drying schedules, while the lumber stack(s) does not move during the drying process.

Air relative humidity is traditionally measured with psychrometers, primarily because of their simplicity and reliability over other methods in monitoring conditions close to moist air saturation (Keey et al. 2000). Most steam-heated kilns use dry- and wet-bulb temperature-sensing bulbs under the overhead coils, one in each plenum, on either side of the heating coils. In kilns divided into two zones along their length, the temperature-sensing bulbs are placed midway in each zone. The temperature-sensing bulbs must be protected from radiation from hot heating coils and be exposed to an air velocity of at least 5 m/s for sensing the psychrometric values correctly. In practice, the error in temperature readings are relatively small and tolerable in most industrial drying operations.

6.5.1.2 Air Venting

As the hot air is passed through the kiln's lumber stack(s), the wet timber absorbs heat, resulting in the movement of moisture to the boards' surfaces, where it is absorbed by the circulating air. When the circulating air becomes saturated with moisture, the air vents (usually included in specific roof panels) open and the excess moisture is vented to the atmosphere. Fresh air is then drawn into the kiln, heated, humidified, and the wood drying process continues.

The air vents are arranged in one- or two-vent lines, depending on the fan arrangement, along the length of the kiln. Each line of air vents is automatically opened and closed by electrically or pneumatically powered motors. Generally, cool dry air is introduced at one end of the kiln while warm moist air is expelled at the other by using the pressure difference generated by the reversible fans. In other words, the air vents on the downstream side draw warmer dryer air into the drying chamber, and the air vents on the upstream side (positive pressure) exhaust cooler moist air. Usually, the air vent system in any kiln exhausts more air than it draws into the kiln, to accommodate the expansion in volume of incoming air as it is heated.

An alternative solution consists in using powered (forced) ventilation for air venting. Even the initial capital investment for the powered venting system is greater, the higher costs could be offset by more efficient drying and lower maintenance costs.

6.5.1.3 Humidification

The requirements for final quality of wood limit the rise to air dry-bulb temperature inside the kiln. Excessively, hot air can almost completely dehydrate the lumber boards' surfaces so that its pores shrink. In order to avoid excessive surface dehydration and board deformation, a humidification process is usually required to compensate for humidity losses due to uncontrolled air leakages and, thus, keep the relative humidity inside the kiln from dropping too low during the drying cycle (Elustondo and Oliveira 2009) and to maintain the kiln humidity at the desired level toward the end of drying schedules.

The control of humidity is essential for satisfactory operation of kiln drying of wood. This may be accomplished by means of opening air vents or through direct steam injection via spray pipes (usually supplied from the same heat source as for drying), or cold water misting (McCurdy 2005) to supply the moisture necessary to maintain the desired relative humidity in a kiln. Steam is injected through special nozzles located in the air stream adjacent to the circulation fans. As with the heating system, steam humidification is regulated by either electrically or, more commonly, pneumatically-controlled valves. Steam injection and water spray are more critical in the case of slow-drying, high-valued hardwoods involving problems with checking, warp, and wastage (Cech and Pfaff 2000).

If high-pressure steam is used for humidification, it may add a considerable amount of heat to the kiln, causing fluctuation and overshoot of the dry-bulb temperature. This is especially critical during the conditioning period when a large volume of steam is added to the kiln to bring up the wet-bulb temperature. Steam pressure should, therefore, be reduced by a pressure regulator or a simple throttling valve. A reduction of pressure increases the time required to attain the wet-bulb setting, but this can be counterbalanced by reducing the dry-bulb temperature, so that the proper conditioning equilibrium moisture content can be achieved at a lower wet-bulb setting.

Once the lumber kiln type has been chosen, designed, and constructed, some of the following aspects should help to evaluate the conditions of the kiln loading and operation (Frank Controls 1986): (i) all dryer fans must work and blow the air in the right direction; (ii) when installing new blades or motors, the fan direction must be easy to reverse because it may seriously affect the airflow through the wood stack(s); (iii) the pitch on the fan blades and amperage on the fan motors should be within normal ranges to ensure that maximum efficiency is obtained without overloading the system; (iv) top and bottom baffles should be in good shape, making contact with the load; overhead baffles must be lowered until all baffles are resting on the top of the load; (v) the height of the load should be the same along the length of the charge; if the top of the loads are uneven, the baffles will not seal the top properly; (vi) if the tracks are two lifts wide, the space or chimney between the loads should be a minimum of 2 inches and a maximum of 6 inches if the lifts are not spaced; (vii) all packages should be butted up as tightly as possible to reduce bypass air cutting across the kiln without going through the load; the uneven ends will allow the most bypass air and should be kept as far from the probes as possible; (viii) the air vents should be inspected for leakage, and properly adjusted; (ix) if leakage from the kiln is severe, the kiln would be leaking at a faster rate than it would vent the kiln; problem areas are vents, doors, and safety windows; a well-sealed kiln will give a much better performance than a kiln that leaks too much, not requiring the vents to open; and (x) wet-bulb temperature sensors must be placed in direct airflow, not around wall obstructions causing turbulence; generally, the further the probes are placed from the wall, the more reliable they will be; the wet bulb must have a constant supply of water and a clean sock; if the sock becomes clogged with resin and dirt, it will begin to read too high, giving a false picture of the relative humidity in the kiln; wet-bulb socks should be regularly checked and replaced for reliable results.

6.5.1.4 Kiln Drying Temperatures

In conventional wood dryers, drying temperatures can be below or above the water boiling temperature at the atmospheric pressure (≈100°C).

6.5.1.4.1 Low and Medium Temperatures

Conventional low-temperature kilns are constructed of materials such as concrete blocks, wood, and steel that are least expensive. They are most advantageous for slow-drying high-value hardwoods, which generally are pre-dried with natural (yard) air to 25% or 30% (dry basis) of moisture content prior entering to conventional kilns.

Low-temperature and medium-temperature dryers operate at standard temperatures between 20°C–45°C and 45°C–82°C, respectively, where interior fans produce air velocities of 2.5–3.6 m/s. They are designed to accommodate large quantities of wood (236–700 m^3) with minimal capital expenditure in building materials and equipment. Most difficult-to-dry hardwoods and easy-to-dry softwoods are typically dried at low dry-bulb temperatures. However, for very difficult-to-dry hardwood species, the drying dry-bulb temperature might not exceed 60°C).

In most of the kilns, air circulates several times through the lumber stack(s), and only a portion is returned to the heating chamber, usually by means of a collecting plenum running along the full length of the kiln. Such drying systems are well suited to medium- and high-temperature kiln drying of softwoods, mainly because the required temperatures are easily achieved and controlled, and any discoloration of wood due to the effect of combustion gases is of little consequence in most softwood operations.

6.5.1.4.2 High Temperatures

To dry large quantities of construction softwood lumber, when color development is not a prime concern, high temperatures (83°C–149°C) are usually used in convective drying kilns to provide relatively fast drying processes with air velocities between the boards higher than 5 m/s.

Refractory hardwood species such as oak, hard maple, and birch cannot be dried satisfactorily at high temperatures directly from the green condition. In these cases, it is recommended to pre-dry the lumber at low temperatures to about 20% (dry basis) moisture content prior to high-temperature kiln drying. Such low-high temperature combination schedules can result in improved lumber quality. However, reductions in drying time of hard-to-dry species are marginal, making the drying of high-density hardwoods in high-temperature kilns economically uncertain.

High-temperature kilns are constructed of insulated aluminum panels whose costs do not differ appreciably compared to those of conventional low-temperature kilns, but, for a given kiln capacity, high-temperature dryers require much higher rates of heat inputs which increases the total cost of the wood drying system. The reduction in drying time for softwood construction lumber, however, is usually sufficient to offset this cost.

In addition to increased heating capacity there are additional requirements for kiln construction materials, insulation, thermal expansion, reliable heat exchange and venting of moisture vapor devices, and maintenance for high-pressure steam boilers. Even the equipment is similar to conventional to conventional low- and medium-temperature dryers, it must be designed for temperatures that may approach 149°C. If kilns are heated by steam, a high pressure boiler is required, and this in turn may require the services of a qualified stationary engineer.

The capital investment per unit capacity of high temperature is higher, but drying times and energy consumption may be reduced by up to 50% compared to conventional wood dryers (Minea 2004). With shorter drying times, fewer or smaller kilns could be required or, alternatively, production can be increased, which may offset the disadvantage of higher investment costs. The reduced energy is primarily the result of the large reduction in drying time to better heat transfer to the wood and improved insulation of the drying enclosure (Cech and Pfaff 2000). The actual savings, however, depend on species, initial moisture content, type of heating system, and drying temperature.

6.6 SPECIALIZED WOOD DRYING METHODS

Drying techniques generally refer to the energy use based on how the drying medium is heated (e.g., by fuel or electricity), how residual energy in exhaust air is recovered, and how the control system is applied to maximize energy utilization, etc.

Specialized drying technologies as direct-fired, radio frequency and microwave, infrared, vacuum, solar (more attractive in remote locations for small kilns), and heat pump-assisted (dehumidification) are usually more expensive and oriented to particular high-value end products. In addition, some of these technologies use higher grade energy (electricity), which, generally, is expensive (Keey et al. 2000; McCurdy 2005; Perré and Keey 2015). Heat pump-assisted dryers may also include auxiliary heating sources, such as electromagnetic radiation in the form of radio frequency, microwave, infrared, and solar energy.

6.6.1 Direct Firing Drying

Direct firing is common in wood drying industry, mainly because this drying technology is more rapid and cost-effective compared, for example, with electric heating methods (Law and Mujumdar 2008).

In direct-fired kilns, the drying medium is the combustion flue from a burner using natural gas (the most common, with calorific gross heating power ranging from approximately 37–45 MJ/m^3, including the heat of vaporization of the water vapor formed by combustion), oil, diesel, and liquefied propane or butane as energy sources, depending on their relative local prices. In the heating chamber, the hot products of combustion are mixed with the circulating air, raising its temperature to the point where subsequent mixing in the kiln will produce drying temperatures required by the drying schedules. The burner, the combustion chamber, and the temperature regulating and safety equipment are commonly located in a kiln control room, and the air is drawn from the kiln, heated, and returned to it by a centrifugal blower and ducts which distribute the hot air to the plenum chamber and to the dryer main circulation fans. Burners commonly have electrically or pneumatically modulated fuel valves which operate in connection with a recording/controlling instrument. The fuel and air supply for combustion is regulated so that the desired kiln temperature is maintained. If the drying air overheats, the temperature limit switches on the inlet and discharge ends of the combustion chamber shut down the burners. Some kilns may use several burner nozzles that can be operated individually or the series can be modulated over a wide operating range. Oil-fired systems may use various oil preheating techniques depending on the grade of oil. Some burners are designed to utilize gas and oil interchangeably.

There are typically two types of direct-fired systems (Cech and Pfaff 2000) (i) those using intermediate heat exchangers to transfer the heat from the burner to the circulating air in order to prevent combustion gases from entering the kiln and, thus, avoiding the danger of lumber discoloration which can result when wood is exposed to combustion gases; however, such wood dryers are likely to contain toxic combustion gases and (ii) those that heat the air directly in mixture with the hot combustion products.

Direct-fired kilns that have no auxiliary steam supply are humidified by hot water spray injected into the kiln by atomizing nozzles located either in the hot air discharge duct or, more commonly, in the air stream opposite the circulation fans. It is highly desirable to heat the spray water since cold water has an appreciable cooling effect in the kiln, causing fluctuation of the dry-bulb temperature and poor control of drying conditions.

Among the advantages of direct firing drying systems using, for example, natural gas as energy source can be mentioned (Cech and Pfaff 2000): (i) ease of installation, (ii) potential for integration into existing control system, (iii) uniform heating, (iv) better control of temperature, and (v) reduced maintenance.

Some limitations of direct drying systems are (i) important energy losses through the exhausted combustion flue and air mixture and (ii) expensive heat recovery solutions from low-temperature exhaust.

6.6.2 Radio Frequency and Microwave Drying

Electric heating (e.g., radio frequency and microwave) of relatively small dryers is an effective method to transfer the energy to the water molecules inside the lumber boards. It can reduce the time needed to dry small pieces of wood, but, generally, is not cost-effective.

Water molecules are dipolar in nature (i.e., they have an asymmetric charge center), and they are normally randomly orientated. The rapidly changing polarity of a radio-frequency and microwave field attempts to pull these dipoles into alignment with the field. As the field changes polarity, the dipoles return to a random orientation before being pulled the other way. This buildup and decay of the field, and the resulting stress on the molecules, causes a conversion of electric field energy to stored potential energy then to random kinetic or thermal energy. Hence, dipolar molecules, such as water, absorb energy in these frequency ranges. The field strength and the frequency are fixed by the equipment, while the dielectric constant, dissipation factor, and loss factor are material dependent. The actual electric field strength is also dependent on the location of the material within the microwave/radiofrequency cavity. The dielectric constant of water is over an order of magnitude higher than most base materials (such as cellulose in wood), moisture is preferentially heated, a process that leads to a more uniformly moist product with time while the overall dielectric constant of most materials is usually nearly proportional to moisture content up to a critical value, often around $0.2-0.3\ \frac{kg_{water}}{kg_{oven\text{-}dry}}$. Hence, microwave and radio-frequency methods preferentially heat and dry wet areas in most materials, processes that tend to give more uniform final moisture contents. For water and other small molecules, the effect of increasing temperature is to decrease the heating rate slightly, hence leading to a self-limiting effect (Schiffmann 1995; Keey et al. 2000). The field's strength and its frequency are fixed by the equipment, whereas the other parameters are material dependent. If the power requirement is over 50 kW, the use of higher power tubes in the radio-frequency range seems to be economically favorable. The least expensive tubes are the microwave oven tubes of which the output power is of the order of 750 W.

Microwave heating operates on the same principles as radio-frequency heating, but with higher frequencies in the range of 300 MHz–300 GHz; thus, the heating thermal power rates can be significantly increased. The cost of drying per unit volume of lumber for microwave drying is influenced by the lumber initial moisture content and density, while for conventional drying, the cost per unit volume is mainly dependent on the length of the drying cycle required to achieve acceptable degrade levels. The cost of the whole system, including the generator, tube from the generator to the dryer, applicator, control system, and conveyor, is much higher, but lower unit costs are associated with higher power equipment (Keey et al. 2000). The microwave drying seems to be appropriate where the high-value hardwood species has a low initial moisture content, causing problems with degrade in conventional drying, and/or is relatively valuable so that capital carrying charges are significant. For example, the microwave drying is economical for Douglas fir with a low initial moisture content of 40%, but not for Estern hemlock at a higher initial moisture content of 86%. On this basis, microwave drying would find most applications for high-value hardwood species (Barnes et al. 1976).

Electromagnetic radiation with wavelengths ranging from the solar spectrum to microwave (0.2 m–0.2 μm) can be combined with conventional air forced convection and/or vacuum wood drying (Mujumdar 2007). Low frequency of electromagnetic radiation (generated by magnetrons and klystrons) cover the range 1–100 MHz, while the high frequencies range from 300 MHz to 300 GHz (Schiffmann 1995; Keey et al. 2000). At radio frequency, the impedance of most moist materials falls dramatically, although they are poor conductors of 50–60 Hz current, thus reducing internal resistance to heat transfer. Energy is absorbed selectively by the water molecules, and as the product gets drier, less energy is used.

Advantages of such electric heating methods include (i) directly supply heat to the product, thus a drying medium is not needed; (ii) possibility for precise drying temperature control; (iii) uniform control of moisture content in a matter of minutes without developing defects, thus avoiding the severe moisture gradients and stresses which may cause checking; (iv) reduced dryer size as chain of heat transfer equipment for direct firing is not required; (v) clean, easy-to-maintain, and no handling of flammable fuel; (vi) better quality, no contamination, and no hazard of combustion; (vii) short start-up and shut-down times; and (viii) can be used to finish the drying of conventionally dried lumber that has not met target moisture contents.

Radio-frequency drying is a rapid drying process whereby the heat necessary to dry wood is generated within the wood itself. Wood containing moisture is subjected to an alternating electric field which causes the dipole water molecules to rotate in response to the changing polarity of the field. This rotation causes molecular friction and, at radio-frequency drying frequencies (1–100 MHz), the frictional heat is sufficient to produce temperatures exceeding the boiling point of water. In the drying apparatus, an alternating field is created between two large, metal-plate electrodes which are connected to a high-frequency generator. Green wood is dried by placing or passing it between these electrodes.

There is also a contribution due to ionic conduction because of the presence of ions in the sap. This mode of heating is not significantly dependent on either the temperature or the frequency of the applied field, but is directly dependent on the charge density and mobility of the ions. Because the heating is internally generated, rather than convectively warmed at the exposed surface of the boards, high and damaging internal pressures can be created in the process.

Materials with poor heat transfer characteristics are traditionally the problem materials when it comes to heating and drying. Radio-frequency heats all parts of the product mass simultaneously and evaporates the water in situ at relatively low temperatures usually not exceeding 82°C. Since water moves through the product in the form of a gas rather than by capillary action, migration of solids is avoided. Warping, surface discoloration, and cracking associated with conventional drying methods are also avoided (Thomas 1996).

In radio-frequency and microwave drying, energy is transferred directly to water molecules throughout the wood so that heating becomes volumetric, resulting in a higher interior temperature, and water would evaporate within the wood. This increases the internal pressure gradient of wood, which pushes the unevaporated water to surface, thus reducing the drying time. However, the capital cost of these electrical drying methods represents a major obstacle to their industrial implementation for drying lumber (Cech and Pfaff 2000; Onn 2007). Because the capital and operating costs are high, the radio-frequency and microwave heating method become economically attractive for small-scale kilns (with power requirements not exceeding 50 kW) for drying high-value hardwood species that are difficult to dry or for final correction of moisture profile, and if the drying rate can be increased fourfold over that for conventional drying means (Perré and Keey 2015). The economics appear to be promising, particularly for new kilns (Schiffmann 1995). For example, the use of radio-frequency heating for the drying of oak in small vacuum kilns of 23 m^3 capacity has lower capital cost but higher energy costs than a conventional dryer for the same duty (Smith and Smith 1994; Mujumdar 2007).

The use of a 300 kW radio-frequency generator for heating, operating at a frequency of 3 MHz in a vacuum kiln (under absolute pressures from 2.7 to 12 kPa) with a 23 m^3 lumber capacity for drying 57 mm square boards of red oak from a moisture content of 85%–8% (dry basis), achieved energy costs almost three times higher than for conventional drying. However, the capital cost was estimated to be 90% of that of a conventional steam-heated system with the same throughput including the boiler (Smith and Smith 1994).

For drying 30 mm thick samples of mountain ash with an initial mass of 0.5 kg from moisture contents of $1.1-0.2\ \frac{kg_{water}}{kg_{oven\text{-}dry}}$, the drying time was of 17 hours with a power input of 50 W. With 100 W power input, the drying time has been reduced at 12 hours. The estimated energy costs for heating were 57% and 76% of the costs of conventional hardwood drying on a small scale for the lower and higher power inputs, respectively (Schaffner 1991), suggesting that the heating cost would not be excessive (Rozsa 1994).

As shown in Table 6.2, the ratios of drying times for the same drying duty using radio-frequency or microwave energy to those from conventional schedules range from 0.25 for white spruce to 0.03 for Douglas-fir, (Keey et al. 2000).

In practice, the use of vacuum in combination with radio-frequency and microwave power inputs is very convenient, especially to avoid generating high internal pressures in the lumber that may develop on drying at atmospheric pressure (Keey et al. 2000). The radio-frequency drying of Australian hardwood and Mountain ash at absolute pressures of 1 kPa compared with 10 kPa increases the initial drying rate twofold at the lower pressure, and there is a rapid rise in the core temperatures at the higher pressure (10 kPa) even at low power inputs (Rozsa 1994). However, for

TABLE 6.2
Examples of Reductions in Drying Times with Radio-Frequency and Microwave Drying

Species (Size of Lumber Board)	Moisture Content (%) Initial	Moisture Content (%) Final	Relative Drying Times (Dielectric vs. Conventional)	Reference
Mixture of western red cedar, western hemlock, and amabilis fir (71–152 mm thick)	80	16	0.12	Avramidis et al. (1994)
Western hemlock (50 mm thick)	86	9–11	0.04	Barnes et al. (1976)
Douglas-fir (50 mm thick)	36	11–13	0.03	
Red oak (25 mm thick)	57–67	7–8	0.06	Harris and Taras (1984)
White spruce (55 mm)	55–65	15	0.25	Miller (1971)
Red oak (57 mm square)	85	8	0.25	Smith and Smith (1994)

power inputs of 1,250 W and, vacuum pressures of up to 10 kPa may need expensive pressure vessels as drying chambers, thus increasing the capital cost (Antti 1992a, b).

Materials that are difficult to dry with convection heating alone (as ceramics and glass fiber) because of poor heat transfer characteristics can be also good candidates for radio frequency assisted heat pump dryers. Such hybrid dryers comprise a mechanical vapor compression heat pump coupled with a pulsed, high-frequency radio-frequency device that generates volumetrically heat within the wet material to heat it and, simultaneously, vaporize the moisture at relatively low temperatures, usually not exceeding 82°C. In the drying chamber, an alternating field is created between two metal-plate electrodes connected to the generator, and the product is dried by placing or passing it between these electrodes. This drying process is rapid because the heat required to dry the wet product is generated within the product itself by applying an alternating polarity of the electric field causing the rotation of dipole water molecules.

A radio frequency-assisted heat pump dryer comprises a vapor compression heat pump retrofitted with a radio frequency generating system capable of imparting radio frequency energy to the drying material at various stages of drying processes (Figure 6.5a). This arrangement can overcome the limitation of heat transfer of conventional hot air drying systems, particularly during the falling period. The radio frequency generator generates heat volumetrically within the wet material by the combined mechanism of dipole rotation and conduction effects. Radio frequency heats all parts of the product mass simultaneously and evaporates the water in situ at relatively low temperatures usually not exceeding 82°C. Since the water moves through the product in the form of a gas rather than by capillary action, migration of solid is avoided. Warping, surface discoloration, and cracking (caused by the stress of uneven shrinkage) associated with conventional drying methods are also avoided. Radio-frequency drying is a rapid drying process whereby the heat necessary to

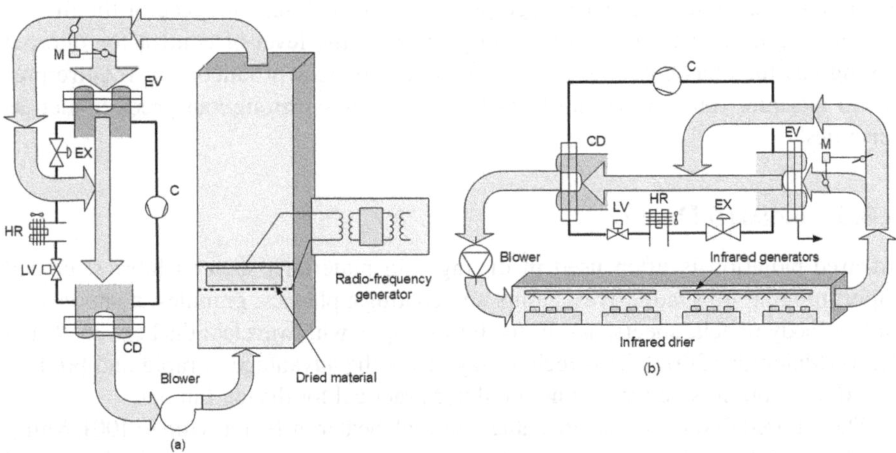

FIGURE 6.5 (a) Radio-frequency; (b) infrared dryers coupled with heat pumps. C, compressor; CD, condenser; D, air damper; EV, evaporator; EX, expansion valve; F, fan; HR, heat rejection coil; M, damper motor.

dry a product is generated within the product itself. Product containing moisture is subjected to an alternating electric field which causes the dipole water molecules to rotate in response to the changing polarity of the field. This rotation causes molecular friction and at radio-frequency drying frequencies (1–30 MHz), the frictional heat is sufficient to produce temperatures exceeding the boiling point of water. In the drying chambers, an alternating field is created between two large metal-plate electrodes which are connected to a high-frequency generator. The material is dried by placing or passing it between these electrodes. The advantage of using radio-frequency for small product items is that they can be dried to uniform moisture content in a matter of minutes without developing defects, especially in the case of wood.

Some high-value hardwood species that are difficult to dry with air convection heated dryers because of poor heat transfer characteristics, particularly during the falling drying rate period, can be candidates for radio frequency heat pump-assisted dryers.

Some characteristics of radio frequency-assisted heat pump-assisted dryers (Kiang and Jon 2007) are as follows: (i) improves the color of the products especially those that are highly susceptible to surface color change since radio-frequency drying starts from the internal to the product surface, minimizing any surface effect; (ii) cracking, caused by the stresses of uneven shrinkage in drying, can be eliminated, even heating throughout the product maintaining moisture uniformity from the center to the surface during the drying process.

The potential for direct application of the radio frequency-assisted heat pump dryers in industry is appreciable for reasons as (Kiang and Jon 2007) follows: (i) simultaneous external and internal drying significantly reduces the drying time to reach the desired moisture content; (ii) good potential for improving the product throughput, for example, by as much as 30% and 40%, for crackers and cookies, respectively, in the bakery industry; (iii) by reducing the moisture variation throughout the thickness of the product, differential shrinkage can be minimized inside materials with high shrinkage properties; (iv) closer tolerance of the dielectric heating frequency may significantly improve the level of control for internal drying and thus has potential in industry that produces products that require precision moisture removal; (v) uniform level of dryness throughout product such as ceramics.

6.6.3 Infrared Drying

Infrared radiation is often used in drying thin materials, such as fabrics; coated paper; printing inks; adhesives; paints and coatings; plastics; granules; coated webs; and car body panels, sheets, and films, for example, with wavelengths between 4 and 8 μm) (Mujumdar 2007). This technology offers the advantage of rapid and predictable drying, but it is neither economical nor practical for drying lumber.

The infrared dryers present advantages as high heat transfer rates (up to 100 kW/m^2), easy to direct the heat source to the surface, and quick response times allowing easy and rapid process control. The radiation penetrates wood only to a slight degree, resulting in high surface temperatures, severe moisture gradients, and casehardening. Only thin slices of wood and veneer can be effectively dried by this process, but even so there are more economical alternatives (Cech and Pfaff 2000). Infrared-assisted heat pump dryers could be used for fast removal of surface moisture during the initial stages of drying, followed by intermittent drying over the rest of the drying process. Heat for drying is generated by radiation from infrared generators (Figure 6.5b). Incorporating infrared into heat pump dryers may ensure faster initial drying rate and offers advantages of compactness, simplicity, ease to control, and low equipment cost. Also, there are possibilities of significant energy savings and enhanced product quality due to reduced residence time in the dryer chamber.

Infrared-assisted heat pump dryers could be used for fast removal of surface moisture during the initial stages of drying, followed by intermittent drying over the rest of the drying process. Heat for drying is generated by radiation from infrared generators. Incorporating infrared into an existing heat pump dryer is simple and capital cost is low. This mode of operation ensures a faster initial drying rate and offers advantages of compactness, simplicity, ease to control, and low equipment cost. There are possibilities of significant energy savings and enhanced product quality due to reduced residence time in the dryer chamber. Infrared-assisted air convective heat pump dryers, where heat for drying is generated by radiation from infrared generators, can also be used for faster removal of surface moisture during the initial stages of drying, followed by intermittent irradiation coupled with convective air drying over the rest of the drying cycle.

In the case of food heat-sensitive products, such a hybrid drying system can reduce the drying time (Zbicinski et al. 1992) and may ensure faster initial drying rate due to the direct transfer of energy from the infrared heating element to the product surface without heating the surrounding air (Jones 1992).

Other advantages of infrared-assisted heat pump dryers are (Paakkonen et al. 1999) as follows: (i) high heat transfer rates (up to 100 kW/m^2); (ii) easy to direct the heat source to the product surface; (iii) quick response times allowing easy process control of the drying process; (iv) system compactness, simplicity, and relatively low cost; and (v) possibility of significant energy savings and enhanced product quality due to reduced residence time in the dryer chamber.

6.6.4 Vacuum Drying

In addition to the previous specialized drying methods that are operated at near atmospheric pressures, vacuum drying (operating at reduced atmospheric pressure), even more expensive, has some industrial practicality, especially for relatively small lumber production of wood heat-sensitive products that must be dried at low temperatures over relatively short periods of time (Perré and Keey 2015).

Vacuum drying requires sealed kilns (Figure 6.6), where the air is evacuated to provide a very low air pressure, so that water can evaporate at much lower temperatures, meaning enhancement of the internal moisture movement and less energy needed (Onn 2007). In order to compensate for the loss of thermal capacity of the air, a slightly higher pressure level (e.g., more than 100 mbar), together with a air velocity (100 m/s m or more), may be very effective.

FIGURE 6.6 View of a typical lumber vacuum dryer (www.bing.com/images/search?q=images+wood+vacuum+dryers. Accessed December 19, 2017).

The energy used to evaporate the water can be supplied by conduction (with aluminum platens heated by electrical resistance or circulation of heated water placed between each layer of boards as the heat source supplied to boards by conduction), radiation (radio-frequency or infrared), cyclic heating (forced air), or a combination thereof. No vent air is required, but some conductive heat losses occur. Compared to drying at atmospheric pressure, vacuum drying reduces the vapor pressure and allows moisture to evaporate faster at the same drying temperature. This process accelerates the drying rate but the capacity of dryers is very low in relation to the cost of equipment. Consequently, the volume of product processed over a given period of time is substantially less in comparison to that obtained from conventional dryers of equal costs.

The vacuum drying reduces the losses of energy through the exhaust air and, thus, is thermally more efficient. The air is eliminated as a heating medium during the vacuum drying, but the diffusion of moisture through wood is extremely slow. Techniques to overcome the inherent poor heat transfer in vacuum dryers include the use of heated plates between the boards or, in larger units, intermittent heating with superheated steam. The water-heated metal platens located between consecutive lumber layers transfer heat directly by contact from the metal to the wood during the vacuum cycle and, as a result, drying rates are substantially accelerated. This technique avoids the discolorations that can sometimes arise with some lumber species when they are dried traditionally in stacks at atmospheric pressure (Keey et al. 2000). However, the heating platens can develop leaks, causing delays and excessive maintenance costs.

Most vacuum drying systems alternate cycles of vacuum drying and heating periods. Discontinuously operating kilns, with two periods of about 1 hour each, consist of a heating period at atmospheric pressure and a drying period at reduced pressure. In other words, the basic process consists of repeated cycles of heating under vacuum followed by short cooling periods during which the evaporated moisture is condensed on the cooled platens, recovering some of the lost moisture, and removed from the drying enclosure (Cech and Pfaff 2000). Tests on drying oak wood at pilot and industrial scales showed that such a discontinuous process is faster, with less susceptibility to mechanical damage of the wood, but the thermal consumption was higher than under continuous vacuum conditions.

Vacuum dryers for drying thick lumber boards present advantages as follows: (i) high-temperature drying without the danger of developing defects with some susceptible species; (ii) reduce drying times by a factor one-half to one-third of that of conventional kilns at atmospheric pressure, depending upon the thickness of the lumber (Hilderbrand 1989) and even five to ten times faster with half to two-thirds of the usual energy consumption when 50 mm-thick oak lumber boards are dried from 40% to 12% moisture content in 4 days (Walker 1993); (iii) accelerate the drying rate (mainly because of the enhanced relative humidity under vacuum) being as rapid as that at a significantly higher temperature at atmospheric pressure; and (iv) produce good quality (e.g., high-quality and heat-sensitive hardwoods which would otherwise be difficult to dry) dried timber without increased drying defects and with minimum color development, and, at the same time, reducing net fuel and electricity consumption by up to 70%.

The disadvantages of vacuum drying are as follows: (i) cost of energy, primarily electrical, and capital costs are high; (ii) the greater specific volume of moisture vapor and air under vacuum reduces the relative space available for the lumber load inside the chamber compared with that in a conventional kiln; (iii) difficulties in getting adequate air circulation through the load to obtain the heat transfer needed (Keey et al. 2000); (iv) may suffer from severe corrosion from released wood acids; therefore, all components of the kiln not made from stainless steel or aluminum are likely to have a service life of less than 2 years; and (v) capacity of vacuum dryers is very low in relation to the cost of equipment; therefore, the volume of product processed over a given period of time is substantially less in comparison to that obtained from conventional dryers of equal cost.

As previously noted, vacuum drying can be combined with radio-frequency or microwave delivery of energy to evaporate water faster.

6.6.5 Solar-Assisted Kilns

To reduce the energy consumption, risks of wood degrade and overall costs, several other processes for drying lumber, such as solar drying, can be considered for small drying operations. Solar drying can result in high quality lumber, primarily because the moisture gradients in the lumber are allowed to equalize at night when drying is not taking place, but drying times vary and could be relatively long. The efficiency of solar dryers is however highly dependent on climate (insolation) and dryer design.

The daily world-average solar radiation (insolation) on a horizontal surface is 3.82 kWh/m^2 (McDaniels 1984) with values in tropical countries being higher (up to 7.15 kWh/m^2) (Imre 1995). For drying of lumber, the solar dryers present attractions in remote locations spread over latitudes from 0° to 50° (Keey et al. 2000).

In most of the tropical and subtropical climates (where solar radiation is generally much higher than the world average), sun drying is only the common method used to preserve agro-food and forest products (Basunia and Abe 2001). In such climates, the average temperature the dryers can easily exceed the average ambient temperature by 17°C with maximum differences of 22°C–33°C.

Under these conditions, air-dry lumber (15%–20% moisture content, dry basis) can be obtained in one-half to one-quarter of the time required or normal air drying, and final moisture content of 10%–12% (dry basis) are possible without unduly extending the drying time. In contrast, solar drying in colder regions (e.g., northern United States and Canada) is only effective for a few months of the year, when both warm temperatures and sunshine prevail. Thus, solar drying is not a commercially viable option in these countries due to the relatively long drying times required.

Solar-heated dryers can be classified into three main groups (Imre 1995): (i) solar natural dryers that use only the sun; (ii) semi-artificial solar dryers with a fan to supply a continuous flow of air through the load; and (iii) solar-assisted artificial dryers, which may use an auxiliary energy source for boosting the heating rate. Solar natural dryers include those in which the airflow is driven by natural convection (passively) and those in which fans driven by solar cells or small wind turbines give some forced convection. These dryers have mainly been used for the drying of loose vegetable materials and grains.

Solar natural dryers may be simple greenhouse kilns where the solar collector(s) is (are) fitted within the structure that holds the load and the airflow is maintained by fans (Langrish and Keey 1992). These structures consist of metal or wood frame structures, sheathed with clear or translucent material such as glass, plastic, or fiberglass. Doors are hung on a solid wall facing away from the sun and the transparent roof is pitched at an angle approximately perpendicular to the mean position of the sun.

Solar kilns are a combination of conventional kilns and heating by solar radiation heating systems including fans and air vents (Figure 6.7), or with heat pump (dehumidification) systems (Figure 6.8a and b). They may have relatively low initial investment costs, but are slower and variable due to the weather variable conditions.

In solar kilns, the thermal solar collectors, constructed of flat or corrugated metal sheets painted in black, filled with either air or water, can be installed on a

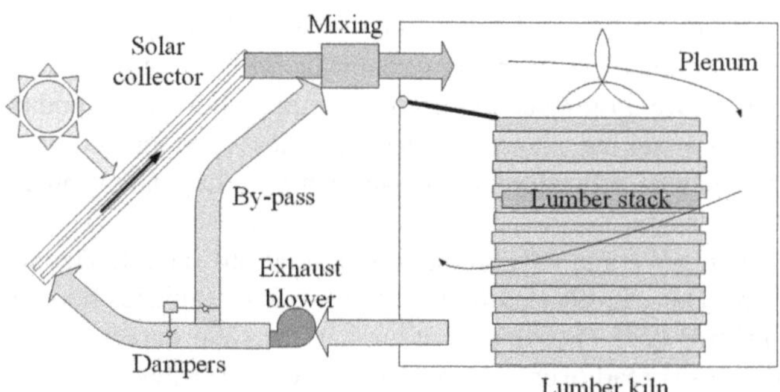

FIGURE 6.7 Simplified representation of a typical conventional solar-assisted lumber kiln.

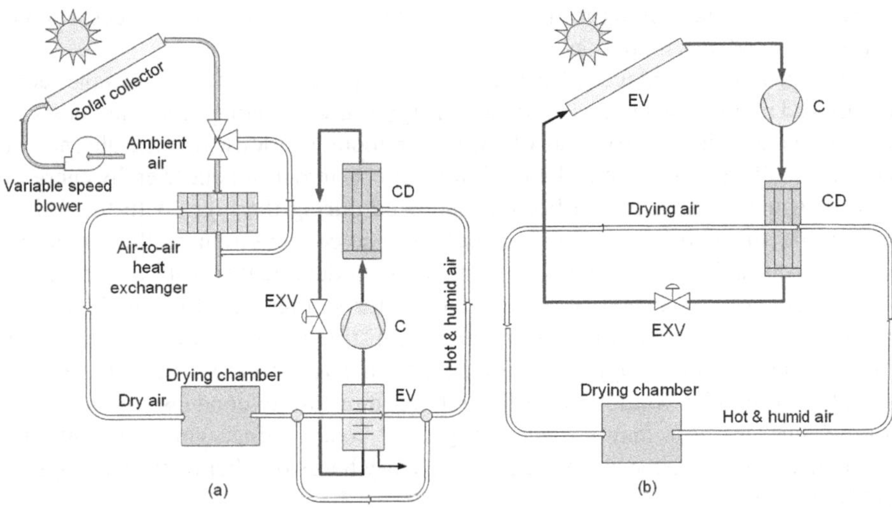

FIGURE 6.8 Schematics of conventional solar-assisted heat pump dryers: (a) with indirect solar collector; (b) with direct solar collector (evaporator). C, compressor; CD, condenser; EV, evaporator; EXV, expansion valve.

separated structure or on the dryer roof, isolated from the lumber drying chamber. Inside the drying chamber, the air circulation is natural or forced by fans mounted in the roof area to circulate air over the solar collectors and through the lumber stack(s), while air venting can be provided by pressure gradients or via powered fans controlled by humidity, temperature, and/or external conditions such as sunlight.

Conventional solar kilns (Figure 6.7) are typically designed and built to keep initial investment costs low. For passive solar kilns that utilize solar radiation for heating air above the temperatures encountered in normal air (Bois 1977; Rice 1987), (i) supplemental energy is necessary to maintain rapid, consistent drying times for all seasons and all locations; (ii) the solar surfaces should be isolated from the dryer during night hours and during periods of low solar influx; (iii) the proper choice of solar cover material and kiln wall insulation is critical for enhanced fuel savings; (iv) the winter months in the north are not practical for solar drying on any scale; (v) the supplemental energy could be direct-fired gas when available; wood waste, while cheaper, might require greater capitalization; electrical energy is too expensive under most circumstances, although capitalization would be low; use of electrical energy in conjunction with dehumidification is possible, but capital costs would be high in this case; (vi) the choice of collector surfaces in practice is restricted to roof and south wall (or sloping roof only); and (vii) the solar kiln, operating as a scheduled dryer, requires that conditioning for stress relief must be provided in some manner at the end of the drying cycle.

A number of extra features have been added in some designs of solar drying systems as (Imre 1995) follows: (i) heat storage systems including rock piles beneath kilns through which hot air from the collectors is pumped during the day and from which hot air is taken at night; (ii) heat storage tanks; and (iii) microprocessor control of vents and air circulation.

Some general advantages of using solar energy are as follows: (i) simple technology and natural, clean, and abundant energy source, although there are costs in collecting and using it; (ii) its availability in remote locations; (iii) the absence of monopoly of its use; (iv) the lack of polluting effects (environment-friendly energy); (v) the air inside the kiln can become saturated during the night with the fall in the outside air temperature; this can lead to some condensation on the lumber to provide a diurnal conditioning which enhances product quality with impermeable wood; (vi) easy conversion of natural energy for storage resulting in significant saving of energy; (vii) easy to implement control strategy; and (viii) when the solar kiln's air vents shut overnight and with the drop in outdoor temperature, the relative humidity in the kiln would rise sufficiently for moisture to condense on the wood boards' surfaces; this may provide a degree of conditioning, which prevents the development of excessive checking in a refractory hardwood being dried (Langrish and Keey 1992).

Among disadvantages of solar energy can be noted (i) intermittent energy source, depending on time, season, and weather; variability of the intensity incident radiation; as well as on average annual sunshine time and annual total quantity of solar radiation; (ii) the relatively low energy flux compared with conventional energy sources; (iii) higher capital costs are incurred for additional solar panels, blowers, storage tank, and valves; (iv) the amount of stored solar energy is greatly subjected to the weather conditions; (v) higher capital costs incurred for additional solar panels, blowers, storage tank, and valves; and (vi) product degradation, spoil, and losses due to rain, wind, dust, birds, and insects.

Compared to conventional kilns, solar drying shows that (i) moisture contents of less than 18% (dry basis) are difficult to attain for most locations; (ii) drying times are considerably higher; (iii) capital, maintenance, and operation are less expensive; (iv) drying degrade cannot be controlled because there is little control over the drying elements; (iv) drying below 60°C do not guarantees killing all the fungi and insects in the wood; (v) if drying by exposing lumber to the sun, the rate of drying may be overly rapid in the dry summer months, causing cracking and splitting, and too slow during the cold winter months (vi) moisture contents of less than 18% are difficult to attain for most locations; and (vii) drying times are considerably higher.

Combining solar energy and heat pump-assisted dryer technology is generally considered an attractive concept able to reduce or eliminate some disadvantages of using solar and heat pump-assisted drying separately.

The solar collector can act as an additional air heater allowing further increase in the air drying temperature prior entering the drying chamber and, thus, improving the overall energy efficiency of the drying system (Figure 6.8a). In such a system, the solar radiation heats the air passing through the solar collector. This heated air then serves to further heat, via an air-to-air heat exchanger, the drying air leaving the heat pump condenser, prior being rejected outside. On the other hand, the hot and humid air leaving the drying chamber passes over the heat pump evaporator where it is cooled and dehumidified. Both sensible and latent heats are absorbed by the evaporating refrigerant and the resulted vapor is compressed by the compressor.

In geographic regions where plentiful sources of solar energy are available, instead of using conventional heating systems to provide for auxiliary heating, the storage of solar energy in phase-change materials for discharging sensible energy to the drying air may lead to cheaper means to provide higher drying temperatures. Such systems also offer the flexibility of operating with heat pumps to further improve the energy efficiency of drying systems (Kiang and Jon 2007).

A typical solar heat pump-assisted dryer mainly comprise a mechanical vapor compression heat pump (evaporator, condenser, compressor, and expansion valve) and a solar collector (Imre et al. 1982; Chaturvedi and Shen 1984; Morrison 1994; Kuang et al. 2003; Imre 2006). The solar collector can act as an additional air heater allowing further increase in the air drying temperature prior entering the drying chamber and, thus, improving the overall energy efficiency of the drying system (Figure 6.8a) or as the heat pump's direct expansion evaporator (Figure 6.8b). In the solar-assisted heat pump dryer with indirect solar collector (Figure 6.8a), the solar radiation heats the air passing through the solar collector. This heated air then serves to further heat, via an air-to-air heat exchanger, the drying air leaving the heat pump condenser, prior being rejected outside. On the other hand, the hot and humid air leaving the drying chamber passes over the heat pump evaporator where it is cooled and dehumidified. Both sensible and latent heats are absorbed by the evaporating refrigerant, and the resulted vapor is compressed by the compressor.

The high-pressure superheated refrigerant vapor condenses inside the condenser by transferring the recovered heat plus the equivalent electrical energy consumed by the compressor to the drying air. After being further heated in this heat exchanger, the hot and dry air passes through the drying chamber and picks up moisture from the dried product. Such a combined system offers the flexibility of operating with the solar collector only, with the heat pump only or, simultaneously, with both systems. Such indirect solar-assisted heat pump dryers may provide advantages as energy savings and higher operating drying temperatures compared to stand-alone heat pump drying systems (Kiang and Jon 2006). However, these systems have higher capital costs required for additional solar panels, blowers, heat exchangers, and controls, while the amount of available solar energy varies significantly throughout the day and/or the year (Chou and Chua 2006).

Experimental research works on solar-assisted heat pump dryer with indirect solar collectors have been conducted in different climate regions (Cervantes and Torres-Reyes 2002; Kuang et al. 2003) in order to dry peanuts (Auer 1980; Baker 1995), rice, and other similar products (Daghigh et al. 2010). For example, in the peanuts solar-assisted dryer system equipped with a mechanical vapor compression heat pump and a solar collector installed on top of the drying chamber and connected to a closed water loop with storage tank (Auer 1980; Baker 1995), part of the hot and moist air leaving the dryer flows through the heat pump evaporator where it is cooled and dehumidified. The sensible and latent heat recovered plus the equivalent heat of the compressor power input are used to heat the intake cold water within the condenser. The hot water produced is stored in a storage tank where it can be further heated with solar energy or used as a heat source to preheat the inlet ambient air (Auer 1980; Imre et al. 1982; Imre 2006).

Direct expansion solar-assisted heat pump dryers (Figure 6.8b) consist of a thermal solar or photovoltaic/thermal collector acting as the heat pump's evaporator and a heat pump (compressor, condenser, expansion valve, etc.). In this case, the refrigerant is directly vaporized inside the solar collector–evaporator due to the solar energy input. Such an experimental 1.5 kW (compressor power input) prototype with HFC-134a as the refrigerant for drying green beans and other agriculture products at 45°C, 50°C, and 55°C has been developed and extensively investigated under different meteorological conditions of Singapore (Hawlader et al. 2003; Hawlader and Jahangeer 2006; Hawlader et al. 2008). Direct expansion solar-assisted heat pump dryers may have advantages such as the elimination of evaporators of traditional heat pump-assisted dryers and higher thermal efficiency of solar collectors. However, even such a concept may reduce the system initial costs, the choice of the working fluid (refrigerant) according to the local ambient temperature conditions, and the control strategy of the integrated hybrid system are very critical issues (Kara et al. 2008). Controlling the air temperature and relative humidity in these kilns controls the rate of drying. Both of the above kilns are similar in that the air is forced through the lumber; the air picks up heat by passing through a heat exchanger (solar collector or steam heated coils) and takes in cooler, dryer ambient air through an intake vent while expelling hot humid air through the exhaust vent. Appropriate monitoring of the temperature and relative humidity of the air, and careful controlling of the vents and temperature of the heat exchanger provide precise control of the drying environment.

Analysis has shown that caution should be exercised in considering solar energy as a means of lowering fuel costs. It is not a universal solution to energy economy in wood drying. Therefore, no one should leap into investing in such technology without carefully considering engineering criteria as well as the overall operating economics.

6.6.6 Superheated Steam Drying

Superheated steam drying is an alternative that has attracted increased interest due to its larger potential for energy recovery. This method would reduce defect in wood such as stresses, crack, warpage, stain, and discolouration and offer faster drying rates to less than 7 days and no risk of fire or explosion. However, the initial cost of this method is relative higher (Onn 2007). The superheated steam drying method relies on the principle of drying in an environment of pure steam (water vapor) which is above its saturation temperature (Keey et al 2000). Superheated steam dryers can operate at low pressure (vacuum), near atmospheric pressure (e.g., fluidized bed dryers for coal), or high pressure (e.g., fluidized bed dryers for pulp and sludge). For that, the internal pressure of the drying chamber is lowered by evacuating air with a vacuum pump, followed by steam being introduced into the drying chamber. At pressures under the atmospheric pressure, steam would be in superheated phase at lower temperatures, usually in the range of 50°C–90°C. In other words, superheated steam drying involves use of superheated steam in direct convective dryers in place of hot air, combustion, or flue gases as the drying medium to supply heat for drying and to carry away the evaporated moisture.

Any direct or direct/indirect (e.g., combined convection/conduction) dryer can be operated as a superheated steam dryer, at least in principle. The technology involved is more complex and, hence, this conversion is not simple, however. Additional criteria must also be considered when selecting a dryer for superheated steam drying operation.

Any convective dryers such as flash, fluid bed, spray, rotary, tray, impingement and conveyor can be constructed and operated as complex superheated steam dryers, but they are not always feasible.

The drying chambers of superheated steam must be leakproof to avoid condensation and energy loss. Exit superheated steam should be recovered to benefit from energy recovery.

Superheat steam as a convective drying medium offers several advantages: (i) higher drying rates under certain conditions, (ii) better quality for certain products, (iii) low-net energy consumption if the excess steam in the dryer is used elsewhere in the process, and (iv) elimination of fire and explosion hazard. When drying with superheated steam, the boards' temperatures correspond to the saturation temperature at the operating pressure, for example, 100°C for steam at 101 kPa. Additional heat is supplied to raise the temperature of the steam above his saturated temperature to obtain superheated steam at the specified pressure. For example, the total evaporation enthalpy of saturated steam at 689 kPa is 2,067.72 kPa, whereas the total enthalpy (latent heat) of evaporation at 101 kPa is 2,256.78 kPa. It can be seen, however, that the total enthalpy (latent heat) of evaporation of steam decreases slightly with a substantial increase in pressure (Cech and Pfaff 2000).

Generally, superheated steam drying could be considered a viable option only if one or more of the following conditions apply in other competitive drying technologies (Mujumdar and Devahastin 2000): (i) energy cost for drying is very high; (ii) product final quality would be superior if dried with superheated steam rather than air; (iii) there are risks of fire, explosion, or other damages; and (iv) high-production capacity required and high quantity of water to be removed.

If applied, some of the advantages of superheated steam drying technology can be summarized as follows (Mujumdar and Devahastin 2000): (i) any direct dryer, in principle, can be converted to a superheated steam dryer (e.g., flash, fluidized bed, spray, impinging jet, and conveyor dryers); dryer's thermal efficiencies can be improved and the unit size is reduced by supplying a part of the heat indirectly (e.g., by conduction or radiation); (ii) superheated steam has heat transfer properties superior to air at same temperature; since there is no resistance to diffusion of the evaporated water in its own vapor, the drying rate in the constant rate period is dependent only on the heat transfer rate; (iii) the vapor evolving from the product may be withdrawn from the chamber, condensed, and the latent heat is recovered; alternatively, the vapor is reheated within the drying chamber by tubular or plate heat exchangers and recirculated as a convective drying medium to enhance the drying rate; such a system is used commercially to dry timber with very attractive results; (iv) if air infiltration is avoided or minimized to an acceptable level, it is possible to recover all of the latent heat supplied to the superheated steam drying by condensing the exhaust steam or by mechanical or thermal compression to elevate its specific enthalpy for reuse in the dryer; since superheated steam drying necessarily produce steam equal in amount to the water evaporated in the dryer, it is necessary to have a useful application for this excess steam in the process plant; if this steam is used elsewhere, the latent heat recovered is not charged to the superheated steam drying, leading to a net energy consumption figure of 1,000–1,500 kJ/kg water removed for superheated steam drying compared with 4,000–6,000 kJ/kg water removed in a corresponding hot air dryer; (v) no oxidative or combustion reactions are possible in superheated steam drying; this means no fire or explosion hazard and often also a better quality product; (vi) higher drying rates are possible in both constant and falling rate periods, depending on the steam temperature; the higher thermal conductivity and heat capacity of superheated steam lead to higher rates of surface moisture removal above the so-called inversion temperature; below the inversion temperature, drying in air is faster; in the falling rate period, the higher product temperature in superheated steam drying (over 100°C at 101 kPa) and lack of diffusional resistance to water vapor lead to faster drying rates; (vii) for products containing toxic or expensive organic liquids that must be recovered, steam drying avoids the danger of fire and/or explosion while allowing condensation of the off-streams in relatively smaller condensers; (viii) superheated steam drying permits pasteurization, sterilization, and/or deodorization of food products; (ix) the effects of superheated steam drying on product quality are all positive compared to air drying or, at most, comparable; (x) fast drying rates resulting in energy savings of about 50% over conventional kilns; (xi) better product quality because staining, mold attack can be avoided, and no oxidative discoloration of wood occur; (xii) minimal danger of corrosion; (xiii) reduced wood stresses, cracking, and warpage; and (xiii) low inventory costs due to several-fold faster drying.

Among the limitations of superheated steam drying the following can be mentioned (Mujumdar and Devahastin 2000): (i) the drying system is more complex; for example, start-up and shutdown are more complex operations than for air convective dryers; (ii) any leak can be allowed since non-condensable gases cause problem with energy recovery by compression or condensation; feeding and discharge of superheated steam drying must not allow infiltration of air; (iii) the product itself may bring in non-condensable gases; (iv) since feed enters at ambient temperature, there is inevitable condensation in the superheated steam drying before evaporation begins; this adds about 10%–15% to the residence time in the dryer; at 1 bar operating pressure, the drying begins at a product temperature of 100°C in the constant rate period when surface water is being removed; alternatively, a preheater is needed for the feedstock; (v) products that may melt, undergo glass transitions, or maybe otherwise damaged at the saturation temperature of steam at the dryer operating pressure cannot be dried in superheated steam even if they contain only surface moisture; (vi) products that may require oxidation reactions (e.g., browning of foods) to develop desired quality parameters cannot be dried in superheated steam; (vii) if the steam produced in the dryer is not needed elsewhere in the sawmill plant, the energy-related advantages of superheated steam drying do not exist; (viii) steam cleaning may not always be a simple task and the chemical composition of the condensate must be carefully evaluated; (ix) costs of feeding, product collection, and exhaust steam recovery systems are much more significant than the cost of the steam dryer alone; (x) in most practical cases, superheated steam drying is a justifiable option only for very large drying facilities, continuously operated systems because of the techno-economics of the ancillary equipment needed; (xi) there is currently limited field experience with superheated steam drying for a smaller range of products; (xii) the autoclaves are much smaller than the conventional hot air kilns; however, since the drying cycles are several-fold shorter, this is not a major limitation; the shorter drying times give the user flexibility in drying different species or sizes of wood while reducing the cost of inventory, especially for wood species requiring several weeks of drying time in air drying; and (xiii) lack of oxygen in the dryer may also help kill microorganisms or insects in wood.

REFERENCES

Anderson, J.O. 2014. Energy and resource efficiency in convective drying systems in the process industry. Doctoral Thesis, Luleå University of Technology, Sweden.

Anderson, J.O., L. Westerlund. 2014. Improved energy efficiency in sawmill drying system. *Applied Energy* 113:891–901.

Antti, A.L. 1992a. Microwave drying of hardwood simultaneous measurements of pressure temperature and weight reduction. *Forest Products Journal* 42(6):49–54.

Antti, A.L. 1992b. Microwave drying of hardwood. Moisture measurements by computer tomograph. In *Proceedings of the 3rd IUFRO International Wood Drying Conference*, Vienna, Austria, August 18–21, pp. 74–77.

Auer, W.W. 1980. Solar energy systems for agricultural and industrial process drying. In *Drying'80*, edited by A.S. Mujumdar, Hemisphere, New York, pp. 280–292.

Avramidis, S., F. Liu, B.J. Neilson. 1994. Radio-frequency/vacuum drying of softwoods drying of thick western red cedar with constant electrode voltage. *Forest Product Journal* 44(1):41–47.

Baker, C.G.J. 1995. *Industrial Drying of Foods*, 1st edition. Springer-Verlag, New York.

Barnes, D., L. Admiraal, R.L. Pike, V.N.P. Mathur. 1976. Continuous system for the drying of lumber with microwave energy. *Forest Products Journal* 26(5):31–42.

Basunia, M.A., T. Abe. 2001. Thin-layer solar drying characteristics of rough rice under natural convection. *Journal of Food Engineering* 47:295–301.

Berit Time. 1988. Hygroscopic moisture transport in wood. Doctoral Thesis, Department of Building and Construction Engineering, Norwegian University of Science and Technology, Trondheim, Norway.

Bian, Z. 2001. Airflow and wood drying models for wood kilns. Master Thesis, Degree of Master of Applied Science in Faculty of Graduate Studies, Department of Mechanical Engineering, the University of British Columbia, Vancouver BC, Canada.

Bois, P. 1977. Constructing and operating a small solar-heated lumber dryer. FPU Technical Report 7. U.S. Department of Agriculture, Forest Service, State and Private Forestry, Forest Products Laboratory, Madison, WI, 12 p.

Cech, M.J., F. Pfaff. 2000. *Operator Wood Drier Handbook for East of Canada*, edited by Forintek Corp., Canada's Eastern Forester Products Laboratory, Québec city, Québec, Canada.

Cervantes, J.G., E. Torres-Reyes. 2002. Experiments on a solar-assisted heat pump and an exergy analysis of the system. *Applied Thermal Engineering* 22:1289–1297.

Chaturvedi, S.K., J.Y. Shen. 1984. Thermal performance of a direct expansion solar assisted heat pump. *Solar Energy* 33(2):155–162.

Chou, S.K., K.J. Chua. 2006. Heat pump drying systems. In *Handbook of Industrial Drying*, edited by A.S. Mujumdar, Taylor & Francis Inc., Boca Raton, FL, pp. 1122–1123.

Daghigh, R., M.H. Ruslan, M.Y. Sulaiman, K. Sopian. 2010. Review of solar assisted heat pump drying systems for agricultural and marine products. *Renewable and Sustainable Energy Reviews* 14:2564–2579.

Elustondo, D.M., L. Oliveira. 2009. Model to assess energy consumption in industrial lumber kilns. *Maderas, Ciencia y Tecnologia* 11(1):33–46.

Frank Controls. 1986. Lumber drying theory (www.frankcontrols.com, accessed June 15, 2015).

Harris, R.A., M.A. Taras. 1984. Comparison of moisture content distribution, stress distribution, and shrinkage of red oak lumber dried by a radio-frequency/vacuum drying process and a conventional kiln. *Forest Product Journal* 34(1):44–54.

Hawlader, M.N.A., K.A. Jahangeer. 2006. Solar heat pump drying and water heating in the tropics. *Solar Energy* 80:492–499.

Hawlader, M.N.A., S.M.A. Rahman, K.A. Jahangeer. 2008. Performance of evaporator-collector and air collector in solar assisted heat pump dryer. *Energy Conversion Management* 49:1612–1619.

Hawlader, M.N.A., S.K. Chou, K.A. Jahangeer, S.M.A. Rahman, K.W.E. Lau. 2003. Solar-assisted heat-pump dryer and water heater. *Applied Energy* 7(1):185–193.

Hilderbrand, R. 1989. *The Drying of Sawn Timber.* Hilderbrand, Maschinenbau, Niirtingen, 240 p.

Imre, L. 1995. Solar drying. In *Handbook of Industrial Drying*, vol. 1, edited by A.S. Mujumdar, Hemisphere, Washington, DC, pp. 373–452.

Imre, L. 2006. Solar drying. In *Handbook of Industrial Drying*, edited by A.S. Mujumdar, Taylor & Francis Inc., Boca Raton, FL, pp. 317–319.

Imre, L., L.I. Kiss, K. Molnar. 1982. Complex Energy Aspects of Solar Agricultural Drying System, In *Proceedings of the 3rd International Drying Symposium.* Wolverhampton, England, pp. 370–376.

Jones, P. 1992. Electromagnetic wave energy in drying processes. In: edited by A.S. Mujumdar, Elsevier Science Publisher BV, Amsterdam, the Netherlands, pp. 114–136.

Kara, O., K. Ulgen, A. Hepbasli. 2008. Exergetic assessment of direct-expansion solarassisted heat pump systems: Review and modeling. *Renewable and Sustainable Energy Reviews* 12:1383–401.

Keey, R.B., J.J. Nijdam. 2002. Moisture movement on drying softwood boards and kiln design. *Drying Technology* 20(10):1955–1974.

Keey, R.B., T.A.G. Langrish, J.C.F. Walker. 2000. *Kiln Drying of Lumber*, Springer Series in Wood Science, edited by T.E. Timell, ISBN-13:978-3-642-64071-1.

Kiang, C.S., C.K. Jon. 2007. Heat pump drying systems. In: *Handbook of Industrial Drying.* 3rd edition, edited by A.S. Mujumdar, CRC Press, New York, pp. 1104–1105.

Kuang, Y.H., R.Z. Wang, L.Q. Yu. 2003. Experimental study on solar assisted heat pump system for heat supply. *Energy Conversion and Management* 44:1089–1098.

Langrish, T.A.G., R.B. Keey. 1992. A solar-heated kiln for drying New Zealand hardwoods. In *Trans IPENZ Chem Elect Mech*, vol. 18, pp. 9–14.

Law, C., A.S. Mujumdar. 2007. Fluidized bed dryers. In *Handbook of Industrial Drying*, 3rd edition, Chapter 8, CRC Press, Boca Raton, FL, pp. 174–193.

Law, C.L., A.S. Mujumdar. 2008. Energy aspects in energy drying. In *Guide to Industrial Drying: Principles, Equipment and New Developments*, Chapter 14, edited by A.S. Mujumdar, IDS2008, Hyderabad, pp. 291–293.

Mackay, J.F.G., L.C. Oliveira. 1989. Kiln operator's handbook for Western Canada, Forintek Canada, SP-31, ISSN 0824-2119.

McCurdy, M.C. 2005. Efficient kiln drying of quality softwood timber. Doctoral Thesis, Degree of Doctor of Philosophy in Chemical and Process Engineering, University of Canterbury, Christchurch, New Zealand.

McDaniels, D.K. 1984. *The Sun, Our Future Energy Source*, 2nd edition. John Wiley, New York, 271 pp.

Miller, D.G. 1971. Combining radio-frequency heating with kiln-drying to provide fast drying without degrade. *Forest Product Journal* 21(12):17–21.

Minea, V. 2004. Heat pumps for wood drying – new developments and preliminary results. In *Proceedings of the 14th International Drying Symposium*, Sao Paulo, Brazil, August 22–25, Volume B, pp. 892–899.

Morrison, G.L. 1994. Simulation of packaged solar heat-pump water heaters. *Solar Energy* 53(3):249–57.

Mujumdar, A.S., L.X. Huang. 2007. Global R&D needs in drying. *Drying Technology* 25:647–658.

Mujumdar, A.S. 2007. An overview of innovation in industrial drying: Current status and R&D needs. *Transport Porous Media* 66:3–18. doi:10.1007/s11242-006-9018-y.

Mujumdar, A.S. 2014. Principles, classification, and selection of dryers. In *Handbook of Industrial Drying*, 4th edition, edited by A.S. Mujumdar, CRC Press, Boca Raton, FL, pp. 4–29.

Mujumdar, A.S., M.L. Passos. 2000. Drying: Innovative technologies and trends in research and development. In *Developments in Drying*, vol. 1, edited by A.S. Mujumdar & S. Suvachittanont. Kasetsart University, Bangkok, pp. 235–268.

Mujumdar, A.S., S. Devahastin. 2000. Fundamental principles of drying. In *Mujumdar's Practical Guide to Industrial Drying*, edited by S. Devahastin. Exergex Corporation, ISBN-10: 9748591395.

Mujumdar, A.S., L.X. Huang. 2007. Global R&D needs in drying. *Drying Technology* 25:647–658.

Nijdam, J.J., T.A.G. Langrish, R.B. Keey. 2000. A high-temperature drying model for softwood timber. *Chemical Engineering Science*. 55:3585–3598.

Onn, L.K. 2007. Studies of convective drying using numerical analysis on local hardwood, University Sains, Malaysia (https://core.ac.uk/download/pdf/32600391.pdf, accessed June 15, 2016).

Paakkonen, K., J. Havento, B. Galambosi, M. Pyykkonen. 1999. Infrared drying of herbs. *Agricultural and Food Science*. 8:19–27.

Pang, S., A.H. Haslett. 1995. The application of mathematical models to the commercial high temperature drying of softwood lumber. *Drying Technology* 13(8&9):1635–1674.

Perré, P., R. Keey. 2015. Drying of wood: principles and practice. In: *Handbook of Industrial Drying*. 4th edition, edited by A.J. Mujumdar. CRC Press, New York, pp. 821–877.

Rasmussen, E.F. 1988. *Dry Kiln Operators Manual*. Forest Products Laboratory, U.S. Department of Agriculture, Hardwood Research Council, Madison, WI.

Rice, R.W. 1987. Solar kiln: A solar heated lumber drying kiln is easy to build, operate, and maintain. *Workbench* Jan.–Feb.: 7.

Rosen, H.N. 1995. Drying of wood and wood products. In *Handbook of Industrial Drying*, 2nd edition, edited by A.S. Mujumdar, Marcel Dekker, New York, pp. 899–920.

Rozsa, A.N. 1994. Dielectric vacuum drying of hardwood. In *Proceedings of the 4th IUFRO International Wood Drying Conference*, Rotorua, New Zealand, August 9–13, pp. 271–278.

Schaffner, R.D. 1991. Drying costs – A brief introduction. In *Australian Timber Seasoning Manual*, Australasian Furnishing Research Development Institute, Launceston.

Schiffmann, R.F. 1995. Microwave and dielectric drying. In *Handbook of Industrial Drying*, vol. 1, 2nd edition, edited by A.S. Mujumdar, Marcel Dekker, New York, pp. 345–372.

Smith, W.B., A. Smith. 1994. Radio-frequency/vacuum drying of red oak energy quality value. In *Proceedings of the 4th IUFRO International Wood Drying Conference*, Rotorua, New Zealand, August 9–13, pp. 263–270.

Tschernitz, J.L. 1986. Chapter 11: Energy in kiln drying (http://www.woodencrates.org/standards/AH188-KD-chapter-11.pdf, accessed May 5, 2017).

Walker, J.C.F. 1993. *Primary* Wood *Processing Principles and Practice*, Chapman & Hall, London, Taylor & Francis Group.

Zbicinski, I., A. Jakobsen, J.L. Driscoll. 1992. Application of infrared radiation for drying of particulate material. In: edited by A.S. Mujumdar, Elsevier Science Publisher BV, Amsterdam, the Netherlands, pp. 704–711.

7 Mechanisms of Lumber Drying

7.1 BOARD INTERNAL DRYING PROCESS

Drying, a complex, thermal, and hydrodynamic process that consists in removing volatile substances (e.g., water) from solid products, is an amalgamation of material science and transport phenomena. According to Mujumdar (2000) and Mujumdar and Huang (2007), our understanding of drying mechanisms at the microscopic level of many solids (including wood and agro-food products) is still rudimentary, in the sense that modeling drying remains a complex and challenging task, and no universal drying theory does exist.

These mechanisms involve transient energy, mass and momentum transfer through porous or nonporous materials with or without phase change, or chemical reactions.

Recent studies have resulted in significant advances in the understanding of the thermodynamics of drying hygroscopic materials, kinetics of drying, vaporization (evaporation) of multicomponent mixtures from porous bodies, behavior of particulate motion in various dryers, etc. In general, the empirical knowledge gained in the past two decades has been of considerable value in modeling, design, and control of industrial dryers (Mujumdar 2015).

Green, fresh-cut timber logs (after harvesting) and sawn lumber boards are hygroscopic, capillary, and porous materials with relatively high initial moisture contents. To artificially dry such materials in batch convective, forced-air kilns, hot and dry air is used to supply heat to the board stack(s) for internal moisture movement, process followed by vaporization at the surface and dissipation of moisture (water vapor) into the bulk of the drying air. The moisture movement inside the wood pieces, as well as moisture vaporization and removal at their surfaces, is thus achieved by supplying thermal energy (heat) by means of conventional or specialized dryers, as well as by unconventional technologies such as heat-pump-assisted dryers (McCurdy 2005).

The mechanisms of convective, forced-air thermal drying of wet, hygroscopic, capillary, porous wood species involve two simultaneous heat and mass transfer processes through the lumber boards to their surfaces, and then to the unsaturated drying air (Meroney 1969; Keey et al. 2000; Bian 2001; Karel and Lund 2003; Mujumdar 2007; Anderson 2014) (Figure 7.1):

i. Internal drying process consisting in moisture (mass) (i.e., water liquid and water vapor) movement (migration and transport) from the interior to the surface within the wood logs or sawn lumber boards; heat is the source from which water molecules in wood boards acquire the kinetic energy necessary for moisture movement toward the surfaces; the movement of

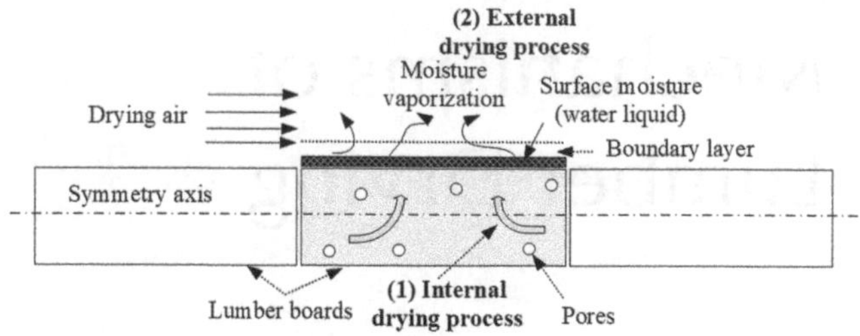

FIGURE 7.1 Main processes of single-lumber board drying.

moisture internally within the lumber boards is a function of the physical nature of the material, the temperature, and its moisture content; it can be described by using properties such as moisture diffusivity; the migration rate of moisture depends on the liquid diffusivity, which is a function of local moisture content and temperature.

ii. External drying process consisting in transfer of heat from the surrounding environment hot drying air to the dried piece of wood as a result of convection, conduction, or radiation and, in some cases, as a result of a combination of these effects, followed by the vaporization of the moisture from the boards' surface and transport away by the drying medium (see Figure 7.5); during forced-convection air drying, heat is supplied by convection, and then by conduction from the surface to the center of lumber boards; heat and mass transfer occur at the lumber and airflow interface through the thermal and moisture concentration boundary layers; the removal of water as vapor from the material surface depends on the external conditions of temperature, air humidity and flow, area of exposed surface, and ambient air pressure.

7.1.1 Mechanisms of Internal Movement of Moisture

Knowing the mechanisms of internal moisture movement within hygroscopic materials upon various operating conditions (as relative humidity and temperatures) may contribute to the appropriate selection of drying methods as well as to appropriate designing and sizing of dryers (Mujumdar 2000).

As previously noted, heat is the source from which the water molecules acquire the kinetic energy necessary for moving inside solids as lumber boards and for the vaporization of moisture at the product's surfaces (Srikiatden and Roberts 2007). The rate of moisture vaporization (evaporation) is dependent upon both the amount of energy supplied per unit of time and the ability of the heating medium to absorb moisture (Keey et al. 2000).

The mechanisms of internal moisture movement during drying of woods (Cech and Pfaff 2000) and other porous, hygroscopic materials, such as foods (Srikiatden and Roberts 2007) are thermally and hydraulically very complex. They depend on the material physical structure and chemical composition, and many other variables that, sometimes, involve phase changes of water (Rosselo et al. 1992).

Generally, the movement of internal moisture from the lumber board core to its surface occurs at the cellular level and is a strong function of the lumber temperature and moisture content, as well as of the temperature, relative humidity, and velocity of the air in contact with the dried board.

The movement of water vapor through void spaces in wood depends on how much water vapor is contained in the air in the voids, or in the air surrounding the wood. If the air surrounding the wood has a low relative humidity, the water vapor will move from the surface of the wet board to the air. Thus, the drying process depends on the surface moisture content of the wood board being dried and whether a difference in moisture content can be developed between the surface and the interior of a board.

This process requires interphase mass transfer from the wet wood board to the hot, dry air that has a strong affinity for moisture.

Several mechanisms (see Table 7.1) can be accounted for moisture movement inside timber logs and lumber boards toward their surfaces during drying. However, the moisture movement occurs through one (predominant) mechanism or combination of two (or more) driving forces at the same time (Choong and Fogg 1968; Skaar et al. 1970; Karel and Lund 2003; Griskey 2006; Srikiatden and Roberts 2007).

As a result of heat transfer to a wet solid, a temperature gradient develops within the solid while moisture evaporation occurs from the surface. This produces a migration of moisture from within the solid to the surface, which occurs through one or more mechanisms, namely, diffusion, capillary flow, and internal pressures set up by shrinkage during drying, and, in the case of indirect (conduction) dryers, through a repeated and progressive occurring vaporization and recondensation of moisture to the exposed surface. An appreciation of this internal movement of moisture is important when it is the controlling factor, as it occurs after the critical moisture content, in a drying operation, carried to low final moisture contents. Variables such as air velocity and temperature, which normally enhance the rate of surface evaporation, are of decreasing importance except to promote the heat transfer rates. Longer residence times, and, where permissible, higher temperatures become necessary. In the case of materials such as ceramics and timber, in which considerable shrinkage occurs, excessive surface vaporization sets up high moisture gradients from the interior toward the surface, which is liable to cause overdrying, excessive shrinkage, and, consequently, high tension, resulting in cracking or warping. In these cases, it is essential to not achieve too high moisture gradients by retarding surface vaporization through the employment of high air relative humidity. The temperature gradient set up in the solid may also create a vapor pressure gradient, resulting in moisture vapor diffusion to the surface simultaneously with liquid moisture movement (Mujumdar 2007).

In the particular case of fresh-cut timber logs and lumber boards, the internal moisture movement from the interior to the surface may occur by one of the following mechanisms of mass transfer (Sherwood 1929a; Choong 1963; Bian 2001; Mujumdar 2007; Berit Time 1998; McCurdy 2005):

a. The magnitude of bulk moisture flow through the interconnected voids under the influence of a pressure gradient as a driving force depends on the wood permeability;

TABLE 7.1
Mechanisms of Internal Moisture Movement in Solids (Srikiatden and Roberts 2007)

Mechanism(s) of Moisture Transfer		
Liquid Water	**Water Vapor**	**References**
Diffusion		Lewis (1921); Sherwood (1929a, b)
Capillary		Ceaglske and Hougen (1937)
Evaporation–condensation		Henry (1948)
Capillary flow	Difference in partial pressure (diffusion)	
Liquid diffusion	Difference in total pressure (hydraulic flow)	Görling (1958)
Surface diffusion	–	
Hydraulic flow	–	
Capillary flow	–	
Knudsen flow	–	Keey (1970)
Hydrodynamic flow	–	
Surface diffusion	–	
Molecular diffusion	–	
Capillary flow	–	
Knudsen flow	–	Briun and Luyben (1980)
Hydrodynamic flow	–	
Surface diffusion	–	
Molecular diffusion within solid	Diffusion in pores	
Capillary flow	Knudsen flow	
Liquid diffusion in pores	Stephan diffusion	Hallström (1990)
Surface diffusion (absorbed water)	Hydraulic flow in pores	
Hydraulic flow in pores	Evaporation–condensation	
Diffusion	Mutual diffusion	
Capillary flow	Knudsen diffusion	
Surface diffusion	Effusion	
Hydrodynamic (bulk) flow	Slip flow	
–	Hydrodynamic (bulk) flow	Waananen et al. (1993)
–	Stephan diffusion	
–	Poiseuille flow	
–	Evaporation–condensation	

b. Movement (diffusion) of liquid (bound) water through the cell walls, if the wet wood is at a temperature below the water boiling point, followed by vaporization of the liquid at the surface and diffusion of the vapor into the surrounding air; a more or less stagnant air film (boundary layer) on the board surface presents a resistance to the passage of moisture from the surface into the air; the diffusion of the bound water could be explained by the fact that the damaged cells at the board surface initially contain water while some of the cells within the wood contain air bubbles; as drying progresses, the moisture is lost from these cells and the air-water meniscus recedes into the bordered pits closest to the surface reducing the radius of curvature and, therefore, increasing the capillary tension in the wood; this tension causes the air bubbles in the wood to enlarge beginning with the largest bubble followed by the smaller bubbles; when the cells have emptied, the tension continues to increase until the meniscus breaks into another cell and empties it (Skaar 1972; Siau 1984).

c. Movement (diffusion) of water vapor through the air inside the wood (through the lumens), if the liquid vaporizes within the wood structure at a point beneath the surface, followed by diffusion of water vapor from that point through the porous wood to the surface and, hence, out into the air; although the vapor diffusion coefficient is much greater than that of the liquid water, the total moisture flow rate appears to be controlled by the latter mechanism (Choong 1963; Kawai et al. 1980).

The moisture movement in wood can be also divided into (Berit Time 1998) (i) the movement of liquid water in the wet zone, that is, above the fiber saturation point; (ii) the movement of bound water and water vapor in the dry zone, that is, below the fiber saturation point (FSP) (McCurdy 2005).

Below fiber saturation point, all the moisture is either in the form of vapor in the lumens or bound water in the cell walls. Above FSP, there is also a point known as the irreducible saturation point where the continuity in the liquid column has broken and liquid flow can no longer occur. Vapor movement in wood below the FSP is more important than movement of water within the cell wall, and it becomes more important as the moisture content and the specific mass of the dry wood decrease.

The migration rate of moisture to the boards' surfaces (i.e., the internal process) should match the moisture removal rate of the dry air (i.e., the external process) in order to ensure cost-effective and safe operation of wood dryers. Therefore, drying conditions (e.g., drying air temperature and relative humidity) need to vary according to boards' actual average moisture content and/or optimum drying time.

Understanding the internal moisture mechanisms from the core of wet lumber boards to the surface, as well as the external mechanism of moisture removal from boards' surfaces, is important to improve and optimize both the production and the economy of drying industry, as well as the quality of many dried products, such as foods (Srikiatden and Roberts 2007) and woods.

Moreover, both internal and external wood drying processes contribute at determining the lumber drying rate and drying time, two technical and economical parameters very important for wood drying industry (Meroney 1969).

In the past, relatively few R&D works focused simultaneously on the internal and external drying processes of lumber boards dried in industrial kilns (Bian 2001).

7.1.2 Passageways for Moisture Movement

Moisture (i.e., water liquid and vapor) movement through the interior of fresh cut green timber logs and sawn lumber boards, which are highly anisotropic materials (Keey et al. 2000), requires passageways (transport spaces and routes) and driving forces (see Section 7.1.4).

The available pathways for moisture movement depend on the wood porosity, permeability, and density (Siau 1984). Depending on these parameters, the spaces available for moisture movement existing in the cellular structure, as well as through the cell walls (Langrish and Walker 1993), are relatively limited, varying from 25% to about 85% of the total volume of wood (Siau 1984).

Porosity (or void fraction) is a measure of the void (i.e., empty) spaces in porous materials. It is a fraction of the volume of voids over the total volume, and, along with the pore structure and distribution, affects moisture diffusivity significantly (King and Margaritis 1971).

Permeability is a wood species dependent parameter that can only exist if the void spaces are interconnected by openings (Keey et al. 2000). It is a measure of the ease with which moisture transported through a piece of wood under the influence of some driving forces, for example, capillary pressure gradient or moisture gradient. For example, softwoods have large permeability and, therefore, they are relatively easy to dry, whereas hardwoods have small permeability and will take longer time to dry (Bian 2001).

The wood density is another important parameter (Keey et al. 2000). Generally, lighter woods (as permeable softwoods) dry more rapidly than do the heavier woods (as impermeable hardwoods). In softwoods, the transport of moisture is often bulk flow (momentum transfer) at high temperature, whereas in hardwoods, the higher density provides increased resistance to diffusion (Siau 1984).

There are five main passageways by which moisture moves through wood depending on the driving force (e.g., moisture content or pressure gradients) (Berit Time 1988; Cech and Pfaff 2000) and variations in wood cellular structure (Langrish and Walker 1993; Onn 2007): (i) within (across) the vessels and cell cavities; (ii) as bound water through fibers and pit chambers from cell to cell by way of the porous pit membrane openings in the cell walls followed by evaporation at the surface on the other side, then across another cell cavity, and so on toward the board's surface; (iii) through the ray cells that represent only 2% of the total volume of wood; (iv) through intercellular spaces, that is, gaps between individual cells where they are not directly connected by the middle lamella representing about 1% of the total volume of wood; and (v) through transitory cell wall passageways that disappear when the liquid is removed by drying.

In softwoods, for example, there is more cell cavity space, thinner walls, and, often, more openings in the cell walls compared to heavy woods, such as the hardwoods. This gives more diffusion space and more passageway area between cells (Erickson 1954).

Because lighter wood species (as softwoods) (e.g., *Pinus radiata*) contain more openings per unit volume, moisture moves through air faster than through wood cell walls, thus they are very permeable (Comstock 1967; Perré and Keey

2015). As a consequence, softwoods dry faster than heavier species (hardwoods) (Langrish and Walker 1993).

According to Stamm and Nelson (1961), the moisture movement through the softwood cells, for example, can be divided into three different pathways and flow types:

i. Pathway 1 (through combination of fiber cavity and cell wall) (also called lumen–cell wall diffusion or cavity wall pathway) in which the water vapor diffuses through the fiber cavities (lumens) in series with the diffusion of bound water through the parts of the cell walls that are discontinuous in the direction of the diffusion; in other words, it is a series combination of bound-water through the cell wall and vapor diffusion across the cell cavity; the pathway 1 accounts for most of the diffusion, although this pathway becomes less important as the basic density of wood increases;
ii. Pathway 2 (through combination of fiber cavity and pit) (also called lumen pit or cavity pit pathway) in which the water vapor diffuses through the cell lumens in series with vapor diffusion through pit chambers and in series as vapor through pit membrane pores, and as bound water through the pit membranes; the diffusion thus occurs through the pit membranes in series with the fiber cavities;
iii. Pathway 3 (through cell walls only) (also called continuous cell wall or continuous cell wall pathway) in which the bound water diffuses through the continuous cell wall; the diffusion takes place through the continuous portion of the cell wall around the fiber cavity.

The proportion of moisture diffusion through each of these pathways depends on wood temperature and basic density. For example, for wood with a basic density of 800 kg/m^3 at 50°C, the proportion of moisture diffusion is 55% (path 1), 22.2% (path 2), and 22.8% (path 3). At 120°C, these percentages become 57%, 24.8%, and 18.2%, respectively.

For a softwood species with a basic density of 200 kg/m^3, dried at 120°C, the proportion of diffusion through path 1 (fiber cavity and cell wall) is of 94%. However, for a softwood species with basic density of 800 kg/m^3, also dried at 120°C, this proportion drops at 57%.

On the other hand, in any piece of wood, the moisture movement takes place in both longitudinal (along the grain) and lateral (across the grain) directions (Cech and Pfaff 2000; Onn 2007).

Although the permeability in the longitudinal direction is many times faster than lateral movement, the geometry of a lumber board is such that the net moisture movement is essentially one-dimensional, normal to the longest face, except near the ends (Chen et al. 1996; Blakemore and Langrish 2007).

Even though the moisture movement is 10–15 times faster along the grain than across the grain, most drying in lumber is by lateral movement (across the grain) because the moisture has to cross a much shorter distance, that is, the lumber board is many times longer than it is thick (Keey et al. 2000; Anderson 2014). Effectively, typical lumber boards are much longer (3–6 m) than in width (100–250 mm) or thickness (40–50 mm). This explains why flat sawn lumber generally dries faster than quarter sawn lumber (McCurdy and Keey 2002).

In other word, for drying, the most important dimension is across the grain, because the cross-sectional area for water movement is much greater than along the grain, even though longitudinal transport is much faster. In softwoods, for example, the longitudinal permeability can be more than 1,000 times greater than the lateral permeability (Banks 1968; Perré and Keey 2015).

Generally, the moisture moves more freely in sapwood than in heartwood because chemical extractives in heartwood plug up passageways. Thus, sapwood generally dries faster than heartwood (Cech and Pfaff 2000; Onn 2007). However, the heartwood of many wood species is lower in moisture content than is the sapwood and can reach final moisture content as fast (Keey et al. 2000).

In sapwood of hardwoods, longitudinal movement of moisture is higher comparatively to lateral diffusion because of the presence of vessels (Walker et al. 1993). The large perforations at each end of the vessels offer comparatively little resistance to longitudinal diffusion relative to the resistance to lateral diffusion through the walls between vessels (Keey et al. 2000). The vessels in hardwoods are sometimes blocked by the presence of tyloses and/or by secreting gums and resins in some other species. The presence of gum veins, the formation of which is often a result of natural protective response of trees to injury, is commonly observed on the surface of sawn boards of most eucalypts (Langrish and Walker 1993).

The influence of longitudinal movement appears to be confined only to the ends that dry out more rapidly than the majority of the board. For this reason, lumber, especially valuable slow-drying hardwoods, is sometimes end-coated to reduce longitudinal moisture movement and this will further decrease its significance (Keey et al. 2000).

7.1.3 Governing Laws of Moisture Movement

Moisture movement within porous materials concern exchanges of momentum (fluid flow), energy (e.g., by heat conduction), and mass (e.g., by molecular diffusion).

All these transport phenomena are described by analogous laws (Table 7.2), that is, (i) Newton's law for fluid momentum, (ii) Fourier's law for heat conduction, and (iii) Fick's laws for mass diffusion (Carrier 1921; Lewis 1921; McCready and McCabe 1933; Sherwood 1929a; b; Tuttle 1925; Welty et al. 1976; Thomas 2000).

TABLE 7.2
Comparison of Transport Phenomena

Transported Quantity	Physical Phenomenon	Equation
Momentum	Viscosity diffusivity (Newton's law)	(7.1)
Heat	Heat conduction (Fourier's law)	(7.2)
Mass	Molecular diffusion (Fick's laws)	(7.4) and (7.9)

To predict the mass transfer by diffusion, the most widely used is the Chilton–Colburn *j-factor* analogy based on experimental data for gases and liquids in both the laminar and turbulent flow regimes, especially over flat plates.

On the other hand, the Reynolds analogy assumes that the molecular diffusivity of momentum (kinematic viscosity, $\nu = \mu/\rho$) and mass (diffusion coefficient, D) are negligible compared to the turbulent diffusivity.

Such analogies require assumptions as (Meroney 1969) (i) Ficks' second law of diffusion in the solid is valid; (ii) diffusion factor (D) is constant; (iii) when the drying process starts, the moisture content is uniform; (iv) moisture movement is normal to the solid's surfaces; (v) surface fibers attain the equilibrium moisture content as soon as drying starts; (vi) thickness of the lumber board does not change during drying; and (vii) equilibrium moisture content remains constant for the drying process.

7.1.3.1 Momentum Transfer

When the drying air is continuously flowing with velocity $u = u_x$ in the x direction parallel to a flat plate (see Figure 7.5a), the air has x-directed momentum, and its specific (mass) flow rate is ρu_x. By random diffusion of molecules, there is an exchange of molecules in the perpendicular direction y. Hence, the x-directed momentum has been transferred in the y-direction from the faster- to the slower-moving layer of the fluid (Incropera and DeWitt 2002).

The Newton's law of viscosity, a simple relationship between the flux of momentum and the velocity gradient, can be written as follows:

$$\tau_{yx} = -\nu \frac{\partial(\rho u_x)}{\partial y} \tag{7.1}$$

where

τ_{yx} is the flux of x-directed momentum in the perpendicular direction y (kg/m·s^2)
$\nu = \mu/\rho$ is the fluid kinematic viscosity (m^2/s)
ρ is the fluid density (kg/m^3)
μ is the fluid dynamic viscosity (kg/m·s or N·s/m^2)
u_x is the fluid velocity component in x-direction (m/s)
y is the direction of momentum transport (m)

Moisture loss during convective hot air drying is a coupled heat and moisture transfer mechanism. Measuring moisture transfer parameters within hygroscopic materials as foods (Srikiatden and Roberts 2007) and wood becomes complicated when both heat and mass transfer processes have to be taken into consideration. Heat transfer should be taken into account when the thermal conductivity is small and the diffusivity is large. Conversely, heat transfer does not need to be considered when the thermal conductivity of the material is large and the diffusivity is small.

Lewis number $\left(Le = \frac{k}{\rho * c_p * D}\right)$, used when a process involves simultaneous heat and mass transfer, is the ratio of the thermal diffusivity $\left(\alpha = \frac{k}{\rho * c_p}\right)$ (where k is the thermal conductivity, in W/mK; ρ is the density, in kg/m^3; and c_p is the specific heat, in J/kg · K) to the mass diffusivity (*D*) and represents the relative effect of temperature and moisture gradients inside a given material (Srikiatden and Roberts 2007).

7.1.3.2 Heat Transfer

For a steady-state thermodynamic system, the heat transfer by conduction, where the molecules are stationary and the transport is done mainly by electrons as a result of random molecular motion caused by a temperature gradient (Crank 1975), is expressed by the Fourier's law of heat conduction:

$$\dot{q} = -k * \frac{dT}{dy} \tag{7.2}$$

where

$\dot{q}$ is the density of heat flux (W/m²)
k is the fluid thermal conductivity (W/mK)
T is the fluid temperature (K)
y is the direction of the heat transfer (m)
dT/dy is the temperature gradient (acting as a driving force for the heat transfer)

Fourier's equation for the heat conduction within the lumber boards is used to predict the internal temperature distribution (Pordage and Langrish 2007; Marinos-Kouris and Maroulis 2014):

$$\frac{\partial T}{\partial t} = \frac{\partial}{\partial t}\left(\frac{k_{wood}}{\rho_{wood} * c_{pwood}} \frac{\partial T}{\partial y}\right) = \alpha \frac{\partial^2 T}{\partial y^2} \tag{7.3}$$

where

k_{wood} is the wood thermal conductivity (W/mK)
ρ_{wood} is the wood density (kg/m³)
c_{pwood} is the wood isobaric specific heat (kJ/kg · K)
$\alpha = \frac{k_{wood}}{\rho_{wood} * c_{pwood}}$ wood thermal conductivity (m²/s)
T is the temperature (K)
t is the time (s)

7.1.3.3 Mass Transfer

Mass transfer within porous material is an unsteady two-phase flow process (Masmoudi and Prat 1991) that cannot by described according to a continuum approach because of large-scale heterogeneities in the fluid distribution (Lenormand et al. 1988) and the material structure containing sufficient pores of different sizes (Maneval and Whitaker 1988).

Most solid materials (such as woods and foods) encountered in industrial drying processes are extensively inhomogeneous (implying that material properties are closely related to the spatial position) and anisotropic (suggesting that the properties depend on the direction).

These conditions mean that the diffusivity (*D*) and permeability (*K*) coefficients, as well as thermal conductivity (*k*), are three-dimensional symmetric tensors.

Lewis (1921) (cited by Sherwood 1931) first recognized that diffusion theory could be applied in the food dehydration process.

Sherwood (1929a, b) proposed a method to quantify moisture loss during the falling rate period and to predict drying times for materials such as woods.

The simplest models describing the mass (moisture) transfer process in drying of porous media at low relative humidity (Comstock 1963; Skaar et al. 1970; Wadso 1993) and relatively low moisture contents are based on both Fick's (Mujumdar 2000) and Darcy's (describing the hydraulic flow of single-phase fluids in porous materials) laws that differ with respect to the solution methodology (Marinos-Kouris and Maroulis 2014).

Fick's laws respectively state that (i) the molar flux due to diffusion is proportional to the concentration gradient (Fick's first law); (ii) the rate of change of concentration at a point in space is proportional to the second derivative of concentration with space (Fick's second law).

Even though the moisture movement in the hygroscopic regions of wood is not governed by diffusion alone, the traditional diffusion equations (i.e., Fick's first and Fick's second laws) are still most often used to describe moisture transport in wood and for evaluation of diffusion coefficients from laboratory experiments (Berit Time 1998).

Simplified analysis methods consist in solving Fick's equations analytically for certain sample geometries (as lumber boards) under assumptions as (i) surface mass transfer coefficient is high enough so that the material moisture content at the surface is in equilibrium with the air drying conditions; (ii) air drying parameters are constant; and (iii) moisture diffusivity coefficient is constant, independent of material moisture content and temperature.

It can be concluded that the potential for heat transfer is the temperature gradient (Section 7.1.3.2), while for the mass transfer, the main potential is the moisture content gradient.

If the fluid thermal conductivity (*k*) and diffusion coefficient (*D*) are constant, the mathematical analogies between the heat and mass transfer can be illustrated as shown in Table 7.3.

However, although the analogy between heat and mass (moisture) transfer is the same mathematically, the actual physical mechanisms are different in porous hygroscopic materials such as foods (Srikiatden and Roberts 2007) and woods.

TABLE 7.3
Analogous Heat and Mass Transfer Equations

Heat Transfer	Mass Transfer
$\dot{q} = -k\frac{\partial T}{\partial y}$	$j_{mass} = -D\frac{\partial(MC)}{\partial y}$
$\frac{\partial T}{\partial t} = \left(\frac{k}{\rho c_p}\right)\frac{\partial^2 T}{\partial y^2}$	$\frac{\partial(MC)}{\partial t} = D\frac{\partial^2(MC)}{\partial y^2}$

7.1.3.3.1 Fick's First Law

To quantify the moisture mass transfer by diffusion, Fick (1855) adopted the Fourier's law (equation 7.2) for heat conduction because of the similarity of these phenomena.

For the one-dimensional steady-state situation (Sherwood 1929a; Berit Time 1998), Fick's first law is formulated as follows: "*diffusion molar (or mass) vapor flux from higher concentration to lower concentration is proportional to the gradient of the concentration of the substance and the diffusivity of the substance in the medium*", and it can be expressed as follows:

$$j = -D * \frac{\partial C}{\partial y} \tag{7.4}$$

where

- j is the mass (kg/m$^2 \cdot$s) or molar (kmol/m$^2 \cdot$s) specific moisture diffusion rates
- D is the diffusion (also termed as transport, diffusivity, or water vapor permeability) coefficient (m^2/s)
- C is the mass (kg/m^3) or molar (kmol/m^3) moisture concentration acting as a driving potential (Berit Time 1998)
- $-\frac{\partial C}{\partial y}$ is the concentration gradient per unit length, that is, the space gradient of the driving potential in direction y

The minus sign indicates that the moisture movement is in the direction of a negative gradient, that is, from high to low concentrations.

Equation 7.4 considers driving force as the concentration (moisture content) gradient and is based on assumptions as follows:(i) liquid concentration throughout the solid at the start is uniform; (ii) moisture mass diffusion is normal to the surface plane; (iii) vaporization takes place at the solid's surface; and (iv) surface resistance to vapor diffusion is negligible, i.e., the liquid concentration on the surface falls to zero immediately after the start of the drying (Sherwood 1929a).

Applied to wood drying processes, where bound water diffusion (in the cell walls) and vapor diffusion (in lumen and pits) are possible pathways to drive moisture from high- to low-content regions, Fick's first law of diffusion shows that the (specific) mass flow rate of the moisture (j_{mass}, kg/m$^2 \cdot$s) through a plane perpendicular to the vertical direction y (see Figure 7.5a) is proportional to the volumetric moisture content gradient (Perré and Keey 2015):

$$j_{mass} = -D * \frac{\partial (MC)}{\partial y} = -D * \nabla (MC) \tag{7.5}$$

where

- D is the diffusion coefficient corresponding to the moisture content as the driving potential (m^2/s)
- MC is the volumetric moisture content, that is, the moisture concentration (kg/m^3)

$-\frac{\partial(MC)}{\partial y}$ is the volumetric moisture content gradient
y is the direction of moisture movement (m)

At the microscopic scale of the transport mechanism, under isothermal conditions (Siau 1984, 1995), the mass flow rates of bound water (j_{bound}) and vapor diffusion (j_{vapor}) can be expressed as follows, respectively (Perré and Keey 2015):

$$j_{bound} = -\rho_b * D_b * \nabla(MC)_b \quad \text{(Bound water flux)} \tag{7.6}$$

$$j_{vapor} = -\rho_v * D_v * \nabla\varepsilon_v \quad \text{(Vapor flux)} \tag{7.7}$$

where

ρ_b and ρ_v are the bound water and vapor mass density, respectively (kg/m^3)
D_b and D_v are the microscopic bound water and vapor diffusion coefficients, respectively (m^2/s)
$(MC)_b$ is the bound moisture content $\left(\frac{kg_{water}}{kg_{dry\text{-}wood}}\right)$
ε_v is the mass fraction of vapor in the gaseous phase (–)

By using the bound liquid diffusivity data of Stamm (1963), Perré and Keey (2015) obtained the following correlation for the bond water diffusion coefficient (D_b) that shows an important increase of the bound water diffusivity as the bound moisture content increases:

$$D_b = \exp\left[-12.82 + 10.90 * (MC)_b - \frac{4{,}300}{T}\right] \tag{7.8}$$

where

T is the temperature (K)

7.1.3.3.2 Fick's Second Law

The moisture transfer in heterogeneous media (in which the heterogeneity of the material is accounted for by the use of the coefficient of diffusivity) can be analyzed by using Fick's second law for homogeneous materials.

Based on the continuity equation $\left(\frac{\partial C}{\partial t} = -\frac{\partial j}{\partial y}\right)$ and if the diffusion coefficient D is constant, Fick's second law of diffusion results as a linear equation where concentration of the chemical species is under consideration, for example, the moisture content in the case of wood drying:

$$\frac{\partial C}{\partial t} = D\frac{\partial^2 C}{\partial y^2} \tag{7.9}$$

Equation 7.9 states that the rate of change of concentration at any point into the diffusion medium in the y direction is proportional to the rate at which the rate of variation of concentration with distance changes is valid. In other words, it describes the rate of accumulation (or depletion) of concentration $\left(\frac{\partial C}{\partial y}\right)$ within the volume as proportional to the diffusivity coefficient (D) and the second derivative of the concentration gradient $\left(\frac{\partial^2 C}{\partial y^2}\right)$.

Applied to the drying processes of hygroscopic, porous materials (such as woods) with constant diffusivity, and by assuming one-dimensional liquid movement (Figure 7.2) with constant moisture content gradient and without internal heat sources, Fick's second law in the falling drying rate period can be written as follows (Meroney 1969; Doe et al. 1996; Mujumdar 2000; Keey et al. 2000):

$$\frac{\partial MC}{\partial t} = \frac{\partial}{\partial y}\left(D\frac{\partial MC}{\partial y}\right) = D\frac{\partial^2 MC}{\partial y^2} \tag{7.10}$$

where

MC is the wood moisture content above equilibrium moisture content $\left(\frac{kg_{water}}{kg_{oven\text{-}dry}}\right)$

D is the constant diffusion coefficient (m^2/s)

y is the distance coordinate perpendicular to the airstream from the core (m)

t is the time (s)

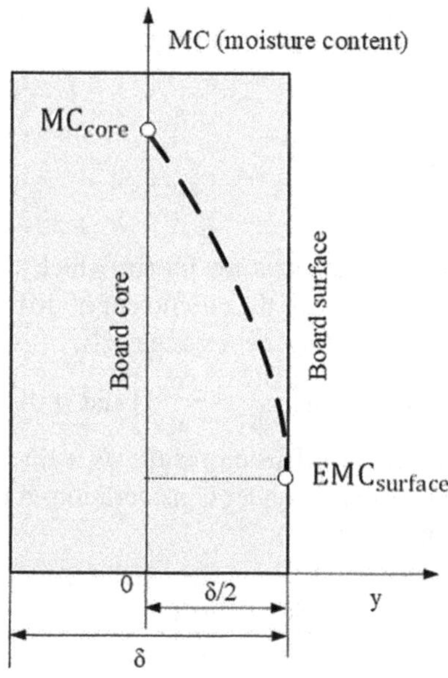

FIGURE 7.2 Typical moisture content profile for moisture movement inside a lumber board. EMC, equilibrium moisture content.

Fick's second law (equation 7.10) describes the time change of the wood moisture distribution within the wood being dried when the controlling mechanism of drying is the diffusion of moisture (Pang and Haslett 1995). It can be used for the calculation of both drying rate (Pakowski and Mujumdar 1987) and drying time (Strumillo and Kudra 1986).

With the coordinate directions (δ is the thickness of the lumber board) and a typical moisture content profile shown in Figure 7.2, the moisture fluxes have negative numerical values when the lumber is dried, since the moisture content falls from a maximum value at the core (MC_{core}) to a minimum value at the surface, equal to the equilibrium moisture content ($EMC_{surface}$).

Fick's second law equation can also be expressed in terms of vapor pressure gradient as the driving force for moisture movement rather than the moisture concentration gradient (Van Arsdel et al. 1973):

$$\frac{\partial p}{\partial t} = \frac{\partial}{\partial x}\left(D_p \frac{\partial p}{\partial x}\right) \tag{7.11}$$

where

p is moisture vapor partial pressure related to moisture content through a sorption isotherm (Pa)

D_p is the moisture diffusion coefficient which relates the rate of moisture movement to a vapor pressure gradient (m^2/s)

7.1.3.3.3 Diffusion Coefficients

In general, moisture movement in solids as woods during drying is a complex process that may involve molecular, capillary, and surface diffusion, and hydrodynamic flow. In addition to being dependent on geometric shapes and physical structure of the dried product, the diffusion coefficient depends as well on the drying conditions.

The diffusion coefficients (D) for various wood species and other hygroscopic materials, such as foods (Burr and Stamm 1956; Comstock 1963; Marshall 1958; Stamm 1964; Srikiatden and Roberts 2007), have been studied both empirically and analytically.

The diffusion of moisture through complex networks of capillaries in wood is analogous to electrical conduction in equivalent resistance circuit. Thus, the diffusion coefficient may be predicted on the basis of a simple electrical conduction analogy (Burr and Stamm 1956; Stamm 1964). They generally vary with drying temperatures, moisture content, and wood basic density during both constant- and falling-rate periods (Collignan et al. 1993; Marinos-Kouris and Maroulis 2014).

The assumption of constant diffusion coefficients is valid within the moisture content range where all drying curves at a given temperature merge into one curve regardless of initial moisture content.

The moisture diffusion coefficient in the tangential direction in softwood varies with absolute temperature as T^{10}, suggesting that it rises fivefold in raising the temperature from 60°C to 120°C. Over the same temperature range, the external mass transfer coefficient for the vaporization from the wood surface increases by only 33%, so the moisture movement within the wood becomes less dominant in restricting the overall drying process than does the moisture vapor transfer in the air. Thus, the convection is much more significant in the drying of permeable softwoods at relatively high temperatures (Luikov 1966).

While the external convection may affect the total time to dry hardwoods, the external mass transfer coefficients influence the moisture content profiles near the wood surface. In turn, the development of such a reduced-moisture zone influences the developments of drying stresses and phenomena such as casehardening, limiting the severity of the drying schedule.

Empirical diffusion coefficients can be obtained only by experiments (Stamm 1959, 1960; Berit Time 1998; Wadso 1993; Meroney 1969) by fitting the experimental data to theoretical drying curves (Tuttle 1925; Sherwood 1929a; Comstock 1963; Stamm 1964). Experimental methods based, for example, on drying curves (the most common method), sorption kinetics, permeation, nuclear magnetic resonance, electron spin resonance, and radiotracer can be used for measuring the effective diffusivity of complex hygroscopic materials as foods (Saravacos and Maroulis 2001; Srikiatden and Roberts 2007). The better direct method able to accurately determine by experiments the diffusion coefficients as functions of local moisture contents varying with time and position during drying of complex materials as foods is the magnetic resonance imaging. Scanning neutron radiography was also used to determine the effective moisture diffusivity as a function of moisture content profiles of porous materials (Srikiatden and Roberts 2007). However, there is no standard method for the experimental determination of diffusion coefficients for foods (Srikiatden and Roberts 2007), wood species, or other similar materials.

By measuring the moisture content gradients of 50.8 mm-thick boards of Sitka spruce (a softwood) dried at 71.1°C under a relative humidity of 75% from an initial moisture content of 0.51 kg/kg (dry basis), a diffusion coefficient of $1.12 * 10^{-9}\,m^2/s$ has been determined (Tuttle 1925).

Diffusion coefficients for Western fir boards ($15 * 15\,mm^2$ with lengths between 40 and 200 mm) dried at temperatures between 30°C and 60°C have been determined in both lateral and longitudinal directions. The radial diffusion coefficients varied from $1.05 * 10^{-11}\,m^2/s$ at 40°C and 5% moisture content to $3.45 * 10^{-11}\,m^2/s$ at 60°C and 15% moisture content, whereas the tangential diffusion coefficients varied from $2.9 * 10^{-9}\,m^2/s$ at 40°C and 5% moisture content to $0.3 * 10^{-9}\,m^2/s$ at 60°C and 15% moisture content (Choong 1963).

The diffusion coefficients below the FSP for softwoods and hardwoods, such as scots and loblolly pine, birch, spruc, douglas-fir, and oak, are lumped parameters which include both bound water and water vapor movement. Above the fiber saturation point, capillary-controlled movement of moisture occurs for permeable species (Stamm 1964, 1967b).

The experimental diffusion coefficients for flat-sawn Eastern hemlock sapwood samples (150 mm long, 50 mm wide and 6.1 mm thick) pre-dried to near the fiber saturation point, then dried at temperatures from 50°C to 120°C ranged from $0.53 * 10^{-9}\,m^2/s$ at 50°C to $3.33*10^{-9}\,m^2/s$ at 120°C (Biggerstaff 1965).

Above the fiber saturation point, for samples ($38.1 * 38.1 * 25.4\,mm^3$) of sapwood and heartwood from hardwood species such as American elm, hackberry, red oak, white ash, sweet gum, and sycamore, the experimental longitudinal diffusion coefficients varied from $2.37 * 10^{-8}$ to $3.43 * 10^{-8}\,m^2/s$, while the tangential coefficients ranged from $1.14 * 10^{-8}$ to $2.16 * 10^{-8}\,m^2/s$. Below the fiber saturation point, the experimental diffusion coefficients were $1.27–2.38 * 10^{-8}\,m^2/s$ (longitudinal) and $3.29–5.74 * 10^{-8}\,m^2/s$ (tangential) (Choong et al. 1994).

For lumber samples (25 mm thick) of Australian hardwoods dried at temperatures from 30°C to 80°C and air velocities around 0.4 m/s, the diffusion coefficients ranged between 10^{-9} and $10^{-10}\,m^2/s$ (Langrish et al. 1997; Langrish and Bohm 1997).

For pre-steamed and untreated samples of Southern beech (an impermeable lumber from New Zealand) dried at 20°C, the diffusion coefficients varied between 10^{-10} and $10^{-11}\,m^2/s$, with the maximum values measured at about 40% moisture content (dry basis) (Grace 1996).

It can be seen that most measured diffusion coefficients lie between 10^{-8} and $10^{-10}\,m^2/s$ (Keey et al. 2000). In softwoods, for example, the longitudinal diffusion coefficients are 55% (Keey et al. 2000) and even 250% (Stamm 1960) higher than the tangential diffusion coefficients. This difference could be due to the combination of a number of features, including the contribution of ray cells to radial diffusion and the concentration of pits on the radial faces relative to those on the tangential faces of the fibers (Stamm and Nelson 1961).

The temperature dependence of diffusion coefficients can be empirically expressed by considering either the activation energy (i.e., least amount of energy needed for a chemical reaction to take place) or the Arrhenius factor (Glasstone et al. 1941; Skaar 1958; Stamm 1964; Choong 1965; King and Margaritis 1971; Wu 1989; Simpson 1993; Mujumdar 2000; Keey et al. 2000; Marinos-Kouris and Maroulis 2014), as follows:

$$D = D_0 * \exp\left(-\frac{E}{\mathcal{R} * T}\right) \tag{7.12}$$

where

D_0 is the Arrhenius (preactivation) factor (m^2/s)
E is the diffusion activation energy (J/mol)
$\mathcal{R}$ is the universal gas constant (8.314 J/mol · K)
T is the absolute temperature (K)

Within the range of moisture contents below the FSP usually encountered in lumber drying, the diffusion coefficient should be proportional to the exponential of the moisture content (Stamm 1967b):

$$D = \exp(C * MC) \tag{7.13}$$

where

C is a constant expressing the effect of moisture content on the diffusion coefficients (see Table 7.4) (Keey et al. 2000)
MC is the moisture content $\left(\frac{kg_{water}}{kg_{oven\text{-}dry}}\right)$

TABLE 7.4
The Effect of Moisture Content on the Diffusion Coefficient (the Parameter *C* Found by Various Authors)

Author	Value of *C*	Species
Choong (1965)	13.5	Western fir
Furuyama et al. (1994)	19.6	Akamatsu; Hinoki
Simpson (1993)	1.15	Desorption, red oak
	1.45	Absorption, red oak
Skaar (1958)	0	Tangential, American beech
	7.27	Radial, American beech
Stamm (1959)	11	Sitka spruce

The diffusion coefficient could also be exponentially related to both temperature and moisture content, if it is assumed that moisture content is the dominant moisture driving force:

$$D = D_0 \exp\left(C * MC + \frac{E}{\mathcal{R} * T}\right) \tag{7.14}$$

Most fitted values for the activation energies (E) are in the range of 23–43 kJ/mol, which are just below those for the evaporation of water (45 kJ/mol) (Keey et al. 2000), while the theoretical model of Stamm (1967b) suggests an activation energy of around 35 kJ/mol.

As the temperature is raised, the increase in the diffusion coefficient appears to be virtually proportional to the increase in vapor pressure of water, a finding which lends weight to the use of the partial pressure of water vapor as the driving force for moisture movement. However, the effect of temperature on the diffusion coefficient persists even at moisture contents for which the vapor pressure is independent of moisture content (above the fiber saturation point), suggesting that the use of a vapor pressure driving force should not be an automatic choice. The use of a chemical potential as driving force results in effective water conductivities which remain strong functions of moisture content, so the use of this driving force does not appear to offer significant advantages over the use of the moisture content as driving force.

If the diffusion coefficient is constant, the moisture content profile would be linear over a material for the steady-state movement of moisture through it. However, the drying of a lumber board is not a steady-state process, and the moisture content profile is non-linear. When the moisture content change occurs over the entire half-thickness of the board, in other words when there is no longer a fully wet region in the core, the moisture content profiles can be shown to be parabolic in shape during drying if the diffusion coefficient is constant.

The effect of moisture content on the diffusion coefficient is less clear than the effect of temperature. Most of the studies showing an effect of moisture content on the diffusion coefficient have been carried out on wood samples which are below the fiber saturation point, for which the parameter C has been in the range of 11–20.

Above the fiber saturation point, the experimental evidence suggests that there is no significant effect of moisture content on the apparent diffusion coefficient (Keey et al. 2000).

Based on the vapor pressure as driving potential, the diffusion coefficient (D_p, kg/m · s · Pa) is dependent of the relative humidity according to the following relation (Berit Time 1998):

$$D_p = a * RH^b \tag{7.15}$$

where

a and *b* are constants given for longitudinal and lateral transport of moisture, respectively

RH is the relative humidity (decimals)

If all phenomena and parameters that influence the moisture movement (diffusion) in solids are combined into one, an effective diffusion coefficient (D_{eff}) can be defined as a material lumped property that does not really distinguish between the transport of water by liquid or vapor diffusion, capillary or hydrodynamic flow due to pressure gradients setup in the material during drying. Such a parameter characterizes the ability of the vapor diffusion through the porous media and relates the moisture content to the vapor pressure as the driving force for diffusion by lumping internal mass transfer mechanisms together (Mujumdar 2000).

Consequently, for hygroscopic porous materials, such as woods and foods, Fick's second law can be expressed by using the effective diffusivity (D_{eff}) as follows (King 1968; Marinos-Kouris and Maroulis 2014):

$$\frac{\partial MC}{\partial t} = \frac{\partial}{\partial y}\left(D_{eff}\frac{\partial MC}{\partial y}\right) = D_{eff}\frac{\partial^2 MC}{\partial y^2} \tag{7.16}$$

where

D_{eff} is the vapor (constant) effective diffusion coefficient (m²/s)

When no free water is present in wood, the value of D_{eff} directly controls the migration of moisture as vapor, while in the case of liquid migration, the effective diffusion coefficient has no influence on drying (Bian 2001).

The measurement of effective moisture diffusivity, which is the most common parameter used in predicting moisture movement in solid products, other than woods (as foods) is still today a challenging task (Srikiatden and Roberts 2007).

7.1.3.3.4 Classical Darcy's Law

Darcy's law is an empirical, phenomenological relationship describing, with a good approximation, the hydraulic flow (movement) of single-phase fluids (e.g., groundwater, moisture) in porous materials (as soil, sand, woods, foods) (Siau 1971; Srikiatden and Roberts 2007; Brown 2016) according to the following relation:

$$\dot{V} = \frac{dV}{dt} = K_h * A * \frac{\Delta h}{L} \tag{7.17}$$

where

$\dot{V} = \frac{dV}{dt}$ is the fluid volumetric flow rate (m^3/s)
K_h is the hydraulic conductivity (m/s)
A is the cross-sectional area to flow (m^2)
Δh is the change in hydraulic head over path L (m)
L is the fluid flow path length (m)

At a specific point in the fluid, the hydraulic head (h) is the sum of the pressure and gravitational (elevation) heads:

$$h = \frac{p}{\rho g + y} \tag{7.18}$$

where

p is the water pressure (N/m^2)
ρ is the water density (kg/m^3)
g is the acceleration of gravity (m/s^2)
y is the elevation head (m)

Under unsaturated conditions, the elevation head (y) is negligible compared to the pressure head. Consequently,

$$h = \frac{p}{\rho g}$$

The hydraulic conductivity can be expressed as a function of permeability as follows:

$$K_h = \frac{k_p * \rho * g}{\mu} \tag{7.19}$$

where

k_p is the permeability (m^2)
μ is the fluid absolute viscosity ($N \cdot s/m$)

Thus, the common form of Darcy's law becomes (Srikiatden and Roberts 2007):

$$\dot{V} = \frac{k_p A}{\mu} * \frac{\Delta p}{L} \tag{7.20}$$

When only one fluid phase migration in wood is present, the classical Darcy's law applies and can be expressed as follows (Perré and Keey 2015):

$$v = -\frac{K_h}{\mu} * \frac{\partial p}{\partial x} \tag{7.21}$$

where

v is the apparent velocity of the fluid through the lumber board specimen (m/s)
K_h is the wood hydraulic permeability (m^2)

μ is the dynamic viscosity of the fluid (Pa · s = kg/m · s)
p is the fluid pressure (Pa = kg/m · s^2)
x is the spatial axis (m)

In equation 7.21, the permeability (K_h), which measures quantitatively the ability of a porous medium to conduct fluid flow, is the most important physical property mainly obtained through experimental measurements.

7.1.3.3.4 Generalized Darcy's Law

When both liquid and vapor phases coexist in wood, the generalized Darcy's law must be used (Bear 1972; Spolek 1981). The volumetric flow rate of each phase is considered to be proportional to the pressure gradient of the corresponding phase. The phenomenological coefficient is the product of the permeability (K) by a function of saturation (called relative permeability, usually supposed to be a function of saturation only) to the considered phase (Perré and Keey 2015).

For the vapor phase,

$$v_g = -\frac{k_p * k_{rg}(S)}{\mu_g}\frac{\partial p_g}{\partial y} \tag{7.22}$$

For the liquid phase,

$$v_f = -\frac{k_p * k_{rf}(S)}{\mu_f}\frac{\partial p_f}{\partial y} \tag{7.23}$$

where
k_{rg} is the vapor relative permeability
k_{rf} is the liquid relative permeability
S is the saturation state (0–1)

Equations 7.22 and 7.23 must be consistent with the classical Darcy's law when one single-fluid phase occupies the porous medium. Consequently, the relative permeability functions fulfill the following conditions:

$$\text{Water liquid only}: f(0) = 0; k_{rf}(1) = 1$$

$$\text{Water vapor only}: k_{rg}(0) = 1; k_{rg}(1) = 0$$

The water liquid pressure (p_f) is related to the water vapor pressure (p_g) through the capillary pressure (p_c) function as follows:

$$p_f = p_g - p_c(S) \tag{7.24}$$

For softwoods, the following functions have been proposed for the relative permeability (Perré et al. 1993).

In the transverse direction,

$$k_{rg}^{lateral} = 1 + (4S - 3)S^4 \quad \text{and} \quad k_{rf}^{lateral} = S^3 \tag{7.25}$$

In the longitudinal direction,

$$k_{rg}^{long} = 1 + (4S - 5)S^4 \quad \text{and} \quad k_{rl}^{long} = S^8 \tag{7.26}$$

7.1.4 Moisture Driving Forces

Drying is a complex thermodynamic process widely depending on the wood initial temperature and moisture content, and the relative air and wood dryness. When green lumber begins to dry, the vaporization of moisture lowers the surface moisture content below that of the interior.

If the surfaces of fresh (cut) harvested green timbers or sawn lumber boards are drier than the interior, the moisture moves (migrates, diffuses) from high to low zones of moisture contents. In other words, moisture tends to distribute itself equally throughout the piece by moving from areas of high to areas of low moisture content. Drying is thus a process of moisture movement (diffusion) from the inner zones of the wood to the surface, followed by vaporization at the surface, driven by the vapor partial pressure difference proportional to the absolute humidity gradient that exists between the surfaces of the lumber boards and the drying air, and dissipation of water vapor into the bulk of drying air (Lewis 1921; Sherwood 1929a, b; Keey and Nijdam 2002).

The moisture is generally driven by affinity (attraction) that both air and wood cell walls have for moisture (water).

The wood's cell walls have an affinity for moisture, thus they will attract replacement moisture in an effort to maintain the FSP condition. They can only absorb moisture from an area of lower affinity, which is any portion of the wood having higher moisture content, that is, the interior zones. Moisture is therefore drawn from the interior of the board, but as it reaches the surface, it is again removed by the dry air because the moisture removal takes place only at the board surface; the drying of wood is essentially a process of moisture movement from the inner zones to the surface; and, in this manner, drying progresses inward. In other words, drying starts from the exterior (surface) of wood boards and moves toward their centers.

The most important factor which governs the rate of moisture movement (whether into, as free and bound water, or out of the wood) consists of the capacity of the surrounding air to absorb water vapor from the surfaces of the wood boards until the vapor pressure of the moisture is equal to the partial pressure of the vapor in the surrounding air (Walker et al. 1993; Cech and Pfaff 2000).

The rate of moisture movement is thus largely dependent upon the relative dryness of the air and wood, and is related to the thickness of the wood boards, and the air temperature. The drying time, for example, increases approximately as the square of the thickness of a lumber (Erickson 1954).

The main dominant driving forces (that can act at the same time) can be classified as follows:

a. Capillary forces (Langrish and Walker 1993; Mujumdar 2007; Anderson 2014);
b. Vapor partial pressure gradients (Skaar 1988; Keey et al. 2000);

c. Moisture content gradients (Erickson 1954; Siau 1980; Skaar 1988; Berit Time 1998; Keey et al. 2000; Cech and Pfaff 2000; Anderson 2014; Mujumdar 2015);
d. Chemical potential (Siau 1980; Skaar 1988; Keey et al. 2000);
e. Osmotic pressures (Skaar 1988).

7.1.4.1 Capillary Forces

Capillarity, defined as the ability of water to form a definite meniscus in a small tube whose sides it can wet, is the main reason why free liquid water flow (or do not flow) inside the wood through cell cavities and small passageways that connect adjacent cell cavities, pit chambers, and pit membrane openings (Erickson 1954).

It consists of the movement of water liquid by molecular attraction between the liquid and the wood body due to simultaneous action of (i) adhesion (i.e., attraction between water molecules and the walls of cells and pit chambers) (Masmoudi and Prat 1988); (ii) cohesion (i.e., attraction of the water molecules to each other); and (iii) porosity (i.e., the ratio of the free space to the total volume of a given piece of wood) (Hougen et al. 1940).

At moisture contents above the fiber saturation point, the water exists as liquid and water vapor in the voids or lumens of wood cells that are long and narrow, hollow in the center, with small orifices or pits that provide fluid paths between adjacent cell lumens.

When green wood starts to dry, the water evaporation from the surface cells exerts a capillary pull on the free water from cell cavities that is progressively removed toward the wood surface.

When the free water has evaporated, the moisture remains in cell cavities as vapor and in cell walls as bound water. That means that at the start of the drying, the capillary action that allows motion of nutrient fluids in the living trees is the main force controlling the movement of the free liquid in wood during drying (Erickson 1954; Anderson 2014).

When the FSP is reached (i.e., when there is no longer any free water in the wood), the capillary force is no longer the main force for the water movement since the free water in the cells no longer forms a continuous system (Erickson 1954; Esping 1992). Capillary forces have important effects on wood quality and transport properties, and influence both natural and artificial drying behaviors (Karathanos et al. 1996; Srikiatden and Roberts 2007).

In permeable wood species, such as softwoods, free water movement through interconnected voids, driven by capillary action, is more important above FSP (Berit Time 1998; Onn 2007). In other words, inside softwood lumber boards, capillary action moves deeper into core and gradually disappears as the moisture content of the core cells approaches the FSP (Cech and Pfaff 2000).

In the latter stages of drying, the vapor diffusion becomes the dominating driving force (Philip and De Vries 1957; Luikov 1966; Berger and Pei 1973; Whitaker 1977).

The mechanism of capillary flow can be described by Darcy's law where capillary pressure gradient is the driving force (Onn 2007; Berit Time 1998). The corresponding equation is mathematically similar to diffusion equation (Perry et al. 1984):

$$\ln\left(\frac{MC - EMC}{MC_{in} - EMC}\right) = \frac{h_s(T - T_s)}{\rho_s * L * h_{fg} * (MC_{in} - EMC)} * t \tag{7.27}$$

where

MC is the moisture content (dry basis) $\left(\frac{kg_{water}}{kg_{oven\text{-}dry}}\right)$

EMC is the equilibrium moisture content $\left(\frac{kg_{water}}{kg_{oven\text{-}dry}}\right)$

MC_{in} is the initial moisture content $\left(\frac{kg_{water}}{kg_{oven\text{-}dry}}\right)$

h_s is the surface heat transfer coefficient ($W/m^2 \cdot K$)

T is the air temperature (K)

T_s is the board surface temperature (K)

ρ_s is the mass density of the solid $\left(\frac{kg_{solid}}{m^3}\right)$

L is the characteristics dimension along the flow path (m)

h_{fg} is the water evaporation (latent heat) enthalpy (kJ/kg)

t is the time (s)

For softwoods as southern pine (Spolek and Plumb 1981) and *P. radiata* (Chen et al. 1996), the capillary pressure can be expressed as a simple algebraic function of saturation (Spolek and Plumb 1981):

$$p_c = A * S^{-B} \tag{7.28}$$

where

$A = 12{,}400$ (Pa)

$B = 0.61$

p_c is the capillary pressure (Pa), a function of the saturation S defined as follows (Spolek and Plumb 1981):

$$S = \frac{Liquid\ volume}{Void\ volume} = \frac{MC - MC_{FSP}}{MC_{max} - MC_{FSP}} \tag{7.29}$$

where

MC is the actual moisture content (%)

MC_{max} is the maximum moisture content of the wood if the entire void structure is filled with water (%)

MC_{FSP} is the moisture content at the FSP when the lumens contain no liquid water (%)

To determine the capillary pressure action in wood, a centrifuge method has been used by Spolek and Plumb (1981) for softwoods and by Choong and Tesoro (1989) for other wood species (Perré and Keey 2015).

7.1.4.2 Vapor Pressure Gradient

When most of the capillary action ceases below FSP (i.e., when no free water exists), the majority of the cell cavities contains only air and water vapor. This establishes a vapor partial pressure gradient that acts as a driving force for water vapor movement through the air in the cells and through the openings in cell walls (Berit Time 1998; Keey et al. 2000; Onn 2007).

During drying, water contained in wood exerts vapor pressure of its own, which is determined by the maximum size of the capillaries filled with water. If the partial pressure of the water vapor in the ambient space is lower than vapor pressure within wood, desorption takes place and water vapor pressure within the wood falls due to osmotic forces. The largest-sized capillaries, which are full of water, empty first. Vapor pressure within the wood falls as water is successively contained in smaller capillaries.

This process occurs because the green wood is a hygroscopic substance and, thus, has the ability to take in or give off moisture in the form of vapor. As the green wood dries under constant atmospheric conditions of temperature and relative humidity, the vapor pressure of the moisture within the wood gradually decreases, that is, the forces which move the moisture from the surface to the surrounding air gradually decrease.

A stage is eventually reached where the vapor pressure of the moisture on the surface of the wood board is equal to the vapor pressure of the surrounding air, that is, the air is no longer able to absorb moisture from the surface of the wood, and further desorption ceases. At this time, the wood still has a moisture gradient, with a dryer surface, and, therefore, the process of moisture diffusion toward the surface continues. The vapor pressure of the surface moisture increases and, again, some of the moisture diffuses into the air. The interior continues to dry until the moisture is distributed uniformly throughout the wood, and the vapor pressure of the moisture in the wood is equal (i.e., in equilibrium) to the partial pressure of the water vapor in the surrounding space. Moisture now exists only in the interior in very small capillaries, bound to the small and large molecules producing considerable lowering of vapor pressure.

The pressure differential (or vapor pressure gradient) between cells, occurring because the amount of vapor in a given volume of air, decreases as the surface of the board is approached, causes moisture in the vapor state to move from areas of high vapor pressure to areas of lower vapor pressure close to the surface of the board, passing through the cell cavities, pit chambers, pit membrane openings, and intercellular spaces (Cech and Pfaff 2000).

Water vapor movement is more significant when the temperature approaches or exceeds the boiling point of water. In this case, more vapor is generated leading to additional partial vapor pressure gradients (Chirife 1983; Van Arsdel 1947; Babbitt 1950; King 1968; Karel and Lund 2003). It is the case of softwoods where, in order to reduce the drying time without decreasing the quality of the dried lumber, the drying temperature is kept above the boiling point of water at normal atmospheric pressure (100°C) (Perré and Keey 2015).

Such conditions ensure that an overpressure exists within the wood pieces, which implies that a pressure gradient drives the moisture (liquid or vapor) toward the exchange surfaces (Lowery 1979; Kamke and Casey 1988).

Consequently, in order to obtain an internal overpressure, the temperature of the porous wood materials must be above the water boiling point during at least one part of the drying process. This is the aim of convective drying at high temperatures with hot air or superheated steam, and a possible aim of microwave or radio-frequency drying.

However, in practice, it is possible to reduce the boiling point of water by decreasing the external pressure and, consequently, to obtain a high-temperature effect with relatively moderate drying conditions (see Figure 7.3). This is the principle of vacuum drying, particularly useful for lumber that would be damaged by high-temperature levels.

The hydraulic flow of vapor driven by the vapor partial pressure gradient can be described by the classical Darcy's law (Nijdam 1998).

If the driving force for moisture movement in wood is the partial pressure of water vapor, the Fick's second law should be modified by using the vapor pressure gradient (δp_v) instead of the wood moisture content (Bramhall 1976a, 1979a,b, 1995; Berit Time 1998; Keey et al. 2000):

$$\frac{\partial(MC)}{\partial t} = \frac{\partial}{\partial y}\left(D\frac{\partial p_v}{\partial y}\right) \tag{7.30}$$

where

p_v is the partial pressure of water vapor at the local temperature and moisture content (Pa)

D is the diffusion coefficient, nearly constant than that obtained by using the moisture content as a driving force (m^2/s)

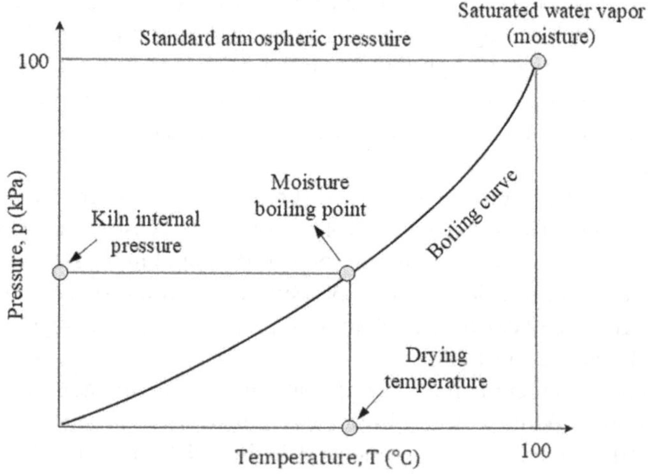

FIGURE 7.3 Variation of moisture boiling temperature with the total pressure of the drying air.

The volumetric mass flow rate of free water $\left(\dot{V}_{free\ water}, \mathrm{kg/m^3}\right)$, occurring due to a pressure gradient in the liquid phase, can be expressed as follows (Keey et al. 2000):

$$\dot{V}_{free\ water} = -E_{water}\frac{\partial p_{water}}{\partial y} = \frac{k_{water} * \rho_{water}}{\mu_{water}}\left(\frac{\partial p_{water}}{\partial y}\right) \tag{7.31}$$

where

E_{water} is the effective permeability to water liquid flow (a measure of the conductance of a porous medium for one fluid phase when the medium is saturated with more than one fluid) that may be related to the water thermal conductivity by the following equation:

$$E_{water} = \frac{k_{water} * \rho_{water}}{\mu_{water}} \tag{7.32}$$

where

k_{water} is the water thermal conductivity (W/mK)
ρ_{water} is the liquid water density (kg/m^3)
μ_{water} is the liquid water dynamic viscosity (kg/s · m)

7.1.4.3 Moisture Gradient

Drying is a coupled heat and mass transfer process consisting of moisture movement from internal zones of the material to the surface (Marinos-Kouris and Maroulis 2011).

The moisture content at the lumber boards' surfaces is controlled by the temperature, relative humidity, and air velocity of the surrounding drying air. The difference between the moisture content inside the lumber board, where the moisture mass transfer process by which moisture moves from one part of the wood board to another as a result of random molecular motion (Geankoplis 1993) in response to moisture content gradients between regions of high moisture concentration and regions of low moisture contents (Srikiatden and Roberts 2007), and the moisture content on the board surface creates a driving force.

A simpler and more practical approach than the vapor pressure gradient or chemical potential to model the moisture movement inside lumber boards particularly for impermeable species is based on the moisture content as a driving force, particularly at higher moisture contents (Perré and Turner 1996).

The drying mechanism based on the moisture content as driving force is slower than that caused by capillary action and, generally, is applied to drying of hardwoods (Berit Time 1998; Keey et al. 2000; Onn 2007).

At low moisture contents, however, there is a little difference between the predictions of drying mechanisms, whether driving forces are based on partial pressure or chemical potential of moisture content gradients.

The moisture content gradient (i.e., the change in moisture content from the board's surface to center) as driving force can theoretically be applied to both bound and free water (Bian 2001), if assumptions as following are applied: (i) fluid viscous dissipation is negligible, (ii) thermal conductivity of each phase is constant, (iii) air and vapor mixture behave as ideal gases, (iv) latent heat of vaporization is constant, (v) enthalpy of the three phases is a linear function of temperature, and (vi) cell wall material is assumed to be rigid above FSP with constant density.

During convective forced air drying, moisture at the lumber surface vaporizes first and, thus, the moisture content drops rapidly there. Then, the moisture existing deeper in wood zones of high moisture contents migrates through the passageways within the cell walls to zones of lower moisture content in an effort to reach the moisture equilibrium throughout the board (Jost 1952) and with the surrounding air (Walker et al. 1993). In other words, since wood has an affinity for moisture, the dryer surface cell walls will absorb moisture from cell walls of higher moisture content, that is, moisture moves from the wetter interior cells to the dryer surface cells to achieve an equilibrium state (Cech and Pfaff 2000; Keey et al. 2000). This is a very complex process.

The greater the temperature of the wood, the faster moisture will move from the wetter interior to the dryer surface. If the relative humidity of air is too low in the early stages of drying, excessive shrinkage may occur, resulting in surface and end checking (Keey et al. 2000). If the temperature is too high, collapse, honeycomb, or strength reduction can occur. While the capillary force only acts on the free water, the diffusion force acts on both the bound and the free water.

Large moisture content gradients may cause a faster water movement, and the moisture then flowing from the inside of the board to the surface is disturbed. As a result, if this happens too fast, case hardening may occur, that is, the surface layers of the board become too dry and the inside more or less wet (Frank Controls 1986). Also, the inhomogeneous water distribution in lumber may result in cracks and deformations, reducing the wood quality. On the other hand, a low water concentration gradient will lead to slow drying.

Contrary, if the moisture content gradient is too small during drying, the resulting drying times are uneconomical. In a kiln, the possibility of controlling the climate exists so that an optimum wood moisture content gradient can be achieved.

The moisture content gradient data is more important than the temperature gradient data. Effectively, since the temperature gradients in lumber dried at less than 100°C are usually very small compared with the moisture content gradients, the lumber is effectively isothermal at any time in the drying process. Even at higher drying temperatures, when a receding evaporative plane may occur, the stresses induced by thermal expansion are usually less than 10% of those induced by shrinkage due to moisture content changes (Keey et al. 2000).

Practically, the rate of moisture movement in wood is governed by molecular diffusion processes as a result of simultaneous combination of cell wall vapor partial pressure and moisture content gradients (Berit Time 1998). This means that a water molecule moves through the cell walls by moisture content gradient, across the cell cavity and through openings by vapor pressure gradient, and, again, through a dryer cell wall by moisture content gradient until it finally reaches the wood surface (Cech and Pfaff 2000; Keey et al. 2000).

Low air relative humidity stimulates diffusion by lowering the moisture content at the surface, thereby steepening the moisture gradient and increasing the moisture movement (diffusion) rate.

The parabolic moisture content profile with distance through a lumber board can be expressed as follows (Keey et al. 2000):

$$\rho_{water} D \frac{(MC - EMC)^2}{2} = \frac{j_{water,\ free}}{\delta}(y - \delta) \tag{7.33}$$

where

ρ_{water} is the water density (kg/m^3)
D is the diffusion coefficient (m^2/s)
MC is the moisture content $\left(\frac{kg_{water}}{kg_{oven\text{-}dry}}\right)$
EMC is the equilibrium moisture content $\left(\frac{kg_{water}}{kg_{oven\text{-}dry}}\right)$
$j_{water,\ free}$ is the free water flow rate (flux) (kg/m · s)
y is the direction of moisture movement (m)
δ is the board thickness (m)

For $MC > 8\%$, the diffusion coefficient in Equation 7.33 was calculated by the following correlation:

$$D = \exp\left[\frac{MC}{0.1792 - 2.553 * MC} - 9.2\right]\left(\frac{0.248}{MC_{FSP}}\right) \tag{7.34}$$

Where the moisture content (MC) and the moisture content at FSP (MC_{FSP}) are expressed as the mass of moisture per unit dry mass of the lumber $\left(\frac{kg_{water}}{kg_{oven\text{-}dry}}\right)$.

The term involving the FSP incorporates a weak temperature dependence through the relationship between the local lumber temperature (T, °C) and the FSP correlated as follows:

$$MC_{FSP} = 0.341 - 0.00133 * T \tag{7.35}$$

where

T is the temperature (°C)

The movement of liquid water is assumed to be proportional to the local gradient in the free water content and to the surface tension. The expression for free water mass flow rate is as follows:

$$j_{water,\ free} = D_c * \gamma * \frac{\delta(MC_{free})}{\delta y} \tag{7.36}$$

where

MC_{free} is the free water moisture content, which is the difference between the local moisture content and that at FSP $\left(\frac{kg_{water}}{kg_{oven\text{-}dry}}\right)$

The effective conductivity coefficient (D_c) in equation 7.36 is fitted by the following relationship:

$$D_c = 2.6 * 10^{-3}(0.145 + 0.0335 * T) * MC_f \tag{7.37}$$

The surface tension γ (N/m) in Equation 7.36 is given by the equation:

$$\gamma = 75.6 - 0.1625 * T \tag{7.38}$$

where
T is the temperature (°C)

In the two-potential coupling model, first developed by Philip and De Vries (1957) and De Vries (1958), the moisture content and temperature gradients were regarded as driving potentials, and the liquid and vapor mass fluxes were defined by Darcy's law and the Stefan diffusion law, respectively.

7.1.4.4 Chemical Potential

Chemical potential is a form of potential energy that can be absorbed or released during a chemical reaction. In a mixture, the chemical potential of a species can be defined as the slope of the free energy of the system with respect to a change in the number of moles of just that species. In other words, it is the partial derivative of the free energy with respect to the amount of the species, all other species' concentrations in the mixture remaining constant at constant temperature. When pressure is constant, chemical potential is the partial molar Gibbs free energy (also known as free enthalpy) (Greiner et al. 1995).

At chemical equilibrium, the total sum of chemical potentials is zero, as the free energy is at a minimum (Berit Time 1998).

In wood, the chemical potential of sorbed water is related to the moisture content gradient. Therefore, a gradient of wood moisture content (e.g., between the board's surface and center) is accompanied by a gradient of chemical potential under isothermal conditions.

However, it is not always easy to relate chemical potential in wood to commonly observable variables, such as temperature, humidity, and moisture content (Keey et al. 2000). The effects of these physical parameters on the mechanism of moisture movement in wood are actually very complex (Cech and Pfaff 2000).

According to Kawai et al. (1980), Siau (1984), Skaar (1988), and Keey et al. (2000), at low moisture contents of wood, the chemical potential is the true driving force for the diffusion of water in both liquid and vapor phases under non-isothermal conditions.

If the true driving force is the chemical potential difference, the moisture transfer will occur in the direction of falling chemical potential rather than decreasing moisture content, that is, moisture may flow from a drier part of a body to a wetter (Keey et al. 2000). That would mean that, when a moisture content gradient occurs in wood, a chemical potential will transport the moisture to redistribute itself throughout the wood piece until it reaches equilibrium (or uniform) chemical potential, resulting in a zero chemical potential gradient. This means that equilibrium will occur when the chemical potential of the wood becomes equal to that of the surrounding air (Skaar 1988). Any departure from equilibrium results in a difference in chemical potential.

The local thermodynamic equilibrium implies that the chemical potential of bound water equals the chemical potential of the water vapor, so that the bound water flow rate may be expressed in terms of the chemical potential of water vapor (Keey et al. 2000).

Because the water conductivity varies with moisture content (within a small temperature dependence), the chemical potential may have limited applicability as a driving force over the entire moisture content range from green to dry wood. However, it does appear to be an appropriate driving force for the movement of the bound water component (Stanish et al. 1986).

In attempting to achieve the equilibrium state under local thermodynamic equilibrium between the vapor and water, moisture flow rate may be assumed to be proportional to the gradient of chemical potential of adsorbed water molecules considered as the driving force $\left(\frac{\partial \mu_b}{\partial y}\right)$ (Keey et al. 2000). It allows calculating the mass rate of bound water migration (a molecular diffusion process) $\left(\dot{m}_{bound\ water}, \frac{kg_{water}}{m^2 s}\right)$, as follows (Stanish et al. 1986; Keey et al. 2000):

$$\dot{m}_{bound\ water} = -B * MC \frac{\partial \mu_b}{\partial y} = -D_b \frac{\partial \left(\rho_{oven\text{-}dry} * MC\right)}{\partial y} \quad (7.39)$$

where

B is a coefficient (–)

MC is the moisture content $\left(\frac{kg_{water}}{kg_{oven\text{-}dry}}\right)$

μ_b is the chemical potential of bound water (J/kg)

y is the dimension in the direction of moisture transfer (m)

D_b is the bound moisture diffusion coefficient, a proportionality constant, independent of moisture concentration if the moisture is unbound (m^2/s)

$\rho_{oven\text{-}dry}$ is the density of the oven-dry wood $\left(\frac{kg_{oven\text{-}dry}}{m^3}\right)$

By using the *effective water conductivity*, similar to the diffusion coefficients (D_b), the concept of the chemical potential as the driving force for moisture movement has been also applied (Cloutier et al. 1992; Cloutier and Fortin 1991, 1993, 1994).

7.1.5 Modeling Internal Moisture Movement

Most of existing models focus separately on the internal moisture movement and external vaporization through and from single-boards and kiln-wide lumber stacks of industrial kilns, respectively). However, today, it is of growing interest to simultaneously model both internal and external processes for lumber drying.

In order to develop models based on effective combination of classical transport theory with solid porous media theory in porous media seems sufficient for dealing with the woods drying processes.

In modeling of wood drying, the wood is assumed to be a porous material with interconnected void space to allow movement of moisture (water liquid and vapor) and air, which depends on the material density, porosity, and permeability allowing the movement of fluids under influence of driving force(s).

The mathematical models of solids' drying generally express the physical mechanism of internal coupled, multiphase heat and mass transfer governed by independent intensive variables such as temperature and pressure (Mujumdar 2000).

Most of the existing mathematic models address only specific drying problems because fundamental data are still lacked due to either understanding of the process or limitation of measurements. Structure properties of the solid matrix, phase equilibrium relationship, transport mechanisms in porous media, as well as the exact interaction between material and dielectric field are still undetermined. Understanding of the material deformation caused by drying induced strain stress is extremely important for constructing a more precise mathematic model for solids drying, including wood. In addition, any mathematic model must be validated on both laboratory and industrial scales, which requires a close cooperation between academia researchers and industrial engineers.

The modeling of wood timber drying includes two interrelated areas: (i) modeling of drying a single board and (ii) modeling of drying a kiln-wide stack of boards (Pang and Haslett 1995).

The existing main drying models are based on diffusion, empirical curve-fitting, transport theories (Bian 2001), and multicomponent concept (Whitaker 1977).

In many existing wood drying models, lumber drying models are based on empirical, diffusion, or multiple-mechanism approaches for predicting average moisture contents and moisture content profiles (Rosen 1987; Keey et al. 2000).

Multiple-mechanism approaches for modeling drying generally include the capillary-driven flow of free liquid, bound water diffusion, and the permeation of water vapor allowing determining the moisture content gradients that help predicting of stress levels (Keey et al. 2000).

Models for detailed internal moisture (and heat transfer) within wood boards have been established by Stanish et al. (1986) and Perré (1996).

Several models exist to describe the isothermal internal movement of moisture inside single lumber boards (Wiley and Choong 1975; Spolek and Plumb 1981; Furuyama et al. 1994; Kamke and Vanek 1994; Hunter 1995). Some of them are based on capillary pressure as a drying force (Perré et al. 1993)

However, few models have been developed for non-isothermal moisture movement in wood (Siau 1983b; Avramidis et al. 1992; Avramidis et al. 1994).

The drying models may be divided into three categories (Kamke and Vanek 1994):

1. Diffusion models (Kayihan 1982; Collignan et al. 1993);
2. Models based on transport properties (Plumb et al. 1985; Nasrallah and Perré 1988; Chen and Pei 1989; Ferguson and Turner 1994);
3. Models based on both the transport properties and the physiological properties of wood related to drying (Pang et al. 1992; Pang et al. 1994).

Simple diffusion models, the earliest attempt to quantify the drying of wood, assume that the moisture migrates by diffusion due to a moisture concentration gradient which can be described by Fick's second law (Rosen 1987; Wu 1989).

The diffusion approach assumes that a single transport mechanism is acting during drying as described by Fick's second law, regardless of the actual moisture transport mechanism (Keey et al. 2000):

$$\frac{\delta(MC)}{\delta t} = \frac{\delta}{\delta y}\left(D * \frac{\delta(\textit{driving force})}{\delta y}\right) \tag{7.40}$$

However, using the diffusion equation to describe overall moisture movement in wood is possibly inexact because in drying each state of moisture (liquid water, bound water, and water vapor) has a different movement mechanism and, thus, it is impossible to describe the overall moisture movement (transport) using a single relation. Therefore, the diffusion model is usually most applicable to impermeable species of wood where the liquid movement is insignificant. In softwood drying, the diffusion model, owing to its relatively simple form (Moren 1989), is often employed when describing the stress development (Pang and Haslett 1995).

The steady-state process of isothermal moisture movement by diffusion (a Brownian movement of water molecules through the noninteracting matrix of cellulose, hemicellulose, and lignin molecules that make up the cell wall) (Booker 1996) and heat transfer within wood material can be simulated based on fundamental laws and mechanistic and/or thermodynamic principles (Siau 1980; Skaar and Siau 1981; Siau 1983a; Stanish 1986; Onn 2007; Avramidis et al. 1992, 1994).

The relative importance of bound water diffusion and vapor movement can vary with the wood permeability and drying temperature. The bound water movement rate can be related to chemical potential difference which in turn can be described by temperature and pressure gradient (Stanish 1986; Pang 1994).

Simultaneous heat and mass transfer models have been developed based on the assumption that moisture movement occurs in both the liquid and vapor states in capillary porous materials (Srikiatden and Roberts 2007).

By assuming that moisture is transferred out of the system by liquid and vapor diffusion simultaneously, the thermal and diffusion properties, and the total pressure are constant. Luikov (1964) introduced the concept of moisture transfer caused by a temperature gradient in addition to a moisture content (or, in general, concentration) gradient.

Based on irreversible thermodynamics, Luikov (1975) proposed a three-potential driving model utilizing three dependent variables (temperature, saturation, and pressure). He derived macroscopic heat and mass transfer governing equations on the basis of flux expressions of phenomenological relationships, without taking into account the contributions of the gas-phase diffusion and liquid bulk flow (Chen and Wang 2005).

For the simple case without total pressure gradient inside the porous material, Luikov (1975) suggested that both moisture content and temperature gradients contribute to the moisture mass transfer flux.

Transient models for moisture transport in wood are based on Fick's law with water vapor pressure and temperature as driving potentials, including a model for hysteresis.

For moisture transport calculations in the hygroscopic region, a Fickian approach, that is, moisture transport by diffusion, is normally used. The selected driving potential often differs and because of this the transport coefficients (diffusion coefficients) themselves differ. Moisture content or air water vapor pressure is such a potential (Berit Time 1998).

The moisture transport in wood is not always obeying Fick's law (Wadso 1993).

Plumb et al. (1985) developed a model for heat and mass transfer in softwood that included liquid transport via capillary action, as well as diffusion of moisture and heat transfer during wood drying. The governing differential equations used to describe heat and mass transfer during wood drying include both diffusive and convective components. According to the author, the model allows predicting the heating and drying rates for typical boards of lumber, as well as the moisture and temperature profiles across the board section.

In Plumb's model, wood in its green state is considered as a three-phase mixture: solid cell wall material (with bound water to the fiber saturation point), free liquid that partially fills the lumens, and gas bubbles containing air and water vapor that occupy the remaining lumen space. Each phase represents a continuum, and its behavior can be predicted through use of the conservation laws. In order to achieve a single set of governing equations that is valid throughout the wood, the phase equations need to be volume averaged to include the effect of each phase on the whole (Plumb et al. 1985), as proposed by Whitaker (1977).

Plumb et al. (1985) assumed that, due to the length of the lumber compared to its thickness, a one-dimensional model for heat and mass (moisture) transfer can be employed. Thus, the transport occurs in the radial direction (i.e., normal to the growth rings) or the tangential direction (i.e., tangential to the growth rings), and not the longitudinal direction, which would be the vertical for a standing tree. The cell wall material was assumed to be rigid above the FSP (Skaar 1972) with constant density. Both liquid- and gas-phase motion are slow, so convective accelerations can be ignored. Within the averaging volume, the thermal conductivity was assumed to be constant, as is the specific heat, so enthalpy is a linear function of temperature. In addition, it was assumed that both reversible work and viscous dissipation are negligible, and that there is no internal heat source.

According to Rizvi (2005), inside highly porous materials, moisture movement occurs mainly in the vapor phase and all evaporation occurs from the interior of the material. Furthermore, in order to supply the heat of vaporization to the interior, heat has to be conducted through both dry solid and pore regions having low thermal conductivity.

Sherwood (1929a,b) was the first to use the moisture content as driving force (potential) based on the following additional assumptions (i) the diffusion coefficient D is constant, (ii) the initial moisture content in the lumber is uniform; and (iii) surface fibers come into equilibrium with the surrounding air instantaneously, so that the resistance of the boundary layer outside the board of lumber is negligible.

Sherwood's (1929a,b) diffusion (transport) mechanism is relatively simple compared with more complex models for moisture movement in limber and, thus, is still in widespread use today (Simpson 1993; Bramhall 1995).

Given the wide variability in diffusion coefficients and permeability, both within and between species (Keey et al. 2000), the simplicity of Sherwood's model is useful when optimizing kiln drying schedules. Moreover, the Sherwood's (1929a,b) diffusion model works well for predicting both average moisture contents and moisture content profiles for some hardwood species, essential for predicting stress levels in lumber (Keey et al. 2000).

Diffusion expressions arise when liquid diffusion is the dominant moisture transport mechanism and capillary forces are dominant.

Wiley and Choong (1975) successfully used this approach when considering the drying of softwood lumber at high moisture contents $\left(> 0.6 \frac{kg_{water}}{kg_{oven\text{-}dry}} \right)$.

The majority of work involving the use of diffusion models has used moisture content driving forces. Some successful applications (mainly impermeable hardwoods) have covered a wide range of moisture contents, for which diffusion (in the strict sense) is unlikely to occur at all moisture contents. Hence, there is some empirical support for the use of moisture content driving forces.

The use of the diffusion model has been more successful with highly impermeable species (particularly, hardwoods). Where the lumber is very permeable, free water movement is probable at much higher moisture contents than the fiber saturation point, and diffusion is unlikely to be the transport mechanism for this process. This suggests that a diffusion model is unlikely to be appropriate for describing the drying of highly permeable species from green, particularly easily dried softwoods, and is more likely to be useful for describing the drying of some hardwood and impermeable softwood species (Keey et al. 2000).

Although describing moisture movement in terms of diffusion has been satisfactory in many practical cases, more attention has been paid in recent years to models in which several transport mechanisms for the movement of moisture are involved (Keey et al. 2000).

The moisture transport models, the most frequently used, are based on potential theories (Luikov 1966) and the water potential concept (Cloutier and Fortin 1991). Some drying studies (Plumb et al. 1985; Stanish et al. 1986; Perré 1987; Chen and Pei 1989; Turner 1990) have described the drying process in lumber by considering different transport mechanisms for bound water, free liquid water, and water vapor.

Fundamental drying transport-based models developed specifically for wood describing it as a hygroscopic porous medium (e.g., Stanish et al. 1986) may be too complicated and computationally too slow for the purposes of process analysis (Kayihan 1993).

Moreover, the derived equations are nonlinear and very complex, and contain coefficients that need to be determined from experiments. Since the bound water is not distinguished from liquid water or water vapor, the models should take into account the effect of sorbed water, particularly when the moisture content is below the FSP (Puiggali and Quintard 1992).

On the other hand, drying models for porous materials as foods (Srikiatden and Roberts 2007) and woods can be classified into three groups (Rosen 1986; Bian 2001; McCurdy 2005):

1. Models involving empirical equations applicable for specific processes;
2. Models based on simultaneous equations for heat and mass transfers;
3. Sophisticated (complicated) models simultaneously associating energy, mass, and momentum transport equations (Sereno and Medeiros 1990) derived based on mechanistic approaches (Philip and De Vries 1957) or from nonequilibrium thermodynamics (Luikov 1966, 1975).

Modeling of drying becomes complicated by the fact that more than one mechanism may contribute to the moisture (mass) total transfer rate and the contributions from different mechanisms may even change during the drying process (Mujumdar 2000). Moreover, detailed experimental data quantifying both drying rates and moisture and temperature distributions within wood during drying are normally not available and the necessary transport properties have received little attention (Plumb et al. 1985). However, when the theoretical development of drying models is insufficient, some parameters are necessary to be determined by experiments and the curve-fitting expressions.

Modeling the mechanism of drying process involves setting up the heat and mass transfer equations correctly, solving the coupled differential equations, and then correlating the prediction with experimental moisture and temperature profiles.

The classical models of heat and mass transfer in drying can be divided into three categories (Zhang et al. 1999): (i) coupled (single, two and three-potential) models; (ii) continuum models mainly based on the fundamental laws of conservation of mass, momentum, and energy; and (iii) combined models.

To develop a simultaneous heat and mass transfer model, the Clausius–Clapeyron equation has been used based on the assumption that mass transfer occurs by liquid capillary flow and vapor diffusion (Berger and Pei 1973).

Generally, for modeling porous materials, such as woods and foods, several models are used as follows (Defraeye et al. 2012a, b): (i) effective penetration depth models (Cunningham 1992); (ii) shrinking core, also called receding front models (Luikov 1975); (iii) macroscopic-scale, continuum models for coupled multiphase heat and mass transfers, using phenomenological (Philip and De Vries 1957; Luikov 1966) and mixture theory (Bowen 1980); and (iv) pore-network and volume-averaging approaches; (Whitaker 1977, 1998; Carmeliet et al. 1999; Blunt 2001; Prat 2002).

Analytical (i.e., the variable separation) method and numerical (i.e., the finite element and difference) method are useful methods to solve the system of differential equations describing the drying behavior in regular-shaped bodies (Sastry et al. 1985; Rossello et al. 1997).

Luikov (1966, 1975) and Whitaker (1973, 1977, 1980, 1984) developed basic models to describe the behavior of solids' drying processes (Plumb et al. 1985).

Luikov's first transport-based model (1966) set of coupled partial differential equations to describe the heat and mass transport in capillary porous media by assuming that the movement of moisture is analogous to heat transfer and that capillary transport is proportional to gradients in moisture content and temperature. Luikov (1966) showed that moisture movement inside a material may result from a significant temperature gradient and, thus, the heat transfer effect should be taken into account along with mass transfer when developing a drying model. The resulting transport coefficients are strong functions of temperature and moisture content, as well as of material properties.

Chen and Pei (1989) and Zhang et al. (1999) proposed heat and mass transfer models during constant and falling rate periods in convective drying of porous, hygroscopic materials, such as woods.

Bian (2001) presented the wood drying model based on simultaneous heat and mass transfer, as developed by Perré and Moyne (1991a, 1991b) for a rigid porous, multiphase, closed, and continuous system (solid, water liquid and vapor, and air) in thermodynamic equilibrium with its surroundings. Using the finite element method, Onn (2007) modeled the heat and mass (moisture) transfer within a rectangular-shaped board of a hardwood (Dark Red Meranti) dried by forced air convection.

Given the complexity of wood structure and the natural variability of the raw material, it may be satisfactory only if a simple correlation of experimental data to estimate the drying time is required. However, most empirical models also give no information on moisture content gradients, information which is required for stress analysis (Keey et al. 2000).

An empirical model, where the driving force is the vapor pressure gradient, can be presented as follows (Bramhall 1976a, b):

$$\frac{d\overline{MC}}{dt} = \frac{p_{vd} - p_{vw}}{R} \tag{7.41}$$

where

$\overline{MC}$ is the average moisture content $\left(\frac{kg_{water}}{kg_{dry\text{-}bone}}\right)$

$p_{v,\,dry}$ is the water vapor pressure at the dry-bulb temperature (Pa)

$p_{v,\,wet}$ is the water vapor pressure at wet-bulb temperature (Pa)

R is the resistance to moisture movement (Pa) found experimentally to be a function of the average moisture content and correlated by one of the following two forms Bramhall (1976a, b):

$$R = c_1 \exp\left(-c_2 \overline{MC}\right) \quad or \quad R = c_3 \overline{MC}^{-c_4} \tag{7.42}$$

where

c_1–c_4 are experimentally fitted constants

Another empirical model correlates the characteristic moisture content curve by the following equation (Tschernitz and Simpson 1979a, b):

$$\bar{\Phi} = \exp\left(\frac{b * t}{l^n}\right) \tag{7.43}$$

where

t is the time (in days)
l is the board thickness (in inches; 1 inch = 25.4 mm)
b and n are empirical parameters

Helmer et al. (1980) used a similar empirical model considering the difference between two absolute humidities as a driving force:

$$\frac{d\overline{MC}}{dt} = \frac{\omega_s - \omega}{R} \tag{7.44}$$

where

ω_s is the saturation absolute humidity at the dry-bulb temperature $\left(\frac{kg_{water}}{kg_{dry\text{-}air}}\right)$
ω is the absolute humidity at the bulk-air temperature $\left(\frac{kg_{water}}{kg_{dry\text{-}air}}\right)$

R is the resistance to moisture movement, stated to be a function of the average moisture content and the dry-bulb temperature (DBT) as follows:

$$R = \overline{MC}^{2.38}(0.289 * DBT - 4.18) \tag{7.45}$$

Because of using uncertain coefficients, different degrees of simplifications and different methods for solving the heat and mass transfer problems, some models did not predict the data well (Kamke and Vanek 1994).

Another fundamental difference between macroscopic models for modeling the internal drying process, that is, the coupled heat and mass transfer in porous materials (Bories 1988; Masmoudi and Prat 1991) consists in the number of independent variables used to describe the process (Perré and Keey 2015; Turner and Perré 2004): (i) moisture content, (ii) temperature, (iii) vapor concentration (or density), and (iv) vapor partial pressure.

Based on this observation, three types of models are possible for drying (i) one-variable model (e.g., moisture content only or an equivalent variable as saturation or water potential) (Keey et al. 2000); as a general rule, the one-variable model based on moisture content only should be avoided (except, possibly, when considering kiln behavior), especially because it is not able to account for the very important coupling that exists during drying between heat and mass transfer (Perré and Keey 2015); (ii) two-variable model (e.g., moisture content and temperature, or an equivalent variable such as enthalpy), an option appropriate for most of the drying conditions encountered in industry when the internal pressure is not the primary concern; and (iii) three-variable model (e.g., moisture content, temperature, and vapor partial pressure or air density) is a quite complex model that should be reserved for processes during which the internal pressure has a significant impact on internal moisture transport, such as those with internal vaporization or vacuum, high-temperature, and radio-frequency drying (Perré and Keey 2015).

To estimate one-dimensional and two-dimensional temperature and moisture content distributions during the preheating and drying of lumber boards, nomographs have been generated (Shubin 1990) in terms of dimensionless parameters based on theory developed by Luikov (1968).

7.1.5.1 One-Dimensional Models

The simple one-variable models are used in drying at the relatively low temperature or ignoring the effect of temperature. They use large board where most of process parameters are independent of the number of sample dimensions and coordinates, and the moisture movement may be assumed to be one-dimensional because the board is much thicker than their width in the air stream direction (Blakemore and Langrish 2007). This kind of a model, more suitable for vacuum drying since the vapor can be regarded as the only component in gaseous phase, should be however avoided because it is not able to account for the very important coupling between mass and heat transfer that exists during drying.

Assuming that the gradient of moisture content (or vapor pressure, or capillary action) is the only mechanism of internal moisture transport in pores, Sherwood (1929a, b, 1931), King (1971), Eckert and Pfender (1978), and Eckert and Faghri (1980) proposed single-variable driving models to describe heat and mass transfer in porous materials.

During the high-temperature drying of *Radiata Pine*, a one-dimensional model of the moisture movement mechanism (used to predict the temperature and average moisture content profiles) showed that the pressure gradient play an important role in heat and mass transfer and that the liquid diffusion is small in comparison to capillary driven moisture convection (Nijdam et al. 2000).

7.1.5.2 Multidimensional Models

Two-dimensional and three-dimensional multidimensional models are needed if the board sample has an irregular geometry or a geometry that is unacceptable for the degradation of the number of dimension or when the porous medium is anisotropic (Turner and Perré 2004). Based on Whitaker's (Whitaker 1977, 1998) models, Perré and Keey (2015) proposed a three-dimensional model allowing a comprehensive geometrical modeling with extensions to account for wood properties and drying with internal overpressure.

7.1.5.3 Multicomponent Models

The multicomponent models seem more rigorous and comprehensive because they (Whitaker 1973, 1977, 1980, 1984) (i) determine at microscopic level the heat and mass transfer conservation equations for a three-phase, rigid structure (solid, liquid, and vapor), such as wood porous bodies, at local thermodynamic equilibrium through a volume-averaged method; (ii) Fick's and Darcy's laws of diffusion are valid; (iii) liquid capillary flow is present, but the bound water flow is absent; and (iv) total pressure gradient of the gaseous phase (water vapor plus air) is considered to be the driving force of the mixture, and the liquid pressure gradient is considered to be the driving force of the liquid phase according to Darcy's law.

7.1.6 Evaporative Front

Each lumber board can be treated as a homogeneous, hygroscopic, porous material (Turner and Ferguson 1995a, b), some new studies have continued to treat wood as homogeneous. During drying, the board's surfaces are first heated by convection, and then the inner body is heated by conduction. The surface vapor pressure, hence the external vapor flow, depends on both temperature and moisture content. To maintain the energy balance, the surface temperature increases as the surface moisture content decreases (Perré and Keey 2015).

Regardless of the mechanism by which the wood is dried, the moisture leaves the wood board by evaporating from the external surfaces and it is in the surface zone where most of the interesting processes occur during drying. Before drying, the surface of a green board has essentially the same moisture content as the core of the board, assuming that there has been no air drying prior to kiln drying. On drying, the moisture at the surface of the board evaporates and the surface layer dries out, while the core moisture content remains the same. It has been observed that the surface dries to below the FSP in the early stage of drying even the core moisture content is very high, mainly as a consequence of pit aspiration in the surface cells during sawing (Pang 1994). As the wet surface from which moisture is evaporating recedes below the surface of the wood, it becomes an evaporative front. Normally the evaporative front recedes to a position 1–2 mm below the surface where it remains until the board approached a moisture content of 50%–60% depending on the drying temperature. During this period of drying, the liquid flows from the core toward the evaporative front due to capillary action generated by moisture gradient (Spolek and Plumb 1981; McCurdy 2005).

Inside the wood, a two-zone process develops: (i) an inner wet zone, where liquid migration prevails and; (ii) a surface dry zone, where both bound water and water vapor diffusion take place (Figure 7.4).

The moisture content at the board's contact surface with the drying air drops very fast (i.e., the thin layer at the surface will be in equilibrium with the airflow very quickly). After a relatively short period of time (e.g., 1 hour) of the

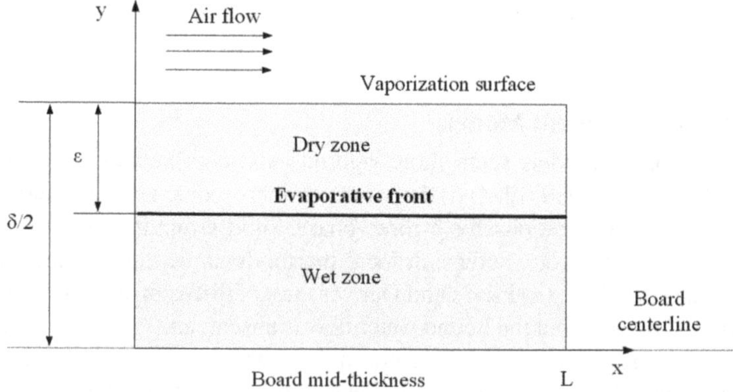

FIGURE 7.4 The evaporative front for drying of a half-width lumber board.

drying process, the moisture content at the board's surface slightly increases due to condensation of the water vapor on the surface when the board temperature is still low.

On the other hand, during the first few hours of the drying process, the pressure inside the lumber increases and the moisture content at the board central line drops slightly, but at a rate less than that near the surface where the moisture is extracted by the capillary force and a slightly low pressure is observed. That means that the liquid water inside the lumber is moving toward the surface at a lower rate than that of the drying rate (Bian 2001). As drying progresses, the low pressure moves inward toward the center, and both high pressure due to the temperature increase near the surface and low pressure due to capillary force are present.

This behavior is explained by a sudden retreat of the moisture level to just below the exposed surface to the narrowest openings in the board where an evaporative front exists for a period, being fed by capillary movement of moisture from the deeper portions of the board.

Under high-temperature drying above 100°C at the atmospheric pressure, the lumber boards reach a point when the liquid is no longer continuous and, thus, the moisture convective flow is interrupted. Further drying will evaporate the remaining liquid, and vapor flow is important.

When the wood reaches the FSP, there is no longer any free moisture in the cell lumens and the moisture movement is by vapor flow and bound water diffusion. Below FSP, further moisture loss from the wood results in dimensional changes in the wood fibers causing them to shrink (Onn 2007).

If the drying temperature is increased, the number of water molecules that evaporate into a cell cavity will increase and the vapor pressure will be greater. Therefore, the vapor pressure in the wet zone will now be much greater than at the dry zone, and the drying rate will be increased because a greater amount of water vapor moves to the surface zone where the amount of water vapor in the cell cavities is kept low by low relative humidity of the outside air.

After a certain time, the pressure reaches a maximum at the center, then it decreases. The increase of the pressure is due to evaporation of the free water and the decrease is due to the loss of the free water because of drying.

High-temperature drying, involving the use of dry-bulb temperatures greater than 100°C, is rapidly becoming the preferred commercial practice (e.g., in New Zealand, Canada, and the United States) for drying softwood timber, such as *P. radiata* (in New Zealand) and southern pine (in North America).

During high-temperature drying of, for example, *P. radiata* (Pang 1994) and Yellow poplar (Beard et al. 1982, 1985), the temperatures at different depths of the board initially reach about 100°C and then progressively start to rise above this temperature. Before rising, the temperature remains at about 100°C while the duration increases with position toward the mid-thickness of the board. This feature of the temperature–time profiles indicates the presence of a receding evaporative front at the boiling point of water at local pressure (Pang and Haslett 1995).

Based on these observations, Pang (1994) and Pang et al. (1992, 1994a) have proposed an evaporative front model in which the physiological properties of green wood and the changes of these properties with drying are considered (Figure 7.4).

The evaporative plane divides the wood board into two parts: a wet zone beneath the plane and a dry zone above it (Bian 2001).

Because in such a process the internal temperatures can remain at the boiling point for some time during the drying, Pang and Keey (1995) propose a model of drying based upon a receding evaporative front until the timber reaches the FSP (Pang and Keey 1995).

Above FSP, when the cell walls are fully saturated, the removal of free moisture from vessels or tracheids is the way in which the porous wood bodies dry out (Keey 1978; Keey et al. 2000). Below the evaporation plane, water exist as unbound water and bound water, whereas at the layer above this plane, water exists as bound water and vapor. Hence, the evaporative plane of wood will slowly recede into the center of the wood, and the moisture movements behave differently for these different layer (Perré et al. 1988; Pang and Haslett 1995; Pang and Keey 1995).

When the moisture content at the lumber board's surface reaches the maximum sorptive value, there is no free water on the surface, and the surface temperature increases rapidly, signaling the start of the second falling rate period. During this period, a receding evaporation front often appears, only a few tracheid thick, dividing the system into two regions: (i) a wet (core) region where no free water exists, all water being in the sorptive or bound water state; in this region, the main mechanisms of moisture transfer are by movement of bound water (sometimes called sorption diffusion) and by vapor transfer with vapor pressure gradient as the main driving force; and (ii) a dry region where the voids contain free water, and the main mechanisms of moisture transfer are capillary flow, and viscous and inertial forces (Strumillo and Kudra 1986; Zhang et al. 1999).

With the sapwood of softwood, there is some evidence for the appearance of an evaporative zone close to the exposed surface, for example, the appearance of a brown stain, particularly at higher kiln temperatures (Pang et al. 1994; Haslett 1998), probably from resinous deposits, at a depth of 0.5–1 mm below the surface (Wiberg and Moren 1999; Keey and Nijdam 2002).

For heartwood, during the first period of drying, the influence of internal resistance to moisture vapor transport gradually dominates over the external resistance; consequently, the drying rate drops sharply. The initial period lasts only 5–15 minutes during which the evaporative front has withdrawn to a short distance from the surface (about 0.5 mm). In the second period, the evaporative front continues to recede into the board at a rate which is not primarily determined by the external conditions. Once the evaporative front reaches the midplane of the board, the drying is controlled by bound water diffusion and water vapor movement. This is the beginning of the third period of drying (Pang and Haslett 1995; Keey and Nijdam 2002).

7.2 BOARD EXTERNAL DRYING PROCESS

The internal drying process of fresh cut green wood, timber logs, or lumber boards consists in migration of moisture (water liquid and vapor) within the wood body toward its surfaces.

On the other hand, the external drying process consists in transfer of heat from the surrounding medium to these pieces of wood, followed by simultaneous removal (e.g., by evaporation or vaporization) and transport of the moisture from the wood surfaces.

During the sawing process of felled wood logs into green boards, dry-out (moisture-denuded) zones (layers) at the damaged board exposed surfaces are created. Here, the moisture content is in equilibrium with the surrounding environment (Pang 1994; Keey and Nijdam 2002; Mujumdar 2007). Therefore, before kiln drying begins, evaporation of moisture occurs naturally by drawing water through the walls of water-filled cells by vacancy diffusion (Booker 1996).

The moisture vaporization process takes place when warm and dry air is forced with a certain velocity over wood exposed surfaces. In this case, the vapor pressure in the slowly moving thin film (boundary layer) of saturated air is substantially higher than in the main air stream. To maintain the desired drying rate, it is essential to remove moisture from this layer as rapidly as possible (Cech and Pfaff 2000; Keey and Nijdam 2002).

In the lumber kilns, forced circulated air supplies sensible and latent heat (enthalpy) of vaporization through the thermal boundary layer to the moisture existing on the surfaces of boards to be dried. Heat is the source from which the water molecules acquire the kinetic energy necessary for moisture vaporization. The air is, thus, cooled while the moisture vaporizes from the boards' surfaces. The water vapor removed by the airflow dissipates into the bulk of the drying agent (air) and is carried away.

The removal of moisture as vapor from the boards' surfaces by vaporization depends on the air temperature, relative humidity, flow rate (or velocity), and the areas of exposed surfaces (Mujumdar 2007).

Both rate of movement of the moisture within the boards (i.e., the internal drying process) and rate of vaporization from their surfaces (i.e., the external drying process), control the drying time of lumber boards in batch lumber kilns.

These rates mainly depend on the total amount of heat supplied per unit of time, the wood structure and transport properties, the air dry- and wet-bulb temperatures, relative humidity, flow rate (or velocity), airflow direction, the ability of the heating medium (air) to absorb moisture, and the relative dryness of the air and wood (Defraeye et al. 2012a, b), as well as on the heat and mass transfer coefficients through the laminar, transient, and turbulent boundary layers. Generally, low relative humidity of the drying air (i.e., low partial vapor pressure) leads to increased moisture vaporization rate.

In other words, the rate of moisture movement in wood and the rate of moisture vaporization are dependent on the capacity of the drying air to absorb moisture, according to the difference in partial vapor pressures between the board's surface moisture layer and the bulk of drying air.

7.2.1 External Flow over a Lumber Board

In kiln drying practice, drying air (essentially, a binary mixture of dry air and water vapor, generally near the atmospheric pressure) circulates over parallel, horizontally stacked boards in their width direction. Neglecting the small, irregular spaces (generally, of order of mm) existing between adjacent boards, it is possible to assimilate each row of boards with a continuous, classical smooth flat plate, as shown in Figure 7.5a. The flow of the drying air over such horizontal, multiple-flat plate surfaces (Figure 7.5b) generates velocity, thermal, and species (moisture) concentration boundary layers where the local heat and mass transfer coefficients vary considerably from one location to another, depending on the nature of the flow (laminar, transient, or turbulent).

7.2.1.1 Flow Regimes

Consider a steady-state, external, parallel flow at zero incidence of a moist air stream (i.e., mixture of dry air and water vapor) of velocity u_∞ (m/s), temperature T_∞ (°C), and moisture concentration (or absolute humidity) $C_\infty \left(\frac{kg_{water}}{kg_{dry\text{-}air}}\right)$ over a succession of multiple, adjacent, stationary, typical lumber boards, theoretically similar to a continuous flat plate of length L, area A_s, and uniform surface temperature $T_s < T_\infty$ (Figure 7.5a). These external thermal and hydraulic parameters, as well as the relative and absolute humidity of the drying air, control the external boundary conditions

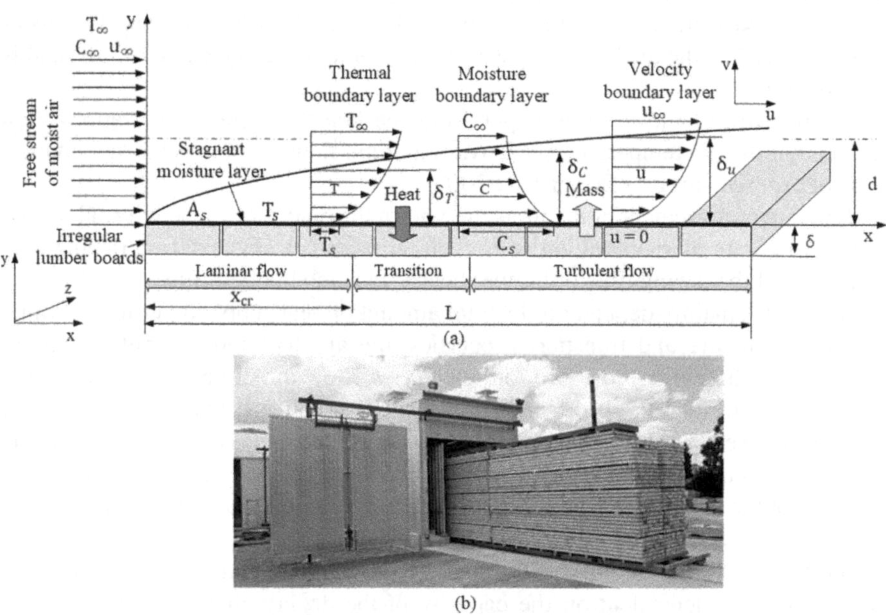

FIGURE 7.5 (a) Schematic representation of thermal, concentration (moisture), and velocity boundary layers developed over one row of horizontal adjacent lumber boards (note: drawing not to scale); (b) view of a typical lumber stack entering a batch kiln (www.google.fr/#q=kiln+drying+pictures. Accessed December 10, 2016). A_s, board's surface area; C, moisture concentration; d, semi-distance between two adjacent layers of boards in a lumber stack; δ, board's thickness; L, width of lumber stack; T_∞, air free-stream temperature; T_s, board's surface temperature; u, velocity component in x direction; v, velocity component in *y* direction; x_{cr}, critical length; δ_C, thickness of concentration boundary layer; δ_T, thickness of thermal boundary layer; δ_u, thickness of velocity boundary layer.

and affect the rate of moisture vaporization (i.e., the board drying rate) (Walker et al. 1993; Keey et al. 2000).

The flow characteristics of the moist air strongly depend on which flow regime exists in the fluid. The flow on a flat plate initially starts in laminar region, but at some distance from the leading edge, small disturbances amplify, and transition to turbulent flow begins to occur. Fluid fluctuations begin to develop in the transition region, and the boundary layer eventually becomes completely turbulent. In the fully turbulent region, fluid motion is highly irregular and is characterized by velocity fluctuations.

As can be seen in Figure 7.5a, a laminar flow begins to develop downstream the board leading edge ($x = 0$), flow characterized by highly ordered air movement and, along a streamline, variable velocity components in both directions x and y (u and v, respectively).

At a certain location (x_{cr}), transient and turbulent flow regimes may occur.

The existence of laminar, transient, or turbulent flows is determined by the local Reynolds number defined as the ratio of inertia forces (associated with the increase in the momentum flux, $F_i \approx \frac{\rho * u_\infty^2}{x}$) to viscous forces $\left(F_v \approx \frac{\mu * u_\infty}{x^2}\right)$ of air moving through the velocity boundary layer:

$$Re_x \approx \frac{Inertia\ \ forces}{Viscous\ \ forces} = \frac{F_i}{F_v} = \frac{\frac{\rho u_\infty^2}{x}}{\frac{\mu u_\infty}{x^2}} = \frac{\rho * u_\infty * x}{\mu} \tag{7.46}$$

where

ρ is the moist air density (kg/m^3)
u_∞ is the velocity of the free stream of moist air (m/s)
μ is the moist air dynamic viscosity (kg/m · s)
v is the moist air kinematic viscosity (m^2/s)
x is the distance from the leading edge of the boards' flat plate (m)

According to equation 7.46, at large *Re* numbers, inertia forces dominate, whereas at small *Re* values, viscous forces dominate.

The location of transient flow (x_{cr}) is determined by the critical Reynolds number that, depending on surface roughness and the turbulence of the free stream, varies between $1 * 10^5$ and $3 * 10^5$ (Incropera and DeWitt 2002):

$$Re_{cr} = \frac{\rho * u_\infty * x_{cr}}{\mu} = \frac{u_\infty * x_{cr}}{v} \tag{7.47}$$

In the fully turbulent flow regime, a larger viscous shear force is observed in the fluid causing a flat velocity profile. Also, the highly irregular movement of drying moist air is characterized by velocity fluctuations and fluid mixing that enhance the heat and mass (moisture) transfer coefficients and rates, while the surface friction significantly increases. In other words, the heat and mass rate transfer to/from the flow and the surface, respectively, are considerably greater in the turbulent than in the laminar regime.

7.2.1.2 Boundary Layers

The airflow over a sharp-edged flat plate is characterized by the development of boundary layers which are zero at the leading edge, gradually becoming thicker with distance.

During external, parallel flow of the drying air over the theoretically isothermal flat plate system shown in Figure 7.5a, three boundary layers (i.e., velocity, thermal, and concentration) will simultaneously develop. The physical processes that occur in these boundary layers affect the surface friction (characterized by the friction coefficient, C_f), the convective heat transfer (characterized by the heat transfer coefficient, h_{heat}), and the convective moisture (mass) transfer (characterized by the mass transfer coefficient, h_{mass}) (Incropera and DeWitt 2002).

In kiln drying of lumber boards, all the three boundary layers are present, but their thicknesses (δ_u, δ_T, and δ_C) at a given location are different.

The velocity boundary layer of variable thickness δ_u occurs because the velocity of air particles in contact with the stagnant water (moisture) layer existing at the boards' surfaces is practically zero. Inside the velocity boundary layer, where the influence of viscosity is observed, the velocity varies from zero on the boards' surfaces to a maximum in the free air stream. At a distance $y = \delta_u$ from the boards' surfaces, the shear stress effect becomes negligible and the air velocity approaches that of free air stream (i.e., $u \approx 0.99 * u_\infty$) (Figure 7.5a).

The surface friction coefficient can be determined as follows (Incropera and DeWitt 2002):

$$C_f = \frac{2\tau_s}{\rho * u_\infty^2} \tag{7.48}$$

where

τ_s is the surface shear stress; for a Newtonian fluid, it may be evaluated from knowledge of the velocity gradient at the surface

$$\tau_s = \mu \left.\frac{\partial u}{\partial y}\right|_{y=o} \tag{7.49}$$

where

μ is the air dynamic viscosity (kg/m · s)

In the drying process of a lumber board, the temperature of the air free stream (T_∞) is higher than that of boards' surfaces (T_s), while the air particles into contact with the stagnant moisture (water) layer are in thermal equilibrium with the boards' surface temperature. The adjoining air layer exchange heat with these particles and, therefore, a thermal boundary layer of thickness δ_T (characterized by temperature gradients and heat transfer) develops. Its thickness is defined (Incropera and DeWitt 2002) as the value of y for which the ratio $\frac{T_\infty - T}{T_\infty - T_s} \approx 0.99$ (Figure 7.5a).

Also, in wood drying, the water vapor (moisture) concentration (equivalent to drying air's absolute humidity) in the free air stream (C_∞) is much lower than that of moisture (water) concentration at the boards' surface (C_s). Therefore, a moisture concentration boundary layer, characterized by moisture concentration gradients and moisture (water vapor) mass transfer, will develop. Its thickness (δ_c) (Figure 7.5a) is defined as the value of y for which $\frac{C_s - C}{C_s - C_\infty} \approx 0.99$ (Incropera and DeWitt 2002).

The mass transfer coefficient is inversely proportional to the thickness of the boundary layer through which the moisture vapor must diffuse to escape into the bulk of the air and be conveyed away. Thus, the mass transfer coefficient as well as the (absolute) humidity potential dwindles in the direction of the airflow. By solving the equations of motion and fitting a polynomial to the velocity distribution in the boundary layer, Pohlhausen (1921) obtained an expression for the local mass transfer coefficient ($h_{x,\,mass}$) as a function of distance x from the leading edge (Perré and Keey 2015).

7.2.1.3 Boundary Layer Equations

For a steady-state, low-speed, two-dimensional forced flow of an incompressible, viscous fluid with temperature-independent properties and no heat generation (as shown in Figure 7.5a), simplified boundary layer equations can be written by assuming that (i) boundary layer thicknesses are very small; (ii) velocity component in x direction (i.e., along the boards' surfaces (u) is much larger than that normal to the boards' surface (v)); (iii) heat and moisture transfer rates in y direction are much larger than those for the x direction; (iv) at the boards' surfaces ($y = 0$), the air velocity in direction y is zero ($v = 0$); and (v) the pressure does not vary in the direction normal (y) to the surface $\left(\frac{\partial p}{\partial y} = 0\right)$; hence, the pressure in the velocity boundary layer depends only on x and is equal to the pressure in the free stream outside the velocity boundary layer.

If, at relatively low-velocity (moderate) air circulation, as generally occurs in batch wood kilns, the pressure gradient in the airflow direction $\left(\frac{\partial p}{\partial x}\right)$ and the viscous dissipation $\left(\left[\frac{v}{c_p}\left(\frac{\partial u}{\partial y}\right)^2\right]\right)$ are neglected, three simplified boundary layer equations can be written (i.e., for momentum, energy, and concentration conservation), in addition to the following continuity equation (Incropera and DeWitt 2002):

$$\frac{\partial u}{\partial x} + \frac{\partial v}{\partial y} = 0 \tag{7.50}$$

where
- u is the air velocity component in x direction (m/s)
- v is the air velocity component in y direction (m/s)

a. Momentum conservation equation in velocity boundary layer in x and y directions:

$$u\frac{\partial u}{\partial x} + v\frac{\partial u}{\partial y} = v\frac{\partial^2 u}{\partial y^2} \tag{7.51}$$

where

ν is the air kinematic viscosity (m²/s)

b. Energy conservation equation in thermal boundary layer:

$$u\frac{\partial T}{\partial x}+v\frac{\partial T}{\partial y}=\frac{k}{\rho c_p}\frac{\partial^2 T}{\partial y^2}=\alpha\frac{\partial^2 T}{\partial y^2} \tag{7.52}$$

where

$\alpha=\dfrac{k}{\rho c_p}$ is the air diffusivity (m²/s)

c. Species (moisture) conservation equation in concentration (moisture) boundary layer (on a molar basis):

$$u\frac{\partial C}{\partial x}+v\frac{\partial C}{\partial y}=D_{s-\infty}\frac{\partial^2 C}{y^2} \tag{7.53}$$

Or, on a mass basis:

$$u\frac{\partial \rho}{\partial x}+v\frac{\partial \rho}{\partial y}=D_{s-\infty}\frac{\partial^2 \rho}{y^2} \tag{7.54}$$

where

C is the moisture molar concentration (mol/m³)

ρ is the moisture density (kg/m³)

$D_{s\infty}$ is the moisture diffusion coefficient in the air (m²/s)

Each of equations 7.50, 7.51, 7.52, and 7.53 includes nonlinear advection terms on the left-hand side (representing convection of momentum and energy) and a diffusion term on the right-hand side.

Because the moisture concentration of the free air stream is lower than that of the surface moisture layer, in boundary layer equations, all thermophysical properties (such as thermal conductivity, dynamic viscosity, and specific heat) must be determined at the board's surface temperature (T_s).

By defining dimensionless independent (x^* and y^*) and dependent (u^*, T^*, and C^*) variables, it is possible to normalize the simplified boundary layer equations (see Incropera and DeWitt 2002 for more details) and conclude the following:

i. The dependence of the velocity boundary layer on the fluid properties (ρ, μ), the air free stream velocity (u_∞), and length scale (L) may be simplified by grouping these variables in the form of a dimensionless group $\dfrac{u_\infty L}{\nu}$ (Reynolds number) over the entire length of the flat plate:

$$Re_L=\frac{u_\infty L}{\nu} \tag{7.55}$$

ii. In the energy conservation equation in the thermal boundary layer, the term $\dfrac{\alpha}{u_\infty L}$ is a dimensionless group that may be expressed as

$$\frac{\alpha}{u_\infty L} = \left(\frac{\nu}{u_\infty * L}\right)\left(\frac{\alpha}{\nu}\right) = \frac{1}{Re_L * Pr} \tag{7.56}$$

where

$\frac{\alpha}{\nu}$ is the reciprocal of Prandtl number $\left(Pr = \frac{\nu}{\alpha}\right)$

$\alpha = \frac{k}{\rho c_p}$ is the air thermal diffusivity (m^2/s)

ν is the air kinematic viscosity (m^2/s)

k is the air thermal conductivity (W/mK)

ρ is the air density (kg/m^3)

c_p is the air specific heat (J/kg·K)

iii. In the conservation equation in the concentration boundary layer, the term $\frac{D_{S\infty}}{u_\infty L}$ is equivalent to $\left(\frac{\nu}{u_\infty * L}\right)\left(\frac{D_{s-\infty}}{\nu}\right) = (Re_L)^{-1}\left(\frac{D_{s-\infty}}{\nu}\right)$, where the ratio $\frac{D_{s-\infty}}{\nu}$ is the reciprocal of the dimensionless Schmidt number:

$$Sc = \frac{\nu}{D_{s-\infty}} \tag{7.57}$$

where

$D_{s-\infty}$ is the moisture diffusion coefficient (m^2/s)

Schmidt number provides a measure of the relative effectiveness of momentum and mass transfer by diffusion in the velocity and concentration boundary layers, respectively.

7.2.1.4 Heat and Mass Transfer

The heat and mass transfer processes at the surface of the boards are of great importance. In kiln drying, the relative bulk motion of the drying air between boards' surfaces is maintained by one or several fans. As a consequence, forced convection heat and mass transfers between the drying air and boards' surfaces occur simultaneously.

During high-temperature drying processes of wood (at >100°C), for example, heat transferred to the lumber board causes the moisture to change from liquid phase to vapor, and then to vaporize at the board surface. In other words, the drying process involves heat transfer from the air to the lumber board, simultaneously with the mass (moisture) transfer from the board's wet surface to the drying air.

In any drying process, the thermal conduction in the flat plate boundary layer and the convective mass transfer are mainly determined by the temperature and moisture concentration differences between the bulk flow and the interface (Mori et al. 1991).

The problem of the simultaneous heat and mass transfer can be simply analyzed by considering that a humid air of temperature T_∞, moisture concentration C_∞, and velocity u_∞ passes over a horizontal flat plate (covered with a very thin and stagnant liquid film) of thickness δ and length L (see Figure 7.5a). The heat and mass transfers through the leading edge (front end) of the plate can be assumed to be negligibly small in comparison with that through the plate surface.

Under such a situation, heat is transferred from the bulk airflow to the interface through the developing boundary layer along the plate surface and the liquid vaporizes and diffuses from the liquid film on the plate surface to the free air stream.

The rates of heat and mass transfers are not independent, but conjugated with each other by the characteristic equilibrium relation being different from that in simple heat and mass transfer problems (Mori et al. 1991).

To simplify the analysis of simultaneous convective heat and mass transfers, some assumptions are considered (Mori et al. 1991): (i) air is an incompressible Newtonian fluid; (ii) all physical properties are constant, being independent of either temperature or concentration; (iii) energy dissipation due to fluid viscosity is negligible; (iv) thermal conduction and mass diffusion in the x-direction in the fluid can be neglected in comparison to those in y-direction; and (v) air at the interface is saturated at the interfacial temperature and its equilibrium relation can be expressed as a linear function of temperature.

The findings of the previous section allow concluding that under certain circumstances, the simplified boundary layer equations for momentum, heat and mass transfer are formally equivalent (similar or analog).

The three similarity parameters (Reynolds - Re_L, Prandtl - Pr and Schmidt - SC) allow applying the results to various geometric configurations (e.g., flat plate lumber boards), and hydraulic and thermal conditions (as the nature and velocity of the fluid) (Meroney 1969).

Consequently, because the equations describing the heat and mass transfer in the external flow over flat plates have analogous forms, it is possible to find solutions for mass convective transfer coefficients (and rates) from the solutions concerning the heat transfer (Masmoudi and Prat 1991), based on the dimensionless parameters Sc and Nu (Incropera and DeWitt 2002) for a wide variety of surface and fluid temperatures, thermal flux rates, and fluid velocities.

Accurate predictions of convective heat and mass transfer at air–porous material interfaces are of essential interest for many engineering applications (Mujumdar 2006; Putranto et al. 2001), one example being optimization of industrial wood convective drying processes with respect to energy consumption and product quality (Erriguible et al. 2006; Younsi et al. 2008; Kowalski 2010; Kowalski and Pawlowski 2011; Defraeye et al. 2012a, b).

In order to select and use correlations, and then determine the average heat and mass transfer coefficients and heat or mass transfer rates for the entire surface (i) an appropriate reference temperature must be chosen; (ii) the fluid properties at that temperature have to be determined; and (iii) Reynolds $\left(Re_L = \frac{u_\infty * L}{\nu}\right)$ and Schmidt $\left(Sc = \frac{\nu}{D_{s-\infty}}\right)$ numbers must be determined for heat and mass transfer, respectively (Incropera and DeWitt 2002).

Heat and mass convective transfer coefficients at the interface between unsaturated porous mediums and external air flows as over flat lumber boards during drying are affected by the spatial nonuniformity of variables as temperature and moisture (species concentration) caused by the leading edge (Masmoudi and Prat 1991).

Thus, under no-slip conditions at the boards' porous surfaces, the prediction of heat and mass transfer coefficients and rates between lumber boards and external airflow is a classical (Crausse et al. 1981) and relatively difficult (Bachmat and Bear 1986; Masmoudi and Prat 1991) coupled heat transfer problem.

By using the Chilton–Colburn analogy applied for fluids flowing over smooth plates and involving heat and mass transfer (Sherwood et al. 1975; Treybal 1968; Welty et al. 1969), Pang (1996a, b, c) derived correlations relating to the heat transfer coefficient to the mass transfer coefficient in kiln drying of timber boards. These correlations allow calculating of the rates of both heat and mass (moisture) transfer to/from the boards' surfaces from/to the air stream, respectively. Table 7.5 shows some of heat and mass transfer coefficients calculated by different authors (Pang 1996a).

As can be seen in Table 7.5, because of variations in drying air temperatures and velocities, the values of external heat and mass transfer coefficients vary significantly (Hunter and Sutherland 1997).

7.2.1.4.1 Heat Transfer

The heat transfer from the drying medium (air) to the board's surface (achieved by forced convection, and then to the interior of board by thermal conduction) can be examined separately from the mass transfer from board' surface to the bulk-airflow, although the two transfer processes are simultaneous, directly related, and described by analogous equations (Keey et al. 2000).

The dominant contribution to the energy transfer by convection between the drying air and the boards surfaces is generally made by the bulk motion of air particles, although the mechanism of diffusion (random motion of air molecules) also contributes to this transfer (Incropera and DeWitt 2002).

TABLE 7.5
Values of Heat and Mass Transfer Coefficients

	Temperature				
Source	Dry-Bulb (°C)	Wet-Bulb (°C)	Velocity (m/s)	Heat Transfer Coefficient (W/m²·K)	Mass Transfer Coefficient (m/s)
Ferguson and Turner (1994)	120–180	80	6	20	0.02
	90 and 125	70	3	17	0.02
Sutherland et al. (1994)	122	70	8	32	0.04
	152	67	8	32	0.04
Perré et al. (1988)	60	n/a	2	19	0.019
Stanish et al. (1986)	75	65	7	58	n/a
Bonneau and Puiggali (1993)	49	38	n/a	33	0.0313
Pang (1994)	120	70	5	35	0.029

n/a, not available.

Forced convective heat transfer coefficient (h_{heat}) depends on the drying air properties (k_f, c_p, μ, ρ), velocity (u_∞), external temperature (T_∞), the board geometry (e.g., length L for flat plates), and surface temperature (T_s) (Mori et al. 1991).

For a flat plate, the Nusselt dimensionless number (that provides a measure of the convection heat transfer) is one universal function of x^*, Re_L, and Pr:

$$Nu = f\left(x^*, Re_L, Pr\right) \tag{7.58}$$

where

$x^* = \dfrac{x}{L}$ is a dimensionless independent variable (–)
x is the distance downstream from the edge of the plate (m)
L is the length of the flat plate (m)
Re_L is the Reynolds number (–)
Pr is the Prandtl number (–)

Over the entire length of a flat plate (L), the average Nusselt number is defined as follows:

$$\overline{Nu} = \frac{\overline{h}_{heat} * L}{k_{air}} = f(Re_L, Pr) \tag{7.59}$$

where

$\overline{h}_{heat}$ is the average heat transfer coefficient (W/m$^2\cdot$K)
L is the flat plate total length (m)
k_{air} is the air thermal conductivity (W/mK)

Thus, the average heat transfer coefficient is:

$$\overline{h}_{heat} = \frac{k_{air}}{L}\overline{Nu} \tag{7.60}$$

When transition from turbulent to laminar boundary layer occurs upstream of the trailing edge $\left(\dfrac{x_{cr}}{L} \le 0.95\right)$, the surface heat transfer coefficients are influenced by conditions in both the laminar and turbulent boundary layers.

However, if transition occurs toward the rear of the plate (in the range $0.95 \le \dfrac{x_{cr}}{L} \le 1$), and by assuming that transition occurs abruptly at $x = x_{cr}$, mixed boundary layer conditions (laminar and turbulent) can be considered, and an average heat transfer coefficient for the entire flat plate can be calculated to a reasonable approximation as follows:

$$\overline{h} = \frac{1}{L}\int_0^L h dx \tag{7.61}$$

Integrating over the laminar region ($0 \le x \le x_{cr}$), and then over the turbulent region ($x_{cr} < x \le L$), this equation may be expressed as

$$\bar{h}_L = \frac{1}{L}\left(\int_0^{x_{cr}} h_{lam}dx + \int_{x_{cr}}^{L} h_{turb}dx\right) \tag{7.62}$$

In situations for which $L \gg x_{cr}\left(Re_L \gg Re_{x_{cr}}\right)$:

$$\overline{Nu_L} = 0.037 Re_L^{4/5} Pr^{1/3} \tag{7.63}$$

Equation 7.63 requires evaluation of the fluid properties at the reference (film) temperature $\left(T_{ref} = \frac{T_\infty + T_s}{2}\right)$.

From knowledge of $\overline{Nu_L}$, the average heat convection coefficient $\left(\bar{h}_{heat}\right)$ can be determined for simplified configurations such as flat plates under assumptions as (Defraeye et al. 2012b) (i) spatial and temporal variations along the surface are not accounted for; and (ii) Chilton–Colburn analogy (Chilton and Colburn 1934) is applied under strict conditions as no radiation, no coupling between heat and mass transfer, and analogous boundary conditions.

The average convective heat transfer coefficient that generally depends upon the nature, geometry, and the degree of roughness of the heat transfer surface, and the drying air velocity, relates the density of the convective heat $\left(\dot{q}_{heat}\right)$ normal to the wall of the air–porous material to the difference between the wall temperature (T_s) and the air bulk air temperature (T_∞) (Defraeye et al. 2012a, b):

$$\bar{h}_{heat} = \frac{\dot{q}}{T_\infty - T_s} \tag{7.64}$$

Since the fluid velocity at the boundary layer adjacent to the lumber board surface is zero, the heat transfer mechanism in this layer is conduction alone and, thus, the heat flux density (W/m^2) can be also expressed by the one-dimension Fourier's law (Anderson 2014; Whitelaw 2016):

$$\dot{q} = -k_{air}\frac{dT}{dy} \tag{7.65}$$

where

$\dot{q}$ is the density of the heat flux transferred (W/m^2)
k_{air} is the thermal conductivity of the air (W/mK)
$\frac{dT}{dy}$ is the temperature gradient in direction y, normal to the boards' surfaces (K/m)
T is the temperature (K)
t is the time (s)
y is the distance measured from the surface in the direction of heat transfer (m)

Equation 6.5 shows that the density of heat transfer across the thermal boundary layer is strongly influenced by the surface temperature gradient $\left(\left.\frac{\partial T}{\partial y}\right|_{y=0}\right)$. Because

$(T_\infty - T_s)$ is supposed constant, independent of x, while the thermal boundary layer (δ_T) increases with increasing x, temperature gradients in the thermal boundary layer must decrease with increasing x. Accordingly, the magnitude of the surface temperature gradient $\left.\frac{\partial T}{\partial y}\right|_{y=0}$ decreases with increasing x, and it follows that $\dot{q}$ and $\bar{h}_{heat}$ decrease with increasing x. Because the temperatures T_∞ and T_s vary with x-distance, it can be difficult to identify a free stream temperature, as well as the heat transfer coefficient, in complex flows as those over dried lumber boards.

Typical values of $\bar{h}_{heat}$ may vary from 10 W/m$^2\cdot$K (forced convection with low speed flow over flat plates) to 100 W/m$^2\cdot$K (forced convection with airflow velocities greater than 5 m/s).

Based on Equation 7.64, the density of the heat flux over unit area of heat transfer surface $\left(\dot{q}, \frac{W}{m^2}\right)$ may then be computed with Newton's law (1701) as follows:

$$\dot{q} = \bar{h}_{heat}\left(T_\infty - T_s\right) \tag{7.66a}$$

where

h_{heat} is the heat transfer coefficient, characteristic of the given airflow and heat transfer surface (W/m$^2\cdot$K)

T_∞ is the bulk air temperature (°C)

T_s is the board surface temperature (°C)

Because at normal drying rates, the effect of the moisture vapor leaving the surface on the net heat transfer is very small and, thus, may be ignored.

For a given bulk velocity (u_∞) and temperature (T_∞) of the air flowing over a horizontal flat plate of area A_s and uniform temperature $T_\infty > T_s$, the total heat transfer rate $\left(\dot{Q}\right)$ for the entire surface may be expressed as follows (Anderson 2014; Marinos-Kouris and Maroulis 2014):

$$\dot{Q} = \bar{h}_{heat} * A_s * \left(T_\infty - T_s\right) \tag{7.66b}$$

where

$\dot{Q}$ is the total rate of heat transfer to the boards' surfaces (W)

$\bar{h}_{heat}$ is the average convective heat transfer coefficient at the board-air interface (W/m$^2\cdot$K)

A_s is the total boards' surfaces area (m^2)

T_∞ is the board's surface temperature in the drying air far from the surface (K)

T_s is the temperature at the board-air interface (K)

7.2.1.4.2 Mass Transfer

During the drying wood process, the dry-bulb temperature of the drying air is higher than the temperature of the lumber. If the drying temperature is kept constant, the rate of vaporization will gradually decrease as the supply of moisture in the wood is diminished and as the vapor pressure of the air is increased. Therefore, to maintain a steady drying rate, the water molecules in the wood must acquire additional thermal energy (heat), or the vapor partial pressure of the kiln atmosphere must be reduced. This is achieved by either increasing the air dry-bulb temperature (i.e., by supplying more heat) or reducing the air relative humidity. That suggests that the rate of moisture movement, whether into or out of the wood boards, is in large part dependent upon temperature and the relative dryness of the air and wood (Cech and Pfaff 2000). The heat and mass transfers occur in opposite directions, they are interlinked and thus, they affect each other.

As for convection heat transfer, similar results may be obtained for convection mass transfer. More specifically, as in convective heat transfer processes, where Prandtl and Nusselt numbers are important, new dimensionless parameters are defined to correlate the convective mass transfer data. Similar to momentum $\left(\nu = \frac{\mu}{\rho}, in\ \frac{m^2}{s}\right)$ and thermal $\left(\alpha = \frac{k}{\rho c_p}, in\ \frac{m^2}{s}\right)$ diffusivity, the mass diffusivity $\left(D_{AB}, in \frac{m^2}{s}\right)$ is defined.

Also, in convective heat transfer, the ratio of the molecular diffusivity of momentum to the molecular diffusivity of heat (thermal diffusivity) was designated as the Prandtl number:

$$Pr = \frac{Momentum\ diffusivity}{Thermal\ diffusivity} = \frac{\nu}{a} = \frac{c_p * \mu}{k} \tag{7.67}$$

In convective mass transfer, Prandtl number $\left(\frac{\nu}{a}\right)$ is replaced by Schmidt number defined as the ratio of momentum diffusivity to mass diffusivity:

$$Sc = \frac{Momentum\ diffusivity}{Mass\ diffusivity} = \frac{\nu}{D_{AB}} = \frac{\mu}{\rho * D_{AB}} \tag{7.68}$$

In convective mass transfer, the Nusselt number $\left(\frac{\overline{h}_{heat} * L}{k_{air}}\right)$ is replaced by the dimensionless parameter termed Sherwood number, defined for mass transfer in a fluid flow over a flat plate where the convection mass transfer coefficient $\left(\overline{k}_{mass}\right)$ depends on the fluid properties ($D_{\infty-S}$, ρ, and μ), velocity (u_∞), and the characteristic length (L). For a flat plate, the Sherwood dimensionless number is universal function of x^*, Re_L, and Sc:

$$Sh = f\left(x^*, Re_L, Sc\right) \tag{7.69}$$

where

$x^* = \frac{x}{L}$ is a dimensionless independent variable (–)
x is the distance from the leading edge (m)
L is the length of the flat plate (m)
Re_L is the Reynolds number based on the plate total length (–)
Sc is the Schmidt number (–)

Thus, in addition to the air velocity, the convective mass transfer coefficient is dependent on the length and the geometry of the vaporization surface, and the direction of the flow (Berit Time 1998).

Sherwood number provides a measure of the convection mass transfer occurring at the board's surface. Sherwood number is therefore to the moisture (concentration) boundary layer what the Nusselt number is to the thermal boundary layer.

Over the entire length of a flat plate (L), the average Sherwood number depends only on Reynolds number (Re_L) and Schmidt number (Sc), and it is defined as follows:

$$\overline{Sh} = f(Re_L,\ Sc) = \frac{\overline{k}_{mass} * L}{D_{s,\infty}} \tag{7.70}$$

where

$\overline{k}_{mass}$ is the average mass transfer coefficient (m/s)
L is the flat plate length (m)
$D_{s,\infty}$ is the moisture (water vapor) diffusion coefficient (m²/s)

The conditions in the concentration boundary layer strongly influence the surface concentration gradient and, also, the convection mass transfer coefficient.

Another dimensionless parameter encountered in processes involving simultaneous convective heat and mass transfer is the Lewis number, defined as the ratio of the thermal diffusivity of heat to the molecular diffusivity of mass:

$$Le = \frac{\textit{Thermal diffusivity}}{\textit{Mass diffusivity}} = \frac{\alpha}{D_{AB}} = \frac{k}{\rho c_p D_{AB}} \tag{7.71}$$

For external forced laminar flow over the entire surface of a lumber board (flat plate), the average Sherwood number can be calculated as follows (for T_{ref} and $0.6 \le Sc \le 50$):

$$\overline{Sh}_L = (0.664 * Re_L^{1/2}) * Sc^{1/3} \tag{7.72}$$

For mixed laminar and turbulent flow for which $L \gg x_{cr}\left(Re_L \gg Re_{x_{cr}}\right)$, the average Sherwood number is (for T_{ref} and $0.6 \le Sc \le 50$)

$$\overline{Sh}_L = (0.037 * Re_L^{4/5} - 871) * Sc^{1/3} \tag{7.73}$$

Equations 7.72 and 7.73 require evaluation of the fluid properties at the reference (film) temperature $\left(T_{ref} = \frac{T_\infty + T_s}{2}\right)$.

Generally, if a fluid of species mass concentration C_∞ flows over a surface at which the species concentration is maintained at some uniform value $C_s > C_\infty$ (Figure 7.5a), mass transfer of the species C_s (typically, water vapor that is transferred into the air stream due to vaporization at the flat plate surface) by convection will occur.

In the particular case of moisture mass transfer between the lumber boards and the drying air, the driving force is the difference between the moisture concentration at the phase boundary (C_s) (at the board surface interface) and the concentration at some arbitrarily defined point in the fluid medium (C_∞).

In other words, similar to the heat transfer, a moisture mass transfer occurs when a moisture concentration gradient exists between the board's surface and the drying air. At steady-state, the specific mass transfer rate is given by the one-dimension Fick's first law in the following form (Incropera and DeWitt 2002; Anderson 2014):

$$j_{mass} = -D * \frac{\partial C}{\partial y} \tag{7.74}$$

where

j_{mass} is the specific mass transfer rate (kg/m$^2 \cdot$ s)
D is the diffusion coefficient (m^2/s)
$\frac{\partial C}{\partial y}$ is the concentration (moisture) mass gradient (kg/m$^3 \cdot$ m)

As for the case of heat transfer, the calculation of the rate at which the mass transfer occurs may be based on the use of a convection mass transfer coefficient. The convection mass transfer is thus introduced by analogy to that of convection heat transfer (Incropera and DeWitt 2002).

Consequently, the moisture specific mass transfer rate between the lumber boards' surfaces and the drying air can be written as follows:

$$j_{mass} = \bar{k}_{mass}(C_s - C_\infty) \tag{7.75}$$

where

j_{mass} is the moisture specific mass transfer rate (kg/m$^2 \cdot$ s)
$\bar{k}_{mass}$ is the average convective mass transfer coefficient (m/s)
C_s is the moisture mass concentration at the board surface (kg/m^3)
C_∞ is the moisture mass concentration at some arbitrary defined point in the drying air bulk flow (kg/m^3)

Similar to Newton's law of cooling, the total convection mass transfer rate from a flat plate surface can be expressed as follows (Anderson 2014):

$$\dot{m}_{mass} = \bar{k}_{mass} * A_s * M * \left(\bar{C}_s - \bar{C}_\infty\right) \tag{7.76}$$

where

$\dot{m}_{mass}$ is the total convection mass transfer rate (kg/s)
$\bar{k}_{mass}$ is the average mass transfer coefficient (m/s)
A_s is the moisture mass transfer surface area (m^2)
M is the moisture (water vapor) molar mass $\left(\frac{kg_{water}}{kmol}\right)$
$\bar{C}_s$ is the average moisture molar concentration at the board (flat plate) surface (kmol/m^3)
$\bar{C}_\infty$ is the average moisture molar concentration in the drying air far from the board surface (kmol/m^3)

In terms of moisture mass density, the total convection mass transfer rate from a flat plate surface can also be expressed as follows:

$$\dot{m}_{mass} = \bar{k}_{mass} * A_s * (\bar{\rho}_s - \bar{\rho}_\infty) \tag{7.77}$$

where

$\dot{m}_{mass}$ is the total convection mass transfer rate (kg/s)

$\bar{\rho}_s$ is the moisture average mass density at the flat plate surface $\left(\frac{kg_{water}}{m^3}\right)$

$\bar{\rho}_\infty$ is the moisture average mass density in the bulk drying air far from the flat plate surface $\left(\frac{kg_{water}}{m^3}\right)$

From Equations 7.76 and 7.77 it can be seen that to calculate the total convection mass transfer rate from a flat plate surface, it is necessary to determine the value of $\bar{C}_s$ or $\bar{\rho}_s$ by assuming thermodynamic equilibrium at the interface between the (wood) solid surface and the moist drying air. Such an equilibrium implies that (i) the temperature of the moisture vapor at the interface is equal to the surface temperature (T_s) and (ii) the moisture vapor is in a saturated state in which case thermodynamic tables for water may be used to obtain its density from knowledge of T_s.

Theoretical analysis has been conducted for vaporization through a laminar boundary layer flow from a flat plate surface, where the convective heat and mass transfers and the two-dimensional thermal conduction in the plate are combined simultaneously. The conjugation was achieved by taking into account the vapor–liquid equilibrium of linear temperature dependence at the interface, the latent heat of vaporization, and the continuities of temperature and heat flux at the interface (Mori et al. 1991).

Numerical calculations have been made for the parallel flow case where both values of the Prandtl and Schmidt numbers are unity.

7.3 LUMBER BOARD DRYING PERIODS

Drying processes of solid materials are generally divided into a number of specific periods in which different physical mechanisms are dominant (Lewis 1921; Sherwood 1929a, b). In general, the drying of hygroscopic materials, such as wood, proceeds in three stages (Fyhr and Rasmuson 1996; Mujumdar and Menon 1995). During the first stage, the wood contains free moisture and has a wet surface maintained by rapid capillary fluid transport.

The drying rate is characterized by the vaporization of free water, and it is constant over time (when the bulk vapor pressure in the drying medium and the turbulent conditions are constant). The second stage occurs when dry spots appear on the surface of the drying material, at the critical moisture content. The movement of water is here characterized by mass diffusion, and the drying rate falls. The third stage begins when the surface film of moisture is completely vaporized. Moisture bound by absorption is removed, and the drying rate is controlled by the net diffusion of water from the bulk of the processed material to its surface.

As for other hygroscopic materials, based on the drying kinetics, the mechanism of convective, forced air drying of timber can be divided into two main, distinct phases: (i) the constant drying rate period and (ii) the falling (or decreasing) drying rate period (Perré and Keey 2015).

In most cases, both these drying periods exist, and for slow-drying wood species (as hardwoods), most of the drying may occur in the falling drying rate period.

The drying periods of a wet lumber board under fixed drying conditions can be graphically represented as variations of the lumber moisture content $\left(\frac{kg_{water}}{kg_{oven\text{-}dry}}\right)$ (Figure 7.6a) and drying rate (kg/s) (Figure 7.6b) or specific evaporative rate $\left(\frac{kg_{water}}{m^2 s}\right)$ (Figure 7.6c) with the drying time, or as the variation of the lumber drying rate (kg/s) with the moisture content $\left(\frac{kg_{water}}{kg_{oven\text{-}dry}}\right)$ (Figure 7.6d).

Figure 7.6a–d shows that, prior beginning the drying cycle, the stack(s) of lumber boards is warmed up (preheated) up to an equilibrium temperature (process 1-2).

Figure 7.6a depicts the drying rate curve, that is, the lumber moisture content loss as a function of time. It can be seen that after an initial period of preheating, the moisture content (dry basis) decreases linearly with time (t), following the start of the vaporization. This is followed by a nonlinear decrease in moisture content with

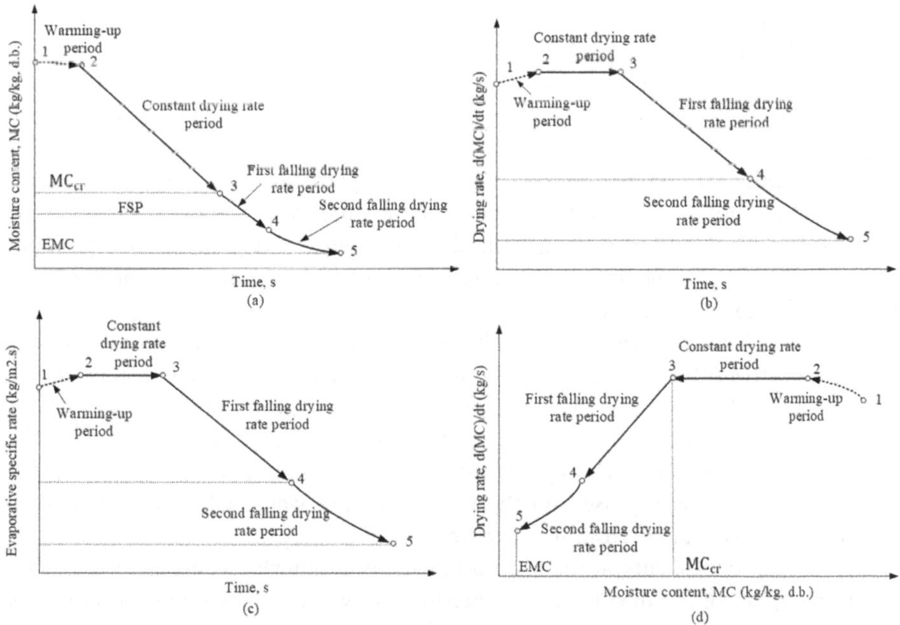

FIGURE 7.6 Wood drying periods represented as (note: graphics not to scale) follows: (a) moisture content vs. time; (b) drying rate vs. time; (c) evaporation specific rate vs. time; (d) drying rate vs. moisture content. MC, moisture content; EMC, equilibrium moisture content; FSP, fiber saturation point; MC$_{cr}$, critical moisture content,; d.b, dry basis.

the time until, after a very long time, the board reaches its *EMC*, and drying stops (Mujumdar 2000).

Figure 7.6b, d shows that at the lumber critical moisture content (MC_{cr}) (point 3 in Figure 7.6a), the drying rate begins to decrease, and then the warming-up period is followed by the *constant drying rate period* (process 2-3).

The period from the moment when the critical moisture content is reached (point 3), until the moment when the lumber is practically *dry*, is called the *falling drying rate period* (process 3-4-5).

During the first critical period after the drying rate is governed by the vaporization of water on the surface wet fraction, and this fraction decreases continuously until the end of this period, the surface is dry.

During the constant speed drying (or critical period), the wood surface is completely covered by a layer of liquid and evaporation which depends only on the speed of diffusion of vapor or intense heat passing through the boundary layer of air (Mujumdar 2015).

In real practice, the initial lumber stack(s) may have high moisture content and the product may be required to have high final (residual) moisture content so that all the drying may occur in the constant rate period.

In other words, during a typical drying process, wet wood boards with initial moisture content high enough being dried under constant drying conditions may experience a constant drying period during which the drying rate is constant.

According to Sherwood (1929b), if the initial moisture content is less than the critical moisture content, no constant rate period appears.

The falling drying rate period is sometimes divided in two parts: a *first falling drying rate period* (3-4) and a *second falling drying rate period* (4-5) (McCready and McCabe 1933; Sherwood 1929a, b; Treybal 1968; Karel and Lund 2003; Rizvi 2005).

The dry-bulb kiln temperature during the drying process is higher than the temperature of the lumber. When the lumber contains free water, the temperature of the wood is approximately the same as that of the wet-bulb. As the supply of free water diminishes and the moisture content approaches FSP, the temperature of the wood begins to rise toward the dry-bulb temperature.

After an initial transition period, the material experiences the constant drying rate period, given that the air–material interface remains wet. The constant drying rate period is characterized by a relative humidity of quasi 100% at the surface, a constant drying rate, and a constant material temperature, which is equal to the wet-bulb temperature (*WBT*) (T_{wb}). In this case, the convective heat supply to the interface is quasi, entirely used for the vaporization of water, which requires latent heat for the phase change from liquid water to vapor.

As vaporization occurs at the air–porous material interface during the constant drying rate period, the drying rate is determined by the airflow conditions and not by the porous material transport properties. Nevertheless, the porous material transport properties do affect the length of the constant drying rate period, since it is dependent on the supply of liquid to the surface. When the material dries out at the interface, the decreasing drying rate period sets in which is characterized by a lower drying rate. During decreasing drying rate period, the liquid water front recedes from the surface and water, once evaporated, must diffuse out via the "dry" outer porous material layer. This dry layer can be seen as an additional resistance to liquid water transport from the inside of the material. Due to this material resistance, in addition to the boundary layer resistance, the drying rate of most porous materials during the decreasing drying rate period is thereby much less sensitive to the convective boundary conditions. This decrease in drying rate is accompanied by a temperature increase since less latent heat is required for the evaporation of water.

7.3.1 Warming-Up Period

During the warming-up period (process 1-2 in Figure 7.6a–d) of typical kiln batch drying schedules, the lumber stack(s) must be heated by air forced convection/conduction until the desired dry-bulb drying temperature (in conventional kilns) or wet-bulb temperature (in heat pump-assisted dryers) are reached, prior starting the first step of the actual drying processes (Anderson and Westerlund 2014).

The stacks of lumber boards are heated, generally between 3 and 8 hours, depending on the wood initial temperature, until the boards are in thermal equilibrium with the drying air, and the board surfaces' temperatures become equal to the wet-bulb temperature of the drying air. The vapor will reach the timber surface by diffusion, and this internal timber moisture diffusion process is governed by intrinsic timber fluid transport properties.

Because lumber boards must be properly heated up as to avoid or create any internal stress, during this period, steam may be added into kiln to decrease the strain in the lumber boards, and thus, prevent cracks and other defects (Erickson 1954).

7.3.2 Constant Drying Rate Period

After the warming-up period, a constant drying rate begins and progresses by increasing the dry-bulb temperature or by decreasing the relative humidity of the drying medium (air).

When the drying process starts up, the moisture content begins decreasing linearly over time, a process followed by non-linear decreasing until the material reach a state of equilibrium.

In kilns, constant rate drying is defined as the period of drying where moisture removal occurs at the surface by vaporization and the internal moisture movement is sufficient enough to maintain the saturated surface, thus the rate of vaporization remains constant.

However, the constant drying rate period, consisting in an initial linear reduction of the average moisture content as a function of time, very common for other porous media, generally occurs for a limited time in wood species having relatively high initial moisture contents (Perré and Martin 1994; Perré and Keey 2015).

In the initial constant rate drying period (drying in which surface moisture is removed), the boards' surfaces take on the wet-bulb temperature corresponding to the air temperature and humidity conditions at the same location.

In the constant draying rate period (Figure 7.6), the moisture migrates inside the wood mostly by capillary forces; vaporization occurs at the exchange surface with a dynamic equilibrium within the boundary layers between the heat and the vapor flows (Perré and Martin 1994).

During the constant drying rate period, the water vaporizes at the surface of wood boards contains free moisture, and the capillary force transports the free water from the cells until the moisture content of the wood surface is equal to a critical moisture content (MC_{cr}). During this period, the drying rate-controlling is the diffusion of the water vapor across the air–moisture interface.

During this period of drying, the drying rate is constant and depends only on the external conditions (air temperature, relative humidity, velocity, and flow configuration). The exposed surface is supplied with liquid water coming from the inside of the board by capillary action or other driving forces.

Surface vaporization is the controlling moisture removal process for the constant drying rate period and the magnitude of the vaporization rate is, in turn, governed by the drying air conditions, specifically, its temperature, relative humidity, and velocity. At the end of the constant rate period, the surface of the board can be assumed dry and vaporization will take place from within the board.

The constant drying rate period lasts as long as the surface is supplied with liquid. Its duration depends strongly on the drying conditions (e.g., air velocity and properties).

During the constant rate period process (2-3 in Figure 7.6), the boards' surfaces remain saturated with liquid water because the movement of water within the boards to surface is equal to the rate of vaporization. Drying takes place by movement of water vapor from saturated surface through a stagnant air film into the main stream of drying air.

At the beginning, the relative humidity of the moisture at the boards' surfaces is practically 100%, and the drying rate and the material wet-bulb temperature of the air at the surface are constant. The convective heat supplied is entirely used for the vaporization of moisture.

During the constant drying rate period, dominated by the convective mass transfer, the temperature need not be very high. Keeping a suitable wet-bulb temperature depression is very important for this period, since this will affect both drying time and drying quality. When free water is not present, the temperature will become the dominant factor in drying.

The liquid migrates from regions with high moisture content (liquid–gas interfaces within large pores) toward regions with low moisture content (liquid–gas interfaces within small pores) according to Darcy's law (permeability gradient of liquid pressure).

The temperature at the surface is equal to the wet-bulb temperature. Moreover, because no energy transfer occurs within the medium during this period, the whole temperature of the board remains at the wet-bulb temperature (Perré and Keey 2015).

The temperature of the liquid film and the adjacent solid surface is the wet-bulb temperature of the moving air (assuming radiation and conduction effects are negligible). High air temperatures can be used in the constant rate period as the solid's temperature will not exceed the wet-bulb temperature.

During the first stage of the drying when free water is present, the temperature of the lumber surface is assumed to be at the wet-bulb temperature. As drying progresses, the surface temperature rises and approaches the dry-bulb temperature.

The dry-bulb temperature is then set to a higher value which makes the temperature difference between the air and the wood relatively constant, thus minimizing drying defects.

During the constant rate period, the vaporization takes place at the surface of wet lumber boards, the rate of drying being limited by the rate of diffusion of water vapor through the surface air film (boundary layer) out into the main body of the air.

In this period, the drying is similar to the vaporization of water from a free liquid surface, and for a sufficiently long constant rate period, the solid may assume a constant equilibrium temperature, just as a free liquid surface is maintained by evaporation at the wet-bulb temperature of the air. The resistance of the surface film to moisture transport is a complex function of the stacking geometry (board spacing), dry-bulb temperature, wet-bulb depression, turbulent intensity level, uniformity of the circulation, surface roughness, and any other factor which controls the effectiveness of the fluid motion to heat and remove the moisture from the lumber surface (Meroney 1969).

During this period, the boards' surfaces are covered with a continuous layer of free water and vaporization takes place mainly at the surface (Srikiatden and Roberts 2007).

The rate of moisture removal is controlled by the rate of vaporization of moisture from the surface to the drying air as well as the rate of heat transfer to the evaporating surface.

The drying rate is determined by external conditions only. If the temperature gradient within the material is negligible, the surface temperature is almost constant, and its value is very close to the wet-bulb temperature of the flowing air (Zhang 1999).

The rate of energy input equals the heat lost during vaporization, so the temperature at the surface is also constant and is lower than the surrounding temperature of the air at a given wet-bulb temperature.

During the constant drying rate period, the exposed surface of the board is still above the FSP. As a result, the vapor pressure at the surface is equal to the saturated vapor pressure, and it is a function of the surface temperature only. Coupled heat and vapor transfer occurs in the boundary layer (Figure 7.7). The heat flux supplied by the airflow is used solely for transforming the liquid water into vapor (Perré and Martin 1994).

In other words, drying in the constant rate period is a surface-based process governed by external conditions such as temperature difference between the dry air and wet surface, area exposed to the dry air, and external heat and mass transfer coefficients (Chirife 1983). The drying in this period is similar to the evaporation of water from a free liquid surface, just as a free liquid surface is maintained at the wet-bulb temperature of the air.

During the constant drying rate period, the moisture vaporization rate is governed by the drying air conditions, specifically its temperature, relative humidity, and velocity. The surface contains free moisture, and vaporization takes place from there. In this stage of drying, the rate-controlling step is the diffusion of the water

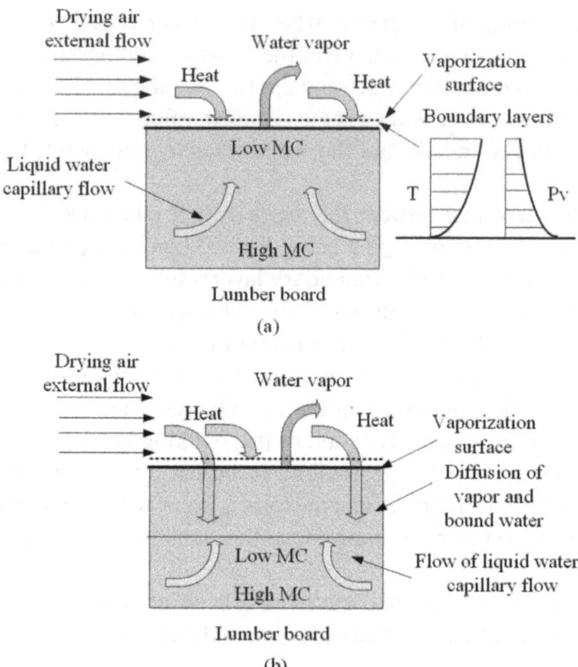

FIGURE 7.7 (a) Constant drying rate period; (b) second (falling) drying rate period. (Adapted and redrawn from Perré and Martin 1994 with permission from Drying Technology.)

vapor across the air–moisture interface and the rate at which the surface for diffusion is removed. During the constant rate period, the wood surface is completely covered by a layer of liquid (moisture).

During the constant drying rate period (i.e., first few hours of a typical drying cycle), the moisture content strongly decreases near the exchange surfaces, the drying rate is high and relatively constant. As the drying progresses, the drying rate decreases. When all the free water is removed, the diffusion of bound water and the convective migration of the vapor ensure the transport of mass.

Also, during this first drying period, the temperature is below the dry-bulb temperature and then increases to the wet-bulb temperature and remains constant at this value because free water is still present.

In the constant rate period, the rate of drying is determined by the rate of vaporization. When all the exposed surface of the solid ceases to be wetted, vapor movement by diffusion and capillarity from within the solid to the surface are the rate-controlling steps.

The rate of drying is dependent on the rate of heat transfer to the drying material and is equal to the rate of mass transfer, thus the temperature of the drying material remains constant. This continues until the critical moisture content is reached at point 3, then the drying rate begins to fall.

During the constant rate period, the drying rate is constant until the critical moisture content, which usually varies with the nature and thickness of the solid material and the external drying conditions (air relative humidity and temperature), is reached (Sherwood 1929b). At this moment, the drying rate starts to decrease and the falling rate period begins (Chirife 1983; Rizvi 2005).

The rate of drying is controlled by vaporization of the surface moisture, which is independent of the nature and internal mechanisms within the solid. The drying rate is governed by the vaporization of water on the wet surface, and it decreases continuously until the surface is dry.

Toward the end of the constant rate period, moisture has to be transported from the inside of the lumber board to the surface by capillary forces and the drying rate may still be constant. The movement of moisture inside the boards is not only maintained by capillary force, but also driven by many other forces, such as liquid–solid matrix interfacial drag, the inertial force of the movement of the liquid film, viscous force, and gravity. At this point, internal moisture transfer to the surface and the vaporization at the surface are in equilibrium, and the free water on the surface will be vaporized steadily and continuously (Zhang 1999).

When the average moisture content has reached the critical moisture content, the surface film of moisture is reduced by evaporation that further drying causes dry spots to appear upon the surface. Since, however, the rate is computed with respect to the overall solid surface area, the drying rate falls even though the rate per unit wet solid surface area remains constant.

At the end of the constant rate period, the surface of the board can be assumed dry and vaporization will take place from within the board. Vapor will reach the timber surface by diffusion, and this internal timber moisture diffusion process is governed by intrinsic timber fluid transport properties and will be largely independent of kiln conditions.

The value of the transition moisture content at which the departure from constant rate drying is first observed is called the critical moisture content, MC_{cr} (Chirife 1983) (see Figure 7.6a, d). The critical moisture content usually varies with the thickness of the material and the external drying conditions (Srikiatden and Roberts 2007).

Vaporization thus depends only on the temperature and the quantity of heat passing through the boundary layer of air. As a consequence, the drying rate during the constant rate period is mostly dependent on the rate of heat transfer to the wood board or stack being dried and, therefore, the maximum achievable drying rate is considered to be heat transfer limited.

In the constant rate period, the rate of drying is determined by the rate of vaporization. When all the exposed surface of the solid ceases to be wetted, vapor movement by diffusion and capillarity from within the solid to the surface are the rate-controlling steps. Whenever considerable shrinkage occurs, as in the drying of timber, it is essential to retard vaporization and bring it in step with the rate of moisture movement from the interior. This could be achieved by increasing the relative humidity of the drying air.

A constant rate drying region is generally concerned with the vaporization of liquid from the solid lateral surfaces of a porous material where resistance to internal diffusion is small as compared to the removal of vapor from the surface. Hence, the drying rate is generally independent of the material being dried.

The rate of vaporization, limited by the relative humidity and the temperature of the drying air, remains constant, while the rate of energy input (i.e., rate of heat transfer to the vaporizing surface) equals the heat lost by moisture vaporization. The temperature at the wood surface is constant and equals to the wet-bulb temperature corresponding to the surrounding air temperature and relative humidity (Sherwood 1929b; Srikiatden and Roberts 2007).

The rate of moisture loss during the constant rate period is limited by the ability of the circulating drying air to remove the water from the lumber surface. At low moisture contents (below 39%), the internal rate of moisture movement (transport) is limited by the ability of the water to diffuse through the wood to its surface and the energy available for sorption, hence the falling rate period (Meroney 1969). Therefore, drying in the constant rate period is a surface-based rate governed by external conditions such as temperature difference between the dry air and wet surface, area exposed to the dry air, and external heat and mass transfer coefficients (Chirife 1983; Geankoplis 1993).

As vaporization occurs, the drying rate is determined by the airflow conditions and not by the wood transport properties. However, the wood transport properties do affect the length of the constant drying rate period, since it is dependent on the supply of liquid to the surface (Defraeye et al. 2012a, b).

7.3.3 Falling Drying Rate Period

If drying is continued after the constant rate period when the lumber dries out at the surface, the falling (decreasing) drying rate period, which is characterized by a lower drying rate, and unsaturated surface drying begins.

The falling rate period (3–4–5 in Figure 7.6a–d), also called critical period post (Mujumdar 2014) or decreasing drying rate period (Perré and Keey 2015) is characterized by the surface of the wood board not being saturated, and the rate of moisture movement from the interior toward the board's surface is less than the rate of vaporization from the surface, and the surface temperature approaches the dry-bulb temperature of the drying medium (Srikiatden and Roberts 2007). Consequently, drying (i.e., moisture movement and moisture vaporization) in the falling rate period is an internally controlled mechanism (Sherwood 1931; Chirife 1983).

The rate of moisture movement from the interior toward the board surfaces is less than the rate of vaporization from the surface; thus, it is an internally controlled drying mechanism (Sherwood 1931).

This decrease in drying rate is accompanied by a temperature increase since less latent heat is required for the vaporization of water (Defraeye et al. 2011, 2012a, b). During this period, the slope of the drying rate curve becomes less steep and, eventually, tends to be nearly horizontal over very long periods of time.

In the falling rate period, the boards' temperatures approach the dry-bulb temperature of the air (Mujumdar 2007).

With woods in which the initial moisture content is relatively low and the final moisture content required is extremely low, the falling rate period becomes important, and dryness times are long. Air velocities will be important only to the extent to which they enhance heat transfer rates. Air temperature, humidity, material thickness, and bed depth all become important. When the rate of diffusion is the controlling factor, particularly when long drying periods are required to attain low moisture contents, the rate of drying during the falling rate period varies as the square of the boards' thickness (Mujumdar 2007).

This stage proceeds until the surface film of liquid is entirely vaporized and the average moisture content has reached the critical moisture content (MC_{cr}).

During the falling rate period, the drying rate slowly decreases until it approaches zero at the *EMC* (point 5 in Figure 7.6a–d) (i.e., when the material comes to equilibrium with the drying air).

During the falling rate period, a conductive heat flux must exist inside the board to increase the temperature and to vaporize the liquid driven by vapor diffusion. The region of liquid migration naturally reduces as the drying progresses, and finally disappears. The process is finished when the temperature and the moisture content attain the outside air temperature and the equilibrium moisture content, respectively (Perré and Keey 2015).

The heat transmission now consists of heat transfer to the surface and heat conduction in the product. Since the average depth of the moisture level increases progressively and the heat conductivity of the dry external zones is very small, the drying rate is increasingly influenced by the heat conduction. During this stage, as the moisture concentration is lowered by drying, the rate of internal movement of moisture decreases. The rate of drying falls even more rapidly than before and continues until the moisture content falls down to the equilibrium value (*EMC*) for the prevailing air humidity and then drying stops (Mujumdar 2007).

During the falling rate period, the surface is completely dry and the drying rate must be evaluated according to the process of removal of moisture from the interior of the solid to the surface, which can be performed by various mechanisms.

In the falling rate period, the rate-controlling step is the transport of moisture to the surface, so the shape of the line is dependent upon the wood species.

In this regime, moisture migration from the product interior to the surface is mostly by molecular diffusion, that is, the water flux is proportional to the moisture content gradient. This means that moisture moves from zones with higher moisture content to zones with lower values. The most common parameter used in predicting moisture transfer in drying of solids is the effective moisture diffusivity, traditionally determined based on Fick's second law diffusion equation (Srikiatden and Roberts 2007). If moisture removal is considerable, the products usually undergo shrinkage and deformation. The drying rate in the falling rate period is controlled by the rate of removal of moisture from the interior of the wood being dried.

In the falling rate period, the rates of internal heat and mass transfer determine the drying rate.

In addition, the process aspects of drying are deterministic, and their distribution affects the size of the drying field. The form of distribution, the mean value, and the standard deviation express the stochastic aspect of the lumber drying process.

Quality of dried lumber boards is measured at the end of drying process. Each board can be evaluated separately before sale.

Although a clear separation cannot be achieved, for any dried material, the falling drying period may be divided in two parts: (i) one, called critical early period post

(3–4 in Figure 7.6a), in which the drying rate varies linearly with moisture from the critical point and (ii) another, called second post-critical period (4–5 in Figure 7.6b) that does not meet a linear variation (Mujumdar 2015).

With solids in which the initial moisture content is relatively low and the final moisture content required is extremely low, the falling rate period becomes important. Air velocities will be important only to the extent to which they enhance heat transfer rates. Air temperature, humidity, material thickness, and bed depth all become important. When the rate of diffusion is the controlling factor, particularly when long drying periods are required to attain low moisture contents, the rate of drying during the falling rate period varies as the square of the material thickness (Defraeye et al. 2011).

The falling rate period can take a far longer time than the constant rate period even though the moisture removal may be less.

7.3.3.1 First Falling Rate Period

When the average moisture content has reached the critical moisture content (MC_{cr}), the surface film of moisture has been so reduced by vaporization that further drying causes dry spots to appear upon the surface. Since, however, the rate is computed with respect to the overall solid surface area, the drying rate falls even though the rate per unit wet solid surface area remains constant.

This gives rise to the second drying stage or the first part of the falling rate period, the period of unsaturated surface drying. This stage proceeds until the surface film of liquid is entirely evaporated (Mujumdar 2007).

On further drying (the second falling rate period), the rate at which moisture may move through the solid as a result of concentration gradients between the deeper parts and the surface is the controlling step.

The first falling rate period (process 3-4 in Figure 7.6a) is the period of the unsaturated surface drying where the surface is dried out and the drying rate decreases as the moisture content decreases due to the additional internal resistance for moisture transfer. This stage proceeds until the surface film of liquid is entirely evaporated. This part of the curve may be missing entirely, or it may constitute the whole falling rate period. In the first falling rate period, the drying rate decreases as the moisture content decreases due to the additional internal resistance for moisture transfer and to the reduction of heat flux into the sample as the surface increases to the heat medium (Karel and Lund 2003; Rizvi 2005).

Some authors suggested that the initial part of the falling rate regime involves a retreat of the vaporization zone into the solid before the free water has been entirely removed from the center portion of the solid (McCready and McCabe 1933).

As drying proceeds, the fraction of wet area at the surface decreases with decreasing surface water content (Zhang et al. 1999). When water passes through randomly distributed paths in a medium, there is a percolation threshold, which usually corresponds to the critical free water content. When the free water content is greater than the critical, the water phase is continuous. Regardless of the rate of internal moisture transfer, as long as the free water content at the surface is less than the critical value (this critical saturation is about 0.3 for most porous materials, De Vries 1958), the surface will form discontinuous wet patches. Thus, the mass transfer coefficient decreases with the surface free water content, and the first falling rate period starts.

In the first falling rate period, a new energy balance will be reached at the surface, accompanied by slowly rising surface temperature. Free water still exists at the surface, the dry patches still contain bound water, and the vapor pressure at the surface is determined by the Clapeyron equation. At this point, the convective mass and heat transfer coefficients at the surface are a function of the surface water content.

7.3.3.2 Second Falling Rate Period

The second falling rate period (process 4-5 in Figure 7.6a) begins when the partial pressure of water throughout the material is below the saturation level. The heat flux from the hot air to the wood boards is very low because of the small temperature gradient between the hot air and the board surface, which is now close to the hot air temperature.

In the second falling rate period, a region in the hygroscopic range develops from the exposed surface. In that region, both vapor diffusion and bound water diffusion act. Vaporization takes place partly inside the medium. Consequently, a heat flux has to be driven toward the inner part of the board by conduction (Figure 7.6b).

The heat transmission now consists of heat transfer to the surface and heat conduction in the product. Since the average depth of the moisture level increases progressively and the heat conductivity of the dry external zones is very small, the drying rate is increasingly influenced by the heat conduction. However, if the dry product has a relatively high bulk density and a small cavity volume with very small pores, drying is determined not so much by heat conduction but by a rather high resistance to diffusion within the product. The drying rate is controlled by diffusion of moisture from the inside to the surface and then mass transfer from the surface. During this stage, some of the moisture bound by sorption is removed. As the moisture concentration is lowered by drying, the rate of internal movement of moisture decreases. The rate of drying falls even more rapidly than before and continues until the moisture content falls down to the equilibrium value (*EMC*) for the prevailing air humidity and then drying stops. The transition from one drying stage to another is not sharp (Mujumdar 2007).

In highly porous materials, or when significant porosity is developed, mass transfer occurs mainly in the vapor phase and all evaporation occurs from the interior of the material (Rizvi 2005). Furthermore, in order to supply the heat of vaporization to the interior, heat has to be conducted through dry solid and pore regions, both of which have low thermal conductivity. The drying rate in this period is extremely slow; therefore, it is not surprising that the time required to remove the last 10% of moisture is almost equivalent to the time required to remove the first 90% of moisture (Srikiatden and Roberts 2007).

During the second falling drying rate period, the water migration is slower and it is done with a combination of the capillary force and a vapor diffusion of bound water in the cell walls until the FSP is reached. Finally, during the last sequence (3), water migration is only done through the diffusion force of bound water in the cell walls.

When vapor pressure of the material becomes equal to the partial vapor pressure of the drying air, no further drying takes place (Rizvi 2005) and the moisture content at this stage is called the equilibrium moisture content (Srikiatden and Roberts 2007).

Drying behavior of some biological and most food materials experience this second falling rate period. When vapor pressure of the material becomes equal to the partial vapor pressure of the drying air, no further drying takes place (Rizvi 2005) and the moisture content at this stage is called the *EMC* (Figure 7.6a).

During the second falling rate period, the rate at which moisture may move through the solid as a result of concentration gradients between the deeper parts and the surface is the controlling step.

REFERENCES

Anderson, J.O. 2014. Energy and resource efficiency in convective drying systems in the process industry. Doctoral thesis, Luleå University of Technology, Luleå, Sweden.

Anderson, J.O., L. Westerlund. 2014. Improved energy efficiency in sawmill drying system. *Applied Energy* 113:891–901.

Avramidis, S., P. Englezos, T. Papathanasiou. 1992. Dynamic non isothermal transport in hygroscopic porous media: Moisture diffusion in wood. *AIChE Journal* 38(8):1279–1287.

Avramidis, S., S.G. Hatzikiriakos, J.F. Siau. 1994. An irreversible thermodynamics model for unsteady-state non isothermal moisture diffusion in wood. *Wood Science and Technology* 28:349–358.

Babbitt, J.D. 1950. On the differential equations of diffusions. *Canadian Journal of Research* 28A:449–474.

Bachmat, Y., J. Bear. 1986. Macroscopic modelling of transport phenomena in porous media: The continuum approach. *Transport Porous Media* 1:213–240.

Banks, W.B. 1968. A technique for measuring the lateral permeability of wood. *Journal of the Institute of Wood Science* 4(2):35–41.

Bear, J. 1972. *Dynamics of Fluids in Porous Media*, American Elsevier, New York.

Beard, J.N., H.N. Rosen, B.A. Adesanya. 1982. Heat transfer during the drying of lumber. In *Proceeding of the 3rd International Drying Symposium,* University of Birmingham, England, Sept. 13–16, Vol. 1, pp. 110–122.

Beard, J.N., H.N. Rosen, B.A. Adesanya. 1985. Temperature distribution in lumber during impingement drying. *Wood Science and Technology* 19:277–286.

Berger, D., D.C.T. Pei. 1973. Drying of hygroscopic capillary porous solids – A theoretical approach. *International Journal of Heat and Mass Transfer* 16:293–302.

Berit Time. 1998. Hygroscopic moisture transport in wood. Doctoral Thesis, Degree of Doctor Engineer, Department of Building and Construction Engineering, Norwegian University of Science and Technology, Trondheim, Norway.

Bian, Z. 2001. Airflow and wood drying models for wood kilns. Master Thesis, Degree of Master of Applied Science in Faculty of Graduate Studies, Department of Mechanical Engineering, the University of British Columbia, Vancouver, Canada.

Biggerstaff, T. 1965. Drying diffusion coefficients in wood as affected by temperature. *Forest Products Journal* 15(3):127–133.

Blakemore, P., T.A.G. Langrish. 2007. Effect of mean moisture content on the steam reconditioning of collapsed *Eucalyptus nitens. Wood Science and Technology* 41(1):87–98.

Blunt, M.J. 2001. Flow in porous media – Pore-network models and multiphase flow. *Current Opinion in Colloid & Interface Science* 6(3):197–207.

Bonneau, P., I.R. Puiggali. 1993 Influence of heartwood-sapwood proportion on the drying kinetics of a board. *Wood Science and Technology* 28:67–85.

Booker, R.E. 1996. New theory for liquid flow in wood. In *Proceedings of the 5th IUFRO Wood Drying Conference*, Québec City, Canada, August 13–17.

Bories, S. 1988. Recent advances in modeling of coupled heat and mass transfer in capillary porous bodies. In *Proceedings of the 6th International Drying Symposium*, Versailles, France, September 5–8, pp. 47–61.

Bowen, R.M. 1980. Incompressible porous media models by use of the theory of mixtures. *International Journal of Engineering Science* 18(9):1129–1148.

Bramhall, G. 1976a. Fick's law and bound-water diffusion. *Wood Science* 8(3):153–161.

Bramhall, G. 1976b. Semi-empirical method to calculate kiln-schedule modifications from some lumber species. *Wood Science* 8(4):213–222.

Bramhall, G. 1979a. Sorption diffusion in wood. *Wood Science* 12(1):3–13.

Bramhall, G. 1979b. Mathematical model for lumber drying I. Principles involved II. The model. *Wood Science* 12(1):14–31.

Bramhall, G. 1995. Diffusion and the drying of wood. *Wood Science Technology* 29:209–215.

Briun, S., K.A.M. Luyben. 1980. Drying of food materials: A review of recent developments. In *Advances in Drying*, edited by A.S. Mujumdar, Hemisphere Publishing Corp., Washington, DC, pp. 155–215.

Brown, G. Darcy's law basics and more. (https://bae.okstate.edu/faculty-sites/Darcy/LaLoi/Basics.htm, accessed March 17, 2016).

Burr, H.K., A.J. Stamm. 1956. Diffusion in wood. Report 1674. Forest Products Laboratory, Madison, WI.

Carmeliet, J., F. Descamps, G. Houvenaghel. 1999. A multi scale network model for simulating moisture transfer properties of porous media. *Transport in Porous Media* 35:67–88.

Carrier, W.H. 1921. The theory of atmospheric evaporation-with special reference to compartment dryers. *Journal of Industrial & Engineering Chemistry* 13:432–458.

Ceaglske, N.H., O.A. Hougen. 1937. The drying or granular solids. *Transactions of American Institute of Chemical Engineering* 33:283–312.

Cech, M.J., F. Pfaff. 2000. *Operator Wood Drier Handbook for East of Canada*, edited by Forintek Corp., Canada's Eastern Forester Products Laboratory, Québec city, Canada.

Chen, G., R.B. Keey, J.C.F. Walker. 1996. Moisture content profiles in sapwood boards on drying. In *Drying '96*, edited by C. Strumillo & Z. Pakowski. Hemisphere, Washington, DC, pp. 679–687.

Chen, P., D.C.T. Pei. 1989. A mathematical model of drying processes. *International Journal of Heat and Mass Transfer* 32(2):297–310.

Chilton, T.H., A.P. Colburn. 1934. Mass transfer (absorption) coefficients. *Industrial and Engineering Chemistry* 26(11):1183–1187.

Chirife, J. 1983. Fundamentals of the drying mechanism during air dehydration of foods. In *Advances in Drying*, Vol. 2 edited by A.S. Mujumdar. Hemisphere Publishing Corp., Washington, DC, pp. 73–102.

Choong, E.T. 1963. Movement of moisture through softwood in the hygroscopic range. *Forest Products Journal* 13:489–498.

Choong, E.T. 1965. Diffusion coefficients of softwood by steady-state and theoretical methods. *Forest Products Journal* 15:21–27.

Choong, E.T., F.O. Tesoro. 1989. Relationship of capillary pressure and water. *Wood Science and Technology* 23(2):139–150.

Choong, E.T., P.J. Fogg. 1968. Moisture movement in six wood species. *Forest Products Journal* 18(5):66–70.

Choong, E.T., Y. Chen, J.D. Mamit, J. Ilic, W.R. Smith. 1994. Moisture transport properties in hardwoods. In *Proceedings of the 4th IUFRO International Wood Drying Conference*, Rotorua, New Zealand, August 9–13, pp. 87–94.

Cloutier, A., Y. Fortin. 1991. Moisture content – Water potential relationship of wood from saturated to dry conditions. *Wood Science Technology* 25(4):263–280.

Cloutier, A., Y. Fortin. 1993. A model of moisture movement in wood based/water potential and the determination of the effective water conductivity. *Wood Science Technology* 27(2):95–114.

Cloutier, A., Y. Fortin. 1994. Wood drying modelling based on water potential a parametric study. In *Proceedings of the 4th IUFRO International Wood Drying Conference*, Rotorua, New Zealand, pp. 47–54.

Cloutier, A., Y. Fortin, G. Dhatt. 1992. A wood drying finite element model based on the water potential concept. *Drying Technology* 10(5):1151–1181.

Collignan, A., J.P. Nadeau, J.R. Puiggali. 1993. Description and analysis of timber drying kinetics. *Drying Technology* 11(3):489–506.

Comstock, G.L. 1963. Moisture diffusion coefficients in wood as calculated from adsorption, desorption and steady state data. *Forest Products Journal* 13:96–103.

Comstock, G.L. 1967. Longitudinal permeability of wood to gases and non-swelling liquids. *Forest Products Journal* 17:41–46.

Crank, I. 1975. *The Mathematics of Diffusion*, 2nd edition. Oxford University Press, London.

Crausse, P., G. Bacon, S. Bories. 1981. Étude fondamentale des transferts couplés chaleur-masse en milieu poreux. *International Journal of Heat and Mass Transfer* 24:991–1004.

Cunningham, M.J. 1992. Effective penetration depth and effective resistance in moisture transfer. *Building and Environment* 27(3):379–386.

De Vries, D.A. 1958. Simultaneous transfer of heat and moisture in porous media. *Transactions – American Geophysical Union* 39:909–916.

Defraeye, T., B. Blocken, J. Carmeliet. 2011. Convective heat transfer coefficients for exterior building surfaces: Existing correlations and CFD modelling. *Energy Correlations and Management* 52(1):512–522.

Defraeye, T., B. Blocken, D. Derome, B. Nicolai, J. Carmeliet. 2012a. Convective heat and mass transfer modelling at air-porous material interfaces: Overview of existing methods and relevance. *Chemical Engineering Science* 74:49–58.

Defraeye, T., B. Blocken, J. Carmeliet. 2012b. Analysis of convective heat and mass transfer coefficients for convective drying of a porous flat plate by conjugate modelling. *International Journal of Heat and Mass Transfer* 55(1–3):112–124.

Doe, P., J.R. Brooke, T.C. Innes, A.R. Olivier. 1996. Optimal lumber seasoning using acoustic emission sensing and real time strain modeling. In *Proceedings of the 5th IUFRO International Wood Drying Conference*, Québec, Canada, August 13–17, pp. 209–220.

Eckert, E.R.G., E. Pfender. 1978. Heat and mass transfer in porous media with phase change. In *Proceedings of 6th International Heat Transfer Conference*, Toronto, Ontario, Canada, August 7–11, vol. 6, pp. 1–12.

Eckert, E.R.G., M. Faghri. 1980. A general analysis of moisture migration caused by the temperature differences in an unsaturated porous medium. *International Journal of Heat Mass Transfer* 23:1613–1623.

Erickson, H.D. 1954. Mechanisms of moisture movement in woods. In *Proceedings of the 6th Annual Meeting of the Western Dry Kiln Clubs at Eureka, CA*, May 14.

Erriguible, A., P. Bernada, F. Couture, M. Roques. 2006. Simulation of convective drying of a porous medium with boundary conditions provided by CFD. *Chemical Engineering Research and Design* 84(2):113–123.

Esping, B. 1992. *Wood Drying: Basics in Drying*. Trätek, Stockholm (in Swedish).

Ferguson, W.J., I.W. Turner. 1994. A two-dimensional numerical simulation of the drying of pine at high temperatures. In *Proceedings of 4th IUFRO Wood Drying Conference*, Rotorua, New Zealand, August 9–13, pp. 415–422.

Fick, A.E. 1855. Poggendorff's Annalen der Physik (in German). *Annals of Physics* 94:59–61.

Frank Controls. 1986. Lumber drying theory (www.frankcontrols.com, accessed June 15, 2015).

Furuyama, Y., Y. Kangawa, K. Hayashi. 1994. Mechanism of free water movement in wood drying. In *Proceedings of 4th IUFRO International Wood Drying Conference*, Rotorua, New Zealand, August 9–13, pp. 95–101.

Fyhr, C., A. Rasmuson. 1996. Mathematical model of steam drying of wood chips and other hygroscopic porous media. *AIChE Journal* 42(9):2491–2502.

Geankoplis, C.J. 1993. *Transport Processes and Unit Operations*. 3rd edition, Prentice Hall, Englewood Cliffs, NJ.

Glasstone, S., K.I. Kaidler, H. Eyring. 1941. *Theory of Rate Processes*. McGraw-Hill, New York, 611 pp.

Görling, P. 1958. Physical phenomena during the drying of foodstuffs. In *Fundamental Aspects of the Dehydration of Foodstuffs*, Society of Chemical Industry, London, pp. 42–53.

Grace, C. 1996. Drying characteristics of Nothofagus truncata heartwood, ME Thesis, University of Canterbury, Christchurch, New Zealand.

Greiner, W., L. Neise, H. Stöcker. 1995. *Thermodynamics and Statistical Mechanics*. Springer-Verlag, New York, Berlin, Heidelberg, p. 101.

Griskey, R.G. 2006. *Transport Phenomena and Unit Operations*. Wiley & Sons, Hoboken, pp. 228–248.

Hallström, B. 1990. Mass transport of water in foods – A consideration of the engineering aspects. *Journal of Food Engineering* 12:45–52.

Haslett, A.N. 1998. *Drying Radiate Pine in New Zealand*. FRI Bulletin 206 NZFRI, Rotonda, New Zealand.

Helmer, W.A., H.N. Rosen, P.Y.S. Chen, S.W. Wang. 1980. A theoretical model for solar dehumidification drying of wood. In *Drying '80*, edited by A.S. Mujumdar, Hemisphere, Washington, DC, pp. 21–28.

Henry, P.S.H. 1948. Diffusion of moisture and heat through textiles. *Discussions of the Faraday Society* 3:243–257.

Hougen, O.A., H.J. McCauley, W.R. Marshal Jr. 1940. Limitations of diffusion equations. *Transactions of the American Institute of Chemical Engineering* 36:183–206.

Hunter, A.J. 1995. Equilibrium moisture content and the movement of water through wood above fiber saturation. *Wood Science and Technology* 29(2):129–135.

Hunter, A.J., J.W. Sutherland. 1997. The evaporation of water from wood at high temperature. *Wood Science and Technology* 31:73–76.

Incropera, F.P., D.P. DeWitt. 2002. *Fundamentals of Heat and Mass Transfer*, 5th edition, John Willey & Sons, New York.

Jost, W. 1952. Fundamental laws of diffusion. In *Diffusion in Solids, Liquids, Gases*, edited by E. Hutchinson. Academic Press Inc., New York, pp. 1–82.

Kamke, F.A., L.J. Casey. 1988. Gas pressure and temperature in the mat during flakeboard manufacture. *Forest Products Journal* 38:41–43.

Kamke, F.A., M. Vanek. 1994. Comparison of wood drying models. In *Proceedings of the 4th IUFRO International Wood Drying Conference*, Rotorua, New Zealand, August 9–13, pp. 1–21.

Karathanos, V.T., N.K. Kanellopoulos, V.G. Belessiotis. 1996. Development of porous structure during air drying of agricultural plant products. *Journal of Food Engineering* 29:167–183.

Karel, M., D.B. Lund. 2003. *Physical Principles of Food Preservation*, 2nd edition. Marcel Dekker Inc., New York.

Kayihan, F. 1982. Simultaneous heat and mass transfer with local three phase equilibria in wood drying. In *Proceeding of the 3rd International Drying Symposium,* University of Birmingham, England, Sept. 13–16, Vol. 1, pp. 123–134.

Kayihan, F. 1985. The process dynamics and the stochastic behavior of batch lumber kilns. *Forest Products* 81(246):104–116.

Kayihan, F. 1993. Adaptive control of stochastic batch lumber kilns. *Computers and Chemical Engineering* 17:265–273.

Kawai, S., K. Nakato, T. Sadoh. 1978. Prediction of moisture distribution in wood during drying. *Mokuzai Gakkaishi* 24(8):520–525.

Kawai, S., K. Nakato, T. Sadoh. 1980. Moisture movement in wood in the hygroscopic range. *Memoirs of the College of Agriculture, Kyoto University*, No. 115: 1–32.

Keey, R.B. 1970. *Drying Principles and Practice*. Pergamon Press, Oxford.

Keey, R.B. 1978. *Introduction to Industrial Drying Operations*. Pergamon, Oxford, 376 pp.

Keey, R.B., J.J. Nijdam. 2002. Moisture movement on drying softwood boards and kiln design. *Drying Technology* 20(10):1955–1974.

Keey, R.B., T.A.G. Langrish, J.C.F. Walker. 2000. *Kiln-Drying of Lumber*, edited by T.E. Timell, Springer Series in Wood Science. Springer, New York (ISBN-13:978-3-642-64071-1).

King, C.J. 1968. Rates of moisture sorption and desorption in porous dried foodstuffs. *Food Technology* 22(502):165–171.

King, C.J. 1971. *Freeze Drying of Foods*. Butterworth, London.

Kowalski S.J. 2010. Control of mechanical processes in drying – Theory and experiment. *Chemical Engineering Science* 65(2):890–899.

Kowalski, S.J., A. Pawlowski. 2011. Intermittent drying of initially saturated porous materials. *Chemical Engineering Science* 66(9):1893–1905.

Langrish, T.A.G., J.C.F. Walker. 1993. Transport processes in wood. In: *Primary Wood Processing*, edited by J.C.F. Walker, Chapman and Hall, London, pp. 121–152.

Langrish, T.A.G., A.S. Brooke, C.L. Davis, H.E. Musch, G.W. Barton. 1997. An improved drying schedule for Australian ironbark timber optimisation and experimental validation. *Drying Technology* 15(1):47–70.

Langrish, T. A. G., N. Bohm. 1997. An experimental assessment of driving forces for drying in hardwoods. Wood Science and Technology 6 (https://www.springerprofessional.de/an-experimental-assessment-of-driving-forces-for-drying-in-hardw/11622450, Accessed January 6, 2018).

Lenormand, R., E. Touboul, C. Zarcone. 1988. Numerical models and experiments on immiscible displacements in porous media. *Journal of Fluid Mechanics* 189:165–218.

Lewis, W.K. 1921. The rate of drying of solid materials. *Industrial & Engineering Chemistry* 13:427–432.

Luikov, A.V. 1966. *Heat and Mass Transfer in Capillary-Porous Bodies*, 1st edition. Pergamon Press, New York.

Luikov, A.V. 1968. *Theory of Drying*, 2nd edition, Energiya Moskva, Moskow.

Luikov, A.V. 1975. Systems of differential equations of heat and mass transfer in capillary-porous bodies. *International Journal of Heat and Mass Transfer* 18:1–14.

Lowery, D.P. 1979. Vapor pressure generated in wood during drying. *Wood Science* 5:73–80.

Maneval, J., S. Whitaker. 1988. Effects of saturation heterogeneities on the interfacial mass transfer relation. In *Proceedings of the 6th International Drying Symposium*, Versailles, France, September 5–8, pp. 499–506.

Marinos-Kouris, D., Z.B. Maroulis. 2014. Transport properties for the drying of solids. In *Handbook of Industrial Drying*, 4th edition, edited by A.S. Mujumdar. CRC Press, Boca Raton, FL, pp. 113–159.

Marshall, W.R., Jr. 1958. Drying – Its status in chemical engineering in 1958. *Chemicial Engineering Progress* 55:213–227.

Masmoudi, W., M. Prat. 1991. Heat and mass transfer between a porous medium and a parallel external flow. Application to drying of capillary porous materials. *International Journal of Heat and Mass Transfer* 34(8):1975–1989.

McCready, P.W., W.L. McCabe. 1933. The adiabatic air drying of hygroscopic solids. *Transactions of the American Institute of the Chemistry Engineers* 29:131–160.

McCurdy, M.C. 2005. Efficient kiln drying of quality softwood timber. Doctoral Thesis, Degree of Doctor of Philosophy in Chemical and Process Engineering to the University of Canterbury, Christchurch, New Zealand.

McCurdy, M.C., R.B. Keey. 2002. The effect of growth-ring orientation on moisture movement in the high temperature drying of softwood boards. *Holz als Roh- und Werkstoff* 60.

Meroney, R.N. 1969. The state of moisture transport rate calculations in wood drying. *Wood and Fiber* 1(1):64–74.

Moren, T.J. 1989. Check formation during low temperature drying on Scots Pine: Theoretical consideration and some experimental results. In *Proceedings of the 2nd IUFRO International Wood Drying Symposium,* Seattle, Washington, DC, July 23–23, pp. 97–100.

Mori, S., H. Nakagwa, A. Tanimoto, M. Sakakibara. 1991. Heat and mass transfer with a boundary layer flow past a flat plate of finite thickness. *International Journal of Heat and Mass Transfer* 34:2899–2909.

Mujumdar, A.S., A.S. Menon. 1995. Drying of solids: Principles classification and selection of dryers. In *Handbook of Industrial Drying*, edited by A.S. Mujumdar, Marcel Dekker, New York, pp. 1–39.

Mujumdar, A.S. 2000. *Mujumdar's Practical Guide to Industrial Drying*, edited by S. Devahastin, Exergex Corp., Brossard, Québec, 187 pp.

Mujumdar, A.S. (Ed.). 2006. *Handbook of Industrial Drying*, 3rd edition. Taylor & Francis Group, Boca Raton, FL.

Mujumdar, A.S. 2007. Principles, classification, and selection of dryers. In *Handbook of Industrial Drying*, 3rd edition, edited by A.S. Mujumdar. Taylor & Francis Group, Boca Raton, FL.

Mujumdar, A.S., L.X. Huang. 2007. Global R&D needs in drying. *Drying Technology* 25:647–658.

Mujumdar, A.S. 2015. *Handbook of Industrial Drying*, 4th edition, edited by A.S. Mujumdar, CRC Press, Boca Raton, FL, 1348 p.Nasrallah, S.B., P. Perré. 1988. Detailed study of a model of heat and mass transfer during convective drying of porous media. *International Journal of Heat and Mass Transfer* 31(5):957–967.

Newton, I. 1701. Scale of degree of heat. *Philosophy Transactions of Royal Society, London* 22:824–829.

Nijdam, J.J. 1998. Reducing moisture-content variations in kiln-dried timber. PhD Thesis, University of Canterbury, Christchurch, New Zealand.

Nijdam, J.J., T.A.G. Langrish, R.B. Keey. 2000. A high-temperature drying model for softwood timber. *Chemical Engineering Science* 55:3585–3598.

Northway, R. 1982. Moisture profiles and wood temperature during very high temperature drying of *Pinus radiata* explain lack of degrade. In *Proceedings of the 2nd IUFRO Wood Drying Conference*, Seattle, July 23–28, pp. 24–28.

Onn, L.K. 2007. Study of convective drying using numerical analysis on local hardwood, University SAINS of Malaysia (https://core.ac.uk/download/pdf/11957610.pdf, accessed August 2, 2016).

Pakowski, Z., A.S. Mujumdar. 1987. Basic process calculations and simulations in drying. In *Handbook of Industrial Drying*, 3rd edition, Chapter 3, edited by A.S. Mujumdar, Taylor & Francis Group, Boca Raton, FL, pp. 53–80.

Pang, S., R.B., Keey, T.A.G. Langrish. 1992. Modelling of temperature profiles within boards during the high-temperature drying of *Pinus radiata* timber. In *Drying '92*, Part A, edited by A.S. Mujumdar, Elsevier Science Publishers BV, Amsterdam, pp. 417–433.

Pang, S., R.B. Keey, J.C.F. Walker. 1994. Modelling of the high-temperature drying of mixed sap and heartwood boards. In *Proceedings of 4th International Wood Drying Conference*, Rotorua, New Zealand, August 9–13, pp. 430–439.

Pang, S., T.A.G. Langrish, R.B. Keey. 1994. Moisture movement in softwood timber at elevated temperatures. *Drying Technology* 12(8):1897–1914.

Pang, S. 1994. The high temperature drying of *Pinus radiata* in a batch kiln. PhD Thesis, University of Canterbury, Christchurch, New Zealand.

Pang, S., A.N. Haslett. 1995. The application of mathematical models to the commercial high-temperature drying of softwood lumber. *Drying Technology* 13(8&9):1635–1674.

Pang, S. 1996a. Moisture content gradient in softwood board during drying: simulation from a 2-D model and measurement. *Wood Science and Technology* 30:165–178.

Pang, S. 1996b. Development and validation of a kiln-wide model for drying of softwood lumber. In *Proceedings of the 5th IUFRO International Wood Drying Conference*, Quebec City, Canada, pp. 103–110.

Pang, S. 1996c. External heat and mass transfer coefficients for kiln drying of timber. *Drying Technology* 14(3&4):859–871.

Perré, P. 1987. The convective drying of resinous woods selection validation and use of a model. PhD Thesis, Paris University, Paris, France.

Perré, P., J.P. Fohr, G. Arnaud. 1988. A model of drying applied to softwood: The effect of gaseous pressure above the boiling point. In *Proceedings of the 6th International Drying Symposium IDS'88*, Versailles, September 5–8, pp. 279–286.

Perré, P., M. Martin. 1994. Drying at high temperature of heartwood and sapwood: Theory, experiment and practical consequence on kiln control. *Drying Technology* 12(8):1915–1941.

Perré, P., C. Moyne. 1991a. Process related to drying: Part I, Theoretical model. *Drying Technology* 9(5):1135–1152.

Perré, P., C. Moyne. 1991b. Process related to drying: Part II, Use of the same model to solve transfers both in saturated and unsaturated porous media. *Drying Technology* 9(5):1153–1179.

Perré, P., M. Moser, M. Martin. 1993. Advances in transport phenomena during convective drying with superheated steam and moist air. *Journal of Heat and Mass Transfer* 36:2725–2746.

Perré, P. 1996. The numerical modelling of physical and mechanical phenomena involved in wood drying: An excellent tool for assisting with the study of new processes. In *Proceedings of the 5th IUFRO International Wood Drying Conference*, edited by A. Cloutier, Y. Fortin & R. Gosselin. August 13–17. Quebec City, Canada, August 13–17, pp. 11–38.

Perré, P., I.W. Turner. 1996. The use of macroscopic equations to simulate heat and mass transfer in porous media. In *Mathematical Modelling and Numerical Techniques in Drying Technology*, edited by I.W. Turner & A.S. Mujumdar. Marcel Dekker, New York, pp. 83–156.

Perré, P., R. Keey. 2015. Drying of wood: Principles and practice. In *Handbook of Industrial Drying*, 3rd edition, edited by A.J. Mujumdar, Dekker, New York, pp. 821–877.

Perry, R.H., D.W. Green, J.O. Maloney. 1984. *Peny's Chemical Engineers' Handbook*. 6th edition, McGraw-Hill, London, pp. 4–14.

Philip, J.R., D.A. De Vries. 1957. Moisture movement in porous materials under temperature gradients. *Transactions of American Geophysical Union* 38(2):222–232.

Plumb, O.A., G.A. Spolek, B.A. Olmstead. 1985. Heat and mass transfer in wood during drying. *International Journal of Heat and Mass Transfer* 28(9):1669–1678.

Pohlhausen, E. 1921. The heat transfer between solid bodies and fluids with small friction and heat conduction. *Angewandte Mathematik und Mechanik* 1:115–121.

Pordage, L.J., T.A.G. Langrish. 2007. Stack-wide behavior for hardwood drying. *Drying Technology* 25:1779–1789.

Prat, M. 2002. Recent advances in pore-scale models for drying of porous media. *Chemical Engineering Journal* 86(1–2):153–164.

Puiggali, J.R., M. Quintard. 1992. Properties and simplifying assumptions for classical drying models. In *Advances in Drying*, Vol. 5, edited by A.S. Mujumdar. Hemisphere, Washington, DC, pp. 109–143.

Putranto, A., X.D. Chen, S. Devahastin, Z. Xiao, P.A. Web. 2001. Application of the reaction engineering approach (REA) for modeling intermittent drying under time-varying humidity and temperature. *Chemical Engineering Science* 66(10):2149–2156.

Reynolds, O. 1895. On the dynamical theory of incompressible viscous fluids and the determination of the criterion. *Philosophical Transactions of the Royal Society of London A* 186:123–164 (doi:10.1098/rsta.1895.0004).

Rizvi, S.S.H. 2005. Thermodynamic properties of foods in dehydration. In *Engineering Properties of Foods*, 3rd edition, edited by M.A. Rao, S.S.H. Rizvi & A.K. Datta, Marcel Dekker Inc., New York, pp. 239–326.

Rosen, H.N. 1986. Recent advances in the drying of solid wood. In *Advances in Drying*, Vol. 4, edited by A.S. Mujumdar, pp. 99–147.

Rosen, H.N. 1987. Recent advances in the drying of solid wood. In *Advances in Drying*, Vol. 4, edited by A.S. Mujumdar, Hemisphere, Washington, DC, pp. 99–146.

Rosselo, C., J. Canellas, S. Simal, A. Berna. 1992. Simple mathematical model to predict the drying rates of potatoes. *Journal of Agricultural and Food Chemistry* 40:2374–2378.

Rossello, C., S. Simal, N. Sanjuan, A. Mulet. 1997. Non-isotropic mass transfer model for green bean drying. *Journal of Agricultural and Food Chemistry* 45:337–342.

Salin, J.G. 1992. Numerical prediction of checking during timber drying and a new mechano-sorptive CREEP model. *Holz als Roh- und Werkstoff* 50:195–200.

Sastry, S.K., R.B. Beelman, J.J. Speroni. 1985. A three-dimensional finite element model for thermally induced changes in foods: Application to degradation of *Agaritine* in canned mushrooms. *Journal of Food Science* 50: 1293–1299, 1326.

Sereno, A.M., G.L. Medeiros. 1990. Simplified model for the prediction of drying rates for foods. *Journal of Food Engineering* 12:1–11.

Siau, J.F. 1980. Non-isothermal moisture movement in wood. *Wood Science* 13(1):11–18.

Siau, J.F. 1983a. Chemical potential as a driving force for non-isothermal movement in wood. *Wood Science and Technology* 17:101–105.

Siau, J.F. 1983b. A proposed theory for non-isothermal unsteady-state transport of moisture in wood. Preliminary Communications. *Wood Science and Technology* 17:75–77.

Saravacos, G.D., Z.B. Maroulis. 2001. *Transport Properties of Foods*. Marcel Dekker, New York.

Skaar, C. 1958. Moisture movement in beech below the fiber-saturation point. *Forest Products Journal* 8:352–357.

Sherwood, T.K. 1929a. The drying of solids – I. *Industrial & Engineering Chemistry* 21(1):12–16.

Sherwood, T.K. 1929b. The drying of solids – II. *Industral & Engineering Chemistry* 21(10):976–980.

Sherwood, T.K. 1931. Application of theoretical diffusion equations to the drying of solids. *Transactions of the American Institute of Chemical Engineers* 27:190–202.

Sherwood, T.K., R.L. Pigford, C.R. Wilke. 1975. *Mass Transfer*. McGraw-Hill, New York.

Shubin, G.S. 1990. *Drying and Heat Treatment of Wood*. Lesnaia Promyshlennost Moskva, Moskow, 154 pp.

Siau, F. 1971. *Flow in Wood*. Syracuse University Press, New York.

Siau, J.F. 1984. *Transport Processes in Wood*, Springer Series in Wood Science. Springer-Verlag, Berlin.

Siau, J.F. 1995. Wood: *Influence of Moisture on Physical Properties*. Virginia Polytechnic Institute and State University, VA.

Simpson, W.T. 1993. Determination and use of moisture diffusion coefficient to characterize drying of northern red oak (Quercus rubra). *Wood Science Technology* 27:409–420.

Skaar, C. 1972. *Water in Wood*. Syracuse University Press, Syracuse, NY.

Skaar, C. 1988. *Wood-Water Relations*. Springer, Berlin Heidelberg New York, 283 pp.

Skaar, C., J.F. Siau. 1981. Thermal diffusion of bound water in wood. *Wood Science Technology* 15:105–112.

Skaar, C., C. Prichananda, R.W. Davidson. 1970. Some aspects of moisture sorption dynamics in wood. *Wood Science* 2(3):179–185.

Spolek, G.A. 1981. A model of convective, diffusive, and capillary heat and mass transport in drying wood. PhD Thesis, Washington State University, Pullman, Washington.

Spolek, G.A., O.A. Plumb, 1981. Capillary pressure in softwood. *Wood Science and Technology* 15:189–199.

Srikiatden, J., J. Roberts. 2007. Moisture transfer in solid food materials: A review of mechanisms, model, and measurements. *International Journal of Food Properties* 10:739–777.

Stamm, A.J. 1959. Bound-water diffusion into wood in the fiber direction. *Forest Products Journal* 9:27–32.

Stamm, A.J. 1960. Bound-water diffusion into wood in across-the-fiber directions. *Forest Products Journal* 10:528–528.

Stamm, A.J. 1963. Permeability of wood to fluids. *Forest Products Journal* 13:503–507.

Stamm, A.J. 1964. *Wood and Cellulose Science*. Ronald Press Company, New York.

Stamm, A.J. 1967a. Movement of fluids in wood – Part 1: Flow of fluids in wood. *Wood Science and Technology* 1(2):122–141.

Stamm, A.J. 1967b. Movement of fluids in wood – Part II – Diffusion. *Wood Science and Technology* 1:205–230.

Stamm, A.J., R.M. Nelson Jr. 1961. Comparison between measured and theoretical drying diffusion coefficients for southern pine. *Forest Products Journal*, November, pp. 536–543.

Stanish, M.A. 1986. The roles of bound water chemical potential and gas phase diffusion in moisture transport through wood. *Wood Science and Technology* 20(1):53–70.

Stanish, M.A., G.S. Schajer, F. Kayihan. 1986. A mathematical model of drying for hygroscopic porous media. *AIChE Journal* 32(8):1301–1311.

Strumillo, C., T. Kudra. 1986. *Drying Principles, Applications and Design*. Gordon and Breach, New York, 448 pp.

Spolek, G.A., O.A. Plumb. 1981. Capillary pressure in softwood. *Wood Science Technology* 15(3):189–199.

Sutherland, J.W., I.W. Turner, R.L. Northway. 1994. A theoretical and experimental investigation of the convective drying of Australian *Pinus radiata* timber. *Drying Technology* 12(8):1815–1839.

Thomas, W.J. 2000. *Introduction to Transport Phenomena*. Prentice Hall, Upper Saddle River, NJ.

Treybal, R.E. 1968. *Mass Transfer Operations*, 2nd edition. McGraw-Hill, New York.

Tschernitz, J.L., W.T. Simpson. 1979a. Drying rate of northern Red oak lumber as an analytical function of temperature, relative humidity and thickness. *Wood Science* 114:202–208.

Turner, I.W. 1990. The modelling of combined microwave and convective drying of a wet porous material. PhD Thesis, University of Queensland, St. Lucia, Brisbane, Australia.

Turner, I.W., W.J. Fergusson. 1995a. An unstructured mesh cell-centred control volume method for simulating heat and mass transfer in porous media. Application to softwood drying. Part I The isotropic model. *Applied Mathematical Modelling* 19:654–667.

Turner, I.W., W.J. Fergusson. 1995b. An unstructured mesh cell-centred control volume method for simulating heat and mass transfer in porous media. Application to softwood drying. Part II The anisotropic model. *Applied Mathematical Modelling* 19:668–674.

Turner, I.W., P. Perré. 2004. Vacuum drying of wood with radiative heating: II. Comparison between theory and experiment. *AIChE Journal* 50:108–118.

Tuttle, F. 1925. A mathematical theory of the drying of wood. *Journal of the Franklin Institute* 200:609–614.

Van Arsdel, W.B. 1947. Approximate diffusion calculations for the falling-rate phase of drying. *Transactions of the American Institute Chemical Engineers* 43(1):13–24.

Van Arsdel, W.B., M.J. Copley, A.I. Morgan. 1973. *Food Dehydration*, Vol. 1. AVI Publishing Co., Westport, CT.

Waananen, K.M., J.B. Litchfield, M.R. Okos. 1993. Classification of drying models for porous solids. *Drying Technology* 11:1–40.

Wadso, L. 1993. Studies of water vapor transport and sorption in wood. PhD Dissertation. Report TVBM-IOI3, Building Materials, Lund University, Sweden.

Walker, J.C.F., B.G. Butterfield, T.A.G. Langrish, J.M. Harris, J.M. Uprichard. 1993. *Primary Wood Processing*. Chapman and Hall, London. 595 p.

Welty, J.R., C.E. Wicks, R.E. Wilson. 1969. *Fundamental of Momentum, Heat and Mass Transfer*, John Wiley & Sons, Inc., New York.

Welty, J.R., C.E. Wicks, R.E. Wilson. 1976. *Fundamentals of Momentum, Heat, and Mass Transfer*, 2nd edition, Wiley.

Wiberg, P., T.J. Moren. 1999. Moisture flux determination in wood during drying above fiber saturation point using CT-scanning and digital image processing. *Holz als Roh-und Werkstoff* 57: 137–144.

Whitaker, S. 1973. The transport equation for multi-phase system. *Chemical Engineering Science* 28:139–147.

Whitaker, S. 1977. Simultaneous heat, mass, and momentum transfer in porous media: A theory of drying. In *Advances in Heal Transfer*, vol. 13, edited by J.P. Hartnett, T.F. Irvine Jr., Academic Press, New York, pp. 119–203.

Whitaker, S. 1984. Moisture transport mechanisms during the drying of granular porous media. In *Proceedings of the 4th International Drying Symposium*, Kyoto, Japan, July 9–12, pp. 31–42.

Whitaker, S. 1998. Coupled transport in multiphase systems: A theory of drying. *Advances in Heat Transfer* 31:1–104.

Whitelaw, J.H. *Convective Heat Transfer*. (http://thermopedia.com/content/660/; DOI:10.1615/AtoZ.c.convective_heat_transfer, accessed May 18, 2016).

Wiley, A.T., E.T. Choong. 1975. An analysis of free-water flow during drying in softwoods. *Wood Science* 7(4):310–318.

Wu, Q.L. 1989. An investigation of some problems in drying of Tasmanian eucalypt timbers. Master Thesis, Degree of Master of Engineering Science, University of Tasmania, Hobart, Australia.

Younsi, R., D. Kocaefe, S. Poncsak, Y. Kocaefe, L. Gastonguay. 2008. CFD modeling and experimental validation of heat and mass transfer in wood poles subjected to high temperatures: A conjugate approach. *Heat and Mass Transfer* 44(12):1497–1509.

Zhang, Z. 1999. Mechanism and mathematical model of heat and mass transfer during convective drying of porous materials. *Heat Transfer-Asian Research* 28(5):337–351.

Zhang, Z., S. Yang, D. Liu. 1999. Mechanism and mathematical model of heat and mass transfer during convective drying of porous materials. *Heat Transfer Asian Research* 28(5):337–351.

8 Kiln-Wide Lumber Drying

8.1 INTRODUCTION

Industrial wood drying processes include operational (such as drying schedule, drying air dry- and wet-bulb temperatures, relative humidity, and velocity), geometrical (such as length of airflow path, stickers length, and number in row), and raw material aspects (such as species origin and density, lumber grade, and initial moisture content).

In industrial batch lumber kilns, thousands of boards are stacked in layers, separated by narrow strips of wood called displacement stickers, which allow air passage along both faces of the boards for drying (see Figure 8.1a). Their scope is to achieve the required final average moisture contents as uniform as possible (Cronin et al. 2002).

The sticker spacing provides pathways of rectangular cross section for the drying air to flow through the lumber stack (Langrish et al. 1993) (see Figure 8.1b). Even the wood stacks are perfectly homogenous and the airflow is totally uniform across their faces, the conditions change progressively through the parallel board layers and throughout the kiln-wide stack(s) in the airflow direction.

In practice, since the wood boards stacks are not of uniform length, and are normally shorter than the width of the stack, the boards are stacked in such a way that the adjacent boards are flush on either side of the stack. This results in possible configurations such as (Sun et al. 1996; Sun and Carrington 1999; Salin 1996) (i) central normal stack of in-line boards; (ii) side aligned stacks of boards, in which the rows of boards and the board-wide gaps are stacked in alignment; and (iii) side staggered stacks of boards, where the rows of boards and the board-wide gaps are stacked in a staggered way.

8.2 IDEALIZED AIRFLOW BETWEEN PARALLEL BOARD LAYERS

As shown in Figure 8.1b, because the horizontal spacing between stickers is much greater than the thickness of these wooden slats, the airflow over each row of boards might be expected to be similar to that between two infinitely wide plates. Even if it is not entirely valid, this statement is a starting point for the analysis of the convection process (Perré and Keey 2015a, 2015b).

Within the lumber stack(s), the main external variables, which control the airflow and heat and mass transfer, are the air dry- and wet-bulb temperatures, relative humidity, and flow rate (or velocity). Higher dry-bulb temperatures provide greater potential for heat supply with lower airflow rates, thus, increasing both rates of internal moisture movement toward the boards' surfaces and surface vaporization (Pang and Haslett 1995). In other words, the stacked lumber is subjected to a sequence of air dry- and wet-bulb temperatures, relative humidity, and flow conditions (air

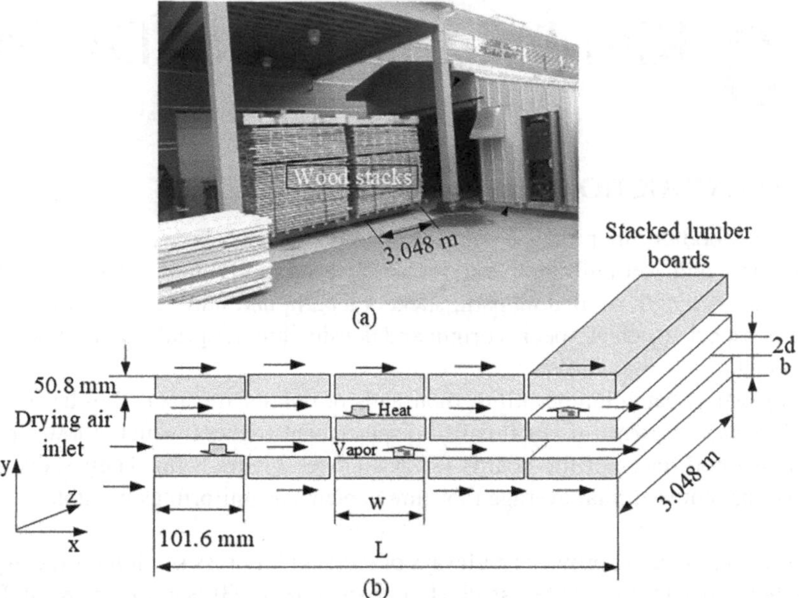

FIGURE 8.1 (a) View of lumber stacks entering a wood dryer (Reproduced with permission from FPInnovations Canada); (b) schematic representation of drying airflow between a series of parallel boards similar to those of batch wood kilns. b, board thickness; d, half of stickers' spacing; L, total (average) width of lumber stacks (notes: schematic not to scale; boards' dimensions given as examples).

velocity, mass rate, etc.). The heating medium (hot, dry air) enters the lumber stack (Figure 8.1b) at one side, flows between parallel board plate layers of average length L, supplies heat to the boards (thus, cools), gains moisture, and, finally, leaves at the other side carrying away the vaporized moisture (water vapor) (Pang 1996a, 1996b; Keey et al. 2000; Pordage and Langrish 2007).

During wood drying, velocity, thermal, and moisture (concentration) boundary layers are formed in the air adjacent to the lumber boards where gradients of velocity, temperature, and moisture concentration are set up. These gradients result in heat and moisture convection between the air and boards' surfaces that depend upon the pattern of airflow (laminar, transient, or turbulent), velocity of the air in the free-air stream (i.e., outside the boundary layer), and temperature and moisture concentration differences across the boundary layers. The flow regimes depend on whether the boards' surfaces are aerodynamically smooth, in transition from smooth to rough, or wholly rough (Hines and Maddox 1985).

Theoretically, at a fixed value of the coordinate y, any two parallel layers of lumber boards can be assimilated with two rectangular, parallel flat plates spaced at $\pm d$ ($2d \ll L$) above and below the centerline of axis symmetry (Figure 8.2).

In kiln lumber drying, the flow of the drying air occurs with negligible gravity, constant density, and viscosity at relatively low Reynolds numbers (which means that viscous forces are much more important than the inertial forces) and low flow

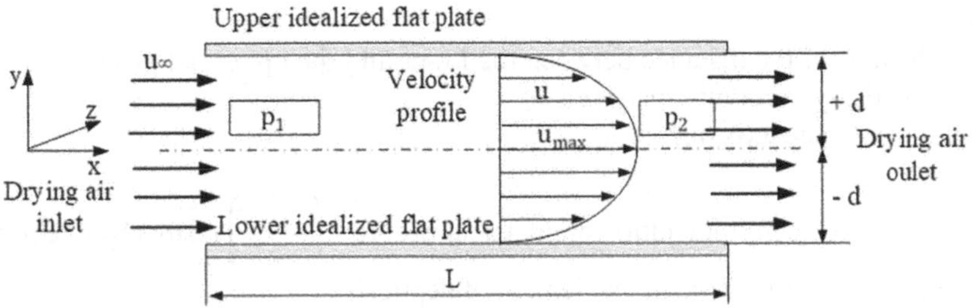

FIGURE 8.2 Idealized flow of the drying air between two parallel layers of lumber boards. p, air pressure; u, air velocity in *x* direction.

rates caused by fans that create a pressure difference between the stack inlet and outlet surfaces.

Generally, the average width of boards is very large (e.g., 3 m) in the z direction when compared to their separation in the y direction (e.g., 50 mm). Therefore, side-wall effects are minor, and the airflow may be considered one dimensional through the stack (Arnaud et al. 1991). In this case, there is only one nonzero velocity component (u) that is constant in the z direction $\left(\frac{\partial u}{\partial z} = 0\right)$, independent of the distance from the stack inlet, and zero in contact with both board layers. Based on these assumptions, it follows that the velocity component is a function of y only [$u = u_x(y)$] (Figure 8.2) (Incropera and DeWitt 2002).

Also, since there is no variation of the pressure in the z direction $\left(\frac{\partial p}{\partial z} = 0\right)$ and $2d \ll L$, relatively small pressure variation in the y direction, as well as negligible gravity, from the momentum balance equation results that the pressure declines linearly (i.e., the pressure gradient is uniform) from the inlet to the exit, according to the following relation (Incropera and DeWitt 2002):

$$-\frac{dp}{dx} = \frac{p_1 - p_2}{L} \tag{8.1}$$

The minus sign is used on the left-hand side of equation 8.1 since *dp*/*dx* is negative. Consequently, the pressure profile becomes

$$p = p_1 - \frac{x}{L}(p_1 - p_2) \tag{8.2}$$

where

p_1 and p_2 are the centerline pressures at the inlet and exit of the air, respectively (Pa)

The velocity profile results by integrating twice the simplified momentum equation $\left(\mu \frac{\partial^2 u}{\partial y^2} = \frac{\partial p}{\partial x}\right)$, as follows:

$$u = \frac{1}{2\mu}\left(-\frac{dp}{dx}\right)\left[d^2 - y^2\right] \tag{8.3}$$

where

d is the half of the distance between the lower and the upper flat plates (m)
μ is the air dynamic viscosity (Pa · s)
y is the ordinate axis (m)

The specific (i.e., per unit width) volumetric flow rate $\dot{V}\left(\frac{m^3}{s \cdot m}\right)$ results by integration of the velocity profile through an element of depth dy:

$$\dot{V} = \int_{-d}^{d} u dy = \int_{-d}^{d} \frac{1}{2\mu}\left(-\frac{dp}{dx}\right)\left[d^2 - y^2\right]dy = \frac{2d^3}{3\mu}\left(-\frac{dp}{dx}\right) \tag{8.4}$$

Since $\dot{V}$ is constant, it follows that dp/dx is also constant, that is, independent of distance x.

The maximum velocity (u_{max}) occurs at the centerline ($y = 0$), and the mean velocity $\bar{u}\left(\frac{m}{s}\right)$, at a given distance x from the air inlet (Incropera and DeWitt 2002):

$$\bar{u} = \frac{\dot{V}}{2d} = \frac{d^2}{3\mu}\left(-\frac{dp}{dx}\right) \tag{8.5}$$

8.3 ACTUAL AIRFLOW BETWEEN PARALLEL LUMBER BOARDS

The variations of lumber moisture content and air external conditions (temperature, relative and absolute humidity, and velocity) across the stack affect the final timber quality (Pang and Haslett 1995). As the air flows over the boards, the dry-bulb temperature decreases and air absolute humidity increases along the airflow direction.

The actual flow of air inside kilns (Figure 8.3) keeps the lumber at the required drying conditions of temperature and relative humidity, in addition of being a carrier for the vaporized moisture. Uniform airflow through the stack is required for

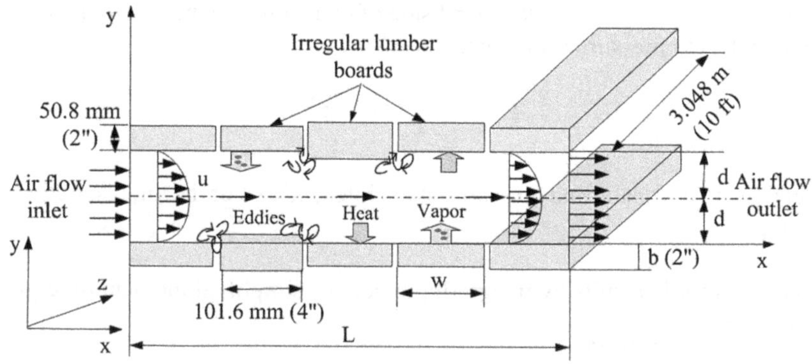

FIGURE 8.3 Schematic representation of airflow between two series of parallel lumber boards (1 feet = 1 ft = 0.3048 m) (1″ = 1 in = 25.4 mm) (note: schematic not to scale).

reaching uniformity in the dryness of the boards at the end of the wood drying schedules (Keey et al. 2000). The uniformity of drying depends on the quality of lumber boards' stacking that can be seen as several sharp-edged parallel plates. But, in reality, the boards at the air inlet face of the stack provide a set of blunt obstacles to the entering airflow.

It would be expected that the incident airflow through a perfectly stacked set of boards would distribute uniformly, but this situation will occur only if the inlet and outlet plenum spaces are infinitely wide (Keey and Nijdam 2002).

The stack height and spaces between boards influence the flow regimes, the boundary layer, and velocity distribution and, finally, determine the mass (volume) rate of drying air (Danckwerts and Anolick 1962; Sorensen 1969; Arnaud et al. 1991). Too thin stickers may lead to a complete fill of the flow passage with the boundary layer, which reduces the moisture-carrying capacity of the drying air and increases the drying time (Milota 1990).

In practice, the flow pattern is significantly altered near the leading and trailing edges. As the air enters the channels between boards, the airflow may detach (separate) near the leading edge due to the abrupt corner causing the flow to shear near the wall, and cause flow nonuniformities that can be reduced by periodic reversing the flow direction through the stack (Bian 2001).

Maldistribution of the air entering the stack from the plenum chamber also induces flow variations inside the spaces between the boards' rows (Keey et al. 2000). The drying air circulates at the boards' surfaces may curl around their leading edges (Dankwerts and Anolick 1962). At high air velocities (e.g., >8 m/s), vortex may appear just after the right-angled bend at the top of the inlet plenum, with the fluid being directed toward the kiln wall. Such vortexes impede the flow into the upper sticker spaces and, if well developed, would induce a reverse airflow (Kroll 1978). The influence of vortexes on the inlet air velocities to the stack appears to become more extensive as the relative width of the plenum chamber narrows (Keey et al. 2000).

It might be expected that strong eddying between the boards at the inlet face leads to greatly enhanced convective heat transfer coefficients from the surface where the boundary layer is thinned. The effect of the leading edge eddy that begins when the Reynolds number (based on the board thickness) is greater than 245 (Sorensen 1969) would diminish away from the stack(s) inlet (Keey et al. 2000).

The airflow through a stack of lumber begins as laminar flow, which is several times less effective in transferring heat and removing moisture than turbulent flow.

Airflow becomes turbulent after about 1.2 m of travel into the stack. This distance depends on the air velocity and whether the individual pieces of lumber have the same size, are smooth from piece to piece, and gaps between each lumber board exist. In the case of drying softwood dimension lumber in high-temperature kilns, the velocity is so high (>5 m/s) that airflow is turbulent soon after the air enters the stack(s). At higher lumber moisture contents (e.g., >30%, dry basis), lower drying times can be achieved at increased velocities of drying air, which provide turbulent airflow, or at lower relative humidity. Below 30% moisture content, the velocity has only a minor or insignificant role in controlling the drying rate and, thus, the drying time. Turbulent flow also appears in kilns where the airflow is periodically disturbed by changes in rotational direction of the fans and when local eddies near the gaps between adjacent boards are produced. The increase of airflow turbulence intensity is higher over wider gaps of 5 mm than at positions over the adjacent boards (Wu et al. 1995; Wu 1989).

Because the velocity boundary layer at the end of the stack grew to about 6.5 mm thick from the board's surface on each side of the flow passage, the sticker thickness during drying of lumber should be at least 13 mm high. A much smaller sticker height would lead to a complete fill of the flow passage with the boundary layer, which may reduce the moisture-carrying capacity of the air (Wu et al. 1989). However, the boards may not be of even thickness everywhere, resulting in surface irregularities in the sticker spaces. Relatively small irregularities in board thickness and board side-gaps can have a remarkable influence on the air motion.

Ideally, the lumber stack(s) must be built with uniform lengths of boards, so that they can be squared off. The sides of lumber packages should be regular from top to bottom, giving all sticker slots equal spaces for airflow. However, in industrial practice, the lumber boards have random lengths, normally placed flush at one side and staggered on the other side. In other words, the stacks are not aligned at each side, leaving gaps within the stack or staggered configurations (Bian 2001). The ends of the stack may not be squared off if random length boards are piled, giving variations in the resistance to the airflow besides providing ends which may cause eddy formation (Keey et al. 2000). Moreover, when alternate boards are missing, some boards will be at least one board-width away from each other in the airflow direction. In addition, the flow spaces between the lumber boards do not have continuous walls. They are built up from boards imperfectly butted-up to each other along their long sides. Thus, the air flows over a series of wall discontinuities at board-width intervals. Therefore, the airflow is disturbed by small gaps (in the range of 0.3–10 mm) between adjacent boards that may open up during kiln drying as the boards shrink and by unevenness due to variation in thickness of the boards (Langrish et al. 1992, 1993; Keey et al. 2000). Gaps between adjacent pieces of lumber may cause airflow variations and can reduce the drying time but also introduce nonuniformity of drying if they are irregular (Bian 2001). Around these gaps, a transient flow is observed accompanied by vortex shedding and airflow variations. With air velocities in the range of 2–6 m/s, this periodicity would take place with gaps as small as 0.3–1 mm, which is highly likely to be present in practice, even with careful stacking.

Even if the lumber boards in a stack are butted together sideways on stacking, inevitably very small board side gaps will develop during the kiln-seasoning process due to the shrinkage of the wood. Large gaps (>5 mm) enhance more local shear stresses than small gaps (≤2 mm) (Kho et al. 1989). Gaps as small as 1 mm might be sufficient to disrupt the flow, with circulation within the gaps themselves. The magnitude of the side gaps in the board influences both the drying rates and the development of drying stresses. Gaps between adjacent boards periodically generate eddies that move upward, deform and, finally, escape over the downstream boards. A circulating flow in the gap between two boards spills over periodically into the boundary layer of the upstream board. Therefore, the air motion over a gap is oscillatory, with air entering the gap, circulating there and, then, being expelled into the boundary layer of the downstream board.

Very small low-frequency oscillations with periods of 0.5–10 seconds may occur at normal kiln-air flow rates, being shorten with increasing air velocity and decreasing gap space (Lee 1990). A period of 2 seconds was estimated for the eddy motion at an air velocity of 6 m/s and a board side gap of 10 mm. At a velocity of 5 m/s and a side gap of 1 mm, a similar period of oscillation has been predicted (Lee 1990; Langrish et al. 1992, 1993). The oscillatory variations in the thickness of the boundary layer in the zone close to the leading edge of each board would lead to enhanced local heat transfer coefficients (Keey et al. 2000). In addition, gaps will reduce the volume of lumber which can be dried in the kiln. Therefore, proper methods, such as fan reversal, have to be introduced to prevent this kind of nonuniformity (Bian 2001). On the other hand, variations in lumber roughness do not result in major variations in drying.

Both the gaps and board irregularities might reduce the drying time for the lumber boards at normal thickness during the first stage of the drying when the free water is present. However, it will also take longer to dry the lumber boards at higher than normal thickness, thus lead to nonuniformity in drying.

It can be concluded that airflow irregularities and variations in surface friction in the airflow direction through horizontal spaces between lumber boards induced by the kiln arrangements and, locally, by the board stacking influence the drying characteristics of lumber kilns and, finally, result in uneven drying between board layers. These variations are superimposed on those which arise from a progressive diminution in drying potential in the direction of the airflow, as well as in lower mass transfer coefficient resulting from the developing moisture concentration boundary layer (Keey and Nijdam 2002).

On the other hand, there are significant differences between the idealized and actual rectangular, horizontal spaces between parallel lumber boards. This is because, in industrial kiln drying practice, most lumber boards are built up from adjacent boards of various thicknesses butted up to each other along their long sides, stacked and then dried (Keey and Nijdam 2002). Due to the existence of many irregularities (or discontinuities), the actual airflow patterns between the parallel lumber boards are more complex compared to idealized pattern previously described (Langrish 1994; Haslett 1998).

8.4 AIRFLOW DISTRIBUTION

The uniformity of kiln wood drying and the quality of dried wood are strongly influenced by the airflow rate and velocity. The distribution of the drying air within the kiln is very important because it influences the drying kinetics as well as the heat and mass transfer process between the air and the lumber stack(s). In order to lower the wood degrade, uniform drying must be achieved both within the kiln-stack and single lumber boards. When the drying air absorbs moisture from the lumber stack(s), it becomes gradually saturated and its potential to remove moisture from the boards placed downstream in the direction of the airflow diminishes. If the distribution of the drying air in the drying chamber is not properly designed, nonuniform drying occurs. Therefore, sections of the lumber stack(s) that first contact the drying air would be drier than other sections that interact with a more saturated moist air.

Inside industrial-scale lumber kilns, the airflow regimes are different in the roof area, in the plenum, and in the flow channels through the wood packages, respectively. Under uniform airflow through the stack, relatively wide plenum spaces may act as infinite reservoirs, with minimal velocity variations and, thus, limited pressure variations (Nijdam 1998; Denig et al. 2000). Some variations in distribution of drying air may be attributed to air leakages if baffles are either missing or not installed. Improper stacking of the wood may also create unwanted air spaces resulting in airflow and drying variations (Langrish et al. 1992; Langrish et al. 1993; Bian 2001).

If the arrangement of axial flow fans in the roof space and the placement of the heating coils ensure an almost uniform airflow distribution as it approaches the end of the roof space to turn into the plenum chamber, velocity variations across the length of the lumber stack will be less than variations down the stack. The variations of airflow must be minimized as much as possible because they strongly affect both the heat and mass transfer rate, although this effect decreases as the wood moisture content drops below fiber saturation point. Usually, the coefficient of variation for the air velocity across the outlet face of a lumber stack should not exceed 12% at velocities between 4.5 and 8 m/s for the drying of softwood species such as *Pinus radiata* under dry-bulb temperature greater than 90°C (Haslett 1998).

8.4.1 Dryer's Fans

Along with the energy consumption for heat production, the electrical energy consumption for air circulation inside the kilns is of major importance, even its capital investment and installation cost are relatively low. Dryer's fans, of which size and number depend on the species being dried and the kiln capacity, minimize both electrical power requirements and operating costs, but the overall labor and maintenance costs increase. In medium- and large-size batch lumber kilns (e.g., with capacities >190 m^3), up to 10 fans can be provided. The operating costs of dryer's fans can be significantly reduced by using speed controllers (frequency inverters). Although such devices may increase the initial capital, the electrical energy within the first year should pay for this additional cost. In wood dryers, it is important to remove moisture from the boundary layer as rapidly as possible to maintain the desired drying rate. This is achieved by controlling the circulation rate of the main air stream, usually through the use of multiple-speed fan motors.

There are two main types of internal fan systems designed for reversible circulation of air (Cech and Pfaff 2000): (i) lineshaft (or longitudinal) and (ii) cross-shaft.

In the line-shaft system, multiblade fans operating at low static pressures and moving large volumes of air at relative low velocities are arranged along a shaft running the full length of the kiln over a false ceiling. The motor is generally in a mechanical room at the end of the kiln, or outdoor. The fans are alternately right hand and left hand, and housed in a baffle system, which deflects the air stream outward and downward into the plenum chamber along the sides of the individual lumber stacks. In this manner, the air is directed through the piles perpendicular to the length of the boards. In the cross-shaft arrangement, three- or four-blade propeller fans are mounted on individual shafts (over a false ceiling or along one side of the kiln) at right angles to the length of the kiln. They are usually driven by individual motors, usually direct-coupled to the fans.

Generally, in conventional kilns, the fans are slightly oversized. During a normal kiln schedule, higher air velocities are supplied during the earlier stages of the schedule because of large quantities of moisture, and gradually reducing in the later stages of drying as the moisture content decreases.

8.4.2 Airflow Velocity

As air passes through the lumber stack it becomes cooler and more humid. Consequently, the airflow rate (or velocity) must be high enough to provide a rapid exchange of air, otherwise the desired drying rate would not be achieved. The role of air in drying is to remove moisture from the lumber surfaces and, at the same time, provide energy to the wood so that vaporization can take place.

The velocity of drying air depends on the lumber permeability and strongly influences the drying rate and drying time (both being reduced at high air velocities, similar to the effect of air relative humidity) (McCurdy and Pang 2007), as well as the economics of the lumber batch dryers (Greenhill 1936; Jain and Sharma 1981). This is because the drying rate of free water is mainly controlled by the external heat transfer coefficient (Langrish and Walker 2006).

In commercial wood drying kilns, air velocity (and, thus, air volumetric flow rate) can be accurately measured with hot wire anemometers, but these devices can be affected by the airflow perturbations. On the other hand, small, reasonably accurate, not expensive, and easy-to-use pressure measurement instruments placed between the sticker spaces could be the best for lumber drying.

Usually, the air free stream mean velocity is measured at the lumber stack(s) exit rather than at the entering side. Unless there are a lot of leaks, the air entering velocity is the same as that at the stack(s) outlet(s) (Bassett 1974; Wu et al. 1995). However, such a calculation usually leads to 15%–23% overestimation of the volumetric airflow rate.

In industrial drying practice, the effect of air velocity in kilns can be estimated by (i) measuring the final moisture content variation across the lumber stack(s); when variations exceed 1% moisture content (dry basis) from the edges to the middle, higher velocities to achieve more uniform drying, especially at moisture contents greater than 30%, should be considered to assure safe drying rates; at 60% moisture content (dry basis), the drying rate increases with increasing velocity, while at 20% moisture content (dry basis) the rate is constant for any velocity; when variations are less than 1%, reductions in velocity should be considered and (ii) measuring the temperatures and relative humidity, then calculating the equilibrium moisture content at each side of lumber load(s); when equilibrium moisture content difference is less than 2% (dry basis), reductions in velocity can be considered; when greater, increases of air velocity may be justified.

Other relative important aspects related to air velocity in kilns are (note: not listed in order of importance and/or relevance) as follows:

a. The airflow velocity (or volume rate) over the surface of lumber boards should be sufficiently high in order to match the rate at which the moisture moves from the interior to the surface of the lumber boards with the vaporization rate at their surfaces;
b. The airflow velocity must be high enough to produce rapid air change inside the lumber stack(s), avoid the formation of death zones, and provide uniform drying;
c. If the air velocity is too low (e.g., lower than 1.8 m/s), the air will rapidly transfer heat to the lumber boards and, at the same time, will pick up the vaporized moisture; therefore, the air will become saturated (100% relative humidity) soon after it enters the lumber stack(s); the edge of the stack(s) will dry very much, while the rest of the load will be very humid;
d. In addition, if airflow velocity is too low, longer times are required for the surfaces of lumber boards to reach moisture content equilibrium, and mold could develop (Bian 2001);
e. Consequently, higher than 1.8 m/s airflow velocities will provide more uniform drying from the inlet edge to the exit edge of the lumber stack(s); higher velocities are beneficial for fast drying species, but the cost of increasing air speed is higher;
f. In practice, to provide more uniform drying from side to side, the direction of airflow must be reversed;
g. At high lumber moisture contents (e.g., >40%, dry basis), the drying rate increases with increasing the air velocity; as the outer shell of the lumber becomes drier, for example, when the average moisture content of the lumber is between 40% and 20% (dry basis), the effect of increasing velocity diminishes; at low moisture contents (e.g., <20%, dry basis), the drying rate becomes practically constant for any air velocity;
h. Higher air velocities can be achieved by increasing the diameter and speed of the dryers' fans, and by adjusting the clearance between the walls and the kiln stack(s); however, electrical energy consumption of the dryers' fans is directly proportional to the third power of the air velocity, thus the operating costs will increase;
i. Higher air velocities are beneficial during the early stages of drying when wood is wet and the requirement for surface vaporization and moisture removal is high;
j. The critical (optimal) air velocity is temperature dependent, becoming higher as the kiln temperature is raised; therefore, maintaining optimum velocity of drying air is important for the total operation costs of wood dryers;
k. With certain species, as the air velocity increases, board surface checking and other drying defects may occur;
l. High air velocity increases the mass transfer coefficients;
m. The air velocity also influences the heat transfer from the heating finned-tube heat exchangers to the air; the higher the velocity through the coils, the

greater the heating rate. This effect would be most important when energy demands for drying are highest, for example, when the lumber is the wettest. As the lumber dries, the heat demand decreases; therefore, somewhat lower velocities across the coils at lower lumber moisture contents are generally not of great concern; poor velocity across the heating coils can be offset by increasing the fin size, or increasing the steam pressure;

n. High air velocity becomes wasteful from an energy standpoint when it has the potential to remove more moisture from the boundary layer than can be supplied by the diffusion of moisture to the wood surface; in other words, air velocity has a diminishing effect on the rate of vaporization as the moisture content of the wood decreases; therefore, air velocity must be adjusted according to air dry- and wet-bulb temperatures and relative humidity, especially during the early stages of drying;
o. As the fiber saturation point is approached, diffusion to the surface becomes the limiting factor in rate of moisture removal, and air velocity should be reduced accordingly;
p. Fan systems are designed to provide reasonably uniform air circulation, generally with or without variable speed motors;
q. At dry-bulb temperatures of 120°C and 140°C, respectively, during drying, for example, *P. radiata*, air velocities of 7 and 6 m/s, respectively, are recommended, instead 4.5–6 m/s under conditions of conventional temperature drying at a dry-bulb temperature of maximum 90°C (Haslett 1998).
r. With the slow-to-dry impermeable species, kiln air velocities as low as 1.5–2 m/s could be sufficient;
s. The choice of air velocity depends on the rate of drying, which in turn depends on the timber permeability and the drying schedule; in low- and medium-temperature drying (below 100°C) the velocity is normally 2–5 m/s, while high-temperature drying at 120°C requires an air velocity of 5–7 m/s, and drying at 140°C needs a velocity of 7–9 m/s (Pang 1996a, 1996b);
t. At increased air velocity, the lumber color changes are reduced (Keey et al. 2000).

The airflow rate (and velocity) also affects the kilns' fans electrical consumption. As velocity increases when fan speeds are increased, electrical consumption increases. Likewise, when speeds are lowered, electrical consumption decreases.

The cost of electricity for running the kilns' fans can be approximately 14%–21% of the total energy costs. Therefore, important cost savings could be achieved by reducing air velocity during some stages of drying. For example, by reducing air velocity from 4 to 2 m/s in the last stage of kiln drying, electricity consumption could be reduced by approximately 30% (Ananias et al. 1992). For high-temperature drying, on the other hand, at a constant air velocity of 7 m/s, the fan energy consumption would be optimum. Little change in cost is achieved for velocities going from 3 to 5 m/s (McCurdy and Pang 2007).

As can be seen in Figure 8.4a, the total kiln operating costs increase as the air velocity increases. This is mainly due to the increased electricity requirements from the fans' motor(s). The relationship between air velocity and fan energy consumption

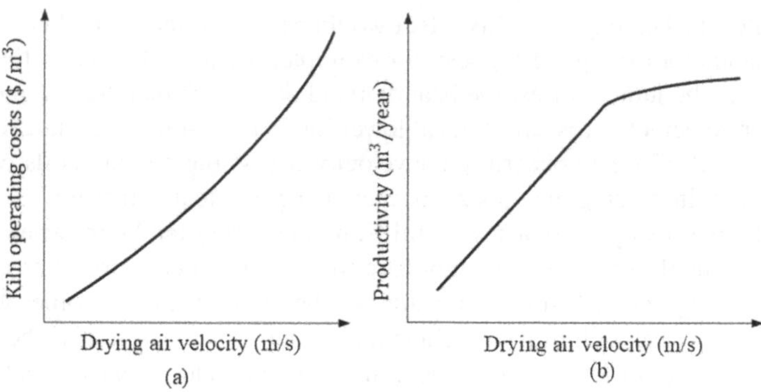

FIGURE 8.4 Kiln operating costs (a) and productivity (b) as functions of drying air velocity.

is nonlinear. The energy consumption increases with the cube of the air speed, thus electricity costs increase rapidly as the air velocity is increased. Eventually, this extra cost outweighs the additional economic benefits in added kiln capacity, and the cost of drying per unit load of lumber rise (Keey et al. 2000).

On the other hand, the kiln productivity increases linearly with kiln velocity as long as the external convection controls the drying rate (Figure 8.4b). At high kiln velocities (e.g., 10 m/s), there would be little gain in productivity by increasing the air velocity further unless very high temperatures are employed (Stoecker 1998).

For most softwood and some easy-to-dry hardwoods, velocities are often over 5 m/s to assure uniform drying. A decrease in air velocity of 20% (all else being kept equal) may reduce energy consumption by approximately 50%. At low moisture contents, the air velocities can be decreased, without a decrease in quality or an increase in drying time.

In conventional wood dryers, heated air is circulated over the lumber stack(s), and the moisture from the boards' surfaces vaporizes, raising the relative and absolute humidity of the drying air. When the absolute humidity of drying air exceeds the level specified by the drying schedule, the warm, moist air is vented to the outside, and cool, drier air is brought in (Keey et al. 2000). Vents thus cause venting from the inside of the kiln to the outside and from the outside inward (note: theoretically heat pump-assisted dryers would not have vents, so this effect would not be present). The kiln fans are generally used to force air through the vents when they are open. The higher the kiln velocity, the higher the venting rate. Variations in velocity will not affect the net amount of air vented but result in changes in the rate of venting. If the vents are properly operating, they will open whenever the conditions in the kiln are too humid and will close when the humidity is at or below the desired level. To reheat the outside fresh air, up to 80% of the total energy to dry lumber may be required.

The venting ratio, that is, the proportion of the vented air to the stream passing through the lumber stack, is the same regardless of airflow direction and can be determined as follows (McCurdy and Pang 2007):

$$\upsilon = \frac{\omega_{air,\ out-stack} - \omega_{air,\ in-stack}}{\omega_{air,\ in-stack} - \omega_{air,\ ambient}} \tag{8.6}$$

where

$\omega_{air,\ out-stack}$ is the air absolute humidity at the lumber stack outlet $\left(\frac{kg_{water}}{kg_{dry\ air}}\right)$

$\omega_{air,\ in-stack}$ is the air absolute humidity at the lumber stack inlet $\left(\frac{kg_{water}}{kg_{dry\ air}}\right)$

$\omega_{air,\ ambient}$ is the ambient air absolute humidity $\left(\frac{kg_{water}}{kg_{dry\ air}}\right)$

8.4.3 Pressure Drops

The geometry of roof and plenum spaces, and lumber stacking (traditionally, with sticker thickness between 1/2″ and 11/4″) (1″ = 25.4 mm) influence the airflow distribution. The airflow distribution is additionally controlled by adding baffles and door strips to minimize the air leakages, and enforcing lumber dimensional stability, as well as by regulating the fans' speeds and changing their reversal schedules (Bian 2001).

The pressure drop through a kiln stack can be expressed as follows (Keey et al. 2000):

$$\Delta p = K\left(\frac{\rho \bar{u}^2}{2}\right) \tag{8.7}$$

where

Δp is the air pressure drop (Pa)
K is the kinetic loss coefficient determined theoretically or from experiments; for example, for air flowing at 10 m/s, a unit velocity head (or $K = 1$) corresponds to a pressure loss of about 50 Pa (Kröll 1978)
ρ is the air density (kg/m^3)
$\bar{u}$ is the average air velocity (m/s)

Theoretically, the kinetic loss coefficient can be calculated as follows (Keey et al. 2000):

$$K = f \times \frac{L}{D_{ch}} \tag{8.8}$$

where

f is the friction factor that depends on the roughness of the surface being swept by the air movement and is a function of the Reynolds number based on the characteristic cross-dimension $\left(\text{Re} = \frac{v \times D_{ch}}{\nu}\right)$
L is the stack's width (m)
D_{ch} is the characteristic cross dimension of the sticker spacing, defined as the ratio of the cross-sectional area of the sticker space to its periphery length (m)

The friction factor f can be calculated using Blasius's equation (1913) derived for flows in smooth pipes under turbulent conditions:

$$\frac{f}{2} = 0.0396 \times Re^{-0.25} \tag{8.9}$$

where

f is the friction factor, that, for developed laminar flow, takes the value $64/Re$ (–)
τ is the hydrodynamic shear stress (N/m^2)
Re is the Reynolds number based on characteristic cross dimension of the sticker spacing (–)

For drying air at 90°C flowing with a velocity of 5 m/s through a stack 2.4 m wide with sticker spacing height of 20 mm, the kinetic loss coefficient and pressure drop were found to be 2.7 and 75 Pa, respectively. At higher drying temperatures, the pressure drop would be greater (Langrish and Keey 1996).

For the air flowing with a velocity of 3 m/s through a row of n (50 × 150 mm) boards stacked with a sticker thickness of 25 mm, the pressure drop was estimated at $(5.02 + 0.99n)$ [Pa]. For a stack of 16 boards wide (corresponding to a 2.4 m-wide stack), the pressure drop for the rows with missing boards was found to be nearly twice that for full rows. However, in practice, the pressure drop across the stack would reach a uniform value, and the air velocities over the rows would redistribute to compensate (Salin and Ohman 1998).

The pressure drops associated with the plenum spaces can be calculated using the following equation (McCurdy and Pang 2007):

$$\Delta p_{plenum} = \left(2K_b + \frac{4f \times L_{plenum}}{D_{H_{plenum}}} \right) \rho_{air} \frac{\overline{u}_{plenum}^2}{2} \tag{8.10}$$

where

K_b is the head loss due to the bend in the plenum (~1.5)
f is the friction factor (–)
L_{plenum} is the swept length of the plenum spaces (m)
$D_{H_{plenum}}$ is the hydraulic diameter of the plenum space (m)
$\overline{u}_{plenum}$ is the average velocity in the plenum space (m/s)

The pressure drop in the headspace can be similarly calculated from the following equation (McCurdy and Pang 2007):

$$\Delta p_{head} = \left(2K_b + K_{fc} + K_{fe} + \frac{4f \times L_{plenum}}{D_{H_{plenum}}} \right) \rho_{air} \frac{u_r^2}{2} \tag{8.11}$$

where

K_{fc} is the head losses due to contractions (~0.4)
K_{fe} is the head losses due expansions through the kiln's fans (~0.7)
ρ_{air} is the air density (kg/m^3)

The pressure drop in heat exchangers can be determined by sizing the heating coils based on the heat transfer requirements and basic heat transfer equations (Coulson et al. 1989; ASHRAE 2009).

8.5 AIRFLOW MALDISTRIBUTION

In conventional and heat pump-assisted commercial kilns, air from the overhead fans has to turn through 180° before entering the lumber stack(s). It is influenced by internal structures such as the false ceiling and baffles, and may not be evenly distributed over the lumber stack(s) (Keey and Nijdam 2002). Although lumber kilns are generally fitted with baffles, hinged plates or curtains at the top and sides of the stack(s), inevitably, at least 10%–25% (Langrish and Keey 1996) or up to 40% (Keey et al. 2000) of the incident flow rate of the drying air bypass around them, even with stacks having well-aligned edges. Air can also bypass through any unblanked spaces between bearers, and in the stacking of random-length boards, any gaps can provide alternative passageways for the flow of air (Keey et al. 2000).

The lumber stacks that have ragged edges or missing board lengths are likely to lead to more airflow maldistribution and, thus, uneven drying. Usually, the plenum width to the roof height ratio should be greater than unity.

If the kiln is designed so that the ratio of the plenum-space width to the ceiling-space height above the stack be at least equal to unity, the airflow maldistribution may be reduced. If this condition is not met, the right-angled turn in the airflow from the ceiling space into the more restricted plenum would result in significant velocity gradients, with the flow directed toward the kiln wall. Moreover, vortex at the top of the inlet plenum may develop, causing a significant maldistribution of the airflow. In industrial practice, the right-angled bend at the top of the plenum causes some flow irregularities and, in extreme cases, vortexes.

On the other hand, if the plenum-space width and the combined sticker-space height have the same extent, the air velocity can be uniformly distributed over the whole cross section of the plenum space. However, if the plenum-space width relative to the combined sticker-spacing height is reduced by 50%, results in a greater flow maldistribution down the stack, with a peak velocity appearing closer to the top. In this case, the air velocity in the sticker space may vary down the stack, for example, from 7 to 4.2 m/s (Nijdam 1998).

Moreover, with equal quantities of air passing through the stack as around it, that is, with 100% bypass, the extent of drying is only 75% of that achieved if all of the delivered air had gone through the stack. Globally, significant bypassing reduces the kiln capacity, but there is no solution to the problem of airflow maldistribution in batch lumber kilns (Nijdam and Keey 1996).

8.6 AIRFLOW REVERSALS

Most industrial kilns are controlled based on the prediction of the variation of moisture content across the board stacks that mainly depends on the air velocity through the stack(s) width(s).

Lumber boards that have final moisture contents above the target value may be unacceptable for their intended use, and may require redrying. Boards with moisture contents below the target value may necessitate conditioning. Both deviations imply an inefficient use of energy and extra handling costs. Hence, reducing as much as possible of the moisture content variability between the dried boards is a challenging task.

8.6.1 Justification

Due to the heterogeneity of the lumber stack(s) and the probabilistic nature of wood drying, no two timber batches dry identically, even for the same type and size of boards, initial moisture content, and nominal operating kiln conditions. In particular, a variable distribution of board initial moisture content and of timber physical properties due to its biological origin, as well as random fluctuations in kiln drying parameters all contribute to dispersion in the final moisture levels.

In wood drying practice, it is not possible to ensure that all the boards have the same initial moisture content. The distribution in the initial moisture content of the wet boards can vary widely depending on the timber species, geographic location, proportion of sapwood and heartwood, elapsed time between harvesting and drying, etc. For example, in the United States, the variation in initial moisture content of southern pine and western hemlock can range from 30% to 200% (dry basis). For green Irish Sitka spruce, a variation in the initial moisture content from 50% up to 180% (dry basis) is possible.

During batch drying, the random variability in the initial moisture content of timber boards provides large dispersions in board drying rates. At the later periods of drying, most of the variability in moisture content between lumber boards is due to dispersion in the drying rate as the influence of variability in initial moisture content diminishes more rapidly. In addition, air dry- and wet-bulb temperatures, relative humidity, and velocity have systematic and random spatial and temporal variation inside the drying chamber (Cronin et al. 2002).

In order to diminish variability in board moisture contents at the specified end of the drying process (generally ranging from 6% to 20% depending on the end-use application), some measures can be applied as follows: (i) reduce the initial moisture variability by sorting the lumber boards and/or timber batches by average moisture contents; (ii) reduce the variability in the lumber stack(s) drying rate (more difficult to achieve) that, generally, arises because of timber thermophysical transport parameters (conductivity, permeability, density, etc.), dispersion in board dimensions and surface finish, randomness in the selection and placing of individual boards in the batch, and board misalignment (Langrish 1994); and (iii) periodically reverse the airflow direction in order to balance the drying at each side of the lumber stack(s) and, thus, achieve more uniform drying (Pang and Haslett 1995); a period of 4 hours is a common industrial practice, for example, for permeable softwoods such as *P. radiata* (Perré and Keey 2015a, 2015b).

As the air flows through the lumber stack, the absolute humidity of the air increases, while the dry-bulb temperature decreases. Therefore, the boards in the front rows at the air inlet edge dry rapidly, whereas those toward the airflow outlet edge of the stack dry more slowly (Keey and Pang 1994). The air that encounters the second board has a lower absolute humidity potential than the air at the start of the stack, which means that the second board does not dry as quickly as the first board. This change in the air parameters and the moisture content of the boards occurs progressively throughout the lumber stack(s) in the airflow direction. If the airflow direction remains the same, the first board in the stream-wise direction will dry more quickly than the last board.

Whenever kiln stacks are built from random-length lumber so that every second board is flush at each end of the stack, variations in openness of the stack result. This gives two different zones: (1) a central zone in which all the available space is filled and (2) two end zones where alternate boards are missing (Salin 2001). This arrangement results in higher within-stack velocities (about 30% higher) in the center than in the end zones, with corresponding implications in the variation in drying behavior.

In practice, the end zones of the lumber stack(s) dry faster than the central, fully filled part. The lower air velocity in the ends is more than compensated by higher heat transfer coefficients associated with the flow disturbance and smaller wood volume. There is a smaller decrease in temperature and increase in absolute humidity along the stack in the airflow direction.

In general, it is expected that the local heat and mass transfer coefficients diminish with distance in the airflow direction due to a thickening of the boundary. The airflow through the lumber stack influences the magnitude of the local air-side mass transfer coefficients, and, thus, the rate of moisture vaporization into the airstream. However, at the higher air velocities used in high-temperature drying, any variations in mass transfer coefficients have a significant effect on the uniformity of drying throughout the stack. Traditionally, the variation of vaporization moisture rates across the lumber stack(s) is lowered by the installation of bidirectional fans and by periodically reversing the airflow direction through the stack(s). This strategy has minimal effect on the drying rates in the center of the stack but reduces the variation in behavior between the two end zones (Perré and Keey 2015a, b).

Among other aspects of air reversal strategy can be mentioned (note: not listed in order of importance and/or relevance) as follows:

a. Airflow reversals at every few hours are used to ensure that the variation in moisture content in the airflow direction is controlled to minimize kiln-wide variations in moisture content from board to board (Pang and Keey 1994);
b. Airflow reversals are employed to counter the progressive humidification of the airflow through the lumber stack(s) with consequential loss of drying potential (Keey et al. 2000);
c. Airflow reversals essentially trim the excessive over- and under-drying of the outer portions of the stack (Ashworth 1977);
d. Once the airflow direction is reversed, the external drying parameters change, resulting in variations of the local drying rates; consequently, the boards near the new air inlet (the old air outlet) dry faster that those close to the new air outlet (the old air inlet); this process is repeated whenever further airflow reversals are performed; soon after the first airflow reversal,

the mid-area in the kiln stack becomes the wettest zone since the external conditions over this region are virtually unchanged even though the airflow may have been reversed several times;

e. Reversing the direction of the airflow has the effect of enhancing the drying of the wetter zone of a stack; therefore, the drying difference between the two outer sides of the stack can be smoothed out to a certain extent (Pang 1994);
f. Optimized airflow reversal can improve the final moisture content deviation from 7.3% to 2.3% (dry basis), reduce the peak moisture content variation and stresses during drying, and, possibly, the residual stresses after drying;
g. The direction and duration of fans' rotation may be achieved manually, but, normally they are controlled through time switches which can be set to stop the fans' motor, reverse the airflow, and restart the fans' motor at desired time intervals, with short periods of time (about 2 minutes) before restarting;
h. Moisture is removed from kilns through roof air vents located on both intake and exhaust sides of the fans; dry fresh air is drawn in on the suction side and moist air forced out on the pressure side of the load; when the direction of rotation of the fans is reversed, the flow of air through the air vents is also reversed;
i. In commercial drying practice, kiln operators select the appropriate reversal strategy for particular drying situations; for example, if minimizing the maximum difference of moisture content is the key factor to be considered, then reversals of overhead fans after 2–6 hours may be suitable (Keey et al. 2000);
j. However, even the airflow reversals reduce the moisture content variability across the kiln within specified limits, there is little benefit in using more than four flow reversals (Pang and Haslett 1995; Keey et al. 2000);
k. Without any flow reversal, the moisture content profiles may show up to 50% variations with time (Ashworth 1977).

8.6.2 Background

Several experimental works studied the important issue of airflow reversals in kilns. Pang (1994), for example, considered the impact of various flow-reversal strategies on the high-temperature drying of *P. radiata* at dry-/wet-bulb temperatures of 120°C/70°C and a kiln-air velocity of 5 m/s through a 2.4 m-wide stack. By drying heartwood without airflow switching, a maximum moisture content gradient across the lumber stack was estimated to be $0.062\frac{kg_{water}}{kg_{oven\text{-}dry}}$. Reversing the flow direction every 8 hours gave very little benefit, but early reversals at 3 and 4 hours after the start are more beneficial, reducing the maximum moisture content difference to $0.04\frac{kg_{water}}{kg_{oven\text{-}dry}}$ (Keey and Pang 1994; Pordage and Langrish 2007). There was essentially no difference in the moisture content variability between making a single flow switch over reversals every 4 hours. Similar trends were found for the drying of sapwood, except that a dual switch over, one after 2 hours and the second after 6 hours, reduces the maximum moisture-content difference to $0.16\frac{kg_{water}}{kg_{oven\text{-}dry}}$ compared with a value of $0.38\frac{kg_{water}}{kg_{oven\text{-}dry}}$ for unidirectional airflow. The greater moisture content variability with sapwood was attributed to the relatively rapid moisture flow within the wood in the early part of the drying cycle.

With different reversal strategies, there are variations in external conditions (air dry- and wet-bulb temperatures and relative and absolute humidity) through the drying schedule and, consequently, variations in local average moisture content through the kiln stack(s) are achieved.

The first reversal is critical, and frequencies of subsequent airflow reversals can be varied depending on the drying temperature used. Frequent airflow reversals are beneficial in reducing the drying stresses, especially commercial kiln drying of softwood lumber (Nijdam and Keey 1996). For kiln drying of *P. radiata* lumber, airflow reversal every 4 hours is recommended. For higher temperature (140°C) or ultra-high temperature (>160°C), airflow direction should be reversed every hour or every half hour.

Concerning the air reversal issue, some additional features have been observed, as (Pang and Haslett 1995, 2015b) follows: (i) the first reversal is critical for reducing the peak value of the maximum differences in moisture content through the kiln-wide stack(s); airflow reversals every 8 hours do not reduce this peak value but can shorten the duration of this peak difference; strategies with the first reversal after 4 hours, or earlier, reduce the peak value of moisture content difference; the strategy of reversing airflow after 2 and 6 hours from the start of drying is supposed to be better than other methods in respect of this consideration; (ii) after 24 hours of drying, airflow reversals every 3 or 4 hours gave virtually the same moisture content distribution; excessive number of airflow reversals may be counter-productive, shortening the technical life of dryer's fans (Wagner et al. 1996); many experimental studies (Keey and Pang 1994; Pang 1994; Pang and Haslett 1995) concluded that two or three reversal at critical drying time may be sufficient and, also, may further reduce drying stresses, particularly critical when the lumber is dried using high- and ultra-high-temperature schedules; (iii) in Scandinavian drying practice, to mitigate the variations in the ease of drying and the development of drying stresses, frequent flow reversals (every 1 or 2 hours, and, sometimes, more frequently) are used in drying small-diameter logs of species such as Norway spruce (Salin and Ohman 1998); and (iv) although manufacturers can supply bidirectional fans with less than a 10% difference between forward and reverse flow throughput (Nijdam and Keey 1996), the forward thrust of the installed fans can be somewhat better than the reverse thrust. The airflow in the reverse direction could be as much as 20% less than that in the forward flow (Keey et al. 2000).

8.7 CONVENTIONAL DRYING SCHEDULES

A well-designed conventional kiln has to maintain reasonably uniform drying conditions throughout its interior by closely controlling them over wide ranges of temperatures and relative humidity, suitable for a given wood species with various lumber board dimensions (Cech and Pfaff 2000).

In industrial heat-and-vent kilns, different softwood and hardwood species are dried from given initial moisture contents to specified average moisture contents suitable for certain value-added items. These items should have upper and lower tolerance limits, and all average moisture contents values should fall within these limits.

There are three environmental variables that control the drying rate, that is, air dry-bulb temperature, relative humidity (or wet-bulb and equilibrium moisture content), and velocity.

In conventional wood dryers, drying schedules (also called indexes or programs) are usually divided into steps (stages) with air dry- and wet-bulb temperatures, relative humidity and velocity, and time durations that mainly depend on the wood species and lumber size. The kiln drying environment (i.e., air dry- and wet-bulb temperatures, and relative humidity) has to be changed according to the actual moisture content of the wood. The drying conditions of each drying step are established based upon moisture content, type of wood, and lumber board dimensions (Cech and Pfaff 2000).

Drying schedules vary by species, thickness, grade, and end use of lumber, and are of two general types: (i) moisture content based and (ii) time based. To avoid most of drying defects and provide the fastest possible drying rates, kiln drying schedules based on lumber moisture content require proper selection of board samples and continuous control of air dry- and wet-bulb temperatures and relative humidity in the kiln because as the moisture content of the lumber stacks drops, the temperature and relative humidity inside the kiln change.

Wood drying schedules represent the time changes of dry- and wet-bulb temperatures, as well as of moisture content and either wet-bulb depression and equilibrium moisture content, or both, during particular drying processes. Because the relative humidity, that is, a measure of the amount of water vapor in the air, expressed as a percentage of the total amount contained in saturated air at a given temperature and pressure does not directly indicate the drying capacity of the kiln atmosphere, this parameter is determined by the use of dry- and wet-bulb temperatures measured by psychrometers.

To sustain the rate of drying as the wood dries out, either the dry-bulb temperature or the wet-bulb depression may be raised, thus providing a greater overall driving force for the drying process. A small wet-bulb depression means relatively high equilibrium moisture content at kiln temperatures, with less surface shrinkage and so smaller differential strains in the early part of the schedule, especially desirable in the drying of less permeable species. However, faster drying rates can lead to steeper moisture content gradients with the risk of excessive strain development and checking. Therefore, most schedules specify relatively small wet-bulb depressions, corresponding to relatively high equilibrium moisture content at kiln temperatures, producing modest surface shrinkage. In conventional low-temperature schedules, the wet-bulb depression is typically about 5°C–10°C, whereas for high-temperature drying schedules, wet-bulb depressions can be 50°C or more (Perré and Keey 2015a).

For wood's given moisture contents, species, and thickness of lumber boards, and for the quality required, there are many sets (or combinations) of air dry- and wet-bulb temperatures, relative humidity (or equilibrium moisture content), and velocity that are intended to dry the lumber without causing degrade.

Drying schedules are thus compromises between the need to dry lumber as fast as possible (i.e., achieving satisfactory drying rates) and avoid drying defects (such as checking, cracking, and discoloration). This means that the stresses that may develop during drying are limiting factors in developing kiln drying schedules. The drying schedules must be developed so that the drying stresses do not exceed the strength of the wood at any given dry- and wet-bulb temperatures, and moisture contents.

Traditionally, the sequences of wood drying schedules have been developed based on empirical trial-and-error studies from small-scale tests involving changes of some pertinent parameters (Meroney 1969). In practice, kiln operators periodically measure the moisture contents a number of representative lumber board samples helping at developing specific time-based schedules. Time-based schedules require the measurement of kiln sample moisture content to determine when finish the drying cycle and, eventually, when equalize and condition the dried lumber.

8.7.1 Basis Rules

To develop lumber drying schedules, some of the following basic rules that must be considered are as follows: (i) type of kiln; (ii) the wood species (because there are many variations in anatomical, physical, and mechanical properties between them); (iii) the thickness of lumber boards (because the drying time is approximately inversely related to thickness and, to some extent, is also influenced by the width of the lumber); (iv) whether the lumber boards are quarter-, flat-, or mixed-sawn (because the sawing pattern influences the distortion due to shrinkage anisotropy); (v) permissible drying degrade (because aggressive drying schedules can cause timber to crack and distort); and (vi) intended use of timber (because the required appearance of the timber surface and the target final moisture contents are different depending on the uses of timber).

In most drying schedules, the air dry-bulb temperature is gradually increased while the air absolute humidity and wet-bulb temperature progressively decrease. Set dry- and wet-bulb temperatures are chosen to achieve a satisfactory drying rate and avoid as much as possible drying defects.

Usually, during stage 1 (at the start of drying, after the preheating step), a relatively low temperature is required to minimize the weakening of wood and subsequent degrade such as checking or honeycomb. The relative humidity is kept high early in drying to minimize surface checking caused by tension stresses that may develop in the outer shell of the lumber. Even under these mild initial kiln conditions, the lumber loses moisture rapidly, and the drying rate is monitored to ensure that drying does not occur too rapidly.

When the lumber losses about 30%–35% (dry basis) of its initial moisture content, stage 2 can start, and, to maintain an acceptable drying rate, the air relative humidity is lowered gradually. When the lumber boards reach 30% average moisture content (dry basis), the third stage begins, during which the air dry-bulb temperature is raised gradually, and the air relative humidity can continually be lowered because of the danger of internal checking (also called honeycomb). As a rule of thumb, when the moisture in lumber at mid-thickness is below 25%–30% (dry basis), it is generally safe to make large increases in dry-bulb temperature to maintain a relatively fast drying rate.

As the dry-bulb temperature increases during drying, some of the phenomena that affect the drying process are (Perré and Keey 2015a, 2015b) as follows: (i) moisture moves faster in wood; (ii) drying is more uniform; (iii) the relative humidity of the air decreases; (iv) wood is weaker; and (v) warp is usually worse, except in high-temperature drying over 100°C.

As the air relative humidity increases during drying, the following changes may occur (Perré and Keey 2015a, 2015b): (i) moisture moves faster; (ii) drying is more uniform; (iii) warp is less; (iv) checking is more likely; (v) staining is less likely to occur; and (vi) with species that are likely to check or crack during drying, the air relative humidity must be quite high, yet not too high so that the risk of staining and warping is increased above tolerable levels; for species that tolerate lower relative humidity, then these low relative humidity should be used to provide brighter and flatter lumber, and faster drying.

Also, as the velocity through a lumber stack of lumber increases, several changes occur that affect drying are as (Perré and Keey 2015a, 2015b) follows: (i) above 40% moisture content, wood dries faster, which increases the risk of checking and cracking but decreases the risk of stain and warp; (ii) below 20% (dry basis) moisture content, wood dries less fast; (iii) between 40% and 20% moisture content (dry basis), the effect of increases in velocity diminishes; (iv) drying is more uniform; and (v) the energy cost of doubling the velocity can increase by 4 times, depending on the electrical energy price; below 20% moisture content (dry basis), it results in significant energy cost savings.

8.7.2 Specific and Modified Schedules

Kiln schedules have been developed to control temperature and relative humidity (via specific sensors) in accordance with the moisture content and stress phenomena within the wood, thus minimizing the drying defects.

Generic kiln drying schedules have been developed in many countries and/or geographic regions to attain objectives as, for example, to produce brighter lumber, or to maintain maximum strength of the lumber for special uses. The generic schedules of kiln drying of Spruce in Europe are totally different to those of southern pines in the United States, radiata pine in Australia and New Zealand, and much-coveted sugi in Japan (Keey et al. 2000).

Most of actual drying schedules used throughout the industry are based on generic (regional) schedules adapted (modified) by sawmills, however, without major changes compared to the conservative schedules developed in the past (Oliveira et al. 2012).

For most of wood species, the modified drying schedules must consider the following specific aspects: (i) depending on wood species, the maximum air dry-bulb temperature prior to equalization and conditioning should be no higher than a given value (e.g., between 71°C and 82°C for softwoods) in order to improve lumber color and machinability, and (ii) the maximum wet-bulb depression (i.e., wet-bulb depression is the temperature difference between dry-bulb and wet-bulb temperatures) should not exceed a given value (e.g., between 25°C and 28°C);

If the lumber is drying too fast, the drying rate can be slowed by (i) lowering the air dry-bulb temperature, (ii) raising the air relative humidity, and (iii) lowering the air velocity. On the other hand, if the wood is drying too slowly, the air relative humidity can be lowered and the velocity increased.

Permeable softwood species can be kiln dried at dry-bulb temperatures above 100°C by using time-based schedules up to final average moisture contents varying between 10% and 20% (dry basis). A typical schedule for eastern Canadian softwoods, for example, begins with a preheating step at maximum 93.3°C dry-bulb temperature, generally for a period of 3–8 hours. At step 1 after the preheating step, the moisture content first decreases linearly with time, a process followed by a nonlinear decrease until the lumber board reaches the scheduled final average moisture content. In New Zealand, where about 95% of the commercial forest harvest derives from plantation softwoods, of which about 90% is *P. radiata* (Keey et al. 2000), comprehensive schedules have been developed for this species, with fast high-temperature schedules for structural grades (Haslett 1998).

Impermeable hardwood species are usually dried at dry-bulb temperatures below 100°C to prevent drying defects. A typical hardwood moisture-based drying schedule might include, for example, 49°C dry-bulb temperature and 80% relative humidity in the beginning (when the lumber is green) and 82°C dry-bulb temperature at the end of the drying cycle when the average moisture content of lumber boards reaches 6%–8% (dry basis) for interior uses such as furniture, cabinets, and flooring. Some basic principles of kiln drying of hardwood are as follows: (i) changes in air dry- and wet-bulb temperatures and relative humidity are based on the average moisture content of the wettest half of the board samples; (ii) the air dry-bulb temperature is controlled to within 0.56°C and air relative humidity within 2%; (iii) in the case of hardwood drying, for high-value uses, such as furniture, millwork, and cabinets, the final moisture content must be controlled to within narrow limits, typically ∓1% moisture content (dry basis); (iv) over-drying any board must be avoided as this will increase shrinkage, cupping, machining defects such as planer splits, and gluing problems; (v) under-drying must also be avoided because the lumber boards of these pieces will shrink and develop end checks; and (vi) because nonuniformity in final moisture content is a result of both resource and dryer variability, in practice, uniformity is achieved by a moisture equalization step.

Finally, because individual identification of wood species (especially those from tropical forests) and separate drying are rarely economic or feasible, economic drying of mixed species involves the grouping of wood species according to criteria such as density, permeability, and extractive content, risks to develop drying defects, and expected total drying times (Keey et al. 2000). A single drying schedule of mixed species (such as spruce–pine–fir) minimizing the wood degrade would be economically feasible.

8.8 KILN-WIDE HEAT AND MASS TRANSFER

In industrial kilns, lumber boards are imperfectly stacked (Kho et al. 1989; Salin 1996). The lack of uniformity in the airflow direction through a lumber stack caused by irregularities, such as different heights of leading, trailing and adjacent boards, as well as by recessions and gaps (that, normally, generate substantial turbulence), strongly influences the flow pattern and shear stresses and, also, increases the heat and mass transfer coefficients. On the other hand, both heat and mass transfer coefficients increase with increasing air velocity but decrease with increasing air temperature (Pang 1996a, 1996b, 2004; Langrish et al. 1997).

Traditionally, the external heat and mass transfer coefficients are evaluated by using empirical correlations.

According to Chilton–Colburn analogy, the heat and mass convective transfers are interrelated to each other by

$$j_{heat} = j_{mass} = \frac{C_f}{2} \tag{8.12}$$

where

j_{heat} is the heat transfer factor (dimensionless) defined as follows:

$$j_{heat} = St_{heat} Pr^{2/3} \tag{8.13}$$

j_{mass} is the mass transfer factor (dimensionless) defined as follows:

$$j_{mass} = St_{mass} Sc^{2/3} \tag{8.14}$$

Cf is the skin friction coefficient defined as follows (dimensionless):

$$C_f = \frac{2\tau_w}{\rho u^2} \tag{8.15}$$

τ_w is the wall shear stress (N/m^2)
ρ is the air density (kg/m^3)
u is the air velocity (m/s)

$St_{heat} = \dfrac{h_{heat}}{c_p \times u \times \rho}$ is Stanton number for heat transfer (–)

$Pr = \dfrac{c_p \times \mu}{k}$ is Prandtl number (–)

h_{heat} is the heat transfer coefficient (W/m^2·K)
c_p is the air specific heat (J/kg·K)
μ is the air dynamic viscosity (Pa·s)
k is the air thermal conductivity (W/m·K)
$St_{mass} = K_{mass}/u$ is Stanton number for mass transfer (–)
K_{mass} is the mass transfer coefficient (m/s)

$Sc = v/D$ is Schmidt number
v is the air kinetic viscosity (m²/s)
D is the moisture diffusion coefficient (m²/s)

By introducing the previous nondimensional criteria, the heat and mass transfer coefficients can be related using Chilton–Colburn analogy (equation 8.12) as follows:

$$\left[\frac{h_{heat}}{c_p \times u \times \rho}\right] \times \left[\frac{c_p \times \mu}{k}\right]^{\frac{2}{3}} = \left[\frac{K_{mass}}{u}\right] \times \left[\frac{v}{D}\right]^{2/3} \tag{8.16}$$

The Chilton–Colburn *j-factors* for mass and heat transfer over blunt boards have been established as follows:

$$j = \left(0.0288 - \alpha \times Re_x^{-\beta}\right) \times Re_x^{-\left(0.2 - \gamma \times Re_x^{-\beta}\right)} \tag{8.17}$$

where
α, β, and γ are positive parameters based on the mass transfer data over blunt boards in a lumber board stack (Kho 1993)
Re_x is the local Reynolds number $\left(Re_x = \frac{u_\infty x}{\upsilon}\right)$
u_∞ is the velocity of the bulk airflow (m/s)
x is the coordinate (m)
υ is the air kinematic viscosity (m²/s)

When the Reynolds number Re_x reaches a critical value of about 500,000, the boundary layer ceases to be fully laminar and becomes turbulent in its outer regions. For a stack width of 2.4 m, an air velocity of 5 m/s, and a kinematic viscosity of $20 * 10^{-6}$ m²/s (corresponding to an air temperature of about 75°C), the maximum Reynolds number is 700,000, so transition to such turbulence is likely within a stack of that width. With even wider stacks, more extensive turbulence would occur with a greater pressure loss. There would also be a greater fall in the mass transfer coefficients from the air inlet to the air outlet face of the stack.

8.8.1 Heat Transfer

Based on the hypothesis that the gaps between adjacent boards reduce the distance from the leading edge, on average, by a factor of 0.845, Salin (1996) calculated the mean board heat transfer coefficients in the center of a 1.5 m-wide stack, with 25 mm sticker widths for a kiln temperature of 60°C, and concluded that the kiln-wide mean heat transfer coefficients vary exponentially with kiln-air velocity according to $u_\infty^{0.65}$. For example, for kiln air velocity of 3 m/s, board dimension 25 × 100 mm and 15 boards, the kiln-wide mean heat transfer coefficient is about 14.6 W/m²·K (Salin 1996).

When the external mass transfer coefficient (K_c) (m/s) is known, the heat transfer coefficient can be predicted as follows:

$$h_{heat} = K_c\left[c_p \times \rho\right]^{1/3} \times \left[\frac{k}{D}\right]^{2/3} \tag{8.18}$$

where

h_{heat} is the heat transfer coefficient (W/m$^2\cdot$K)
c_p is the air specific heat (J/kg$\cdot$K)
ρ is the air density (kg/m^3)
k is the air thermal conductivity (W/m$\cdot$K)
D is the mass diffusion coefficient (m^2/s)

By measuring the weight variations of a very wet sample board (i.e., with surface at the wet-bulb temperature) in a stack dried at dry-bulb temperature of 60°C and wet-bulb temperature of 50°C, with air velocities varying from 0.46 to 2.74 m/s, Stevens et al. (1956) suggested the following empirical correlation to predict the average external heat transfer coefficient (W/m$^2\cdot$K):

$$h_{heat} = 10.79 + 8.94 \times u_\infty \tag{8.19}$$

where

u_∞ is the average air velocity at the lumber stack inlet (m/s)

At low drying temperatures, the average heat transfer coefficients calculated from this correlation are higher than those estimated from the Chilton–Colburn analogy using the experimental data of Kho et al. (1989, 1990). This is mainly because the moist air has relatively low values of specific heat due to lower values of relative humidity. However, at high temperatures, where the air absolute humidity is high, these two methods are in agreement. For example, for air at dry-/wet-bulb temperatures of 120°C/70°C and with air velocity of 3 m/s, the heat transfer coefficient from equation 8.21 is 37.61 W/m^{2}s while the value over the front board from the analogy is 35.52 W/m^{2}s. Over successive boards, the heat transfer coefficients predicted from the Chilton–Colburn analogy are however about 50% lower than this value (22.2 W/m$^2\cdot$K).

It can be noted that in a lumber stack, the heat transfer coefficients may reach a limiting asymptotic value after several rows, considerably higher than that predicted from the equation for transfer from a sharp-edged plate (Sparrow et al. 1982).

Due to the boundary layer separation, reattachment, and redevelopment of flow, the measured heat transfer data over blunt-edged flat plates reach a maximum at the point of reattachment and pass through a minimum at a point between the leading edge and the reattachment point (Sørensen 1969; Kho 1993).

For drying temperatures of 60°C and 100°C, and air velocity of 3 m/s, the heat transfer coefficients calculated with equation 8.18 are 26 and 23 W/m$^2\cdot$K, respectively.

Under dry-/wet-bulb temperatures of 149°C/82°C and 121°C/55°C with the air velocity of 11 m/s, the measured heat transfer coefficient values were $47.1 \mp 5.6\ \text{W/m}^2\text{K}$ and $53.4 \mp 10.4\ \text{W/m}^2\text{K}$, respectively (Beard et al. 1985).

When the rate of sublimation from naphthalene-coated boards is used, and the reference conditions are taken at a temperature of 41°C and an air velocity of 5 m/s, the external mass transfer coefficient (K_c, m/s) is reduced to the form as follows:

$$K_{mass} = 0.0418\left[\frac{u}{5}\right]^{0.78} \times \left[\frac{T}{314.15}\right]^{1.17} \times \left[\frac{v}{1.794\times10^{-5}}\right]^{-0.45} \quad (8.20)$$

where

K_{mass} is the external mass transfer coefficient (m/s)
u is the air velocity (m/s)
T is the air temperature (K)
v is the air kinetic velocity that can be estimated as a function of temperature using the following relation (Perry and Chilton 1984):

$$v = \left(1.02\times10^{-5}T^2 + 3.31\times10^{3}T - 0.3157\right)\times10^{-5} \quad (8.21)$$

Under these conditions, for water vapor as a diffusate, the value of mass transfer coefficient (K_{mass}) is 0.0418 m/s (Pang 1996c).

At drying temperatures of 60°C and 100°C and air velocity of 3 m/s, the experimentally measured values of the external mass transfer coefficient calculated with equation 8.20 are 0.028 and 0.029 m/s, respectively (Milota and Tschernitz 1994; Keey et al. 2000), in agreement with the experimentally measured values.

8.8.2 Mass Transfer

As the air flows through the lumber stack(s), it is humidified because the moisture moves from the boards into the air causing the air to become cooler and more humid in the stream-wise direction (Pordage and Langrish 2007). This can be related to the difference in the humidity just above the timber surface and that in the bulk air. Given that the initial (inlet) air absolute humidity (ω_{air}^{inlet}) can be measured, then the absolute humidity (ω_{air}) above any board can be estimated from the distance into the stack and the drying rate of the upstream boards.

The change in air temperature from the inlet of the stack to any position within the lumber stack(s) can be related to the absolute humidity at that point by the equation (Keey and Pang 1994):

$$T_{air}^{in} - T_{air} = \frac{h_{fg}}{(1+\alpha_H)c_{p,\,vapor}}\ln\left[\frac{c_{p,\,air} + \omega_{air}\times c_{p,\,vapor}}{c_{p,\,air} + \omega_{air}^{in}\times c_{p,\,vapor}}\right] \quad (8.22)$$

where

T_{air}^{inlet} is the air temperature at the inlet to the stack (°C)
T_{air} is the actual air temperature corresponding to the actual air absolute humidity $\left[\omega_{air} = \omega_{air}^{inlet} + \sum\Delta\omega_{air}\right]$ at a given point in the stack (°C)
h_{fg} is the enthalpy (latent heat) of vaporization of water (kJ/kg)
α_H is the ratio of radiative to convective heat transfer (>0.2 above 100°C and ≤0 below 80°C)

$c_{p,\,vapor}$ is the specific heat of water vapor $\left(\frac{kJ}{kg_{water}K}\right)$

$c_{p,\,air}$ is the specific heat of dry air $\left(\frac{kJ}{kg_{dry\ air}K}\right)$

ω_{air}^{in} is the air absolute humidity at the inlet to the stack $\left(\frac{kg_{water}}{kg_{dry\ air}}\right)$

As for the convective heat transfer, the board irregularities and the small gaps between boards increase the local mass transfer rates near the leading edges as well as throughout the entire length of stack (Kho et al. 1989). In addition, the enhancement of the mass transfer coefficients far from the stack inlet is considerable, going up to 83% (with naphthalene sublimation from a test board 640 × 100 × 20 mm at 41°C) (Table 8.1) (Langrish et al. 1993). If a board is missing within a package, an increase of up to 40% in mass transfer rates may occur with larger increases occurring at lower velocities (Sparrow et al. 1982).

The turbulence levels at the stack's inlet also increases the moisture mass transfer rates. For example, for a flat plate, increasing the turbulence level of the airflow from 1% to 8% increases the mass transfer rate by 55%. However, the turbulence intensity may decrease inside wood packages (Bian 2001).

In conclusion, over the first five boards of a lumber stack, the variations of mass transfer coefficient with distance from the air inlet present important features such as (i) the maximum value occurs over the first board 10–30 mm away from the leading edge; (ii) the maximum values over successive boards occur at positions close to the leading edge of each board, then reach an asymptotic value from 30 to 40 mm onwards; this phenomenon can be explained by the influence of eddy formation in the board gap, outward motion, and forward movement; as the deformed eddies move over the board, they vanish into the airstream, hence the effect is diminished; and (iii) from the second board along the airstream direction, the mass transfer coefficient profiles over each board are identical for a fixed air velocity, with higher values for higher air velocities (Pang and Haslett 1995).

Experimentally, the moisture mass transfer coefficients for the vaporization of moisture from a wood surface have been obtained in two ways (Keey et al. 2000): (i) by using the rate of sublimation from naphthalene-coated boards as a measure of the vapor transfer rate through the boundary layer (Kho et al. 1989, 1990) and (ii) by using a sorption method, measuring the weight change of a sample following a step change in relative humidity (Choong and Skaar 1972a, 1972b).

TABLE 8.1
Mass Transfer Coefficients in a Pilot-Plant Kiln

Velocity between Boards (m/s)	Turbulence Intensity before Stack (%)	Measured Mass Transfer Coefficient (m/s)	Enhancement Over Flat Plate Value (%)
3	16	0.0078	66
5	19	0.0115	62
7	23	0.0167	83

8.9 KILN-WIDE DRYING MODELING

The drying models are fundamental to understanding the drying phenomena and to predict variables such as drying curves, moisture content gradient, temperature profiles, and drying stress within the single lumber boards.

Modeling of the commercial kiln-wide drying processes, essential for improving the design and analysis of wood drying equipment (Ashworth 1977; Keey and Ashworth 1979), is generally based on equations aiming at relating the lumber moisture content to drying rate and drying time.

In a drying process occurring inside a batch lumber kiln, a complex coupling exists between the single-board and stack-wide scales. In such a drying system, the time constant of the whole process is driven by the time constant of one single board. Therefore, the single boards cannot be assumed to be at thermodynamic equilibrium (Perré and Remond 2006).

Drying models at low and high temperatures (Pang and Haslett 1995; Hukka 1997; Perré and Turner 1999a, 1999b; Nijdam et al. 2000; Kocaefe et al. 2009) for the entire kiln lumber stack (Salin 2001; Awadalla et al. 2004) used single-board models based on the one-dimensional/one variable diffusion equation (Kho 1993; Pang 1994), while the drying air leaving the first board was corrected for the next board in the layer.

Several high-temperature drying models have been presented in the literature. Pang and Haslett (1995), for example, simulated the drying process of a kiln stack of 2.4 m width comprising 100 × 50 mm *P. radiata* boards by assuming dry/wet-bulb temperatures of 120°C/70°C and an air velocity of 5 m/s in only one direction. They observed, after about 26 hours of drying, when the wood was relatively dry, the air no longer gains moisture, so the air absolute humidity everywhere was similar to that at the air inlet.

In their dynamic kiln-wide wood drying model, Sun et al. (2000) solved the dynamic unsteady-state mass, momentum, and energy balance equations for both the airflow and the wood boards in normal, aligned, and staggered stacks of *P. radiata* sapwood, respectively, dried under typical dehumidifier kiln conditions. Transient profiles of the air humidity, temperature, pressure, and velocity, and the wood temperature and moisture content have been obtained along the airflow direction within the stack. For simplicity, the airflow model was one dimensional, but two parallel airflow streams were used, and it was assumed that local thermal and phase equilibria are satisfied within the wood boards. The local pressure of moist air inside the wood boards was assumed to be equal to the local pressure of the gas flow. Since the wood drying system is comprised of two parts, moist air and moist wood, the balance equations have been established for each subsystem.

Sun and Carrington (1999) and Sun et al. (2000) have developed a dynamic kiln-wide wood drying model based on the characteristic drying curve of Sun et al. (2003, 2005) by solving the mass, momentum, and energy balance equations for both the airflow and wood boards in drying three types of stacks: normal, aligned, and staggered. For the airflow inside a stack, the model solved the integral form of the unsteady-state mass, momentum, and energy balance equations. For simplicity, the airflow model was one dimensional, but two parallel air streams for the central stack and side stack were used.

Also for simplicity, the characteristic drying curves (see Section 8.10) of low-temperature drying developed by Sun et al. (1996) previously were adopted for the description of the internal moisture transfer processes inside the wood boards (Sun et al. 2000). The scope was to predict the distribution of the average mass fractions, temperature, pressure and velocity of the air stream, and the average moisture content of the wood boards and their temperature.

Perré and Rémond (2006) proposed a dual-scale computational model of wood drying in a batch lumber kiln containing a stack with 100 boards arranged in layers. The model simulates each single board of the stack using one module of a comprehensive computational model for heat and mass transfer in porous media (TransPore), while the timber variability was taken into account by a Monte-Carlo method (Perré and Turner 1996, 1999a, 1999b; Perré and Remond 2006). The simulation of the coupling between the lumber board (micro scale) and the kiln (stack) wide (macro scale) was a two-step process consisting in calculation of drying conditions (temperature and vapor pressure) throughout the stack from the heat, vapor, and moisture fluxes leaving each faces of the boards at time t, as well as the air velocity and the sticker thickness; (ii) calculation of heat, vapor, and moisture fluxes at time $t + dt$ by using the TransPore module for each single board. This computational model confirmed that increasing the air velocity and the stick thickness dramatically increases the uniformity of drying conditions within the stack (Perré and Remond 2006).

McCurdy and Pang (2007) developed an integrated kiln (stack)-wide modeling concept to simulate the drying process (i.e., the lumber moisture content, humidity, and temperature change through the stack during drying based on given schedule conditions), energy use, and wood color change in kiln drying of softwood timber. The model included two kiln configurations: (i) the first, where the humid air is exhausted before the heating coil; and (ii) the second, where the humid air is exhausted after two heating coils. The first kiln configuration, containing one heating coil located on one side of the kiln, seemed particularly useful for high-graded timber because drying airflow can be directed in both directions, and the air conditions at the stack inlet side are controlled at set points. In the second kiln configuration, the heating elements were installed on both sides of the lumber stack(s). The airflow can also be in either direction by controlling the operation (on/off) of the heating elements, depending on the airflow direction.

Practical operation sequences have been applied for dry-bulb temperatures ranging from 50°C to 70°C, air velocities from 3 to 9 m/s, and wet-bulb depression of 15°C–20°C. The computer models were run for a number of schedules with a dry-bulb temperature ranging from 50°C to 70°C and an air speed range of 3–9 m/s with reversals every 6 hours. The stack was 2.4 m wide, 3.6 m high, and 7 m long; thus, the stack volume was 60.5 m^3. The board thickness was 40 mm, with initial moisture content of 120% and basic density of 400 kg/m^3 (van Meel 1958; Pang and Haslett 1995; Pang and Keey 1995; Nijdam 1998; McCurdy and Pang 2007).

8.10 CHARACTERISTIC DRYING CURVE

Coupling the process of moisture movement through a single lumber board with that, more complex, of moisture transport within the entire kiln in order to determine the kiln-wide drying behavior of stack(s) of lumber is a difficult approach mainly because of very large variations in thermodynamic parameters within the kiln (Keey and Nijdam 2002). In other words, coupling the equations developed for single-board drying with those describing the kiln-wide conditions may generate equation sets, which may be complicated and slow to compute (Kayihan 1993).

8.10.1 Generalities

To overcome such a situation (i) extensive laboratory tests have been conducted to obtain sufficient data to generate simplified drying kinetics for kiln-wide analysis (Ashworth 1977; Keey 1978; Keey et al. 2000); (ii) separate experiments have been performed in order to validate the single lumber board's mathematical models; and (iii) if the single-board model can describe the true picture of timber drying, experimental and simulation results may be used to produce drying curves (termed as characteristic drying curves) for kiln-wide design (van Meel 1958; Keey 1978; Pang 1994; Pang and Keey 1995; Keey and Pang 1994; Pang and Haslett 1995).

Although few comprehensive models of heat and mass transfer inside one single board do exist, simplified approaches at the board level, such as those derived from the concept of characteristic drying curve, are usually adopted when the kiln (stack)-wide level is considered (van Meel 1958; Cronin et al. 2003).

The concept of the characteristic drying curve is based on the premise that the relative drying rate is only a function of some averaged (or characteristic) moisture content under constant external convective conditions. In other words, at each volume-averaged, free moisture content (above the equilibrium moisture content), there is a corresponding specific drying rate (relative to the unhindered drying rate) that is independent of the external drying conditions. It provides a basis to determine the influence of the various external factors on kiln-drying behavior.

The characteristic curves are normally obtained from (i) laboratory oven tests with small specimens of the same thickness of the lumber of interest (McCurdy et al. 2005) or (ii) deduced from more detailed models of moisture movement (Pang and Keey 1994).

The original concept of the characteristic curve (Van Meel 1958) assumed that the airflow is uniform and the local mass transfer coefficients for moisture vaporization do not vary with position. When applied to the kiln-wide drying of lumber boards, the concept is rough, not only due to variations in drying behavior between boards but also due to embedded assumptions in the concept itself. For that, the concept has been criticized for being inexact. However, it provides reasonable approximations for kilns stacked with uniformly thick lumber and if the relative transfer resistance between the wood and the environment does not change greatly (Keey and Nijdam 2002). Nevertheless, it is a sufficient representation of drying behavior to determine the effect of kiln parameters, such as the uniformity of the airflow, the number of airflow reversals, the velocity, dry- and wet-bulb temperatures, and relative humidity settings, on the drying cycle (Perré and Keey 2015a, 2015b).

The concept of the characteristic drying curve has been successfully used in order to assess kiln (stack)-wide behavior for highly permeable softwoods such as radiata pine under high-temperature drying conditions for which the final moisture content variations and endpoint determination are of most concern (Keey and Suzuki 1974; Pang and Haslett 1995; Langrish 1999; Pordage and Langrish 2007).

The endpoint of typical kiln-wide stack drying cycle has been determined by using physiological transport properties and a single-board drying model to generate characteristic drying curves of mixed sap/heartwood boards (e.g., highly permeable softwood species such as *Pinus*) (Pang and Haslett 1995).

Even the concept of the characteristic drying curve was useful in studying the complex process of kiln drying, it is not applicable, for example, for drying hardwood timbers such as Australian iron bark. Effectively, for the drying of impermeable species, for which the surface moisture content reaches equilibrium quickly, there is unlikely to be any significant connection between the volume-averaged and the surface moisture contents, so the concept is unlikely to apply under these conditions (Pordage and Langrish 2007).

In addition, irrespective of the external of convective batch drying conditions, the geometrically similar single characteristic drying curves can be drawn for given species of wood being dried over specific ranges of drying conditions (van Meel 1958, 1959; Keey 1978; Pang and Haslett 1995; Keey et al. 2000; Pordage and Langrish 2007).

8.10.2 The Concept

For a given wood species, the drying rate curves determined over a given range of conditions appear to be geometrically similar. If these curves are normalized with respect to average moisture content and the initial drying rate, then all the curves could often be approximated to a single, "characteristic" drying curve.

If the board's drying behavior is to be described by the characteristic curve, then its properties must satisfy the following two criteria (adapted from Mujumdar 2007): (i) the critical moisture content (MC_{cr}) is invariant and independent of initial moisture content and external conditions; and (ii) all drying curves for a specific wood species are geometrically similar so that the shape of the curve is unique and independent of external conditions.

These criteria are restrictive, and it is quite unlikely that any wood species will satisfy them over an exhaustive range of conditions. However, the concept is widely used and, often, utilized for interpolation and prediction of dryer performance.

Conventionally, the characteristic drying curve describes a functional relationship between a relative (characteristic) drying rate (f) (the ordinate axis) and the normalized volume-averaged (characteristic) moisture content Φ (Figure 8.5) (Keey 1992; Pang and Haslett 1995; Pordage and Langrish 2007).

Figure 8.5 compares the two-stage quasi-linear characteristic falling rate curve of both high- and low-temperature drying. For low-temperature drying, the characteristic drying curve is concave up and for higher temperatures, it is concave down. In Figure 8.5, the normalized variables are defined by the following expressions (Pang and Haslett 1995; Keey et al. 2000; Mujumdar 2007; Pordage and Langrish 2007).

$$f = \frac{\dot{j}_{mass}}{\dot{j}_{mass,initial}} = \frac{\dot{j}_{mass}}{\dot{j}_{mass,critic}} = \frac{\dot{j}_{mass}}{\dot{j}_{mass,sat}} \tag{8.23}$$

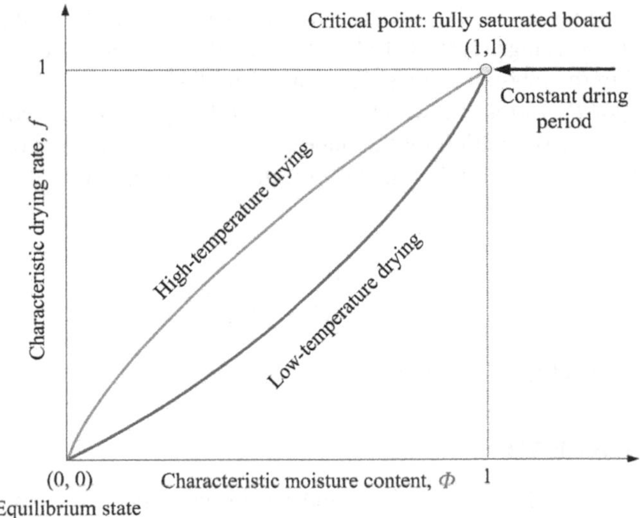

FIGURE 8.5 Typical characteristic drying curves for low- and high-temperature drying processes (note: graphic not to scale).

and

$$\Phi = \frac{\overline{MC} - EMC}{MC_{cr} - EMC} \tag{8.24}$$

where

- f is the relative drying rate (–)
- j_{mass} is the specific drying rate (for a unit surface) at any location and at any moisture content MC (kg/m²s)
- $j_{mass,\ initial} = j_{mass,\ critic} = j_{mass,\ sat}$ is the initial (critical or saturated) drying rate when the board is fully saturated (i.e., in the first constant drying period, above the critical moisture content MC_{cr}) (kg/m²s)
- $\overline{MC}$ is the board volume-averaged moisture content $\left(\frac{kg_{water}}{kg_{oven\text{-}dry}}\right)$
- EMC is the equilibrium moisture content under the prevailing temperature and relative humidity $\left(\frac{kg_{water}}{kg_{oven\text{-}dry}}\right)$
- MC_{cr} is the critical moisture content marking the end of drying entirely controlled by the moisture vapor transport in the air $\left(\frac{kg_{water}}{kg_{dry\text{-}bone}}\right)$

The relative drying rate (f), a function of normalized moisture content (Φ), is the most important parameter. Equation 8.24 has been used extensively as the basis for understanding the behavior of industrial drying plants owing to its simplicity. Only after it is determined can the kiln-wide calculation be performed to predict variations of moisture content, temperature, and drying rate. As this relationship (f)

varies with factors such as the properties of the material and the configuration of a dryer (Keey 1978; Langrish et al. 1992), it is usually correlated from experimental data obtained in the laboratory on small board samples.

For drying of *P. radiata* sapwood, Sun et al. (1996) proposed an empirical characteristic drying curve relation for the internal moisture movement. For the constant rate period ($MC \geq MC_{cr}$), $f = 1$. For the falling rate period ($MC < MC_{cr}$), $f = \Phi^{A-B\times\Phi_{FSP}}$, where

$$\Phi_{FSP} = \frac{\overline{MC} - MC_{FSP}}{MC_{cr} - MC_{FSP}} \tag{8.25}$$

MC_{FSP} is the moisture content at fiber saturation point (%)

8.11 DRYING RATE

When the humid wood boards are dried under stable conditions, the moisture content decreases at first linearly. In the case of pure convective drying, the wood surface is always saturated with water, and, consequently, the boundary layer reaches the air wet-bulb temperature. This first step is followed by a nonlinear process where the moisture content decreases until the wood board attains hygroscopic equilibrium.

If the temperature of the drying air is constant, the rate of evaporation gradually decreases. To maintain a steady drying rate, the water molecules in the wood must acquire additional energy, or the vapor partial pressure of the drier atmosphere must be reduced. This is achieved by either increasing the air dry-bulb temperature (more energy) or reducing the relative humidity (lower vapor pressure) of the drying air. When the vapor partial pressure of the moisture in the wood is equal to the vapor pressure of the surrounding air and neither gains nor loses moisture at a given temperature and relative humidity, the equilibrium moisture content is achieved.

The drying rate of a given wood species (Langrish and Walker 1993; Walker et al. 1993; Keey et al. 2000; Pordage and Langrish 2007) is generally a function of (i) external drying conditions (e.g., kiln type, dimensions, and thicknesses of lumber boards); (ii) physical characteristics of wood (species, grain pattern, initial moisture content); and (iii) structure of drying schedule (dry- and wet-bulb temperatures, air velocities) (Meroney 1969; Berit Time 1988; Onn 2007).

The drying rate of wood is controlled by external transport rates and, thus, is a complex function of the stacking geometry, the dry-bulb temperature, the wet-bulb depression, the diffusion mechanism within the wood boards, the air velocity. In any given situation, it is to be expected that there exists some optimum combination of these states, which provides the fastest drying rate compatible with low lumber degrade (Meroney 1969).

The drying rate of lumber boards is proportional to (Walker 1993; Keey et al. 2000) (i) (lumber density)$^{-1}$ (i.e., more cell wall material is traversed per unit distance, which offers resistance to diffusion); (ii) (board thickness)$^{-1}$ (i.e., during the later stages of drying, the gradient in moisture content is shallower for thicker lumber); and (iii) saturation vapor partial pressure which corresponds to a higher equilibrium moisture content.

8.11.1 Influence of Drying Temperature

If the relative humidity of drying air is kept constant, the higher the dry-bulb temperature, the higher the drying rate. Both dry-bulb temperature (defined as the temperature of a vapor mixture determined by inserting a thermometer with a dry-bulb) and wet-bulb temperature (defined as the temperature reached by a small amount of liquid evaporating in a large amount of an unsaturated air–vapor mixture) influence the drying rate by increasing the moisture-holding capacity of the air, as well as by accelerating the diffusion rate of moisture through the lumber boards. Higher dry-bulb temperature means more energy available to break the bonds between water molecules and wood and to evaporate water.

The rate of moisture movement to the lumber board's surface, as well as the moisture-carrying capacity of the air, increases at higher temperatures (at dry-bulb temperatures above 100°C) reducing the amount of humid air that needs to be vented (Keey et al. 2000).

During the drying process, the lumber boards' temperature approaches the air stream dry-bulb temperature, and the rate of heat transfer diminishes. As drying progresses, the rate of heat transfer is further reduced and the dry-bulb temperature must be increased to maintain an acceptable drying rate.

Both dry-bulb temperature and drying rate influence the development of drying defects because the wood cell walls are weaker at high temperatures than at low temperatures, and they are also weaker at high moisture contents than low moisture contents. The strength of wood cells during drying is directly related to the development of defects as the wood dries. For example, wood at a high moisture content exposed to a high dry-bulb temperature may experience enough strength loss in the cell walls that the stresses of drying cannot be resisted, and the cells may collapse. This complex interaction between the air dry-bulb temperature and relative humidity, and the lumber moisture content and drying rate is the basis for controlling the wood drying (Keey et al. 2000).

8.11.2 Influence of Air Relative Humidity

The relative humidity of air is defined as the partial pressure of water vapor divided by the saturated vapor pressure at the same temperature and total pressure. If the dry-bulb temperature is kept constant, lower relative humidity results in higher drying rates due to the increased moisture gradient in lumber boards.

8.11.3 Influence of Airflow Rate

The timber drying rate and quality, everything else being equal, are affected by the airflow rate and velocity. It is essential to remove moisture from the lumber boards' boundary layers as rapidly as possible to maintain the desired drying rate, mainly by controlling the air velocity inside the dryer.

At constant dry-bulb temperatures and relative humidity, the highest possible drying rate is obtained by rapid circulation of the drying air across the lumber stack(s), giving rapid removal of moisture vaporized from the lumber boards.

Higher circulation rates are beneficial during the early stages of drying when lumber boards are wet and the requirements for surface forced vaporization and moisture removal are high.

However, high drying rates are not always desirable, particularly for impermeable hardwoods, because higher drying rates develop greater stresses that may cause the timber to crack or distort. At very low fan speeds (e.g., less than 1 m/s), the airflow through the stack is often laminar flow, and the heat transfer between the timber surface and the moving air stream is not particularly effective (Walker et al. 1993). The low (externally) effectiveness of heat transfer is not necessarily a problem if internal moisture movement is the key limitation to the movement of moisture, as it is for most hardwoods (Pordage and Langrish 1999).

During the constant drying period of permeable woods (above the fiber saturation point), the rate of drying is controlled by the rate of heat transfer from the air to the lumber. An increase in both the air velocity and the wet-bulb depression enhances the heat transfer and, thus, the evaporation rate. The forced vaporization rate increases with increasing air velocity, and they are proportional to wet-bulb depression. In other words, as long as the moisture can move from the interior to the surface at a fast enough rate to keep the surface moist, the drying rate can be increased if the surface vaporization rate is increased, notably by increasing the airflow rate across the surface of the wood, by increasing the temperature of the air surrounding the wood, or by reducing the relative humidity of the air (Walker 1993; Keey et al. 2000).

8.11.4 Estimation of Drying Rate

The rate of heat transfer is primarily affected by the dry-bulb temperature differential (air stream and wood surface), which to some extent is affected by the rate of moisture vaporization. If velocity is constant, the rate of heat transfer from a moving air stream to a wood surface is more or less proportional to the dry-bulb temperature difference. Initially, when the wood is wet, the temperature difference between air stream and wood surface will be equal to the wet-bulb depression. Large quantities of heat are required to vaporize the free water brought to the wood surface, and the rate of heat transfer is at a maximum. Later, as the fiber saturation point is reached at progressively deeper levels in each lumber board, the wood temperature approaches the air stream temperature, and the rate of heat transfer diminishes. As drying progresses below the fiber saturation point, the rate of heat transfer is further reduced, and, since this affects the drying rate, the dry-bulb temperature must be increased to maintain an acceptable drying rate. At this stage, high air velocity has little effect on drying rate and its continuation serves no useful purpose (Cech and Pfaff 2000). In the last stage of drying, the drying rate is controlled by moisture diffusion, and air velocity can be reduced without considerably affecting drying time (Ananias et al. 1992; Breitner et al. 1990).

An accurate prediction of the lumber overall drying rate requires a comprehensive description of the drying process (lumber dimensions, dry-bulb temperature, wet-bulb temperature, and air velocity) and wood physical properties (heat capacity, basic density, thermal conductivity, gas permeability, diffusion coefficient, and capillarity) (Meroney 1969).

The dispersion of the drying rate at high moisture contents will tend to arise from variations in kiln air conditions and irregularities in the geometry of timber loading. Drying rate distribution at lower moisture contents will be more sensitive to the inherent biological variability of wood macro- and microstructure. Variability in any of these parameters will produce dispersion in the magnitude of the drying rate constant (*k*).

The drying rate (i.e., the rate of change of the moisture content *MC* with respect to time *t*) is proportional to how far the moisture content (*MC*) of a wood sample is from its equilibrium moisture content (*EMC*), which is a function of the temperature *T* and relative humidity (*RH*) (Simpson 1993):

$$\frac{d(MC)}{dt} = -\frac{MC - EMC}{\tau(T,L)} \tag{8.26}$$

where

MC is the moisture content $\left(\frac{kg_{water}}{kg_{oven\text{-}dry}}\right)$

EMC is the equilibrium moisture content $\left(\frac{kg_{water}}{kg_{oven\text{-}dry}}\right)$

τ is a function of the temperature *T* and a typical wood dimension *L* called time constant, having units of time (*s*)

The solution to equation 8.26 is:

$$\frac{MC - EMC}{MC_{in} - EMC} = e^{-t/\tau} \tag{8.27}$$

where

MC_{in} is the initial moisture content $\left(\frac{kg_{water}}{kg_{oven\text{-}dry}}\right)$

For red oak lumber, for example, the time constant τ can be expressed as follows:

$$\tau = \frac{L^n}{a + b \times p_{sat}(T_{sat})} \tag{8.28}$$

where

a, *b*, and *n* are empirical constants (–)

$p_{sat}(T)$ is the water vapor partial pressure (Pa) at saturation temperature T_{sat} (°C)

By assuming that the drying rate is a linear function of the difference between instantaneous board moisture content (*MC*) and the equilibrium moisture content (*EMC*) (Lewis 1921; Sherwood 1929), the drying rate at any time of the drying process can be expressed as follows (Cronin et al. 2002):

$$\frac{d(MC)}{dt} = -k(MC - EMC) \tag{8.29}$$

where

MC is the moisture content (dry basis) $\left(\frac{kg_{water}}{kg_{oven\text{-}dry}}\right)$
EMC is the equilibrium moisture content (dry basis) $\left(\frac{kg_{water}}{kg_{oven\text{-}dry}}\right)$
k is the drying rate constant (h^{-1})
t is the time (h)

Integrating this expression and knowing that the initial moisture content (MC_{in}) gives

$$MC(t) = (MC_{in} - EMC)e^{-kt} + EMC \tag{8.30}$$

If dry- and wet-bulb temperatures remain constant, the equilibrium moisture content is assumed to be invariant with respect to time. The magnitude of the equilibrium moisture content, *EMC* can be estimated from the wet-bulb depression ($\Delta T_{depression}$) with an approximate formula sufficiently accurate for wet-bulb temperatures in the range from 35°C to 65°C (Malmquist 1991):

$$EMC = \frac{0.24}{1 + \frac{\Delta T_{depression}}{5.3}} \tag{8.31}$$

There are a number of alternative expressions for the drying rate constant. In the diffusion approach to timber drying, the board can be considered as a flat plate with a moisture gradient in the vertical direction only and vaporization occurring from the top and bottom faces only (Simpson 1993).

For long drying periods of flat lumber boards with moisture gradients in the vertical direction only and moisture vaporization occurring from both the top and bottom faces (Simpson 1993), the drying rate constant (*k*) will have the form:

$$k = \frac{\pi^2 D}{b^2} \tag{8.32}$$

where

D is the diffusion coefficient (m^2/s)
b is the board thickness (m)

Another empirical expression for the drying rate constant is given as follows:

$$k = \alpha_D \left(\frac{0.025}{b}\right)^{1.25} \times \frac{T}{65} \tag{8.33}$$

where

α_D is a coefficient which depends on wood species, moisture content, and kiln design (varying around 0.048 for softwoods)
b is the board thickness (m)
T is the board temperature (°C)

Both equations 8.31 and 8.32 suggest that the drying rate constant is inversely related to board thickness though the exact nature of the relationship differs.

When applied to the whole drying process, the diffusion coefficient (D) will no longer be the actual theoretical water-in-timber diffusivity but rather an effective (or lumped) empirical coefficient taking into account a number of internal and external thermophysical drying mechanisms and properties. Therefore, if lumber drying is controlled by moisture diffusion, the lumber drying rate can be calculated as an effective diffusion coefficient (D_{eff}) multiplied by the difference between actual moisture content (MC) and equilibrium moisture content (EMC). Such a diffusion model could practically be applied to the entire drying process and can be expressed as follows (Elustondo and Oliveira 2009):

$$\frac{d(MC)}{dt} = -D_{eff}\left(MC_{eff} - EMC\right) \tag{8.34}$$

where

MC_{eff} is the effective moisture content equal to the actual lumber moisture content ($MC_{eff} = MC_{actual}$) below the fiber saturation point and $MC_{eff} = MC_{FSP}$ above the fiber saturation point $\left(\frac{kg_{water}}{kg_{oven\text{-}dry}}\right)$.

EMC is the apparent equilibrium moisture content that results from assuming that the experimental drying rate under diffusion control reduces approximately linearly with the reduction of moisture content $\left(\frac{kg_{water}}{kg_{oven\text{-}dry}}\right)$.

The relationship between the effective diffusion coefficient (D_{eff}) and temperature (Siau 1995) is equal to a constant (D_0) multiplied by an exponential function of the energy of activation ($\Delta h = 34{,}150\,\text{kJ/kmol}$), the ideal gas constant ($R = 8{,}314\,\text{kJ/kmol}$), and the absolute temperature (T) (Siau 1995):

$$D_{eff} = D_0 e^{(-\Delta h/RT)} \tag{8.35}$$

Under constant parameters of the drying air (dry- and wet-bulb temperatures, pressure, relative humidity, and velocity), the specific drying rate (i.e., the amount of moisture removed from the unit area of a dried surface) can be expressed as follows (Mujumdar 2007):

$$\dot{m}_{water} = \frac{m_{oven\text{-}dry}}{A}\left[-\frac{d(MC)}{dt}\right] \tag{8.36}$$

where

$\dot{m}_{water}$ is the specific drying rate $\left(\frac{kg_{water}}{m^2 \cdot s}\right)$

$m_{oven\text{-}dry}$ is the mass of dry solid ($kg_{oven\text{-}dry}$)

A is the surface area exposed (m^2)

MC is the moisture content at time t $\left(\frac{kg_{water}}{kg_{oven\text{-}dry}}\right)$

t is the time (s)

$-\frac{d(MC)}{dt}$ is the infinitesimal variation of moisture content with regard to time $t\left(\frac{kg_{water}}{kg_{oven\text{-}dry}\cdot s}\right)$

When the surface of the moist lumber board is no longer saturated with moisture, the characteristic drying curve concept can be used to relate the unhindered drying rate from the saturated surface to the diminished drying rate (Ashworth 1977; Keey 1992; Pang 1994).

8.12 DRYING TIME

Dimensional lumber is dried from its green condition to achieve structural stability and to avoid shrinkage and stresses. To procure suitable lumber in an economically convenient time, it is artificially dried in kilns (Meroney 1969).

In general, the drying time is shorter when the velocity and dry-bulb temperature of air are high and the relative humidity is low. However, an excessively low relative humidity may produce a surface zone with low moisture content, thus reducing moisture migration close to the surface (Perré and Keey 2015a, 2015b).

If the wood is dried too fast, serious cracking, warping, and fiber collapse may occur. The phenomenon of shrinkage reduces the surface area for heat and mass transfer, which may have an effect on the overall drying rate. Contrary, if the drying operation lasts too long, excessive energy and time costs decrease productivity of the plant (Plumb et al. 1985).

The drying of any wood is rarely controlled by a single process. Empirically, the drying time of lumber is proportional to the quantity of moisture to be removed and to the square of the specimen's density and thickness, as well as inversely proportional to the water saturation vapor pressure, air velocity, and air wet-bulb depression (Hildebrand 1989).

Drying time of lumber also depends on the air velocity. At very low air velocities (e.g., less than 1 m/s), the airflow through the lumber stack(s) is often laminar, and the heat transfer between the timber surface and the moving air stream is not too effective (Walker et al. 1993; Pordage and Langrish 1999).

For drying temperatures under 100°C, the following equation can be used to estimate the drying time t (in days) for different board thicknesses l (mm) and temperatures T(°C):

$$t = \frac{1}{\alpha}\ln\left(\frac{MC_{in}}{MC_{fin}}\right)\left(\frac{l}{25}\right)^{1.25}\frac{65}{T} \quad (8.37)$$

where

MC_{in} and MC_{fin} are the initial and final moisture contents, respectively $\left(\frac{kg_{water}}{kg_{oven\text{-}dry}}\right)$

α is a coefficient that depends on the wood species, initial moisture content, humidity, kiln design, and other factors; it can be 0.048 for softwoods and 0.027 for hardwoods.

The total drying time of a drying cycle is a sum of drying times in the succeeding drying periods. If the drying rate curve is known, the total drying time required to reduce the solid (wood) moisture content from MC_{in} (initial) to MC_{fin} (final) can be calculated by

$$t_{drying} = -\frac{m_{oven\text{-}dry}}{A} \int_{MC_{in}}^{MC_{fin}} \frac{d(MC)}{\dot{m}_{water}} \tag{8.38}$$

where

$m_{oven\text{-}dry}$ is the mass of bone-dry solid ($kg_{oven\text{-}dry}$)

A is the total exposed drying surface of the solid material (m^2)

MC_{in} is the initial average moisture content $\left(\frac{kg_{water}}{kg_{oven\text{-}dry}}\right)$

MC_{fin} is the final average moisture content $\left(\frac{kg_{water}}{kg_{oven\text{-}dry}}\right)$

$\dot{m}_{water}$ is the specific drying rate (per unit surface of dried area) $\left(\frac{kg_{water}}{m^2 \cdot s}\right)$

The negative sign in equation 8.38 indicates that the moisture content decreases with time.

8.13 KILN OVERALL DRYING EFFICIENCY

Drying of wood is energy intensive, because, among other factors, a high amount of energy is required to move the moisture inside the lumber boards and then vaporize it from the boards' surfaces. In addition, depending upon the type of equipment used to dry wood, the thermal efficiency of the operation may require up to 4 times the energy actually needed to evaporate the water. Therefore, typical convective dryers, even if well designed and well operated, can be less than 50% efficient.

Green wood to be dried may contain as much as 75% water (by mass). Even when optimally operated, the kiln drying efficiency is affected by environmental factors. Certain operating practices and maintenance procedures may reduce (or increase) the dryer efficiency (Tschernitz 1986; McCurdy 2005).

Generally, the inefficiency of conventional dryers is due to one or more of the following factors (Law and Mujumdar 2008): (i) most conventional dryers are convective type where drying medium has to be in contact with the material for heat and mass transfer; hence, contacting efficiency between the drying medium and the materials determines the drying performance in terms of thermal efficiency; (ii) in convective drying, significant loss of energy occurs when enthalpy in exhaust air is vented without partial recycle or heat recovery, especially if the flow rate of exhaust air is high; and (iii) poor insulation dryers causes appreciable loss energy to the environment.

The kiln energy efficiency can be defined as the ratio of total energy used for moisture vaporization to total energy supplied for drying:

$$\eta_{kiln,\ energy} = \frac{E_{vap}}{E_{total}} < 1 \tag{8.39}$$

where

E_{vap} is the energy used for moisture vaporization (J)
E_{total} is the total energy supplied for drying (J)

At low dry-bulb temperature and relative humidity, the thermal energy (heat) is proportional to temperature gradients, and the kiln energy efficiency can be approximated to thermal efficiency:

$$\eta_{kiln,\ thermal} = \frac{T_{in} - T_{out}}{T_{in} - T_{amb}} \tag{8.40}$$

where

T_{in} is the air dry-bulb temperature entering the lumber stack(s) (°C)
T_{out} is the air dry-bulb temperature leaving the lumber stack(s) (°C)
T_{amb} is the ambient medium dry-bulb temperature (°C)

Kiln thermal efficiency could also be expressed in terms of the actual absolute humidity change over the lumber stack(s) dryer compared to the saturated absolute humidity change (i.e., the maximum amount of moisture the air can absorb) (Anderson 2014):

$$\eta_{kiln,\ thermal} = \frac{\omega_{out} - \omega_{in}}{\omega_{sat} - \omega_{out}} \tag{8.41}$$

where

ω_{in} is the air absolute humidity at the lumber stack(s) inlet $\left(\frac{kg_{water}}{kg_{dry\ air}}\right)$

ω_{out} is the air absolute humidity at the lumber stack(s) outlet $\left(\frac{kg_{water}}{kg_{dry\ air}}\right)$

ω_{sat} is the absolute humidity of the air at saturation $\left(\frac{kg_{water}}{kg_{dry\ air}}\right)$

The maximum kiln efficiency is achieved when the drying air leaves the lumber stack(s) at (saturated) wet-bulb temperature. Thus, it can be expressed as the ratio of the maximum temperature difference between the air temperatures at the inlet and outlet of the lumber stack(s) to temperature difference between the inlet temperature and ambient temperature:

$$\eta_{kiln,\ \max} \frac{T_{in} - T_{wet\text{-}bulb}}{T_{in} - T_{amb}} \tag{8.42}$$

where

T_{in} is the air temperature entering the lumber stack(s) (°C)

$T_{wet\text{-}bulb}$ is the wet-bulb temperature of the air leaving the lumber stack(s) (°C)

T_{amb} is the ambient temperature (°C)

The performance level of a given dryer should not be generalized because it is dependent, among many other factors, on its geographical location, ambient and operating conditions, maintenance, etc.

Among the measures allowing improving the kiln energy and thermal efficiency, the following can be mentioned (Law and Mujumdar 2008): (i) proper selection of dryer by using models and software; (ii) use intermittent drying to allow to internal moisture to move to the evaporating surfaces during off cycles; (iii) provide appropriate kiln maintenance to minimize or reduce air leakage, uncontrolled heat losses, and reduce the humidification requirements; and (iv) use efficient heat recovery methods (e.g., heat pumps) and/or alternate primary energy sources (solar, biomass, etc.)

8.14 ENERGY REQUIREMENTS

Drying technologies (such as air-convective batch drying of lumber), used to remove moisture under controlled conditions of temperature and humidity, are essential to add value, achieve adequate quality, and lower shipment costs of many products. However, they can be very expensive operations in terms of energy usage (Comstock 1976).

In practice, drying time and quality of dried timber have sometimes higher priority than the energy use, which limits the possibilities to achieve high energy efficiency of industrial sawmills.

Most of the wood drying technologies in use today were designed when conventional energy (fuel, electricity) and biomass prices were relatively low. At that time, the energy use was not a critical issue, and relatively few efforts have been deployed to increase the energy efficiency of lumber kilns (Anderson 2014).

Today, with higher energy prices and production costs, sawmills must decrease their energy consumption and increase the heating system efficiency of lumber kilns. One of the most efficient situations for wood product mills is still to use their own wood waste to fire the boilers, thus reducing the global energy costs.

The total amount of energy requirement (or consumed) depends on the following factors: (i) those affecting consumption of energy by the wood being dried (including species density, initial and final moisture contents, and actual drying schedule used); the material factors have the greatest influence on energy requirements; if, for example, both wood initial moisture content and density are doubled independently, the total energy requirements may increase by 82% and 61%, respectively; also, by increasing the lumber final moisture content from 15% to 19%, energy savings up to 5% could be realized; (ii) those affecting consumption by structural and external components (including kiln type, size and construction materials, environmental conditions, type of heating, air temperature and velocity, venting, and control systems); since structural losses account for 25%–30% from the total energy consumed, thermal conductivity values of kiln construction materials must be carefully selected; if heat losses through the kiln structure double, the energy consumption may increase by 17%, mainly because of wet insulation, leaks through and/or around vents, doors, and cracks; regular maintenance of kilns is one of the best solutions for keeping external influences to a minimum; (iii) those affecting the drying time; kiln energy consumption can be minimized by keeping drying times to a minimum; for example, when drying time is doubled and keeping everything else constant, about 30% more energy could be consumed; this may be attributable to the influence of structural losses, which may change in time; and (iv) those affecting the steam spray consumption to humidify incoming air; if, for example, 25% of the incoming air is humidified instead of 5%, the total energy requirement may increase by about 11%; changes in the relative humidity of the outside air may also affect the energy consumption for incoming air humidification.

It can be concluded that (i) the energy consumed by industrial lumber dryers is usually very significant; (ii) energy requirements vary with kiln size; smaller kilns require more energy per unit of dried lumber; (iii) material variables (such as wood density, initial and final moisture contents) have the greatest influence on energy requirements; (iv) the kiln total energy consumption can be minimized by optimized operating procedures; and (v) regular maintenance practices of kilns are among the best methods to reduce the structural heat losses over technical life of industrial wood dryers.

8.14.1 World Energetic Context

The total world energy supply (consumption) is today more than 150,000 TWh (2008) of which oil and coal combined represent over 60% (Figure 8.6a) (IEA 2014). With the current consumption rates, most of the nonrenewable energy resources could be consumed during the next 100 years (IEA 2014). Industrial users (agriculture, mining, manufacturing, and construction) consume about 37%; personal and commercial transportation—20%; residential (heating, lighting, and appliances)—11%; and commercial buildings (lighting, heating, and cooling)—5% of the total world energy supply. The other 27% of the world's energy is lost in energy transmission and generation.

In most countries, electricity is generated from fuels, and the primary energy consumption can be about 3 times higher than the final energy delivered (Kemp 2012).

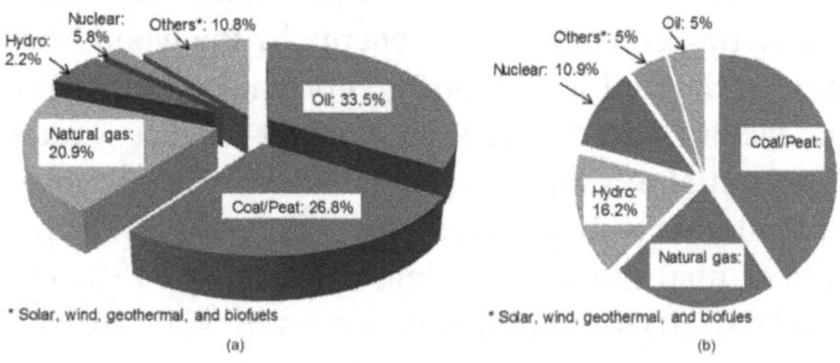

FIGURE 8.6 (a) Total world's energy supply (consumption); (b) world's electricity generation by fuel (2008).

On the other hand, the world electricity generation exceeds 21,000 TWh (2008), of which more than 60% is produced by using coal and natural gas as primary energy sources (Figure 8.6b). Refrigeration, heat pump, and air conditioning industries consume about 10%–15% of the total available electrical energy (IEA 2015).

Global CO_2 emissions come from electrical power generation (40%), as well as from industry (17%), buildings (14%), and transport (21%) energy consumptions.

This world energy context opens up opportunities for developing alternative renewable and clean energy sources, such as solar, wind, hydrogen, water hydrokinetic, nuclear, ambient air, and geothermal. In addition, such considerable global energy use and CO_2 emissions could be reduced, especially in industry, if best available technologies were to be developed and worldwide applied.

8.14.2 Energy Consumption in Wood Drying Industry

Between 9% and 25% of national energy of the developed countries is consumed for industrial thermal drying processes. Particularly in wood sawmills of developed countries, drying processes, including material processing, handling, sawmilling, and lumber drying, as well as services such as compressed air, space heating, and lighting of premises (Vidlund 2004), may consume as much as 70% of the plant's total required energy for primary transformation of woods (fossil, electrical, and renewable) (Mujumdar 1995). Material handling features as a major consumer of electricity accounting for some 15%–28%. The energy needs for services such as lighting, space heating, hot water or steam systems, compressed air, and workshops facilities are quite significant. During the last three decades, national energy consumption for industrial drying operations ranged from 10%–15% for the United States (1600 * 10^9 MJ/year), France (168 * 10^9 MJ/year), and the United Kingdom (128 * 10^9 MJ/year) to 20%–25% for Denmark and Germany (Baker and Reay 1982; Larreture and Laniau 1991; Law and Mujumdar 2008). In particular, energy consumption in drying ranged from a low value of under 5% for chemical process to 35% for papermaking operations.

In Canada, from 1990 to 2011, energy use in the wood products industry increased by 14% (Nyboer and Bennett 2013). During the same period, gross output rose by 24% and gross domestic product, by 27% (Statistics Canada 2013).

In Sweden, sawmill industry produces 16.4 Mm^3 timber annually, and the drying process uses between 78% and 83% of the total heat consumed in sawmills (Stridberg and Sandqvist 1984; Vidlund 2004).

On a national scale, the proportion of energy consumed in the wood industry in relationship to industry as a whole varies according to the degree of its importance to the national economy. For example, in Canada and the United States, the energy consumption in the wood industry typically represents 1.3% and 0.9%, respectively, versus that of the total industry (www.fao.org/docrep/t0269e/t0269e04.htm, Accessed August 11, 2016).

In Sweden, the specific electrical energy requirements for drying species such as softwoods and hardwoods vary from 20 to 30 GJ/m^3, respectively (for air-dried saw timber) and from 1.5 to 2.5 GJ/m^3 (for kiln-dried saw timber), respectively (Table 8.2) (Vidlund 2004). Also in Sweden, the specific thermal energy requirements for drying

TABLE 8.2
Energy Specific Consumption within Mechanical Wood-Based Industry in Sweden

	Electrical (kWh/m³)	Thermal (GJ/m³)	Fuel (L/m³)
Sawn timber (air-dried)			
Hardwood	30	–	5
Softwood	20	–	4
Sawn timber (kiln-dried)			
Hardwood	75	2.5	5
Softwood	45	1.5	4

species such as softwoods and hardwoods vary from 0.06 to 0.20 GJ/m^3, respectively (for air-dried saw timber) and from 45 to 75 kWh/m^3, respectively (for kiln-dried saw timber) (Table 8.2) (Vidlund 2004).

Expressed in terms of specific energy required for moisture vaporization, species such as southern yellow pine may require between 3.7 and 5.1 MJ/kg moisture evaporated, douglas-fir, from 4.7 to 7.0 MJ/kg, and refractory lumbers such as white oak, more than 14 MJ/kg (Rosen 1995).

Table 8.3 shows, as typical examples, the electricity and heat specific consumptions for lumber production processes in Sweden (Anderson 2012; Anderson and Westerlund 2014). It can be seen that the drying process accounts there for about 77 kWh/m^3 of the total energy consumed in sawmills, and the remaining part is used for building (office) heating.

The specific electricity (kWh/m^3), purchased or on-site generated by diesel generators or cogeneration plants, is most used for kiln's fans (31%), sawing (23%), and grinding (13%).

Today, as in the future, energy savings is an important issue due to the limited energy resources and ever-increasing demand for the primary transformation of wood (Comstock 1976; Department of Energy Buildings 2015). Improving the

TABLE 8.3
Breakdown of Specific Electricity and Heat Usage in Swedish Lumber Production

Process	Electricity (kWh/m³)	Heat (GJ/m³)	Temperature (°C)
Barking	4	–	–
Sawing	23	10	30
Sorting	2	5	30
Drying	31	299	75
Dry handling	4	5	30
Grinding	13	5	30
Building (offices)	–	15	30
Total	77	339	–

energy efficiency of drying equipment in order to reduce and/or recover a part of energy losses, mostly due to the moist air venting, is one of the most relevant objectives of R&D activities throughout the world.

The potential reduction in energy use by dry kilns in Canada, for example, would be 5.5 PJ/year (1 PJ = 10^{15} J), which corresponds to 355 kT/year (1 kT = 103 MT) in carbon dioxide emissions. Furthermore, it is estimated that CO_2 emissions could be reduced by an additional 90 kT/year through a decrease in the amount of downgraded lumber.

Historically, the increase in the worldwide energy demand along with cheap fossil fuels has contributed to an increase in greenhouse gas (GHGs) emissions and global warming. However, in Canada, for example, while energy use increased by 14% over the 1990–2010 period, direct GHGss are today (2016) down 41% from 1990 levels. That is because the industry became increasingly dependent on wood waste (carbon-neutral) and electricity (no direct emissions) to meet their energy needs. The contribution of fossil fuels to total energy use declined from 37% in 1990 to just 19% in 2011, while the production of softwoods and hardwoods was in 2011 nearly 37% lower than the peak production year of 2004 (source: Energy Data-ICE, Emission Factors-NRCAN, Environment Canada. Accessed December 15, 2017). In Canada, more than 54% (2004) of wood residues (biomass) are employed to produce 25.6% of electricity and 14.4% of natural gas used in the wood products manufacturing industry, including drying processes (Nyboer and Bennett 2013).

GHGs per unit of gross output were in 2011 over 50% below 1990 levels, mainly because of using biomass and electricity instead of fossil fuels in a context where, over the time, the contribution of wood and electricity increases, while the contribution of fossil fuels declines. However, although the use of electricity does not produce direct GHGs, it causes indirect emissions in Canadian provinces (such as Alberta, eastern Canada) where electricity is, in majority, generated from fossil fuels.

8.14.3 Energy Sources and Costs

Among the economic criteria for choosing the energy sources for drying wood, the following can be mentioned: (i) the specific cost, (ii) the continuity availability of supply, (iii) the capital cost of the equipment needed to convert the energy source into useful thermal energy, and (iv) the efficiency of converting the energy source into useful thermal energy.

In regard to all these criteria, direct-fired kilns, where the combustion gases enter directly the drying chamber, are up to 100% efficient since any steam boiler and heat transfer coils are needed and, thus, require less capital investment.

Industrial sawmills typically use heat generated from fossil (mainly, oil, natural gas, coal, or propane), wood residues (biomass), and electrical energy sources.

Table 8.4 shows the relative use of fuels (%) of total energy use in the wood products manufacturing industry in Canada (NAICS 321. 2012).

Worldwide, within the renewable energy sector, which represents 13% of the total energy production, the use of waste and biofuels has increased drastically during the last decade accounting for 75.2% of the renewable energy supply (IEA 2014). The

TABLE 8.4
Relative Use of Energy Sources versus Total Energy Use in the Wood Product Manufacturing Industry in Canada

Energy Type	%
Wood (biomass)	54.5
Electricity	25.6
Natural gas	14.2
Steam	0.9
Propane	0.7
Others	4.1

subsidies for biomass, together with the rising energy prices, have made biomass a desirable product on the wood energy market. This has led to higher biomass prices and an increased interest in improving the resource and energy efficiency associated with biomass production.

Biofuel is another substitute for fossil fuels to decrease the GHGs. One challenge with biofuels is to find sufficient amounts of biomass since the foresting is already close to its maximum sustainable capacity. In Canada, for example, electrical (4560 GWh) and thermal (940 GWh) annual production using wood waste represent 6.2% of the country's renewable energy capacity (2009), about 8% of total energy generation (estimated) (Statistics Canada 2013). Even the available biomass from foresting residues is not sufficient to replace the nonrenewable resources, it could be better used by increasing the overall resource and energy generation efficiency.

In sawmills, fossil and biomass (produced by themselves as by-products from their own lumber production or otherwise purchased from nearby biomass industries) and energy sources are burned in boilers (furnaces) with an annual average efficiency of about 80% in order to produce steam or hot water (Kemp 2014).

Sawmills are important suppliers to the biomass market. In Sweden's sawmills, for example, about half (47% of the entering mass of timber) is transformed into lumber for final transportation to the market, and the remainder (around 53%, wt) is considered as by-product (biomass), for example, wood chips (26%), bark (19%), and sawdust (8%) (Figure 8.7). A significant part of the biomass is used for internal heat production, mainly for forced drying of lumber in kilns, via boilers (Anderson and Westerlund 2014).

Relative costs of energy sources, such as fossil fuels, wood residues, and electricity, vary from one country to another. In many countries, the cost of electricity is typically higher than that of fuel. The exception comes where electricity is mainly generated as hydroelectricity, in which case the fossil fuel use is almost zero and the generating cost is also very low (e.g., Canada/Québec, Norway, and South Korea). The charge made for hydroelectric power is usually based on amortization of the high capital costs, with a small amount for operating costs (principally, labor and maintenance). Likewise, nuclear power tends to have relatively low fuel cost but high capital charges, which should also allow for final decommissioning costs (Kemp 2014).

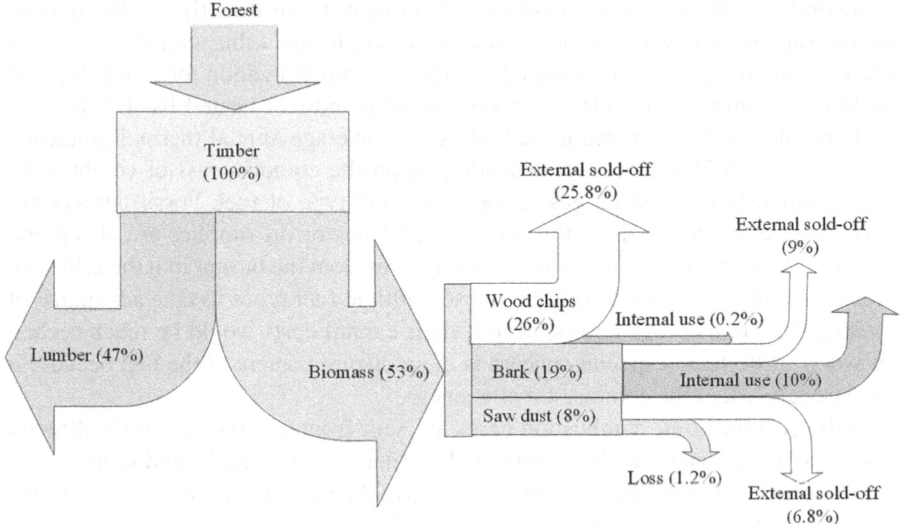

FIGURE 8.7 Example of lumber and biomass production from timber in Sweden.

Table 8.5 shows, as typical example, the specific costs of current fossil and electrical energy sources in the United States in 2017 (in 2016 US$) (source: www.eia.gov/outlooks/aeo/pdf/electricity_generation.pdf. Accessed February 6, 2018).

8.14.4 Heat Carrier Fluids

The heat produced by means of electricity, fossil fuels, or biomass energy sources is transported to conventional and heat pump-assisted kilns by thermal carriers (fluids) in order to heat the wood load and auxiliary equipment, and vaporize the moisture extracted from the lumber stack(s).

Usually, steam (one of the most efficient heat carriers), combustion gases (used in direct-fired kilns), and hot water and thermal oil (generally used in small-scale kilns) serve as heating thermal agents in forced-air convective wood drying facilities.

Most of these conventional heat sources can be combined with other nonconventional heat sources such as solar and electrical energies (microwave, radio frequency, infrared, and heat pumps).

TABLE 8.5
Prices of Energy Sources in the United States (2017)

Energy Type (Unit)	Price
Electricity (2016 US$/kWh)	0.10
Distillate fuel oil (2016 US$/GJ)	17.47
Natural gas (2016 US$/GJ)	5.24
Propane (2016 US$/GJ)	14.00

Steam is supplied to heating coils (indirect heating) or directly to kiln as saturated or superheated water vapor. Although indirect heated kilns with steam are less efficient than direct-fired kilns, the advantage of steam utilization for equalizing and conditioning must be considered. In conventional indirect heated batch kilns, the most common sources of steam are boilers with average annual thermal efficiency varying between 70% and 80% depending upon the completeness of combustion, excess combustion air, stack gas temperature, and type of fuel. Fossil burners are particularly efficient for dry kilns when natural gas or oil supplies are cheap and available. A special concern consists in ash flyover from the burner into the kiln. The most desirable boilers are those able to use multiple fuel types to take advantage of changing fuel markets and costs, even if their capital costs would be much higher. For wood waste fuel, a special concern is the moisture content of the fuel because it may strongly affect the combustion efficiency.

In direct-fired kilns, combustion gases are sent from a burner assembly directly into the dryer chamber. Such systems use less fuel than indirect-heated kilns.

For sawmills that use large quantities of thermal energy, the cogeneration systems supplying high-pressure steam for production of electricity can be considered, resulting in 8%–20% savings in energy costs (Tschernitz 1986).

8.14.5 Energy Requirements for Batch Wood Dryers

The average specific electrical energy consumption (per m^3 of dried wood) of modern kiln dryers is about 285 kWh/m^3 in Norway (Vidlund 2004). In Canada, for drying wood species such as fir, it may reach about 300 kWh/m^3 (Clément and Fortin 1996). For drying radiata pine in New Zealand (Ananias et al. 2012), where this lumber is predominantly sapwood with high moisture contents of approximately 150% (Sun et al. 2003), and Chile (Meil et al. 2010), the average specific thermal (heat) energy is about 3 GJ/m^3. In Canada, the thermal consumption for drying mixed spruce, pine, and fir is approximately 1 GJ/m^3 (Garraham et al. 2008).

This section identifies and quantifies the main heat requirements of conventional, convective, forced-air industrial batch dryers during the warming-up (preheating) stage, as well as during the subsequent steps of the entire drying process.

In a simple, open air-convective, continuous drying system (schematically shown in Figure 8.8a), the dried material enters the drying chamber at state *a* with total mass $\dot{M}_{dry} + \dot{M}_{wet}$, initial moisture content MC_{in}, and dry-bulb temperature T_a. After drying, the dried material leaves the drying chamber (state *b*) at temperature T_b, mass $\dot{M}_{dry}$, and final moisture content MC_{fin}.

On the other hand, the drying air (state 1) with flow rate $\dot{m}_{air}$ is heated at constant absolute humidity (process 1-2), then enters the dryer as hot and dry air at state 2 and, finally, absorbs moisture from the dried material (process 2-3).

The thermal power provided for heating the drying air is given as follows:

$$\dot{Q}_{12} = \dot{m}_{air}\left(h_2 - h_1\right) \tag{8.42a}$$

where

$\dot{Q}_{12}$ is the thermal power provided for heating the drying air (W)
$\dot{m}_{air}$ is the mass flow rate of drying air (kg/s)

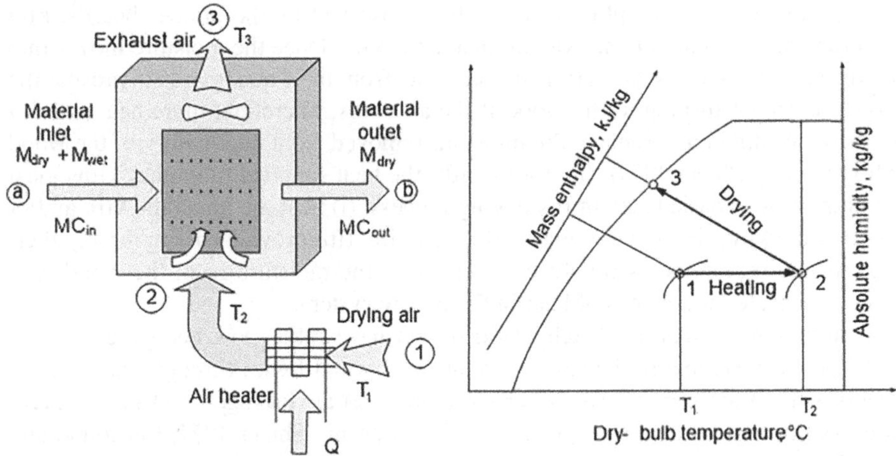

FIGURE 8.8 (a) Schematic representation of an open, continuous drying system; (b) thermodynamic process of the drying air in the Mollier diagram; $\dot{M}_{dry}$, mass of dry material leaving the dryer; $\dot{M}_{wet}$, mass of wet material entering the dryer; MC_{in}, material initial moisture content; MC_{fin}, material final moisture content; $\dot{Q}$, heat rate input; T, dry-bulb temperature.

h_1 is the specific enthalpy of drying air at the inlet of air heater (J/kg)
h_2 is the specific enthalpy of drying air at the outlet of air heater (J/kg)

The maximum rate of moisture (water) (kg_{water}/s) extracted from the material is given as follows:

$$\dot{M}_{water} = \dot{m}_{air}(\omega_3 - \omega_2) = \dot{m}_{air} \tag{8.42b}$$

where

ω_1, ω_2, and ω_3 are the absolute humidity of the drying air at states 1, 2, and 3 during the process 1-2-3 $\left(\frac{kg_{water}}{kg_{dry\ air}}\right)$.

If $\bar{c}_p$is the average specific heat of the material, the thermal power balance (kW) of the simplified drying process represented in Figure 8.8b is given as follows:

$$\dot{m}_{air} \times h_3 - \left(\dot{m}_{air} \times h_2 + \dot{M}_{water} \times c_p \times T_a\right) = \dot{M}_{dry} \times c_p \times (T_a - T_b) \tag{8.43}$$

For wood batch drying kilns, the total amount of energy demand depends on the following factors: (Shottafer and Shuler 1974; Comstock 1976; Elustondo and Oliveira 2009) (i) kiln location, type, size, and building materials; (ii) wood density, initial and final moisture contents, and actual drying schedule used; (iii) environmental conditions (winter, summer); and (iv) type of heating, air circulation, venting, and control systems.

The energy requirements of typical conventional batch wood dryers obtained from burning fossil fuels or wood residues and electrical energy (Simpson 1991) can be divided into (Clément and Fortin 1996; Anderson 2014) (i) thermal (95% or more) and (ii) electrical (5%, mainly for circulation fans) (see Figure 8.8).

A part of the heat supplied to the kiln is absorbed by the lumber boards, kiln structure, and the equipment existing inside the kiln. Once the moisture moves into the drying air, the moist air is transported away from the wood boards to prevent the moisture from returning to the wood if the air cools. Therefore, more heat and dry air must be supplied to replace the moist air removed from the vicinity of the wood (Shottafer and Shuler 1974). In other words, the heat supplied to lumber kilns must accomplish four basic tasks in the drying process: (i) heat the air in the kiln and/or entering the kiln, (ii) heat any material in the kiln, (iii) provide energy for any thermodynamic requirements involved in removing the moisture from the wood, and (iv) compensate for any loss of heat in the drying system.

Conventional industrial batch wood dryers have rather low energy efficiency, often below 50%, mainly because of multiple thermal losses through and/or associated with (i) air venting, (ii) dryer structure, (iii) air leakages, (iv) heat generation system, and (v) heat transportation (Shottafer and Shuler 1974; Elustondo and Oliveira 2009; Anderson 2014).

During the first stages of drying, the energy efficiency is usually high because free water evaporates easily from the surfaces of lumber boards. However, energy efficiency is reduced toward the end of drying when absorbed water migrates slowly through the wood by diffusion. The energy consumption can be optimized according to the stages of drying (Strumillo et al. 1995), for example, by setting small wet-bulb depressions during warm-up to reduce the vapor condensation and the energy required to maintain the air relative humidity (Perré and Keey 2006).

Figure 8.9 schematically represents the main entering and leaving heat fluxes in a typical conventional, forced-air convection kiln as well as their approximate

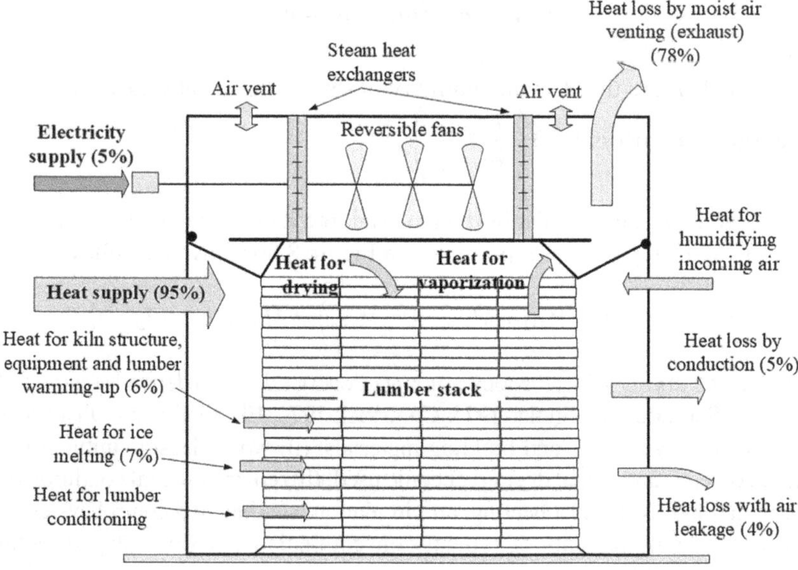

FIGURE 8.9 Schematic representation of a typical conventional kiln showing the approximate percentages of heat requirements.

percentages of the total energy demand (Elustondo and Oliveira 2009; Anderson and Westerlund 2014).

It can be seen that the energy supplied to such conventional batch wood dryers (i) heats the kiln structure and equipment, and lumber stack(s); (ii) heats the air in the kiln and/or air entering the kiln; (iii) helps the moisture to move inside the boards by overcoming the moisture retention forces until the equilibrium moisture content is reached; (iv) provides heat to remove the moisture from the wood boards by forced vaporization; and (v) compensates for any loss of heat in the system (Shottafer and Shuler 1974).

With more detailed words, the conventional batch wood dryers require energy to (Tschernitz 1986) (i) compensate the heat losses of the heat production and delivery systems; (ii) compensate heat losses during the kiln start-up, shutdown, and off-load periods; (iii) preheat the lumber and kiln structure and equipment from the initial temperatures to first set drying temperature; (iv) move moisture inside the lumber boards and vaporize it from the lumber boards' exposed surfaces; (v) compensate the heat losses through dryer's structures (walls, ceiling, doors, and floor) by conduction along the supporting framework, and by convection and radiation from the outer surfaces, especially in cold climates; (vi) compensate the heat losses associated with air venting used to remove water vapor from the dryer; (vii) compensate the heat losses associated with dryer air leakages; and (viii) energize the dryer's fans for air circulation.

The largest heat losses of wood dryers (about 78% of the total energy requirement) are associated with the air venting losses (Figure 8.8) (Keey et al. 2000; Anderson and Westerlund 2011; Anderson 2014). The high amount of air venting (evacuation) losses gives opportunities to recover a part of loosed energy, usually by using air-to-air heat recovery heat exchangers or phase change systems.

Other energy requirements of batch wood dryers are attributed to the following (Figure 8.8): (i) heat losses by thermal conduction through the kiln envelope (walls, roof, and floor) (5%); if the kiln cannot reach the set dry-bulb temperatures, these losses may cause water vapor condensation, thus increasing the heat required to maintain the air humidity inside the kiln (Perré et al. 2007); if the kiln insulation is effective, using high temperatures for drying can reduce energy consumption (Elustondo and Oliveira 2006; Menshutina et al. 2004; Kudra et al. 2009); (ii) heat losses with air leakages that mainly arise when the kiln is open during lumber loading lumber (4%); (iii) heat to warming-up the lumber stack(s), dryer structure (walls, doors, floor, etc.), and equipment at the beginning of the drying processes (6%); (iv) heat to melt ice when the lumber has been stored below 0°C (7%) (Anderson and Westerlund 2014); (v) heat for humidifying the incoming air, when required (Shottafer and Shuler 1974); and (vi) heat for lumber conditioning, when required.

It can be seen that heat losses from conventional batch kilns, other than those associated with air venting, vary typically from 5% to 10% from the total energy consumption. These heat losses can be greater if the dryers are poorly insulated, especially in cold climates.

The detailed knowledge of all energy requirement (consumption) components of wood dryers, as well as energy efficiency of the production systems, allows to better manage the energy generation and distribution and determine the global, overall energy requirements for the entire wood drying facility (Clément and Fortin 1996). The main components of energy requirements could be described and evaluated by a simple, practical step-by-step method (Shottafer and Shuler (1974).

8.14.5.1 Warming-Up (Preheating) Step

The warming-up (preheating) step is required for both low- and high-temperature wood drying processes in order to raise the temperature of wood substance and of kiln envelope, and equipment from their initial values up to the set dry-bulb temperature of the (next) first drying step.

8.14.5.1.1 Warming-Up the Wood Substance

During the preheating step, sensible heat is required to warming-up the kiln-wide stack(s) of green wood and other woody materials (mainly, stickers and timber sleepers) from their initial temperature at the dryer inlet to the dry-bulb set temperature of the first drying step. The quantity of heat absorbed depends upon the total mass of each material, the specific heat of the material, and the dry-bulb temperature change from initial to final states. This preheating process is more critical when the dryer is charged with cold or frozen lumber during winter in cold climates (Shottafer and Shuler 1974; Cech and Pfaff 2000).

8.14.5.1.1.1 Initial Temperature >0°C If the initial temperature of the woody materials at the dryer inlet is higher than 0°C, the sensible heat (kJ) required for warming-up the lumber kiln-wide stack(s) and auxiliary materials can be expressed as follows (Shottafer and Shuler 1974):

$$Q_{1a} = \sum_{i=1}^{3} m_i \times c_i \times \left(T_{set,1} - T_{in,i}\right) \tag{8.44}$$

where

$i = 1$ for lumber stack(s)
$i = 2$ for woody stickers
$i = 3$ for timber sleepers
m_i is the anhydrous mass of green wood boards, stickers, and sleepers, respectively (kg)
c_i is the specific heat of green wood boards, stickers, and sleepers, respectively (KJ/kg·K)
$T_{set,1}$ is the set dry-bulb temperature of the first drying stage (°C)
$T_{in,i}$ is the initial temperature of green wood boards, stickers, and sleepers, respectively (°C)

It can be noted that the specific heat of wet wood is greater than that of dry wood and varies with the moisture content. For example, at 100% of moisture content, $\overline{c}_{p,wet\ wood} = 5.552\ \text{kJ/kg} \cdot \text{K}$, while at 10% it becomes $\overline{c}_{p,wet\ wood} = 1.786\ \text{kJ/kg} \cdot \text{K}$.

Exercise E8.1

A green lumber stack having the total mass of 10,000 kg and specific heat of 3 kJ/kg·K enters a batch wood dryer at T_{in} = 20°C and is heated up to the first set dry-bulb drying temperature $T_{set,1}$ = 90°C. Calculate the sensible heat (Q_{sens}) (kJ) required to warm-up the wood stack only.

Solution

The sensible heat (kJ) required to warm-up the wood stack can be calculated as follows (equation 8.44):

$$Q_{sens} = m_{wood} \times \overline{c}_{p,wood} \times \left(T_{set,1} - T_{in}\right) = 10{,}000 \times 3 \times (90 - 20) = 2.1 \times 10^6 \text{ kJ}$$

where

m_{wood} is the mass of green wood stack (kg)
$\overline{c}_{p,wood}$ is the average specific heat of wood (kJ/kg·K)
T_{in} is the initial temperature of green wood boards (°C)
$T_{set,1}$ is the set dry-bulb temperature of the first drying stage (°C)

8.14.5.1.1.2 Initial Temperature <0°C If the lumber initial temperature is lower than 0°C, special considerations must be applied depending upon the wood species to dry and their optimum drying temperatures. This is because wood is a good insulating material, and, during the initial preheating period, the temperature of the wood surface rises rapidly in comparison to the core.

Particularly, easy-to-dry softwoods, dried at high temperatures, dry rapidly, and, consequently, the core may still be frozen as the shell begins to dry. This establishes severe moisture gradients with comparatively dry shell in which tensile stresses due to shrinkage are usually sufficient to cause surface checking. Consequently, the softwood lumber stack(s) must be thawed prior entering the first stage of the drying cycle. This is accomplished by running the kiln with both wet- and dry-bulb set points at the same temperature of about 60°C for 5–10 hours, depending on the species and lumber thickness. On the other hand, the slow-drying hardwoods, normally dried at low temperatures, do not pose a problem in this respect since the initial drying conditions are relatively mild, allowing the lumber to thaw before the drying takes place (Cech and Pfaff 2000).

If the lumber enters the kiln at an initial temperature below 0°C, the thermal energy (sensible and latent heat) required for melting ice and, then, heating the resulting water up to the first dry-bulb set temperature of the first drying step can be calculated with the following equation (Shottafer and Shuler 1974):

$$Q_{1b} = m_{ice}\left[\overline{c}_{p,\,ice} \times \left(0 - T_{in,\,ice}\right) + h_{melt}\right] + m_{water}\left[\overline{c}_{p,\,water} \times \left(T_{set,1} - 0\right)\right] \quad (8.45)$$

where

Q_{1b} is the thermal energy (sensible and latent heat) required for melting ice and, then, heating the resulting water up to the first dry-bulb set temperature of the first drying step (kJ)
m_{ice} is the mass of ice initially present in the lumber stack(s) (kg)

$\overline{c}_{p,\, ice}$ is the average specific heat of ice (kJ/kg · K)
$T_{in,\, ice}$ is the initial temperature of ice (°C)
h_{melt} is the enthalpy (latent heat) of melting (kJ/kg) of water (for water, it is about 334.94 kJ/kg)
m_{water} is the mass of water resulted from the melting of ice (kg)
$\overline{c}_{p,\, water}$ is the average specific heat of water (kJ/kg)
$T_{set,\, 1}$ is the set dry-bulb temperature of the first drying step (°C)

8.14.5.1.2 Warming-Up the Kiln Structure and Equipment

The kiln envelope (i.e., structure, foundation, walls, roof, doors, etc.) and auxiliary equipment (e.g., fans, motors, baffles, temperature sensors, controls) must be preheated prior each new drying cycle from their initial temperatures up to the set dry-bulb temperature of the first drying step.

The sensible thermal energy (sensible heat) required for warming-up the dryer envelope (Q_{env}) and equipment (Q_{equip}) can be calculated for each kiln's component as follows:

$$Q_2 = Q_{env} + Q_{equip} = \left[\sum_{env}^{j} m_j \times c_{p,j} + \sum_{equip}^{k} m_k \times c_{p,k}\right] \times \left(T_{set,1} - T_{in}\right) \tag{8.46}$$

where

Q_2 is the sensible thermal energy (sensible heat) required for warming-up the dryer envelope and equipment (kJ)
m_j is the mass of the envelope component j (kg)
m_k is the mass of the auxiliary equipment k (kg)
$c_{p,\, j}$ is the specific heat of the envelope component j (kJ/kg · K)
$c_{p,\, k}$ is the specific heat of the auxiliary component k (kJ/kg · K)
$T_{set,\, 1}$ is the set dry-bulb temperature of the first drying stage (°C)
T_{in} is the initial temperature of the kiln's envelope and auxiliary equipment (°C)

By neglecting the masses and specific heats of all auxiliary equipment, the overall heat capacity of the kiln structure (mainly concentrated in the walls, roof, and floor) can be approximately calculated as follows:

$$\overline{C}_{p,\, kiln} \approx m_{kiln} \times \overline{c}_{p,\, kiln} \tag{8.47}$$

where

$\overline{C}_{p,\, kiln}$ is the overall heat capacity of the kiln structure (kJ/K)
m_{kiln} is the approximate mass of the kiln's structure (kg)
$\overline{c}_{p,\, kiln}$ is the approximate average specific heat of kiln calculated between the initial and final (maximum) drying temperatures (kJ/kg · K)

By using the overall heat capacity ($C_{p,\, kiln}$) of the kiln structure, the total sensible heat (kJ) required to warm-up the kiln structure could be approximated as follows (Elustondo and Oliveira 2009):

$$Q_2 \approx \bar{C}_{p,\,kiln} \times \left(T_{set,\,1} - T_{in}\right) \tag{8.48}$$

If, for the entire kiln structure of each particular kiln, the overall heat transfer coefficient $\left(\bar{U}_{kiln},\ \mathrm{W/m^2K}\right)$ and the heat capacity ($C_{p,\,kiln}$) can be experimentally estimated based on overall calibration parameters, the approximate total sensible thermal power (or heat transfer rate) required for warming-up the kiln structure (envelope) and equipment could be calculated as follows (Elustondo and Oliveira 2009):

$$\dot{Q}_{kiln} \approx \bar{U}_{kiln} \times A_{kiln} \times \left(T_{set,\,1} - T_{in}\right) \tag{8.49}$$

where

$\dot{Q}_{kiln}$ is the total sensible thermal power required for warming-up the kiln structure (envelope) and equipment (kW)

A_{kiln} is the approximate total heat transfer area of the kiln in contact with the ambient air and ground (m^2)

8.14.5.2 Energy Requirements during Drying Steps

Energy requirements during the next effective drying steps can be calculated based on the kiln thermal efficiency and/or on its overall energy balance.

The kiln thermal efficiency can be expressed by the following relation:

$$Eff_{therm} = \frac{h_{fg} + h_{sorption}}{N} \tag{8.50}$$

where

Eff_{therm} is the thermal efficiency of kiln (%)

h_{fg} is the enthalpy (latent heat) of vaporization of water (e.g., $2310\,\frac{kJ}{kg_{water}}$ or $0.641\,\frac{kWh}{kg_{water}}$ at 80°C)

$h_{sorption}$ is the wood's enthalpy of sorption, that is, the heat required to mechanically remove the water from wood (e.g., for spruce, $h_{sorption} = 130\,\frac{kJ}{kg_{water}}$ at 15% moisture content and $h_{sorption} = 340\,\frac{kJ}{kg_{water}}$ at 8% moisture content)

N is the total quantity of energy consumed by the dryer to remove and evaporate 1 kg of moisture (water) from wood $\left(\frac{kJ}{kg_{water}}\right)$.

The thermal efficiency is characteristic of each particular drying operation and depends on variables such as wood species to be dried, type of dryer and drying method, drying schedule, initial and final moisture contents, outdoor climates, etc. For example, for drying at medium temperatures (50°C < T_{drying} < 100°C) of resinous species of 50 mm thick, the thermal efficiency (Eff_{therm}) may vary from 45% to 60%. For larger thickness of lumber boards, the thermal efficiency will be lower, while the drying at high temperatures (T_{drying} > 100°C) may result in higher thermal efficiency (Clément and Fortin 1996).

The thermal (sensible and latent heat) and electrical energies required at each stage (step) of a typical wood batch drying cycle can be identified step by step (Shottafer and Shuler 1974). The electrical energy consumption is relatively easy to calculate (and measure) because it is required only by the kiln's fans and in the case of heat pump-assisted kilns by the heat pump compressor(s) and blower(s) (Clément and Fortin 1996).

8.14.5.2.1 Overcome the Moisture Retention Forces

To maintain constant drying rates, the water molecules inside the lumber boards must absorb sensible heat in order to increase their kinetic energy and, thus, overcome hygroscopic (retention) forces.

The sensible heat required for breaking molecular bonds between wood fiber and absorbed water, that is, to overcome the moisture retention (hygroscopic) forces, can be calculated at each drying step l as follows (Shottafer and Shuler 1974):

$$Q_{3,\,retention} = \sum_{l} (m_{wood,\,l} \times h_{desorption}) \tag{8.51}$$

where

$Q_{3,\,retention}$ is the sensible heat required to overcome the moisture retention (hygroscopic) forces (kJ)

$M_{wood,\,l}$ is the anhydrous mass (oven-dry weight) of wood under drying process at each drying step l (kg)

$h_{desorption}$ is the latent heat (enthalpy) of desorption (or sorption) of wood at final moisture content of each drying step l (kJ/kg)

To overcome hygroscopic forces whenever 1 kg of wood substance if dried from green condition to oven-dry state ($MC_{final} = 0$), about 79 $\frac{kJ}{kg_{oven\text{-}dry}}$ is required (see Figure 8.10). However, if the wood is dried from green condition to, for example, $MC_{final} = 10\%$, only 30.23 $\frac{kJ}{kg_{oven\text{-}dry}}$ is required. Thus, the enthalpy (or latent heat) of

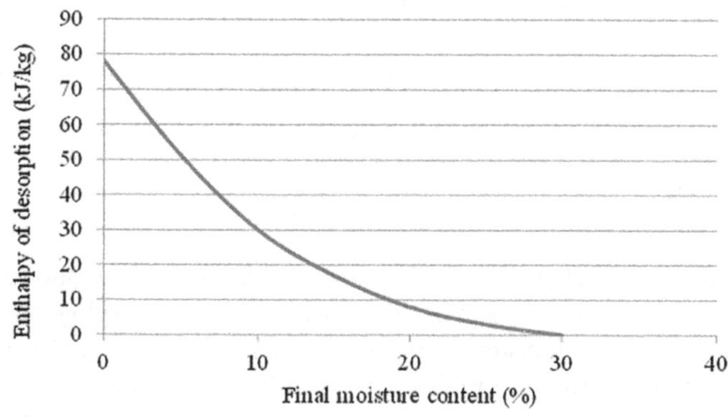

FIGURE 8.10 Enthalpy of desorption required to overcome hygroscopic forces as a function of final moisture content. (Adapted and redrawn from Shottafer and Shuler 1974).

desorption drops off rapidly as higher final moisture contents are desired (Shottafer and Shuler 1974).

8.14.5.2.2 Raise the Temperature and Evaporate the Moisture Removed from the Wood

The moisture extracted from the lumber boards must be heated and, then, vaporized by forced convection. These processes that require considerable amounts of energy, being function of the initial and final moisture contents, wood species, density, temperature, and total volume of wood in the dryer, can be achieved by supplying heat from the drying hot air to the wood board surfaces.

The total (sensible and latent) heat required for heating and vaporizing the moisture extracted from wood boards at each step n of the drying schedule is the sum of the heat used at all the drying steps (Shottafer and Shuler 1974):

$$Q_4 = \sum \left(m^n_{moisture} \times \bar{c}^n_{p,\,moisture} \times \Delta T_n\right) + \left(m^n_{moisture} \times \bar{h}_{fg}\right) \tag{8.52}$$

where

Q_4 is the total (sensible and latent) heat required for heating and vaporizing the moisture extracted from wood boards (kJ)

$m^n_{moisture}$ is the mass of moisture removed at step n (kg)

$\bar{c}^n_{p,\,moisture}$ is the average specific heat of water vapor (moisture) at step n (kJ/kg·K)

$\Delta T_n = T_{set,\,n} - T_{in,\,n}$ is the temperature change of water at step n to attain the saturation state (°C)

$T_{set,\,n}$ is the set dry-bulb temperature at step n (°C)

$T_{in,\,n}$ is the initial dry-bulb temperature of wood at step n (°C)

$\bar{h}_{fg}$ is the average enthalpy (latent heat) of vaporization of wood at step n (kJ/kg)

For a given wood species, the enthalpy (latent heat) of vaporization (i.e., the energy consumed at constant temperature for phase change from liquid to vapor) depends upon the particular temperature at which the process occurs. Consequently, the enthalpy (latent heat) of vaporization (h_{fg}) must be calculated for each drying step occurring at different dry-bulb temperatures according to the drying schedule:

$$\bar{h}_{fg} = \left(606.5 - 0.695 \times DBT_n\right) \times 4.18 \tag{8.53}$$

where

DBT_n is the set dry-bulb temperature at the drying step n

Enthalpy (latent heat) of vaporization may represent 20%–60% of the total energy consumed within a wood kiln. It is dependent only upon the temperature and the mass of water vaporized, and is slightly decreasing with the temperature. For example, at 0°C, the latent heat of vaporization is 2,501 kJ/kg; at 20°C, 2,400 kJ/kg; and at 100°C, 2,256 kJ/kg.

For continuous wood drying processes, the total sensible and latent heat (kJ) required for heating and vaporizing the moisture extracted from the wood at each drying step n can also be calculated using the following equation (Shottafer and Shuler 1974):

$$Q_4 = \sum \left\{ m_{oven\text{-}dry}^{n} \times \frac{\Delta(MC)_n}{100} \times \left[\bar{c}_{p,\,moisture}^{\,n} \left(T_{set,\,n} - T_{in,\,n} \right) + \bar{h}_{fg} \right] \right\} \tag{8.54}$$

where

$m_{oven\text{-}dry}^{n}$ is the anhydrous mass (oven-dry) of the wood stack at step n (kg)
$\Delta(MC)_n$ is the average wood moisture content variation at step n (%, dry basis) $\left(\frac{kg_{moisture}}{kg_{oven\text{-}dry}} \right)$
$\bar{c}_{p,\,moist}$ is the average specific heat of water at step n (kJ/kg · K)
$T_{set,\,n}$ is the set dry-bulb temperature at step n (°C)
$T_{in,\,n}$ is the initial dry-bulb temperature of wood at step n (°C)
$\bar{h}_{fg}$ is the average enthalpy (latent heat) of evaporation of water at step n (kJ/kg)

The minimum amount of latent heat required for moisture vaporization only during the wood drying process is given as follows:

$$Q_{latent} = Q_{evap}^{\min} = m_{moisture} \times \bar{h}_{fg} \tag{8.55}$$

where

Q_{latent} is the minimum amount of latent heat required for moisture vaporization only (kJ)
m_{evap} is the mass of water (moisture) removed from the wood stack (kg)
$\bar{h}_{fg}$ is the average enthalpy (latent heat) of evaporation of water (kJ/kg)

For continuous wood drying processes, expression 8.54 becomes:

$$Q_{evap}^{\min} = m_{oven\text{-}dry} \times \left(MC_{in} - MC_{fin} \right) \times \bar{h}_{fg} \tag{8.56}$$

where

$m_{oven\text{-}dry}$ is the anhydrous (oven-dry) mass of green wood ($kg_{oven\text{-}dry}$)
MC_{in} is the initial moisture content $\left(\frac{kg_{water}}{kg_{oven\text{-}dry}} \right)$
MC_{fin} is the final moisture content $\left(\frac{kg_{water}}{kg_{oven\text{-}dry}} \right)$

The latent heat component can also be determined by using the board stack's volume and lumber density, and expected percentage of moisture change (dry basis) (Tschernitz 1986):

$$Q_{latent} = \rho_{wood} \times V_{wood} \times \bar{h}_{fg} \times \frac{MC_{in} - MC_{fin}}{100} = Q_{free} + Q_{bound} \tag{8.57}$$

where

Q_{latent} is the total amount of latent heat required to evaporate water from wood boards (kJ)

ρ_{wood} is the density of wood (kg/m^3)

V_{wood} is the volume of wood (m^3)

$\bar{h}_{fg}$ is the average enthalpy (latent heat) of vaporization of water $\left(\frac{kJ}{kg_{water}}\right)$ (at moisture contents > 20%)

M_{in} is the initial moisture content of wood (%)

M_{fin} is the final moisture content of wood (%)

Q_{free} is the energy needed to evaporate free water per drying cycle (kJ)

Q_{bound} is the energy needed to evaporate bound water per drying cycle (kJ)

ρ_{water} is the density of water (kJ/m^3)

For final moisture contents >20%,

$$Q_{bound} = 0$$

Thus, in this case, $Q_{latent} = Q_{free}$.

For final moisture contents <20%, Q_{bound} can be calculated from equation 8.57 where volume V_{wood} is total green volume charged to the kiln (m^3) and the wood density based on oven-dry green volume of wood $\left(\frac{kg}{m^3_{green}}\right)$.

When removing bound moisture, extra latent heat (Δh_{extra}) is required to break the bonds between the water and the wood boards. This appears as an increase above the normal enthalpy (latent heat) of evaporation, becoming greater as the moisture content falls (Kemp 2012):

$$h'_{fg} = h_{fg} + \Delta h_{extra} \tag{8.58}$$

Finally, the total heat transfer rate (or thermal power) required to evaporate the moisture from the surfaces of wood boards can be expressed as the sum of heat transfer rates required at each drying step n:

$$\dot{Q}_4 = \sum_n \left[\bar{h}_s \times A \times \left(T_{set,\,n} - T_{surf,\,n}\right)\right] \tag{8.59}$$

where

$\dot{Q}_4$ is the heat transfer rate (or thermal power) required to evaporate the moisture from the surfaces of wood boards during the entire drying process (kW)

$\bar{h}_s$ is the wood board's surface average heat transfer coefficient (W/m$^2\cdot$K)

A is the total heat transfer area of the dried wood boards (m^2)

$T_{set,\,n}$ is the air set dry-bulb temperature at each drying step n (°C)

$T_{s,\,n}$ is the temperature at the surface of dried boards at each drying step n (°C)

For a minimum amount of heat required, the minimum rate of latent vaporization (thermal power) required for a given drying process can be expressed as follows:

$$\dot{Q}_{evap}^{min} = \dot{Q}_{latent} = \dot{m}_{moisture} \times \overline{h}_{fg} \tag{8.59a}$$

where

$\dot{Q}_{evap}^{min}$ is the minimum rate of latent vaporization (thermal power) required (kW)

$\dot{m}_{moisture}$ is the moisture vaporization rate (kJ/s)

Exercise E8.2

In a continuous drying process of softwood (see Figure 8.8), the outside air at 10°C (state 1) and initial absolute humidity $\omega_{in} = \omega_1 = \omega_2 = 0.0075 \frac{kg_{water}}{kg_{dry\ air}}$ is heated up to 45°C (state 2) before entering the lumber stack. If the average moisture evaporation rate ($\dot{m}_{moisture}$) and average enthalpy (latent heat) of vaporization (h_{fg}) of water during the air humidification process 2-3 are 0.2622 kg/s and 2,400 kJ/kg, respectively, calculate

1. Minimum latent thermal power (latent vaporization rate) required;
2. Minimum airflow rate required.

Solution

The air heating and humidifying processes represented in the Mollier diagram (Figure E8.2) allow determining the final (saturated) absolute humidity of drying air $\omega_{fin} = \omega_3 = 0.017 \frac{kg_{water}}{kg_{dry\ air}}$.

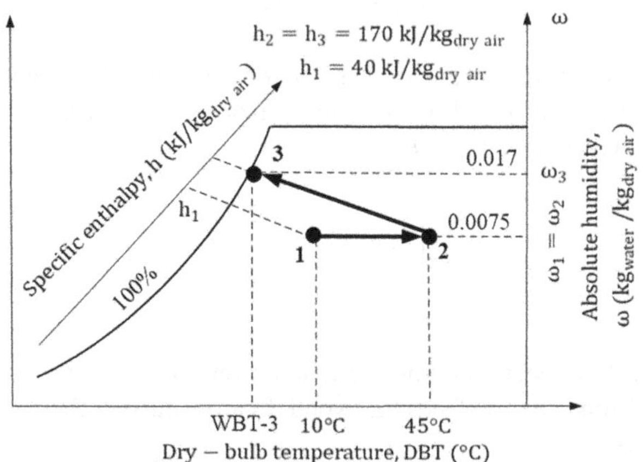

FIGURE E8.2 Heating and humidifying process of drying air inside an open wood dryer represented in a Mollier psychrometric diagram.

1. Because the moisture evaporation rate (i.e., air humidification rate) is known, equation 8.55 allows calculating the minimum latent thermal power (latent vaporization rate) of the drying system:

$$\dot{Q}_{evap}^{min} = \dot{Q}_{latent} \approx \dot{m}_{moisture} \times \bar{h}_{fg} = 0.2622 \times 2400 = 629.28 \text{ kW}$$

2. By assuming negligible air leaks and knowing the average moisture evaporation rate, equation 8.42 allows calculating the minimum air mass flow rate required:

$$\dot{m}_{air} = \frac{\dot{m}_{moisture}}{\omega_{fin} - \omega_{in}} = \frac{\dot{m}_{moisture}}{\omega_3 - \omega_2} = \frac{0.2622}{0.017 - 0.0075} = 2.76 \ \text{kg/s} \approx 3.31 \ \text{m}^3/\text{s}$$

In summary, the main steps for calculating the heat required to raise the temperature of the water removed from the wood and to evaporate it are

a. By using thermodynamic tables or software, determine the enthalpy (latent heat) of vaporization (h_{fg}, kJ/kg) of water at each dry-bulb temperature of the drying schedule stages;
b. Using the kiln drying schedule, determine the moisture content change during each drying step (%) and express it as decimals:

$$\Delta(MC)_{step\ i} = MC_{start,\ step\ i} - MC_{end,\ step\ i} \tag{8.60}$$

where

MC_{start} is the moisture content at the start of the drying step i $\left(\frac{kg_{water}}{kg_{oven\text{-}dry}}\right)$

MC_{end} is the moisture content at the end of the drying step i $\left(\frac{kg_{water}}{kg_{oven\text{-}dry}}\right)$

3. Determine the temperature variations (ΔT_i) from the initial temperature ($T_{in,\,i}$) to the scheduled dry-bulb temperature of each drying stage i (DBT_i):

$$\Delta T_i = DBT_i - T_{in,\ i} \tag{8.61}$$

4. Calculate the heat required at each drying step (in kJ) as follows:

$$Q_{4_i} = m_{oven\text{-}dry}^{wood} \times \Delta(MC)_{step\ i} \times \left[c_p^{water} \times \Delta T_i + \bar{h}_{fg,\ i}\right] \tag{8.62}$$

5. Calculate the total heat required to raise the temperature of the water removed from the wood, and to evaporate it:

$$Q_4 = \sum Q_{4_i} \tag{8.63}$$

Exercise E8.3

The initial temperature and moisture content of a lumber stack are 15°C and 100% (dry basis), respectively. The lumber stack is kiln dried in three successive steps according to the schedule parameters given in Table E8.3. Knowing that the total oven-dry mass is 50,000 kg and the water specific heat at constant pressure is constant $\left(c_p^{water} = 4.18\,\text{kJ/kg}\right)$, determine the total (sensible and latent) heat required to raise the temperature of the water removed from the wood and, then, to evaporate it during the whole drying cycle (kJ).

Solution

By applying equation 8.62 at each of the three successive drying steps, the following results are obtained (see Table E8.3.1):

$$Q_{4_{i=1}} = m_{oven\text{-}dry}^{wood} \times \Delta(MC)_{step\ 1} \times \left[c_p^{water} \times \Delta T_{step\ 1} + \bar{h}_{fg,\ step\ 1}\right]$$

$$= 50{,}000 \times 0.7 \times \left[4.18 \times (95) + 2372\right] = 96.9 \times 10^6\,\text{kJ}$$

$$Q_{4_{i=2}} = m_{oven\text{-}dry}^{wood} \times \Delta(MC)_{step\ 2} \times \left[c_p^{water} \times \Delta T_{step\ 2} + \bar{h}_{fg,\ step\ 2}\right]$$

$$= 50{,}000 \times 0.1 \times \left[4.18 \times (100) + 2{,}358\right] = 13.88 \times 10^6\,\text{kJ}$$

$$Q_{4_{i=3}} = m_{oven\text{-}dry}^{wood} \times \Delta(MC)_{step\ 3} \times \left[c_p^{water} \times \Delta T_{step\ 3} + \bar{h}_{fg,\ step\ 3}\right]$$

$$= 50{,}000 \times 0.13 \times \left[4.18 \times (110) + 2{,}380\right] = 18.56 \times 10^6\,\text{kJ}$$

The total heat (sensible plus latent) required is given as follows:

$$Q_4 = \sum Q_{4_i} = 129.34 * 10^6\,\text{kJ}$$

TABLE E8.3
Given Data for Example E8.3

	Parameter				
Drying step	Lumber Oven-Dry Mass ($kg_{oven\text{-}dry}$)	Moisture Content Variation (% [d.b.])	Dry-Bulb Temperature Change versus Initial Temperature (°C)	Enthalpy of Vaporization of Water $\left(\frac{kJ}{kg_{water}}\right)$	Heat Required Q_{4_i} (kJ)
1	50,000	70	95	2,372	$96.9 * 10^6$
2	50,000	10	100	2,358	$13.88 * 10^6$
3	50,000	13	110	2,380	$18.56 * 10^6$
Total	–	–	–	–	$129.34 * 10^6$

$Q_4 = \sum_i Q_{4_i}$

8.14.5.2.3 Heat the Remaining Water in Wood after Drying

Usually, kiln operations do not dry wood to zero moisture content. Therefore, at the end of any kiln drying cycle, there is still some water remaining in the wood, and some of the heat supplied to the kiln is used to raise the temperature of the water.

The heat requirement to compensate the energy absorbed by the residual water can be determined as follows (Shottafer and Shuler 1974):

$$Q_5 = m_{water}^{residual} \times \overline{c}_{p,\ water} \times \Delta T \tag{8.64}$$

where

Q_5 is the heat required to compensate the energy absorbed by the residual water (kJ)
$m_{water}^{residual}$ is the mass of residual water (kg)
$\overline{c}_{p,\ water}$ is the average specific heat of water (kJ/kg · K)
ΔT is the temperature change of the residual water (°C)

The amount of water remaining in the wood (neglecting stickers and timber sleepers) at the final moisture content can be calculated as follows:

$$m_{water}^{residual} = m_{oven\text{-}dry} \times MC_{fin} \tag{8.65}$$

where

$m_{water}^{residual}$ is the amount of water remaining in the wood at the final moisture content (kg_{water})
$m_{oven\text{-}dry}$ is the green wood's anhydrous (oven-dry) mass ($kg_{oven\text{-}dry}$)
MC_{fin} is the final moisture content of wood expressed in decimal $\left(\frac{kg_{water}}{kg_{oven\text{-}dry}}\right)$

The specific heat of water is not constant but is dependent upon the drying temperature of wood. For practical considerations, however, the thermal capacity of water can be considered to be 1.055 kJ/kg · K.

Exercise E8.4

For the lumber dryer described in exercise E8.3, determine:

a. The mass of water that remains in the lumber charge at the end of the drying cycle;
b. The heat required to raise the temperature of the water remaining in the lumber charge.

Solution

a. The amount of water remaining in the wood charge can be calculated by using equation 8.65:

$$m_{water}^{residual} = m_{oven\text{-}dry} \times MC_{fin} = 50{,}000 \times 0.07 = 3{,}500 \text{ kg}$$

b. The heat required to raise the temperature of this quantity of water remaining in the lumber charge is:

$$Q_5 = m_{water}^{remaining} \times \bar{c}_{p,\,water} \times \Delta T = 3{,}500 \times 4.18 \times 110 = 1.6 \times 10^6 \text{ kJ}$$

8.14.5.2.4 Raise the Temperature and Humidify the Incoming Vent Air

In conventional batch drying kilns, as moisture (water vapor) is removed from the wood, it is absorbed by the air circulating within the kiln and becomes saturated with moisture (water vapor). When the absolute humidity of the air inside the kiln is too high or saturated, a part of the kiln hot and humid air is evacuated via air vents, and simultaneously, new, fresh air from outside, that already contains some moisture, is introduced into the kiln for replacing it. The incoming air is then further humidified by moisture from the wood, and, thus, its absolute humidity is elevated to the level existing in the kiln. In addition, the dry-bulb temperature of the fresh air must be elevated to the kiln's temperature.

Since kiln controls are not sufficiently precise to allow the exact amount of necessary air into the kiln, a certain amount of excess air is usually admitted. The excess air must be heated and partially humidified by the moisture removed from the wood. In practice, the complete humidification of the excess air is achieved by the direct spraying of steam into the kiln.

Globally, the total heat required to raise the dry-bulb temperature and humidify the incoming vent air is dependent on the volume and the mass of water vapor contained in the vent air brought into the kiln, the specific heats of the air and the water vapor, the temperature change, and the enthalpy (latent heat) of vaporization needed for the production of the humidifying vapor (Shottafer and Shuler 1974). This means that, energetically, the air venting operation involves substantial heat losses. Currently, up to 78% of the total energy consumption of the conventional wood dryers is required for heating and humidification of the incoming air.

The total volume of the vented (evacuated) moist air (i.e., dry air plus water vapor) from the dryer can be expressed as follows:

$$V_{moist\,air}^{evacuated} = V_{dry\,air}^{evacuated} + V_{water\,vapor}^{evacuated} \tag{8.66}$$

where

$V_{moist\,air}^{evacuated}$ is the total volume of vented moist air (m^3)
$V_{dry\,air}^{evacuated}$ is the volume of vented dry air (m^3)
$V_{water\,vapor}^{evacuated}$ is the volume of vented moisture (water vapor) (m^3)

Also,

$$m_{dry\,air}^{evacuated} = \rho_{dry\,air}^{evacuated} \times V_{dry\,air}^{evacuated} \tag{8.67}$$

$$m_{water\,vapor}^{evacuated} = \rho_{water\,vapor}^{evacuated} \times V_{water\,vapor}^{evacuated} \tag{8.68}$$

where

$\rho_{dry\,air}^{evacuated}$ is the density of evacuated dry air (kJ/m^3)
$\rho_{water\,vapor}^{evacuated}$ is the density of evacuated moisture (water vapor) (kg/m^3)

From expressions 8.67 and 8.68,

$$V_{moist\ air}^{evacuated} = \frac{m_{dry\ air}^{evacuated}}{\rho_{dry\ air}^{evacuated}} + \frac{m_{water\ vapor}^{evacuated}}{\rho_{water\ vapor}^{evacuated}} = \frac{m_{dry\ air}^{evacuated}}{\rho_{dry\ air}^{evacuated}} + \frac{\omega_{water\ vapor}^{evacuated} \times m_{dry\ air}^{evacuated}}{\rho_{water\ vapor}^{evacuated}} \tag{8.69}$$

where

$\omega_{water\ vapor}^{evacuated}$ is the absolute humidity of vented water vapor $\left(\frac{kg_{water}}{kg_{oven\text{-}dry}}\right)$

The total heat lost through the air vents can be calculated as follows:

$$\begin{aligned} Q_{vent\ air}^{loss} &= Q_{dry\ air}^{loss} + Q_{water\ vapor}^{loss} \\ &= m_{dry\ air}\left[\overline{c}_{p,\ dry\ air} \times (T_2 - T_1) + \omega_{water\ vapor}^{evacuated} \times \overline{c}_{p,\ water} \times (T_2 - T_1)\right] \end{aligned} \tag{8.70}$$

where

$Q_{vent\ air}^{loss}$ is the total heat lost through the air vents (kJ)
T_1 is the dry-bulb temperature of the ambient outdoor air (°C)
T_2 is the dry-bulb temperature of the kiln indoor air (°C)
$\overline{c}_{p,\ dry air}$ is the dry-air average specific heat (kJ/kg·K)
$\overline{c}_{p,\ water}$ is the average specific heat of water (kJ/kg·K)

The total (sensible plus latent) heat required to heat the incoming fresh air could be evaluated as follows:

$$Q_6 = m_{fresh} \times (h_{out} - h_{in}) = \rho_{fresh} \times V_{fresh} \times (h_{out} - h_{in}) \tag{8.71}$$

where

Q_6 is the total (sensible plus latent) heat required to heat the incoming fresh air (kJ)
m_{fresh} is the mass of the new fresh air entering the kiln (kg)
h_{out} is the specific enthalpy of the evacuated moist air (kJ/kg)
h_{in} is the specific enthalpy of the incoming air (kJ/kg)
ρ_{fresh} is the density of the incoming air (kg/m^3)
V_{fresh} is the volume of the incoming air (m^3)

8.14.5.3 Compensate the Heat Losses through the Kiln Envelope

Wood kilns are generally structures constructed above grade, thus exposed to outdoor and ground thermal conditions. Structure envelopes of lumber kilns mainly refer to the walls, roof, floor, and doors.

Both outdoor and indoor thermal conditions influence the kiln heating load and calculations for sizing the heating system components. In cold climates, weather conditions that vary considerably from location to location, from season to season, and from year to year may be critical for industrial drying processes, especially for poor insulated dryers. Kiln indoor design conditions are determined for each wood species according to general or specific drying schedules.

At indoor and outdoor solid surfaces of walls and doors in contact with indoor and outdoor air, respectively, thermal free (induced by air density differences, usually confined to a boundary layer near the heated surfaces) and forced convection (induced by the air displacement with fans), which can take place independently or in combination, are the most important modes of heat transfer.

Inside the kiln, thermal convection is associated with drying air in motion along the wall surfaces. The convective heat transfer coefficients mainly depend on the drying air velocity and the wall shape or orientation. Many correlations exist for predicting the convective heat transfer coefficients under specific thermal and geometric conditions.

Along outdoor walls, roofs, and doors of lumber kilns, forced convection is due to wind of which both the direction and magnitude are very unpredictable.

Heat required for compensating the convective and conductive losses through the dryer structure (walls, floor, ceiling, doors, including thermal insulations) is proportional to (i) the temperature difference between the inside and the outside of the kiln (the greater the temperature difference, the greater the heat loss); (ii) total surface area of each element; (iii) time of kiln operation (the longer the drying time at a given temperature, the greater the heat loss); and (iv) to a lesser extent, increased air velocity within and outside the dryer.

The amount of conductive heat losses through the kiln structure depends upon the thermal conductivities of the various materials of the kiln walls, roof, floor, and doors, and the temperature difference between the inside and the outside of the dryer. The heat transfer through each element is also dependent upon the surface area in question, the materials of construction, and the time of dryer operation for any batch run (Shottafer and Shuler 1974).

The kiln's walls and roof are complex assemblies of materials, and floors are in contact with the ground. Because of these conditions, precise calculation of heat transfer rates is difficult, but experience and experimental data make reliable estimates possible. To solve the problem of heat transfer in kiln structures, the concept of thermal resistance is very useful (McQuiston et al. 2005).

The existence of several materials in single components (such as walls, roofs, and floors) makes it difficult to accurately evaluate the average thermal conductivity. Therefore, overall heat transfer coefficients are used for each structural component as a whole unit, generally provided by the kiln construction companies.

The general procedure for evaluation of heat losses of a kiln structure during the worst thermal conditions that can occur during a drying cycle includes the following (ASHRAE 2008): (i) selection of outdoor design parameters (e.g., temperature, relative humidity, and wind direction and speed); (ii) selection of kiln thermal design conditions to be maintained during typical drying cycles; (iii) selection of surface convective heat transfer coefficients; (iv) determination of materials' thermal conductivities; (v) calculation of heat losses through walls, floors, ceilings, and doors; generally, in practice, the heat losses through the kiln floor is negligible compared to the heat loss through the kiln walls; (v) calculation of heat losses through air venting and exfiltration; and (vi) summation of all previously identified heat losses.

A simple equation to express the heat transfer rate by convection is given as follows:

$$\dot{Q}_{conv} = \bar{h} \times A \times \left(T - T_{surf}\right) \tag{8.72}$$

where

$\dot{Q}_{conv}$ is the convective heat transfer rate from the indoor or outdoor air to or from walls, ceiling, and doors' surfaces (W)

$\bar{h}$ is the average convective heat transfer coefficient ($W/m^2 \cdot K$)

A is the walls, ceiling, or doors heat transfer areas (m^2)

T is the surrounding air dry-bulb temperature (°C)

T_{surf} is the wall, ceiling, or door surface temperature (°C)

Equation 8.72 may also be expressed in terms of thermal resistance:

$$\dot{Q} = \frac{T - T_{surf}}{\dfrac{1}{h \times A}} \tag{8.73}$$

Or

$$\dot{q}_{conv} = \frac{T - T_{surf}}{\dfrac{1}{h}} = \frac{T - T_{surf}}{R} \tag{8.74}$$

where

$\dot{q}_{conv} = \dfrac{\dot{Q}}{A}$ is the density of convective heat transfer (W/m^2)

$R = 1/h$ is the surface thermal resistance ($m^2 \cdot K/W$)

The free (natural) convective heat transfer coefficients are low (e.g., about 6 $W/m^2 \cdot K$), while (at air velocity higher than about 6 m/s) the forced convective heat transfer coefficients are higher than 35 $W/m^2 \cdot K$.

Thermal conduction is the mechanism of heat transfer between parts of continuum solids due to the transfer of energy between particles or groups of particles at the atomic level. The Fourier equation expresses steady-state conduction in one dimension from a high to a low temperature through a medium (McQuiston et al. 2005):

$$\dot{Q}_{cond} = -k \times A \times \frac{dT}{dx} \tag{8.75}$$

where

$\dot{Q}_{cond}$ is the conductive heat transfer rate (W)

k is the thermal conductivity ($W/m \cdot K$)

A is the normal area to heat flow (m^2)

dT/dx is the temperature gradient (K/m)

In equation 8.75, the negative sign indicates that $\dot{Q}_{cond}$ is transferred in positive direction of x when dT/dx is negative (see Figure 8.11).

Consider a flat wall where uniform temperatures T_1 and T_2 are assumed to exist on each surface. If the thermal conductivity, the heat transfer rate, and the area are constant, equation 8.75 may be integrated to obtain

$$\dot{q} = \frac{\dot{Q}}{A} = \frac{-k \times (T_1 - T_2)}{\delta} = \frac{-(T_1 - T_2)}{\frac{\delta}{k}} = \frac{-(T_1 - T_2)}{R} \tag{8.76}$$

Where, for a plate wall, $R = \delta/k$ is the unit thermal resistance ($m^2 \cdot K/W$)

Based on equation 8.76, the thermal conductivity of a given material can be defined as follows:

$$k = \frac{\dot{q} \times \delta}{(T_1 - T_2)} \tag{8.77}$$

where

k is the thermal conductivity of a given material ($W/m \cdot K$)

$\dot{q}$ is the density of heat transferred (w/m^2)

$(T_1 - T_2)$ is the temperature difference between hot and cold parts of the wall (°C)

δ is the thickness of the given material (m)

During a steady-state heat transfer process, the convective heat transfer from the kiln indoor air and indoor wall surface can be expressed as follows (see Figure 8.11):

$$\dot{Q}_{indoor,\ conv} = h_{indoor,\ conv} \times A \times (T_{indoor} - T_1) \tag{8.78}$$

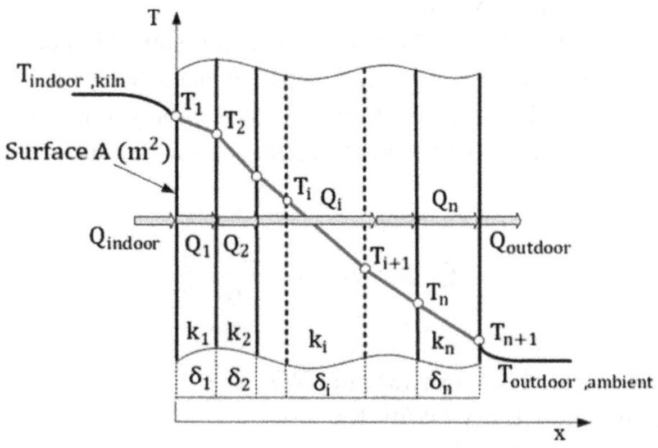

FIGURE 8.11 Schematic representation of a multiple kiln external wall; δ_i, thickness of material layer i; k_i, thermal conductivity of the material of layer i.

where

$h_{indoor,\ conv}$ is the convective heat transfer coefficient of the surface air layer inside the kiln (W/m^2·K)

A is the indoor heat transfer surface (m^2)

T_1 is the temperature of the wall indoor surface (°C)

Thus,

$$(T_{indoor} - T_1) = \frac{\dot{Q}_{indoor}}{h_{indoor} \times A} \tag{8.79}$$

In the same manner, it can be written (see Figure 8.10) as follows:

$$\dot{Q}_1 = A \times \left(\frac{T_1 - T_2}{\dfrac{\delta_1}{k_1}} \right) = \frac{k_1 \times A}{\delta_1}(T_1 - T_2) \tag{8.80}$$

Thus,

$$(T_1 - T_2) = \frac{\dot{Q}_1 \times \delta_1}{k_1 \times A} \tag{8.81}$$

In the same manner,

$$(T_5 - T_6) = \frac{\dot{Q}_5 \times \delta_5}{k_5 \times A} \tag{8.82}$$

and

$$\dot{Q}_{outdoor,\ conv} = h_{outdoor,\ conv} \times A \times (T_5 - T_{outdoor}) \tag{8.83}$$

where

$h_{outdoor}$ is convective heat transfer coefficient of the surface air layer outside the kiln (W/m^2·K)

The difference between the kiln indoor temperature and the outside air temperature can be written as follows:

$$\begin{aligned} T_{indoor} - T_{outdoor} &= (T_{indoor} - T_1) + (T_1 - T_2) + (T_2 - T_3) + (T_3 - T_4) \\ &\quad + (T_4 - T_5) + (T_5 - T_6) + (T_6 - T_{outdoor}) \end{aligned} \tag{8.84}$$

Or by using equations 8.79–8.83,

$$\begin{aligned} T_{indoor} - T_{outdoor} &= \frac{\dot{Q}}{h_{indoor} \times A} + \frac{\dot{Q} \times \delta_1}{k_1 \times A} + \frac{\dot{Q} \times \delta_2}{k_2 \times A} + \cdots + \frac{\dot{Q} \times \delta_5}{k_5 \times A} + \frac{\dot{Q}}{h_{outdoor} \times A} \\ &= \frac{\dot{Q}}{A}\left(\frac{1}{h_{indoor}} + \frac{\delta_1}{k_1} + \frac{\delta_2}{k_2} + \cdots + \frac{\delta_5}{k_5} + \frac{1}{h_{outdoor}} \right) \end{aligned} \tag{8.85}$$

The convective heat transfer rate is given as follows:

$$\dot{Q} = \frac{A \times (T_{indoor} - T_{outdoor})}{\left(\frac{1}{h_{indoor}} + \frac{\delta_1}{k_1} + \frac{\delta_2}{k_2} + \dots + \frac{\delta_5}{k_5} + \frac{1}{h_{outdoor}}\right)} \tag{8.86}$$

If the thermal conductivities (k_1, k_2, … k_i) and layers' thicknesses (δ_1, δ_2, …, δ_i) are known, the steady-state overall heat transfer coefficient $\left(\bar{U}, \frac{m^2K}{W}\right)$ can be calculated as follows:

$$\bar{U} = \frac{1}{\frac{1}{h_{indoor}} + \frac{\delta_1}{k_1} + \frac{\delta_2}{k_2} + \dots + \frac{\delta_5}{k_5} + \frac{1}{h_{outdoor}}} = \frac{\dot{Q}_{cond}}{A \times (T_{indoor} - T_{outdoor})} \tag{8.87}$$

For practical considerations of energy losses in dry kiln operation, the overall heat transfer coefficient U is sufficient for determining heat losses.

The overall heat transfer coefficient may vary from 0.4 to 0.6 W/m^2·K (Holman 1992; McCurdy and Pang 2007).

The coefficient U is very often expressed as an R value, where $R = 1/U$. The reason for expressing heat loss in this form is that when the characteristics of a given construction are known, the heat loss can be calculated simply by knowing the area and the difference in high (inside) and low (outside) temperatures.

The total heat transfer rate through the kiln wall(s) is given as follows:

$$\dot{Q}_6 = A \times \bar{U} \times (T_{indoor} - T_{outdoor}) \tag{8.88}$$

where

$\dot{Q}_6$ is the total heat transfer rate through the kiln wall(s) (W)
A is the kiln's total heat transfer area (m^2)

Practically, the dryer heat losses through structural sections (walls, ceiling, floor) can be calculated as follows:

$$\dot{Q}_6 = \sum \frac{A_n \times (T_{indoor} - T_{outdoor})}{\bar{R}_n} = \sum \bar{U}_n \times A_n \times (T_{indoor} - T_{outdoor}) \tag{8.89}$$

where

$\dot{Q}_6$ is the total heat transfer rate through the kiln wall(s) (W)
A_n is the heat transfer area of the dryer section i (m^2)
T_{indoor} is the indoor air dry-bulb temperature (°C)
$T_{outdoor}$ is the outdoor air dry-bulb temperature (°C)

$\bar{R}_n$ is the overall heat transfer resistance of the dryer section n (m^2·K/W) calculated for each construction element n with the following equation:

$$\bar{R}_n = \frac{1}{\bar{U}_n} = \left[\frac{1}{h_{indoor}} + \frac{\delta_n}{k_n} + \frac{1}{h_{outdoor}}\right] \tag{8.90}$$

where

$\bar{U}_n$ is the average overall heat transfer coefficient of construction element n (W/m²·K)

h_{indoor} is the convection heat transfer coefficient at the interior surface (m²·K/W)

δ_n is the thickness of the section n (m)

k_n is the thermal conductivity of the section n (W/m·K)

$h_{outdoor}$ is the convection heat transfer coefficient at the exterior surface of section n (W/m²·K)

If it is assumed that the radiation heat loss is transferred to the immediate surroundings at the same ambient temperature used for the external convection calculations and, by considering that the heat loss from the kiln walls is mainly by convection, the total thermal power heat loss can be calculated with the following expression (McCurdy and Pang 2007):

$$\dot{Q}_6 = \frac{T_{indoor} - T_{outdoor}}{\dfrac{1}{h_{indoor,conv} \times A_{outdoor}} + \sum \dfrac{\delta_i}{k_i A_{outdoor}} + \dfrac{1}{h_{outdoor,conv} A_{outdoor,conv}}} \tag{8.91}$$

where

T_{indoor} is the interior temperature of wall (°C)

$T_{outdoor}$ is the external temperature of wall (°C)

$h_{indoor,conv}$ is the indoor convective heat transfer coefficient (W/m²·K)

$A_{outdoor}$ is the external surface area (m²)

δ_i is the thickness of the wall panel i (m)

k_i is the thermal conductivity of the wall panel i (W/m·K)

$h_{outdoor,conv}$ is the external convective heat transfer coefficient (W/m²·K)

h_r is the external radiation heat transfer coefficient (W/m²·K)

For drying cycles of a given drying time (t), the thermal heat loss can be expressed as the sum of heat loss through various kiln surfaces at different times in the drying schedule:

$$Q_{loss} = \sum \bar{U}_i A_i \left(T_{inside} - T_{outside}\right) \times t_i \tag{8.92}$$

where

Q_{loss} is the heat loss through walls, floor, doors, etc. (kJ)

U_i is the overall heat transfer coefficient of individual dryer structural components (W/m²·K)

A_i is the surface area of walls, ceiling, floor, and doors (m²)

T_{inside} is the dryer dry-bulb temperature (°C)

$T_{outside}$ is the exterior or ambient dry-bulb temperature (°C)

t_i is the drying time (s)

Equation 8.92 shows that, since the temperature of the dryer varies with time according to the drying program, the time t broken into time step (t_1, t_2, …, t_i). Also, the U values of walls, ceiling, and floors will differ, that is, U_1, U_2, …, U_n.

The outside surface temperature ($T_{outside}$) varies from night to day, and the ground temperature is higher than the outside air. The value T_1 will have seasonal variation for any one location and will vary according to the local climate. Wind may be a factor.

Exercise E8.5

A lumber kiln operates at an average dry-bulb temperature of 80°C. Knowing the average outdoor (0°C) and ground (5°C) temperatures, as well as the overall heat transfer coefficients ($\bar{U}$) and heat transfer areas for the various structural components of the kiln, determine the heat loss through each of these structural components of the kiln. Average outdoor and ground temperatures are 0°C and 5°C, respectively (Table E8.5).

Solution

The heat loss through each structural component of the kiln can be evaluated by using equation C (see Table E8.5):

$$\dot{Q}_{6,\,walls} \approx \bar{U}_{walls} \times A_{walls} \times \left(T_{indoor} - T_{outdoor}\right) = 0.67 \times 300 \times 80 = 16.08 \text{ kW}$$

$$\dot{Q}_{6,\,roof} \approx \bar{U}_{roof} \times A_{walls} \times \left(T_{indoor} - T_{outdoor}\right) = 0.65 \times 200 \times 80 = 10.4 \text{ kW}$$

$$\dot{Q}_{6,\,door} \approx \bar{U}_{door} \times A_{walls} \times \left(T_{indoor} - T_{outdoor}\right) = 1.25 \times 20 \times 80 = 2.0 \text{ kW}$$

$$\dot{Q}_{6,\,floor} \approx \bar{U}_{floor} \times A_{walls} \times \left(T_{indoor} - T_{ground}\right) = 0.51 \times 200 \times 75 = 7.65 \text{ kW}$$

TABLE E8.5
Data and Results for Exercise E8.5

Structural Component	Overall Heat Transfer Coefficient ($\bar{U}$) (W/m²·K)	Heat Transfer Area (A) (m²)	Weighted Average Kiln Dry-Bulb Temperature (°C)	Outdoor and Ground Average Temperature (°C)	Heat Rate Loss (W)
Walls	0.67	300	80	0	16.08
Roof	0.65	200	80	0	10.4
Door	1.25	20	80	0	2.0
Floor	0.51	200	80	5	7.65
Total	–	–	–	–	36.13

8.14.5.4 Compensate the Kiln Air Leakage Heat Losses

Although the wood kilns are generally well sealed, some moisture exfiltration will occur because the kiln moisture vapor pressure is generally greater than the ambient moisture vapor pressure. Uncontrollable air leakages account for all the fresh air that enters into the kiln through openings other than the kiln vents. They occur mainly when the dryer is open during wood loading, but there can also be leakage through cracks, dryer poor insulation, etc.

Warm air flowing out of the kiln via air leakage contains moisture vaporized from the lumber, as well as steam supplied by the humidification system. The heat losses associated with such uncontrolled air leakages are usually small but strongly depends on the quality of kiln construction. They can increase the humidification requirements but also may be useful means of controlling excessive kiln temperature increases. The amount of air leakage cannot be calculated theoretically. Usually, it can be estimated experimentally through an energy and mass balance over the entire wood dryer.

The following equation may relate the rate of moisture exfiltration ($\dot{m}_{ex}$) to the difference in absolute humidity between the kiln and outdoor air:

$$\dot{m}_{ex} = K_{ex} \times (\omega_{indoor} - \omega_{outdoor}) \tag{8.93}$$

where

$\dot{m}_{ex}$ is the rate of moisture (water vapor) exfiltration $\left(\frac{kg_{water}}{s}\right)$

K_{ex} is a dry air exfiltration experimental coefficient $\left(\frac{kg_{dry\,air}}{s}\right)$

ω_{indoor} is the indoor air average absolute humidity $\left(\frac{kg_{water}}{kg_{oven\text{-}dry}}\right)$

$\omega_{outdoor}$ is the outdoor air average absolute humidity $\left(\frac{kg_{water}}{kg_{oven\text{-}dry}}\right)$

The rate of enthalpy evacuated out of the kiln by moisture exfiltration can be expressed as follows:

$$\dot{H}_{ex} = \dot{m}_{ex} \times \overline{c}_{ex} \times T_{ex} \tag{8.94}$$

where

$\dot{H}_{xf}$ is the rate of total enthalpy of moisture exfiltration (W)

$\dot{m}_{ex}$ is the rate of moisture (water vapor) exfiltration $\left(\frac{kg_{water}}{s}\right)$

$\overline{c}_{ex}$ is the average specific heat of moisture exfiltration (J/kg·K)

T_{ex} is the temperature of moisture exfiltration (°C)

Uncontrolled air leakages must be compensated by fresh air flowing into the kiln that requires to be heated.

8.14.5.5 Losses in Energy Generation and Heat Distribution Systems

In industrial-scale wood dryers, steam is the most frequent heat carrier medium from boiler to the heat exchangers.

The total thermal consumption of steam boilers can be calculated as follows:

$$Q_{boiler} = \frac{D_b\left(MC_i - MC_f\right)\times V_{green}\times N}{100\times\eta} \tag{8.95}$$

where

Q_{boiler} is the boiler thermal energy requirement (consumption) (kJ)

D_b is the basic mass density of green wood $\left(\frac{kg_{oven\text{-}dry}}{m^3_{green}}\right)$

MC_i is the initial moisture content of wood (%, dry basis) $\left(\frac{kg_{water}}{kg_{oven\text{-}dry}}\right)$

MC_f is the final moisture content of wood (%, dry basis) $\left(\frac{kg_{water}}{kg_{oven\text{-}dry}}\right)$

V_{green} is the volume of the green wood dried $\left(m^3_{green}\right)$

N is the energy (heat) required to remove and vaporize 1 kg of moisture (water) extracted from the wood $\left(\frac{kJ}{kg_{water}}\right)$

η is the global efficiency of the heat production device and distribution system (expressed in decimals)

Even if the heat losses of a pressurized steam distribution system (e.g., steam provided at 150°C) and the condensate return system are of 5% and 10%, respectively, so that most condensate is returned above 100°C, the average overall system efficiency might be about 33.5%. Usually, for indirect-heated convective dryers, the overall thermal efficiency (expressed as latent heat of evaporation compared to gross calorific value of fuel) is never greater than 50% (Kemp 2012).

Some boilers in sawmills do not have a pressurized condensate return system, which limits the boiler feed water return temperature to about 85°C–90°C before boiling occurs.

Because of multiple heat losses, the net energy delivered to the kilns represents generally between 75% and 85% of the total energy produced by the fuel-burned (e.g., oil, propane, natural gas) boilers. This means that, even the flue gases leave at 120°C–150°C, the maximum boiler efficiency (defined as the ratio between heat transferred to the process fluid and the heat released from combustion of the fuel) is about 75%–85%. Lower flue gas temperatures would provide condensation in the exhaust gas, which can lead to corrosion due to acid gases, even with relatively clean fuels such as natural gas.

Heat losses occur in the steam generation (e.g., high fuel moisture content, incomplete combustion in fuel burner, losses in boiler tubing, high stack gas temperature) in the steam distribution network to the heating coils (about 7%), as well as in the condensate return pipes to the boiler (about 13% for condensate at 90°C).

Modern boilers recover as much heat as possible from the flue gas, for example, by economizers which heat boiler feed water and the incoming air.

8.14.5.6 Electrical Energy Requirements

In wood dryers, electrical energy is mainly required for fans (centrifugal, vane-, or axial flow) used to move the drying air through the lumber stack(s) and for the blowers of steam boilers.

Centrifugal fans are the most widely used, because they can efficiently move large quantities of air over a wide range of pressures, while the blades of a rotating impeller may be forward or backward curved, or radial. Vane-axial fans produce axial flows of the air, and guide vanes are provided before and after the wheel to reduce rotation of the airstream. Axial-flow fans are similar to the vane-axial fans but do not have the guide vanes. They are not able to produce pressures as high as those of the centrifugal fan but can move large quantities of air at low static pressure.

There are several simple relationships between fan capacity, pressure, speed, and power which are referred to as the fan laws (McQuiston et al. 2005). The most useful first three fan laws assuming constant air density are as follows: (i) fan capacity (i.e., total airflow, or volume per unit time) is directly proportional to the fan speed; (ii) fan pressure head (static, total, or velocity) is proportional to the square of the fan speed; and (iii) fan's required power is proportional to the cube of the fan speed. For example, if the fan speed is reduced by 50%, the air velocity will be reduced 50%, the pressure head will drop by 75%, and the shaft power by 87.5%. The other three fan laws are as follows (McQuiston et al. 2005): (i) fan pressure and power are proportional to the density of the air at constant speed and capacity; (ii) fan speed, capacity, and power are inversely proportional to the square root of the density at constant pressure; and (iii) fan capacity, speed, and pressure are inversely proportional to the density, and the power is inversely proportional to the square of the density at a constant mass flow rate.

For wood dryers, the actual electrical power demand for circulating the drying air varies with wood species, stack(s) width, board roughness, sticker thickness, drying air temperature and velocity, and drying rate. The increase of drying air velocity and stack(s) width, and a decrease in sticker thickness may increase the required fan power input requirement as well as associated energy consumption (Figure 8.12).

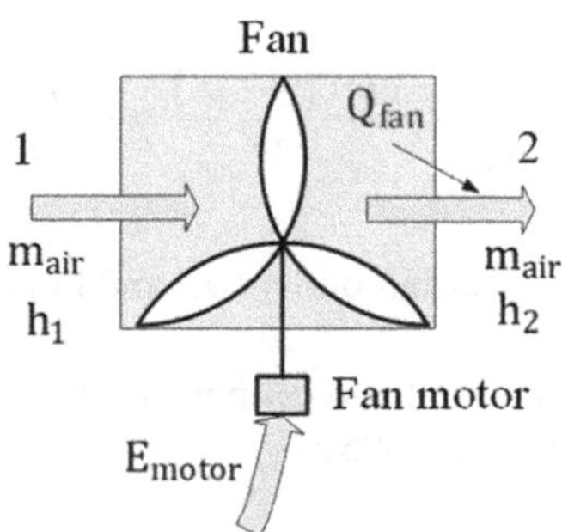

FIGURE 8.12 Schematic of air moving through a fan.

The fan' performances are generally given in the form of a graph showing pressure, efficiency, and power as a function of capacity. The energy transferred to the air by the impeller results in an increase in static and velocity pressure. The sum of the two pressures gives the total pressure.

The electrical power required at the fan shaft (Figure 8.11) must account for fan inefficiency, which may vary from 50% to 70%. It can be determined from (McQuiston 2005)

$$\dot{W}_{shaft} = \frac{\dot{W}_{total,\ air}}{\eta_{total}} \tag{8.96}$$

where

$\dot{W}_{shaft}$ is the electrical power required at fan shaft (kW)

$\dot{W}_{total,\ air}$ is the total power imparted to the air (kW)

η_{total} is the total fan efficiency (–)

It follows that the total fan efficiency (η_{total}) is the ratio of total power imparted to the air $\left(\dot{W}_{total,\ air}\right)$ to the shaft power input $\left(\dot{W}_{shaft}\right)$:

$$\eta_{total} = \frac{\dot{W}_{total,\ air}}{\dot{W}_{shaft}} = \frac{\dot{m}\left(p_{02} - p_{01}\right)}{\bar{\rho} \times \dot{W}_{shaft}} = \dot{V}\frac{\left(p_{02} - p_{01}\right)}{\dot{W}_{shaft}} \tag{8.97}$$

where

$\dot{m}$ is the mass flow rate of air (kg/s)

$p_0 = p + \frac{\rho u^2}{2}$ is the total pressure of air at fan outlet (p_{02}) and inlet (p_{01}) (Pa)

p is the static pressure of air (Pa)

$\frac{\rho u^2}{2}$ is the velocity pressure of air (Pa)

$\bar{\rho}$ is the average density of air (kg/m^3)

$\dot{V} = \frac{\dot{m}}{\bar{\rho}}$ is the volumetric flow rate of air (m^3/s)

Similarly, the fan static efficiency can be defined as the ratio of the static air power to the shaft power input and expressed as follows:

$$\eta_{static} = \frac{\dot{W}_{static}}{\dot{W}_{shaft}} = \frac{\dot{m}\left(p_2 - p_1\right)}{\bar{\rho} \times \dot{W}_{shaft}} = \dot{V}\frac{\left(p_2 - p_1\right)}{\dot{W}_{shaft}} \tag{8.98}$$

where

p is the static pressure of air at fan outlet (p_2) and fan inlet (p_1) (Pa)

From equation 8.97, it follows that the total power imparted to the air can be calculated as follows (McQuiston et al. 2005):

$$\dot{W}_{total,\ air} = \dot{m}_{air}\frac{p_{02} - p_{01}}{\bar{\rho}_{air}} = \dot{V}\left(p_{02} - p_{01}\right) \tag{8.99}$$

Similarly, the fan static power is the part of the total power that is used to produce the change in static pressure (McQuiston et al. 2005):

$$\dot{W}_{total,\,air} = \dot{m}_{air}\frac{p_2 - p_1}{\bar{\rho}_{air}} = \dot{V}\left(p_2 - p_1\right) \quad (8.100)$$

The power required at the fan's motor input may be determined as follows:

$$\dot{W}_{motor} = \frac{\dot{W}_{shaft}}{\eta_{motor} \times \eta_{drive}} \quad (8.101)$$

where

$\dot{W}_{motor}$ is the electrical power required at input to fan's motor (kW)
$\dot{W}_{shaft}$ is the electrical power required at fan's shaft (*kW*)
η_{motor} is the fan motor efficiency, generally, varying from 80% to 95% (–)
η_{drive} is the belt drive efficiency accounting for belt drive (generally, 3% of the fan power) (–)

Fans that circulate air through drying kilns add sensible heat to the system that is equivalent to kinetic energy developed and to energy dissipated by the fans' electrical motors and drives inefficiency.

Almost all the energy required by fans to generate airflow and static pressure is dissipated heat within the dryer chamber.

Electrical energy is converted to thermal energy within the dryer in two ways. If the motors are external to the dryer, then only the work done in air movement is converted to heat by friction (air and bearing friction) minus the work of venting. The heat generated within the motor is lost to the external environment (approximately 10% of power input). If the motors are within the dryer compartment, then all the electrical consumption is converted into heat. Between 10% and 20% of the fan electrical power is however lost inside the fan motors (Tschernitz 1986).

The portion of fan heat released to the airstream depends on the location of the fan motor and drive. If they are within the airstream, all the energy input to the fan motor is released to the airstream. If the fan motor and drive are outside the airstream, the energy is split between the airstream and the room housing the motor drive.

To calculate heat generated by the fan motor and drive located inside the drying chamber, the following relation can be used:

$$\dot{Q}_{fan,\,airstream} = \dot{W}_{fan,\,motor} \quad (8.102)$$

where

$\dot{Q}_{fan,\,airstream}$ is the thermal power released to airstream (kW)
$\dot{W}_{fan,\,motor}$ is the motor electrical power input (kW)

For low pressure rises (<2.5 kpa), the temperature rise may be found by the following (ASHRAE 2008):

$$\Delta T = \frac{\Delta p}{\bar{\rho} \times \bar{c}_p \times \eta_{fan}} \quad (8.103)$$

where

ΔT is the temperature rise across fan (°C)
Δp is the pressure rise across fan (kpa)
ρ is the air average density (kg/m^3)
$\overline{c}_p$ is the air average specific heat (kJ/kg)
η_{fan} is the fan efficiency (decimal)

The cost of electrical power and energy consumption can be reduced if one can control the air velocity over wide limits. The combined motor-shaft-drive system may reduce the power savings.

Actually, a properly designed electronic speed control fan motor unit will come close to obeying the cubic law—all other factors being constant. There will always be bearing losses, but for the most part, these are small. With adequate speed control of the motor drive system, power savings approaching 87.5% can be realized with a 50% reduction in fan speed. The efficiency of a fan motor installation, expressed as m^3 of delivered air per kW power, will vary with actual fan design even though the cubic law still applies.

Traditionally, velocity was reduced economically only by using two-speed motors or variable mechanical drives. For AC induction motors operating at constant voltage, the speed can be changed by varying the frequency and current (i.e., variable frequency drive system) (Tschernitz 1986).

8.14.5.7 Total Kiln Energy Requirement

Based on the previous energy requirement components, some factors have to be considered in practice: (i) over 90% of the heat consumed is used to heat and evaporate the water, heat and humidifying the vent air, and compensate the kiln heat losses; (ii) as the outside air temperature increases, kiln heat requirements decrease; (iii) for the same moisture content (%), higher specific gravity woods contain more water and result in higher energy consumption; and (iv) for a particular kiln charge, all the elements of heat consumption, except one (i.e., heat loss which is time dependent), are essentially fixed values since they are dependent primarily upon the quantity of wood and the amount of water contained; the effect of drying time may be adjusted by either using lower temperature for longer times or using higher temperatures for shorter times.

8.15 CONTROL OF WOOD DRYERS

Drying is a critical step of most wood products manufacturing processes. The methods (single, multiple, or combined) used and their proper control are factors key to achieve appropriate production levels, quality, and costs.

The aim of industrial wood drying processes is to produce solid products of desired quality at minimum cost and maximum throughput while utilizing the least amount of energy and avoiding wood drying defects. Good quality implies that the product corresponds to a number of technical, chemical, and biological parameters, each within specified limits. While wood quality can be defined differently by its various users (such as industrial wood processors or consumers), certain aspects of quality remain similar, such as minimizing warp, checks, and splits, and discoloration, and maintaining or enhancing lumber mechanical properties.

Product quality, along with drying efficiency, is thus the most important aspect in operating a drying process (Law and Mujumdar 2008; Perré and Keey 2015a, 2015b).

Upon the maximum operating temperature, the conventional industrial kilns can be classified as (i) conventional (or low temperature), (ii) medium temperature, and (iii) high temperature.

Control of the drying process generally varies with (i) the type of kiln used, (ii) species being dried, and (iii) temperatures used in the process. It usually involves measurements of the moisture content of the wood being dried.

Because a great number of parameters affect product quality, the control of wood dryers is a challenging task. Any control system for dryers requires the consideration of factors such as (i) critical process parameters; (ii) process dynamics; (iii) number of process variables that are to be controlled and/or monitored; (iv) operating ranges of air dry- and wet-bulb temperatures, pressure, and relative/absolute humidity; (v) product throughputs; (vi) airflow rates; (vii) initial and final moisture contents; (viii) data acquisition system; (ix) type of control calibration and sensing devices; and (x) system reliability.

In addition to product quality and optimum energy consumption, other objectives and performance indicators for industrial drying operations are as follows (Onn 2007; Oliveira et al. 2012; Yumah et al. 2015): (i) maximization of throughput at minimum drying time, energy and maintenance costs, and desired final moisture contents; (ii) avoidance of overdrying and under-drying that may cause thermal damage especially to heat-sensitive products; (iii) reduction of defective lumber boards and particle emission; and (iv) adequate control strategy improving the process reliability, providing uniform drying within the lumber stack(s) boards and reducing drying defects that can adversely affect the serviceability and economics of the dried lumber boards.

The simplest control systems of kilns include discrete programmable controllers, timers, and recorders for temperature and energy (e.g., natural gas via steam) supply. More sophisticated systems are computer software based on control kiln conditions during specific schedules (Keey et al. 2000). In the past, the lack of direct, online, reliable methods for sensing product moisture content leads to relatively little progress in the field of kilns' control strategies (Robinson 1992, 2000).

Any control system of industrial dryers requires (Marchant 1985; Yumah et al. 2015) (i) accuracy, that is, the exit product moisture content must be close to the desired value; (ii) stability, that is, the system must avoid large fluctuations in output moisture content; (iii) quick response to any disturbances or changes, for example, in input moisture contents in order to provide reasonable recovery times; and (iv) robustness, that is, the control system should be able to operate successfully over a wide range of process parameters.

The control strategies of industrial dryers are based on classical (using transfer functions) or modern (using differential equation) control theories. Both classical and modern control theories are difficult to apply due to one or more of the following reasons (Yumah et al. 2015): (i) the drying process is complex, time variant, and nonlinear; (ii) it is not possible to adequately represent all parameters that exhibit interacting behavior; (iii) some drying variables (e.g., product quality and color) cannot be measured directly; other measurements (e.g., moisture content) may be inconsistent, imprecise, incomplete, or not totally reliable; (iv) dryer models are generally approximations to the real process and may require large computing time; and (v) difficulties are experienced when the process operates over a wide range of conditions.

In wood drying industry, the control strategy is usually based on anticipation, practical knowledge of drying process, and operators' personal experience.

The development of new methods for controlling the drying process focuses on new ways to measure moisture content or moisture content variation, dry- and wet-bulb temperatures drops across the lumber stack(s), and drying stresses.

8.15.1 Basic Variables and Control Systems

The basic variables associated with industrial drying processes are of two categories (Courtois 1997) (i) input variables, which denote the effect of the surroundings on the drying process and (ii) output variables (see Figure 8.12).

The input variables can be (i) manipulated (e.g., airflow and heating rates) and (ii) load (e.g., drying air temperature and relative humidity, product moisture content, and composition) that cannot be adjusted by the control system.

Among the most desirable output variables, but often difficult to measure directly on line at reasonable costs, the following can be noted (Yumah et al. 2015) (i) moisture content of dried product, (ii) exhaust air temperature and relative humidity, and (iii) product quality (e.g., defects, color).

The purpose of any kiln drying control systems is to provide desired output variable data by changing the manipulated variables in order to compensate for changes in the main load variables.

8.15.1.1 Manual Control

For small-scale batch dryers, drying small volumes of sensitive and/or valuable materials and manual control strategies, requiring less expertise than the automatic control systems, can be used. The manual control consists in simple sequences (Yumah et al. 2015): (i) turn on the dryer; (ii) set the initial volume (mass) of material to be dried; (iii) measure the output moisture content and compare with the desired value; and (iv) based on the difference between the desired and the measured moisture content value, make adjustments to the manipulated variables (e.g., energy input, feed rate) to maintain the desired moisture content.

The manual control of large batch wood drying would require much higher labor costs per unit product throughput.

8.15.1.2 Automatic Control

Most large-scale industrial dryers are provided with automatic control of drying air dry- and wet-bulb temperatures, relative and absolute humidity, and air velocity in order to achieve effective drying processes (Cech and Pfaff 2000; Onn 2007). Automatic control, including online measurement of the product's moisture content, allows improving the dryer operation and energy efficiency. Typical automatic batch dryer control systems use the exhaust-air temperature as the controlled variable to determine when to end the drying process.

The main features of automatic controls are as follows: (i) ability to control the kiln dry- and wet-bulb temperatures within 0.5°C and relative humidity within 1% or 2%, (ii) change the kiln drying parameters very smoothly, (iii) determine the drying rate of individual boards as well as of the entire lumber stack(s), and (iv) accurately measure the moisture contents above and below saturation fiber point within 5% error and 0.5% repeatability.

The major benefits of automatic control via preprogrammed temperature or airflow of batch drying processes are as follows: (i) increased safety, (ii) increased production through a reduction of drying time, and (iii) increased consistency of product quality.

The control strategy of dryers aims at holding the main controlled variables at their target set points by (i) proportional (that actuates the manipulated variables proportionally to the error signals), (ii) integral (that actuates the manipulated variable based on the time integral of the error), and (iii) derivative (that forecasts fast changes in the error signal) actions. In industrial drying applications, these three control actions are integrated within (i) proportional, (ii) proportional–integral, and (iii) proportional–integral–derivative (PID) controllers (Luyben 1990; Keey 1992).

Typically, an automatic control system (see Figure 8.13) receives a measured signal of the controlled output variable (i.e., moisture content) and compares it with the set point value, which generates an error signal. The value of the error is supplied to the main controller that, in turn, changes the value of the manipulated variable in such a way to reduce the magnitude of the error. The controller corrects the manipulated variables via a final control element (e.g., a control valve, motor, fan, or heater) in order to adjust them back to the desired set points (Yumah et al. 2015).

Modern microprocessor-based control systems may implement various control algorithms in order to achieve additional tasks such as (Garcia and Morari 1982; Prett and Garcia 1988; Luyben 1900; Coughanour 1991; Boseley et al. 1992; Luyben and Luyben 1996; Morari et al. 1997) (i) servicing several control loops; (ii) controlling parameters from different process instruments via a central computer (e.g., moisture content and relative/absolute humidity); (iii) providing mathematical functions, data acquisition, and storage for different measured parameters, such as air temperature, flow rate, pressure, relative humidity, and product moisture contents; and (iv) providing planning, supervision, optimization, quality control, and control of mode of operation.

As can be seen in Figure 8.13, the computer system collects data from the process measurements, calculates the values of the manipulated variables, and implements the control action on the process, based on the control algorithm that is already programmed and stored in the memory of the computer. Signals are converted by digital to analog and analog to digital converters. The operator communicates with the control system with a keyboard, a monitor, and a printer or plotter. Improvements in dryer controls will be performed because of the development of better sensors and analyzers (Liptak 1999).

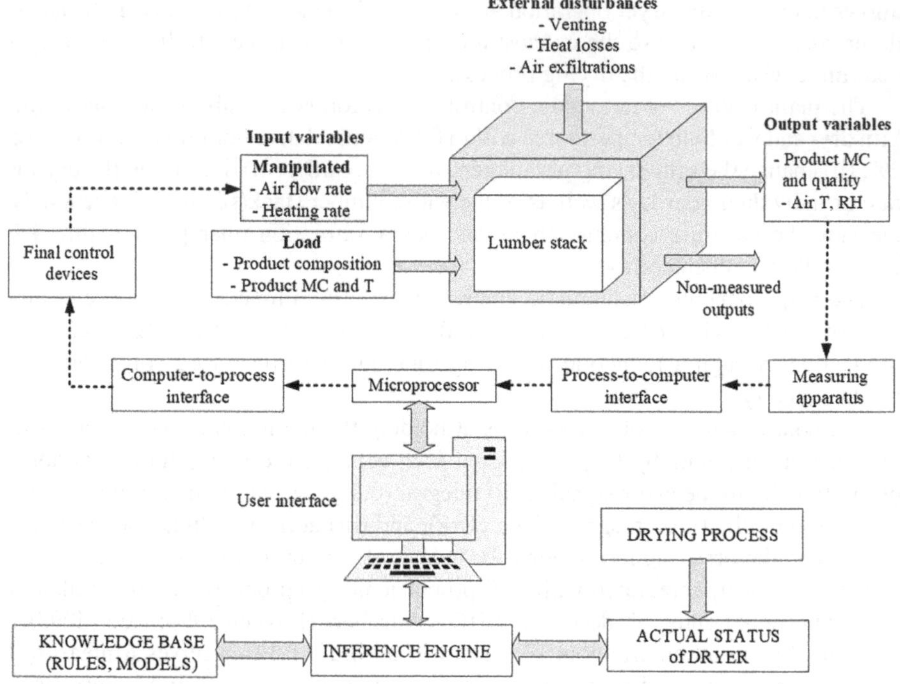

FIGURE 8.13 Simplified diagram of a typical drying process variables and structure of an expert control system; MC, moisture content; RH, relative humidity; T, temperature.

Smaller lumber drying installations having one or two kilns might have a simple electronic control system with discrete programmable controls, timers, relays, and a chart recorder for temperature and steaming control. Other options include a preset timer and endpoint control based on the temperature drop across the load. More sophisticated systems for a medium-sized facilities might have a low-cost, computer-based systems running with software to give visual readouts of kiln parameters. These systems might also incorporate programmable logic controllers to enable the operator to change the drying schedules while the kilns are working and ramp set points up and down. Advanced computer-based kiln control systems are available to provide centralized supervision in a larger installation comprising several kilns, together with ancillary plant such as steaming chambers and heating units (Keey et al. 2000).

The installation of process control systems incorporating programmable logic controllers may increase the overall lumber grade recovery from 70.7% to 81.9%, with reductions in energy costs of 43% for steam and 10% for electricity, with up to 1.2-year payback on the capital investment.

8.15.1.3 Advanced Control Systems

During the next years, since the dryer performance is highly nonlinear and difficult to predict with simple mathematical models, it is expected that industrial dryers utilize emerging, real-time, intelligent control technologies based on (Garcia and Morari 1982; Quantrille and Liu. 1991) (i) expert systems (useful for reasoning about process parameters such as temperatures and pressures, and product structure); (ii) artificial (biological neural) neurocontrollers useful for reasoning about process trends; and (iii) fuzzy logic controllers (Astram and McAvoy 1992).

An intelligent, computerized expert control systems consist of the following (Quantrille and Liu 1991; Gevarter 1987; Bernard 1988): (i) knowledge base of the drying process; (ii) availability of a control structure using inference procedures to draw conclusions and to infer the correct control action based on the stored information and the current state of the process; (iii) availability of a global database for keeping track of the system status, the input data, and the relevant history of the process; and (iv) availability of an user interface to provide communication between the user and the drying program.

Expert systems can be employed in many process controls (Prett and Garcia 1988; Quantrille and Liu 1991; Bernard 1988; Tzouanas et al. 1988): (i) process management and optimization, trend analysis, alarm processing, control system design, and adaptive control; (ii) enhancement of classical controller performance by sensor failure identification, valve saturation, and process constrains; (iii) system status based on global database; (iv) fault detection, diagnosis, and troubleshooting; and (v) supervisory control and start-up or shutdown procedures.

Artificial neutral network control (biological neural) systems consist of interconnected processing elements (also called neurons or nodes) and have the potential to treat many problems that cannot be handled by traditional control techniques. The strength of the connection among these neurons is characterized by its assigned weight adjusted with an algorithm in order to reach a desired input/output (Yumah et al. 2015). Artificial neutral network control can be used in many potential applications related to industrial dryers' design, operation, and control (Bhat and McAvoy 1990; Psichogios and Unger 1991): (i) process modeling; (ii) qualitative interpretation of process data for the purpose of control, extraction of control rules for fuzzy logic controllers; (iii) detection of sensor failure; (iv) provision of inferred values for signals that are difficult to measure in practical situations; and (v) estimation of model parameters (e.g., mass diffusivity) from experimental data.

Fuzzy logic control systems are knowledge-based control strategies that use fuzzy linguistic variables to manage the uncertainty in drying systems. They are used to convert linguistic variables into precise numerical control actions (Zadeh 1972; Mamdani 1974, 1977; Mamdani et al. 1975).

8.15.2 Control of Lumber Kilns

The kilns' control systems mainly focus on changes in lumber average moisture content in order to obtain the final moisture content as close as possible to the optimum value. The optimum final values of lumber moisture content depends on multiple factors such as the kiln capacity and dynamics, the variability of the physical properties of lumber boards, initial moisture content, changes of air parameters as it passes through the stack(s), etc. In other words, the lumber moisture content is the main control variable that must be well measured and compared with the set point (target) values (Léger and Amazouz 2003). Thus, the design of control systems of lumber kilns requires the knowledge (measurement) of optimum change in moisture content.

In control systems based on measurement of the mean moisture content of lumber, the measured output moisture content values are compared with set values, and, once it reaches those values, the drying advances to the next step, and new set points are established for dry- and wet-bulb temperatures. However, it is difficult to measure moisture content as an output signal because the moisture content varies from one board lumber to another and even within a single board, and, furthermore, it can be measured only for a limited range of values. It is therefore impossible to compare a measured value with the set point value. Therefore, the control systems based on moisture content measurement are not reliable in industrial practice.

The simplest control systems are time based, but they are not sensitive to changes in conditions inside the kiln. The kiln operators must estimate the drying rate and, accordingly, the time the lumber will take to reach a mean moisture content. Based on the control system shown in Figure 8.14, the kiln operators usually set and adjust the drying schedules by a trial-and-error approach in order to determine where enhancements can be made and change the drying schedule accordingly. This control strategy operates in successive, preprogrammed steps established by the operators on the basis of their knowledge of the drying process, and lumber load properties (Cech and Pfaff 2000).

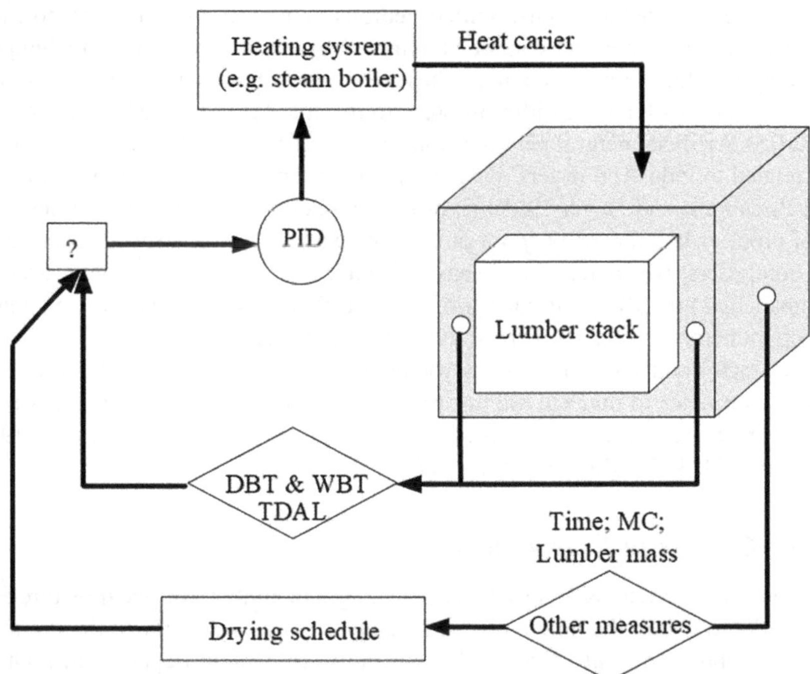

FIGURE 8.14 Simplified diagram of a typical PID drying process control; DBT, dry-bulb temperature; MC, moisture content; PID, proportional–integral–derivative; TDAL, temperature drop across the load; WBT, dry-bulb temperature.

A typical kiln control system that acts on air dry- and wet-bulb temperatures and relative/absolute humidity on the basis of the difference between the desired temperature or humidity and measured values is schematically represented in Figure 8.14.

Such a control system measures air dry- (with thermocouples, Resistance Temperature Detectors-RTDs, thermistors, etc.) and wet-bulb (with electronic probes, psychrometers, etc.) temperatures, or Temperature Drops across the Load (*TDAL*), and compare them with respective set point temperatures and, then, acts on the process to reduce any error is detected.

Other variables in such a control system are the air temperature, velocity and direction, or lumber stack(s) actual mass. Systems based on lumber mass measurement give the total mass of part of the load, and the moisture content of the load can be calculated from its estimated oven-dry mass. Depending on the degree of precision of the scale, the mean drying rate of the load can also be estimated (Léger and Amazouz 2003).

The heating system ensures that the desired temperatures exist in the kiln. It must offset energy losses due to evaporation, increases in lumber temperature, the heating up of the fresh air intake required to adjust the moisture content in the kiln, and energy lost through kiln walls and doors. Its ability to offset energy loss depends on its power and on air velocity. The indirect heating systems (steam—commonly used as an energy source, hot oil, hot water) via heat exchangers generally provide the most effective distribution of heat and best local area control.

The control system shown in Figure 8.13 contains a PID (proportional, integral, and derivative) regulator that controls auxiliary parameters such as the temperature of drying air, moisture content and mass of lumber stack(s), drying time, Temperature Drop across the Load (*TDAL*), etc. Depending on the detected errors, the PID controller regulates by raising or lowering the heat carrier fluid output (e.g., steam) from the heating source. Such a control system can achieve results that are accurate, fast, stable and reliable drying cycles. The PID control loop compares the actual air dry- and wet-bulb temperatures, or *TDAL* measurements with their respective set points, and, according to the proportional action, opens and/or shutoff a tap proportionally to the observed discrepancy between the actual parameters in the kiln and the desired values. When a set value (e.g., dry- or wet-bulb temperatures, air relative or absolute humidity, dew point, lumber equilibrium moisture content, *TDAL*, or elapsed time) is reached, the PID controller changes the temperatures in the kiln via the heating air system. The integral action is a continuous reaction in the event that the observed error persists in spite of the proportional action on the tap. The derivative action is an amplified reaction directed to the tap once the temperature discrepancy occurs, generating a predictive action and offsetting the slow response of some systems to new demands (e.g., for more energy) (Léger and Amazouz 2003).

The moisture content remains the main output parameter estimated based on periodical measurements of board sample moisture contents or by measurements of *TDAL* or of the stack actual mass during the drying cycle, or simply on time. Based on such measured parameters, the control system will determine when to move to the next stage in the drying program. In this way, new air temperature settings are activated throughout the drying schedule. In other words, in a control of the drying process, the change in moisture content exerts no direct effect on control except to activate the next stage in the drying schedule. The *TDAL* is a modern method that indirectly generates an estimate of the drying rate and the moisture content of the load measurements needed to run the kiln. It produces changes in the kiln temperature settings and may help adjust the drying schedule.

8.15.3 Drying Software

Over the past few decades, considerable efforts have been devoted to the development of properly designed software programs applicable to drying of various materials in order to optimize the thermal drying operations (Kemp 2007; Kemp et al. 2004; Menshutina and Kudra 2001; Marinos-Kouris et al. 1996). However, few commercial drying software packages have been developed and/or well accepted by the drying industry (Gong and Mujumdar 2008).

According to Devahastin (2006), the development of drying-related process simulation software has almost been completely neglected over the past decades. The apparent progress has been disappointingly slow, especially in terms of commercially available software packages.

However, one might expect that, given the huge developments in computing power over recent years, many new drying software programs would have appeared, giving increasing accurate and comprehensive simulations of dryers.

The relatively low number of software for drying processes can be attributed to the following reasons (Kemp et al. 2004; Kemp 2007): (i) complexity and variability of drying kinetics, (ii) difficulties in modeling hygroscopic materials such as woods and foods, and (iii) limited market and lack of replicability.

In solids drying, there has often been a great difference between academic theory and industrial design practice. Traditionally, practical dryer design has tended to be based on simple correlations and scale-up from pilot-plant tests, rather than on rigorous theoretical models. Moreover, subjects such as selection and troubleshooting were largely neglected. This dichotomy has been reflected in software programs. Dryer manufacturers and users have developed in-house programs based on simple correlations, sometimes including other aspects such as mechanical design. On the other hand, numerous computer-based models have been developed in academia research projects but have very rarely been tested on industrial data or used for practical improvement of industrial dryers. Moreover, few commercial software has been developed. The reason for this is that drying processes, like all processes involving solids, are much more difficult to model than fluid-phase (liquid and gas) processes. Physical properties of fluids can be obtained easily from data banks and are uniquely defined for given temperatures and pressures, while the system is controlled by equilibrium thermodynamics (Kemp 2007).

In contrast, physical properties of solids vary considerably with solids structure, and drying kinetics can differ for the same material depending, for example, on particle size and porosity governed by highly nonlinear equations. Moreover, parameters such as diffusion coefficients in solids are difficult to measure or predict accurately, and, thus, the uncertainty is high, and simple lumped parameter models such as the characteristic drying curve concept can be more effective and reliable. As a result, it is difficult to evaluate the parameters required for dryer models and software with sufficient accuracy for practical use.

Today, and in the future, user-friendly software is needed to improve the energy efficiency of drying a particularly energy-consuming operation and reduce the carbon footprint of drying products. They may be very cost-effective in the design, analysis, troubleshooting, as well as control and optimization of drying systems.

Available commercial drying software is limited for various reasons (Kemp 2007). Drying programs fall into types such as (Kemp 2007) (i) numerical (calculation) programs, such as numerical models at various levels of complexity for dryer design, performance rating, and scale-up; (ii) process simulators such as Aspen Plus, HYSYS, and Batch Plus; (iii) expert systems that can help with selection and other aspects of design without numerical calculations; (iv) on-line information (currently, scattered or inadequately exploited) such as the process manual; and (v) auxiliary calculations (e.g., drawing psychrometric charts or processing experimental data).

Computer software can be helpful in the following ways (Gong and Mujumdar 2008): (i) process simulation and control of drying process that can lead to optimized design and operation, (ii) dryer design, including heat and mass balance calculations, and (iii) simulation of drying kinetics that may predict the transient coupled heat and mass diffusion within the material and determine the drying time.

Three commercial software packages specifically intended for drying have been identified (Gong and Mujumdar 2008): (i) Simprosys, (ii) dryPAK, and (iii) DrySel.

Simprosys is a Windows-based process simulator used for design and simulation of dryers and evaporation systems (Devahastin 2006; www.simprotek.com. Accessed December 12, 2017).

dryPAK is a dryer design software package developed on the DOS platform that includes heat and mass balances and drying kinetics modeling by combining the equilibrium and the characteristic drying curve methods (www.tandfonline.com/doi/abs/. Accessed January 6, 2018). Drying kinetics is based on Fick's diffusion equation for three basic geometries (plate, cylinder, and sphere) and two types of boundary conditions for isothermal or adiabatic case. Mass transfer coefficients and other kinetic data can be entered to calculate the dryer length.

DrySel is an expert system for selection and comparison of over 50 different types of dryers (e.g., batch or continuous, contact, or convective heating) based on their respective advantages and disadvantages, material properties, specified throughput and moisture content, and safety and environmental issues. Dryers may be ranked in order of merit score, and both graphical and numerical displays are provided (Gong and Mujumdar 2008).

Any dryer software should be based on a theoretical model that is capable of useful practical application to real industrial-scale dryers. The calculations can be used to (i) design a new dryer for a given duty, (ii) determine the performance for an existing dryer under a different set of operating conditions, and (iii) scale-up from laboratory-scale or pilot-plant experiments to a full-scale dryer.

For solids drying, key factors such as drying kinetics and internal moisture transport within a solid cannot be predicted from first principles but only measured by experiment. Hence, scale-up calculations have been found to be more reliable than design based only on thermodynamic data. The experimental data are used to verify the theoretical model and find the difficult-to-measure parameters; the full-scale dryer can then be modeled accurately.

The calculations are also of levels of complexity, such as (Kemp and Oakley 2002) (i) level 1 based on simple heat and mass balances directly gives information about, for example, the mass flow rates, moisture content, and relative humidity; however, no information is provided about the required equipment size or the dryer; (ii) level 2 based on simple data, assumptions, and approximate calculations gives rough sizes and throughputs for dryers. In the case of batch dryers, it calculates the required size of the dryer and estimates the drying time; (iii) level 3 gives overall sizes and performances of dryers by scaling-up drying curves from small-scale or pilot-plant experiments; the characteristic drying curve concept is used in the integral model; and (iv) level 4 methods require more input data and much more complex modeling techniques to track the local conditions of the solids and water vapor during drying; they include incremental models, in one or more dimensions, and *CFD* (computational fluid dynamics).

It can be noted that higher levels of complexity can potentially give more precise results but also require more data and may be more susceptible to cumulative error. Therefore, the most detailed model may not be the most appropriate for a given application.

Process simulators, including Aspen Plus, HYSYS, and Batch Plus, allow the dryer to be modeled as part of the overall process flowsheet or recipe.

However, the major simulators for continuous processes, such as Aspen Plus, HYSYS, and Prosim, have had extremely limited capabilities for solids, as they were originally designed for fluid processes, and improvement has been relatively slow. Batch Plus, which was designed from the outset to handle pharmaceuticals and similar processes, handles solids more effectively.

HYSYS does not include a dryer unit. Aspen Plus includes a dryer unit which appears to be too simplistic to be of much practical use. Popular process simulators such as Hysys, Aspen Plus, and ProSim were designed mainly for materials of very well-defined chemical compositions. Their fundamental calculations are based on components' liquid–vapor equilibrium which is calculated according to the gas state equation.

Almost all simulators use simplified models at level 1 or 2 that are in fact adequate to study the main interactions between dryers and the rest of the process, which need to be considered in the earliest stages of process design. However, these models will not be adequate for detailed design or analysis of dryers (Kemp 2004).

In the cases where the requirement is for decision-making rather than equipment sizing or performance simulation, expert systems, which are decision-making tools, may be appropriate for dryer selection procedures of great practical importance in drying industry (Van't Land 1984, 1991; Kemp and Bahu 1995; Kemp 1999).

An obvious candidate for computer assistance is the processing of data from drying experiments. In particular, the handling of large numbers of data points and the interconversion between the different types of drying curve (moisture–time, rate–time, and rate–moisture) are fairly easily accomplished using a spreadsheet.

It can be seen that there is potential to apply software in all aspects of drying operations, from initial process development and equipment selection and design through to commissioning, optimization, debottlenecking, and troubleshooting of existing plants. It can also be used by all departments within a company, from research and development to operations.

Although many drying programs have been written to perform simulations and calculations in design mode, this may not match well to industrial needs. Design and selection tools are only required when a new dryer is being installed, which for end-user companies will only be at intervals of many years. The main users of design and scale-up programs are equipment manufacturers, who usually have their own in-house methods. Simulation and performance rating tools are useful on existing plants, and a dryer is only designed once but will then operate for maybe 20–50 years. Even so, when a model has been developed for a particular dryer, it does not need to be rerun often. In contrast, troubleshooting needs may arise without warning at any time, so if an effective program were available, it might be quite widely used.

For dryers there has been little software available commercially, and it has covered only a few types and a relatively small range of calculations. Among the main reasons, the following can be mentioned (Kemp 2007): (i) complexity of the calculations (with its multiple nonlinear equations, drying analysis presents a fascinating intellectual challenge but is a nontrivial programming problem; moreover, the many different factors involved, such as heat and mass transfer, vary between different dryer types, so that the corresponding models are also complex and different); (ii) difficulties in modeling solids (the numerous parameters involved, their variability, and the difficulty in measurement all make modeling of drying much harder than their gas- and liquid-phase equivalents); (iii) limited market and lack of replicability because dryers are relatively specialized equipment; the different types of dryers need to be modeled in very different modes; and (iv) changes in operating system software (e.g., from DOS to Windows).

In conclusion, relatively slow progresses have been achieved in implementation of software for dryers' modeling. Various in-house and research software has been written, but only a handful of commercial programs have been successfully launched and updated. The barriers to successful development of drying software are as much commercial as scientific. In particular, the market size is inadequate to pay for the development of complex programs, and interactive user interfaces need frequent and costly rewriting as operating systems change (Kemp 2007).

Because of the complexity of drying and solids processing operations, and the difficulty in determining key parameters, detailed and rigorous mathematical models are not necessarily the best way to analyze dryers. Simplified methods, consistent with the level of data available and the objective of the calculation, may be more appropriate. Spreadsheets often provide an effective way to implement these simplified models. Many dryer models work better in scale-up mode, fitted to even limited experimental data, than for design from scratch. Software should reflect this and be usable for scale-up calculations, including input of existing process conditions.

Decision-making algorithms and tools for selection and problem-solving can be useful, and in-depth qualitative information is essential to develop a clear understanding of the key factors in the process.

The dryer must also be considered in the context of the entire process flowsheet, not just as a unit operation. This needs to be done in the early stages of development, and simulators are valuable for this, preferably with data links to programs for individual unit operations which can be used for later optimization.

Future developments seem likely to be in low-cost models such as spreadsheet implementations and in detail improvements to generic simulators to allow them to handle solids better. Market considerations make it unlikely that detailed design models will become available as commercial programs, unless either industry is prepared to make a very substantial investment or a reduction from current standards in user-friendliness and debugging is accepted.

Among available software specially designed for wood drying, including heat pump-assisted dryers, the following can be mentioned:

a. MRS SA 2000 Microsoft Visual Basic-based software (www.mrsdeshumidification.com/mrs. Accessed June 15, 2015) that includes a simple, user-friendly, visual interface, and may program up to 12 drying steps per schedule; preprogrammed drying modes are as follows: preheating, conventional, dehumidifying, hybrid, cooling, and balancing; essential parameters such as actual drying step, elapsed time, dry- and wet-bulb temperatures, compressor running and vent air opening status, compressor cumulative running time, fans' direction of rotation, and alarms or critical shutdowns are graphically displayed;
b. MEC Control Software (www.sechoirmec.com. Accessed October 14, 2016) that allows controlling the wood drying cycle in PID or standard mode according to the progression of the wood moisture content; the drying schedule can be adapted according to various wood species and ranges of moisture content, by choosing operating parameters of fans, air vents, and humidification and heating devices in different stages of drying; the operating system, Windows XP or 2000, can be connected on any type of company network allowing a best follow-up as well as a better breakdown service by MEC dry kiln; dry- and wet-bulb wood temperature and outside temperature are measured with precision sensors RTD resistant to temperatures up to 200°C; up to 16 board moisture contents can be read and corrected according to wood dried species via an analog multiplexer moisture transmitter; if a failure occurs, 48 messages of critical and no-critical alarms are available for safety operation of the kiln; critical alarms stop the fans, as well as the heating, air venting, and humidification systems, and immediately show a message on the screen;
c. dTOUCH—Touch Screen Kiln Controller (www.logica-hs.com/. Accessed October 14, 2016) includes a drying program database for over 400 timber species that together with additional information received from the sensors and inputs from the dryer allows defining the most suitable drying schedule for each particular case; customized programs using dTOUCH can automatically identify some critical conditions (e.g., frozen and casehardened wood) and modify the drying schedule in order to correct the identified problem;

d. vacuTOUCH—Vacuum kilns controller (www.logica-hs.com. Accessed November 2, 2017) is equipped with special time-based, multifunction devices provided with wireless sensors for the timber temperature and moisture measurement; it may automatically or manually adjust the parameters and functions of different continuous or discontinuous vacuum kilns with single and/or double chambers, and with or without humidification/dehumidification systems; up to 30 steps for each drying schedule can be programmed;
e. SIEMENS software (http://w3.siemens.com. Accessed July 15, 2016)is a PC-based device fully independent of the drying system during operation. The programs and engineering handling of the software controller is compatible with other standard controllers;
f. TIMBERplus software (www.timberplus.com. Accessed January 15, 2016) has been developed for the veneer production and trade;
g. Brunner-Hildebrand software (www.brunner-hildebrand.com. Accessed June 12, 2016) uses platforms such as Windows XP professional or WinXP to customize the control system; this WinXP-based software is easy to install on customer's own PCs and provides networking monitoring and remote controlling with modem; Brunner-Hildebrand software allows to control (i) time-based drying schedule in function of temperature and dryer ambient climate, without moisture content measurement; (ii) heating of conventional and vacuum kilns with up to six timber temperature probes; (iii) steaming with (in vacuum kilns) or without (in conventional kilns) measurement of the inner timber temperature measurement for a light/bright coloration; (iv) reconditioning; and (v) variable speed drive systems for energy savings and a tender timber drying;
h. Nyle NSC100 Advanced Control System (http://lumber.nyle.com/controls. Accessed July 19, 2016) is available for any dry kiln, new or retrofit, allows kiln operators to dry lumber using a specific drying method, sets an automated schedule, and monitors/controls the drying chamber; the NSC100 controller does not require a computer for data logging or control because data logs are stored entirely within the Programmable Logic Controller (PLC); the standard package includes (i) color touch screen; (ii) built-in data logging; (iii) schedule control; (iv) multiple schedules of drying; (v) web server for remote access via LAN or mobile phone; (vi) wireless access; (vii) printer; (viii) works with unlimited number of kilns; (ix) automatic recording during drying cycle (temperatures, moisture contents, electrical power, etc.); (x) control set points and climate during heat treating cycle; (xi) calculate average moisture content for readings in the chamber; and (xii) automatically turn off the equipment when heat treating has been completed;
i. KISS, kiln interface software system (http://hotkilns.com/kiss. Accessed October 14, 2016). Remotely monitors and controls up to ten kilns and connects up to ten DynaTrol controllers to a personal Windows-based computer;
j. SKC, Single Kiln Temperature Controller Software (www.temperature-controlwiki.com. Accessed January 15, 2017) includes components such as temperature controllers and PC software;

k. TAP kiln controllers (www.kilncontrol.com/explore/. Accessed January 6, 2018) utilizes a PID control algorithm to ensure accurate schedule following with the fastest response, minimal overshoot, and limited steady-state error; it provides access to many features and settings within the controller such as extensive logging capabilities, virtually unlimited number of schedule segments, on-the-fly schedule modifications, real-time graphical display, built-in diagnostics, preventative maintenance, and semiannual software updates; diagnostics provide a real-time look into the usage of critical kiln components. From the diagnostics screen, you can view the actual number of relay actuations, the amount of time the heating elements have been on, and the amount of time the thermocouples have been exposed to high temperatures; users have virtually unlimited storage of schedules and segments.

8.16 ENVIRONMENTAL ISSUES

Conventional and heat pump-assisted wood dryers are concerned by current global environmental aspects mainly because of greenhouse gas (GHGs) emissions due to usage of fossil fuels and electricity, and release of wood volatile organic compounds (VOCs).

8.16.1 Carbon Footprint of Energy Sources

Carbon footprint varies between different fossil fuels because of different ratios of carbon to hydrogen and calorific values. Typical values for CO_2 produced per kWh of energy for some energy sources are as follows: (i) natural gas: 0.184 $\frac{kg_{CO_2}}{kWh}$, (ii) diesel and fuel oil: 0.25 $\frac{kg_{CO_2}}{kWh}$, and (iii) coal: 0.324 $\frac{kg_{CO_2}}{kWh}$. For oil and coal, these numbers vary with grade and, for natural gas, depend on the proportion of other hydrocarbons and gases mixed with the main constituent, methane.

The values for electrical energy depend on how it is generated. It is assumed that electricity requires 3 times as much primary energy per kWh as heat for a country which generates all its electrical energy from fossil fuels, the expected values would be approximately 0.55 $\frac{kg_{CO_2}}{kWh}$, 0.75 $\frac{kg_{CO_2}}{kWh}$, and 0.95 $\frac{kg_{CO_2}}{kWh}$ if the main energy sources are natural gas, oil, and coal, respectively.

In practice, typical values are $0.4-0.6$ $\frac{kg_{CO_2}}{kWh}$ for Europe (e.g., 0.458 $\frac{kg_{CO_2}}{kWh}$ in Germany and 0.541 $\frac{kg_{CO_2}}{kWh}$ in the United Kingdom), 0.6 $\frac{kg_{CO_2}}{kWh}$ for North America (e.g., 0.613 $\frac{kg_{CO_2}}{kWh}$ in the United States), and $0.8-1.0$ $\frac{kg_{CO_2}}{kWh}$ for developing countries (e.g., 0.836 $\frac{kg_{CO_2}}{kWh}$ in China and 0.924 $\frac{kg_{CO_2}}{kWh}$ in India). The highest values are for countries using a high proportion of coal, such as Australia $\left(0.953 \frac{kg_{CO_2}}{kWh}\right)$.

As expected, the carbon footprint values are substantially lower for countries which generate much of their electricity from renewable sources (e.g., hydroelectricity, wind) or nuclear, all having virtually zero CO_2 emissions. For example, in France (where over 70% of electricity is nuclear) (2010), the average carbon footprint is $0.088 \frac{kg_{CO_2}}{kWh}$. In Québec (eastern Canada), where almost 100% of electricity is hydro-electric, the electricity carbon footprint is only $0.00122 \frac{kg_{CO_2}}{kWh}$ (source: National Energy Foundation, UK, 2010. http://nef.org.uk/. Accessed November 17, 2017).

8.16.2 Global GHGs

Global carbon emissions from fossil fuels have continuously increased after 1900. Since 1970, the CO_2 emissions have increased by about 90%, with emissions from fossil fuel combustion and industrial processes contributing about 78% of the total GHGs. Agriculture, deforestation, and other land-use changes have been the second largest contributors (IPCC 2014a, 2014b).

After Montreal (1987) and Kyoto (1997) protocols, Paris Climate Conference (2015) agreed to limit global average temperature rise below 2°C, mainly caused by global GHGs.

Figure 8.15 shows the key greenhouse gases emitted by human activities at the global scale based on global emissions (2010) (IPCC 2014a).

It can be seen that carbon dioxide (CO_2) is the major greenhouse gas contributor with 76%, mainly from burning fossil fuels, but, also, from direct human-induced impacts on forestry and other land use, such as deforestation, land clearing for agriculture, and degradation of soils. Methane (CH_4) emissions (16%) come from agricultural activities, waste management, and energy use, and nitrous oxide (N_2O) (6%) from agricultural activities, such as fertilizer use and fossil fuel combustion. Finally, fluorinated gases (F-gases) are emitted by various industrial processes, refrigeration and heat pumps, and the use of a variety of products such as hydrofluorocarbons (HFCs), perfluorocarbons, and sulfur hexafluoride (SF_6).

Figure 8.16 represents the global GHGs by economic sectors (2010) (IPCC 2014a)

It can be seen that, electricity and heat production account for 25% of global GHGs, mainly from burning of coal, natural gas, and oil for electricity and heat

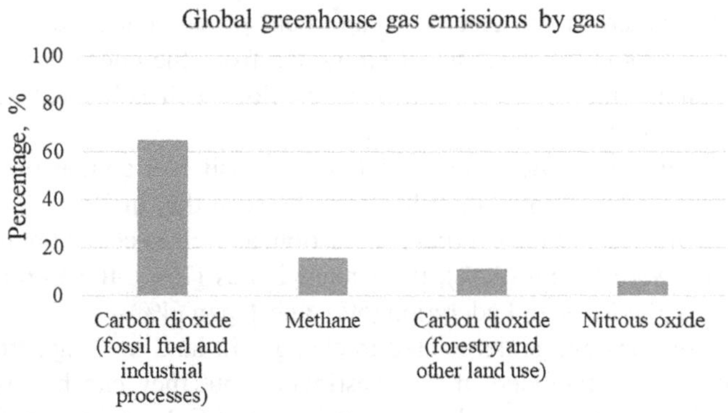

FIGURE 8.15 Global greenhouse gas emissions by gas (*source*: www.epa.gov/ghgemissions/global-greenhouse-gas-emissions-data. Accessed November 17, 2017).

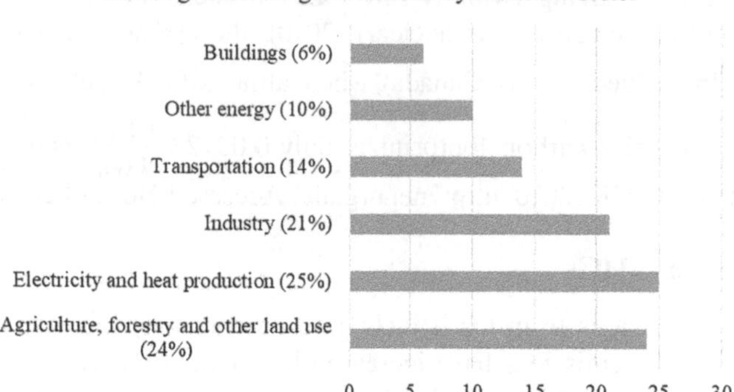

FIGURE 8.16 Global greenhouse gas emissions by the economic activities (*source*: www.epa.gov/ghgemissions/global-greenhouse-gas-emissions-data. Accessed November 17, 2017).

generation. Industry (with 21% of 2010 global GHGs) involves fossil fuels burned on-site at facilities for energy production, including emissions from chemical, metallurgical, and mineral transformation processes not associated with energy consumption, as well as emissions from waste management activities. However, this percentage does not include emissions from industrial electricity use that are covered in the electricity and heat production sector. Agriculture (mainly from cultivation of crops and livestock), forestry (mainly from deforestation), and other land use account for 24% of 2010 global GHGs. These estimates do not include the CO_2 that ecosystems remove from the atmosphere by sequestering carbon in biomass, dead organic matter, and soils, which offset approximately 20% of emissions from this sector (FAO 2014). Global GHGs from transportation (14% in 2010) primarily involve fossil fuels burned for road, rail, air, and marine transportation. Almost all (95%) of the world's transportation energy consumption comes from petroleum-based (gasoline and diesel) fuels. GHGs from buildings (6%) come from on-site energy generation and burning fuels for heat in buildings or cooking in homes. As for industry, the emissions from electricity use in buildings are excluded being instead covered in the electricity and heat production sector. GHGs from other energy sectors (10%) refer to all emissions from the energy sectors are not directly associated with electricity or heat production, such as fuel extraction, refining, processing, and transportation.

Figure 8.17 represents the global GHGs from fossil fuel combustion and some industrial processes by country in 2014. It can be seen that in 2014, the top carbon dioxide (CO_2) emitters from fossil fuel combustion, as well as cement manufacturing and gas flaring, were China (30%), the United States (15%), the European Union (9%), India (7%), the Russian Federation (5%), and Japan (4%).

Even the emissions and sinks related to changes in land use (e.g., from agriculture, forestry) are not included in these estimates, but they can be important. In 2014, these emissions were over 8 billion metric tons of CO_2 equivalent (Boden et al. 2017), or about 24% of total global GHGs (IPCC 2014b). In areas such as the United States and Europe, changes in land use associated with human activities have the

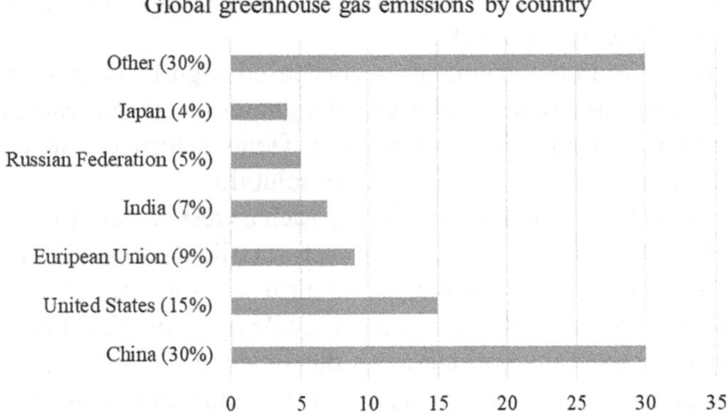

FIGURE 8.17 Global greenhouse gas emissions by country (*source*: www.epa.gov/ghgemissions/global-greenhouse-gas-emissions-data. Accessed November 17, 2017).

net effect of absorbing CO_2, partially offsetting the emissions from deforestation in other regions.

In 2013, Canada's total GHGs were 726 million ton of carbon dioxide equivalent (Mt CO_2 eq.) mainly in economic sectors such as oil and gas (25%), electricity production (12%), transportation (23%), emission-intensive and trade-exposed industries (11%), buildings (12%), agriculture (10%), and waste and others (7%). In Canada, 79% of electricity is generated from non-emitting sources. Some current policies encourage further development of hydroelectricity and other forms of renewable electricity generation. However, GHGs are projected to be higher by 5.8% in 2020 and by 12.2% in 2030 compared to 2013.

8.16.3 Volatile Emissions from Wood

The harmful emissions produced by forests and sawmills (including wood dryers) can be significant. In kilns, the higher the operating temperature, the larger the amount of emissions are produced per kg of moisture removed.

The amount and chemical composition of VOCs depend on species of tree where they act as nutrient reserve, plant hormones, protective and conserving substances.

In many countries, sawmill operations are subject to environmental audit and must comply with limitations to emitted VOCs, condensable organic vapor, and particles less than 10 µm in diameter (as in the United States).

VOCs, primarily monoterpenes (Granström 2005), terpenes, and other hydrocarbons naturally emitted by forests' woods, cause damage to economically important trees and crops during drying of wood, and may have certain environmental impacts.

VOCs can be classified into hazardous air pollutants, such as formaldehyde (coming from the thermal degradation of the hemicelluloses and the lignin) and substances that react in sunlight to produce ozone.

In the presence of nitrogen oxides and sunlight, VOCs generated during storage and drying of wet wood (e.g., fungal particles and extractives) contribute to the formation of photo-oxidants that are harmful to humans' respiratory tract and sensitive parts of the lungs and, also, disturb photosynthesis, causing damage to forests and crops (Berghel and Renström 2004).

When the moisture in the drying air is condensed, significant amounts of monoterpenes can be found as a separate phase on the condensate (Strömvall and Petersson 1999). In the absence of solubility-enhancing surfactants, terpenes are unlikely to be found in the condensate due to their low water solubility.

Volatile compounds emitted during drying, such as resin acids, are found as vapor in the drying air at high temperature. When such volatile compounds leave the dryer, they cool in the ambient air and condense to form aerosols that, if the water vapor does not condense, can be visible as a blue haze (Bridgwater et al. 1995).

The majority of the VOCs emitted to air during drying in sawmills is composed of monoterpenes (Johansson and Rasmuson 1998), while emissions of biogenic volatile organic substances come from trees and wood processing due to machining, logging, chipping, drying, barking, sawing, and boards' production. These emissions are largely variable when fresh wood is sawn (Conners et al. 2001).

The content of monoterpenes of different species (such as conifers) varies between different trees and even within individual tree. The composition of monoterpenes changes in response to external conditions such as herbivore or pathogen attacks (Raffa 1991). In sawmills, the toxic effects on humans of monoterpenes may occur.

As can be seen in Table 8.6 (Granström 2005), the monoterpene emissions from the drying of wood lumber range from $50\frac{mg}{kg_{oven\text{-}dry}}$ for Norway spruce, $380\frac{mg}{kg_{oven\text{-}dry}}$ for Scots pine, $120\frac{mg}{kg_{oven\text{-}dry}}$ for radiata pine, $210\frac{mg}{kg_{oven\text{-}dry}}$ for Douglas-fir, to $1490\frac{mg}{kg_{oven\text{-}dry}}$ for Ponderosa pine. These emissions depend not only on the wood species but also on the drying temperature and time, initial moisture content, and

TABLE 8.6
Emissions of VOC during Drying of Wood Boards and Timber as Measured in the Exhaust Moist Air

References	Species	Dimensions (mm)	T_{max} (°C)	Monoterpenes $\left(\frac{mg}{kg_{oven\text{-}dry}}\right)$	Other VOC $\left(\frac{mg}{kg_{oven\text{-}dry}}\right)$
Broege et al. (1996)	Norway spruce	24;–:–	60	50	10
Broege et al. (1996)	Scots pine	30;–:–	65	380	50
Lavery and Milota (2000)	Douglas-fir	41; 165;1,120	n/a	210	790
Lavery and Milota (2001)	Ponderasa pine	40; 120–300; 4900	n/a	1490	2,210
McDonald and Wastney (1995)	Radiata pine	35; 205; 600	120	120	110

Notes: Dimension numbers mean thickness, width, and length; n/a, indicates no information available; T_{max}, maximum temperature of the drying medium; MC_{fin}, final moisture content of the dried wood; VOCs, volatile organic compounds.

the size of the lumber boards. It can be seen that Scots pine emits considerably more monoterpenes during drying than does Norway spruce (Broege et al. 1996; Englund and Nussbaum 2000).

The concentration of monoterpenes in emission plumes caused by anthropogenic activities is typically 10–1,000 times higher than the background level in conifer forests. Due to the short atmospheric life span of monoterpenes, the highest photo-oxidant concentrations can be expected within 5 hours after the emission takes place, and within a distance of 50 km (Strömvall and Petersson 1999).

In Sweden, the concentration levels in air for monoterpenes emitted during forestry operations are around 1.0–1.5 mg/m^3 near during logging of Scots pine and Norway spruce. During barking of timber, the concentration of monoterpenes in air is 5–20 mg/m^3. On the other hand, in the Swedish sawmills, the average terpene concentrations in air vary between 50 and 550 mg/m^3. In a production hall of Scots pine boards, the average emission of terpenes during sawing was of 153 mg/m^3. Also in Sweden, the average exposure limits for monoterpenes are 150 mg/m^3 over an 8 hour workday, and 30 mg/m^3 over a 15 minute period (Granström 2005).

An increase in monoterpene emissions in the presence of nitrogen oxides and sunlight by a factor of 2 may raise local tropospheric ozone production (Litvak et al. 1999).

Volatile terpenes, the major components of resin, are particularly produced by conifers and released in air by trees more actively in warmer weather. Their production is affected by the availability of nitrogen and water, and the tree' growth stadium (Lerdau et al. 1995). During drying in direct-fired dryers, terpenes follow the used drying air out of the dryer. In this case, the drying air is often led to an incinerator, where the terpenes are combusted. Flue gas recycling is favorable both in terms of emission control and in terms of energy efficiency. In indirect dryers, terpenes are found in the inert gases and can easily be combusted. Emissions of terpenes in combination with emissions of nitrogen oxides increase ozone levels considerably.

On the other hand, the exposure to terpenes at concentrations of 100–200 mg/m^3 in air during the processing of wood may cause acute toxicological effects in the respiratory tracts and difficulties and irritation to mucous membranes of sawmill workers (Ruppe 1929; Falk et al. 1990; Dahlqvist 1992, 1996; Halpin et al. 1994; Eriksson et al. 1996). Terpenes may cause allergies and give over-sensibility of the skin (Matura et al. 2003), while turpentine can cause allergic and nonallergic contact dermatitis (Cronin 1979). In addition, terpene oxidation products may cause an increase in the sensitivity of respiratory tract and a decrease in the lungs' gas exchange ability (Malmberg et al. 1996).

Terpene emissions from dryers, primarily a work environment issue, become a problem when mixed with polluted air. However, they can be measured with good accuracy, despite problems with diffuse emissions and high moisture content of the drying medium.

Exposure to sawmill fumes has also been shown to cause acute decreases in carbon monoxide lung diffusing capacity (Eriksson et al. 1996; Dahlqvist and Ulfvarson 1994). However, as wood dust causes irritant effects not observed during exposure to terpenes, some problems could be caused by oxygenated terpenes formed during the processing of wood or by terpenes reaching the lungs through adsorption to particles of wood dust (Malmberg et al. 1996; Dahlqvist et al. 1996).

The corrosive nature of the volatile substances released during kiln drying particularly with certain species (such as oaks that have both high tannin and phenolic content (Arni et al. 1965a, 1965b) is derived from the presence of free acetic acid in the wood dried at high temperatures.

The acetic acid can attack metals, unprotected concrete and wood, the wall linings of older kilns and stickers. However, the corrosion is avoidable through the extensive use of aluminum, plastic pipework replacing copper, masonry or fibreglass construction, and by eliminating features that trap condensation (Barton 1972).

Batch drying of softwoods generally yields greater VOC emissions than hardwoods because they are dried at medium or high temperatures. Table 8.7 lists, as an example, the concentrations of volatile emissions arising from drying of *P. radiata* at two different high-temperature levels (McDonald and Wastney 1995).

The specific (per kg of water removed) VOC emissions show a high initial peak followed by linear decrease, and, then, a broad peak that begins when the wood is almost dry (Banerjee et al. 1995; Ingram et al. 2000; Conners et al. 2002). The emissions from dried wood is driven by vapor pressure (Banerjee et al. 1998).

When drying is achieved with air at 140°C than at 120°C, 60%–70% larger VOC emissions occur (McDonald and Wastney 1995). However, no difference in VOC emissions occurs when drying at between 82°C and 118°C, although emissions per unit of time are larger at the higher temperature (Ingram et al. 1996).

During very-high-temperature drying processes, wood thermal degradation (that is dependent on the wood species being dried and may begin at 130°C or 200°C) (Broege et al. 1996; Bridgwater et al. 1995) also rises the emissions of substances such as formic acids, alcohols, aldehydes, and carbon dioxide (Bridgwater et al. 1995; Fengel and Wegener 1984). Some volatile substances, such as formaldehyde, come from the thermal degradation of the hemicelluloses and lignin (Perré and Keey 2015a).

Hardwoods produce lower volatile emissions than the resinous softwoods, particularly as kiln temperatures are normally lower than those for most softwoods (Granström 2005).

TABLE 8.7
Concentrations of Volatile Emissions Arising from Drying *P. radiata* at Two High-Temperature Levels

	Concentration (g/m^3)	
	Dry/Wet-Bulb Temperatures (°C)	
Compound	120/70	140/90
Formaldehyde	19.5	31.0
Acetic acid	21.7	38.2
Monoterpenes	34.8	66.4
Hydroxylated monoterpenes	12.6	10.7
Condenser residues (resins and fatty acids)	16.4	16.4

Drying installations may cause air pollution by emission of dust and gases. In some geographic areas, even plumes of clean water vapor are unacceptable, particulates below the range 20–50 mg/m^3 of exhaust air are a common requirement (Packman 1960; Perré and Keey 2015a).

8.16.4 Global Impact of Heat Pumps

In many industrial processes, heat pumps are applied to recover and reuse process waste heat, as in heat pump-assisted wood dryers.

The impacts of heat pumping technology on the global energy policy generally concern (Hitchin 2006) (i) energy access and supply security; even the heat pumps cannot directly influence energy supply risks, their use reduces energy demand and, therefore, the magnitude of exposure to such risks; and (ii) reduction of GHGs that affect the stratospheric ozone layer and earth's global warming.

Because of negative influence of chlorofluorocarbon (CFC) and hydrochlorofluorocarbon (HCFC) refrigerants on the environment, strong international legislations such as Montreal (1987) and Kyoto (1997) protocols have been implemented to substitute them by new HFC, natural, and other low-global warming potential (GWP) fluids.

Tremendous efforts have also been made to develop and design refrigeration, air conditioning and heat pump systems with lower refrigerant charges, and less electrical energy consumptions.

It was estimated (2005) that, by assuming that average household's heating demand is $15{,}000\frac{kWh}{year}$ and 50% of replacements are natural gas-fired boilers and 50% oil boilers, residential heat pumps only may save more than 157 million tonnes of CO_2 per year corresponding to 0.7% of total CO_2 emissions. Globally, a 30% market penetration of heat pumps into existing heating markets would reduce total global CO_2 emissions by up to 8% in 2030 (IEA 2014). This target could be further increased by (i) improving the heat pumps' efficiency allowing them to be comparable with other renewable technologies such as biomass, wind, and solar; and (ii) lowering the carbon emissions from each kilowatt hour of electricity consumed.

The cost of saving a tonne of CO_2 emissions through heat pumping technology varies from one country to another depending on climate, nature of the electricity supply systems, and local costs of alternative heating systems and combustibles.

8.16.4.1 Life Cycle Climate Performance

The environmental impacts of heat pumps in general can be achieved by evaluating their specific life cycle climate performances (LCCPs) (UNEP/TEAP 1999; Papasavva et al. 2010; Beshr 2014; Lee et al. 2016), an evaluation method by which cooling, dehumidifying, and heating can be evaluated based on global warming impact over the course of systems' complete life cycle.

Even subject to a certain amount of uncertainty, the LCCP methodology is flexible and can be applied to mechanical vapor compression cycles powered by electricity from the electricity grids. LCCP calculations are dependent on a number of technical data such as (i) system characteristics and performances, (ii) manufacturing emissions, and (iii) energy generation emissions.

LCCP index (expressed in $\frac{kg_{CO_2eq}}{kWh}$) is calculated as the sum of direct and indirect emissions generated over the lifetime of the heat pump:

$$LCCP = Direct\ emissions + Indirect\ emissions \tag{8.104}$$

Direct emissions include direct leakages of refrigerant into the atmosphere during the lifetime of the system (including annual refrigerant loss from gradual leaks, large losses during operation of the heat pump, and catastrophic leaks), as well as losses during improper disposal of old heat pump equipment and atmospheric reaction (degradation) products created by the refrigerant when it decomposes in the atmosphere.

Indirect emissions include emissions generated by manufacturing of materials used to build the heat pump, manufacturing of the refrigerant, and energy consumption since power plants indirectly emit CO_2 when they generate electricity thus contributing to global warming. The indirect factors depend on the climate and the nature of the electricity supply systems. For example, in the Canadian eastern province of Québec, where almost 100% of electricity is hydroelectric, the indirect emission factor is $0.00122\frac{kg_{CO_2}eq}{kWh}$, while in Alberta (eastern Canada) where electricity is mostly generated based on fossil fuels, it is $0.92527\frac{kg_{CO_2}eq}{kWh}$ (source: www.canada.ca/en/environment-climate-change/services/climate-change/greenhouse-gas-emissions.html. Accessed March 2, 2016).

For mechanical vapor compression heat pumps, powered by electricity from the local electricity grids, the contribution of direct emissions can be calculated using the rate of refrigerant leakage multiplied by the refrigerant charge and the GWP of the refrigerant (Lee et al. 2016; IIR 2016):

$$Direct\ emissions = m_{refr}\left(t_{life} \times ALR \times EOL\right) \times \left(GWP + GWP_{degr}\right) \tag{8.105}$$

where

m_{refr} is a refrigerant charge (kg)
t_{life} is the average lifetime of equipment (years)
ALR is the annual leakage rate (% of refrigerant charge)
EOL is the end of life refrigerant leakage (% of refrigerant charge)
GWP is the refrigerant global warming potential $\left(\frac{kg_{CO_2}}{kg}\right)$
GWP_{degr} is the GWP of atmospheric degradation products of the refrigerant when it decomposes in the atmosphere $\left(\frac{kg_{CO_2}}{kg}\right)$

Assuming that the heat pump being evaluated uses 100% of the electrical energy from the local grid, the indirect emissions can be expressed as follows (IIR 2016):

$$Indirect\ emissions = t_{life} \times AEC \times EM + \sum MM \times m_{heat\ pump} + \sum RM \times m_{recycle} + m_{refr} \times \left(RFM + t_{life} \times ALR \times RFM\right) + RFD \tag{8.106}$$

where

t_{life} is the average lifetime of equipment (years)
AEC is the annual energy consumption (kWh/year)
EM is the equivalent mass of CO_2 emissions per electrical kWh generated $\left(\frac{kg_{CO_2}eq}{kWh}\right)$
MM is the equivalent mass of CO_2 emissions per unit mass of heat pump manufacturing material $\left(\frac{kg_{CO_2}eq}{kg}\right)$
$m_{heat\,pump}$ is the mass of heat pump (kg)
RM is the equivalent mass of CO_2 emissions per unit mass of recycled material $\left(\frac{kg_{CO_2}eq}{kg}\right)$
$m_{recycle}$ is the mass of recycled material (kg)
m_{refr} is the mass of refrigerant charge (kg)
RFM is the equivalent mass of CO_2 emissions per unit mass of refrigerant manufacturing $\left(\frac{kg_{CO_2}eq}{kg}\right)$
ALR is the annual leakage rate (in % of refrigerant charge)
RFD is the equivalent mass of CO_2 emissions per unit mass of refrigerant disposal $\left(\frac{kg_{CO_2}eq}{kg}\right)$

The emissions generated from heat pump annual energy consumption (AEC) (measured in $\frac{kg_{CO_2}eq}{kg}$) are the largest factor in the LCCP equation 8.106.

Material manufacturing equivalent CO_2 emissions (MM) (also expressed in $\frac{kg_{CO_2}eq}{kg}$) are given by various industry sources such as trade associations, governmental departments, and published research works. Many materials are manufactured with a mixture of virgin and recycled materials. The emission values for recycled materials were then taken and weighted to develop the mixed manufacturing emissions (Table 8.8) (Kemp 2014).

TABLE 8.8
Material Manufacturing Emissions

	Virgin Manufacturing Emissions $\left(\frac{kg_{CO_2}eq}{kg}\right)$	Percentage of Recycled Materials in Mixed Materials (%)	100% Recycled Material Manufacturing Emissions $\left(\frac{kg_{CO_2}eq}{kg}\right)$	Mixed Manufacturing Emissions $\left(\frac{kg_{CO_2}eq}{kg}\right)$
Steel	1.8	29	0.54	1.43
Aluminum	12.6	67	0.63	4.5
Copper	3.0	40	2.46	2.78
Plastics	2.8	7	0.12	2.61

Finally, material disposal (end of life) equivalent CO_2 emissions (*RFD*) include all emissions up to the production of recycled material. For metals $\left(0.07\,\frac{kg_{CO_2}eq}{kg}\right)$ and plastics $\left(0.01\,\frac{kg_{CO_2}eq}{kg}\right)$, this includes the shredding of the material. For refrigerants, equivalent CO_2 emissions include energy required to recover the refrigerant and may be included in the manufacturing emissions of the materials produced from recycled materials.

8.16.4.2 Life Cycle Costs

The life cycle cost (LCC) is an economic method that evaluates present and future costs from investing, operating, and maintaining of a heat pump project over its life cycle. To calculate the LCC, it is necessary to compute the present value of all costs over the life of the project that generally are of two types: (i) single costs (such as investment and repair costs that may occur one or more times during the life cycle); and (ii) annually recurring costs (such as energy and annual maintenance costs that occur regularly every year).

The present values of single costs can be calculated as follows:

$$PVSC_n = \frac{SC_n}{\left[1+(i-p)\right]^n} \tag{8.107}$$

where

$PVSC_n$ is the present value of a single cost SC_n after n years (\$)
SC_n is the single cost after n years (\$)
n is the number of years after year 0 (–)
i is the discount rate (decimals)
p is the annual inflation (decimals)

The present value of annually recurrent costs should be calculated as follows:

$$PVARC_n = ARC_n \times \frac{1-\left[1+(i-p)\right]^{-n}}{(i-p)} \tag{8.108}$$

where

$PVAR_n$ is the present value of an annually recurrent cost ARC_n (\$)
ARC_n is the annually recurrent cost (\$)
n is the number of years of the study period (–)
i is the discount rate (decimals)
p is the annual inflation (decimals)

The total LCC may also be evaluated as follows (IIR 2016):

$$LCC_{total} = PVIC + LCC_{energy} + LCC_{operating} + LCC_{environment} + LCC_{others} \tag{8.109}$$

where

$PVIC$ is the present value of investment costs that may include, for instance, costs of products, installation, administration, etc.
LCC_{energy} is the present value of annually energy costs ($)
$LCC_{operating}$ is the present value of nonfuel operating, maintenance, and repair costs ($)
$LCC_{environment}$ is the present value of environmental costs ($)
LCC_{others} is the present value of other costs ($)

8.16.4.3 Total Equivalent Warming Impact

The total equivalent warming impact (TEWI) index takes into account not only the direct warming effects due to refrigerant losses but also the indirect effects due to the plant efficiency and CO_2 release by utility companies supplying power to a given device. It allows to make comparison of large size heat pump drying systems in relation with their impact on the environment. The TEWI index combines the direct emissions of CO_2 due to refrigerant leakage and the indirect emissions of CO_2 associated with energy consumption. The TEWI index is one of the useful tools aiming at comparing the impact of heat pump drying systems on the global warming.

The TEWI calculation of a heat pump system is based on the following relation:

$$TEWI = m_{refrig}^{leaks} \times t \times GWP_{refrig} + RIEF \times AEC \times t \quad (8.110)$$

where

m_{refrig}^{leaks} is the mass of refrigerant annually leak (kg/year)
t is the time of equipment operation (years)
GWP_{refrig} is the refrigerant global warming potential (–)
$RIEF$ is the regional indirect emission factor of CO_2 per unit of energy delivered $\left(\frac{kg_{CO_2}}{kWh}\right)$
AEC is the annual energy consumption of the equipment (kWh/year)

The regional indirect emission factor ($RIEF$) of CO_2 per unit of energy delivered varies from $0.00 \frac{kg_{CO_2}}{kWk}$ (Norway) and $0.00122 \frac{kg_{CO_2}}{kWk}$ (Quėbec, eastern Canada) to $0.84 \frac{kg_{CO_2}}{kWk}$ (Denmark) and $0.04 \frac{kg_{CO_2}}{kWk}$ (Sweden).

8.16.5 Emissions from Heat Pump-Assisted Wood Dryers

The main polluted emission from heat pump-assisted wood dryers is the condensed (liquid) moisture. In Canada, for example, for industrial-scale wood dryers of average capacity of 354 m^3 (Minea 2008), the volume of water (condensate) extracted from softwood stacks with high-temperature heat pumps is about 15 m^3/day. This water is acidic (with pH generally between 3.3 and 4.3) and contains low concentrations of organic loads such as total biochemical oxygen demand (BOD) (<320 mg/L) and total chemical oxygen demand (COD) (<454 mg/L) (Table 8.9).

TABLE 8.9
General Characteristics of Condensed Water

Parameter	Effluent	Protection of Aquatic Life	
		Acute Toxicity	Chronic Effect
Flow (m^3/day)	5–15		
Total BOD (mg/L)	73–320	<5.0 and >9.5	<6.5 and >9.0
Total COD (mg/L)	106–454		
pH	3.3–4.3		

On the other hand, condensates produced by wood heat pump-assisted dryers do not contain suspended particles but contain toxic quantities of VOCs including problematic pollutants such as formaldehyde (which can exceed the acute toxicity limit by up to 5.7 times) and, and to a lesser degree, acetaldehyde (which can exceed the acute toxicity limit by up to 1.3 times). Even formaldehyde and acetaldehyde are both fully biodegradable, they are harmful, for example, to trout and daphnia (Table 8.10).

Formaldehyde is an organic molecule normally found in the air at concentrations lower than 1 μg/m^3 where it has a half-life of 4.1 hours when submitted to direct solar radiation and in rainwater at concentrations approaching 10 μg/m^3. At higher concentrations, it quickly becomes toxic. Formaldehyde is biodegradable and its molecules degrade through photolysis. When dissolved in water under natural conditions, it is not volatile and forms methylene glycol.

The condensate treatment options may be selected taking into consideration factors as the capacity to reduce the organic contaminant content in water in low concentrations (effluent COD concentration of less than 500 ppm), as well as the availability and ability of equipment for treating wastewater volumes below 15 m^3/day. Consequently, the wood-drying companies that are not connected to municipal sewer systems have to use water treatment technologies in order to reduce the concentration of toxic substances to levels below the limits set by national regulations. Fees for collecting and transporting water to the municipal waste disposal site must be applied. If the kiln site is connected to a sewer system, the municipality may accept the condensate but charge to intercept and treat the effluent.

Among the options to consider for effluent treatment are the activated carbon adsorption, natural mitigation, biofiltration, and reverse osmosis. Advanced

TABLE 8.10
Problematic Pollutants of Condensed Water Extracted from Softwood

Parameter	Concentration (μg/L)	Protection of Aquatic Life	
		Acute Toxicity Limit (μg/L)	Chronic Effect Limit (μg/L)
Formaldehyde	34–5,700	1,000	120
Acetaldehyde	200–1,500	1,200	n/a

oxidation techniques such as beta radiation, UV-catalyzed chemical oxidation, ozonation, and UV photolysis can also be considered. Analysis based on the capacity to attain standard norms, low investment and operating costs, and quick start-up capability shows that the optimum technology would be the *UV peroxidation*, followed by *natural mitigation.* In terms of water treatment performance, the UV peroxidation process is similar to ozonation, but it offers the advantage of significantly lower investment and operating costs because it uses a stable oxidant manufactured at lower cost in a factory. The process can be easily restarted after long shutdowns, and the peroxide decomposes quickly without leaving residues. The *natural mitigation* is one of the most inexpensive processes in terms of investment and operating costs. Purification is performed over a period of 70 days in a series of watertight basins. In general, three basins are used: the first one reduces the organic load as much as possible, while the other two act as buffer zones in the event that the first basin overflows. The purification mechanisms combine solar UV radiation, input of atmospheric oxygen and surface algae photosynthesis, aerobic digestion at the surface, and anaerobic digestion on the bed of the basin. More than 85% of the COD can be eliminated from the liquid phase at the lagoon outflow, a part of which is left in the sludge that settles on the bed of the basin. This technology is less effective in winter, but requires long treatment periods and significant surface areas, for example, up to 1,050 m^2 to treat 15 m^3/day of liquid effluent. Natural mitigation method makes use of resistant microorganisms, but they remain subject to toxic shocks and temperature variations, and restarting can be a slow process.

For treatment capacities of 15 m^3/day, the investment costs average US$123,000 for *natural mitigation* and US$68,500 for *UV peroxidation* (2005) (Minea 2008). The electrical energy consumption is zero for natural mitigation and of about 6.6 US$ for UV peroxidation. The annual operating costs (maintenance and labor) of such systems generally represent 5% of the total investment costs. This does not include electrical, fuel, or chemicals costs, where required. Table 8.11 shows the final concentrations and percentages of reduction for both recommended treatment technologies, that is, *natural* mitigation and *UV peroxidation.*

Based on the previous ecological issues related to heat pump-assisted drying of softwood, *natural mitigation* and *UV peroxydation* technologies could be recommended as the best in terms of ability to adequately treat condensates at the low investment and operation costs, and low electricity consumption.

TABLE 8.11
Summary of Final Concentrations of the Condensed Water

Technology	Final Concentration					
	COD		BOD		CH_2O	CH_3CHO
	(mg/L)	% of reduction	(mg/L)	% of reduction	(µg/L)	(µg/L)
Natural mitigation	150	67	16	95	<285	<75
UV peroxidation	91	80	36	89	<647	<170

The investment costs of treatment equipment depend on treatment capacities and can be calculated as follows:

$$IC = RIC \times \left(\frac{ATC}{RTC} \right)^{a} \tag{8.111}$$

where

IC is the investment cost to be estimated ($)
ATC is the actual treatment capacity (m^3)
RTC is the reference treatment capacity (m^3)
RIC is the reference investment cost ($)
a is the scaling factor (~0.60)

The installation and annual operating (maintenance and labor) costs may represent about 30% and 5% of the total investment costs, respectively. These percentages not include electrical, fuel, or chemicals costs, if required. Table 8.12 resumes the costs of chosen treatment technologies for two treatment capacities (7.5 m^3/day and 15 m^3/day), and Table 8.13 gives the final concentrations and percentages of reduction.

High-temperature heat pump-assisted dryers used for softwood drying reduce the oil consumption and, consequently, the emissions of CO_2 greenhouse gas. In Canada, for example, the average conversion factor for electricity production ranges from 0.92527 kg CO_2/kWh (in Alberta) to 0.00122 kg CO_2/kWh (in Quebec) with country average value of around 0.3 kg CO_2/kWh. That means that, in Canada, each additional 1,000 kWh electrical energy consumption corresponds to about 300 kg of

TABLE 8.12
Cost Summary of Chosen Technologies for the Treatment of Condensed Water

Technology	Dose	Investment Cost		Operating Cost (C$/m^3)	Electrical Consumption (kWh/m^3)
		C$	C$/(m^3/day)		
7.5 m^3/day capacity					
Natural mitigation with non-aerated lagooning	–	45,193	6,026	0.64	0.0
UV peroxidation	0.60 kg/kg$_{COD}$	98,549	13,140	3.64	6.6
UV peroxidation	0.27 kg/kg$_{COD}$	98,549	13,140	3.33	6.6
15.0 m^3/day capacity					
Natural mitigation	–	68,500	4,567	0.48	0.0
UV peroxidation	0.60 kg/kg$_{COD}$	123,022	8,201	2.48	6.6
UV peroxidation	0.27 kg/kg$_{COD}$	123,022	8,201	2.18	6.6

Note: Costs for 1 C$ = 0.85 US$ (2007).

TABLE 8.13
Summary of Final Wastewater Concentrations

		Final Concentration					
		COD		BOD			
Technology	Dose	(mg/L)	% of reduction	(mg/L)	% of reduction	CH_2O (μg/L)	CH_3CHO (μg/L)
Natural mitigation with non-aerated lagooning	–	150	67	16	95	<285	<75
UV peroxidation	0.60 kg/kg_{COD}	45	90	18	94	<323	<85
UV peroxidation	0.27 kg/kg_{COD}	91	80	36	89	<647	<170

CO_2 equivalent emission. The very low regional conversion factor in the province of Quebec is attributable to the fact that almost 98% of the province's energy is hydroelectric electricity, while in Alberta province the electricity production is almost 100% fossil based.

Despite additional electrical energy consumption for driving the heat pump compressors and blowers, by using heat pump-assisted dryers with high-temperature heat pumps, the annual CO_2 emissions may by substantially reduced in Canada compared to conventional drying systems using oil as primary energy. For example, if the softwood drying market penetration would attain 5% in Canada during the next 5 years, the net reduction in CO_2 may attain 36,600 tons/year (Minea 2008).

REFERENCES

Ananias, R., P. Steinhagen, C. Mujica. 1992. Study of air velocity, drying time, and lumber degradation in industrial conventional drying of radiata pine. In *Proceedings of the Segundo Simposio Pinus radiate Investigacion en Chile*; Universidad Austral: Valdivia, Chile, pp. 265–272 (in Spanish).

Ananias, R.A., J. Ulloa, D.M. Elustondo, C. Salinas, P. Rebolledo, C. Fuentes. 2012. Consumption in industrial drying of radiata pine. *Drying Technology* 30:774–779.

Anderson, J.O. 2014. Energy and resource efficiency in convective drying systems in the process industry. Doctoral Thesis, Luleå University of Technology, Sweden.

Anderson, J.O., L. Westerlund. 2011. Surplus biomass through energy efficient kilns. *Applied Energy* 88:4848–4853 (doi:10.1016/j.aplenergy.2011.06.027).

Anderson, J.O., L. Westerlund. 2014. Improved energy efficiency in sawmill drying system. *Applied Energy* 113:891–901.

Arnaud, G., J.P. Fohr, J.P. Garier, C. Ricolleau. 1991. Study of the air flow in a wood drier. *Drying Technology* 9(1):183–200.

Arni, P.C., G.C. Cochrane, J.D. Gray. 1965a. The emission of corrosive vapor by wood—Part 1: Survey of the acid-release properties of certain freshly felled hardwoods and softwoods. *Journal of Applied Chemistry* 15:305–313.

Arni, P.C., G.C. Cochrane, J.D. Gray. 1965b. The emission of corrosive vapours by wood—Part 2: The analysis of the vapours emitted by certain freshly felled hardwoods and softwoods by gas chromatography and spectrophotometry. *Journal Applied Chemistry* 15:463–468.

ASHRAE. 2008. *ASHRAE Handbook*. HVAC Systems and Equipment. SI Edition. American Society of Heating, Refrigerating and Air-Conditioning Engineers, Atlanta, GA.

ASHRAE. 2009. *Handbook Fundamentals*, SI edition, ASHRAE Inc., Atlanta, Georgia.

Ashworth, J.C. 1977. The mathematical simulation of batch-drying of softwood timber. Ph.D. Thesis, University of Canterbury, NZ, 2 vols.

Astram, K.J., T.J. McAvoy. 1992. Intelligent control: An overview and evaluation, in *Handbook of Intelligent Control*, edited by D.A. White and D.A. Sofge, Van Nostrand Reinhold, New York.

Awadalla, H.S.F., A.F. El-Dib, M.A. Mohamad, M. Reuss, H.M.S. Hussein. 2004. Mathematical modelling and experimental verification of wood drying process. *Energy Conservation and Management* 45:197–207.

Baker, C.G.J., D. Reay. 1982. Energy usage for drying in selected U.K. industrial sectors. In *Proceedings of the 3rd International Drying Symposium*, Vol. 1, pp. 201–209.

Banerjee, S. 1998. Wet line extension reduces VOCs from softwood drying. *Environmental Science & Technology* 32(9):1303–1307.

Banerjee, S., M. Hutten, W. Su, L. Otwell, L. Newton. 1995. Release of water and volatile organics from wood drying. *Environmental Science & Technology* 29(4):1135–1136.

Barton, G.M. 1972. How to prevent dry kiln corrosion. *Canadian Forest Industries* 92(4):27–29.

Bassett, K.H. 1974. Sticker thickness and air velocity. In *Proceedings of the Western Dry Kiln Clubs of the 25th Annual Meeting*, School of Forestry, Oregon State University, Corvallis, OR, pp. 40–45.

Beard, J.N., H.N. Rosen, B.A. Adesanya. 1985. Temperature distribution in lumber during impingement drying. *Wood Science and Technology* 19:277–286.

Berghel, J., R. Renström. 2004. Controllability of product moisture content when nonscreened sawdust is dried in a spouted bed. *Drying Technology* 22(3):507–519.

Berit Time. 1988. Hygroscopic moisture transport in wood. A Thesis presented for the degree of Doctor Engineer of the Norwegian University of Science and Technology, Department of Building and Construction Engineering.

Bernard, J.A. 1988. Use of rule-based system for process control. *IEEE Control Systems Magazine* 8(5):3–13.

Beshr, M. 2014. ORNL life cycle climate performance – V1.0 (http://lccp.umd.edu/ornllccp/. Accessed October 15, 2017).

Bhat, N., T. McAvoy. 1990. Use of neural nets for dynamic modeling and control of chemical processes. *Computers & Chemical Engineering* 14:573–583.

Bian, Z. 2001. Airflow and wood drying models for wood kilns. A Thesis submitted in partial fulfilment of the requirement for the Degree of Master of Applied Science in Faculty of Graduate Studies, Department of Mechanical Engineering, University of British Columbia.

Blasius, H. 1913. The similarity law for friction processes in fluids. Forsch VDI 131.

Boden, T.A., G. Marland, R.J. Andres. 2017. National CO_2 emissions from fossil-fuel burning, cement manufacture, and gas flaring: 1751–2014. Carbon Dioxide Information Analysis Center, Oak Ridge National Laboratory, U.S. Department of Energy, doi: 10.3334/CDIAC/00001_V2017.

Boseley, J.R., T.E. Edgar, A.A. Patwardhan, G.T. Wright. 1992. Model-based control: A survey, in *Advanced Control of Chemical Processes*, edited by K. Najim and E. Dufour, IFAC Symposia series, No. 8. Pergamon Press, Oxford, New York.

Breitner, T., S. Quarles, D. Huber, D. Arganbright. 1990. Steam and electrical consumption in a commercial scale lumber dry kiln. In *Proceedings of the 41st Dry Kiln Association*, Convallis, OR, May 9–11, pp. 83–94.

Bridgwater, A. 1995. The nature and control of solid, liquid and gaseous emissions from the thermochemical processing of biomass. *Biomass and Bioenergy* 9:325–341.

Broege, K., K. Aehlig, M. Scheithauer. 1996. *Emissionen aus Schnittholztrocknern*, Institut fur Holztechnologie, Dresden.

Cech, M.J., F. Pfaff. 2000. Operator wood drier handbook for east of Canada, edited by Forintek Corp., Québec city, Canada.

Choong, E.T., C. Skaar. 1972a. Diffusion and surface emissivity in wood drying. *Wood Fiber* 492:80–86.

Choong, E.T., C. Skaar. 1972b. Diffusion and surface emissivity in wood drying. *Wood Fiber* 492:89–96.

Clément, C., Y. Fortin. 1996. Calcul de la consommation énergétique au cours du séchage du bois. Ateliers-Conférences sur le séchage. Deuxième édition, Association des Manufacturiers de Bois de Sciage du Québec, Québec, 25–26 avril.

Comstock, G.L. 1976. Heat requirement for drying: Lumber, veneer, particles. In *FPRS Proceedings—Wood Residue as an Energy Source*. Paper No. P-75-13, Madison, WI.

Conners, T.E., L.L. Incram, W. Su, S. Banerjee, A.T. Dalton, M.C. Templeton, S.V. Diehl. 2001. Seasonal variation in southern pine terpenes. *Forest Products Journal* 51(6):89–94.

Conners, T.E., H. Yan, S. Banerjee. 2002. Mechanism of VOC release from high-temperature southern pine lumber drying. *Wood and Fiber Science* 34(4): 666–669.

Coughanour, D.R. 1991. *Process Systems Analysis and Control*, 2nd Edition. McGraw-Hill, New York.

Coulson, J.M., I.F. Richardson, R.K. Sinnot.1989. *Chemical Engineering: An Introduction to Chemical Engineering Design*, Pergamon Press, Oxford.

Courtois, E. 1997. Automatic control of drying processes. In *Computerized Control Systems in the Food Industry*, edited by G.S. Mittal, Marcel Dekker, New York.

Cronin, E. 1979. Oil of turpentine: A disappearing allergen. *Contact Dermatitis* 5:308–311.

Cronin, K., K. Abodayeh, J. Caro-Corrales. 2002. Probabilistic analysis and design of the industrial timber drying process. *Drying Technology* 20(2):307–324.

Cronin, K., P. Baucour, K. Abodayeh, A. Barbot Da Silva. 2003. Probabilistic analysis of timber drying schedules. *Drying Technology* 21(8):1433–1456.

Dahlqvist, M. 1992. Lung-function and precipitating antibodies in low exposed wood trimmers in Sweden. *American Journal of Industrial Medicine* 21(4):549–559.

Dahlqvist, M. 1996. Acute effects of exposure to air contaminants in a sawmill on healthy volunteers. *Occupational and Environmental Medicine* 53(9):586–590.

Dahlqvist, M., U. Ulfvarson. 1994. Acute effects on forced expiratory volume in one second and longitudinal change in pulmonary-function among wood trimmers. *American Journal of Industrial Medicine* 25(4):551–558.

Danckwerts, P.V., C. Anolick. 1962. Mass transfer from a grid packing to an air stream. *Transactions of the Institution of Chemical Engineers* 40:203–213.

Denig, J., E.M. Wengert, W.T. Simpson. 2000. *Drying Hardwood Lumber.* Forest Products Laboratory, United States Department of Agriculture, Forest Service.

Department of Energy Buildings, Energy Data Book. 2015. http://ledsgp.org/resource/buildings-energy-data-book/?loclang=en_gb, Accessed September 26, 2016.

Devahastin, S. 2006. Software for drying/evaporation simulations: Simprosys. *Drying Technology* 24:1533–1534.

Elustondo, D.M., L. Oliveira. 2006. Opportunities to reduce energy consumption in softwood lumber drying. *Drying Technology* 24:653–662.

Elustondo, D.M., L. Oliveira. 2009. Model to assess energy consumption in industrial lumber kilns. *Maderas, Ciencia y Tecnologia* 11(1):33–46

Englund, F., R.M. Nussbaum. 2000. Monoterpenes in Scots pine and Norway spruce and their emission during kiln drying. *Holzforschung* 54(5): 449–456.

Eriksson, K.A., N.L. Stjernberg, J.O. Levin, U. Hammarstrom, M.C. Ledin. 1996. Terpene exposure and respiratory effects among sawmill workers. *Scandinavian Journal of Work Environment & Health* 22(3):182–190.

Falk, A., A. Löf, M. Hagberg, E. Wigaeus-Hjelm, W. Zhiping. 1990. Uptake, distribution and elimination of a-pinene in man after exposure by inhalation. *Scandinavian Journal of Work Environment & Health* 16:372–378.

FAO. 2014. *Agriculture, Forestry and Other Land Use Emissions by Sources and Removals by Sinks.* Climate, Energy and Tenure Division, FAO. http://www.fao.org/docrep/019/i3671e/i3671e.pdf, Accessed July 15, 2017.

Fengel, D., G. Wegener. 1984. *Wood: Chemistry, Ultrastructure, Reactions*, Walter de Gruyter, Berlin and New York, 613 pp.

Garcia, C.E., M. Morari. 1982. Internal model control. 1-An unifying review and some new results. *Industrial & Engineering Chemistry Process Research and Development* 21:308.

Garraham, P., G. MacKay, L. Oliveira, M. Savard, D. Elustondo, L. Jozsa. 2008. *Drying Spruce-Pine-fir Lumber.* Special publication SP-527E, FPInnovations, Forintek, Vancouver, Canada.

Gevarter, W.B. 1987. Introduction to artificial intelligence. *Chemical Engineering Progress* 83(9):21–37.

Gong, Z.-X., A.S. Mujumdar. 2008. Software for design and analysis of drying systems. *Drying Technology* 26(7):884–894

Granström, K. 2005. *Emissions of Volatile Organic Compounds from Wood.* Karlstad University Studies, Sweden. Division for Engineering Sciences, Physics and Mathematics, Department of Environmental and Energy Systems.

Greenhill, W.L. 1936. The measurement of the flow of air through timber seasoning stacks. *Journal of the Council for Scientific and Industrial Research* (Australia) 9:128–134.

Halpin, D.M.G., B.J. Graneek, M. Turnerwarwick, A.J.N. Taylor. 1994. Extrinsic allergic alveolitis and asthma in a sawmill worker: Case report and review of the literature. *Occupational and Environmental Medicine* 51(3):160–164.

Haslett, A.N. 1998. Drying radiata pine in New Zealand: Research and commercial aspects. *FRI Bulletin* 206, NZ Forest Res. Inst. Rotorua, NZ, 24 pp.

Hines, A.L., R.N. Maddox. 1985. *Mass Transfer: Fundamentals and Applications.* Prentice-Hall, Englewood Cliffs, NJ, 542 pp.

Hitchin, R. 2006. The potential impact of heat pumps on energy policy concern: Position paper. IEA Heat Pump Programme. http://www-v2.sp.se/hpc/publ/HPCOrder/ViewDocument.aspx?RapportId=388. Accessed November 18, 2017.

Holman, J.P. 1992. *Heat Transfer*, 7th Edition. McGraw Hill Book, Singapore, 702 pp.

Hukka, A. 1997. Evaluation of parameter values for a high temperature drying simulation model using direct drying experiments. *Drying Technology* 15(4):1213–1229.

IEA. 2014. Key world energy statistics. http://www.iea.org/publications/freepublications (accessed May 15, 2016).

IEA. 2015. Heat pump program annex 35. Industrial energy-related systems and technologies, Application of Industrial Heat Pumps, Final report, Part 1.

IIR. 2016. International institute of refrigeration. The 32nd Information Note on Refrigeration Technology, October.

Incropera, F.P., D.P. DeWitt. 2002. *Fundamentals of Heat and Mass Transfer*, 5th Edition. John Wiley & Sons, New York, 981 p.

Ingram, L., F. Taylor, M. Templeton. 1996. Volatile organic compound emissions from southern pine kilns: Drying Pacific northwest species for quality control. *Forest Products Society*, 41–45.

Ingram, L.L., R. Shmulsky, A.T. Dalton, F.W. Taylor, M.C. Templeton. 2000. The measurement of volatile organic emissions from drying southern pine lumber in a laboratory-scale kiln. *Forest Products Journal* 50(4):91–94.

IPCC. 2014a. Climate change 2014: Mitigation of climate change. Contribution of Working Group III to the Fifth Assessment Report of the Intergovernmental Panel on Climate Change. Cambridge University Press, Cambridge, UK and New York, NY.

IPCC. 2014b. Climate change 2014: Synthesis report. Contribution of Working Groups I, II and III to the Fifth Assessment Report of the Intergovernmental Panel on Climate Change. IPCC, Geneva, Switzerland, 151 pp.

Jain, P.K., S.N. Sharma. 1981. Studies on air circulation in a new design of side-mounted fans kiln. *Journal of the Timber Development Association of India* 17(3):12–25.

Johansson, A., A. Rasmuson. 1998. The release of monoterpenes during convective drying of wood chips. *Drying Technology* 16(7):1395–1428.

Kayihan, F. 1993. Adaptive control of stochatic batch lumber kilns. *Computers Chemical Engineering* 17(3):265–273.

Keey, R.B. 1978. *Introduction to Industrial Drying Operations*. Pergamon, Oxford, UK, 376 pp.

Keey, R.B. 1992. *Drying of Loose and Particulate Materials*. Hemisphere Publishing, New York.

Keey, R.B., J.C. Ashworth. 1979. The Kiln Seasoning of Softwood Timber Boards. *The Chemical Engineering Journal* 347(8):593–598.

Keey, R.B., J.J. Nijdam. 2002. Moisture movement on drying softwood boards and kiln design. *Drying Technology* 20(10):1955–1974.

Keey, R.B., S. Pang. 1994. The high-temperature drying of softwood boards a kiln-wide model. *Transactions of Industrial & Chemical Engineering* 72(A):741–753.

Keey, R.B., M. Suzuki. 1974. On the characteristic drying curve. *International Journal of Heat and Mass Transfer* 17:1455–1464.

Keey, R.B., T.A.L. Langrish, J.C.F. Walker. 2000. *The Kiln Drying of Lumber*, Springer Series in Wood Science. Timell, T.E. (Editor) (ISBN-13:978-3-642–64071-1), Berlin, Germany.

Kemp, I.C. 1999. Progress in dryer selection techniques. *Drying Technology* 17;1667–1680.

Kemp, I.C. 2004. Drying in the context of the overall process. *Drying Technology* 22:377–394.

Kemp, I.C. 2007. Drying software: past, present and future. *Drying Technology* 25(7):1249–1263.

Kemp, Ir. I.C. 2012. Fundamentals of energy analysis of dryers. Introduction. Editors Prof. Evangelos Tsotsas & Prof. Arun S. Mujumdar. Published by Wiley-VCH Verlag GmbH & Co. KGaA, Weinheim, Germany. Wiley Online Library. https://doi.org/10.1002/9783527631681.ch1, Accessed January 15, 2017.

Kemp, I.C. 2014. Fundamentals of energy analysis of dryers, in *Modern Drying Technology*, edited by E. Tsotsas and A.S. Mujumdar (doi: 10.1002/9783527631728.ch21).

Kemp, I.C., R.E. Bahu. 1995. A new algorithm for dryer selection. *Drying Technology* 1995, 13, 1563–1578.

Kemp, I.C., D.E. Oakley. 2002. Modelling of particulate drying in theory and practice. *Drying Technology* 20:1699–1750.

Kemp, I.C., N.J. Hallas, D.E. Oakley. 2004. Developments in aspen technology drying software. In *Proceedings of the 14th International Drying Symposium (IDS)*, Sao Paulo, August, Volume B, pp. 767–774.

Kho, P.C.S. 1993. Mass transfer from in-line slabs: Application to high temperature kiln drying of softwood timber boards. Ph.D. Thesis, Canterbury University, New Zealand.

Kho, P.C.S., R.B. Keey, J.C.F. Walker. 1989. Effects of minor board irregularities and air flows on the drying rate of softwood timber boards in kilns. In *Proceedings of IUFRO Wood Drying Conference*, pp. 150–157.

Kho, P.C.S., R.B. Keey, J.C.F. Walker. 1990. The variation of local mass-transfer coefficient in stream wise direction over a series of inline blunt slabs. In *Proceedings of the Chemeca'90 Conference*, Auckland, New Zealand, pp. 348–355.

Kocaefe, D., J.L. Shi, D.Q. Yang, M. Bouzara. 2008. Mechanical properties, dimensional stability, and mold resistance of heat-treated jack pine and aspen. *Forest Products Journal* 58(6): 88–93.

Kroll, K. 1978. *Dryers and Drying Processes*, 2nd Edition, Springer, Berlin Heidelberg New York, 654 pp.

Kudra, T. 2004. Energy aspects in drying. *Drying Technology* 22(5):917–932.

Kudra, T., R. Platon, P. Navarri. 2009. Excel-based tool to analyze the energy performance of convective dryers. *Drying Technology* 27:1302–1308.

Kyoto. 1997. https://en.wikipedia.org/wiki/Kyoto_Protocol, Accessed February 15, 2017.

Langrish, T., J. Walker. 2006. Drying of timber. In *Primary Wood Processing: Principles and Practice*, 2nd Edition, Springer, Dordrecht, The Netherlands.

Langrish, T.A.G., J.C.F. Walker. 1993. Transport Processes in Wood. In *Primary Wood Processing*, edited by J.C.F. Walker. Chapman and Hall, London, pp. 121–152.

Langrish, T.A.G. 1994. Assessing the variability of mass transfer coefficients in stacks of timber with the aid of a numerical simulation. In *Proceedings of the 4th IUFRO International Wood Drying Conferences*, Rotorua, NZ, pp. 150–157.

Langrish, T.A.G. 1999. An assessment of the use of characteristic drying curves for the high-temperature drying of softwood timber. *Drying Technology* 17(4&5):991–998.

Langrish, T.A.G., R.B. Keey. 1996. The effects of air bypassing in timber kilns on fan power consumption. In *Proceedings of the CHEMICA '96*, Sydney, Australia, Vol. 2, pp. 103–108.

Langrish, T.A.G., P.C.S. Kho, J.C.F. Walker. 1992. Experimental measurement and numerical simulation of local mass transfer coefficients in timber kilns. *Drying Technology* 10:753–781.

Langrish, T.A.G., R.B. Keey, P.C.S. Kho, J.C.F. Walker. 1993. Time-dependent flow in arrays of timber boards: Flow visualization, mass-transfer measurements and numerical simulation. *Chemical Engineering Science* 48(12):2211–2223.

Langrish, TA.G., A.S. Brooke, E.L. Davis, H.E. Musch, G.W. Barton. 1997. An improved drying schedule for Australian ironbark timber: Optimisation and experimental validation. *Drying Technology* 15(1):47–70.

Larreture, A., M. Laniau. 1991. The state of drying in French industry. *Drying Technology* 9(1):263–275.

Law, C.L., A.S. Mujumdar. 2008. Energy aspects in energy drying. In *Guide to Industrial Drying: Principles, Equipment and New Developments*, Chapter 14th, edited by Arun S. Mujumdar, IDS, Hyderabad, India.

Lee, H.S. 1990. *Flow Visualization on High-temperature Drying*. B.E. Res. Rep. CAPE, University of Canterbury, Christchurch, New Zealand.

Lee, H.S., S. Troch, Y. Hwang, R. Rademacher. 2016. LCCP evaluation on various vapor compression cycle options and low GWP refrigerants. *International Journal of Refrigeration* 70:128–137.

Léger, F., M. Amazouz. 2003. Evaluation of wood kiln control practices. Prepared by CANMET Energy Technology Centre – Varennes for Forintek Canada Corp.

Lerdau, M., P., Matson, R. Fall, R. Monson. 1995. Ecological controls over monoterpene emissions from Douglas-fir. *Ecology* 76(8):2640–2647.

Lewis. W.K. 1921. The rate of drying of solid materials. *Industrial & Engineering Chemistry* 13:427–432.

Liptak, B.G. 1999. *Optimization of Industrial Unit Processes*, 2nd Edition, CRC Press, Boca Raton, FL.

Litvak, M.E., S. Madronich, R.K. Monson. 1999. Herbivore-induced monoterpene emissions from coniferous forests: Potential impact on local tropospheric chemistry. *Ecological Applications* 9(4):1147–1159.

Luyben, M.L., W.L. Luyben. 1996. *Essentials of Process Control*, McGraw-Hill, New York.

Luyben, W.L. 1990. *Process Modeling. Simulation, and Control for Chemical Engineers*, 2nd Edition, McGraw-Hill, New York.

Malmberg, P.O. 1996. Increased bronchial responsiveness in workers sawing Scots pine. *American Journal of Respiratory and Critical Care Medicine* 153(3):948–952.

Malmquist, L. 1991. Lumber drying as a diffusion process. Halz als Roh und Werkstoff 49.

Mamdani, E.H. 1974. Application of fuzzy algorithms for simple dynamic plant. *Proceedings of the Institution of Electrical Engineers* 121(12):1585–1588.

Mamdani, E.H. 1977. The application of fuzzy control systems to industrial processes. *Automatica* 13(3):235–242.

Mamdani, E.H., S. Assilian. 1975. An experiment in linguistic synthesis with a fuzzy logic controller. *International Journal of Man-Machine Studies* 70:1–13.

Marchant, J.A. 1985. Control of high temperature continuous flow dryers. *Agriculture Engineering* 40:145–149.

Marinos-Kouris, D., Z.B. Mroulis, C.T. Kiranoudis. 1996. Computer Simulation of Industrial Dryers. *Drying Technology* 14(5):971–1010.

Matura, M. 2003. Patch testing with oxidized R-limonene and its hydro-peroxide fraction. *Contact Dermatitis* 49(1):15–21.

McCurdy, M.C., S. Pang. 2007. Comparison of colour development in sugar–amino acid solutions with colour changes in wood duringdrying. *Asia-Pacific Journal of Chemical Engineering* 2(1):30–34.

McCurdy, M.C. 2005. Efficient kiln drying of quality softwood timber, Thesis submitted in fulfilment of the requirement for the Degree of Doctor of Philosophy in Chemical and Process Engineering to the University of Canterbury.

McCurdy, M.C., S. Pang. 2007. Optimization of kiln drying for softwood, through simulation of wood stack drying, energy use, and wood color change. *Drying Technology* 25:1733–1740.

McCurdy, M.C., S. Pang, R.B. Keey. 2005. Measurement of colour development in Pinus radiata sapwood boards during drying at various schedules. *Maderas: Ciencia y Technologia* 7(2):79–85.

McDonald, A., S. Wastney. 1995. Analysis of volatile emissions from kiln drying of radiata pine. In *Proceedings from 8th International Symposium on Wood and Pulping Chemistry*, Helsinki, pp. 431–436.

McQuiston, F.C., J.D. Parker, J.D. Spitler. 2005. *Heating, Ventilating, and Air Conditioning Analysis and Design*, 6th Edition, John Wiley & Sons, New York.

Meil, J., L. Bushi, P. Garrahan, R. Aston, A. Gingras, D. Elustondo. 2010. *Status of Energy Use in the Canadian Wood Products Sector.* Canadian Forest Service, Ottawa.

Menshutina, N.V., M.G. Gordienko, A.A. Voynovskiy, T. Kudra. 2004. Dynamic analysis of drying energy consumption. *Drying Technology* 22(10):2281–2290.

Menshutina, N.V., T. Kudra. 2001. Computer aided drying technologies. *Drying Technology* 19(8):1825–1850.

Meroney, R.N. 1969. The state of moisture transport rate calculations in wood drying. *Wood and Fiber* 1(1):64–74.

Milota, M.R. 1990. For quality drying, see if your kiln works properly. *Forest Industries* 117(5):26–29.

Milota, M.R., J.L. Tschernitz. 1994. Simulation of drying in a batch lumber kiln from single-board tests. *Drying Technology* 12(8):2027–2055.

Milota, M.R. 2000. Warp and shrinkage of hem-fir stud lumber dried at conventional and high temperatures. *Forest Products Journal* 50(11/12):79–84.

Minea, V. 2008. Energetic and ecological aspects of softwood drying with high-temperature heat pumps. *Drying Technology* 26(11):1373–1381.

Montreal. 1987. https://en.wikipedia.org/wiki/Montreal_Protocol, Accessed February 15, 2017.

Morari, M., E. Zafiriou, M. Zafiriou. 1997. *Robust Process Control*. Prentice-Hall, Englewood Cliffs, NJ.

Mujumdar, A.S. 2007. *Handbook of Industrial Drying*, 3rd Edition, CRC Press – Taylor & Francis Group, Boca Raton, FL.

Mujumdar, A.S., A.S. Menon. 1995. Drying of solids. In *Handbook of Industrial Drying*, 2nd Edition, edited by A.S. Mujumdar, pp. 1–46, Marcel Dekker, New York.

NAICS 321. 2012. Wood product manufacturing. https://www.census.gov/econ/isp/sampler/. Accessed August 18, 2017.

Nijdam, J.J. 1998. Reducing moisture content variation in kiln-dried timber. Ph.D. Thesis, University of Canterbury, Christchurch, New Zealand.

Nijdam, J.J., R.B. Keey. 1996. Influence of local variations of air velocity and flow direction reversals on the drying of stacked timber boards in a kiln. *IChemE Journal* 74(A):882–892.

Nijdam, J.J., R.B. Keey. 2000. Impact on kiln-drying softwood boards. *Bulletin of the Polish Academy of Sciences Technical Sciences* 48(3):301–314.

Nijdam, J.J., T.A.G. Langrish, R.B. Keey. 2000. A high-temperature drying model for softwood timber. *Chemical Engineering Science* 55:3583–3598.

Nyboer, J., M. Bennett. 2013. Energy use and related data: Canada wood products and industry, 1990, 1995–2011, Forest Products Association of Canada, Canadian Industry Program for Energy Conservation of the Canadian Industrial Energy End-use Data and Analysis Centre. http://www2.cieedac.sfu.ca/Report_2012, Accessed August 11, 2016.

Oliveira, L., D. Elustondo, A.S. Mujundar, R. Ananias. 2012. Canadian developments in kiln drying. *Drying Technology* 30(15):1792–1799.

Onn, L.K. 2007. Study of convective drying using numerical analysis on local hardwood, University SAINS of Malaysia, https://core.ac.uk/download/pdf/11957610.pdf. Accessed August 2, 2016.

Packman, O.F. 1960. The acidity of wood. *Holzforschung* 14(6):178–183.

Pang, S. 1994. The high temperature drying of Pinus radiata in a batch kiln. Ph.D. Thesis, University of Canterbury, Christchurch, New Zealand.

Pang, S. 1996a. External heat and mass transfer coefficients for kiln drying of timber. *Drying Technology* 14(3&4):859–871.

Pang, S. 1996b. Development and validation of a kiln-wide model for drying of softwood lumber, Cloutier, A., Fortin, Y. and Gosselin, R. (Eds.). In *Proceedings of the 5th IUFRO International Wood Drying Conference*, August 13–17, Quebec City, Canada, pp. 103–110.

Pang, S. 1996c. Moisture content gradient in softwood board during drying: simulation from a 2-D model and measurement. *Wood Science and Technology* 30:165–178.

Pang, S. 2004. Optimizing airflow reversals for kiln drying of softwood timber by applying mathematical models, Maderas. *Ciencia y tecnología (Chile)* 6(2):95–108.

Pang, S., A.N. Haslett. 1995. The application of mathematical models to the commercial high temperature drying of softwood lumber. *Drying Technology* 13(8&9):1635–1674.

Pang, S., R.B. Keey. 1994. The high-temperature drying of softwood boards: a kiln-wide model. *IChemE Journal* 72(A):741–753.

Pang, S., R.B. Keey. 1995. Drying kinetics of Pinus radiata boards at elevated temperatures. *Drying Technology* 13(5–7):1395–1409.

Papasavva, S., W.R. Hill, S.O. Andersen. 2010. GREEN-MAC-LCCP: A tool for assessing the life cycle climate performance of MAC systems. *Environmental Science and Technology* 44(19):7666–7671.
Paris Climate Conference. 2015. https://en.wikipedia.org/wiki/2015_United_Nations_Climate_Change_Conference, Accessed February 15, 2017.
Perré, P., A. Degiovanni. 1990. Simulation par volumes finis des transferts couples en milieux poreux anisotropes: Séchage du bois à basse et à haute température. *International Journal of Heat and Mass Transfer* 33(II):2463–2478.
Perré, P., I. Turner. 1999a. A 3D version of transpore: A comprehensive heat and mass transfer computational model for simulating the drying of porous media. *International Journal of Heat and Mass Transfer* 42(24):4501–4521.
Perré, P., I. Turner. 1999b. TransPore: A generic heat and mass transfer computational model for understanding and visualizing the drying of porous media. *Drying Technology* 17(7–8):1273–1290.
Perré, P., R. Keey. 2006. Drying of wood: Principles and practice, in *Handbook of Industrial Drying*, 3rd Edition, edited by A.J. Mujumdar, pp. 821–877. Dekker, New York.
Perré, P., R. Keey. 2015a. Drying of wood: Principles and practice, in *Handbook of Industrial Drying*, 3rd Edition, edited by A.J. Mujumdar, pp. 821–877. Dekker, New York.
Perré, P., R. Keey. 2015b. Drying of wood: Principles and practice, in *Handbook of Industrial Drying*, 4th Edition, edited by A.J. Mujumdar, pp. 821–877. CRC Press, Taylor & Francis, Dekker, New York,.
Perré, P., R. Remond. 2006. A dual-scale computational model of kiln wood drying including single board and stack level simulation. *Drying Technology* 24:1069–1074.
Perry, R.H., C.H. Chilton. 1984, *Handbook of Chemical Engineers*, 5th Edition, Hemisphere, New York.
Plumb, O.A., G.A. Spolek, B.A. Olmstead. 1985. Heat and mass transfer in wood during drying. *International Journal Heat Mass Transfer* 28(9):1669–1678.
Pordage, L.J., T.A.G. Langrish. 2007. Stack-wide behavior for hardwood drying. *Drying Technology* 25:1779–1789.
Prett, D.M., C.E. Garcia. 1988. *Fundamental Process Control*, Butterworth-Heinemann, Boston, MA.
Psichogios, D., L. Unger. 1991. Direct and indirect model based control using artificial neural networks. *Industrial & Engineering Chemistry Research* 30:2564–2573.
Quantrille, T.E., Y.A. Liu. 1991. *Artificial Intelligence in Chemical Engineering*, Academic Press, San Diego, CA.
Raffa, K. 1991. Phytochemical induction by herbivores, in *Phytochemical Induction by Herbivores*, edited by M.J. Raupp, pp. 245–276, Wiley, New York.
Robinson, J. 1992. Improve dryer control. *Chemical Engineering Progress* 88(12):28–33.
Robinson, J. 2000. Improved moisture content control saves energy. http://www.process-hcating.com. Accessed October 12, 2016.
Rosen, H.N. 1995. Drying of wood and wood products, in *Handbook of Industrial Drying*, 2nd Edition, edited by A.S. Mujumdar. Marcel Dekker, New York, pp. 899–920.
Ruppe, K. 1929. Diseases and functional disturbances of the respiratory tract in workers of the woodworking industry. *Z. Gesamte Hyg.* 19:261–264.
Salin, J.G. 1996. Prediction of heat and mass transfer coefficients for individual boards and board surfaces. A review, in *Proceedings of the 5th IUFRO International Wood Drying Conference*, August 13–17, edited by A. Cloutier, Y. Fortin, and R. Gosselin, pp. 49–58, Quebec City, Canada.
Salin, J.G. 2001. Global modelling of kiln drying, taking local variations in the timber stack into consideration. In *Proceedings of the 7th International IUFRO Wood Drying Conference*, Tsukuba, Japan, July 9–13.

Salin, J.G., G. Ohman. 1998. Calculation of drying behaviour in different parts of timber stack. In *Proceedings of the 11th International Drying Symposium IDS '98, Halkidiki, Greece*, Vol. B, pp.1603–1610.

Sherwood. T.K. 1929. The drying of solids - II. *Industrial & Engineering Chemistry* 21(10):976–980.

Shottafer, J.E., C.E. Shuler. 1974. *Estimating Heat Consumption in Kiln Drying Lumber, Life Science and Agriculture Experiment Station*, Technical Bulletin 73, University of Main, Orono, Main, September.

Siau, J.F. 1995. *Influence of Moisture on Physical Properties.* Virginia Polytechnic Institute and State University, New York.

Simpson, W. 1991. *Agricultural Handbook. Dry Kiln Operator's Manual.* USDA Forest Service, Madison, WI.

Simpson, W.T. 1993. Determination and use of moisture diffusion to characterize drying of northern Red Oak. *Wood Science & Technology* 27:409–420.

Sorensen, A. 1969. Mass transfer coefficients on truncated slabs. *Chemical Engineering Science* 24:1445–1460.

Sparrow, E.M., J.E. Niethammer, A. Chaboki. 1982. Heat transfer and pressure drop characteristics of arrays of rectangular modules encountered in electronic equipment. *International Journal of Heat and Mass Transfer* 31:961–973.

Statistics Canada 2013. https://www.statcan.gc.ca/eng/start. Accessed November 14, 2017.

Stevens, W.C, D.O. Johnson, G.H. Pratt. 1956. An investigation into the effect of air speed on the transference of heat from air to water. *Timber Technology* 64(2208):537–539.

Stoecker, W.F. 1998. *Industrial Refrigeration Handbook*, McGraw-Hill, New York.

Stridberg, S., I. Sandqvist. 1984. Sawmills energy equilibrium (In Swedish), Atu-rapport 82–4287, 82–5736, ISBN 91–7850-028–1.

Strömvall, A.M., G. Petersson. 1999. Volatile terpenes emitted to air, in *Pitch Control, Wood Resin and Deresination*, edited by L. Allen, Tappi Press, Atlanta.

Strumillo, C., P. Jones, R. Zylla. 1995. Energy aspects in drying, in *Handbook of Industrial Drying*, 2nd Edition, edited by A.S. Mujumdar, pp. 1241–1275, Marcel Dekker, New York.

Sun, Z.F., C.G. Carrington, C. Davis, Q. Sun, S. Pang. 2005. Mathematical modelling and experimental investigation of dehumidifier drying of Radiata pine timber. *Maderas. Ciencia y Tecnología* 7(2):87–98. http://dx.doi.org/10.4067/S0718-221X2005000200003. Accessed November 16, 2016.

Sun, Z.F., C.G. Carrington, C. McKenzie, P. Bannister, B. Bansal. 1996. Determination and application of characteristic drying rate curves in dehumidifier wood drying, in *Proceedings of the 5th IUFRO International Wood Drying Conference*, August 13–17, edited by A. Cloutier, Y. Fortin, and R. Gosselin, pp. 495–503, Quebec City, CA.

Sun, Z.F., C.G. Carrington. 1999. Effect of stack configuration on wood drying processes, In *Proceedings of the 6th IUFRO International Wood Drying Conference,* January 25–28, edited by H. Vermaas, and D. Steinmann, pp. 89–98, Stellenbosch, South Africa.

Sun, Z.F., C.G. Carrington, P. Bannister. 2000. Dynamic modelling of the wood stack in a wood drying kiln. *IChemE Journal* 78(A):107–117.

Sun, Z.F., C.G. Carrington, S. Davis, Q. Sun, S. Pang. 2003. Drying radiata pine timber under dehumidifier conditions: Comparison of modelled results with experimental results. In *Proceedings of the 8th International IUFRO Wood Drying Conference*, Brasov, Romania, August 24–29.

Szafran, R.G., A. Kmiec, W. Ludwig. 2005. CFD modelling of a spouted-bed dryer hydrodynamics. *Drying Technology* 23(8):1723–1736.

Tschernitz, J.L. 1986. Solar energy for wood drying using direct or indirect collection with supplemental heating and a computer analysis. Res. Pap. FPL-RP-477. Madison, WI: U.S. Department of Agriculture, Forest Service, Forest Products Laboratory, 81 p.

Tzouanas, Y.K., C. Geogakis, W.L. Luben, L.H. Unger. 1988. Expert multivariable control. *Comput. Chem. Eng.* 12:1065.
UNEP/TEAP report. 1999. The implications to the Montreal Protocol of the inclusion of HFCs and PFCs in the Kyoto Protocol
van Meel, D.A. 1958. Adiabatic convection batch drying with recirculation of air. *Chemical Engineering Science* 9:36–44.
Van't Land, C.M. 1984. Selection of industrial dryers. *Chemical Engineering* 91:53–61.
Van't Land, C.M. 1991. *Industrial Drying Equipment: Selection and Application*, Marcel Dekker, New York.
Vidlund, A. 2004. Sustainable production of bio-energy production in the sawmill industry. Licentiate Thesis, Stockholm, Sweden.
Wagner, F.G., T.M. Gorman, R.L. Folk, H.P. Steinhagen, R.K. Shaw. 1996. Impact of kiln variables and green weight on moisture uniformity of wide grand-fir lumber. *Forest Products Journal* 46(11/12):43–46.
Walker, J.C.F. 1993. *Primary Wood Processing Principles and Practice*, Chapman and Hall, London, 595 pp.
Walker, J.C.F., B.G., Butterfield, T.A.G. Langrish, J.M. Harris, J.M. Uprichard. 1993. *Primary Wood Processing*. Chapman and Hall, London. 595 p.
Wu, Q. 1989. An investigation of some problems in drying of Tasmanian Eucalypt timbers. Research Report, Department of Civil and Mechanical Engineering, University of Tasmania, Hobart, Tasmania, Australia, 237 pp.
Wu, Q., A.R. Oliver, P.E. Doe. 1995. A study of the boundary layer flow through a lumber drying stack. *Drying Technology* 13(8&9):2011–2026.
Yumah, R.Y., Mujumdar, A.S., V.G.S. Raghavan. 2015. Control of industrial drying, in *Industrial Drying Handbook*, 4th Edition, CRC Press – Francis & Taylor, Boca Raton, FL.
Zadeh, L.A. 1972. A rational for fuzzy control. Journal of Dynamic Systems, Measurement and Control 3–4, March, 1972.

9 Heat Pump-Assisted Wood Dryers

9.1 INTRODUCTION

Industrial wood drying is an energy-intensive consuming operation. Consequently, the improvement of energy efficiency of convective forced-air batch kilns by recovering a part of sensible and latent heat losses, mostly due to the moist (humid) air venting, is one of the most relevant objectives of R&D activities throughout the world (Chua et al. 2002). Heat recovery in kiln lumber drying is common particularly in moderate and cold climates where winter temperatures frequently drop below 0°C.

In addition to conventional heat recovery devices (e.g., air-to-air and desiccant heat exchangers) aiming at reducing global (electrical and fossil) energy consumption in the wood drying industry, there are heat pump-assisted dryers (also known as dehumidification kilns).

These systems, that are among the most efficient technical options available today, generally integrate air-to-air low (up to 55°C), moderate (up to 80°C), and high-temperature (>100°C) subcritical mechanical vapor compression heat pumps, acting as simultaneous cooling and dehumidifying, and heating devices, coupled with slightly modified conventional drying enclosures (kilns, chambers) (Law and Mujumdar 2008).

Heat pump-assisted dryers reduce the thermal energy consumption by incorporating one or more heat pump(s), which recover heat by cooling the kiln air below its dew point and recycling the latent heat of condensation (Minea 2012). Most of the moisture is removed from the kiln as liquid rather than by venting moist, warm air. However, some venting is required only for control and safety purposes (Mujumdar 2015).

Heat pump-assisted dryers use supplementary energy sources for preheating and as backup heating, as fossil fuels (e.g., oil, propane, natural gas) or renewables (e.g., biomass) (supplied via boilers), and electricity (supplied via resistive elements).

Theoretically, heat pump-assisted batch-type dryers are totally closed systems, which use suitably designed air-to-air heat pumps to condense moisture vaporized from the wood in order to rewarm a part of the recirculated drying air. Single-stage heat pump-assisted dryers are extensively used because of limited temperature differences that can be achieved inside most of wood drying chambers.

The basic construction of heat pump-assisted dryers is similar to that of conventional kilns (see Chapter 8), but require more thermal insulation, especially in cold climates.

9.2 BACKGROUND

In the 1970s and 1980s, in countries such as the United States, France, Canada, and New Zealand, the kiln drying industry promoted heat pump-assisted drying concepts.

Since the 1980s, wood drying has been one of the largest markets for heat pump-assisted drying in countries with cold and/or moderate climates, such as the United States, Canada, New Zealand, Finland, Sweden, and Norway.

However, despite the intense promotion by kiln drying industry under the rise of fossil energy prices and new legislation on environmental pollution, the performance of heat pump-assisted drying was sometimes disappointed mainly due to inappropriate kiln structures and/or dryer/heat pump integration, low performance of available equipment, deficient airflow control strategy, inadequate dehumidifying capacity, and use of not environmentally friendly chlorofluorocarbons (CFCs) as working fluids. The reliability of heat pump-assisted dryers was often low, and equipment suppliers did not provide enough information on the actual performance of their systems (Minea 2016).

Because of their high energy efficiency and economic performances, heat pump-assisted (dehumidifying) kilns are still considered as valuable options for drying various products as woods and agro-foods (Mujumdar 1996, 2002, 2004, 2006, 2007; Kiang and Jon 2007; Mujumdar and Wu 2008; Minea 2013a, b, 2014a, b).

In spite of the large number of heat pump-assisted wood drying installed throughout the world, relatively few R&D works have been published in the specialized literature.

A medium-temperature heat pump operating at 50°C dry-bulb temperature and 90% relative humidity, and using HFC-134a as a refrigerant, passive evaporator, and liquid subcooling, achieved a maximum specific moisture extraction rate ($SMER_{heat\ pump}$) of $5.11\frac{kg_{water}}{kWh_{heat\ pump}}$ (Carrington et al. 1995).

The performance of a commercial-scale heat pump-assisted timber dehumidifier with a 5 kW compressor and HFC-134a as a refrigerant, designed for drying at dry-bulb temperatures up to 60°C, has been affected by excessive quantities of vented air and heat losses through the kiln envelope (Bannister et al. 1998). However, the energy efficiency of such a system should be greater than a comparable, conventional air-vented kiln dryer by a factor of the order of two (Bannister et al. 1999).

A heat pump-assisted kiln for drying 40 mm boards of green red beech from New Zealand (a hard-to-dry timber) allowed providing the following valuable conclusions: (Carrington et al. 1999; Carrington et al. 2002) (i) during a drying cycle of 100 days, the average compressor power input was approximately 0.7 kW per cubic meter of dried material; (ii) the drying period was extended to 200 days without incurring too high energy costs; (iii) on-site energy use was relatively small compared to that of a conventional heat-and-vent kiln; (iii) there was a need to size the fans correctly in order to avoid high power demands; (iv) very low air velocities (e.g., 1 m/s) through the timber stack may be used without extending the drying time significantly; (v) kiln chamber must be well sealed against leaks to prevent undesired venting and, also, well insulated; (vi) dry-bulb temperature control may be achieved using venting or supplementary heating, when required, without incurring major additional energy costs, or by switch on/off the heat pump; (vii) good drying conditions can also be obtained without modulating on/off the heat pump, if it is correctly sized; (viii) improved control of the timber seasoning process may lead to reduced drying time, less operating costs, and higher value of hard-to-dry timbers; and (ix) control limits for the evaporating and condensing temperatures must be well matched to minimize the drying time, as well as overall efficiency.

Based on Bannister et al. (1999) concept, Chua et al. (2002) proposed a dual purpose, entirely recirculatory heat pump (dehumidifier) dryer for drying high-value seed pine cones and pine pollen catkins. During the first 30 hours of operation, the drying rate varied from 3.45 (at the beginning) to $1.2\frac{kg_{water}}{h}$ (at the end of the drying cycle).

By using a new characteristic drying curve (Sun et al. 1996), based on an isothermal diffusion model developed by Davis (2001) and using the experimental data obtained by Pang (1999) under heat pump drying conditions with low-to-medium temperatures, Sun et al. (2005) assessed the kiln-wide wood drying model previously developed by Sun and Carrington (1999a, b) and Sun et al. (2000).

The warming-up (preheating) of lumber stack(s) prior starting the drying process with electric-driven heat pumps of different softwood and hardwood lumber species is one of the highest important issues (Fernandez-Golfin Seco et al. 2004).

A number of studies indicate that the high-temperature drying of woods, such as, for example, southern yellow pine, in less than 24 hours at both 115.6°C (Koch 1971, 1974) and 132.2°C (Price and Koch 1980), can not only increase the drying rate but also, potentially, improve the wood's final quality, as reducing warp and enhancing color, as compared to the product dried at 82.2°C (Cai and Oliveira 2008).

Drying experiences at two high temperatures (104°C and 110°C) indicated that (Cai and Oliveira 2008) (i) the drying rates increase by 2.2–3.5 times, respectively, in comparison to a conventional drying schedule; (ii) both high-temperature drying schedules with steam result in greater internal stresses in the dried lumber compared to conventional kiln drying; (iv) warp is, in general, reduced by using the high-temperature drying schedules except at 110°C without steam conditioning; less crook was observed for both high-temperature drying levels; and (v) reductions in modulus of elasticity and rupture were found for both high-temperature drying schedules.

By initially fast increasing the drying temperature during the high-temperature drying schedule, lower residual stresses have been achieved (Vansteenkiste et al. 1997; Chen et al. 1997; Keey et al. 2000). However, even southern yellow pine crook is reduced by the high-temperature drying schedule, bow and twist are not significantly reduced (Wu and Smith 1998).

By drying hem-fir stud lumber at medium- (82°C) and high-temperature (116°C–132°C) schedules, bow and crook can be slightly reduced (Milota 2000). By drying a mixture of spruce and jack pine according to both 52 hours medium- (82°C) and 24 hour high-temperature schedules (115°C), the losses due to lumber degrades have been reduced slightly during the high-temperature drying at 115°C, and the modulus of rupture was reduced by about 16.5% in Jack pine and 10% in Spruce (Cech and Huffman 1974).

Approximately 90% of checks form in the first 6 hours of drying of *Pinus radiata* at a temperature of 120°C and a wet-bulb depression of 50°C, as the average dry-basis moisture content fell from 146% to 72% (Keey and Nijdam 2002).

Due to high permeability of relatively permeable species such as radiata pine, they can be kiln dried at dry-bulb temperatures ranging from 40°C to 180°C, without loss of structural strength, as in Australia and New Zealand, generally, followed by a process of steam conditioning (Pang et al. 2001).

The schedule for drying a radiata pine lumber stack consisting of 40 (thickness) × 100 (width) mm boards can be based on some of the following recommendations (McCurdy and Pang 2007): (i) high air speed reduces color change, though very high air speeds probably reduce returns and drying costs; (ii) in order to achieve the maximum economic benefits, a dry-bulb temperature of 60°C–70°C is recommended with a wet-bulb temperature depression of 15°C–20°C; (iii) air speed of 9 m/s is beneficial; and (iv) accurate calculations are possible once the market data are available as well as the kiln capital, depreciation, and labor costs, energy prices, and product prices at different grades.

High-temperature (120°C–130°C) schedules could dry mixtures of spruce–pine–fir lumber satisfactorily because the drying times would be 3–4 times shorter compared to conventional kiln drying (70°C–80°C) when the lumber is dried, for example, from initial moisture content of 80% to a final moisture content of 10%–12% (Aleon et al. 1988).

In spruce lumber, the internal mass transfer resistance is small when unbound moisture is present and the external mass transfer controls the moisture loss rate. Thus, all of the unbound moisture can be removed after 20 hours within a total drying time of 50 hours, a result that suggests that the external resistance is a significant factor for approximately 40% of the total drying time (Perré and Moyne 1991; Keey et al. 2000).

For softwood (resinous) lumbers dried with high-temperature heat pump-assisted batch dryers (Lewis 2003), the heat pump's average specific moisture extraction rate ($SMER_{heat\ pump}$) was 2.35 $\frac{kg_{water}}{kWh}$ (with white spruce) and 1.5 $\frac{kg_{water}}{kWh}$ (with balsam fir), while the heat pump's average coefficients of performance varied from 3.0 (at the end) up to 4.6 (at the beginning of drying cycles) (Minea 2004, 2014a, b).

The wet-bulb depression may affect the level of color change and the moisture content variation. The effect of the wet-bulb depression on the color change depends on the drying time. In the early stages of drying, lower wet-bulb depression (i.e., higher wet-bulb temperature at a given dry-bulb temperature) means that the wood surface temperature is higher, and, thus, the color change rate is higher. During the later stages of drying, this effect is relatively less significant (McCurdy and Pang 2007).

9.3 ADVANTAGES AND LIMITATIONS

Some of the advantages of using heat pump technology in wood drying operations can be summarized as follows (note: not listed in priority and/or relevance order):

1. Air-to-air heat pumps can be adapted to any conventional drying enclosures (kilns) which is reasonably airtight, moisture-resistant, and well insulated (Cech and Pfaff 2000);

2. Waste heat recovery and heat demand occur simultaneously;
3. Recovery of sensible heat and enthalpy (latent heat) of vaporization from the drying air reduces the overall (global) energy (electrical and fossil) consumption for each unit (kg) of moisture removed, provide efficient and cost-effective drying of both low- and high-grade wood, and decrease the wood drying industry's vulnerability to fluctuations in the fossil fuel prices (Kiang and Jon 2006);
4. May save up to 56% of operating (energy) costs compared to conventional drying methods (Barker and Guttridge 1988), resulting in relatively short payback period; for example, on-site measurement of energy usage for softwood drying with a heat pump-assisted kiln from 140% to 12% moisture content (dry basis) at a maximum dry-bulb temperature of 70°C showed that the specific total energy consumption (fuel and electricity) was of 0.6 GJ/m^3, much lower compared to that of a conventional gas-fired kiln (2.5 GJ/m^3) (Carrington et al. 1998);
5. Ability to control the air drying temperature (within a wide range of drying conditions, typically from 20°C to 120°C, generally with supplementary heating) and relative humidity (between 15% and 80%) (Mujumdar 2006; Chou and Chua 2006; Kiang and Jon 2007) in ways not possible for vented warm-air dryers (Claussen et al. 2007); condensation of moisture occurring at the heat pump evaporator reduces the relative humidity of the drying air and, thus, increases the driving force for moisture movement inside the lumber boards; at low drying temperatures, the drying potential of the air can be maintained by further reduction of the air relative humidity (Kiang and Jon 2007);
6. Operated at temperatures above the moisture freezing temperature in both continuous and intermittent drying modes, new or retrofit heat pump-assisted forced-air convection dryers may provide drying times similar to conventional kilns, reduce the overall energy consumption, and improve the final quality of dried lumber products;
7. If properly operated with accurate control of the drying parameters, there is possibility to achieve efficient control of product moisture contents, and air dry- and wet-bulb temperatures; this preserves and/or improves the quality of heat-sensitive (e.g., agricultural and biomedical products) and high-value dried products (especially at high temperatures), generally, with similar drying rates and drying times to those of traditional air-vented, steam, or electrically-heated kilns resulting in reduction of physical defects and enhanced productivity (Prasertsan et al. 1997; Soponronnarit et al. 1998; Bannister et al. 1999; Strommen et al. 1999; Islam and Mujumdar 2008; Pal and Khan 2008); improvement of product final quality (sometimes, more important than energy efficiency for many sawmills and users) mainly occurs since the drying parameters are practically unaffected by ambient weather thermal conditions; drying is achieved in almost closed kilns where much lower amount of air is exhausted outside the drying enclosure to control internal relative humidity, compared to conventional kiln dryers (Fernandez-Golfin Seco et al. 2004);

8. Potential for achieving lower operating energy costs, with specific moisture extraction ratios often in the range of 1.0–4.0 $\frac{kg_{water}}{kWh_{heat\ pump}}$ since heat can be recovered from the moist (humid) air, which should compensate for the higher initial capital costs (Kiang and Jon 2007);
9. Although the capital costs of heat pump-assisted kilns are normally higher than that of equivalent direct-fired units run at the same temperature, the operating costs may be less unless fuel at a discounted or nil value (such as waste wood) is used; however, if very low kiln temperatures are required, costs may still favor the use of a heat pump-assisted kiln; a reduction in kiln temperature from 70°C to 50°C, for example, will cause the energy demand of the heat pump-assisted kiln to increase by approximately 7% compared with 60% for a conventional air-vented kiln (Carrington et al. 1998); (Carrington et al. 2000)
10. The economics of heat pump-assisted drying systems depends on how the heat pump is integrated in the process; identification of feasible installation alternatives (e.g., with split or compact heat pumps) is of crucial importance;
11. When high quantities of moisture are removed from large volumes of lumber, the efficiency of kiln drying processes increases, and shorter payback periods can be achieved compared to traditional air-forced drying methods;
12. In large sawmills, heat pump-assisted kilns help use low-pressure and low-capacity boilers for steam generation, thereby eliminating the requirement of one or more stationary engineers or for additional boilers;
13. Possibility for retrofitting conventional kilns without major additional investment for new boilers is generally very expensive;
14. Potential for increased throughput by using air with lower relative humidity to enhance drying rate with little impact on drying-induced stresses;
15. Potential for reducing industrial emissions (greenhouse gases, fumes, odors, and chemical pollutants) into the atmosphere under strict national and international environmental regulations;
16. Easier management of condensate's environmental impacts compared to conventional kilns (Perera and Rahman 1997; Minea 2004); however, volatile organic chemicals, normally removed with the vented moist air, now appear in the condensate stream, which potentially could be sent to a separate unit for chemical recovery (Minea 2010a, b; Perré and Keey 2015); moreover, as moisture is removed as condensed liquid rather than vapor in warm discharged air, the associated thermal losses are avoided;
17. High annual factors of utilization and high energy efficiency due to low temperature lifts.

Among some practical limitations of heat pump-assisted kilns compared to conventional air-vented dryers, the following can be mentioned:

1. May incur higher initial capital and maintenance costs compared to conventional dryers due to additional refrigeration components (heat pump compressor and blower, heat exchangers, refrigerant filters);

2. Operating costs are highly dependent on species, lumber dimensions, size of kiln, and local climate (if the kiln is located outdoors);
3. More complex operation and, thus, requirement for competent design engineers and operators/technicians;
4. Additional floor space and/or building required for the heat pump;
5. Need for temperature resilient materials and fluids (refrigerants, oils, belts, etc.), and regular maintenance;
6. Need for regular maintenance (including refrigerant charge checking and periodical recharging);
7. Risk of refrigerant leakage;
8. Use of electricity, a high-grade energy source, as the main energy source for driving the heat pump compressor(s) and blower(s) instead fossil or renewable energies;
9. Higher consumption of electricity due mainly to the heat pump compressor(s) and blower(s) (Anderson and Westerlund 2014);
10. In some installations, especially in high-temperature drying systems operating in moderate and cold climates, it may be a requirement for an auxiliary (supplementary) and backup heating source to maintain effective drying temperatures (Cech and Pfaff 2000);
11. To prevent excessive heat losses during cold winter months and reduction in drying temperature resulting in an increase of drying time, heat pump-assisted kilns must be well insulated;
12. Compared to conventional kilns, drying may be slower; thicker lumber boards require a proportionately longer drying time, especially in species which are tolerant of rapid drying conditions at higher temperatures;
13. The maximum dry-bulb temperature of the drying process is limited due to limits on the safe working temperature of the heat pump compressor and other components;
14. In countries where the price of electricity is much higher, for example, compared to that of biomass, there will be a negative effect on the global economic profitability (Anderson and Westerlund 2014);
15. Compared with air convective vented dryers, heat pump systems are more vulnerable to heat transfer mechanisms over large temperature differences;
16. Poor control of air circulation through the entire lumber stack(s) and of heat recycled through the heat pump dryer may create zones of low humidity, reducing the drying rate and increasing the global energy consumption;
17. As in the case of conventional dryers, uneven drying conditions along the length of the kiln inevitably accentuate the overall variation of the final moisture content; the heat input and venting systems must be correctly designed with the use of zonal sensing and control along the kiln to minimize temperature and humidity variation; to ensure an even air velocity, all of the space around the stack must be completely baffled (Pang and Haslett 1995).
18. Because the input energy to a heat pump-assisted dryer is relatively small, even minor air leaks or heat losses can cause the dryer temperature to fall

uncontrollably with adverse consequences for the system performance, while such losses would, normally, have little effect on the performance of warm-air-vented dryers;

19. If all basic parameters of kiln schedule are not properly controlled, some negative impacts such as shrinkage, surface splits, end checks and splits, and strength properties of softwoods, usually dried at high temperatures (>80°C–100°C) with drying times of 48–60 hours, can be achieved (Pang and Haslett 1995).

The heat pump-assisted drying technology is not widespread in industrial applications as it should be, particularly in the field of drying agro-food and wood sectors. Among others, reasons for their neglect still are as follows: (i) uncertainty by potential users as to heat pump reliability, (ii) lack of good hardware in some types of potential applications, (iii) lack of experimental and demonstration installations in different types of industries, (iv) lack of required knowledge engineering of heat pump technology in target industries, and (v) relative cost of electricity and fossil fuels affecting the commercial viability of heat pump-assisted dryers.

9.4 OUTLOOK FOR R&D WORKS ON HEAT PUMP DRYERS

The majority of industrial dryers are air convective dryers using fossil fuels, biomass, or electricity as their primary energy sources.

Most heat losses are due to the exhausted moist air and the poor thermal insulation of the drying enclosures. Because of energy losses, attention has focused on reducing and/or recovering the wasted heat, as part of the global effort to reduce energy consumption and control greenhouse gas emissions.

Among other heat recovery devices, the heat pumps assisting the air convective dryers have the potential to save up to 50% or more of the primary energy used. The heat pump drying principle is based on the conservation of energy, the system being energetically closed (Strumillo 2006).

In heat pump-assisted dryers, the process air is heated up to the required drying temperature, typically between 30°C and 57°C (Strumillo 2006). The increased temperature enhances the heat transfer rate into the drying materials and increases the moisture diffusion rate. The low relative humidity of the drying process air finally helps remove moisture from the dried products (Chua et al. 2002).

Heat pump drying operation is based on the dehumidifying principle and no warm air loaded with moisture is discharged into the environment, but goes back to the dryer. This process includes two major steps: energy conservation through reheating and dehumidification of the drying air. Warm, dry, and unsaturated air is led over the surface of the product to be dried. This air takes up the surface water on the product, the produced vapor condensates in the heat pump evaporator, and finally, is warmed in the condenser.

This section reviews a number of published R&D works on drying materials other than woods, for example, agro-foods. It focuses on two of the most important issues of the drying technology in general, that is, integration and efficient control of electrically driven, mechanical vapor compression heat pumps coupled with air convective dryers. Some of observations and comments may apply to industrial heat pump-assisted wood drying technology.

The expression "efficiently controlled heat pump dryers" mainly refers to strategies aiming at matching water thermal dewatering (moisture vaporization) capability of the dried material with the heat pump dehumidification capacity in order to ensure safe operating parameters for the heat pump and provide the best final quality of the dried product (Minea 2013a, b).

The drying industry is highly diversified because of the huge number of products that needs drying, as woods and agro-foods. For the majority of these products, the application of heat pump-assisted drying technology has been extensively studied, mainly at the academic level, and several more or less innovative concepts have been proposed for dryer–heat pump integration and control.

A number of published works recommended relatively complex concepts. In most of them, for example, methods aiming at matching the heat pump dehumidification capacity with the product thermal dewatering (moisture vaporization) rate were rarely employed, while external heat rejection devices were frequently used, even heat pump-assisted drying systems that reject heat cannot be enough energetically efficient. Instead, they have to be physically and energetically isolated from the ambient environment, and heat may be added only in order to reach the set dry- and wet-bulb temperatures, and/or compensate enclosure losses, but not rejected from the heat pump thermodynamic cycle itself.

Relatively complicated configurations reported involved, for example, two-stage series evaporator drying heat pump systems with two series subcoolers and heat pipe-type economizers, two-cycle heat pump dryers, two-stage heat pump/two-stage drying chamber systems, and multistage fluidized bed heat pump dryers. Dryers have also been combined with ground-source heat pumps and thermal loops or air-to-air heat exchangers without proving that such concepts could help increase overall system energy efficiency. It can be concluded that conventional drying heat pumps are already complex systems because of the interdependency of many thermodynamic parameters. Increasing their complexity cannot contribute to any further industrial deployment, in spite of their technical, economic, and environmental advantages. Most of the improvements have to focus on the simplification of the concepts in order to increase the reliability and dehumidification efficiency of industrial drying applications.

Another problem consisted in eventual deficient correlation between the data provided and the results reported by several published R&D works, especially on agro-food and unconventional products dried with heat pumps (Minea 2014a, b). It was found that relatively few of them provided complete input information about the experimental and/or theoretical studies achieved. This situation suggests that, for each material or group of similar products, pertinent drying parameters have to be well defined and controlled in order to make the heat pump-assisted drying systems more reliable and efficient.

On the other hand, drying schedules were not always provided in terms of set dry- and wet-bulb temperatures, temperature depression in relation with the air relative and absolute humidity, and flow rate. Drying curves of the dried products were sometimes provided without specifying whether their moisture content was measured and how (oven, continuously or intermittently) it was measured. Preheating and supplementary (backup) heating were rarely used, especially when drying heat pumps were installed outdoors.

In several published studies, essential data, such as the input/output quantities and initial/final moisture contents of the dried materials, the heat pump dehumidification capacity and/or compressor rated input power, the condenser heating and heat rejection capacity, the heat pump pressures and temperatures throughout the drying cycles were missing. Heat pump malfunctions, including compressor short cycles, excessive or insufficient refrigerant pressures and temperatures, and evaporator frosting, were rarely reported. Without such information, industrials and/or young researchers may be disappointed, and academic R&D reports may also compromise the credibility and future industrial applications of the drying heat pump technology.

A number of R&D studies reported the fabulous advantages of heat pump dryers compared to conventional convective drying methods without providing any accurate measured and/or calculated parameters. Much academic R&D work focused on the miraculous impact of heat pump dryers on the final structure, color, and nutritional quality of dried products, while the experimental setup and drying methods were sometimes highly questionable. Some published studies also claimed that a specific heat pump dryer having a given dehumidification capacity could work miracles: it could efficiently dry different products (solids, liquids), in different drying modes (batch, continuous, intermittent), with different drying mediums (air, inert gases, CO_2), and at negative or positive temperatures, without modifying and/or adjusting the dryer–heat pump integration and/or control strategies. However, any change made in one aspect of the heat pump drying system will inevitably influence many others (Minea 2014b).

To be able to validate the reported performances, future published R&D studies on heat pump-assisted drying processes have to provide at least the following information (Minea 2014b):

a. The dryer's main characteristics, such as the location (hot or cold weather, indoor, outdoor), volume (or mass) capacity, airflow configuration through both the dryer and heat pump (fans, blowers, air vents), and percentage of dryer heat leakage rates (conduction, air exfiltration and infiltration);
b. The drying schedules for each type of dried material, including the product initial and required final moisture contents, as well as input and output masses (or volumes), preheating requirements and type of supplementary (backup) heating (if applicable), and for each drying step, air setting dry- and wet-bulb temperatures, or air temperature depression, and air relative humidity;
c. The heat pump's critical operating parameters, for example, type of refrigerant used, maximum and minimum allowed discharge or suction pressures and temperatures, drying airflow rate and pattern through both the dryer and heat pump heat exchangers, and heat transfer rates of the heat pump evaporator (dehumidification capacity), condenser (heating capacity), and heat rejection device (when required).

9.4.1 Dryer–Heat Pump Coupling

Dryers and heat pumps are both complex thermodynamic systems. Consequently, systems integrating dryers and heat pumps are much more complex than each of these components separately. Therefore, they should not be analyzed independently due to the complex interaction between the air drying process and the heat pump (refrigerant) thermodynamic cycle. While the fundamentals of the drying process (i.e., heat and mass transfer theories) and the thermodynamic cycles of heat pumps are well known, the interaction between these two thermodynamic systems requires careful theoretical and experimental approaches (Minea 2013a, b).

9.4.2 Schematic Representation of the Principle

Some published research work schematically represented the heat pump drying principle as shown in Figure 9.1 (Chua et al. 2002). This schematic drawing suggests that the same air flows through the heat pump evaporator and condenser. The problem in this case is that the dehumidification capacity of the heat pump as well as the relative humidity of the process air entering the drying chamber cannot be efficiently controlled. Generally, for purposes of economical design and efficient dehumidification control, the drying airflow going through the condenser has to be constant and higher than the one going through the evaporator, while that circulating through the evaporator must be variable (Minea 2013a).

9.4.3 Using Heat Rejection Devices

Several research studies used parallel condensers (Figure 9.2) (Alves-Filho and Tokle 1999; Alves-Filho et al. 2008a) and refrigerant desuperheaters (Alves-Filho and Eikevik 2008). These heat exchangers were meant to discharge the excess heat outside of the drying system. The main drawbacks of the heat pump drying systems shown in Figure 9.2 are twofold. First, there is no air bypassing the evaporators. Second, the three-way valve cannot operate as a flow modulating valve to direct

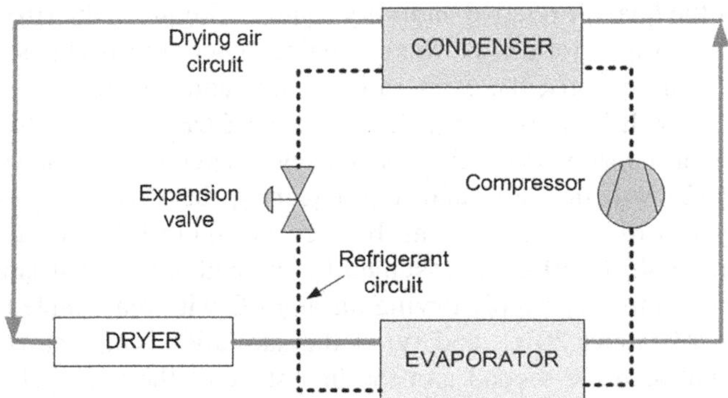

FIGURE 9.1 Schematic representation of the heat pump drying principle (Redrawn and reproduced from Chua et al. (2002) with permission from *Drying Technology*).

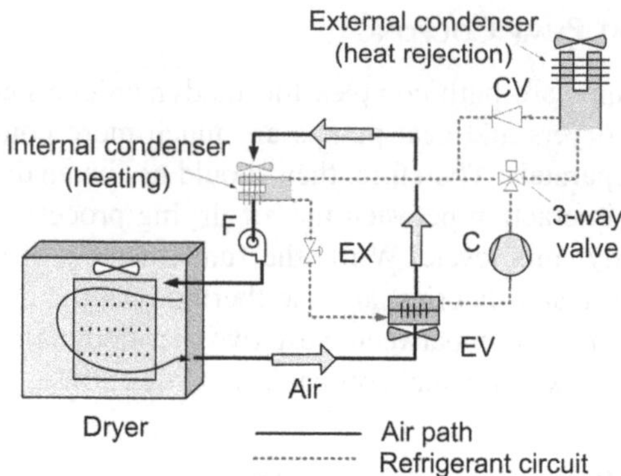

FIGURE 9.2 Heat pump batch dryer with external condenser as a heat rejection heat exchanger (Redrawn and reproduced from Alves-Filho and Tokle (1999) and Alves-Filho et al. (2008a) with permission from *Drying Technology*). C, compressor; CV, check valve; EV, evaporator; EX, expansion valve; F, fan.

the whole refrigerant flow to the internal condenser to heat the drying chamber or to the external condenser to discharge the excess heat outside. When the external condenser is operating, its blower and the heat pump compressor are running, and, thus, electrical power is used while no drying air is being dehumidified. Heat rejection with the refrigerant desuperheater could be a better solution but, depending on the cooling air temperature, it may reject sensible heat (15%–20% of the total), or latent heat. In the latter case, the heating capacity of the internal condenser could be affected because these two heat exchangers operate in series.

9.4.4 Using Two-Stage Heat Pumps for Drying

The complex two-cycle heat pump dryer concept shown in Figure 9.3a (Lee et al. 2008) may raise questions regarding (i) the control of the dehumidification rates of evaporators 1 and 2 according to the material actual moisture vaporization; (ii) the control of the final air relative humidity (state S) (Figure 9.3b); (iii) the hot and humid air enters evaporator 2 at a lower dry-bulb temperature after passing through evaporator 1; consequently, the evaporating temperature of the first-stage refrigerant HFC-134a will be lower, and, therefore, the $COP_{heat\ pump,1}$ of the first cycle will decrease; at the same time, the condensing temperature of the second-stage refrigerant R-124 will increase, and, consequently, the $COP_{heat\ pump,2}$ of the second series cycle will decrease, as can be seen in Figure 9.3c; (iv) if the drying process occurs at 45°C–50°C, as assumed by the authors in their simulations, a one-stage heat pump may supply drying air at 75°C with reasonable temperature lifts between 25°C and 30°C; and (v) if the saturation temperature difference ($T_{cond} - T_{evap}$) between the second and the first stage of the two-cycle heat pump increases, the question is if the drying chamber will be able to provide equivalent temperature drops through the dried material without any control of the drying airflow rates.

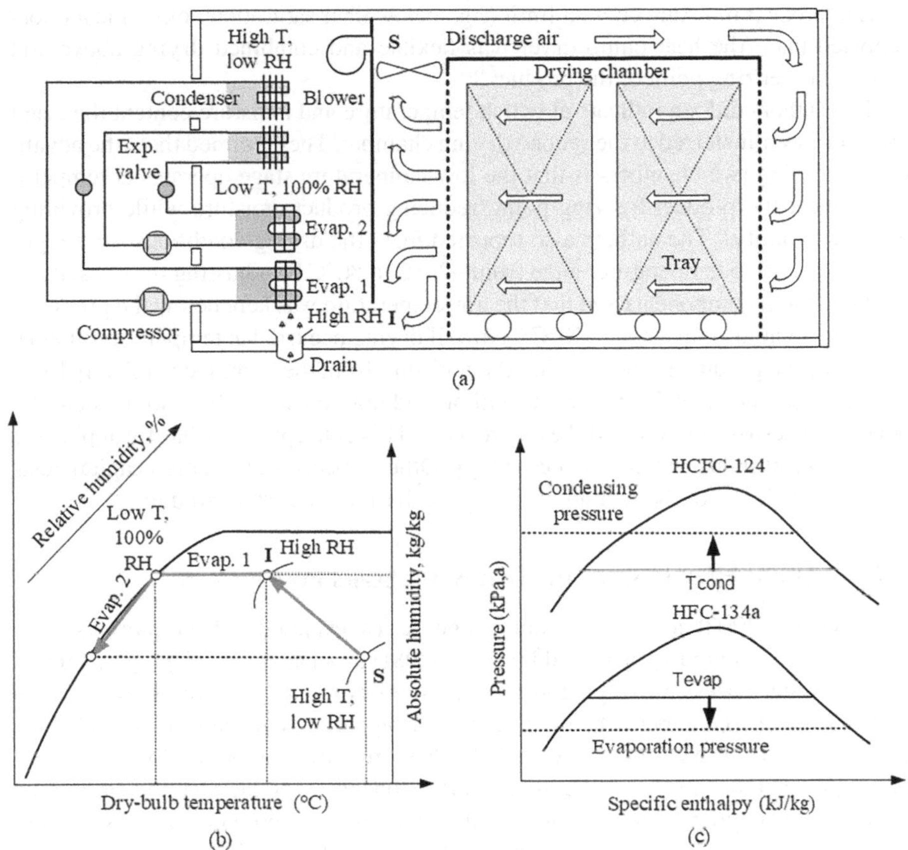

FIGURE 9.3 (a) Schematic of the proposed two-cycle heat pump dryer (Redrawn and reproduced from Lee et al. (2008) with permission from *Drying Technology*); (b) Mollier diagram of the drying air process; (c) p–h diagram of the refrigerant evaporating and condensing; I, state of drying air entering evaporator 1, S, state of drying supplied to the dryer.

All these observations suggest that using two-cycle heat pumps for drying seems less relevant while intending to improve drying energy efficiency and/or product final quality.

9.4.5 Using Two-Stage Fluidized Bed Heat Pump Dryer

A multi (two)-stage fluidized bed heat pump dryer with two independent drying loops, two refrigerant circuits, and external condensers with bypasses around the evaporators was proposed (Alves-Filho and Tokle 1999). Installed after the low-temperature drying chamber, the high-temperature drying chamber receives the same flow of material as the first drying stage, but at lower moisture contents. Consequently, the second drying chamber may operate with much lower $SMER_{heat\ pump}$ performances. Even though heat pump dryers are generally less efficient when operating below a certain product moisture content limit (e.g., below fiber saturation point), the authors reported that "the heat pump dryer was flexible and combined drying above and below the freezing point of the product."

The authors did not indicate at which temperature and moisture content the dried product was transferred to the second drying chamber. They claimed that "the advantage of having two chambers is that the low-temperature stage operates at temperatures under the product freezing point, reducing product moisture while providing excellent quality." The authors also reported that "the drying conditions were regulated over a wide temperature range from –20°C to 80°C by adjusting the capacity of the heat pump components and that the air temperature was kept below the product's freezing point, usually above –10°C." Now, if drying occurred at temperatures below or around the product freezing point, the moisture from the humid air probably froze on the evaporator fins. However, the authors did not give any information about the periodical defrosting cycles, if they were used. This concept seems hardly applicable as an efficient industrial drying technology. Other aspects, such as high capital costs and low efficiency of the second drying stage, have to be further studied.

9.4.6 Using Heat Pumps with Two-Stage Series Evaporators

Chua et al. (2002) analyzed a multistage series evaporator heat pump system (Figure 9.4a), previously proposed by Perry (1981) as an advanced drying system.

This system uses two (high and low pressure) evaporators on the same air path from the drying chamber. The high-pressure evaporator was used for sensible heat cooling up to the air dew point, and the low-pressure evaporator for latent heat removal. The most important drawback of this concept is the fact that the air coming from a single drying chamber flows through two series evaporators. That is because larger air pressure losses will occur and the practical control of dehumidification rates within the two series evaporators operating with the same drying airflow will be more complex. Since with one evaporator, the dehumidification rate control

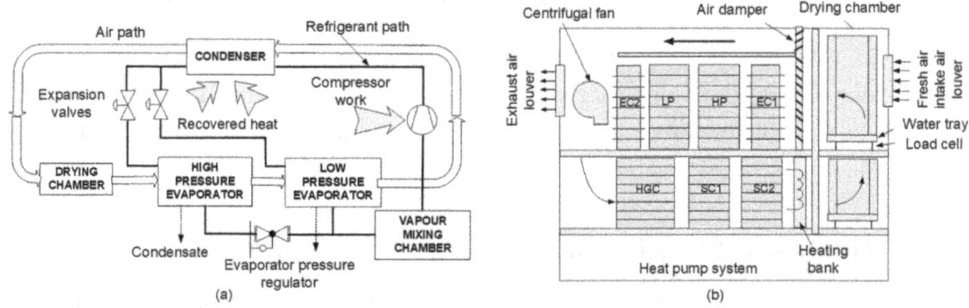

FIGURE 9.4 (a) Two-stage series evaporator heat pump drying system coupled with a drying chamber (Redrawn and reproduced from Perry (1981) and Chua et al. (2002) with permission from *Drying Technology*); (b) two-stage modular heat pump dryer (Redrawn and reproduced from Chua (2000), Chua et al. (2002, 2005), and Chua and Chou (2005) with permission from *Drying Technology*); EC, economizer; HP, high-pressure evaporator; LP, low-pressure evaporator; HGC, hot gas condenser; SC, subcooler.

would be much easier, using two series evaporators, with two different evaporating temperatures and the same drying air stream, seems a very complicated concept.

Perry's concept (Figure 9.4b) was later developed, built, and tested by Chua et al. (2000, 2002, 2005) for drying agricultural products. In their system, the drying air comes from a single drying chamber, and the high-pressure evaporator surface temperature was controlled by adjusting the spring knob of the back-pressure regulator. However, such control seems difficult to apply because it is not clear how the spring knob of the back-pressure regulator can be continuously set during actual drying cycles where the material thermal dewatering (moisture vaporization) rate is variable. In this concept, there are also two series refrigerant subcoolers, an unnecessarily complicated approach for drying heat pump systems. Using two series subcoolers does not make practical sense. Moreover, the economizers shown in Figure 9.4b seem to be part of a heat pipe system with a circulating pump and an unidentified thermal carrier. This heat recovery system precools the air before entering the evaporators and preheats it before entering the condenser. However, by decreasing the evaporating temperature and increasing simultaneously the condensing temperature, a lower coefficient of performance can be achieved, while the initial cost of the heat pump dryer may increase because of additional materials and controls. In spite of this, the authors claimed that "the installation of the economizer resulted in the improvement in COP and SMER of about 3% and 4%, respectively," without providing experimental results validating this performance.

Ho et al. (2001) presented a two-stage heat pump dryer similar to the system shown in Figure 9.4b. The authors reported that "improvements in the total heat recovered can be up to 106%" when drying sensitive products (potato slices). This conclusion seems highly questionable because with heat pump drying systems the purpose is not necessarily to recover higher and higher amounts of latent heat, but rather to correlate the heat pump latent heat extraction capacity with the material thermal dewatering (moisture vaporization) rate. On the other hand, multistage evaporator systems are complex and involve greater initial capital costs.

9.4.7 Using Supercritical CO_2 Heat Pumps

Studies on using carbon dioxide (CO_2) heat pump dryers first focused laundry drying systems.

Schmidt et al. (1998) compared the thermodynamic behavior of two dehumidification heat pump cycles: (i) a subcritical process with HFC-134a and (ii) a supercritical process with CO_2. Their simulation results showed that both cycles are equivalent in terms of energy efficiency. However, better compression efficiency was expected with CO_2, as well as improved energy efficiency.

Klöcker et al. (2001) studied the feasibility of using CO_2 as a working fluid for laundry heat pump dryers and compared it with a heat pump using HFC-134a as a refrigerant. They found that the CO_2 heat pump did not need more energy than the HFC-134a heat pump, and that the CO_2 supercritical cycle was suitable for heat pump dryers due to the good environmental characteristics and thermal properties of CO_2. The authors compared the energy performance of two optimized CO_2 heat pumps with those of a conventional electrically heated dryer, and concluded that if the drying time is not the key requirement, energy savings of about 65% can be achieved. Also, if a high water extraction rate is sought, energy savings of about 53% can be achieved, including fan energy consumption, but with much shorter drying times than that of the conventional system.

Klöcker et al. (2002) built a laboratory prototype laundry dryer using CO_2 as a working fluid. Their experiments showed that the heat pumps used in laundry dryers at 50°C–60°C exhibited a significant energy savings potential (between 53% and 65%) compared to conventional high temperature (130°C) drying methods. However, the drying cycles with trans-critical CO_2 heat pumps in batch mode were relatively short (about 54 minutes). The reported experimental results showed that the end of the drying cycles was dependent on the maximum air temperature entering the evaporator (40°C) rather than the material final moisture content. The authors also showed that up to 35% of each drying cycle was a transient process with air temperatures leaving the gas cooler (entering the drum) being below the desired set points (50°C–60°C). Now, during transient periods, the heat pump drying performance declines. Despite this, the reported *SMERs* were relatively high $\left(1.54 - 2.05 \frac{kg_{water}}{kWh_{heat\ pump}}\right)$, depending on the type of compressor CO_2 used. Moreover, because the drying airflow rate through the evaporator and the gas cooler was the same, the gas cooler heat transfer surface (118 m^2) was overdesigned compared to the evaporator heat transfer surface (30.1 m^2). That generally implies large CO_2 charges and finned-tube heat exchangers.

Sarkar et al. (2006) validated their simulation model with the experimental results on CO_2 heat pump-assisted laundry dryer reported by Klöcker et al. (2002). The authors reported that dryer efficiency, recirculation air ratio, ambient temperature, and air mass flow rate have a significant impact on system performance. It was observed that "by increasing the bypass air ratio, the evaporator dehumidification rate increased and the moisture extraction rate continuously increased. At 0.3 bypass air ratio, the SMER attained a maximum." In spite of this sound observation, the authors surprisingly concluded that the "bypass of air was not a very effective mode of system control. With increase in bypass air ratio at 0.5, the moisture extraction ratio ($MER_{heat\ pump}$) $\left(\frac{kWh_{heat\ pump}}{kg_{water}}\right)$ increased by only 3.5%, and the maximum value of Specific Moisture Extraction Rate ($SMER_{heat\ pump}$) $\left(\frac{kg_{water}}{kWh_{heat\ pump}}\right)$ at 0.3 bypass air ratio was only 4% more than that at air bypass ratio of 0."

Honma et al. (2008) presented an experimental study on a compact heat pump dryer using CO_2 as a refrigerant for domestic clothes washers/dryers. The authors mentioned the requirement to balance the amount of heat provided by the air with the heat supplied by the gas cooler and to adapt the refrigerant cycle based on the progress of the drying cycle. They controlled the expansion valve to keep the superheat constant in order to maintain the targeted heating capacity at a constant level of 2.7 kW and reduce the drying time. The prototype did not include any other control means, such as air bypassing the evaporator. On the other hand, the gas cooler had a heat transfer surface about 40% higher than that of the evaporator for the same airflow rate. The heat pump CO_2, defined as the gas cooler heating capacity divided by the compressor input power, was estimated at 4.07. The tests performed revealed that the prototype was able to reduce electric power consumption by 59.2% and the drying time by 52.5% in comparison with electrical heater drying systems. Furthermore, the authors estimated that the drying time can be further reduced by 3% by keeping the refrigerant superheating temperature at 6°C–10°C.

Fornasieri et al. (2010) compared a trans-critical CO_2 process with a subcritical HFC-134a cycle in terms of energy efficiency, and the experimental results achieved with a CO_2 prototype confirmed the theoretical analysis. The authors concluded that CO_2 operation in trans-critical conditions is a *viable alternative to the traditional technology in the heat pump dryers.* Tests performed on a CO_2 heat pump dryer prototype showed a slight decrease of the electrical power input, with a very limited (+3.8%) increase in the drying time, in comparison with a conventional HFC-134a heat pump dryer. About this work, it can be noted that (i) the authors did not indicate whether a preheating step was required; their graphical results showed that the heat pump required a long stabilization period prior to reaching its permanent running speed; however, it is well known that drying heat pump efficiency is drastically affected at nonstationary running speeds; (ii) graphical results also showed that air temperature at the drum outlet (or at the CO_2 evaporator inlet) reached up to 55°C toward the end of the cycle; such inlet temperatures may result in evaporating temperatures being equal to or higher than the CO_2 critical point (31°C), with suction pressures as high as 75 bars and compressor discharge pressures exceeding 130 bars; (iii) other graphs showed that the same drying airflow rate goes through the evaporator and the gas cooler; now, in this case, the heat transfer area of the gas cooler has to be much higher than that of the evaporator. It may cause practical issues, such as the dryer physical dimensions and cost; and (iv) finally, the authors discussed neither the risk of evaporator and gas cooler clogging with clothes fibers, dust, and other materials nor the associated cleaning issue.

Alves-Filho et al. (1998b) presented a "new, innovative carbon dioxide heat pump" dryer for high-quality dried products. The schematic representation of this innovative heat pump dryer showed the CO_2 condenser (gas cooler) and evaporator operating with the same airflow. The authors reported that the drying circuit was designed to handle other gases than air, such as nitrogen and helium, but did not indicate whether modifications and/or adaptations of the drying system are required when the drying medium is changed. Moreover, they claimed that "the CO_2 heat pump dryer can operate at temperatures ranging from –30°C to 110°C to provide the conditions for heat sensitive products which require temperatures below or above its freezing point." Operating the CO_2 heat pump at evaporating temperatures as low as –30°C may provide drying processes at temperatures far below 0°C. In this case, moisture frosting may occur and evaporator defrosting cycles have to be periodically initiated. On the other hand, to provide drying air at 110°C, the heat pump compressor discharge temperature has to be around 130°C, and the drying airflow rate through the gas cooler has to be relatively small. These high drying temperatures imply return air temperatures entering the heat pump at 60°C probably, leading to evaporation temperatures above the CO_2 critical point (≈31°C). The $COPs_{heat\ pump}$, $SMERs_{heat\ pump}$, and heat pump refrigeration capacity computations for conventional and supercritical CO_2 cycles were not validated by any valid experiments. Moreover, without specifying either the quantities of dried materials versus the heat pump dehumidification capacity or the initial and final material moisture contents, the authors reported $COPs_{heat\ pump}$ as high as 11 and $SMERs_{heat\ pump}$ as high as 8 $\frac{kg_{water}}{kWh_{heat\ pump}}$ based only on a given drying air temperature and a given variation of the drying air relative humidity. Finally, the study did not demonstrate why the carbon dioxide drying heat pump has a greater impact on the quality of the dried product.

Eikevik et al. (2005b) built a prototype fluidized bed heat pump dryer using CO_2 as the working fluid. The diameter of the fluidized bed varied from 10 to 35 cm, and the airflow varying from 100 to 800 m^3/h. The measured specific moisture ratios ranged from 1.65 to 3.73 at a temperature range from –20°C to 90°C. The maximum CO_2 pressure was 130 bars. As previously noted, when drying with air at –20°C, moisture frosting may occur and evaporator defrosting cycles have to be periodically initiated. On the other hand, if the drying air enters the heat pump evaporator at, let us say, 60°C and leaves at 50°C, the evaporating temperature will probably be higher than the CO_2 critical temperature (≈31°C), the evaporating pressure above 73 bars and the discharge pressure in excess of 150 bars.

9.4.8 Using Ground-Source Heat Pumps for Drying

Colak and Hepbasli (2005, 2009) proposed a ground-source heat pump for drying apples, a typical example of the inefficient use of heat pumps in drying processes. First, because the ground-source heat pump has to operate in the heating mode only, no drying air dehumidification is provided. For relatively small quantities of dried materials, such a ground-source heat pump could well operate during relatively short running cycles in order to supply heat to the drying chamber at temperatures between 35°C and 45°C. In this range of drying temperatures, air-to-air heat pumps may be technically and economically more competitive. However, for large-scale industrial systems with longer drying cycles (dozens of hours, or days), the concept – as proposed – could be inapplicable without supplying auxiliary heat to the drying chamber. In continuous drying processes, the temperature and thermal potential (capacity) of the ground will progressively decline. The heat pump will cool the ground excessively, while the evaporating temperature will drop unduly, as well as the $COPs_{heat\ pump}$. After a certain period of time, the heat will be extracted from the ground at evaporating temperatures as low as –15°C and even lower. For batch dryers, it may be worse because the time between two consecutive charges is short and the heat pump will operate practically in continuous mode but with many shut-off and transient cycles. Finally, this paper did not provide any long-term validation of the simulated results.

9.4.9 Using Heat Pumps Coupled with Thermal Loops

The concept proposed by Phaphvangwittayakul et al. (2007) combined a heat pump dryer with a thermal loop (heat pipe) heat recovery system. The authors reported that "the loop thermosyphon reduced the specific energy consumption from 12 to 20%, depending on the operating conditions, and that the payback period was of about 3 years." These energy consumption reductions seem unlikely first because the thermal loop simultaneously reduces the heat pump's evaporating temperature and increases the condensing temperature. Consequently, the heat pump compression ratio will increase, the heat pump coefficient of performance will decrease, and the compressor input power and energy consumption will be higher. In practice, the payback period cannot be improved by adding an additional expensive device (the heat pipe) to the heat pump drying system, because the system's total energy consumption increases.

9.4.10 Using Atmospheric Freeze-Drying Fluidized Bed Heat Pumps

Conventional vacuum freeze-drying, involving the removal of water from a frozen product by sublimation, is used for drying sensitive food products. Atmospheric freeze-drying is another convective drying process at temperatures below the product freezing point. It occurs at higher temperatures compared to vacuum freeze-drying, typically in the range of –3°C to –10°C. Integrating of heat pumps in atmospheric freeze-drying fluidized beds (Figure 9.5a) has been proposed as a new application with the possibility of using air or inert gases as drying mediums (Strommen and Kramer 1994).

As published, the system schematic, as well as the drying process shown in a Mollier diagram (Figure 9.5b), may do not efficiently allow controlling the heat pump's dehumidification rate, while providing unbalanced air flows through the evaporator and the condenser. The authors reported that all the drying air is recirculated through the heat pump and that no air (or heat) is discharged from the drying chamber. No useful data were provided about other essential parameters, such as the heat pump cooling and dehumidification capacity or, at least, the compressor input power, and the initial and final material moisture contents. Also, the authors did not provide any information about the material drying curves, preheating requirements (temperature, time), dehumidification cycle duration, control of the dryer dry- and wet-bulb temperatures, quantity of water extracted, heat pump refrigerant operating parameters (pressures, temperatures, flow rates, etc.), and energy consumptions (compressor, blower, and, if required, additional heat).

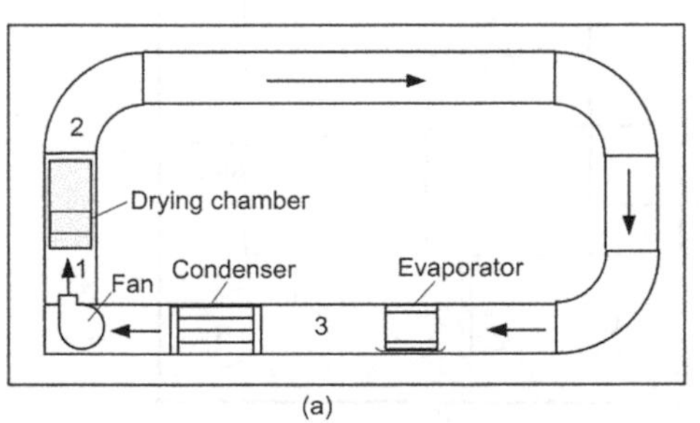

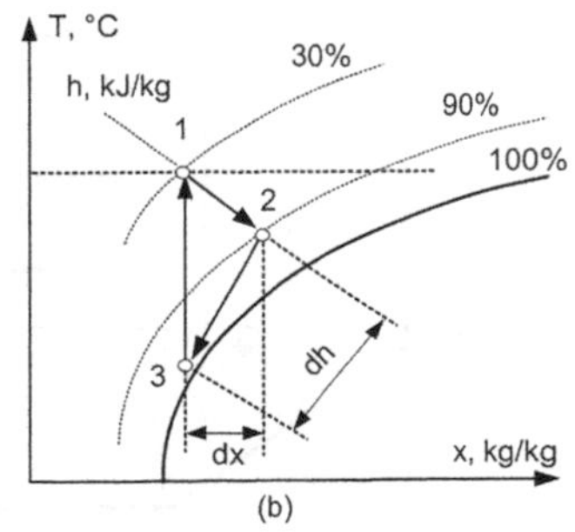

FIGURE 9.5 (a) Experimental heat pump-assisted fluidized bed dryer plan shown schematically; (b) drying process in the Mollier diagram (Redrawn and reproduced from Strommen and Kramer (1994) with permission from *Drying Technology*).

In this work, drying experiments with *air* and *inert gases* were conducted at freeze-drying temperature of –5°C. At that drying temperature, the refrigerant evaporating temperature was around –15°C, and the air moisture probably froze on the evaporator fins, a highly undesirable situation in heat pump-assisted drying technology.

In their review, Claussen et al. (2007) presented a fluidized bed heat pump drying unit with external condenser (Figure 9.6). This schematic representation did not show any means to control the heat pump's dehumidification rate according to the moisture vaporization rate of the dried material, but indicates that the excess heat is rejected outside by an external, parallel condenser. When the external condenser is operating its blower as well as the heat pump compressor run, and, thus, electrical power is consumed while no drying air is being dehumidified. Normally, the compressor excess input energy must be used to compensate part of drying chamber heat losses and air leakages. If not, the drying process becomes almost an electrical heated process, not a dehumidification process with heat pump.

Alves-Filho et al. (2004) reported that their fluidized bed heat pump-assisted dryer under atmospheric freeze-drying conditions received wet materials and discharged dried products continuously. The authors did not provide any representations of the heat pump fluidized bed system nor any information about the heat pump dehumidification capacity and/or compressor rated input power. Other missing data included the refrigerant used, drying airflow rates, and heat pump main operating parameters (pressures, temperatures). The authors did not specify either how the control of the actual temperatures of the dry- and wet-bulb drying air has been achieved, and how the drying temperatures were kept constant successively at –10°C, –5°C, and 25°C. Without that basic information, it was impossible to validate whether the heat pump dehumidification capacity balanced the material capacity to provide moisture to the

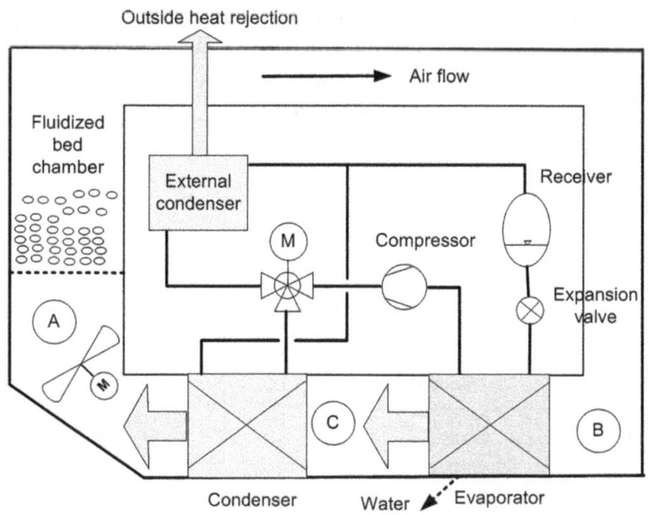

FIGURE 9.6 Schematic of a fluidized bed heat pump dryer (Redrawn and reproduced from Claussen et al. (2007) with permission from *Drying Technology*).

drying air, and if the drying heat pump was efficient or not. The initial moisture content of the material (green peas) was 73% (wet basis), and the drying time was 9 hours for freeze-drying at –5°C and –10°C. If humidity removal occurred at –5°C and –10°C, the water vapor probably frosted on the evaporator finned surface.

Eikevik et al. (2005a) studied the effect of operating conditions on atmospheric freeze-dried codfish. Their fluidized bed heat pump-assisted dryer was identical to the experimental fluidized bed dryer plant used in a previous paper. The authors did not provide the initial quantity of dried product, the heat pump dehumidification capacity, and/or the compressor rated input power. They reported that "the drying chamber outlet air was dehumidified and cooled in the evaporator, and that the air temperature was controlled by the heat pump three-way valve redirecting refrigerant flow partly into condensers 1 and 2." At this point, it may be noted that when controlling the air dry-bulb temperature in this manner, it may be difficult to achieve efficient *SMER* ratios in the drying process. The authors also reported that "the drying time at –10°C only was in the order of 300 minutes and that with an initial drying time period of 10 hours at –5°C, the SMER was reduced by 67% compared to drying at 30°C only, and the energy consumption was 7.5 times lower compared to drying at 30°C only." It can be seen that the heat pump operated during the first 10 hours with drying air entering the evaporator at –5°C, while the refrigerant evaporating temperature was probably as low as –15°C. In this case, water vapor existing in the moist drying air has probably froze on the evaporator heat transfer external surface. Other drawbacks of this study include a lack of information on the heat pump operating parameters during the drying process, above and below the freezing point.

In two separate papers, Alves-Filho et al. (2008a, b) reported results for "protein dried in a laboratory fluidized bed heat pump dryer." Neither study provided information about the heat pump (capacity, etc.) and drying cycles (quantity of dried material, moisture contents, drying schedules, etc.). The first drying stage occurred at a constant temperature (–5°C) during 2, 3, 6, or 8 hours, respectively, the second, at a constant temperature (25°C) during 2 hours. It can be understood that, between –5°C and 25°C, the heat pump was running continuously to remove moisture from the drying air. During the drying cycle at –5°C (2–8 hours), the drying air entered the heat pump evaporator at –5°C and vaporized, let us say, at –15°C. At that evaporating temperature, the water vapor from the drying air condensed and froze on the evaporator finned heat transfer surface. In that case, the process probably required defrosting cycles, but the authors did not discuss this issue. Their graphical results did not show any defrosting periods during the drying process at temperatures below the moisture freezing point. In spite of such undesirable situations, the authors reported that "the refrigeration coefficient of performance ranged from 3.66 to 4.19 (first stage) or from 4.32 to 4.82 (second stage). The heat pump coefficient of performance varied from 4.66 to 5.19 (at –5°C drying temperature) and from 5.32 to 5.82 (at 25°C drying temperature)," without providing experimental parameters validating such high performance levels, especially during drying at –5°C.

Alves-Filho and Eikevik (2008) reported a "newly built laboratory scale heat pump fluidized bed dryer." Their drying heat pump used a desuperheater to reject the excess heat outdoors, but any estimates were provided about the actual amount of heat rejected outdoors by the desuperheater (composed from the heat recovered from the dried material and the equivalent heat of the compressor input power), as well as about the additional energy consumption of the desuperheater's fan.

9.4.11 Quantity of Dried Material versus Heat Pump Size

A fundamental issue for heat pump-assisted dryers consists in the correlation between the initial quantity (and, thus, the moisture content) of the dried material and the heat pump size (i.e., its nominal dehumidification or cooling capacity). This aspect is important because any heat pump-assisted dryer must be a closed thermodynamic system in which, with appropriate control strategies, the heat pump must, practically, run continuously in order to optimize its cooling and dehumidification performance.

Unfortunately, several R&D academic works published in the past reported relatively high dehumidification performances of heat pump dryers (most of them, at the laboratory scale), while the quantities of dried products (e.g., agro-food, biological) were too small versus the sizes of heat pumps used during experimental tests. In other words, the dried materials did not contain enough moisture (water, humidity) able to run the heat pump under normal parameters. Such a situation may, also, provide to too low compressor suction pressures and, consequently, may lead to the moisture freezing on evaporator coils.

9.4.12 Information Provided versus Results Reported

Several past research studies reported good and very good drying performances of heat pump-assisted dryers, especially for agro-food and biological materials, however, without providing complete (or enough) data allowing the readers to understand and/or validate themselves the published results. Obviously, without sufficient technical data, it is difficult to improve the drying technology in general and, also, implement the new academic R&D findings in industry.

A number of features of drying processes using heat pumps are missing from many published R&D studies. The most frequent missing data is the initial (and/or final) quantity (mass or volume) of dried products. This information, when available, allows readers to validate essential results such as the drying time and the heat pump drying (dehumidification) performance.

9.4.12.1 Agro-Food Products

The drying characteristics and product quality (especially the change in color) of traditional *Chinese medicinal mushroom* dried continuously and intermittently were investigated by Chin and Law (2002). First, no information about the heat pump rated capacity versus the initial quantity of dried product was provided to determine whether the desuperheating coil was able to supply the required thermal power for heating the water and, then, the drying air prior to entering the drying chambers. Second, the quantity of dried material was 1 kg with an average initial moisture content of 203% (dry basis), but the article did not provide information about the heat pump capacity versus the material mass), as well as any heat pump energy consumption measurements in the continuous and/or intermittent drying modes. Third, the study reported that the material was dried over 30- to 40-hour periods, while two graphical representations showed that the moisture ratios rapidly decreased by up to 70% of the initial values within the first 3–4 hours for all experimental intermittencies, whether the air heater was activated or not. Fourth, the heat pump operating parameters, when operated at drying temperatures of 40.6°C and 28.4°C, respectively, were not shown prior to analyzing the final color of the dried product.

Rectangular potato and apple slices forming composite samples were dried inside a batch heat pump-assisted dryer (Rahman et al. 2007).The publication did not provide any information about the heat pump capacity (e.g., rated compressor input power or nominal dehumidification capacity) and initial quantities of dried samples. Even though initial and final moisture contents as well as drying times and temperatures were given, no heat pump performance parameters, such as $SMER_{heat\ pump}$ or $COP_{heat\ pump}$, were reported.

Codfish was dried using a single-stage ammonia heat pump at –5°C, 10°C, 25°C, 45°C, and 75°C, and 40% relative humidity (Eikevik et al. 1999). Without providing any information about the initial product mass and initial/final moisture contents, heat pump capacity, and moisture extraction rates, the frosting phenomenon probably occurred at drying temperatures below 10°C, the authors reported $COPs_{heat\ pump}$ higher than 5, and heat pump Specific Moisture Extraction Ratios (SMERs) as high as 3 and 5.5 $\frac{kg_{water}}{kWh_{heat\ pump}}$, respectively.

Strommen et al. (2003) also published results for drying codfish pieces with heat pump. The authors reported $SMERs_{heat\ pump}$ higher than $3.5-4\ \frac{kg_{water}}{kWh_{heat\ pump}}$ without indicating the heat pump capacity/size, and the initial quantity of dried material.

According to Alves-Filho and Eikevik (2007), cheese drying tests with heat pumps were successfully performed at 0°C, 4°C, 8°C, and 12°C, and a relative humidity of 50%, over a period of 5 days (120 hours). It should be noted that if the drying process were performed at temperatures as low as 0°C, 4°C, and 8°C, the moisture removed from the dried cheese probably froze, impacting the heat pump dehumidification performance dramatically. In spite of this technical drawback, experimental cheese drying curves as a function of the drying chamber inlet air temperatures were reported without specifying the heat pump dehumidification capacity nor, at least, the compressor rated input power. Also, no performance data, such as the moisture extraction rate, $COP_{heat\ pump}$, or $SMER_{heat\ pump}$, were provided.

Many academic research studies have been published on drying vegetables with heat pumps, but several of them failed to provide the data required to support all or a part of the reported results. For example, even though the initial mass of the dried material (vegetable seeds) and the initial/final moisture contents were provided, the dehumidification capacity of heat pumps used (single- and two-stage supercritical CO_2) as well as the drying time were not. On the other hand, with a dried material initial mass as small as 200 g, overall $COPs_{heat\ pump}$ and $SMERs_{heat\ pump}$ as high as 6.5 and 4.48 $\frac{kg_{water}}{kWh_{heat\ pump}}$, respectively, were obtained by (theoretically) varying the air mass flow rates and intermediate pressures. Such unlikely performances, achieved with theoretical drying concepts involving both single- or two-stage series supercritical CO_2 heat pumps and two series drying chambers, have little credibility for the academic R&D community and industrial drying industry.

Contrary to the previous study, Jinjiang and Yaosen (2007) provided a lot of useful information in their paper. The initial mass of the dried products (paddy and shredded radish) seemed well correlated with the compressor input power and initial/final moisture contents. The results reported, such as drying times, total water volumes removed and moisture extraction rates, specific electrical energy consumption, and average $SMER_{heat\ pump}\left(2\ \frac{kg_{water}}{kWh_{heat\ pump}}\right)$, help validate the reliability of the experimental work, as well as the overall performance of the heat pump-assisted drying system. These authors made a commendable effort to provide enough credible data about their experimental work to better promote the drying heat pump technology.

Drying raw, cold-extruded cranberry and potato–turnip mixtures for conversion into high-quality instant products with a CO_2 drying heat pump was investigated by Alves-Filho (2002). The author did not provide any information about the initial and final moisture contents of the dried materials, the heat pump capacity and operating parameters, and the quantity (or rate) of water removed. On the other hand, the results showed that air entered the drying chamber at –13°C and left it at about –10°C. If the inlet air relative humidity was 50%, the CO_2 evaporating temperature would have to be as low as –18°C and thus the moisture probably froze on the external surface of the evaporator. In spite of the relatively high performance reported (i.e., $COP_{heat\ pump}$ of 4 and $SMERs_{heat\ pump}$ up to 2.73 $\frac{kg_{water}}{kWh_{heat\ pump}}$), the author did not mention whether moisture freezing on the evaporator finned surface had any negative impacts.

Batch and continuous bed dryers with two parallel heat pumps for drying crops (chopped alfalfa) were modeled and experimented by Adapa et al. (2002a). The authors provided useful information about the heat pump refrigerant (HFC-134a), the heat pump (compressor) rated input power, the evaporating and condensing pressures and temperatures (0.97 and 2.03 MPa and 40°C and 68°C, respectively), initial and final dried product masses, initial and final moisture contents, drying times, maximum and minimum temperatures, and air relative humidity, as well as the mass of water extracted. However, the experimental tests provided poor $SMERs_{heat\ pump}$, which varied between 0.5 and 1.02 $\frac{kg_{water}}{kWh_{heat\ pump}}$. The authors explained this disappointing performance by the lesser drying rate of hygroscopic alfalfa compared to other nonagricultural products. Additional causes may include the fact that the heat pump evaporators did not have any external fins, and that the temperature inside the drying chamber was maintained at a constant level by exhausting part of the air from the cabinet, resulting in a loss of latent and sensible heat. Obviously, drying chopped alfalfa with heat pumps down to a moisture content as low as 10% (wet basis) seems inappropriate.

Queiroz et al. (2004) used heat pumps to dry tomatoes without providing any information about the product initial mass and heat pump capacity. Even if the refrigerant was identified (HCFC-22), the heat pump capacity, the material initial mass and initial/final moisture contents were not provided. The authors reported that "energy savings of about 40% with drying heat pump were achieved," but no information on the heat pump operating parameters, drying schedules, and times was provided. The heat pump $COPs_{heat\,pump}$ (compressor only and compressor air fan) at drying temperatures of 40°C, 45°C, and 50°C was calculated, but no information about the heat pump dehumidification performance ($SMER_{heat\,pump}$) was reported.

Fatouh et al. (2006) investigated some of the drying characteristics of different herbs (Jew's mallow, spearmint, and parsley) using a heat pump-assisted dryer with HFC-134a as a refrigerant. The authors provided information about the initial mass and moisture contents (initial and final) of the dried products, and the drying times for each test, but none about the heat pump capacity and/or energy dehumidification performance ($COP_{heat\,pump}$, $SMERs_{heat\,pump}$). The effects of dried product quantities, drying temperatures, air velocities, herb size and type, and dryer productivity and specific energy consumption were presented and discussed.

Ginger was dried in a modified atmosphere (normal air, nitrogen, carbon dioxide) with heat pump and other (freezing and vacuum) drying methods (Hawlader et al. 2006). Despite their nice conclusions concerning the increase of *6-gingerol retention and flavor retention*, the experimental methods and the results obtained with the drying heat pump seemed less relevant. First of all, only 100 g of ginger slices approximately seem to have been used in each heat pump-assisted drying process. The drying cycles lasted 8 hours, but no information on the heat pump capacity (e.g., compressor input power or evaporator dehumidification capacity), as well as the product initial and final moisture contents, the quantity of water removed and/or associated drying curve, and drying performance ($COPs_{heat\,pump}$, $SMER_{heat\,pump}$), was provided. Second, the authors did not indicate whether changes in the heat transfer process at the heat pump evaporator and condenser level were observed when normal air was replaced with nitrogen and carbon dioxide as drying mediums.

Ginger drying at 40°C, 50°C, and 60°C in a conventional tray dryer was also reported by Phoungchandang et al. (2009). This publication reported that untreated ginger (probably, 200 g) was dried with a heat pump over periods of 2 hours (at 60°C) and 3.16 hours (at 40°C), without providing any information about the heat pump size or dehumidification capacity and/or the water quantities extracted. The authors concluded that "the heat pump dryer provided shorted drying times as percent of 29.63, 12.5 and 7.69 for drying temperature of 40, 50 and 60°C, respectively, due to lower relative humidity of heat pump dryer during drying" without indicating how the drying air temperature and relative humidity were controlled and kept constant.

Contrary to other published studies, Lee and Kim (2007) provided much more useful information about the dried product (shredded radish) inlet mass and initial/final moisture contents, as well as about the drying time, moisture extraction rate, duration of the drying cycles and dehumidification performance ($SMER_{heat\,pump}$). The authors found that batch heat pump drying took 1.2–1.5 times longer than convection hot air drying (i.e., 25 vs. 15 hours). This result was attributed to the *insufficient capacity of the heat pump system used* versus the initial mass and humidity content of the dried material. In spite of this, the authors concluded that "the heat pump dryer showed considerable improvement in energy saving" because the $SMER_{heat\,pump}$ $\left(1.5\ \frac{kg_{water}}{kWh_{heat\ pump}}\right)$ was about 3 times higher than that of the hot air dryer using diesel fuel as a heat source, even if the drying time was longer with the heat pump-assisted drying method than with the conventional hot air drying method.

Apparently, very small quantities (25 ± 2 g) of green sweet pepper thin layers were dried in a heat pump-assisted dryer at hot air temperatures ranging from 30°C to 45°C, and relative humidity from 19% to 50% (Pal et al. 2008). The most important missing data was the size of the drying heat pump (compressor rated input power or evaporator dehumidification capacity). The tests reduced the initial moisture content of the material by up to 10% (dry basis) with drying air temperatures of 30°C and 45°C during 36 and 17 hours, respectively. As a result, very low $SMERs_{heat\ pump}$ ($0.55 \frac{kg_{water}}{kWh_{heat\ pump}}$ at 30°C and $0.93 \frac{kg_{water}}{kWh_{heat\ pump}}$ at 40°C) were achieved. In spite of this poor dehumidification performance, the authors recommended to "dry green sweet peppers at 35°C with a heat pump dryer" in order to "obtain acceptable product quality attributes of dehydrated products."

Olive leaves were dried in a pilot-scale heat pump-assisted conveyor dryer using HFC-407C as a refrigerant (Erbay and Icier 2009). In this publication, except for the drying temperature and cycle duration, almost all other required data, such as the quantity of dried materials, initial and final product moisture contents, heat pump size, capacity and energy consumption, and dehumidification performance ($SMERs_{heat\ pump}$), were missing. In spite of this, several less relevant aspects, such as the experimental uncertainties and exergetic efficiency analysis, were largely presented, while most of the article conclusions had little to do with its actual content.

A relatively small quantity (0.5 kg) of saffron stigma was dried with "a hybrid PVT solar dryer and a heat pump unit arranged to create flow air in a closed (or open) cycle current" (Mortezapour et al. 2012). In spite of some data which were provided about the heat pump and dryer (e.g., refrigerant used, mass and initial/final moisture contents of the dried material, fresh air maximum relative humidity), the work raised serious questions about its technical relevance. First, the ambient air entered the photovoltaic (PV) solar collector at temperatures between 25°C and 38°C from 9 AM to 3 PM, but the outlet temperatures were not indicated. After the PV solar collector, the air was mixed *with the dryer air to drop its temperature and relative humidity* prior to entering the heat pump condenser. After the condenser, an auxiliary electrical heater reheated the air up to the desired set drying temperature. A logical question could be why does the system have to cool down the air coming from the solar PV collectors (by mixing it with fresh air) and, then, heat it again with the heat pump condenser and/or electrical heater. Second, according to the authors, the heat pump dryer input power was entirely provided by the PV collectors (~120 W) during only one cycle per day (i.e., about 30 minutes). However, the drying system was intended to *dry the daily harvest from a farm in the area where the experiments were conducted* by using power from the grid for the additional drying cycle. Unfortunately, the authors did not indicate how many additional cycles were required each day. And, if this number was relatively high, how could they justify the economic efficiency of the proposed hybrid drying system, which provided a maximum $SMER_{heat\ pump}$ of $1.16 \frac{kg_{water}}{kWh_{heat\ pump}}$ at 60°C during only one cycle per day?

A study on the intermittent heat pump-assisted drying of salak fruits was published by Ong et al. (2012). The dehydration process was divided into three distinct phases, that is, the initial (5 hours at 37°C), intermediate (intermittent ratio of 0.75 during 120 minutes), and final stages, but the schematic diagram of the experimental setup was not provided. The authors indicated "the mass of salak slices (100 g) arranged in a thin layer on a drying tray without staking," but, unfortunately, not the initial mass of the dried product nor the initial/final moisture contents and heat pump dehumidification capacity have been reported. The authors reported that "the heat pump-assisted drying process at 37°C took the shortest time to achieve a moisture content of 0.63 gram of water per gram of dry solid compared to conventional hot air drying at 50°C."

The chlorophyll retention of ivy gourd leaves by using a heat pump-assisted dehumidification air dryer was theoretically studied (Potisate and Phoungchandang 2010). None of the operating parameters of the drying heat pump were provided to help academia and industry validate the relevance of the performance claimed, and a small quantity of material (35 g) was dried using a heat pump equipped with a 250 W compressor during 1–1.3 hours. There probably was an unbalanced ratio between the initial quantity of dried material and the heat pump dehumidification capacity, which led to extremely low $COPs_{heat\,pump}$ (≈1.2).

About 800 g of grapes were dried with a "pilot-scale plant, comprising a closed-circuit, hot air convection chamber with heat pump" (Velazquez et al. 1997). Even though the heat pump refrigerant was identified (HFC-134a), the heat pump size or capacity was not. For all dried grape samples, moisture content–time curves were determined at specified dry- and wet-bulb temperatures. This information, generally missing in many other studies, helped understand that drying at different relative humidity levels means drying at different dry- and wet-bulb temperature combinations. The influence of grape variety and relative humidity of the drying air on the product moisture content (expressed in $\frac{kg_{water}}{kg_{oven\ dry}}$) as well the effects of the grape pretreatment temperature, an essential step in the drying process, were studied and results were provided. This study also provided experimental conditions (i.e., severe pretreatment combined with drying at high temperatures and low relative humidity) making it possible to reduce the drying time from 40 (normally required for sun drying) to 24 hours. Finally, although the heat pump dehumidification performance ($SMER_{heat\,pump}$, etc.) was not reported, it could be estimated that this paper was one of the most relevant and complete to have been published, providing a good understanding of the research method and the results reported.

A pilot-scale heat pump dryer for measuring and modeling the moisture desorption isotherms of peas was used by Rahman et al. (1998). Data on the initial mass of the dried material as well as the air drying temperature and relative humidity ranges under study were provided, but the material initial and final moisture contents and the heat pump capacity and drying times were not provided. Finally, the system start-up operation was not clearly explained, and, consequently, the reader may be led to understand that the heat pump reached the operating parameters without any product inside the drying chamber.

Green peas were also dried "in a heat pump-assisted dryer under atmospheric freeze-drying conditions (−6°C, −3°C and 0°C) and at 25°C, operating either at constant or combination of temperatures" (Alves-Filho et al. 2004). First, this paper did not provide any information about the refrigerant used and the heat pump operating parameters (pressures, temperatures) nor the initial quantities and final moisture content of the dried materials, the moisture extraction rates, and the heat pump energy ($COPs_{heat\ pump}$) or dehumidification performance ($SMERs_{heat\ pump}$). Second, the authors reported that "the drying air passed through the bed of particles and removed moisture from the wet material. Then, the moistened air flowed through the filter and air cooler (evaporator) where it was cooled below its dew point leading to condensation of water vapour. The evaporator absorbed latent heat of condensation of moisture to boil the fluid inside the tubes." These statements are generally right but, when drying occurs at temperatures near or below 0°C, the refrigerant evaporating temperatures may be much lower than the air dew points corresponding to air temperatures at normal atmospheric pressure. Consequently, in such circumstances, the moisture removed from the dried material may freeze on the surface of the evaporator fins, and, therefore, complex and costly defrosting procedures have to be initiated. The authors did not mention whether this phenomenon occurred during the 9 hour freeze-drying process and how they achieved the evaporator defrosting cycles, if any.

The effect of multistage heat pump fluidized bed atmospheric freeze-drying and microwave vacuum drying on the drying kinetics, moisture diffusivities, microstructure and physical parameters of green peas was evaluated by Zielinska et al. (2012a, b). The initial moisture content of green peas was about $3.23 \mp 0.02 \frac{kg_{water}}{kg_{oven\ dry}}$), and the material was dried under different drying conditions to an equilibrium moisture content of $0.09 \mp 0.04 \frac{kg_{water}}{kg_{oven\ dry}}$. According to the authors, the samples were placed in a freezer at –20°C before the tests were performed with a heat pump at –5°C, which is below the freezing point of the frozen green peas, and the inlet air relative humidity was kept at both a low (20%) level and a high (55%) level. However, the authors did not explain how they kept these parameters constant during each drying cycle. No information was provided on the initial quantity of dried products and the heat pump dehumidification capacity nor on the heat pump performance results. These authors reported that with the combined method (i.e., fluidized bed followed by heat pump drying), the initial drying rate was about 0.04 L/min, but the use of microwave vacuum drying increased the drying rate to 0.08 L/min. The drying rates of green peas dried by microwave vacuum drying and hot air convective drying were 0.59 and 0.20 L/min, respectively. Even though many input data were missing, the authors reported results on product drying kinetics, final morphological parameters, texture properties, and color. It would be impossible to validate such complex results without any fundamental experimental data about the combined drying systems studied.

Ceylan et al. (2007) used a heat pump-assisted dryer for drying an unknown quantity of tropical fruit (kiwi, avocado, and banana) at 40°C during 6 hours. Even though the initial and final product moisture contents (expressed as $\frac{kg_{water}}{kg_{oven\ dry}}$) were provided, the heat pump size and dehumidification performance ($COP_{heat\ pump}$, $SMER_{heat\ pump}$) as well as the moisture extraction rate or quantity of water removed were not provided.

According to Sunthonvit et al. (2007), the best systems for preserving the volatile compounds of nectarines are heat pump-assisted dryers. In this publication, the drying temperatures and the material final moisture content were provided, but not its initial mass and moisture content. Moreover, the heat pump size, moisture extraction rate, and drying performance ($COP_{heat\ pump}$, $SMER_{heat\ pump}$) were not provided. As a consequence, the reader may not understand very well nor validate the author's conclusion stating that the heat pump dryer "followed by cabinet and tunnel dryer was the best system for preservation of volatile compounds" of tropical fruits.

A small quantity of single-layer whole figs (400 g) was dried by Xanthopoulos et al. (2007) with a 2 kW (compressor rated input power) heat pump using HCFC-22 as a refrigerant. The initial and final moisture contents (wet basis) of the dried product were provided as well as the drying times. Missing information included the drying schedules used (with or without preheating steps), the water removal time profiles, and the heat pump energy consumption and dehumidification performance ($COP_{heat\ pump}$, $SMER_{heat\ pump}$).

Mature kaffir lime leaves were dried at 40°C, 50°C, and 60°C with a heat pump dehumidified dryer in order to determine their desorption isotherms (Phoungchandang et al. 2008). According to the data provided by the authors, tests were conducted on 30 g of product during periods of 1–8 hours, without indicating the initial and final moisture contents, the heat pump size or dehumidification capacity, and its energy performance ($COP_{heat\ pump}$, $SMER_{heat\ pump}$).

Plum slices were dried at 45°C and 50°C in a heat pump-assisted conveyor dryer, but no experimental results were provided (Hepbasli et al. 2010). Even though the initial and final product moisture contents were provided, the initial mass of the dried product, the heat pump size or capacity, operating parameters, and the drying times and heat pump dehumidification performance ($SMER_{heat\ pump}$) were not provided. In addition, the authors did not explain how the air drying temperatures and relative humidity were kept constant. In spite of such essential missing results, the authors performed a superfluous exegetic analysis of the heat pump cycle without any apparent link with the heat pump-assisted system used to dry plums in a particular hot and dry climate.

9.4.12.2 Biological Materials

Without indicating the product initial mass and the heat pump rated dehumidification capacity, Strømmen et al. (2007) reported some results on biological material (e.g., blood, sick tissues, and rat liver) drying processes at temperatures between −10°C and 20°C. Some experimental results, such as the drying curve and dehumidification performance for rat liver drying as a function of air inlet temperature, were reported. However, the dehumidification performance achieved at drying temperatures as low as −5°C to 10°C during relatively long periods of time did not seem relevant because the moisture freezing process provided low $SMERs_{heat\ pump}$ $\left(0.52\text{–}0.87\ \frac{kg_{water}}{kWh_{heat\ pump}}\right)$.

Alves-Filho et al. (2006) applied an atmospheric sublimation and evaporative concept to dry bovine intestine with a fluidized bed heat pump at temperatures as low as −10°C and above the material freezing point (up to 25°C). In this publication, no information was provided about the material initial/final mass and moisture contents, and the heat pump dehumidification capacity. Moreover, the authors did not describe the heat pump drying system, and no drying schedule was provided. Tests were conducted at constant temperatures of –10°C, –5°C, and 5°C with drying times of 21, 20, and 5 hours, respectively; no moisture freezing phenomena occurring at drying temperatures below 0°C were mentioned.

Similar unlikely results were reported by Senadeeva et al. (2012) in a study on bovine intestine drying using an atmospheric two-stage fluidized bed heat pump-assisted drying system. In this publication, the heat pump dehumidification capacity, the refrigerant used, and the initial mass of the dried products were not specified. In spite of this missing data, the authors reported moisture removal with a single-stage heat pump operating at –10°C and –5°C drying temperatures for more than 20 hours, without indicating any possible occurrence of moisture freezing and/or low heat pump efficiency ($COP_{heat\ pump}$, $SMER_{heat\ pump}$) at such low drying temperatures. The author's main conclusion stating that "two-stage fluid bed heat pump drying of bovine intestines is an efficient … technology that has the potential to improve moisture removal, keeping improved product quality at reduced costs," seems unlikely.

The dehydration of lactic acid bacteria for starter cultures within a heat pump dryer was studied as an alternative to freeze-drying without providing any information about the custom-made heat pump (operating parameters, dehumidification and/or energy performance, etc.) (Cardona et al. 2002). Even though no schematic diagram of the integrated system was provided, the evaporator air bypass process was described. The material was dried at seven temperatures varying between 10°C and 40°C, and the moisture content initial value and variations were provided when the activity of the dried material was measured. The authors concluded that *Lactococcus lactis* can be successfully dried at 20°C–25°C using a heat pump dryer to a similar level of activity and viability as that obtained with a freeze dryer, provided adequate dehydro-protectants are added. Also, heat pump drying is more economical than freeze-drying, both in terms of capital and running costs. The lower operating costs were attributed by the authors to reusing the evaporation latent heat, which requires about 20% of the freeze dryer energy.

9.4.12.3 Unconventional Products

Strømmen et al. (2004) reported a number of results on sulfate and sulfite cellulose dried at temperatures varying between –15°C and 20°C with a two-stage ammonia drying heat pump. Without providing any minimum required data, such as the product initial mass, its initial and final moisture contents, the drying system operating parameters, the drying cycle duration, the volume of water removed, and the heat pump capacity, these authors reported poor dehumidification performance ($SMERs_{heat\ pump}$). According to them, the first stage of the ammonia heat pump-assisted dryer operated with an air inlet drying temperature and a relative humidity of –15°C and 40%, respectively, and the heat pump evaporating temperature and outlet relative humidity were –2°C and 80%, respectively. The authors reported that system energy consumption at –15°C (during 7 hours) was 14.5 times greater than drying at 20°C, with $SMER_{heat\ pump}$ values of 0.28 and 4.05 $\frac{kg_{water}}{kWh_{heat\ pump}}$, respectively. The authors did not explain the poor performance achieved and how the air drying inlet temperature was kept constant at –15°C during 7.5 hours without moisture freezing and defrosting cycles (if any). In spite of these simple observations, the authors surprisingly concluded that "a considerable improvement in tensile index and water retention value was achieved at –15°C and 20°C compared to conventional industrial dried" cellulose.

An experimental study on sludge dehumidification with heat pumps was published by Guang et al. (2007). The authors provided a lot of useful information, such as the refrigerant used (HFC-236fa), the compressor, electrical heater and fans rated input powers, the initial weight of the dried material, the initial and final product moisture contents, the drying cycle duration, and the evaporator air inlet and outlet temperatures. This paper was one of the rare documents addressing the importance of the preheating stage in heat pump drying technology. A study was also conducted to determine optimum heat pump usage in relation with the sludge actual moisture content. From this publication, it can be seen that the drying heat pump has to be used only when the moisture content is between 46% and 19% approximately. With moisture contents below 19%, the drying heat pump cannot provide acceptable energy performance, and, therefore, it must be switched off, and conventional heat sources such as electrical energy or fossil fuels have to be used for further drying.

9.5 COUPLING CONVENTIONAL WOOD DRYERS WITH HEAT PUMPS

Most of the conventional air convective batch lumber dryers can be coupled with compact- or split-type heat pumps allowing partial recirculation of drying air in order to recover sensible and latent heat, thus achieving high thermal efficiencies.

The integration of conventional batch lumber dryers with electrically (or gas)-driven air-to-air (or other types) heat pumps is a challenging task.

As in conventional kilns (see Chapter 8), the lumber stack(s) are placed inside the batch heat pump-assisted dryer on trays and removed once the desired wood moisture content is reached (Perera and Rahman 1997; Kiang and Jon 2007). The dry and hot drying air flows through the lumber stack(s). In the case of occasional overheating of the drying chamber, as well as when the dryer's fans shut down for few minutes during periodical change of their sense of rotation, some inlet and outlet air vents automatically open. The last operation aims at avoiding the inward collapse (or implosion) of the drying enclosure due to the sudden drop of the interior air pressure if the air vents remain closed. By neglecting such inevitable, but relatively small heat losses, heat pump-assisted dryers can be practically considered as closed systems Minea (2014a).

9.5.1 COMPACT-TYPE HEAT PUMPS

In the case of compact heat pumps, all components (compressor, evaporator, condenser, expansion valve, refrigerant piping, fans, and controls) are installed outside the drying chamber in a mechanical room (Figures 9.7 and 9.8). The hot and humid air from the drying enclosure enters the evaporator EV and the condenser CD. A mixing process using motorized air dampers and constant (or variable) speed blower (B) is achieved prior reheating the air through the condenser CD. Heating coils, installed inside the drying chamber, supply heat for the warming-up (preheating) of lumber stack(s) and, when required, supplement the heat that the drying air receives

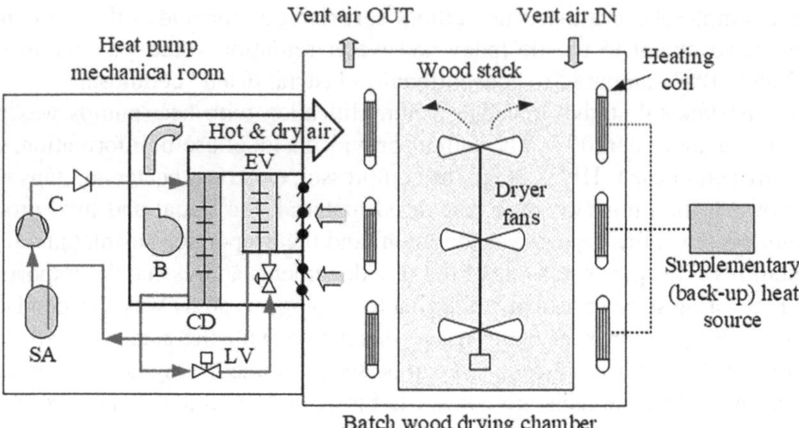

FIGURE 9.7 Schematic plan representation of a heat pump-assisted dryer with compact-type heat pump (note: schematic not to scale); B, blower; C, fixed or variable speed compressor; CD, condenser; EV, evaporator; EX, expansion valve; LV, liquid valve; SA, suction accumulator.

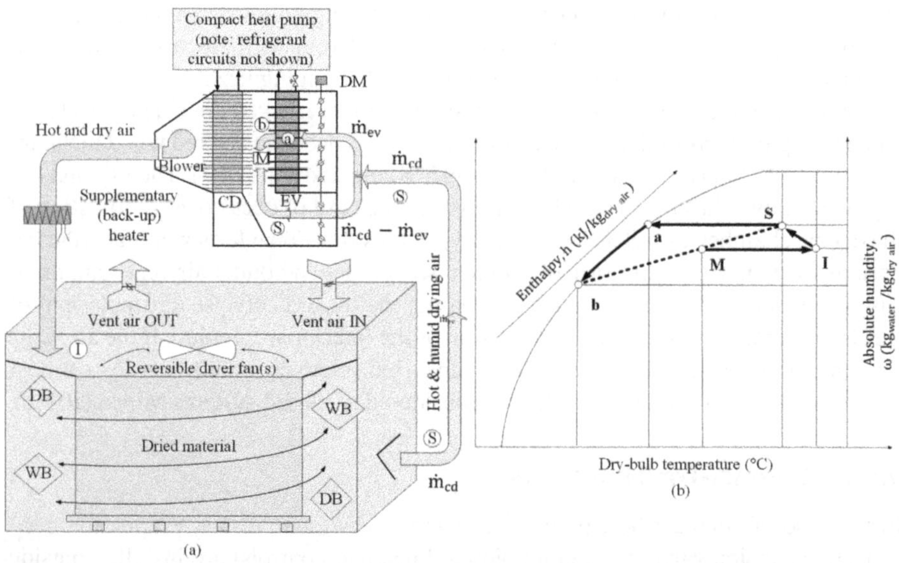

FIGURE 9.8 (a) Schematic front representation of a heat pump-assisted dryer with compact-type heat pump (note: not to scale); (b) air thermodynamic cycle in the Mollier diagram. CD, condenser; DB, dry-bulb temperature sensor; EV, evaporator; HEX, heat exchanger; $\dot{m}$, air mass flow rate; WB, wet-bulb temperature sensor (for the rest of the legend, see Figure 9.7) (notes: schematic not to scale; positions of air drying inlets and outlets, in and from dryer, do not consider air reversal operations).

from the heat pump's condenser. When the dryer dry-bulb temperature drops under its setting point, heating coils raise the air temperature before it is returned to the wood stack(s) being dried. The supplementary (backup) heat may be provided by natural gas, bark or oil-fired boilers, or by electrical coils (elements).

Multiblade fans circulate the air through the wood stack inside the drying chamber. The fans' rotation direction periodically changes, for example, every 3 hours at the beginning, and every 2 hours at the end of the drying cycles. As noted, the air vents open when the dryer fan changes rotation direction in order to avoid air implosion hazards, and, also, when the dryer dry-bulb temperature exceeds its setting point, in order to prevent excessive superheating.

Figure 9.8a shows, in a different manner, the principle of a lumber dryer coupled with a compact-type air-to-air heat pump, and Figure 9.8b represents in the Mollier diagram the thermodynamic process of the drying air through the integrated lumber dryer and compact-type heat pump drying system. The thermodynamic process of the drying air is explained in Section 9.6.1.

When a variable speed compressor is chosen, the evaporator bypass may be eliminated, and the heat pump dehumidification capacity matched with the material thermal dewatering (moisture vaporization) capacity achieved by varying the speed of heat pump's compressor and/or blower.

9.5.2 Split-Type Heat Pumps

In the case of split heat pumps (Figures 9.9–9.11), there are also two chambers thermally isolated from one another: (i) a drying enclosure (chamber) where the

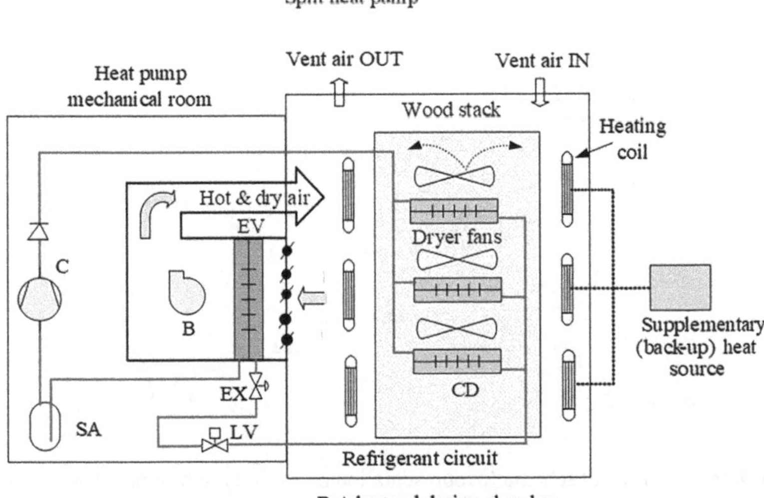

FIGURE 9.9 Schematic plan representation of a heat pump-assisted wood dryer with split-type heat pump (note: schematic not to scale). EX, expansion valve (for the rest of the legend, see Figure 9.7).

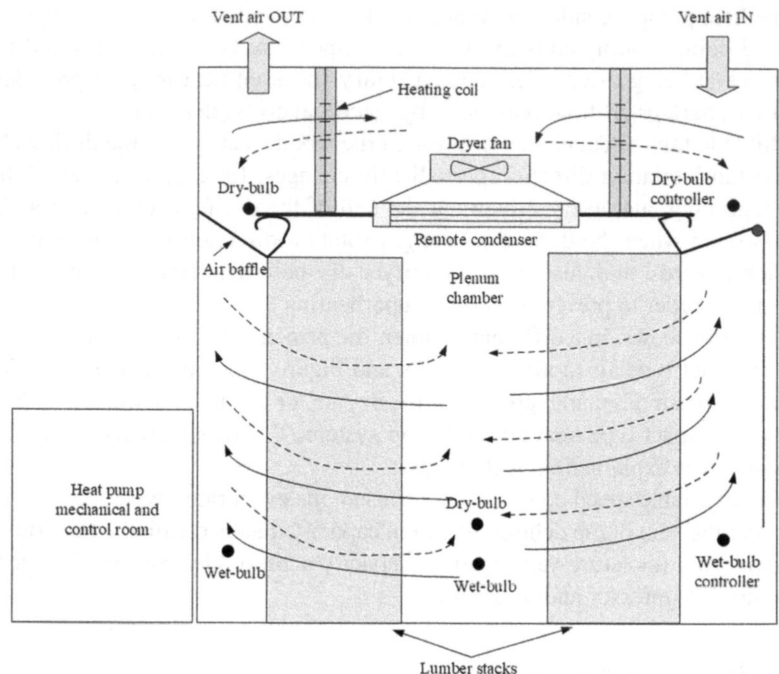

FIGURE 9.10 Schematic front representation of a heat pump-assisted dryer with remote condenser. Adapted from Lewis (1981, 2003). Redrawn and reprinted with permission from Nyle Systems LLC.

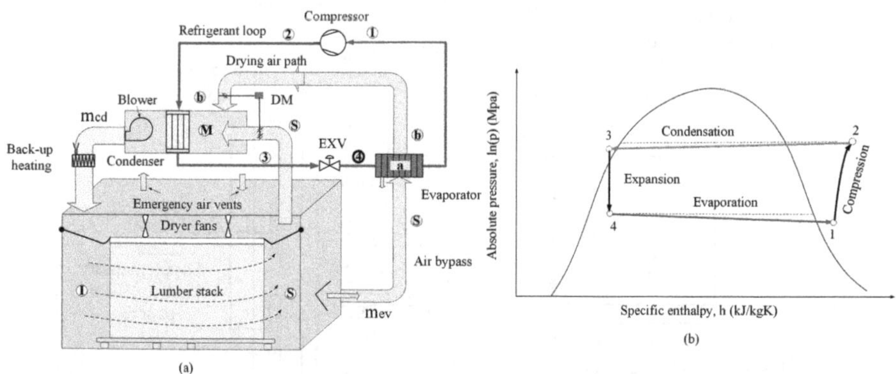

FIGURE 9.11 Additional schematic representation of a split-type heat pump-assisted dryer; (a) drying air and refrigerant loops; (b) thermodynamic cycle of refrigerant represented in ln(p)–h diagram. EXV, expansion valve; HEX, heat exchanger; DM, damper motor (for the rest of the legend, see Figure 9.7; for the thermodynamic process of drying air, see Figure 9.8b) (notes: schematic not to scale; positions of air drying inlets and outlets, in and from dryer, do not consider air reversal operations).

heat pump remote condenser CD is installed; and (ii) a mechanical room where all temperature-sensitive components (compressor, evaporator, variable speed blower, and expansion valve) and air ducts are installed. The airflow through the evaporator EV is controlled by using a variable speed blower (B). If a variable speed compressor is chosen, the heat pump dehumidification capacity could be matched with the material thermal dewatering (moisture vaporization) rate by combined variation of the compressor and blower speeds, and/or dampers.

Figure 9.11b presents the thermodynamic cycle of the refrigerant in ln(p)–h diagram, explained in Section 9.6.1.

9.5.3 Multiple Split-Type Heat Pumps

Heat pump-assisted wood dryers can be coupled with multiple split-type heat pumps (Figure 9.12). Such a configuration allows matching in a different manner (i.e., by varying the number of heat pumps running at the same time) the cooling and dehumidifying capacity of heat pumps with the thermal dewatering (moisture vaporization) rate of the lumber stacks during successive steps of drying cycles.

9.5.4 Parallel Heat Pumps

In the case of sawmills that use parallel drying chambers operating simultaneously at different drying temperatures and humidity conditions, a useful concept may include

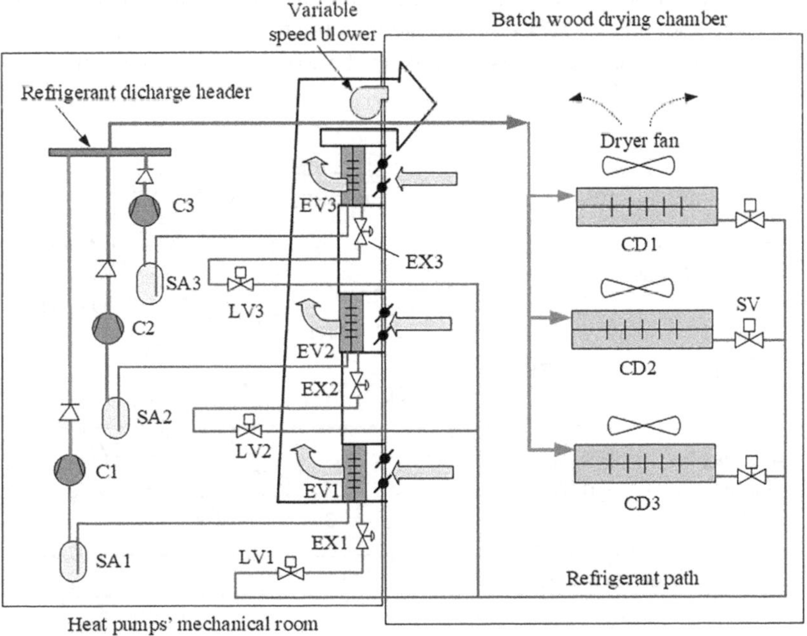

FIGURE 9.12 Schematic of a heat pump-assisted dryer with multiple split-type heat pumps (note: not to scale). SV, back pressure solenoid valve (for the rest of the legend, see Figure 9.7).

two air streams coming from two separate drying chambers, each passing through two separate evaporators of the same heat pump (Figure 9.13). In such a concept, the subcooled refrigerant liquid is split into two streams at the exit of the heat pump condenser. One stream enters the expansion valve EX1 of the high-temperature evaporator EV1, and the second, the expansion valve EX2 of the low-temperature evaporator EV2. A two-temperature valve in the suction line keeps the low-side pressure of the refrigerant in evaporator EV2 at a higher pressure than in the evaporator EV1. A check valve, located in the suction line coming from the colder evaporator EV1, prevents the warmer, higher pressure low-side vapor from entering the colder evaporator EV1 during the off cycles. The vaporized refrigerant is returned to the compressor where it is compressed and becomes high-pressure, high-temperature superheated vapor (Minea 2010b).

9.6 WORKING PRINCIPLE OF HEAT PUMP-ASSISTED DRYERS

Heat pump-assisted wood dryers use the cold side of the heat pump (i.e., the evaporator, having its surface temperature below the dew point of the ambient drying air) in order to condense the moisture removed from the wood stack(s). In other words, instead of venting outside a part of the humid air, as in conventional kilns, moisture is condensed, and, thus, the recovered sensible and latent heats are returned to the drying process. In this way, industrial heat pump-assisted dryers can provide efficient and cost-effective drying of lumber over conventional heat-and-vent kilns, particularly where quality is a key issue (Bannister et al. 1999).

Based on such a simple principle, heat pump-assisted kilns are capable of drying most wood species at low (35°C–50°C), medium (up to 75°C–85°C), and high (practically, up to 110°C) temperatures, and lower the wood's average moisture contents up to as low as 5%–6%.

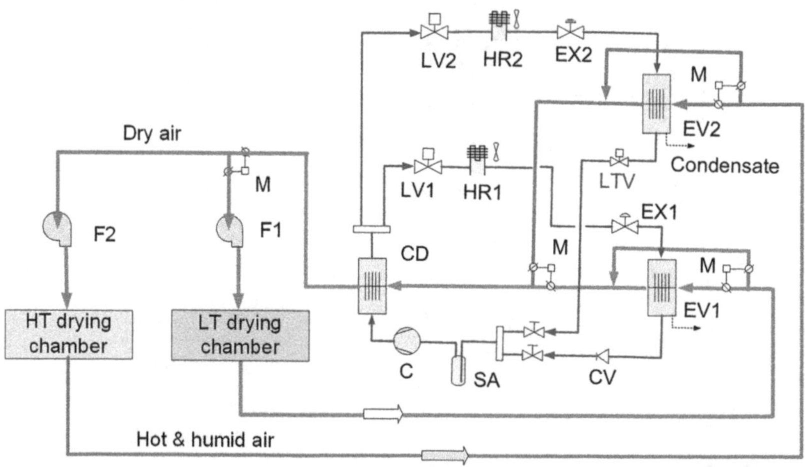

FIGURE 9.13 Schematic representation of a heat pump-assisted wood drying system with two parallel drying chambers. CV, check valve; F, fan (for the rest of the legend, see Figure 9.7).

9.6.1 Drying Air Thermodynamic Process

In both heat pump-assisted batch dryers coupled with compact (Figures 9.7 and 9.8) or split-type (Figures 9.9–9.11) heat pumps, a part (i.e., a variable mass flow rate $\dot{m}_{ev} < \dot{m}_{cd}$) of the hot and humid air leaving the dryer at state S and mass flow rate $\dot{m}_{cd}$ passes through the heat pump evaporator (see Figure 9.8a) where it is first cooled (process S-a) and, then, dehumidified (releasing about 2,300 $\frac{kJ}{kg_{water}}$) when condensation of moisture occurs as the air temperature goes below its dew point (process a-b) (see Figure 9.8b). The sensible heat as well as enthalpy (latent heat) of condensation removed from the drying air are transferred to the refrigerant, causing it to vaporize and superheat, while the condensed moisture is drained out from the system.

9.6.2 Refrigerant Thermodynamic Process

The cooled and dehumidified air at state b is mixed with the rest of the drying airflow coming from the drying chamber at state S. The mixed air at state M is further heated by the heat pump's condenser up to state I. Inside the drying chamber, the moisture vaporized from the lumber stack(s) is removed through an isenthalpic process I-S.

Within the heat pump, the working fluid (refrigerant) at low pressure and temperature is vaporized and superheated in the evaporator (moisture condenser) (process 4-1) by heat drawn from the drying air with variable mass flow rate $\dot{m}_{ev}$ (Figure 9.11a and b). In the evaporator, the refrigerant at state 4 recovers sensible and latent heat from the warm moist air that is cooled and, then, dehumidified. The superheated refrigerant at state 1 enters the compressor which quasi-adiabatically raises the enthalpy and pressure of the refrigerant (process 1-2) and discharges it as superheated vapor (state 2) at temperatures as high as 120°C. The superheated refrigerant vapor then enters the condenser where it condenses (process 2-3). Through this process, the sensible and latent heat recovered from the drying air at the evaporator, plus the equivalent heat corresponding to the compressor electrical energy input. In other words, heat is removed from the condensing refrigerant and transferred to the drying air. After passing through the condenser (process 2-3), the refrigerant subcooled liquid (at state 3) is throttled to a low pressure through the expansion valve EX (isenthalpic process 3-4), prior entering the evaporator to complete the cycle. In this way, heat is recycled to maintain the drying conditions in the kiln, whereas in conventional kiln-drying systems a considerable part of heat is exhausted to the atmosphere through the air venting system.

9.7 ENERGY REQUIREMENT

A part of energy requirement for heat pump-assisted dryers, mainly needed to overcome the wood's moisture retention forces, and heating and evaporating the water extracted from the wood, comes from the sensible and latent heat recovered by the heat pump from the wood stack(s). The rest of the required energy is provided by the heat pump compressor and blower equivalent electrical energy consumption, as well as by the supplementary (backup) (electrical or fossil) heating coil(s).

Figure 9.14 shows that the sensible and latent heat provided by cooling and condensing moisture from the drying air passing through the heat pump's evaporator must be added the equivalent electrical energy consumed by the heat pump compressor and blower, and dryer's fans. In addition, for preheating the wood stack, stickers and timber sleepers, the dryer envelope (structure, insulation, slab, foundation) and other auxiliary equipment, and for preventing the heat losses through walls and floor of the dryer, supplementary (backup) heating energy is required for most of heat pump-assisted wood dryers.

Heat pump-assisted batch kilns may use significantly less energy (usually, up to 50%), compared to two conventional heat vent kilns (Carrington et al. 1998; Minea 2004) and, also, may improve the final quality of dried lumber. Such benefits can be partially offset in some countries by relatively higher costs of electricity compared with the cost of fossil energy (Carrington et al. 2003).

9.7.1 Energy Balance

To correctly design the complex coupled system dryer–heat pump, it is mandatory to calculate the system's energy requirements that, globally, are similar to those of conventional kilns (see Section 8.14.5).

The total energy required (excluding the heat consumption for the kiln preheating) for wood drying with air-to-air heat pumps may be expressed as follows (Minea 2007) (Figure 9.15):

$$Q_{tot} = (\chi + 1)\sum\left(E_{fan} + Q_{backup}\right) + E_{heat\ pump} - Q_{rec} \tag{9.1}$$

where

Q_{tot} is the total energy required by the heat pump-assisted dryer (kWh)

χ is the heat losses' parameter (usually, between 0.1 and 0.3) accounting for uncontrollable losses of poorly insulated and/or leaky dryers (–)

$E_{heat\,pump}$ is the total electrical energy consumption of heat pump compressor(s), blower(s), and controls (kWh)

E_{fan} is the dryer's fan(s) electrical energy consumption (kWh)

$Q_{backup} = Q_{loss\text{-}venting} + Q_{loss\text{-}others}$ is the backup heat consumption to compensate the venting ($Q_{loss\text{-}venting}$) and other heat losses (walls, doors, floor, air leakages, etc.) ($Q_{loss\text{-}others}$) (kWh)

Q_{rec} is the sensible (Q_{sens}) and latent heat (Q_{lat}) recovered by the heat pump from the dried lumber stack(s) (kWh)

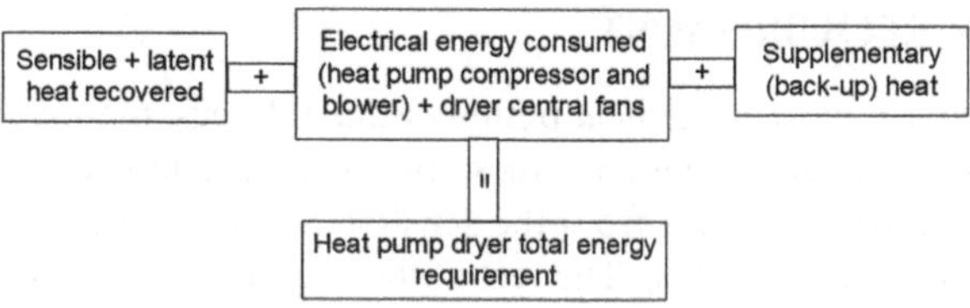

FIGURE 9.14 Main components of energy required by heat pump-assisted wood dryers.

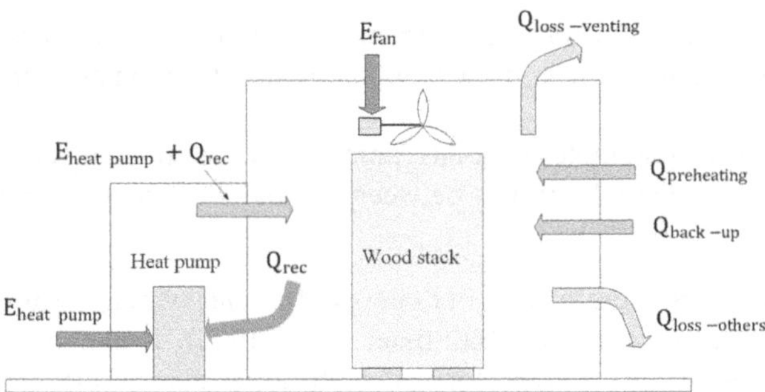

FIGURE 9.15 Schematic of energy balance of a typical heat pump-assisted wood dryer. E_{fan}, dryer fan(s) electrical energy consumption; $E_{heat\ pump}$, heat pump electrical energy consumption (compressor + blower + controls); Q_{backup}, backup heat consumption; $Q_{loss\text{-}others}$, other heat losses (walls, doors, floor, air leakages, etc.); $Q_{preheating}$, preheating (warming-up) heat consumption; $Q_{loss\text{-}venting}$, venting heat losses; Q_{rec}, sensible and latent heat recovered.

9.8 PERFORMANCE OF HEAT PUMP-ASSISTED DRYERS

9.8.1 Heat Pump Only

Energy performances of heat pumps used as cooling, dehumidifying, and reheating devices in lumber drying processes can be characterized by their coefficients of performance $\left(COP_{heat\ pump}^{heating}\right)$ defined as the total heat supplied to the dryer by the heat pump's condenser (Q_{cond}) (kWh) divided by the compressor and blower electrical energy consumption ($E_{compr+blower}$) (kWh) during a given drying cycle:

$$COP_{heat\ pump}^{heating} = \frac{Q_{cond}}{E_{compr+blower}} = \frac{Q_{evap} + E_{compr+blower}}{E_{compr+blower}} = 1 + \frac{Q_{evap}}{E_{compr+blower}} \tag{9.2}$$

where

Q_{evap} is the total heat recovered by the heat pump's evaporator (kWh)

Generally, to be energetically and economically acceptable, the average $COP_{heat\ pump}^{heating}$ in industrial drying processes must be equal or higher than 4, preferably, nearer 10 when the temperature lifts are relatively low.

The performances of heat pumps used as dehumidification devices in drying operations can be defined as follows:

$$SMER_{heat\ pump} = \frac{m_{water}}{E_{compr+blower}^{heat\ pump}} \tag{9.3}$$

where

$SMER_{heat\,pump}$ is the specific moisture extraction rate $\left(\frac{kg_{water}}{kWh_{heat\ pump}}\right)$

m_{water} is the total mass of moisture (water) extracted by the heat pump from the dried material (kg_{water})

$E^{heat\ pump}_{compr+blower}$ is the total energy consumed by heat pump's compressor, blower, and controls during the entire heat pump-assisted drying cycle ($kWh_{heat\ pump}$)

Such a performance parameter depends, among other parameters, on the heat pump running time, and generally ranges between 1.5 and 4, with average values around $2.5\ \frac{kg_{water}}{kWh}$.

Heat pump dryers operate at higher energy efficiency when the amount of water removed increases (Table 9.1) (Bannister et al. 1999). It can be observed that the heat pump-assisted dryer performed with better energy efficiency (measure by *SMER*) when more water is removed, which can result in shorter payback periods.

9.8.2 Coupled Heat Pump-Assisted Dryer

The dehumidification performance of a coupled heat pump–dryer can be similarly described by the dryer's specific moisture extraction ratio ($SMER_{dryer}$) defined as the ratio between the amount of moisture (water) extracted from the dried lumber and the total energy (electrical and fossil) consumed by the dryer, including the dryer (fans, supplementary heat, controls) and the heat pump (compressor, blower, controls) energy consumptions:

$$SMER_{dryer} = \frac{m_{water}}{E^{dryer}_{input,\ total}} \tag{9.4}$$

where

$SMER_{dryer}$ is the dryer's specific moisture extraction ratio $\left(\frac{kg_{water}}{kWh_{dryer}}\right)$

m_{water} is the total mass of moisture (water) extracted by the heat pump from the dried material (kg_{water})

$E^{dryer}_{input,\ total}$ is the total energy consumed by the dryer (fans, warming-up and backup heat, controls) and the heat pump (compressor, blower, controls) (kWh_{dryer})

TABLE 9.1
Energy Performance of a Heat Pump Dryer for Timber Drying versus the Mass of Water Condensed

SMER $\left(\frac{kg_{water}}{kWh_{heat\ pump}}\right)$	Water Removed (kg)
2.2	1,300
2.6	1,750
2.9	5,200
3.2	7,200

This dehumidification performance indicator should be compared with $SMER_{dryer}$ of conventional air convective dryers that, generally, ranges between 0.12 and 1.28 $\frac{kg_{water}}{kWh_{dryer}}$ (Table 9.2).

Specific energy consumption of heat pump only $\left(SEC_{heat\ pump}, \frac{kWh_{heat\ pump}}{kg_{water}}\right)$ and of heat pump-assisted dryers $\left(SEC_{dryer}, \frac{kWh_{dryer}}{kg_{water}}\right)$ is defined as reciprocals of $SMER_{heat\ pump}$ and $SMER_{dryer}$, respectively.

9.9 SIMPLE PAYBACK PERIOD

The total cost of heat pump-assisted dryers generally consists of the following: (i) total fixed costs (*TFC*) unrelated to the amount of moisture removed from the dried product including interests and the maximum demand charges for the electricity power supply; and (ii) total variable costs (*TVC*) that include the energy used and, possibly, the dryer maintenance expenses, depending on the energy cost and the effectiveness of the dryer; the variable costs progressively increase as the dryer operates.

The total specific cost for removing 1 kg of moisture during a given drying cycle can be expressed as follows (Kiang and Jon 2007):

$$Cost_{1\ kg} = \frac{TFC + TVC}{\text{Total moisture removed}} \tag{9.5}$$

where

$Cost_{1kg}$ is the total specific cost for removing 1 kg of moisture (US$/kWh)

If the heat pump-assisted kiln operating hours are long (e.g., 8,000 hours/year), the total cost of removing a liter of water from the product is significantly lower. Several factors are expected to influence the overall economic viability of a heat pump dryer.

For typical drying conditions with air temperature and relative humidity in the respective range of 25°C–65°C and 40%–100%, a heat pump dryer designed for a drying capacity of 200 kWh has a payback period of 2–3 years. The payback period for initial investment is generally reduced if more product moisture is available for

TABLE 9.2
Comparison of Heat Pump Dehumidifiers with Vacuum and Hot-Air Drying

Parameter	Unit	Hot Air Drying	Vacuum Drying	Heat Pump Drying
SMER	kg/kWh	0.12–1.28	0.72–1.2	1.0–4.0
Drying efficiency	%	35–40	<70	95
Temperature range	°C	40–90	30–60	10–65 (10–100)
Relative humidity	%	Variable	Low	10–65
Capital cost	–	Low	High	Moderate
Operating cost	–	High	Very high	Low

heat recovery. Also, the payback period is sensitive to the operating pressure of the evaporator and condenser. The minimum effectiveness factor should be more than 0.55 if the payback period is less than 3 (Kiang and Jon 2007).

The simple payback ($SPB_{investment}$) period of a heat pump-assisted dryer can be defined as the total initial capital investment cost (*TIC*, expressed, for example, in US$) divided by the annual net energy cost savings (*ANSEC*, in US$/year):

$$SPB_{investment} = \frac{TIC}{ANES} \tag{9.6}$$

The initial investment cost, very site specific but decreasing with the dryer size, generally includes the costs of system design, procurement, drying chamber fabrication, installation and commissioning, compressor, refrigerant/lubricant, blower and fans, heat exchangers, ductwork, air filters, controls, etc. To increase the heat pump-assisted dryer's Simple Payback (*SPB*), the total investment cost must be reduced as much as possible, namely the cost of the heat pump itself that may represent more than 25% of the system total cost. Higher $SMER_{dryer}$ or $SMER_{heat\ pump,}$ both strongly influenced by the electricity price and the heat pump *COPS*, provides lower operating costs and, consequently, shorter simple payback periods. For drying industry, acceptable $SPB_{investment}$ values are normally between 2 and 3 years. If it is assumed that the useful heat generated by heat pump partially replaces heat from an existing boiler with thermal efficiency η_{boiler}, the simple payback period can be expressed using the following equation (Mujumdar 2008):

$$SPB_{boiler} = \frac{1}{\left(\frac{FEP}{\eta_{boiler}} - \frac{EEP}{COP_{heat\ pump}}\right) \times 8760 - AMC} \tag{9.7}$$

where

FEP is the fuel energy price (US$/kWh)
EEP is the electrical energy price (US$/kWh)
$COP_{heat\ pump}$ is the heat pump's coefficient of performance (–)
AMC is the heat pump's annual operating and maintenance costs (US$/kWh · year).

It can be seen that SPB_{boiler} is sensitive to fuel prices and heat pump's operating and maintenance costs. It may decrease at higher fuel prices and lower electricity costs.

9.10 DRYING SCHEDULES FOR HEAT PUMP-ASSISTED DRYERS

During the wood drying process, as the moisture is removed from the board surfaces, the moisture migrates from inside the wood boards to the surface because of moisture content gradients and other driving forces. The migration rate of moisture depends on the liquid diffusivity, which is a function of local moisture content and temperature. For cost-effective operation of the dryer, the migration rate of moisture to the drying surface should match the moisture removal rate of the drying air. Therefore, drying conditions need to vary in time according to the actual moisture content of the wood stack(s) for cost-effective and safe operation of the heat pump-assisted dryer. An appropriate initial quantity of wood may help attaining high overall drying performances.

In heat pump-assisted dryers, a fraction of the kiln air, usually controlled by means of bypass dampers to maintain the evaporation pressure at a specified value, is circulated through the heat pump evaporator where it is cooled and dehumidified, then circulated through the heat pump condenser where it is reheated and returned to the kiln.

The scope is to remove the excess of moisture from the drying air and condense it in order to recover sensible heat and latent heat of vaporization.

In order to reduce the energy consumption per unit of removed moisture and the drying time to obtain the desired product moisture content, drying schedules involving parameters such as air dry- and wet-bulb temperatures, relative humidity, and velocity must be used.

Generally, drying schedules that imply time-varying supply of thermal energy can be classified into the following categories (Kiang and Jon 2007): (i) intermittent supply of drying air (that has the most significant influence on the product drying kinetics and quality parameters) or intermittent regulation of the air temperature and relative humidity; (ii) dry aeration, which is a combination of high-temperature drying and tempering periods; (iii) air reversal drying; and (iv) cyclic drying.

Some of such time-dependent drying schemes may lead to the following (Kiang and Jon 2007): (i) thermal energy savings, (ii) shorter drying times, (iii) higher rates of moisture removal, and (iv) higher product quality (e.g., reduced shrinkage, cracking, and brittleness).

Drying process by dehumidification with heat pumps is quasi-identical to that used in conventional air convective heated kilns. For both conventional and heat pump-assisted drying methods, high-quality record and control of kiln's temperature and relative humidity (wet-bulb depression, wet-bulb temperature, or equilibrium moisture content), and reverse fan direction every several hours, are essential.

Basically, schedules for lumber drying in heat pump-assisted dryers are similar to those employed in conventional kilns. According to the heat and mass transfer between the air and the lumber boards, drying schedules must determine the air conditions suitable for drying via the control systems that act directly on air dry- and wet-bulb temperatures, relative humidity, and velocity (Léger and Amazouz 2003).

The drying schedules and thermal conditions of each step are established based upon the type and quality of wood species, temperature and initial moisture content, board dimensions, etc. in conformity with wood drying schedules (also called programs or indexes) established for specific geographic regions.

The typical (basic) drying programs, generally known for each category (or group) of wood species, are set prior beginning each drying cycle. However, kiln's operators can analyze and, then, modify them using a variety of parameters such as actual (measured) average lumber moisture contents.

The measured dry- and wet-bulb temperatures, as well as the mean average moisture contents of the lumber stack(s) or the time elapsed, are the principal set points for any drying schedule. The methods used to change air thermal and hydraulic parameters may vary from one to another commercial control systems currently available on the market.

As those for conventional air-vented kilns, drying schedules of heat pump-assisted wood dryers are divided into steps (stages), generally based on dry- and wet-bulb set temperatures, and dryer and heat pump running times.

Some steps depend on the drying process itself, while others are related to lumber response during drying.

The first step in a drying schedule is a preheating step when the lumber temperature (and the absolute humidity of the air inside the drying enclosure) rises to a specified value using a supplementary heat source (fossil or electric). With wood species dried at low temperatures, maximum preheating temperatures may attain 66°C because of the limits imposed by operating temperatures of the compressor. However, with softwood (such as white spruce and balsam fir), the preheating periods require up to 93°C temperatures for periods of 6–8 hours prior the heat pump starts at the first step of the drying cycle. Such relatively high temperatures also help at destroying microorganisms responsible for sapwood discoloration (Minea 2004). In cold climates, when the kiln, the air and the lumber have to be heated up in winter, this also can involve melting the snow and thawing the lumber, stickers, etc. During the preheating step, the heat pump is not running. The heat pump is called to run when the wet-bulb temperature of the air inside the drying chamber reaches its preset value corresponding to a relatively high absolute humidity. At the end of the preheating step, that is, when the kiln internal absolute humidity exceeds the set point of the selected schedule, the heat pump starts.

When the heat pumps start up at step 1, the lumber moisture content first decreases linearly with time, a process followed by a nonlinear decrease until the lumber boards reach the final moisture content. The heat pump must recover and, then, supply a part of the necessary heat to continue the drying process at low (limited to about 49°C), medium (≈75°C), or high temperature (up to 110°C). The next steps (generally, up to six), when the heat pump is operating, form the actual drying process. Each of these drying steps must ensure appropriate thermal and hydraulic conditions for removing moisture from the lumber stack(s) at rates able to avoid defects, above as well as below fiber saturation point. Each intermediate step of drying schedules ends when the final preset moisture content value is reached, value usually measured based on the oven-dry method.

Generally, in heat pump-assisted dryers, the heat pump running is mainly controlled by the absolute humidity of the drying air. When the condensing capacity of the heat pump is exceeded, or when the internal temperature rises above the value considered in the software, the venting system opens and the kiln operates as an open circuit for short periods of time. The rest of the time the kiln operates as a closed circuit. If the heat pump is correctly designed and the kiln construction is very tight, all air vents are closed. However, the opening of vents in any intermediate step of drying process with heat pump in operation may be required if the dry-bulb temperature inside the drying chamber increases over the maximum admitted value in order to preserve safe operating parameters for both heat pump and dryer (Minea 2004; Fernandez-Golfin Seco et al. 2004).

Supplementary heat may be required during the entire drying process if the dry-bulb temperature inside the dryer descends below the set point of each drying step indicating that there is not enough moisture inside the kiln to be condensed by the heat pump under optimum refrigerant operating parameters. This situation may occur when the heat pump is not correctly designed according to the quantity and properties of the dried wood stack(s).

However, if the kiln's heat losses are too high (notably in cold climates with old, poor insulated dryers), supplementary heat must be supplied because, in this case, the total heat requirement may exceed the heat generated by the heat pump.

If the moisture condensation rate is high enough to permit the constant running of the heat pump, the external (auxiliary, supplementary) heating system switches off.

At the end of heat pump-assisted drying, when the wood is almost completely dried (generally, at a final average moisture content of 8%–20%) and the demand of energy inside the kiln is still very high, the internal humidity is very low and the external heating system usually runs again to increase the kiln's internal temperature. Eventually, internal humidifiers have to be used during the final drying steps in order to control the air's relative humidity that can drop below the set point and, thus, produce internal stresses inside lumber boards.

9.11 DESIGN OF HEAT PUMP-ASSISTED WOOD DRYERS

As for many other industrial dryers, no single design procedure exists for conventional and heat pump-assisted dryers because of variety of product size and porosity, drying time, production capacities, drying temperatures, operating pressure, and heat transfer modes (Mujumdar 2000).

In the particular case of the heat pump-assisted wood drying process, several more of less simplified design methods have been proposed in the past and experimentally validated for representative wood species, such as softwoods and hardwoods, at both laboratory and industrial pilot scales.

For appropriate design of heat pump-assisted wood dryers, both wood drying and heat pump specialists must relatively well know: (i) wood's basic drying phenomena (Figure 9.16), such as moisture migration inside the lumber boards and moisture vaporization from the board surfaces (see Chapter 7) and from the kiln-wide stacks (see Chapter 8), and moisture removal from the drying air (by condensation); and the heat pump basic thermodynamic processes (e.g., refrigerant evaporation, compression, and condensation).

In industrial heat pump-assisted wood drying practice, two situations generally may occur: (i) an existing, conventional kiln, of which the maximum/minimum wet (humid, green) wood drying capacities (expressed in *MBF*, m^3, etc.) (1 MBF = 2.36 m^3) are known, must be equipped (retrofitted) with a mechanical vapor compression heat pump as an add-on dehumidification device; in this case, the nominal capacity of the heat pump must be determined; and (ii) a given mechanical vapor compression heat pump, of which capacity (e.g., the nominal compressor power input or cooling and dehumidification nominal capacity) is known, must be coupled with a new wood dryer; in this case, the maximum/minimum wet (green humid) wood drying capacities (expressed in *MBF*, m^3, etc.) of the new dryer must be determined.

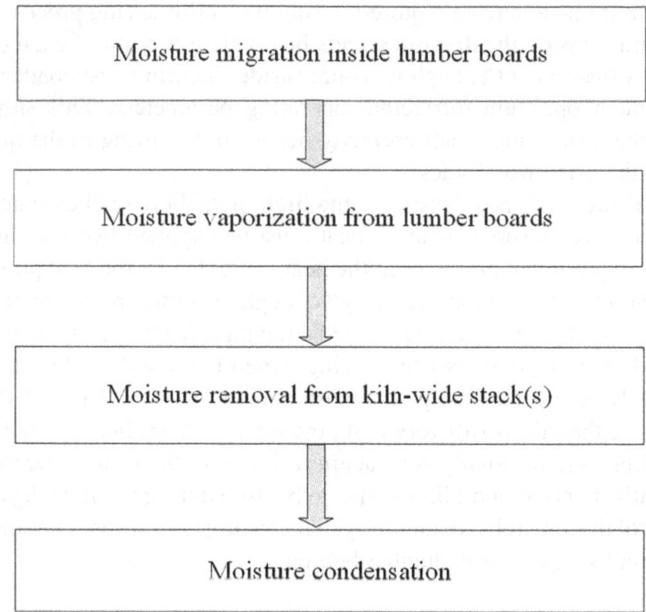

FIGURE 9.16 Main processes involving moisture transport in heat pump-assisted kilns.

9.11.1 Basic Principles

Moisture movement and removal from lumber boards is a complex thermodynamic process depending, among many other parameters, on the wood species, initial temperature, and moisture contents, etc.

The main role of heat pump is to recover sensible and latent heat from the drying air that passes through the evaporator that acts as an air cooler and dehumidifier. The recovered heat, plus the equivalent heat of the electrical energy consumed by the heat pump compressor and blower, serves to overcome the water retention forces inside the wood pieces and heat losses of kiln's structure and equipment, as well as to remove moisture from the boards' surfaces.

These two simultaneous thermodynamic processes, along with the moisture condensation on the heat pump evaporator, must be matched as much as possible in order to ensure the system thermal equilibrium, provide efficient dehumidification performances, and avoid heat pump troubleshooting and/or compressor failure.

In other words, for efficient operation, heat pump-assisted dryers coupled with low-, medium-, or high-temperature heat pumps must ensure adequate correlation between the wood thermal dewatering (moisture vaporization) rate that depends on the boards' internal moisture movement and surface moisture vaporization processes, and the heat pump cooling and dehumidification capacity during each step of drying cycles. Errors in such a correlation may compromise the heat pump operation and, thus, the successful application of heat pump-assisted wood drying in industry (Minea 2008, 2009).

Heat pumps cannot accurately operate as dehumidifiers if the total sensible and latent heat extraction is not large enough to provide appropriate operating parameters, such as compressor suction temperatures and pressures. In other words, the heat pump optimum coefficient of performance cannot be reached if the evaporator heat extraction capacity is too low compared to the material moisture vaporization rate. This means that sufficient quantities of dried materials have to be supplied to provide the required thermal input to the evaporator. Insufficient quantities of wood mean small quantities of moisture and heat extraction, too low evaporating and compressor suction pressures and temperatures, risks of humidity freezing on the evaporator fins, short drying cycles, too high compressor discharge pressures and temperatures, compressor mechanical failures, and, finally, poor dehumidification performances (Minea 2010a, b).

9.11.2 General Heat Pump Sizing Rule

Mechanical vapor compression air-to-air heat pumps include basic components such as compressors, evaporators, condensers, and expansion devices. For heat pump-assisted dryers, bypass factor, conditions of air entering the evaporator and leaving the condenser, and the performance of heat pump system must be accurately determined. Heat pumps coupled with dryers are complex systems due to the interdependency of all components and operating parameters. Any change occurring in one component or parameter will inevitably influence the others, as well as the heat pump dryer's overall performance. Therefore, the nominal capacity of a heat pump dedicated to drying porous and hygroscopic material such as woods has to be carefully sized. Sizing the heat pump nominal capacity for drying purposes implies a good comprehension of the typical lumber drying curve, that is, the profile of moisture content variation during the entire drying cycle.

Figure 9.17 schematically shows the main thermodynamic processes involved in heat pump-assisted dryers on both air and refrigerant sides.

Over- or undersized heat pumps may drastically penalize both the energy and drying performance of heat pump-assisted dryers. To achieve as high as possible

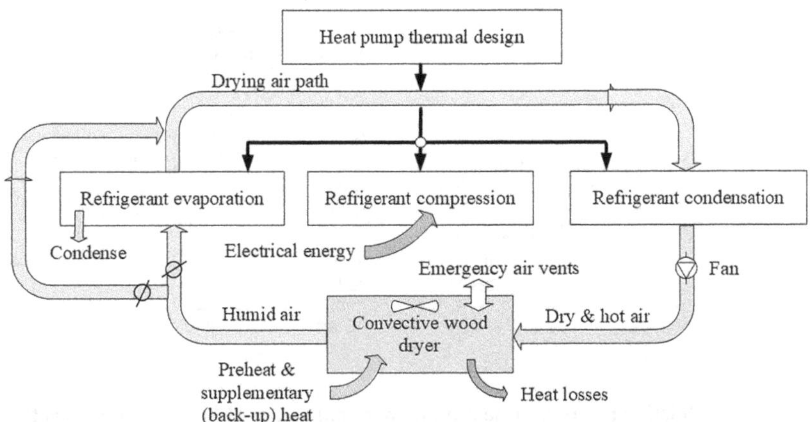

FIGURE 9.17 Schematic representation of main thermodynamic processes and energies involved in heat pump-assisted drying.

$COP_{heat\ pump}$ (defined as the ratio between the thermal energy supplied to the dryer and the electrical energy consumed by the heat pump compressor and blower, preferably higher than 4) and $SMER_{heat\ pump}$ (defined as the ratio of total moisture removed and the total electrical energy consumed by the heat pump compressor and blower, preferably higher that 2.0 $\frac{kg_{water}}{kWh}$), the dried lumber must give off enough moisture to be able to provide sufficiently high absolute humidity to the drying air at the heat pump evaporator inlet. This will ensure enough moisture condensing enthalpy (latent heat) to be removed. Such a challenge could be reached if the nominal heat pump's cooling and dehumidification capacity is sized as low as possible in order to match the quantity of moisture that can be removed from the product stack. Such an approach may help keep the heat pump compressor in continuous running mode while avoiding excess heat rejection outdoor.

Figure 9.18a and b suggests the optimum location (note: graphics not to scale) of the heat pump nominal design point for batch-type dryers. This point must be determined for each dried material (such as woods and agro-foods), and, among the parameters previously mentioned, according to the material's initial and final moisture contents (Minea 2015).

More precisely, in the case of wood drying, for example, the heat pump nominal design point must be at or just above the wood fiber saturation point (*FSP*), defined as the point in dried wood at which all free moisture has been removed from the cells themselves, while the cell walls remain saturated with absorbed moisture (Figure 9.18a). At moisture contents (*MC*) above *FSP*, the heat pump $COP_{heat\ pump}$ and $SMER_{heat\ pump}$ could be relatively high because the absolute humidity of the drying air can be kept high enough, and there is practically no change in volume (shrinkage or swelling), while *MC* still decreases. However, as the *FSP* approaches, surface moisture diffusion becomes the limiting factor of the moisture removal rate. As a consequence, the $COP_{heat\ pump}$ and $SMER_{heat\ pump}$ decrease, and, at the *FSP*

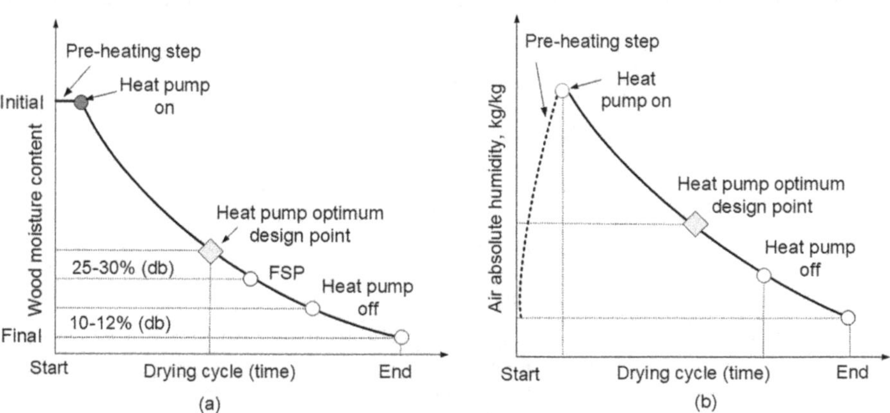

FIGURE 9.18 Relative position of heat pump nominal sizing (design) point versus the drying curves as a function of (a) wood's actual moisture content; (b) air drying actual absolute humidity. FSP, fiber saturation point (note: key point positions are not shown to scale) (Reprinted from Minea (2015) with permission from International Refrigeration Institute).

or at a certain absolute humidity of the drying air, the heat pump must be shut down in order to save energy and preserve wood quality. To prevent frequent heat pump shutdowns, the dehumidification rate must be progressively decreased using different methods in order to match the material thermal dewatering (moisture vaporization) rate.

On the other hand, Figure 9.18b suggests the relative location of the heat pump optimum design point (also, not shown to scale) versus the actual absolute humidity of the drying air. When the absolute humidity of the drying air drops below a certain value, generally corresponding to wood's *FSP*, the evaporator recovers less heat and the heat pump $COP_{heat\ pump}$ and $SMER_{heat\ pump}$ decrease, even if the compressor consumes less electrical energy.

It can be concluded that, if sized at an optimum nominal capacity, the heat pump may operate within a certain dehumidifying rate range (max, min) corresponding to the product moisture contents and/or the absolute humidity of the drying air, each varying, like the heat pump dehumidification rate, between their maximum and minimum values.

In other words, an optimum ratio between the maximum quantity of moisture that can be removed from the wood stack(s) and the heat pump's maximum and minimum dehumidification capacity has to be provided, while keeping the heat pump in quasi-continuous running mode and without rejecting excess heat outdoors.

9.11.3 Heat Pump Nominal Capacity

If an existing (conventional) kiln (see Chapter 8) has to be equipped (retrofitted) with a mechanical vapor compression heat pump as a dehumidification device, the nominal size of the heat pump, expressed in terms of cooling and dehumidification capacity $\left(\dot{Q}_{evap}\right)$ or compressor input power $\left(\dot{W}_{comp}\right)$, must be first determined.

In this case, the capacity (i.e., lumber volume or mass) of the existing dryer enclosure and, thus, the initial green (humid, wet, saturated, swollen) volume of the wood $\left(V_{in}^{green}\right)$ (expressed in m^3_{green}) that could be introduced inside the kiln are known.

The green volume is defined as the solid volume of a wood when it is in equilibrium with the relative humidity of the surrounding medium. For initial moisture contents (MC_{in}) higher than 30% (dry basis), this volume is constant.

Based on the initial volume of the green wood, it is possible to determine the wood initial dry (anhydrous) mass with the following relation:

$$m_{in}^{dry} = \rho_{basic} \times V_{in}^{green} \tag{9.8}$$

where

m_{in}^{dry} is the wood initial dry (anhydrous) mass (kg_{dry})

ρ_{basic} is the wood basic density $\left(\frac{kg_{dry}}{m^3_{green}}\right)$ defined as the oven-dry mass of a wood sample divided by its green (swollen) volume.

The wood basic density indicates the maximum amount of moisture that a wood species may contain and, thus, is species dependent (Perré and Keey 2015). It can vary from 50 $\frac{kg_{dry}}{m^3_{green}}$ for some balsa wood to 1,400 $\frac{kg_{dry}}{m^3_{green}}$ for lignum vitae, corresponding to maximum moisture contents of 19 and 0.05$\frac{kg_{water}}{kg_{oven\ dry}}$, respectively. However, most commercial species have densities in a narrower span from 350 to 800 $\frac{kg_{dry}}{m^3_{green}}$ (Walker 1993). In eastern Canada, for example, the average basic densities of wood species to dry are 340 $\frac{kg_{dry}}{m^3_{green}}$ for the balsam fir, 390 $\frac{kg_{dry}}{m^3_{green}}$ for the red pine, and 350 $\frac{kg_{dry}}{m^3_{green}}$ for the white spruce (Cech and Pfaf 2000).

The initial moisture content (%, dry basis) can be defined as follows $\left(\frac{kg_{water}}{kg_{oven\ dry}}\right)$:

$$MC_{in} = \frac{m_{in}^{green} - m_{in}^{dry}}{m_{in}^{dry}} = \frac{m_{in}^{green}}{m_{in}^{dry}} - 1 \tag{9.9}$$

where

m_{in}^{green} is the initial mass of the green (humid) wood entering the kiln (kg_{green})

In eastern Canada, for example, MC_{in} is of about 40% for white spruce and red or white pine, and 88% for balsam fir.

From equation 9.9, the initial mass of the green (humid) entering the kiln can be expressed as follows:

$$m_{in}^{green} = (1 + MC_{in}) \times m_{in}^{dry} \tag{9.10}$$

if MC_{in} is expressed in decimals (e.g., 0.75), or

$$m_{in}^{green} = \left(1 + \frac{MC_{in}}{100}\right) \times m_{in}^{dry} \tag{9.11}$$

if MC_{in} is expressed in percentage (e.g., 75%)

Similarly, based on the definition of the final moisture content (%, dry basis) $\left(\frac{kg_{water}}{kg_{dry}}\right)$:

$$MC_{fin} = \frac{m_{fin}^{green} - m_{in}^{dry}}{m_{in}^{dry}} = \frac{m_{fin}^{green}}{m_{in}^{dry}} - 1 \tag{9.12}$$

the final mass of the green (humid) wood leaving the kiln (kg_{green}) can be calculated as follows:

$$m_{fin}^{green} = (1 + MC_{fin}) \times m_{in}^{dry} \tag{9.13}$$

if MC_{fin} is expressed in decimals (e.g., 0.15), or

$$m_{fin}^{green} = \left(1 + \frac{MC_{fin}}{100}\right) \times m_{in}^{dry} \quad (9.14)$$

if MC_{fin} is expressed in percentage (e.g., 15%)

In eastern Canada, for example, the optimum final moisture content for almost all softwood species must be 18% (Cech and Pfaff 2000).

Based on equations 9.11 and 9.14, the theoretical, maximum (total) mass of moisture (water) (kg_{water}) that can be removed from the initial mass (volume) of the green lumber stack(s) during the entire drying cycle is

$$m_{moisture}^{removed} = \left(1 + \frac{MC_{in}}{100}\right) \times m_{in}^{dry} - \left(1 + \frac{MC_{fin}}{100}\right) \times m_{in}^{dry} = \left(\frac{MC_{in} - MC_{fin}}{100}\right) \times m_{in}^{dry} \quad (9.15)$$

From this theoretical (total, maximum) mass of moisture (water) removed from the lumber stack(s), it may be assumed that the heat pump will recover (condense) between 70% and 85%:

$$m_{moisture}^{condensed} = \varepsilon \times m_{moisture}^{removed} = \varepsilon \times \left(\frac{MC_{in} - MC_{fin}}{100}\right) \times m_{in}^{dry} \quad (9.16)$$

where

$$\varepsilon = 0.7 - 0.85$$

The rest of water removed from the wood stack(s) (15%–30%) includes moisture losses by ventilation (vent air openings, exfiltration, etc.) and various uncontrolled condensations (on walls, floor, equipment, etc.) and leakages, all of them depending on the quality of the insulation of the drying envelope and the efficiency of the system control of operation.

Knowing the mass of moisture condensed by the heat pump's evaporator, the heat pump latent (dehumidification) capacity (Q_{evap}^{lat}, expressed in kJ) is given as follows:

$$Q_{evap}^{lat} = \eta_{evap} \times m_{moisture}^{condensed} \times h_{fg} \quad (9.17)$$

where

η_{evap} is the thermal efficiency of the heat pump evaporator (0.85–0.9)

$m_{moisture}^{condensed}$ is the mass of moisture condensed (kg)

h_{fg} is the water evaporation (latent heat) enthalpy (kJ/kg)

If it is assumed that the sensible heat removed by the heat pump evaporator $\left(Q_{evap}^{sens}\right)$ from the drying air represents about 15% of the latent heat, the total heat removed (recovered) by the heat pump evaporator (expressed in kJ) is given as follows:

$$Q_{evap}^{total} = Q_{evap}^{lat} + Q_{evap}^{sens} = 1.15 \times \eta_{evap} \times m_{moisture}^{condensed} \times h_{fg} \quad (9.18)$$

Based on the total (sensible plus latent) heat recovered and the heat pump total running time $\left(t_{heat\ pump}^{run}\right)$, the evaporator average thermal power ($\dot{Q}_{evap}^{avrg}$, in kW) can be determined as follows:

$$\dot{Q}_{evap}^{avrg} \approx \frac{1.15 \times \eta_{evap} \times m_{moisture}^{condensed} \times h_{fg}}{t_{heat\ pump}^{run}}$$

$$= \frac{1.15 \times \eta_{evap} \times \varepsilon \times \left(\dfrac{MC_{in} - MC_{fin}}{100}\right) \times \rho_{basic} \times V_{in}^{green} \times h_{fg}}{t_{heat\ pump}^{run}} \tag{9.19}$$

where

$t_{heat\ pump}^{run}$ is the heat pump's total running time (s)

If the heat pump's average heating coefficient of performance ($COP_{heat\ pump}$), based on the compressor and blower electrical power consumptions (kWh), is assumed to be known or approximated, and using equation 9.20, the required nominal electrical input power of the heat pump compressor $\left(\dot{W}_{comp},\ \text{kW}\right)$ will be approximated as follows:

$$\dot{W}_{comp} = \frac{1.15 \times \eta_{evap} \times \varepsilon \times \left(\dfrac{MC_{in} - MC_{fin}}{100}\right) \times \rho_{basic} \times V_{in}^{green}}{\left(COP_{heat\ pump} - 1\right) \times t_{heat\ pump}^{run}} \tag{9.20}$$

From equations 9.19 and 9.20, the condenser average thermal power ($\dot{Q}_{cond}^{avrg}$, kW) can be finally estimated as follows:

$$\dot{Q}_{cond}^{avrg} \approx \dot{Q}_{evap}^{avrg} + \dot{W}_{comp} \tag{9.21}$$

If the specific enthalpies of the drying air at the evaporator inlet and outlet (h_i and h_b) are known (see Figure 9.19), the average mass flow rate of the drying air through the evaporator (kg/s) is given as follows:

$$\dot{m}_{ev}^{air} \approx \frac{\dot{Q}_{evap}^{avrg}}{h_i - h_b} \approx \frac{1.15 \times \eta_{evap} \times m_{moisture}^{condensed} \times h_{fg}}{t_{hp}^{run}\left(h_i - h_b\right)} \tag{9.22}$$

Similarly, if the specific enthalpies of the drying air at the condenser inlet and outlet (h_m and h_d) are known (see Figure 9.19), the average mass flow rate of the drying air through the condenser (kg/s) could be estimated as follows:

$$\dot{m}_{cond}^{air} \approx \frac{\eta_{cond} \times \dot{Q}_{cond}^{avrg}}{h_m - h_d} \approx \frac{\eta_{cond} \times \left(\dot{Q}_{evap}^{avrg} + \dot{W}_{comp}\right)}{\overline{c}_p(T_m - T_d)} \tag{9.23}$$

where

η_{cond} is the thermal efficiency of the heat pump condenser (0.85–0.9)
h_m is the specific enthalpies of the drying air at the condenser inlet (kJ/kg)
h_d is the specific enthalpies of the drying air at the condenser outlet (kJ/kg)

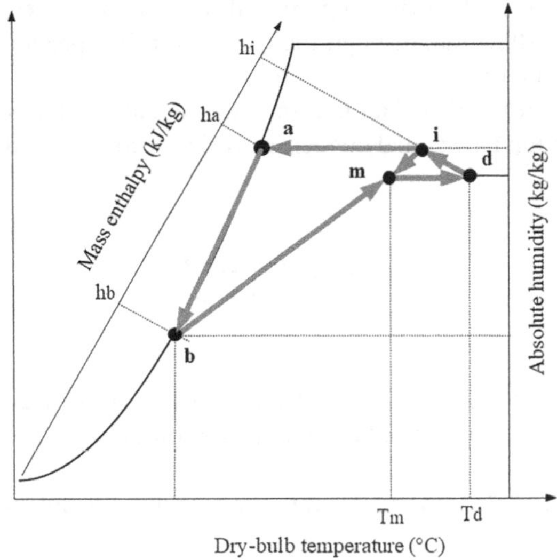

FIGURE 9.19 Typical theoretical thermodynamic processes of the drying air represented in the Mollier diagram.

$\overline{c}_p$ is the average specific heat of the drying air passing through the condenser (kJ/kg·K)

T_m is the dry-bulb temperature of the drying air entering the heat pump condenser (°C)

T_d is the dry-bulb temperature of the drying air leaving the heat pump condenser (°C)

Because $\dot{m}_{air}^{cond} > \dot{m}_{air}^{evap}$, an air bypass circuit around the evaporator is required (Figures 9.7 and 9.8). Such a bypass allows mixing the air at the evaporator outlet (at 80%–90% of relative humidity) with a part of the air leaving the dryer. The air bypass allows controlling the evaporator dehumidification rate during the whole drying process when the wood actual moisture content continuously decreases.

This process also supplies a sufficient high airflow rate through the condenser to avoid excessive compressor discharge pressures and temperatures without overdesigning the condenser heat transfer area.

The correlation between the evaporator airflow rate $\left(\dot{m}_{air}^{evap}\right)$ and the condenser airflow rate $\left(\dot{m}_{air}^{cond}\right)$ can be expressed as follows:

$$\dot{m}_{air}^{evap} = \varepsilon \times \dot{m}_{air}^{cond} \tag{9.24}$$

where

ε is the bypass factor $(0 < \varepsilon < 1)$

In the case of split-type heat pumps (see Figures 9.9–9.11), the heat pump evaporator may be provided with variable speed blower(s) to control the dehumidification rate.

The combined action of the bypassing air and the variable speed blower may also provide optimum airflow rate through the evaporator, independently of the airflow rate through the condenser.

At airflow conditions other than nominal, the actual volumetric flow rate (m^3/s) through the heat pump blower and condenser varies according to the following fan's low:

$$\dot{V}_{air,\ cond}^{actual} = \dot{V}_{air}^{nominal}\left(\frac{v_{actual}}{v_{nominal}}\right)^{0.5} \tag{9.25}$$

where

ν_{actual} and $\nu_{nominal}$ (m^3/kg) are the specific volumes of the air leaving the blower at the actual and the nominal thermal conditions, respectively.

Consequently, the actual air mass flow rate (kg/s) is given as follows:

$$\dot{m}_{air,\ cond}^{actual} = \bar{\rho}_{air} \times \dot{V}_{air,\ cond}^{actual} \tag{9.26}$$

where

$\bar{\rho}_{air}$ is the average density of drying air through the heat pump condenser (kg/m^3)

Exercise E9.1

The green volume of a Canadian white spruce lumber stack at the inlet of a high-temperature heat pump-assisted dryer is of $335\ m_{green}^3$, and the average initial moisture content is of 40% (dry basis). The average final moisture content at the dryer's outlet is of 18% (dry basis).

Determine:

i. the initial dry (anhydrous) mass of the lumber stack;
ii. the initial green (humid) mass of the lumber stack;
iii. the final green (humid) mass of the lumber stack;
iv. the theoretical (maximum, total) mass of moisture (water) (kg_{water}) removed from the lumber stack;
v. the theoretical (total, maximum) mass of moisture (water) condensed by the heat pump's evaporator.

Solution

i. The Canadian white spruce's basic density (ρ_{basic}) is 350 $\frac{kg_{dry}}{m_{green}^3}$ (Cech and Pfaff 2000), and the initial dry (anhydrous) mass (kg_{dry}) of the lumber stack is given as follows:

$$m_{in}^{dry} = \rho_{basic} \times V_{in}^{green} = 350\ \frac{kg_{dry}}{m_{green}^3} \times 335\ m_{green}^3 = 117{,}250\ kg_{dry}$$

ii. The initial green (humid) mass of the lumber stack can be calculated using equation 9.11:

$$m_{in}^{green} = \left(1 + \frac{MC_{in}}{100}\right) \times m_{in}^{dry} = \left(1 + \frac{40}{100}\right) \times 117,250 = 164,150\ kg_{green}$$

iii. The final green (humid) mass of the lumber stack can be calculated using equation 9.14:

$$m_{fin}^{green} = \left(1 + \frac{MC_{fin}}{100}\right) \times m_{in}^{dry} = \left(1 + \frac{18}{100}\right) \times 117,250 = 138,355\ \ kg_{green}$$

iv. The theoretical (maximum, total) mass of moisture (water) (kg_{water}) removed from the lumber stack is given as follows (equation 9.16):

$$m_{moisture}^{removed} = \left(\frac{MC_{in} - MC_{fin}}{100}\right) \times m_{in}^{dry} = \left(\frac{40-18}{100}\right) \times 117,250 = 25,795\ kg_{water}$$

v. If 20% of the moisture (water) removed from the lumber stack are lost, the theoretical (total, maximum) mass of moisture condensed by the heat pump's evaporator is given as follows:

$$m_{moisture}^{condensed} = 0.8 \times m_{moisture}^{removed} = 0.8 \times 25,795 = 20,636\ kg_{water}$$

Exercise E9.2

The total mass of moisture (water) removed by the evaporator (moisture condenser) of a heat pump-assisted softwood batch dryer during 65 hours of continuous running with an average coefficient of performance is about 20,000 kg. If the enthalpy (latent heat) of water condensation at 60°C is 2,400 kJ/kg, determine:

a. theoretical total thermal energy (heat) removed from the condensed moisture;
b. compressor average input power.

Solution

The total thermal energy (heat) removed from the condensed moisture (1 kWh = 3,600 kJ) is given as follows:

$$Q_{evap} = 20,000\ \text{kg} \times 2,400\ \text{kJ/kg} = 48,000,000\ \text{kJ} = 13,333\ \text{kWh}$$

Thus:

$$\dot{Q}_{evap} = \frac{Q_{evap}}{65\ \text{h}} = \frac{13,333\ \text{kWh}}{65\ \text{h}} \approx 205\ \text{kW}$$

Compressor average input power:

$$\dot{W}_{compr} = \frac{\dot{Q}_{evap}}{COP_{heat\ pump} - 1} = \frac{205}{4-1} \approx 68.37\ \text{kW}$$

9.12 CONTROL OF HEAT PUMP-ASSISTED DRYERS

The thermodynamic processes involved in softwood and hardwood heat pump-assisted drying with heat pumps are complex and highly nonlinear. The energy efficiency of dryers comes from their ability to recover sensible and latent heat and to control air drying temperatures and relative humidity.

Efficient running of heat pump-assisted dryers is based, among many other elements, on the knowledge of the mechanisms of moisture movement inside the wood species, moisture vaporization and removal at/from boards' surfaces, moisture condensation on the heat pump's evaporator finned coil, and energy requirements and drying schedules. Therefore, apart from the optimum integration and design of wood batch dryers and heat pumps, appropriate control strategies are needed to provide optimum and safe thermodynamic parameters for both dryer and heat pump, and good final quality of the dried wood.

In the case of medium- or high-temperature split-type heat pumps (as represented in Figures 9.9–9.11), the most difficult, but fundamental task of optimum operation may be achieved using the following design and control basic principles: (i) size the heat pump nominal cooling and dehumidification rate according to the dryer capacity (volume) and the product initial and final moisture contents; insufficient quantity of material to be dried means small quantities of moisture and heat to be extracted, excessively low evaporating and compressor suction pressures and temperatures with associated risks of humidity freezing on the evaporator finned tubes, and short compressor running cycles; conversely, if the heat pump dehumidification capacity is oversized, the equivalent heat corresponding to the electrical energy consumed (compressor and blower) will progressively build up inside the adiabatic drying enclosure and, thus, must be periodically rejected outdoor by means of external parallel air- or water-cooled condensers, or refrigerant subcoolers; such heat rejection must be avoided as much as possible in order to optimize the heat pump-assisted dryer energy performance; and (ii) continuously match the heat pump dehumidification capacity with the product thermal dewatering (moisture vaporization) rate.

In addition, the design of heat pump must consider the following other elements: (i) specific thermophysical properties and flow control strategies for the refrigerant, as the evaporator superheating amount; (ii) adequate protection of remote condensers installed inside the drying chambers against thermal shocks; and (iii) management of refrigerant migration because of relatively high temperature differences existing between the dryer chambers and the heat pump mechanical rooms.

If all or even a part of such design and control requirements is not met in practice, frequent compressor cycling caused, for example, by unbalances between thermal dewatering rate of the dried product and the heat pump dehumidification capacity, excessive hot air entering the remote condenser at sudden and excessive high temperatures, and/or refrigerant liquid entering the compressor suction line at each start-up, may occur, leading to the compressor premature wear, failure, and, even, mechanical destruction.

The control of heat pump (e.g., reciprocating, scroll) compressors in response to the actual cooling and dehumidifying loads is probably the most important aspect of controlling the complete heat pump-assisted drying system. The compressors can be gradually unloaded, and, eventually, be stopped when the moisture load is too low. With large-capacity compressors, this cooling and dehumidification operation may lead to possible flooding of the compressor by liquid refrigerant if the pump-down sequence is not activated.

9.12.1 Control of Heat Pump Dehumidification Capacity

The heat pump cooling and dehumidification capacity must continuously match with the product thermal dewatering (moisture vaporization) rate, for example, by varying the airflow (and, thus, velocity) through the evaporator (by using evaporator bypassing dampers or variable speed blowers), by varying the compressor refrigerating capacity (with variable speed or multiple-rack compressors), or by using intermittent drying (Chua et al. 2002) strategies in the case of relatively small drying facilities (Minea 2006).

9.12.1.1 Variable Air Velocity through the Evaporator

By neglecting the cooling (sensible) thermal capacity, the heat pump (evaporator) dehumidification capacity could be approximated as follows:

$$\dot{Q}_{evap}^{latent} \approx \dot{m}_{water} \times h_{fg} = \dot{m}_{dry\ air} \times \Delta\omega_{air} \times h_{fg} = \rho_{dry\ air} \times A_{evap} \times \bar{u}_{air} \times \Delta\omega_{air} \times h_{fg} \quad (9.27)$$

where

$\dot{Q}_{evap}$ is the evaporator dehumidification capacity (kW)

$\dot{m}_{water}$ is the mass flow rate of moisture condensed $\left(\frac{kg_{water}}{s}\right)$

h_{fg} is the water condensing specific (latent heat) enthalpy $\left(\frac{kJ}{kg_{water}}\right)$

$\dot{m}_{dry\text{-}air}$ is the mass flow rate of dry air $\left(\frac{kg_{dry\text{-}air}}{s}\right)$

$\Delta\omega_{air}$ is the variation of the absolute humidity of the air through the evaporator $\left(\frac{kg_{water}}{kg_{dry\text{-}air}}\right)$

$\rho_{dry\text{-}air}$ is the density of dry air (kg/m^3)

A is the evaporator air-side heat transfer area (m^2)

$\bar{u}_{air}$ is the air average velocity through the evaporator external heat transfer area (m/s)

It can be seen that the mass of the removed moisture from a kiln strongly depends on the airflow rate and velocity across the heat pump evaporator. Because the evaporator heat exchange surface (A_{evap}) is constant, it is necessary to vary the airflow rate to match the actual rate of moisture vaporization with the heat pump dehumidification capacity.

From equation 9.27, by assuming both $\rho_{dry\text{-}air}$ and h_{fg} also constant, it can be seen that the absolute humidity of the air ($\Delta\omega_{air}$) varies approximately inversely proportional with the air velocity through the evaporator:

$$\Delta\omega_{air} \approx \frac{\dot{Q}_{evap}}{\rho_{dry\text{-}air} \times A \times \bar{u}_{air} \times h_{fg}} \tag{9.28}$$

Equation 9.28 suggests that the evaporator dehumidification capacity can be effectively matched with the lumber actual moisture vaporization rate by varying the air velocity (or flow rate) through the heat pump evaporator.

In drying, such a control is required because at the beginning of a drying cycle, the lumber stack(s) is very humid, and, thus, it is able to provide by vaporization high quantities of moisture resulting in drying air with relatively high absolute humidity. As a result, the variation of absolute humidity of the air ($\Delta\omega_{air}$) through the evaporator also will be relatively high. For that, according to equation 9.28, at the beginning of the drying cycle, the average velocity of the drying air through the heat pump evaporator must be relatively low in order to condense more moisture and, thus, achieve a maximum evaporator dehumidification capacity $\left(\dot{Q}_{evap,\max}\right)$.

Toward the end of the drying cycle, the moisture content of the dried lumber stack(s) progressively decreases, as well as the absolute humidity of the air and its variation through the evaporator. In other words, the lumber thermal dewatering (moisture vaporization) rate decreases. In order to match the lumber decreasing thermal dewatering rate with the heat pump evaporator dehumidification capacity up to a minimum value $\left(\dot{Q}_{evap,\min}\right)$, according to equation 9.28, the air velocity through the evaporator must me progressively increased.

In this way, the heat pump will practically operate continuously, without frequent on/off cycles, even the quantity of moisture removed from the drying air decreases toward the end of the drying cycle. In other words, the heat pump evaporator (moisture condenser) thermal capacity will vary between the maximum and minimum values $\left[\dot{Q}_{evap} \in \left(\dot{Q}_{evap,\min}, \dot{Q}_{evap,\max}\right)\right]$ without cycling on/off the compressor because the actual wet-bulb temperature inside the dryer is continuously maintained higher than its set points.

In the case of compact-type heat pumps (see Figures 9.7 and 9.8), if the drying air flows at the same rate through the evaporator and the condenser, a number of operating troubleshooting problems may occur.

First, if the airflow rate through the evaporator is constant and too high, the quantity of moisture extracted may be low, and the heat extracted insufficient for efficient drying. In addition, the heat pump dehumidification capacity cannot be continuously matched with the material thermal dewatering (moisture vaporization) rate.

Second, if the airflow rate through the condenser is too low, the heat pump compressor may trip off on its high pressure control, while the temperature and relative humidity of the drying air at the heat pump condenser outlet may be not adequate for efficient drying purposes.

A standard method for adapting the heat pump cooling and dehumidification rates to the actual moisture vaporization rate of the lumber stack(s) consists in varying the air velocity through the heat pump evaporator (moisture condenser) by using bypass dampers (ASHRAE 2009; Lewis 2003), as shown in Figure 9.20a This method is based on the fact that lower airflow rates through the evaporator help to increase the moisture condensation rate from the dried lumber stack(s), and vice versa.

A temperature sensor controls both primary and bypass dampers so that the temperature of the air leaving the evaporator is the same as the temperature of the refrigerant leaving the evaporator. The airflow through the evaporator is thus controlled to maintain the evaporation pressure at a value corresponding to the actual quantity of moisture available inside the dryer, and, also, reduce the airflow rate through the evaporator in order to avoid excessive refrigerant condensation pressure. The bypass factor (i.e., the ratio of the mass flow rate through the evaporator to the mass flow rate through the condenser) can be controlled by the heat pump evaporation pressure and can theoretically vary between 0 and 1.

However, in industrial applications, the air bypass control is a relatively imprecise method for wood drying mainly because of low accuracy of temperature readings and relatively low reliability of air dampers.

Another, more reliable method to vary the airflow rate through the heat pump evaporator consists in using a variable speed blower (Figure 9.20b). However, this control method can be used only when the heat pump is of split type (see Figures 9.9–9.11), that is, when the refrigerant condenser is remote installed inside the drying chamber (see Figures 9.9–9.11).

When variable speed blowers are employed, empirical time-based transfer functions (as that shown by equation 9.29) can be used to provide variable air speeds of the drying air circulating through the heat pump evaporator in order to achieve high dehumidification rates at the beginning of the drying cycle and progressively lower rates toward the end of the drying process (Minea 2010a, 2010b):

$$v = K \times t + 25 \tag{9.29}$$

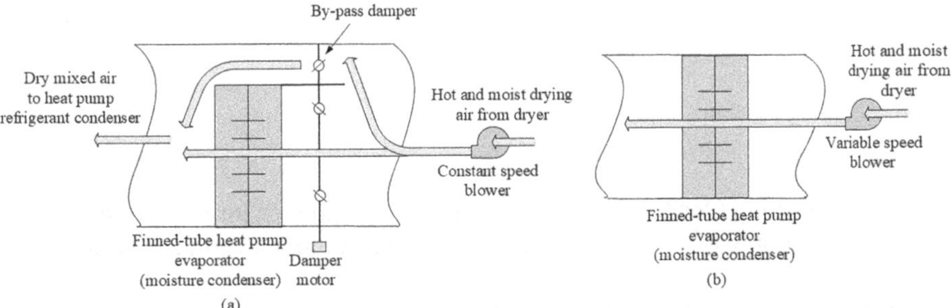

FIGURE 9.20 (a) Principle of heat pump evaporator air bypass with constant speed blower applied to compact-type heat pump-assisted wood dryers; (b) heat pump evaporator with variable speed blower applied to split-type heat pump-assisted wood dryers.

where

- K is an empirical (experimental) constant depending on wood dried species (–)
- t is the drying time measured from the beginning of the dehumidification process with heat pump (hours)
- υ is the frequency of the electrical current supplying the heat pump blower (Hz)

For example, for softwood, such as white spruce and pine (eastern Canada), the experimental constant K may vary from 0.2 (for high dehumidification rates) to 0.5 (for low dehumidification rates).

9.12.1.2 Intermittent Drying

In industrial drying, great care is taken to avoid defects of dried materials, such as cracking and shrinkage phenomena (mostly at the end of the constant drying rate period or at the beginning of the falling drying rate period, especially if the material is thick and the drying rate is high), caused by the induced stresses (Kowalski and Pawlowski 2008).

For that, intermittent drying (Chua et al. 2002; Minea 2006) strategy could be well adapted to capillary porous materials such as woods (and ceramics), allowing the heat pump dehumidification capacity to better match with the wood vaporization rates. This method avoids cracking and shrinkage by slowing down the drying rate to enable the moisture inside the capillary pores to be consistent throughout the material and, thus, diminish the risks.

Lowering the drying rate, on the other hand, increases the drying time, which is economically unprofitable. Therefore, the drying rate has to be increased again when the moisture distribution has become sufficiently consistent.

A high drying rate is not risky at the beginning of the drying process because the material is fully saturated and the capillary transportation of moisture from the interior is sufficient enough to preserve the liquid film on the surface. During this period, the drying proceeds as from an open water surface. When the water on the surface disappears, the surface begins to shrink and drying-induced stresses appear. That is the moment for periodically increasing the air humidity by stopping the heat pump compressor.

The heat pump's intermittency ratio is defined using the following equation:

$$\alpha = \frac{t_{on}}{t_{on} + t_{off}} \tag{9.30}$$

where

- t_{on} is the running ("on") time of the heat pump compressor (minutes or hours)
- t_{off} is the stopping ("off") time of the heat pump compressor (minutes or hours)

9.12.1.3 Variable Speed Compressors

The heat pump cooling and dehumidification capacity can also be adjusted by varying the compressor speed that is directly proportional to the compressor volumetric displacement rate. The compressor speed could, for example, be varied it within a range of 50%–150% of specified displacement rate. The compressor motor performance factor is assumed less than that for heat pumps with constant speed compressors because of the inverter efficiency. In addition, the speed variation of the compressor is not always adequate to properly control the heat pump over a wide range of operating conditions as permitted by, for example, the variation of the bypass factor.

9.12.1.4 Multiple Compressors

Instead of varying the airflow rate through the heat pump evaporator only by means of bypass dampers, variable speed blowers, and/or variable speed compressors, the heat pump cooling and dehumidification capacity could be additionally adjusted according to the wood actual moisture vaporization rate by using multiple compressor racks (see Figure 9.12).

At the beginning of the drying cycle, when the available quantity of vaporized moisture is relatively high, all heat pump compressors can simultaneously run. Toward the end of the drying process, when less quantity of moisture to condense is available inside the drying chamber, one or several compressors must shut down.

9.12.2 Safe Operating Parameters for Heat Pump Dryers

Heat pump-assisted wood dryers are subject to particular requirements related to their safe operation because of (i) refrigerant thermophysical properties, (ii) particular location of heat pump refrigerant condensers (compact, i.e., installed close to the evaporators, or remote), (iii) temperature differences existing between the dryer chambers and the heat pump mechanical rooms, (iv) efficiency of strategy employed to set the wet-bulb temperature of the drying air, and (v) strategy adopted to control the amount of heat supplied for preheating and supplementary (and backup) heating.

If such requirements are not addressed in practice, frequent compressor cycling (caused, e.g., by the refrigerant liquid entering the compressor suction line), excessive hot air entering the remote condenser, and/or unbalances between the wood moisture vaporization rate and the heat pump cooling and dehumidification capacity may occur and, thus, contribute to the compressor premature wear, failure, and even mechanical destruction (Minea 2010).

9.12.2.1 Refrigerant Superheating

Typically, the low-pressure refrigerant vapor is superheated inside the heat pump evaporator prior to entering the compressor suction line. Refrigerant superheating is required to avoid the saturated vapor–liquid mixtures, and even saturated liquid from entering the compressor suction line. The amount of refrigerant superheating required for preventing the liquid from entering the compressor suction line is controlled by the expansion valve based on the saturated evaporating pressure and temperature, and the compressor actual suction temperature.

For most common refrigerants, a suction superheating (process A-B in Figure 9.21a) of about 5°C is sufficient to prevent saturated vapor from entering the compressor (isentropic process B-C). This is achieved due to the slope of the isentropic (adiabatic) curves ($ds = 0$) in the superheated vapor zone versus the position of the saturated vapor curve. If the compressor sucks up superheated vapor at state

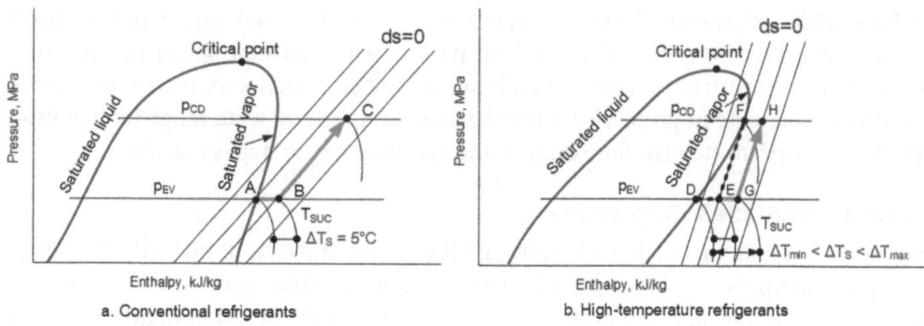

FIGURE 9.21 Comparison of adiabatic compression processes in ln(p–h) diagram; (a) for most common refrigerants; (b) for some new and/or unconventional refrigerants. s, specific entropy.

B during the quasi-adiabatic compression process B-C, the isentropic curve does not touch the saturated vapor line.

However, with other refrigerants, the isentropic curves (E-F and G-H compression processes in Figure 9.21b) could be more vertically oriented compared to those of most common refrigerants. This suggests that, if the superheating amount (process D-E) is too small, the compression adiabatic curve could cross the refrigerant saturation line. In this case, the thermodynamic state of the refrigerant leaving the compressor (F) can be located on the saturated line. To avoid such a situation, higher amounts of suction superheating have to be provided within a safe, optimum range:

$$\Delta T_{\min} \leq \Delta T_{superheating} \leq \Delta T_{\max} \tag{9.31}$$

With such higher superheating amounts, the thermodynamic state of the refrigerant vapor (state H in Figure 9.21b) will be far enough from the vapor saturated curve, as can be seen in the superheated vapor region of ln(p–h) diagram. This will ensure safer operating conditions, especially for reciprocating refrigerant compressors.

For high-temperature refrigerants, such as HFC-236fa and HFC-245fa, the optimum amount of superheating is between 15°C and 20°C at any saturated suction pressure (Minea 2010).

To automatically keep the compressor suction superheating within the optimum range, simple control sequences may be used. For that, both compressor suction pressure (evaporating pressure) (p_{EV}) and suction temperature (T_{SUC}) sensors send proportional signals to the heat pump process controller. It automatically may calculate the actual evaporating temperature according to the refrigerant saturated pressure–temperature correlation:

$$T_{EV} = f\left(p_{EV}\right) \tag{9.32}$$

Then, for each pair of suction and evaporating temperatures, the refrigerant actual suction superheating is calculated and periodically adjusted as follows:

$$\Delta T_{\min} \leq \Delta T_{superheating} = T_{SUC} - T_{EV} \leq \Delta T_{\max} \tag{9.33}$$

According to the superheating amount, the process controller will automatically modulates the opening of the electronic expansion valve to keep the superheating within the preset optimum range. If the actual superheating is, for example, 0.5°C bellow the lower limit (ΔT_{min}), the expansion valve will slightly close in order to increase the suction superheating. If the actual suction superheating is, for example, 0.5°C higher than the upper limit (ΔT_{max}), the expansion valve will slightly open to decrease the suction superheating.

9.12.2.2 Wet-Bulb Temperature

In heat pump-assisted dryers, the operation of heat pump's compressor is controlled according to the dryer actual wet-bulb temperature. Toward the end of the preheating step (during which the lumber stack and dryer auxiliary equipment are heated, but heat pump is not running), when the air wet-bulb temperature gets near the set point of the first drying step, the compressor starts, and the first stage of the heat pump-assisted drying process begins. When the air wet-bulb temperature inside the dryer becomes equal or lower than its set point, the compressor shuts down in order to allow the dried lumber stack(s) getting additional moisture. When the air wet-bulb temperature reaches the first set point, the heat pump restarts.

On the other hand, if the heat pump nominal cooling and dehumidification capacity is higher than the wood dewatering (moisture vaporization) rate, the heat pump compressor will run with frequent and short on/off cycles each time the air wet-bulb temperature reaches its set points. However, too short on/off cycles and associated transient regimes contribute to the compressors wear and lower the system overall efficiency.

To avoid such a situation, the dryer wet-bulb temperature could automatically be set according to the actual measured values, as explained below (Minea 2010).

It is well known that prior starting a new wood drying cycle, the air wet-bulb temperature (WBT_{set}) is preset for each step of the cooling and dehumidification process. Also, the air actual wet-bulb temperature (WBT_{actual}) is measured inside the dryer and transferred to the system process controller automatically.

At the end of the preheating step, the actual wet-bulb temperature inside the dryer $\left(WBT_{actual}^{preheat}\right)$ becomes equal to the preset value of the first drying step $\left(WBT_{set}^{1}\right)$, and, thus, the heat pump compressor starts.

During the first step of the heat pump-assisted drying process, because the heat pump compressor is running and the dryer is a relatively tight enclosure with all air vents closed, the difference between the actual air wet-bulb temperature and the first preset wet-bulb temperature $\left(WBT_{set}^{1}\right)$ gradually decreases over time. When this difference reaches, for example, 5°C, the system controller may automatically lower the set point of the air wet-bulb temperature, for instance, by 2.5°C. This is the second set point for the air wet-bulb temperature $\left(WBT_{set}^{2}\right)$.

After having modified the first set point of the air wet-bulb temperature, the heat pump compressor does not stop but continues to run in cooling and dehumidification mode. Consequently, the difference between the actual (measured) wet-bulb temperature and the second set wet-bulb temperature $\left(WBT_{set}^{2}\right)$ continues to gradually decrease over time. When the difference between the actual air wet-bulb temperature and the second set air wet-bulb temperature reaches 5°C once again, the system controller may again automatically lower the second set air wet-bulb temperature by 2.5°C. This is the third set point for the air wet-bulb temperature $\left(WBT_{set}^{3}\right)$.

Such an automatic control sequence could be applied several times depending on the physical characteristics of the dried lumber stack(s) and many other parameters.

Globally, the wet-bulb temperature is set according to a transfer function as follows:

$$WBT_{set}^{i} = WBT_{set}^{i-1} - 2.5°C \tag{9.34}$$

where

i is the number of successive wet-bulb temperature settings during the entire drying cycle

9.12.2.3 Refrigerant Charge and Migration

The heat pump refrigerant charge mainly consists of condenser and evaporator, and suction, discharge, and liquid line quantities. The heat pump total charge also includes the liquid receiver minimum operating and some amounts existing in other (incontrollable) parts of the system.

Proper refrigerant charging contributes to normal energy efficiency of heat pump dryers, while improperly charged systems work with poor heat transfer coefficients and reduced efficiency. For example, a 15% under-charge can result in a 25% decrease in heat pump efficiency. As the charge drops further, efficiency decreases faster, to 50% or more. Similar losses of efficiency also result from overcharging the heat pump. Either over- or undercharging finally increases operating costs and maintenance expenses. Moreover, a poorly charged heat pump will not properly cool or lubricate the compressor and, thus, may significantly shorten its technical life.

Another major danger for the heat pump compressors comes from the liquid entering the suction port that represents an additional significant danger for the mechanical integrity of the compressor.

To manage the refrigerant migration and prevent the liquid from entering the compressor suction line, especially in small-scale heat pump dryers, a current method consists in installing a liquid valve LV between the condenser CD and the liquid receiver LR, as shown in Figure 9.22a. The liquid valve LV controls the refrigerant migration during the compressor on/off cycling and standby periods. When the compressor shuts down, the liquid valve LV immediately closes, but the compressor continues to run for additional few seconds reducing the pressure in the system between the solenoid valve and the suction side of the compressor, while compressed hot vapor is condensed and stored in the liquid receiver. The refrigerant from the evaporator EV is also pumped down and stored inside the condenser CD and the liquid receiver LR. When the suction pressure reaches the set point of the low-pressure sensor, the controller disconnects the power to the compressor. At the end of this sequence, the evaporator is practically free of any refrigerant, and the condenser and liquid receiver store the majority of the refrigerant charge. Later, when the drying control system calls for cooling and dehumidifying, the solenoid valve opens and refrigerant flows toward the compressor with a consequent rise in pressure. When the pressure reaches the low-pressure switch set point, the controller starts the compressor. When the

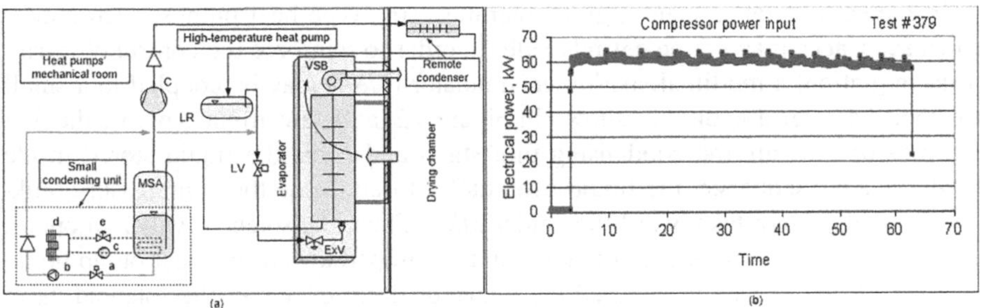

FIGURE 9.22 (a) Means of managing liquid migration in large- and very-large-scale industrial heat pump-assisted dryers (Reprinted from Minea (2015) with permission from International Refrigeration Institute); (b) example of a continuous compressor running profile achieved by preventing the refrigerant liquid from entering the suction port; a, solenoid valve; b, refrigerant liquid pump; c, small compressor; d, small air-cooled condenser; e, expansion valve; C, heat pump compressor; LR, liquid receiver; LV, liquid valve; VSB, variable speed blower.

compressor restarts, the liquid valve LV first opens for approximately 5–8 seconds, while the compressor is still off. During this period of time, a certain amount of liquid flows from the condenser to the inlet port of the expansion valve, and then the compressor restarts. At this moment, the oil pressure difference will be higher than the minimum differential limit, and the compressor will not shut down. In relatively small capacity heat pump dryers, the liquid valve is able to totally address the problem.

During the off cycle, there is only a small amount of refrigerant in the system on the suction side of the compressor, and since the low-pressure switch is set for a relatively low pressure, the compressor starts under a relatively light load. After it starts, the unloading system takes over control of the compressor.

In the case of high-temperature split heat pump dryers, there is a significant temperature difference (up to 100°C) between the drying enclosure and the mechanical room during the preheating step and heat pump off and standby periods.

In the case of large- and very-large-capacity high-temperature split heat pumps, the control of refrigerant migration may not be entirely safe for the compressor mechanical integrity only by using the pump-down strategy. When high-capacity wood drying systems are not running for relatively long periods of time, the refrigerant may flow in both directions through the check valve CV1 (Figure 9.22a). Because the compressor C and the suction accumulator SA are metallic masses at lower temperatures, the refrigerant naturally migrates and condenses inside them. When the compressor starts at the first dehumidification step, it may suck up the stored liquid refrigerant.

Because the liquid is incompressible, the pressure differential will be too low and the large-scale compressor will automatically shut down. After the pressure levelling, the compressor will try to start-up again. Several such on/off cycles may occur until the whole quantity of liquid refrigerant is pumped out and the set pressure differential is restored.

In spite of conventional protections that may be used, however, too frequent on/off cycles reduce the useful life of reciprocating compressors and may damage their moving parts (pistons, connecting rods) and cylinders.

In the case of large-scale high-temperature split-type heat pumps, such pump-down sequences may not be entirely safe. In order to reinforce the control of refrigerant migration, a modified suction accumulator (*MSA*) may be coupled to a small air-cooled condensing unit, as shown in Figure 9.22a. A few minutes before the first compressor start-up, the condensing unit starts and cools the liquid stored inside *MSA*. As a consequence, the liquid eventually stored inside the compressor crankcase naturally migrates toward the colder *MSA*. Then, the solenoid valve (a) opens, the liquid pump (b) starts-up, and the liquid is pumped toward the heat pump liquid receiver *LR*. After having pumped the whole quantity of liquid, the solenoid valve (a) closes, the liquid pump (b) shuts down, and the heat pump compressor C starts up safely. Figure 9.22b represents an improved, continuous compressor running profile achieved by preventing the refrigerant liquid from entering the compression suction port (Minea 2015).

9.12.2.4 Thermal Shocks

Heat pump-assisted drying systems need smaller quantities of backup energy compared to most conventional kilns. The reason is that the heat pump recovers sensible and latent heat from the dryer warm and humid air, and returns them to the dryer enclosure. Moreover, the compressor input electrical energy is added to the recovered heat in the form of equivalent thermal energy. Consequently, the additional quantity of heating energy required by dryers using heat pumps as dehumidifiers is much smaller than that of conventional wood drying systems. Additional heat is, however, needed only to compensate for any lack of heat pump heating capacity, dryer heat losses, air leakages, and other uncontrollable energy losses.

With both compact and split drying heat pumps, the backup heating supplying control is a prominent issue. Appropriate strategy helps avoid useless material stresses and, especially, saves energy.

It is well known that multiblade dryer fan circulate the air through the lumber stack(s) (Figure 9.23a). Its rotation direction periodically changes every 3 hours at the beginning, and every 2 hours at the end of the heat pump dehumidification process (Figure 9.23b). A number of air vents open when the dryer fan changes rotation direction in order to avoid air implosion hazards, and, also, when the dryer dry-bulb temperature exceeds its setting point, to avoid excessive superheating. When the dryer fans turn in direction 1, the dryer warm and humid air first passes through the steam heat exchangers where it is heated, if necessary, then through the heat pump condenser. When the dryer blower turns in the direction 2, the dryer warm air first passes through the heat pump condenser, then through the backup heat exchanger and, finally, through the material stack.

In the case of rotation direction 2, sudden and intense thermal shocks on the remote condenser may provide unusual high compressor discharge (and condensing) pressures and temperatures. To avoid such an abnormal operation condition, adequate backup heating strategies have to be used especially for large-scale split heat pump dryers.

In large-scale conventional heat pump dryers, large-capacity solenoid valves (SV) are used (Figure 9.24a) to direct the heat carrier (e.g., steam) at full flow rates

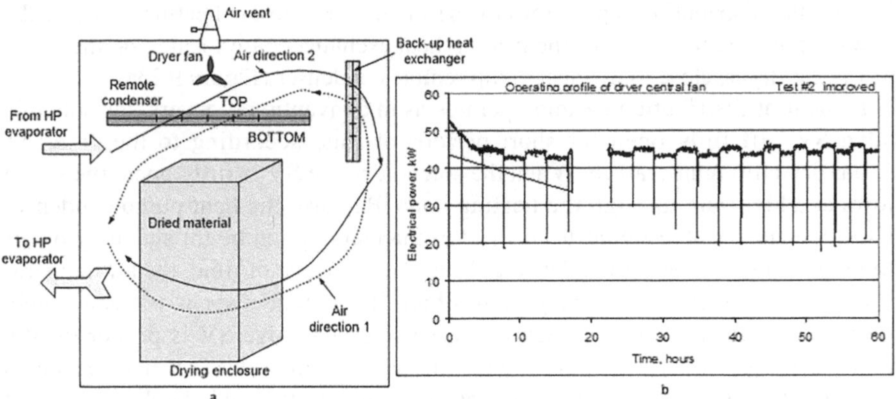

FIGURE 9.23 Softwood dryer with HT heat pump; (a) rotation directions of the dryer central fan; (b) operating profile of the dryer central fan(s).

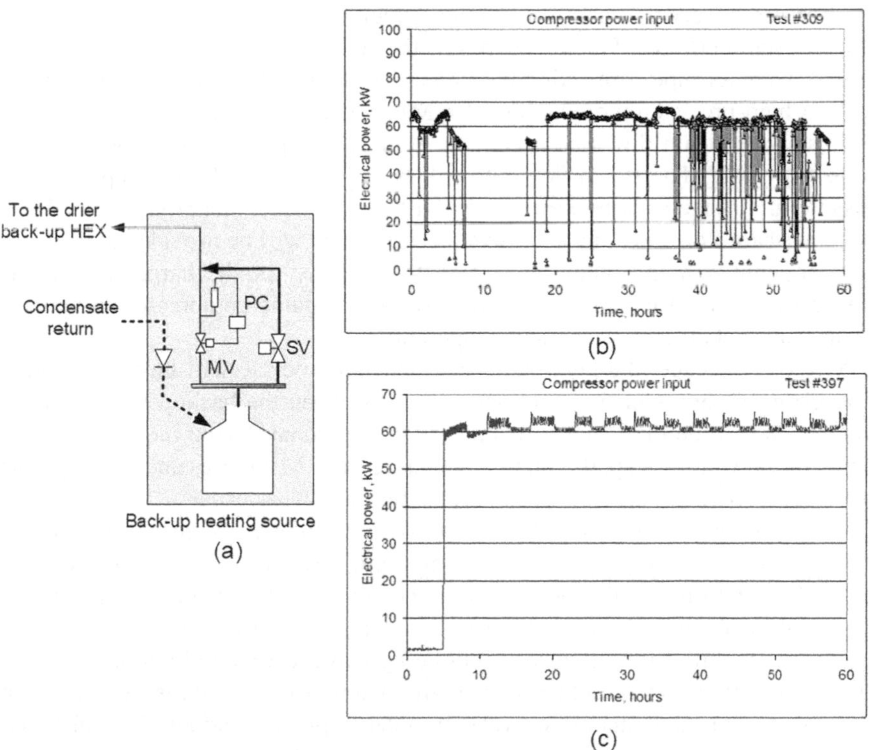

FIGURE 9.24 Improved back-up heating control with small modulating valve (MV) and pressure sensor (PS); (a) configuration; (b) abnormal compressor operating profile with large steam solenoid valve SV; (c) improved, normal compressor operating profile with small steam modulating valve MV.

through the distribution piping toward the dryer steam-to-air heating coils. In the case of split-type heat pumps, the heating heat exchangers are located on the upper part of the dryer, close to the heat pump remote condenser (Figure 9.23a).

If the heat distribution system operates as in conventional dryers, the solenoid valve SV will fully open for short periods of time according to the dryer air actual dry-bulb temperature. When the solenoid valve SV is fully open, the steam flows at a high rate through the heating coils. Because the heat pump condenser is located close to the steam-to-air heat exchangers, it can be hit suddenly by air flowing at high temperatures. During these short periods of time, the air temperatures at the condenser inlet may be too high if the condenser is not adequately thermally protected. Even though the large solenoid valve SV is proportionally modulated, as shown in Figure 9.24a, the heating coils will receive too much quantity of heat, a situation that may produce sudden thermal shocks at the heat pump's condenser top inlet that excessively increase the compressor discharge pressure and temperature, and may break down the compressor lubricating oil and materials.

Exceeding the discharge pressure limit also allows the compressor operating with short on/off cycles that may mechanically damage it. In order to address these safety issues, a second steam distribution circuit on the steam distribution header may be installed (Figure 9.24a). It bypasses the main solenoid valve SV and includes a small-diameter pipe connecting the steam manifold to the steam distribution network. A high-precision modulating valve MV, activated by the pressure sensor PS, modulates the steam flow rate according to the steam pressure leaving the modulating valve MV. The steam flow rate will be adjusted in order to keep the outlet pressure in a range of approximately 5%–10% of the inlet steam pressure. With a constant steam flow rate, any excessive thermal shock will be provided at the top of the heat pump remote condenser (Figure 9.24a). Thus, the discharge pressure and temperature will not increase suddenly, and the heat pump compressor will operate at constant discharge pressures and temperatures.

As a consequence, the heat pump thermodynamic cycle will be stable, and no unnecessary compressor on/off cycles will occur when the backup heating system operates. When the dryer has to operate in conventional drying modes, and, also, during the preheating step, the small modulating valve MV closes and the large solenoid control valve SV controls the steam heating supply, as usual.

During the preheating step, the steam-to-air heating coils installed inside the drying enclosure are supplied with high-pressure steam via a large on/off solenoid valve SV that fully (100%) opens. After the heat pump start-up, the quantity of additional heat required is much lower compared to that of the preheating step. Therefore, a small, high precision modulating valve, activated by a pressure sensor, must installed in parallel with the main solenoid valve SV. This valve will supply small but constant steam flow rates in order to prevent sudden thermal shocks at the remote condenser aïr inlet and, thus, ensure discharge pressure and temperature within normal, safe operating ranges. As can be seen in Figure 9.24b, if the large on/off solenoid valve is used to intermittently supply additional (backup) heating steam, excessively high compressor discharge pressures can be achieved (Minea 2015).

9.12.2.5 Avoid Compressor Failure

A major issue for industrial heat pump-assisted wood dryers consists in increasing the useful life of the compressors. Inadequate amount of refrigerant superheating (see Section 9.12.2.1), inappropriate setting of wet-bulb temperature (see Section 9.12.2.2), and deficient management of refrigerant migration (see Section 9.12.2.3), able to provide thermal shocks on remote condenser(s) (see Section 9.12.2.3), may contribute at the heat pump compressor failure and even at its destruction.

The most significant dangers come from the liquid refrigerant entering the compressor suction line, from the partial vapor condensation during the actual compression process, because non-superheated vapor entering the compressor may produce serious mechanical damages to the compressor.

Figures 9.24b and 9.25a, b show some abnormal high compressor discharge pressures and frequent cycling during wood drying processes with a high-temperature heat pump.

Figure 9.25c shows a number of moving parts (cylinders, pistons, and connecting rods) of a destroyed compressor as a consequence of frequent abnormal operation conditions.

Usually, the heat pump compressors are provided with devices for protection against abnormal operating conditions. Among them the following can be mentioned (ASHRAE 2008): (i) a high-pressure cutout (a pressure sensor that sends a signal to the switch to cut power to the compressor motor); (ii) a high-to-low-side internal relief valve (e.g., to limit maximum pressure in hermetic compressors not

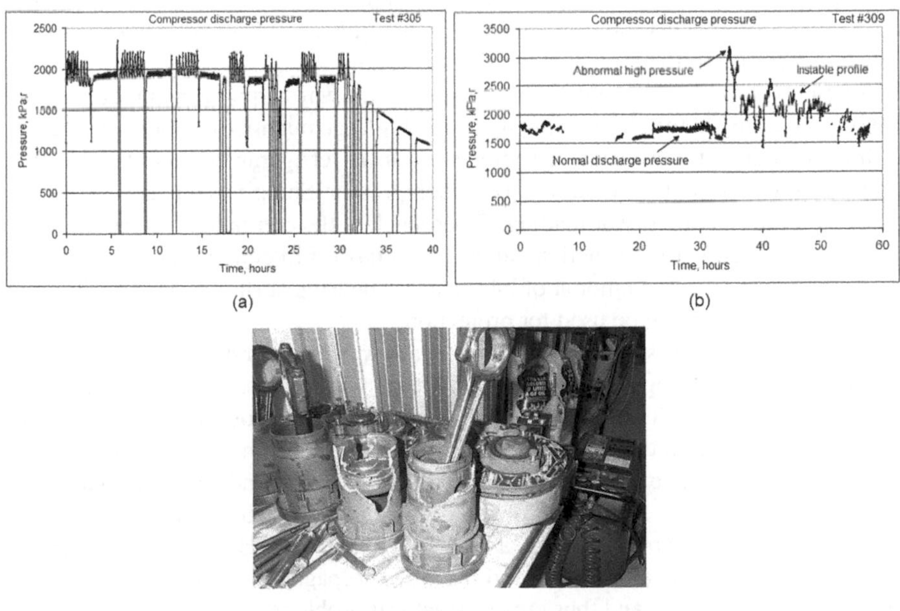

FIGURE 9.25 (a) and (b) abnormal profiles of compressor discharge pressures; (c) view of damaged compressor components

equipped with other high-pressure control devices) and an external relief valve of which differential pressure settings depend on the refrigerant used and operating conditions; (iii) a relief valve assembly on the oil separator of screw compressors; (iv) high-temperature temperature control sensor to protect against overheating the compressor motor and refrigerant and oil breakdown; sometimes, they are used to stop the compressor when discharge temperature exceeds safe values via a switch that may be placed internally (near the compression chamber) or externally (on the discharge line); (v) on large-capacity compressors, lubricant temperature may be controlled by cooling with a heat exchanger or direct liquid injection, or the compressor may shut down on high lubricant temperature; (vi) here lubricant sump heaters are used to maintain a minimum lubricant sump temperature, and a thermostat may be used to limit the maximum lubricant temperature; (vii) low-suction pressure protection set to limit the compression ratio or evaporator freeze-up; (ix) forced-feed lubrication systems based on lubricant pressure, minimum-flow, or minimum-level protectors to prevent the compressor from operating with insufficient lubricant pressure; (x) time delay or lockouts with manual resets prevent damage to both compressor motor and contactors from repetitive rapid starting cycles; fixed-speed compressor motors experience a significant inrush electrical current during start-up; this current can reach the level of locked rotor amps; if the motor restarts rapidly, without adequate cooling, overheating damage can occur; time delays should be set for an appropriate interval to avoid this hazard; and (xi) suction line inlet strainer to remove any dirt that might be in suction line piping, normally required in all field-assembled systems.

Liquid is essentially incompressible, so damage may occur when a compressor is handling liquid. This damage depends on the quantity of liquid, frequency with which it occurs, and the type of compressor.

Slugging, flood-back, and flooded starts are three ways liquid can damage a compressor. Slugging is the short-term pumping of a large quantity of liquid refrigerant and/or lubricant. It can occur just after start-up if refrigerant accumulated in the evaporator during shutdown returns to the compressor. Slugging can also occur with quick changes in compressor loading. Flood-back is the continuous return of liquid refrigerant mixed with the suction vapor. It is a hazard to compressors that depend on maintaining a certain amount of lubricant for bearing surfaces. A properly sized suction accumulator can be used for protection.

Flooded start occurs when refrigerant is allowed to migrate to the compressor during shutdown. Compressors can be protected with crankcase heaters and automatic pump-down cycles. Suction pulsation occurs because of the sudden flow and slight pressure drop in the suction line at the end of the reverse portion of the cycle and during suction. The frequency of this pulsation corresponds to the frequency of the compression cycle. Amplitude of the pulsation can reach 10 kPa, especially if the compression chamber serves as a resonator. Suction pulsations may affect the compressor volumetric efficiency. Moreover, the propagation of compression waves may generate vibrations and thus create structural problems.

At the locked rotor conditions, the temperature increase must be kept low enough to prevent excessive motor temperature. In other words, the maximum temperature under these conditions should be held within the limits of the materials. With better protection (and improved materials), higher temperatures can be tolerated, and less expensive motors can be used.

In the case of heat pump-assisted dryers, one or more of previous mentioned factor may lead to heat pump compressor damage and even mechanical failure.

REFERENCES

Adapa, P.K., G.J. Schoenau, S. Sokhansanj. 2002a. Performance study of a heat pump system for specialty crops – Part 2: Model verification. *International Journal of Energy Research* 26:1021–1033.

Aleon, D., G. Negrie, J. Perez, O. Snieg. 1988. *Séchage de sapin et de pin sylvestre à haute température.* Centre technique du bois et de l'ameublement, Paris, France.

Alves-Filho, O., T. Tokle. 1999. Heat pumps dry food products in Norway. *IEA Heat Pump Centre Newsletter* 17(4):4–5.

Alves-Filho, O. 2002. Combined innovative heat pump drying technologies and new cold extrusion techniques for production of instant foods. *Drying Technology* 20(8):1542–1557.

Alves-Filho, O., P.G. Pascual, T.M. Eikevik, I. Strommen. 2004. Dehydration of green peas under atmospheric freeze-drying conditions. In *Proceedings of the 14th International Drying Symposium (IDS2004)*, Sao Paulo, Brazil, August 22–25, Vol. C, pp. 1521–1528.

Alves-Filho, O., M.L. Garcia, I. Strømmen, T.M. Eikevik. 2006. Heat pump atmospheric sublimation and evaporative drying of bovine intestines. In *Proceedings of the 15th International Drying Symposium (IDS2006)*, Budapest, Hungary, August 20–23.

Alves-Filho, O., T.M. Eikevik. 2007. Cheese drying and ripening kinetics in applications of heat pump technologies. In *Proceedings of the 5th Asia Pacific Drying Conference*, Hong Kong, August 13–15.

Alves-Filho, O., T.M. Eikevik. 2008. Hybrid heat pump drying technologies for porous materials. In *Proceedings of the 16th International Drying Symposium (IDS2008)*, Hyderabad, India, November 9–12.

Alves-Filho, O., O. Guzev, S. Goncharova-Alves. 2008. Kinetics and energy aspects of heat pump drying of protein. In *Proceedings of the 16th International Drying Symposium (IDS 2008)*, Hyderabad, India, November 9–12.

Alves-Filho, O., E. Thorbergen, I. Strommen. 1998a. A component model for simulation of multiple fluidized bed heat pump driers. In *Proceedings of the 11th International Drying Symposium*, Vol. A, pp. 94–102, August 19–22.

Alves-Filho, O., T. Lystad, T. Eikevik, I. Strommen. 1998b. A new carbon dioxide heat pump dryer – An approach for better product quality, energy use and environmentally friendly technology. In *Proceedings of the Symposium on Drying Technologies*, St-Hyacinthe, Québec, Canada, September 3–5.

Anderson, J.O., L. Westerlund. 2014. Improved energy efficiency in sawmill drying system. *Applied Energy* 113:891–901.

ASHRAE. 2008. *ASHRAE Handbook.* HVAC Systems and Equipment. SI Edition, American Society of Heating, Refrigerating and Air-Conditioning Engineers, Inc., Atlanta, GA.

Bannister, P., B. Bansal, C.G. Carrington, Z.F. Sun. 1998. Impact of kiln losses on a dehumidifier drier. *International Journal of Energy Research* 22:515–522.

Bannister, P., G. Carrington, G. Chen, Z.F. Sun. 1999. Guidelines for operating dehumidifier timber kilns. Energy Group's Heat Pump Dehumidifier Research Programme Report, EGL-RR-02.

Bannister, P., G. Carrington, G. Chen, Z. Sun. 1999. Energy Group's Heat Pump Dehumidifier Research Programme, Report EGL-RR-02, 1st edition.

Barker, D., J. Guttridge. 1988. Hardwood drying at La-Z Boy Canada Ltd. sharing experiences. In *Seminar on Industrial Applications for Heat Pumps*, Ontario Research Foundation, Mississauga, Ontario, October 1–2, pp. 103–108.

Cai, L., L.C. Oliveira. 2008. A simulation of wet pocket lumber drying. *Drying Technology* 26(5):525–529.

Cardona, T.D., R.H. Driscoll, J.L. Paterson, G.S. Srzednicki, W.S. Kim. 2002. Optimizing conditions for heat pump dehydration of lactic acid bacteria. *Drying Technology* 20(8):1611–1632.

Carrington, C.G., P. Bannister, Q. Liu. 1995. Performance analysis of a dehumidifier using HFC-134a. *International Journal of Refrigeration* 18:477–488.

Carrington, G., Z. Sun, P. Bannister. 1998. Dehumidifier driers – What to expect and pitfalls to avoid. In *Proceedings of the Wood Technology Workshop*, Christchurch, New Zealand, October 23, pp. 6–12.

Carrington, C.G., Z.F. Sun, Q. Sun, P. Bannister, G. Chen. 1999. Dehumidifier dryers for hard-to-dry timbers. In *Proceedings of the 20th International Congress of Refrigeration*, Sydney, Australia, September 19–24.

Carrington, C.G., C.M. Wells, Z.F. Sun, G. Chen. 2002. Use of dynamic modeling for the design of batch-mode dehumidifier dryers. *Drying Technology* 20(8):1645–1657.

Cech, M.Y., D.R. Huffman. 1974. *High-Temperature Drilling of Mixed Spruce. Jack Pine and Balsam Fir*. Eastern Forest Products Lab, Ottawa, Ontario.

Cech, M.J., F. Pfaff. 2000. *Operator Wood Drier Handbook for East of Canada*, edited by FORINTEK Corp., Canada's Eastern Forester Products Laboratory, Québev city, Québec, Canada.

Ceylan, I., M. Aktas, H. Dogan. 2007. Mathematical modelling of drying characteristics of tropical fruits. *Applied Thermal Engineering* 27:1931–1936.

Chen, G., R.B. Keey, J.C.F. Walker. 1997. The drying stress and check development on high-temperature kiln seasoning of sapwood *Pinus radiata* boards. *Holz als Roh-und Werkstoff* 55:169–173.

Chin, S.K., S.L. Law. 2002. Product quality and drying characteristics of intermittent heat pump drying of *Ganoderma tsugae murrill*. *Drying Technology* 27:1457–1465.

Chua, K.J. 2000. Dynamic modelling, experimentation, and optimization of heat pump drying for agricultural products. PhD Thesis, National University of Singapore, Department of Mechanical Engineering, Singapore.

Chua, K.J., S.K. Chou, J.C. Ho, N.A. Hawlader. 2002. Heat pump drying: Recent developments and future trends. *Drying Technology* 20(8):1579–1610.

Chua, K.J., S.K. Chou. 2005. A modular approach to study the performance of a two-stage heat pump system for drying. *Applied Thermal Engineering* 25:1363–1379.

Chou, S.K., K.J. Chua. 2006. Heat pump drying systems. In *Handbook of Industrial Drying*, edited by A.S. Mujumdar, Taylor & Francis Inc., Boca Raton, FL, pp. 1122–1123.

Claussen, I.C., T.S. Ustad, I. Strommen, P.M. Walde. 2007. Atmospheric freeze drying – A review. *Drying Technology* 25:957–967.

Colak, N., A. Hepbasli. 2005. Exergy analysis of drying of apple in a heat pump dryer. In *Proceedings of the 2nd International Conference of the Food Industries & Nutrition*, Division on Future Trends in Food Science and Nutrition, pp. 145–158.

Colak, N., A. Hepbasli. 2009. A review of heat pump drying: Part 2 – Applications and performance assessments. *Energy Conversion and Management* 50:2187–2199.

Davis, C. 2001. Dehumidifier drying of *Pinus radiata* boards. PhD Thesis, Physics Department, Otago University, New Zealand.

Eikevik, T.M., I. Strommen, O. Alves-Filho. 1999. Design and dimensioning criteria of heat pump dryers. In *Proceedings of the 20th International Congress of Refrigeration*, Sydney, September 19–24.

Eikevik, T.M., I. Strommen, O. Alves-Filho, A.K.T. Hemmingen. 2005a. Effect of operating conditions on atmospheric freeze dried codfish. In *Proceedings of the 3rd Inter-American Drying Conference (IADC2005)*, Montreal, Canada, August 21–23.

Eikevik, T.M., I. Strommen, O. Alves-Filho. 2005b. Heat pump fluidized bed dryer with CO_2 as refrigerant – Measurements of COP and SMER. In *Proceedings of the 3rd Nordic Drying Conference*, Karlstad, Sweden, June 15–17.

Erbay, Z., F. Icier. 2009. Optimization of drying of olive leaves in a pilot-scale heat pump dryer. *Drying Technology* 27:416–427.

Fatouh, M., M.N. Metwally, A.B. Helali, M.H. Shedid. 2006. Herb drying using a heat pump dryer. *Energy Conversion Management* 47:2629–2643.

Fernandez-Golfin Seco, J.I., J.J. Fernandez-Golfin Seco, E. Hermoso Prieto, M. Conde Garcia. 2004. Evaluation at industrial scale of electric-driven heat pump dryers (HPD). *Holz Roh Werkst* 62:261–267; doi: 10.1007/s00107-004-0483-0.

Fornasieri, E., F. Mancini, S. Minetto. 2010. Theoretical analysis and experimental test of a CO_2 household heat pump dryer. In *Proceedings of the 9th IIR Gustav Lorentzen Conference*, Sydney, Australia, April 12–14.

Guang, Z.B., L. Liang, Z.Y. Dong, F.Y. Fei. 2007. Experimental study on sludge drying with heat pump dehumidifier. In *Proceedings of the 5th Asia-Pacific Drying Conference*, Hong Kong, August 13–15.

Hawlader, M.N.A., C.O. Pereira, M. Tian. 2006. Comparison of the retention of 6-gingerol in drying under modified atmosphere heat pump drying and other drying methods. *Drying Technology* 24:51–56.

Hepbasli, A., N. Colak, E. Hancioglu, F. Icier, Z. Erbay. 2010. Exergoeconomic analysis of plum drying in a heat pump. *Drying Technology* 28:1385–1395.

Ho, J.C., S.K. Chou, A.S. Mujumbar, M.N.A. Hawlader, K.J. Chua. 2001. An optimization framework for drying of sensitive products. *Applied Thermal Engineering* 21:1779–1798.

Honma, M., T. Tamura, Y. Yakumaru, F. Nishiwaki. 2008. Experimental study on compact heat pump system for clothes drying using CO_2 as a refrigerant. In *Proceedings of the 7th IIR Gustav Lorentzen Conference on Natural Working Fluids*, Trondheim, Norway, May 29–31.

Islam, M.R., A.S. Mujumdar. 2008. Heat pump-assisted drying. In *Guide to Industrial Drying: Principles, Equipment, and New Development*, edited by A.S. Mujumdar, International Drying Symposium, Hyderabad, India.

Jinjiang, Z., W. Yaosen. 2007. Experimental study on drying high moisture content paddy by super-conducting heat pump dryer. In *Proceeding of the 5th Asia-Pacific Drying Conference*, Hong Kong, August 13–15.

Keey, R.B., T.A.G. Longish, J.C.F. Walker. 2000. *Kiln-Drying of Lumber*, Springer Verlag, Berlin, Heideberg.

Keey, R.B., J.J. Nijdam. 2002. Moisture movement on drying softwood boards and kiln design. *Drying Technology* 20(10):1955–1974.

Kiang, C.S., C.K. Jon. 2006. Heat pump drying systems. In *Handbook of Industrial Drying*, edited by A.S. Mujumdar, Taylor & Francis Inc., Boca Raton, FL, pp. 1104–1105.

Kiang, C.S., C.K. Jon. 2007. Heat pump drying systems. In *Handbook of Industrial Drying*, 3rd edition, edited by A.S. Mujumdar, CRC Press, Taylor & Francis, New York, pp. 1104–1132.

Klöcker, K., E.L. Schmidt, F. Steimle. 2001. Cabon dioxide as a working fluid in drying heat pump. *International Journal of Refrigeration* 24(2):100–107.

Klöcker, K., E.L. Schmidt, F. Steimle. 2002. A drying heat pump using carbon dioxide as working fluid. *Drying Technology* 20(8):1659–1671.

Koch, P. 1971. Process for straightening and drying southern pine 2 by 4's in 24 hours. *Forest Products Journal* 21(5):17–23.

Koch, P. 1974. Serrated kiln sticks and top load substantially reduce warp in southern pine studs dried at 240°F. *Forest Products Journal* 24(II):30–34.

Kowalski, S.J., A. Pawlowski. 2008. Drying in non-stationary conditions. *Chemical Process Engineering* 29:337–344.

Lee, K.H., O.J. Kim. 2007. Experimental study on the energy efficiency and drying performance of the batch-type heat pump dryer. In *Proceeding of the 5th Asia-Pacific Drying Conference*, Hong Kong, August 13–15.

Lee, K.H., O.J. Kim, J.R. Kim. 2008. Drying performance simulation of a two-cycle heat pump dryer for high temperature drying. *Drying Technology* 28:683–689.

Law, C.L., A.S. Mujumdar. 2008. Energy aspects in energy drying. In *Guide to Industrial Drying: Principles, Equipment and New Developments*, Chapter 14, edited by A.S. Mujumdar, IDS2008, Hyderabad, India.

Léger, F., M. Amazouz. 2003. Évaluation des pratiques de control et de commande des séchoirs à bois (in French). Centre de la technologie de l'énergie de CANMET – Varennes, April.

Lewis, D.C. 1981. Lumber conditioning kiln. US Patent # 4250629 A.

Lewis, D.C. 2003. High temperature dehumidification systems. US Patent # 20030208923 A1.

Marnoto, T., E. Sulistyowati, M.M. Syahri. 2012. *The Characteristic of Heat Pump Dehumidifier Dryer in the Drying of Red Chili (Capsium annum L)*. Chemical Engineering Department, Faculty of Industrial Technology, UPN Veteran Yogyakarta, Indonesia.

McCurdy, M.C., S. Pang. 2007. Optimization of kiln drying for softwood, through simulation of wood stack drying, energy use, and wood color change. *Drying Technology* 25:1733–1740.

Minea, V. 2004. Heat pumps for wood drying – New developments and preliminary results. In *Proceedings of the 14th International Drying Symposium*, Sao Paulo, Brazil, Vol. B, pp. 892–899, August 22–25.

Minea, V. 2006. Hardwood drying with low temperature heat pumps. In *Proceedings of the 15th International Drying Symposium*, Budapest, Hungary, Vol. C, pp. 1757–1762, August 20–23.

Minea, V. 2007. Energetic and ecological aspects of wood drying with heat pumps. In *Proceedings of the 5th Asia-Pacific Drying Conference*, Hong Kong, Vol. 1, pp. 434–442, August 13–15.

Minea, V. 2008. Design and control optimizations of drying heat pumps. In *Proceedings of the 16th International Drying Symposium (IDS2008)*, November 9–12, Hyderabad, India.

Minea, V. 2009. Avoiding failures of large-scale high-temperature drying heat pumps. In *Proceedings of the 6th International Drying Symposium* (*ADC2009*), Bangkok, Thailand, October 19–21, pp. 179–187.

Minea, V. 2010a. Improvements of high-temperature drying heat pumps. *International Journal of Refrigeration* 33(1):180–195.

Minea, V. 2010b. Industrial drying heat pumps. In *Refrigeration: Theory, Technology and Applications*. Nova Science Publishers, Hauppauge, NY, pp. 1–70.

Minea, V. 2012. Efficient energy recovery with wood drying heat pumps. *Drying Technology* 30:1630–1643.

Minea, V. 2013a. Review – Part I: Drying heat pumps – System integration. *International Journal of Refrigeration* 36:643–658.

Minea, V. 2013b. Review – Part II: Drying heat pumps – Agro-food, biological and wood products. *International Journal of Refrigeration* 36:659–673.

Minea, V. 2014a. Overview of heat pump-assisted drying systems. Review – Part I: Integration, control complexity, and applicability of new innovative concepts. *Drying Technology* 33(5):515–526. doi:10.1080/07373937.2014.952377.

Minea, V. 2014b. Overview of heat pump-assisted drying systems. Review – Part II: Data provided vs. results reported. *Drying Technology* 33(5):527–540. doi:10.1080/07373937.2014.952378.

Minea, V. 2015. High-temperature heat pump-assisted softwood dryer: Sizing and control requirements & energy performances. In *Proceedings of the 24th International Congress of Refrigeration (ICR2015)*, Yokohama, Japan, August 16–22.

Minea, V. 2016. Advances in heat pump-assisted drying technology. In *Advances in Drying Technology*, edited by V. Minea, Francis & Taylor, CRC Press – Taylor & Francis Group, Boca Raton, FL, pp. 1–116.

Metaxas, A.C., R. Meredith. 1983. *Industrial Microwave Heating*. Peter Peregrinus Ltd, London.

Mortezapour, H., B. Ghobadian, S. Minaei, M.H. Khoshtaghaza. 2012. Saffron drying with a heat pump-assisted hybrid photovoltaic-thermal solar dryer. *Drying Technology* 28:560–566.

Mujumdar, A.S. 1996. Innovation in drying. *Drying Technology* 14(6):1459–1475.

Mujumdar, A.S. 2000. Fundamental principles of drying. In *Mujumdar's Practical Guide to Industrial Drying, Principles, Equipment and New Developments*, edited by S. Devahastin, Thananuch Business Ltd. Publication, Bangkok, Thailand, pp. 1–56.

Mujumdar, A.S. 2000. *Mujumdar's Practical Guide to Industrial Drying, Principles, Equipment and New Developments*, edited by S. Devahastin, Thananuch Business Ltd. Publication, Bangkok, Thailand.

Mujumdar, A.S. 2008. Guide to industrial drying. Principles, equipments & new developments.

Mujumdar, A.S., Z.H. Wu. 2008. Thermal drying technologies – Cost-effective innovation aided by mathematical modeling approach. *Drying Technology* 26(1):146–154.

Mujumdar, A.S. 2015. Principles, classification, and selection of dryers. In *Handbook of Industrial Drying*, 4th edition, edited by A.A. Mujumdar, Taylor & Francis.

Ong, S.P., C.L. Law, C.L. Hii. 2012. Optimization of heat pump-assisted intermittent drying. *Drying Technology* 30:1676–1687.

Pal, U.S., M.K. Khan, S.N. Mohanty. 2008. Heat pump drying of green sweet pepper. *Drying Technology* 26:1584–1590.

Pal, U.S., M.K. Khan. 2008. Calculation steps for the design of different components of heat pump dryers under constant drying conditions. *Drying Technology* 26:864–872.

Pang, S., A.N. Haslett. 1995. The application of mathematical models to the commercial high-temperature drying of softwood lumber. *Drying Technology* 13(8&9):1635–1674.

Pang, S. 1999. Drying of Radiata pine lumber under dehumidifier conditions: Experiments and model simulation. Research Report, Forest Research, Rotorua, New Zealand.

Pang, S., I.G. Simpson, A.N. Haslett. 2001. Cooling and steam conditioning after high-temperature drying of *Pinus radiata* board: Experimental investigation and mathematical modelling. *Wood Science & Technology* 35:487–502.

Perera, C.O., M.S. Rahman. 1997. Heat pump dehumidifier drying of food. *Trends in Food Sciences Technology* 8:75–79.

Perré, P., C. Moyne. 1991. Processes related to drying. Part II. Use of the same model to solve transfers both in saturated and unsaturated porous media. *Drying Technology* 9:1153–1179.

Perré, P., R. Keey. 2015. Drying of wood: Principles and practice. In *Handbook of Industrial Drying*, 4th edition, edited by A.J. Mujumdar, CRC Press, Taylor & Francis Group, Dekker, New York, pp. 821–877.

Perry, E.J. 1981. Drying by cascade heat pumps. *Institute of Refrigeration Management* 1:1–8.

Phaphvangwittayakul, W., G.S.V. Raghavan, P. Terdtoon. 2007. Loop thermosyphon application for heat pump drying. In *Proceedings of 3rd Inter-American Drying Conference (IADC2007)*, Montréal, Canada, August 21–23.

Phoungchandang, S., W. Srinukroh, B. Leenanon. 2008. Kaffir Lime Leaf (*Citrus hystric DC*) drying using tray and heat pump dehumidified drying. *Drying Technology* 26:1602–1609.

Phoungchandang, S., S. Nongsang, P. Sanchai. 2009. The development of ginger drying using tray drying, heat pump-dehumidified drying and mixed-mode solar drying. *Drying Technology* 27:1123–1131.

Potisate, Y., S. Phoungchandang. 2010. Chlorophyll retention and drying characteristics of ivy gourd leaf (*Coccinia grandis voigt*) using tray and heat pump – assisted dehumidified air drying. *Drying Technology* 28:786–797.

Prasertsan, S., P. Seansaby, G. Prateepchaikul. 1997. Heat pump dryer. Part 3: Experiment verification of the simulation. *International Journal of Energy Research* 21:1–20.

Price, E.W., P. Koch. 1980. Kiln time and temperature affect shrinkage, warp, and mechanical properties of southern pine lumber. *Forest Products Journal* 30(8):41–47.

Queiroz, R., A.L. Gabas, V.R.N. Telis. 2004. Drying kinetics of tomato by using electric resistance and heat pump dryers. *Drying Technology* 22(7):1603–1620.

Rahman, M.S., C.O. Perera, C. Thebaud. 1998. Desorption isotherm and heat pump drying kineticsof peas. *Food Research International* 30(7):685–491.

Rahman, S.M.A., M.R. Islam, A.S. Mujumbar. 2007. Study of a coupled heat and mass transfer in composite food products during convective drying. *Drying Technology* 25:1359–1368.

Sarkar, J., S. Bhattacharyya, M.R. Gopal. 2006. Transcritical CO_2 heat pump dryer: Part 2. Validation and simulation results. *Drying Technology* 24:1593–1600.

Schmidt, E.L., K. Klöcker, N. Flacke, F. Steimle. 1998. Applying the trans-critical CO_2 process to a drying heat pump. *International Journal of Refrigeration* 21(3):202–211.

Senadeeva, W., O. Alves-Filho, T. Eikevik. 2012. Influence of atmospheric sublimation and evaporation on the heat pump fluid bed drying of bovine intestines. *Drying Technology* 30:1583–1591.

Soponronnarit, S., A. Nathakaranakule, S. Wetchacama, T. Swasdisevi, P. Rukprang. 1998. Fruit drying using heat pump. *International Energy Journal* 20:39–53.

Strommen, I., K. Kramer. 1994. New applications of heat pumps in drying processes. *Drying Technology* 12(4):889–901.

Strommen, I., T.M. Eikevik, O. Alves-Filho. 2003. Operational modes for heat pump drying – New technologies and production of a generation of high quality dried fish products. In *Proceedings of the 21st International Congress of Refrigeration*, Washington, DC, August 17–22.

Strømmen, I., T.M. Eikevik, O. Alves-Filho, K. Syverud. 2004. Heat pump drying of sulphate and sulphite cellulose. *Drying 2004*. In *Proceedings of the 14th International Drying Symposium (IDS2004)*, Sao Paulo, Brazil, August 22–25, Vol. B, pp. 1225–1232.

Strømmen, I., T. Eikevik, I.C. Claussen. 2007. Atmospheric freeze drying with heat pumps – new possibilities in drying of biological materials. In *Proceedings of the 5th Asia-Pacific Drying Conference*, Hong Kong, August 13–15.

Strommen, I., T.M. Eikevik, A.F. Odilio. 1999. Optimum design and enhanced performance of heat pump dryers, In *Proceedings of the ADC'99*, pp. 66–80. Bandung, Indonesia. .

Strumillo, C. 2006. Perspectives on development in drying. *Drying Technology* 24:1059–1068.

Sun, Z.F., C.G. Carrington, C. McKenzie, P. Bannister, B. Bansal. 1996. Determination and application of characteristic drying-rate curves in dehumidifier wood drying. In *Proceedings of the 5th IUFRO International Wood Drying Conference*, edited by A. Cloutier, Y. Fortin & R. Gosselin. Quebec City, Canada, August 13–17, pp. 495–503.

Sun, Z.F., C.G. Carrington. 1999. Dynamic modelling of a dehumidifier wood drying kiln. *Drying Technology* 17(4&5):711–729.

Sun, Z.F., C.G. Carrington, P. Bannister. 2000. Part A – Dynamic modelling of the wood stack in a wood drying kiln. *Transactions of I. Chem. E.* 78:107–117.

Sun, Z.F., C.G. Carrington, C. Davis, Q. Sun, S. Pang. 2005. Mathematical modelling and experimental investigation of dehumidifier drying of *Radiata pine* timber. *Maderas Ciencia y Tecnología* 7(2):87–98. (http://dx.doi.org/10.4067/S0718-221X2005000200003, accessed November 16, 2016).

Sunthonvit, N., G. Srzednicki, J. Craske. 2007. Effect of drying treatments on the composition of volatile compounds in dried nectarines. *Drying Technology* 25:877–881.

Vansteenkiste, D., M. Stevens, J. Van Acker. 1997. High temperature drying of fresh sawn poplar wood in an experimental convective dryer. *Holz als Roh-und Werkstoff* 55:307–314.

Velazquez, G., F. Chenlo, R. Moreira, E. Cruz. 1997. Grape drying in a pilot plant with a heat pump. *Drying Technology* 15:899–920.

Wu, Q., W.R. Smith. 1998. Effects of elevated and high-temperature schedules on warp in southern yellow pine lumber. *Forest Products Journal* 48(2):52–56.

Xanthopoulos, G., N. Oikonomou, G. Lambrinos. 2007. Applicability of a single layer drying model to predict the drying rate of all figs. *Journal of Food Engineering* 81:553–559.

Zielinska, M., P. Zapotoczny, O. Alves-Filho, T.M. Eikevik, W. Blaszczak. 2012a. Combined heat pump fluidized bed atmospheric freeze drying and microwave vacuum drying of green peas. In *Proceedings of the 18th International Drying Symposium (IDS2012)*, Xiamen, China, November 11–15.

Zielinska, M., P. Zapotoczny, O. Alves-Filho, T.M. Eikevik, W. Blaszczak. 2012b. Microwave vacuum assisted drying of green peas using heat pump and fluidized bed. *Proceedings of the 18th International Drying Symposium (IDS2012)*, Xiamen, China, November 11–15.

10 Heat Pump Systems for Drying

10.1 INTRODUCTION

Heat pumps are thermodynamic devices allowing for the recovery of energy in the form of sensible and/or latent heat from relatively low-temperature heat reservoirs (e.g., ambient air, soil, groundwater, surface water, or industrial waste heat) and, simultaneously, supplying sensible heat to higher-temperature heat reservoirs (e.g., air or water) for residential, commercial, or industrial usage.

Because the heat is lifted from a lower temperature level to a higher temperature level, this thermodynamic process is called heat pumping where the user is interested in the heat delivered at the high-temperature level. The steady-state thermodynamic processes of heat pumping are governed by the first (conservation of energy) and second (that states that heat cannot be transferred from a lower to a higher temperature without the consumption of energy) laws of thermodynamics (Sauer and Howell 1983; Radermacher and Hwang 2000).

10.2 BASIC THERMODYNAMIC VARIABLES

The first law of thermodynamics uses concepts such as internal energy, heat, work, enthalpy, and specific heat.

10.2.1 Internal Energy

For any thermodynamic system of a given mass m (kg), the total energy (E) generally represents the sum of all energies of the system in a given state: (i) energy associated with the atoms' structure, (ii) kinetic energy associated with molecules' movement, (iii) potential energy associated with molecules' position and attractive forces (i.e., due to gravitational position of the system and the attractive forces existing between molecules), and (iv) any other form of energy (chemical, electrical, etc., due to electrical charges):

$$E = E_{kinetic} + E_{potential} + E_{chemical} + E_{electrical} + \cdots \tag{10.1}$$

It can be noted that changes in the average velocity of the molecules are indicated by temperature changes of the system, while variations in position are denoted by changes in phase of the system.

In thermodynamics, it is common to regroup the kinetic and potential energies into one unique variable called internal (or thermal) energy (U, in J). Thus,

$$E = U + E_{chemical} + E_{electrical} + \cdots \tag{10.2}$$

where

$$U = E_{kinetic} + E_{potential} = \frac{V^2}{2g} + g * z \tag{10.3}$$

where

V is the velocity of the molecules and atoms (m/s)
z is the system elevation (m)
g is the gravitational acceleration (m^2/s)

While conserving the total energy of the system, the potential energy can be converted into kinetic energy.

The internal energy refers to the invisible microscopic energy on the atomic and molecular scale. It is thus defined as the energy associated with the random, disordered/disorganized motion of molecules and atoms.

Internal energy depends only on the state of the system and does not depend on how the system got to that state. In other words, ΔU is independent of the path in the thermodynamic process.

Internal energy is an intensive variable since it depends upon the mass of the system and can be stored within it (Sauer and Howell 1983):

$$U = m * u \tag{10.4}$$

where

U is the total internal energy of a given mass of a substance (J)
m is the mass of the substance (kg)
u is the specific internal energy (or internal energy per unit mass) (J/kg)

The internal energy of ideal gases (defined as a gas at sufficiently low density that intermolecular forces and the associated energy are negligibly small), is a function of temperature only, mainly because changes in the average velocity of the molecules are indicated by temperature changes of the system. This means that an ideal gas at a given temperature has a certain definite specific internal energy (u), regardless of the pressure. At low pressures, real gases approach the ideal behavior and the ideal gas state equation ($pv = RT$) provides good approximations, being useful in most of industrial theoretical and application studies because of its simplicity.

The relation between the specific internal energy (u) and the temperature (T) for ideal gases can be established by using the definition of constant-volume specific heat (see Section 10.2.4):

$$du = c_v dT \tag{10.5}$$

For ideal gas mixtures ($A + B + C + \cdots$), according to Gibbs–Dalton law, "internal energy of a gas mixture is equal to the sum of total internal energy (expressed in J) for each component when it occupies the total volume by itself":

$$U_{mix} = U_A + U_B + U_C + \cdots \tag{10.6a}$$

or, in terms of specific internal energy (expressed in j/kg):

$$u_{mix} = u_A + u_B + u_C + \ldots \tag{10.6b}$$

For two-phase, saturated vapor (i.e., mixture of saturated liquid with saturated vapor in equilibrium), the specific internal energy is expressed as follows:

$$u = u'_f + x\left(u''_g - u'_f\right) \tag{10.7}$$

where

u'_f is the specific internal energy of saturated liquid (J/kg)

x is the vapor title $\left(x = \dfrac{m_{vapor}}{m_{vapor} + m_{liquid}}\right)$ (–)

u''_g is the specific internal energy of saturated vapor (J/kg)

10.2.2 Heat and Work

In thermodynamic systems, the heat and work that can lead to changes in the system's internal energy are equivalent variables (Joule 1850; Glassley 2015).

The specific (mass) heat (q, J/kg) is defined as the microscopic (internal) energy naturally in transit from a high-temperature body to a low-temperature object via a heating process.

If a thermodynamic system is fully insulated from the outside environment, it is possible to have a change of state in which no heat is transferred into or to the system. Such a process which does not involve heat transfer is an adiabatic process.

On the other hand, the specific (mass) work (w, J/kg) done by the unit mass (kg) of a working fluid (e.g., an ideal gas) crossing the boundary of a thermodynamic system is defined as follows:

$$w = \int_1^2 p dv \tag{10.8}$$

where

p is the system pressure (Pa)

dv is the change of the fluid specific volume (m^3/kg)

Both heat and work are not functions of state, and can be measured and quantified, but not stored or conserved independently since they depend on the process. Heat and work are forms of energy in transit, that is, energy that enters a system as heat may leave as work or vice versa. That means any change in the energy of a thermodynamic system must result in a corresponding change in the energy of the surroundings outside the system.

10.2.3 Enthalpy

The specific enthalpy, a useful state variable of thermodynamic systems, is defined as follows:

$$h = u + pv \tag{10.9}$$

where
- h is the specific enthalpy (J/kg)
- u is the specific internal energy (J/kg)
- p is the system pressure (Pa)
- v is the specific volume (m^3/kg)

For a given ideal gas, the specific enthalpy is

$$h = u + pv = u + RT \tag{10.10}$$

where
- R is the ideal gas constant (J/kgK)
- T is the absolute temperature (K)

Since R is a constant and u is a function of temperature only, it follows that the specific enthalpy (h) of an ideal gas is also a function of temperature only.

The enthalpy is an extensive variable, thus the total enthalpy is

$$H = m * h = U + pV = m * h + pV \tag{10.11}$$

where
- H is the system total enthalpy (J)
- m is the system's mass (kg)
- h is the system specific enthalpy (J/kg)
- p is the system pressure (Pa)
- V is the system total volume (m^3)

For many substances, the enthalpy, given in properties tables, diagrams, or properties software, is relative to an arbitrary reference value. For example, for water vapor, the reference state is the saturated liquid at 0°C where the enthalpy is zero. For ammoniac, the reference state is the saturated liquid at −40°C, the enthalpy being zero at this temperature. It is thus possible to have negative values for enthalpy.

For a thermodynamic process 1-2 occurring at constant pressure (p = constant), the finite variation of enthalpy represents the heat exchanged, that is, added to or removed from the system:

$$\Delta h = q_{12} \tag{10.12}$$

According to Gibbs–Dalton law for an ideal gas mixture (e.g., $A + B + C + \cdots$), enthalpy of a gas mixture is equal to the sum of total enthalpy (expressed in J/K) of each component when it occupies the total volume by itself:

$$H_{mix} = H_A + H_B + H_C + \cdots \tag{10.13a}$$

or, in terms of specific enthalpy (expressed in J/kgK):

$$h_{mix} = h_A + h_B + h_C + \cdots \tag{10.13b}$$

The enthlapy of a two-phase saturated liquid and vapor mixture is as follows:

$$h = h'_f + x\left(h''_g - h'_f\right) \tag{10.14}$$

where

h'_f is the specific enthalpy of saturated liquid (J/kg)
x is the title of two-phase saturated mixture (–)
h''_g is the specific enthalpy of saturated vapor (J/kg)

10.2.4 Specific Heat

The specific heat (J/kgK) is defined as the heat required to raise the temperature of a unit mass of substance by a unit temperature.

Because the changes in pressure and volume (i.e., how a process is carried out) of any thermodynamic system affect the specific heat, the specific heat is expressed either at constant pressure (c_p) or at constant volume (c_v). Therefore, both constant-volume and constant-pressure specific heats are thermodynamic properties of a substance (Sauer and Howel 1983; Tsilingiris 2008).

The constant-pressure specific heat (c_p) is the amount of heat (δq) required to increase the temperature of a closed thermodynamic system of constant composition by dT when the system is heated in a reversible transformation of state at constant pressure with fully specified path (Glassley 2015):

$$c_p = \left(\frac{\delta q}{dT}\right)_{p,\ rev} \tag{10.15}$$

where

δq and dT are differential changes in heat and temperature, respectively

For a thermodynamic process occurring at constant pressure $\delta q = dh$, it follows that the constant-pressure specific heat (c_p) can also be defined by the following relation:

$$c_p = \left(\frac{dh}{dT}\right)_{p,\ rev} \tag{10.16}$$

Similarly, by definition, the constant-volume specific heat (c_v) is the amount of heat required to increase the temperature by dT when the system is held at constant volume:

$$c_v = \left(\frac{\delta Q}{dT}\right)_{v,rev} \tag{10.17}$$

For ideal gases, since the internal energy is not a function of volume, regardless of what kind of process is considered, the constant-volume specific heat can be expressed as

$$c_v = \left(\frac{du}{dT}\right)_{v,rev} \tag{10.18}$$

The heat capacities at constant pressure and constant volume (J/K) are defined, respectively, as

$$C_p = m * c_p \tag{10.19}$$

$$C_v = m * c_v \tag{10.20}$$

The ratio of constant-pressure and constant-volume specific heats $\left(k = \frac{c_p}{c_v}\right)$ represents the isentropic expansion factor for ideal gases or adiabatic exponent for real gases.

For ideal gases, the relation between the specific heat at constant pressure and the specific heat at constant volume is as follows:

$$c_p - c_v = R \tag{10.21}$$

where

R is the gas specific constant (J/kgK)

Exercise E10.1

A superheated water vapor undergoes a transformation of state at constant pressure of 0.5 MPa from T_1 = 350°C (state 1) to T_2 = 400°C (state 2). Determine the average specific heat at constant pressure associated with this process.

Solution

With the given pressure and temperatures can be determined from thermodynamic properties tables and/or ln(p) – h diagram of water vapor the following properties:

The specific enthalpy at state 1 (0.5 MPa and 350°C):

$$h_1 = 3{,}167.7 \text{ kJ/kg}$$

The specific enthalpy at state 2 (0.5 MPa and 400°C):

$$h_2 = 3{,}271.9\ \text{kJ/kg}$$

According to equation (10.16), the specific heat at constant pressure is as follows:

$$c_p \approx \left(\frac{\Delta h}{\Delta T}\right)_p = \frac{h_2 - h_1}{T_2 - T_1} = \frac{3{,}271.9 - 3{,}167.6}{400 - 350} = 2.084\ \text{kJ/kgK}$$

10.3 FIRST LAW OF THERMODYNAMICS

The first law of thermodynamics (also known as the law of energy conservation) states that energy can neither be created nor be destroyed, and that it can only be transferred (or changed) from one physical body form to another.

In the frame of this law, the thermodynamic systems can be (i) closed (with no exchange of matter but some exchange of heat and/or work; and (ii) open (with exchange of matter heat and work, as, for example, in the case of heat pumps' compressors, evaporators, and condensers).

10.3.1 Closed Systems

The energy balance for any thermodynamic system can be formulated as "net amount of energy added to system is equal to the net increase in stored energy of system" or "entering energy minus leaving energy are equal to the increase in energy in system" (Sauer and Howell 1983).

If the potential, kinetic, and chemical energies of a thermodynamic system with constant mass m (kg) are constant, or not important, and the total heat (Q, J) and total work (W, J) are added in a thermodynamic cycle, the change in the total internal energy (U, J) can be expressed as follows according to the first law of thermodynamics:

$$\Delta U = |Q| + |W| \tag{10.22}$$

where
$|Q|$ is the heat added to the system (J)
$|W|$ is the energy (or equivalent work) added to the system (J)

or, in a differential form:

$$Q = dU + W \tag{10.23}$$

For a change from state 1 to state 2 it is as follows:

$$Q_{12} = U_{12} + W_{12} \tag{10.24}$$

where

Q_{12} is the total heat exchanged (J)
W_{12} is the total work exchanged (J)

10.3.2 Cycles

Closed thermodynamic systems can undergo cycles by adding work (W, expressed in J) from the surrounding medium to state 1 and by transferring equivalent heat (Q, expressed in J) to the surrounding in order to return the system to state 1 (Figure 10.1). In this case, the amount of the work added is always proportional to heat supplied.

If the thermodynamic cycle involves several heat and work (absorbed or rejected) quantities, the first law of thermodynamics for the closed system undergoing a cycle can be expressed as follows:

$$\left(\sum W\right)_{cycle} = \left(\sum Q\right)_{cycle} \tag{10.25}$$

This means that the algebraic sum of heats exchanged with the surrounding medium is equal to the algebraic sum of mechanical works exchanged with the surrounding medium, that is, net energy transferred by heat for the cycle equals the net energy transferred by work for the cycle. The work absorbed by the system is thus related to the amount of the heat absorbed and/or rejected.

Applied to heat pumps, the first law of thermodynamics, as expressed by Equation 10.25, states that for transferring heat from one temperature level to a higher one, an expenditure of energy in the form of work(absorbed or rejected) or heat (as in the case of absorption heat pumps, desiccant systems and engine-driven heat pumps) is required, and that the amount of energy rejected at the higher temperature level (Q_{warm}) is the sum of the heat removed on the lower temperature level

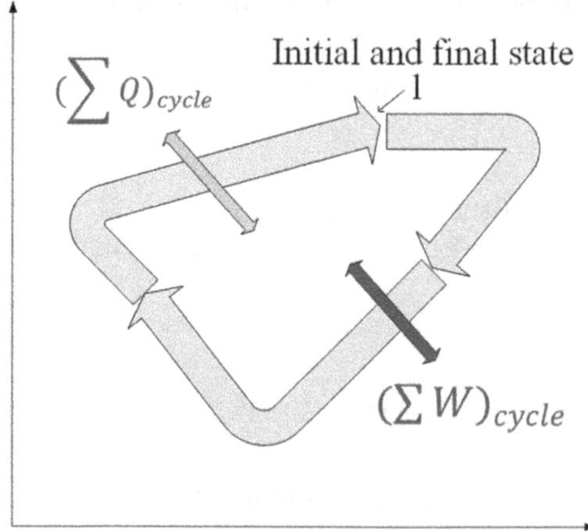

FIGURE 10.1 Thermodynamic cycle of a closed system with multiple heat and work exchanges with the surrounding medium.

(Q_{cold}) plus the added energy (or equivalent work) (W) to accomplish the energy lift (Radermacher and Hwang 2000).

Since at the end of a heat pump thermodynamic cycle the system returns to the same state as at the beginning, the change in the system's total internal energy must be zero.

Thus,

$$0 = |Q| + |W| \tag{10.26}$$

where

$$|Q| = |Q_{warm}| + |Q_{cold}| \tag{10.27}$$

By convention, the total heat added to a heat pump thermodynamic system ($Q_{added} = Q_{cold}$) (e.g., at the heat pump evaporator) is considered negative, and if extracted (removed and supplied) from the cycle ($Q_{extracted} = Q_{warm}$) (e.g., at the heat pump condenser) is assigned as positive. Also, the work added (or equivalent heat) (W) to a thermodynamic cycle is assigned as negative.

Consequently, with careful attention to what term is positive and what is negative, equation 10.26 becomes:

$$0 = |Q_{warm}| + |Q_{cold}| + |W| = Q_{warm} - Q_{cold} - W \tag{10.28}$$

or

$$Q_{warm} = Q_{cold} + W \tag{10.29}$$

Q_{cold} is the total heat extracted from the (cold) heat source (J)
W is the total work (or equivalent energy) added to the cycle (J)
Q_{warm} is the total heat removed from the cycle and transferred to the (warm) heat source (J)

The thermal efficiency of the heat pump cycle (called coefficient of performance, COP) is the ratio of the heat rejected to the absorbed work:

$$COP_{heat\ pump} = \frac{Q_{warm}}{W} = \frac{Q_{cold} + W}{W} \tag{10.30}$$

10.3.3 Open Systems

Applied to open systems with fluid flowing in permanent regime (e.g., heat pumps' compressors, condensers, and evaporators), the first law of thermodynamics can be expressed as follows:

$$\dot{Q}_{12} - \dot{W}_{12} = \dot{m}\left[(h_2 - h_1)\frac{1}{2}(V_2^2 - V_1^2) + g(z_2 - z_1)\right] \tag{10.31}$$

or

$$q_{12} - w_{12} = (h_2 - h_1)\frac{1}{2}(V_2^2 - V_1^2) + g(z_2 - z_1) \tag{10.32}$$

where

$\dot{Q}_{12}$ is the total thermal power exchanged by the system with ambient medium (W)

$\dot{W}_{12}$ is the total (e.g. electrical to thermal) power exchanged by the system with ambient medium (W)

$\dot{m}$ is the fluid mass rate (kg/s)

h_1 and h_2 are the specific enthalpy of working fluid entering and leaving the open thermodynamic system, respectively (J/kg)

V_1 and V_2 are the velocities of working fluid at the inlet and outlet of the open thermodynamic system, respectively (m/s)

z_1 and z_2 are the elevations of working fluid at the inlet and outlet of the open thermodynamic system, respectively (m)

$q_{12} = \frac{\dot{Q}_{12}}{\dot{m}}$ is the specific (mass) heat exchanged by the system with ambient medium (J/kg)

$w_{12} = \frac{\dot{W}_{12}}{\dot{m}}$ is the specific work exchanged by the system with ambient medium (J/kg)

10.4 SECOND LAW OF THERMODYNAMICS

The first law of thermodynamics (see Section 10.3) states that "*energy can neither be created nor destroyed*", but does not provide information about the direction of processes and does not determine the final equilibrium state, even if it is known that energy is transferred from high-temperature to low-temperature bodies.

On the other hand, the second law of thermodynamics states that heat cannot be transferred from a body at lower temperature to another body at higher temperature without the expenditure of energy. In other words, "*heat will not spontaneously flow from a lower temperature to a higher temperature*," or, according to Clausius (1867), "*no process is possible whose sole result is the removal of heat from a reservoir at one temperature and the absorption of an equal quantity of heat by a reservoir at a higher temperature.*"

The second law of thermodynamics introduces the notion of entropy as a measure of the molecular disorder of a system:

$$S = m * s \tag{10.33}$$

where

S is the total entropy of the system (J/K)

m is the mass of the system (kg)

s is the specific entropy of the system (J/kgK)

By application of the theory of probability to thermodynamic systems of enormous number of molecules, there is a simple relationship between the entropy of molecules and the probability of the occurrence of a given state, according to the following relationship (Boltzmann 1877a, 1977b):

$$s = k_B * \ln\left(\hat{W}\right) \quad (10.34)$$

where

k_B is the Boltzmann constant ($1.38 * 10^{-23}$ J/K)

$\hat{W}$ is the thermodynamic probability (i.e., the number of ways the atoms or molecules of a thermodynamic system can be arranged).

Boltzmann's equation is thus a probability equation relating the specific entropy (s) of an ideal gas to the thermodynamic probability $\hat{W}$.

The more shuffled any system is, the greater is its entropy. Conversely, an orderly or unmixed configuration is one of lower entropy.

Entropy is thus an absolute quantity, and if a substance reaches a state in which all randomness has disappeared, it should then have zero entropy (Sauer and Howell 1983).

By definition, at a given absolute temperature (T, in K), any infinitesimal heat transfer between the thermodynamic system and its surroundings ($\delta q = c_p dT$) (in J) will lead to an increase of the system entropy:

$$ds = \frac{\delta q}{T} \quad (10.35)$$

Enounced as the principle of the increase of entropy, the second law of thermodynamics states that in any process, whatever between two equilibrium states of a system, the sum of the increase in entropy of the system and the increase in entropy of its surroundings is equal to or greater than zero.

In other words, entropy cannot be destroyed, but it can be created. Therefore, the entropy of any isolated system always increases, and the system spontaneously evolve toward thermal equilibrium, that is, the state of maximum entropy. The second law expresses thus the improbability of the spontaneous passage of the system from a highly probable state (random or disordered) to one of lower probability.

In the case of heat pumps' closed thermodynamic cycles, the second law of thermodynamics states that, for a given delivered heat (Q_{warm}), better performance means a smaller input work (W) (or equivalent heat), but the smallest work (W) will never be zero. However, the second law of thermodynamics does not say anything about an upper limit for the work. The heat pump system may be thus very inefficient and operate irreversibly if the input work becomes much larger than necessary. That means that there is no upper limit on the degree of irreversibility for the heat pumping operation.

For ideal gas mixtures ($A + B + C + \cdots$), "entropy is equal to the sum of entropy for each component when it occupies the total volume by itself" (Gibbs–Dalton law):

$$S_{mix} = S_A + S_B + S_C \tag{10.36a}$$

or

$$s_{mix} = s_A + s_B + s_C \tag{10.36b}$$

10.5 IDEAL HEAT PUMP CYCLES

10.5.1 Carnot Cycle

The Carnot cycle (Carnot 1824) is an imaginary, idealized thermodynamic cycle consisting of two isothermal and two adiabatic, completely reversible (i.e., equilibrium) processes (see Figure 10.2), for which particular working medium do not need to be specified (McQuiston et al. 2005). That means that even the Carnot cycle is based on nearly ideal gas behavior, but no matter what the medium is, the efficiency for all Carnot cycles operating between the same two temperatures is the same.

Figure 10.2a shows the principle of a single-stage heat pumping process which removes the total heat $Q_1(J)$ from a thermal reservoir at constant temperature level T_1

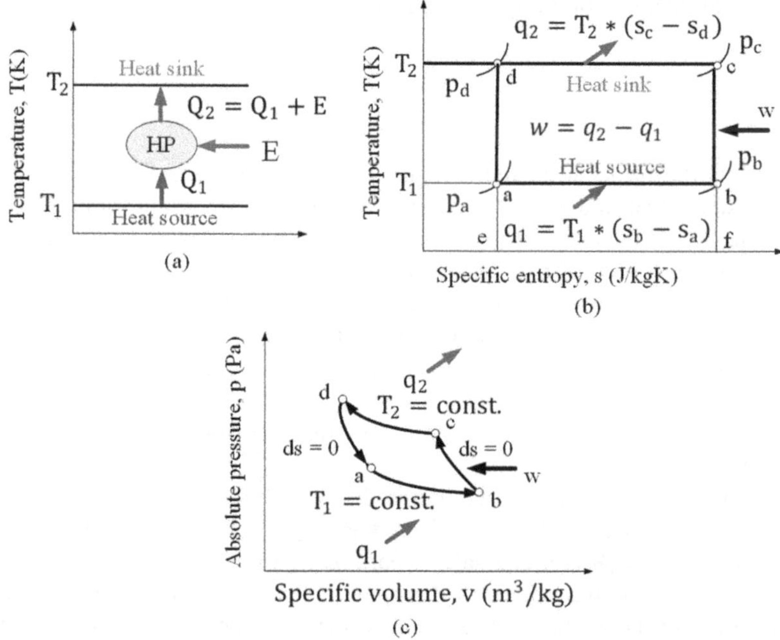

FIGURE 10.2 (a) Theoretical thermodynamic principle of heat pumps; (b) reverse Carnot cycle represented in the temperature–entropy (T–s) diagram; and (c) reverse Carnot cycle represented in the pressure–volume (p–v) diagram.

and releases the total heat $Q_2(J)$ to another thermal reservoir at higher constant temperature level T_2 ($T_2 > T_1$). To perform such a process, a mechanical, electrical, or thermal energy input E(J) is required.

In other words, according to the first law of thermodynamics, the amount of heat rejected q_2(in J/kg) at temperature T_2 is the sum of the specific work input w (in J/kg) required by the heat pump and the amount of heat q_1 (in J/kg) removed at the temperature $T_1 < T_2$:

$$q_2 = q_1 + w \tag{10.37}$$

In terms of thermal rate and mechanical powers, it can be expressed as

$$\dot{Q}_2 = \dot{Q}_1 + \dot{W} \tag{10.38}$$

where

$\dot{Q}_2 = \dot{m} * q_2$ is the total thermal power supplied at the high-temperature source T_2 (W)
$\dot{Q}_1 = \dot{m} * q_1$ is the total thermal power removed from the low-temperature source $T_1 < T_2$ (W)
$\dot{W} = \dot{m} * w$ is the total mechanical (electrical or thermal) input power (W)
$\dot{m}$ is the refrigerant mass flow rate (kg/s)

Figure 10.2b and c represents the reverse Carnot heat pumping process (*a-b-c-d*) (that can be explained based on the evolution of an ideal gas inside a cylinder with insulated piston) in diagrams *p–v* and *T–s*, respectively.

It can be seen that heat Q_1 is added to the working fluid at constant temperature T_1 along the process *a-b*. The fluid is then compressed adiabatically (i.e., with no friction and no transfer of heat) (reversible process *b-c*), heat Q_2 is rejected at constant temperature T_2. Finally, the fluid is expanded by the reversible adiabatic/isentropic process *d-a* with no friction and no change in temperature.

In the *T–s* diagram (Figure 10.2b), the areas under isothermal lines represent the heats transferred during the respective reversible processes. Thus, the amount of heat absorbed at temperature T_1 is represented by the area *a-b-f-e* and the amount of heat rejected at temperature T_2 by area *c-d-e-f*. The difference between areas *c–d–e–f* and *a–b–f–e* (area *a–b–c–d*) represents the amount of work input required for the cycle (Radermacher and Hwang 2000; McQuiston et al. 2005).

The efficiency of the ideal heat pump Carnot reversible cycle involving heat and work exchanges is the highest possible and can be mathematically expressed as (Glassley 2015)

$$COP_{heat\ pump}^{Carnot} = \frac{|q_2|}{|w|} = \frac{|\dot{Q}_2|}{|\dot{W}|} \tag{10.39}$$

where

q_2 is the amount of heat output (J)
w is the amount of input work (J)

Based on areas represented in diagram *T–s* (Figure 10.2b), the upper limit of the COP of a heat pump Carnot cycle (that is always larger than unity) operating between the temperatures $T_{min} = T_1$ and $T_{max} = T_2$ can be expressed as follows (Radermacher and Hwang 2000; McQuiston et al. 2005):

$$COP_{heat\ pump}^{Carnot} = \frac{|q_2|}{|w|} = \frac{T_2\left(s_c - s_d\right)}{\left(T_2 - T_1\right)\left(s_c - s_d\right)} = \frac{T_2}{T_2 - T_1} = \frac{1}{1 - \dfrac{T_1}{T_2}} = \frac{1}{1 - \dfrac{T_{min}}{T_{max}}} \quad (10.40)$$

where

$T_1 = T_{min}$ is the absolute (constant) temperature of the heat source (K)
$T_2 = T_{max}$ is the absolute (constant) temperature of the heat sink (K)

The heat pump reversible cycles operated between the same source (T_1) and sink (T_2) temperatures have identical coefficients of performance. However, because complete reversible processes require that no pressure or temperature gradients develop during the operation of a heat pump, the Carnot heat pump cycle cannot be realized in practice. Moreover, any heat pump cycle cannot have a higher COP than that of a reversible cycle operating between the same source and sink temperatures.

The Carnot cycle is indispensable as a reference frame, that is, as a means for understanding the relationship between heat, work, and cycle efficiencies (Stoecker 1998; Glassley 2015).

Exercise E10.2

A heat pump operates according to the ideal Carnot cycle with air as a working fluid and the following basic parameters (see Figure 10.2):

The heat source lowest temperature $T_{min} = T_1 = 323$ K
The heat sink highest temperature $T_{max} = T_2 = 473$ K
The lowest specific entropy $s_a = s_d = 6.947$ kJ/kgK
The highest specific entropy $s_b = s_c = 7.092$ kJ/kgK

Calculate the following:

i. The specific heat extracted from the heat source reservoir,
ii. The specific heat rejected to the heat sink reservoir,
iii. The specific work consumed,
iv. The COP of the Carnot cycle.

Solution

It can be seen in Figure 10.2b that

$$T_1 = T_a = T_b = 323 \text{ K}$$

$$T_2 = T_c = T_d = 473 \text{ K}$$

i. The specific heat extracted from the heat source reservoir is represented by the area under line *a–b*:

$$q_{extr} = T_1(s_b - s_a) = 323(7.092 - 6.947) = 46.8 \text{ kJ/kg}$$

ii. The specific heat rejected to the heat sink reservoir is represented by the area under line *c–d*:

$$q_{rej} = T_2(s_c - s_d) = 473(7.092 - 6.947) = 68.6 \text{ kJ/kg}$$

iii. The specific work consumed is represented by the area delimited by the state points *a–b–c–d*:

$$w = q_{rej} - q_{extr} = (T_2 - T_1)*(s_b - s_a) = (473 - 323)*(7.092 - 6.947) = 21.75 \text{ kJ/kg}$$

iv. The COP of the ideal Carnot Cycle will be:

$$COP = \frac{T_2}{T_2 - T_1} = \frac{473}{473 - 323} = 3.15$$

Exercise E10.3

A mechanical vapor compression heat pump operates according to the theoretical (ideal) Carnot cycle, as represented in diagram *T–s* (see Figure 10.2b). Heat is extracted from a heat source reservoir at constant temperature 50°C (323 K) and pressure $p_b = p_{min}$ = 100 kPa, and rejected to a heat sink reservoir at constant temperature 100°C (473 K) and $p_d = p_{max}$ = 2,240.8 kPa. The working fluid is an ideal gas (air) with gas constant R = 287 kJ/kg and isentropic exponent k = 1.4.

Calculate the following:

a. The specific heat extracted,
b. The specific work required,
c. The COP of the ideal reversible Carnot cycle.

Solution

a. The ideal Carnot cycle consists in four reversible processes, that is, two isotherms (*a-b* and *c-d*) and two isentropes (*b-c* and *d-a*) (see Figure 10.2b and c).

With the air as an ideal gas (working) fluid, the specific heat extracted during the isothermal process *a-b* can be calculated as follows (Ciconkov 2001):

$$|q_{a-b}| = RT * \ln \frac{p_a}{p_b}$$

where

R = 287 J/kgK is the ideal air gas constant

Pressure p_a can be calculated from the isentropic expansion process *d-a*:

$$p_a = p_d * \left(\frac{T_1}{T_2}\right)^{\frac{k}{k-1}} = 2240.8 * \left(\frac{323}{473}\right)^{\frac{1.4}{1.4-1}} = 2240.8 * (0.788)^{3.5}$$

$$= 2240.8 * 0.4343 = 973.179 \text{ kPa}$$

Thus,

$$|q_{a-b}| = RT * \ln \frac{p_a}{p_b} = 0.287 * 323 * \ln\left(\frac{973.179}{100}\right)$$

$$= 287 * 323 * 2.2753 = 210.92\,\text{kJ/kg}$$

ii. For the isentropic processes *b-c* and *d-a*, it can be write, respectively, as follows:

$$\frac{p_d}{p_a} = \left(\frac{T_2}{T_1}\right)^{\frac{k}{k-1}}$$

$$\frac{p_c}{p_b} = \left(\frac{T_2}{T_1}\right)^{\frac{k}{k-1}}$$

It can be seen from these relations that $\frac{p_d}{p_a} = \frac{p_c}{p_b}$, thus $\frac{p_d}{p_c} = \frac{p_a}{p_b}$.

The specific heat supplied to the heat sink reservoir during the isothermal process *c-d* can be calculated as follows:

$$|q_{c-d}| = RT * \ln \frac{p_d}{p_c} = RT * \ln \frac{p_a}{p_b} = 0.287 * 473 * \ln \frac{973.179}{100}$$

$$= 287 * 473 * 2.2753 = 303.874\ \text{kJ/kg}$$

b. Specific work required will be as follows:

$$w_{b-c} = |q_{c-d}| - |q_{a-b}| = 303.874 - 210.920 = 97.954\ \text{kJ/kg}$$

iii. COP of the reversible Carnot cycle is as follows:

$$COP_{Carnot} = \frac{|q_{c-d}|}{w_{b-c}} = \frac{303.874}{97.954} = \frac{1}{1 - \frac{T_1}{T_2}} = \frac{1}{1 - \frac{323}{473}} = 3.10 - 3.15$$

Exercise E10.4

A large industrial-scale ammonia mechanical vapor compression heat pump operates according to the ideal Carnot cycle shown in Figure E10.4a, and b. The refrigerant flow rate is 3 kg/s, the condensing temperature is 70°C, and the evaporating temperature is 40°C.

1. Represent in both ln(*p*)–*h* and *T*–*s* diagrams the Carnot cycle;
2. Determine
   2a. The total thermal power extracted from the heat source;
   2b. The total compressor input power required;
   2c. The total thermal power supplied to the heat consumer;
   2d. The total thermal power extracted from the heat source;
   2e. The theoretical COP.

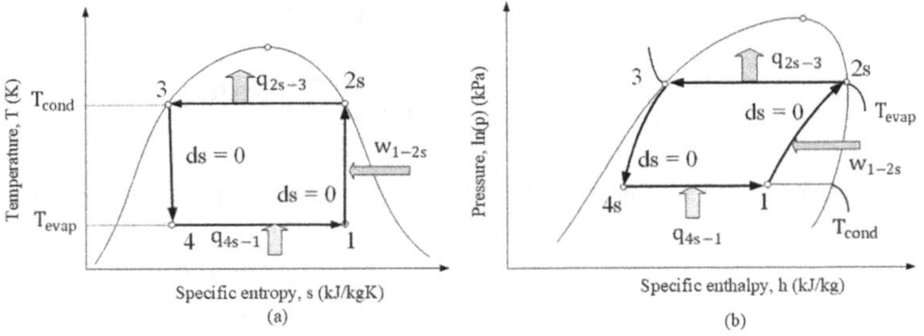

FIGURE E10.4 Ideal Carnot cycle represented in (a) T–s diagram; (b) ln(p)–h diagram.

Solution

1. Figure E10.4a and b represents the ideal Carnot cycle in *T–s* and ln(*p*)–*h* diagrams, respectively.

2a. State 2*s* of ammonia at the end of the isentropic compression process 1-2*s* is saturated vapor at the condensing temperature 70°C.

At this temperature, because the theoretical compression process 1-2*s* is isentropic, the specific entropy of the saturated vapor at state 1 will be $s_1 = s_{2s}$ = 4.84 kJ/kgK (from ammonia thermophysical properties tables or from an ammonia properties software).

Also, at the condensing temperature 70°C,

$$h_{2s} = h_g = 1483.94 \text{ kJ/kg}$$

$$h_3 = h_f = 545.04 \text{ kJ/kg}$$

State 1 of the refrigerant (ammonia) is a mixture of saturated liquid and saturated vapor at the evaporating temperature of 40°C.

At this saturated temperature, from ammonia thermophysical properties tables or from ammonia properties software, the saturated pressure can be determined: 155.54 kPa (absolute).

The specific entropies of saturated liquid and saturated vapor corresponding to states 1 and 4 are, respectively,

$$s_f = 1.644 \text{ kJ/kgK}$$

$$s_g = 5.155 \text{ kJ/kgK}$$

Also, from ammonia thermophysical properties tables or from an ammonia properties software, the specific enthalpies of saturated liquid and saturated vapor corresponding to states 1 and 4 are, respectively,

$$h_f = 390.64 \text{ kJ/kg}$$

$$h_g = 1498.91 \text{ kJ/kg}$$

The vapor quality (title) of ammonia at state 1 will be:

$$x_1 = \frac{s_1 - s_f}{s_g - s_f} = \frac{s_{2s} - s_f}{s_g - s_f} = \frac{4.84 - 1.644}{5.155 - 1.644} = 0.91$$

On the other hand, the specific enthalpy of ammonia at state 1 is defined as follows:

$$x_1 = \frac{h_1 - h_f}{h_g - h_f} = \frac{h_1 - 390.64}{1498.91 - 390.64} = 0.91$$

Thus,

$$h_1 = 1{,}399.2 \text{ kJ/kg}$$

State 3 of ammonia at the end of the isotherm process 2s-3 is saturated liquid at the condensing temperature 70°C. At this temperature, the specific entropy of the saturated liquid at state 3 will be $s_3 = s_f = 2.1$ kJ/kgK (from ammonia thermophysical properties tables, or from an ammonia properties software).

The vapor quality (title) of ammonia at state 4 will be

$$x_4 = \frac{s_4 - s_f}{s_g - s_f} = \frac{s_3 - s_f}{s_g - s_f} = \frac{2.1 - 1.644}{5.155 - 1.644} = 0.13$$

On the other hand, the vapor quality at state 4 is defined as follows:

$$x_4 = \frac{h_4 - h_f}{h_g - h_f} = \frac{h_4 - 390.64}{1{,}498.91 - 390.64} = 0.13$$

Consequently, the specific enthalpy at state 4 will be

$$h_4 = 534.71 \text{ kJ/kg}$$

The total thermal power extracted from the heat source can be calculated as follows:

$$\dot{Q}_{4s-1} = \dot{m}(h_1 - h_4) = 3 * (1{,}399.2 - 534.71) = 864.5 \text{ kW}$$

2b. The total compressor input power required is as follows:

$$\dot{W}_{1-2s} = \dot{m} * (h_{2s} - h_1) = 3 * (1{,}483.94 - 1{,}399.2) = 254.2 \text{ kW}$$

2c. The total thermal power supplied to the heat sink is as follows:

$$\dot{Q}_{2s-3} = \dot{m} * (h_{2s} - h_3) = 3 * (1{,}483.94 - 545.04) = 2{,}816.7 \text{ kW}$$

2d. The total thermal power extracted from the heat source is as follows:

$$\dot{Q}_{4-1} = \dot{Q}_{2s-3} - \dot{W}_{1-2s} = 2{,}816.7 - 254.2 = 2{,}562.5 \text{ kW}$$

2e. The theoretical COP is as follows:

$$COP_{Carnot} = \frac{\dot{Q}_{2s-3}}{\dot{W}_{1-2s}} = \frac{2816.7}{254.2} \approx 11.1$$

or

$$COP_{Carnot} = \frac{T_2}{T_2 - T_1} = \frac{70+273}{(70+273)-(40+273)} = \frac{343}{30} \approx 11.4$$

10.5.2 Lorenz Cycle

The Carnot cycle is based on the assumption that the heat pump recovers and supplies heat at constant temperature levels, as shown in Figure 10.2.

However, in many heat pump applications (Figure 10.3a), heat is recovered from (process 1-2) and supplied (process 3-4) to fluids with variable temperatures, even the evaporating (process *a-b*) and condensing (process *c-d*) temperatures of a given pure refrigerant are constant (Radermacher and Hwang 2000). In these cases, evaporating (process *a-b*) and condensing temperatures (process *c-d*) are fixed by the lowest cooled fluid (state 2) and highest cooling fluid (state 4) temperatures, respectively. Pinches (*a*-2) and (4-*c*) (Figure 10.2a) show where the temperature differences (approaches) between the cold fluid and the evaporating refrigerant, and between the condensing refrigerant and the heated fluid, respectively, become very small, as well as the associated heat transfer rates due to very small approach temperatures.

In contrast, between the state points *b* and 1 and, also, between *d* and 3, the temperature differences are large, as well as the heat transfer rates.

While the temperature difference at pinch points approaches zero, the temperature differences at the other end of the heat exchangers stay finite. In evaporation

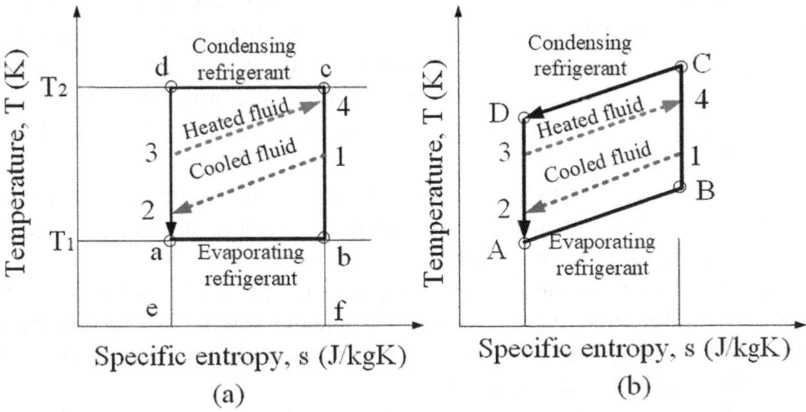

FIGURE 10.3 (a) Carnot cycle for heat pumping with a pure refrigerant represented in the temperature–entropy (T–s) diagram; (b) Lorenz cycle for heat pumping with a refrigerant mixture represented in the temperature–entropy (T–s) diagram.

and condensing processes, the highest temperature of the evaporating refrigerant and the lowest temperature of the condensing refrigerant are fixed by the respective water temperatures. In turn, their pressures are fixed, and thus the compressor power (Radermacher and Hwang 2000).

As a consequence, for a given heat exchanger, in order to facilitate heat exchange, the available heat transfer surface area has to be increased.

In this case, even when the heat transfer area approaches infinity, the performance improvement of the heat exchanger, as well as that of the entire heat pump, is limited. The Lorenz cycle addresses this issue. When working fluids as refrigerant mixtures that change their temperatures during the course of evaporation and condensation processes (i.e., fluids with evaporation and condensation temperature glides) are available, the schematic of Figure 10.3a can be modified to become the Lorenz cycle (Figure 10.3b) (Alefeld 1987; Herold 1989; Rice 1993; Radermacher and Hwang 2000).

For the ideal case shown in Figure 10.3b, it can be seen that the change in temperature from the beginning to the end of the evaporation process matches that of the cooled fluid, both process curves become parallel and the pinch point is eliminated. With an increase in heat transfer area, the temperature difference between the two fluids continues to shrink everywhere in the exchanger, and thus the overall entropy production can theoretically approach zero. Even for the same heat exchange area, the Lorenz cycle better matches the source and sink glides, resulting in higher overall cycle efficiency; mainly it operates at improved mean temperatures compared to the corresponding pure refrigerant (Rice 1993; Radermacher and Hwang 2000).

10.6 THEORETICAL HEAT PUMP CYCLE

The mechanical vapor compression cycle is the cycle of choice for a wide range of heat pump applications, mainly because (i) the latent heat during the liquid–vapor phase change process is used as the mechanism to transfer heat to and from the working fluid which greatly reduces the mass flow rate in the cycle and (ii) the expansion process is conducted almost entirely in the liquid phase and, since the liquid is essentially incompressible, only a minim amount of work is lost compared to the expansion of a gas within other cycles (Radermacher and Hwang 2000).

Subcritical mechanical single-stage vapor compression heat pumps use electrical- or gas-driven compressors to compress synthetic, natural, or mixed refrigerants according to the reverse Rankine thermodynamic cycle.

To accomplish such a cycle, the heat pump uses the physical properties of a pure, volatile working fluid (refrigerant), as well as some amount of external electrical (or thermal/fossil) primary energy to run the compressor and auxiliary equipment as blowers and/or fluid circulating pumps.

Figure 10.4a shows the theoretical (standard) cycle of heat pumps working with pure refrigerants to transfer heat from a low-temperature heat source to a higher temperature (warmer) heat (sink) source.

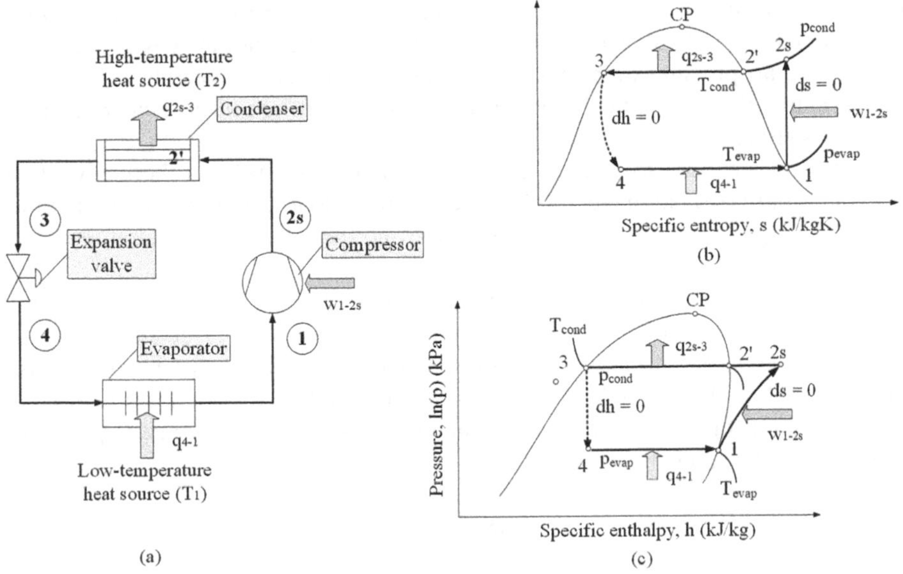

FIGURE 10.4 Theoretical thermodynamic cycle of single-stage mechanical vapor compression heat pumps: (a) main mechanical components; (b) cycle representation in T–s diagram; and (c) cycle representation in ln(p)–h diagram.

The main mechanical components are (i) compressor (reciprocating, scroll, screw, etc.), (ii) expansion device (thermostatic or electronic valve, orifice or capillary tube), and (iii) two heat exchangers (condenser and evaporator).

The thermodynamic process of the heat pump theoretical mechanical vapor compression cycle is represented in both temperature–entropy (Figure 10.4b) and pressure–enthalpy (Figure 10.4c) diagrams.

The thermodynamic cycle consists of processes as (i) saturated vapor (state 1) is leaving the evaporator and enters the compressor as a saturated vapor (title $x = 1$); (ii) the compression process 1-2*s* is assumed to be reversible and adiabatic ($\delta q = 0$) and, therefore, isentropic ($ds = 0$); it is continued until the condenser pressure (p_{cond} at state 2*s* located in the superheated vapor region) is reached; (iii) the process 2*s*-2′ consists in sensible cooling of superheated vapor at constant pressure p_{cond} (i.e., no pressure drop exists in the condenser and connecting piping) with temperature of the vapor decreasing until the saturated vapor condition is reached (state 2′) (iv) the condensation process (2′-3) occurs at constant pressure (p_{cond}) and temperature (T_{cond}); both vapor cooling and condensation occur by transferring heat (q_{2s-3}) to the heat pump heat sink; at state 3, the refrigerant leaves the condenser as a saturated liquid (title $x = 0$). (v) saturated liquid leaving the condenser at state 3 is expanded through an expansion (throttling) valve, where partial evaporation occurs as the pressure drops across the valve; in this process, the enthalpy at exit and at inlet of expansion valve are equal ($h_3 = h_4$), that is, the expansion process is isenthalpic; because the expansion process 3-4 is irreversible, with an increase in entropy, the process is shown as a dashed line in Figure 10.4b and c; and (vi) the two-phase refrigerant with a title $0 < x < 1$ (state 4) is flowing through the evaporator at constant pressure (p_{evap}) (i.e., no pressure drop exists in the evaporator and connecting piping) and temperature (T_{evap}) where the liquid phase is vaporized by absorbing heat (q_{4-1}) from the heat source.

It can be seen that to plot the heat pump theoretical cycle on thermodynamic diagrams, only two parameters have to be specified, that is, the saturation pressure (p_{evap}) or temperature (T_{evap}) for the evaporator, and the saturation temperature (T_{cond}) or pressure (p_{cond}) for the condenser.

During both condensing and evaporating processes, the specific (mass) enthalpy, entropy, and volume of the two-phase refrigerant with a given quality (title) (x), defined as the ratio of the mass of saturated vapor (designed by letter g) to total mass of saturated liquid (designed by letter f) plus saturated vapor when the refrigerant is in a saturation state, can be computed using the following expressions, respectively (see Figure 10.4b and c):

$$h = xh_g + (1-x)h_f \tag{10.41a}$$

$$s = xs_g + (1-x)s_f \tag{10.41b}$$

$$v = xv_g + (1-x)v_f \tag{10.41c}$$

Specific internal energy can then be obtained from the definition of enthalpy as $u = h - pv$.

According to relation 10.41a, the specific enthalpy is the sum of the saturated vapor [xh_g] and saturated liquid [$(1 - x)h_f$] enthalpy. The same procedure is followed for determining the specific entropy (relation 10.41b) and specific volume (relation 10.41c) for given qualitics.

In the superheat region, the refrigerant thermodynamic properties must be obtained from tables or plots of the thermodynamic properties.

If the mass flow rate of the refrigerant is known, the following thermal powers can be determined from the pressure–enthalpy diagram:

$$\dot{Q}_{extracted} = \dot{Q}_{evap} = \dot{m}_{refr}\left(h_1 - h_4\right) \tag{10.42}$$

$$\dot{W}_{required} = \dot{W}_{compr} = \dot{m}_{refr}\left(h_{2s} - h_1\right) \tag{10.43}$$

$$\dot{Q}_{supplied} = \dot{Q}_{cond} = \dot{m}_{refr}\left(h_{2s} - h_3\right) \tag{10.44}$$

where

$\dot{m}_{refr}$ is the refrigerant mass flow rate (kg/s)

$Q_{extracted} = Q_{evap}$ is the theoretical thermal power extracted from the heat source at the evaporator (W)

$W_{required} = W_{compr}$ is the theoretical electrical (or thermal) power required at the compressor (W)

$Q_{supplied} = Q_{cond}$ is the theoretical thermal power supplied to the heat sink at the condenser (W)

h is the refrigerant specific enthalpy in the points of the theoretical cycle represented in Figure 10.4b, and c (J/kg)

The theoretical heating COP of the heat pump cycle is defined as follows:

$$COP_{heat\ pump}^{heating} = \frac{Useful\ heat\ pump\ effect}{Net\ energy\ input} = \frac{h_{2s} - h_3}{h_{2s} - h_1} \tag{10.45}$$

Although simple and quasi-practical, this theoretical, but nonideal cycle does not have a COP as high as that of Carnot cycle mainly because (i) the flow through the expansion valve (process 3-4) is an irreversible expansion process where an opportunity to produce useful work is lost and (ii) the heat rejection (process 2*s*-3) does not occur at constant pressure.

It is however a good model for understanding the basic features of real vapor compression heat pump cycles (McQuiston et al. 2005).

Exercise E10.5

A single-stage mechanical vapor compression heat pump operates according to the theoretical cycle schematically represented in Figure 10.4a–c. The compression process 1-2*s* is adiabatic ($\delta q = 0$) and the expansion process 3-4 is isenthalpic ($dh = 0$). The refrigerant is the HFC-134a, and the heat pump operates in a lumber kiln with an average evaporating temperature of 20°C and an average condensing temperature of 70°C.

Determine

1. The specific cooling and dehumidifying effect (i.e., the specific heat extracted from the drying air flowing through the evaporator);
2. The specific compressor required work;
3. The specific heating effect (i.e., the specific heat supplied to the lumber dryer by the condenser);
4. The heat pump theoretical COP.

Solution

The evaporating and condensing saturated pressures can be found as functions of the given evaporating (20°C) and condensing (70°C) saturated temperatures from a pressure–enthalpy diagram or from a thermophysical properties table for the refrigerant HFC-134a, or by the aid of appropriate refrigerant properties software:

$$p_{evap} = 571.71\ \text{kPa}, a$$

$$p_{cond} = 2116.8\ \text{kPa}, a$$

As can be seen in Figure 10.4b and c, states 1 and 3 are located on the vapor and liquid saturated curves, respectively.

Consequently, their specific enthalpies can be determined from a pressure-enthalpy diagram or from the thermos-physical properties table for the refrigerant HFC-134a, or by the aid of appropriate refrigerant properties software:

$$h_1 = 409.75 \text{ kJ/kg}$$

$$h_3 = 304.28 \text{ kJ/kg}$$

The simplest way to find the enthalpy at state 2*s* (h_{2s}) is to use a pressure–enthalpy diagram for the refrigerant HFC-134a. State 2 is located at the intersection of adiabatic compression curve 1–2*s* with the horizontal constant pressure line p_{cond} = 2,116.8 kPa, *abs*. Thus,

$$h_{2s} = 438 \text{ kJ/kg}$$

Because the expansion process 3-4 is theoretically isenthalpic, $h_4 = h_3 = 304.28$ kJ/kg

1. The specific cooling and dehumidifying effect is as follows:

$$q_{4-1} = h_1 - h_4 = 409.75 - 304.28 = 105.47 \text{ kJ/kg}$$

2. The specific compressor work is as follows:

$$w_{1-2s} = h_{2s} - h_1 = 438 - 409.75 = 28.25 \text{ kJ/kg}$$

3. The specific heating effect is as follows:

$$q_{2s-3} = h_{2s} - h_3 = 438 - 304.28 = 133.72 \text{ kJ/kg}$$

4. The heat pump theoretical COP is as follows:

$$COP_{heat\ pump}^{theoretical} = \frac{q_{2s-3}}{w_{1-2s}} = \frac{133.72}{28.25} = 4.73$$

10.7 REAL HEAT PUMP CYCLE

Compared to the ideal reverse Carnot (Section 10.6.1) and theoretical (Section 10.6.2) cycles, the real (actual, practical) heat pump mechanical vapor compression cycle has inherent thermodynamic losses, mainly associated with single-phase vapor compression (that results at high discharge refrigerant temperature and high compression work) and isenthalpic expansion (that results at large throttling losses and low heating capacity).

Real heat pumping cycles can be realized with both pure and mixed refrigerants.

10.7.1 Pure Refrigerants

Real (actual, practical) heat pumps used for industrial wood drying work in the heating mode only (Figure 10.5a) according to a nonreversible (real) subcritical Rankine reverse thermodynamic cycle (Kiang and Jon 2007; Minea 2016).

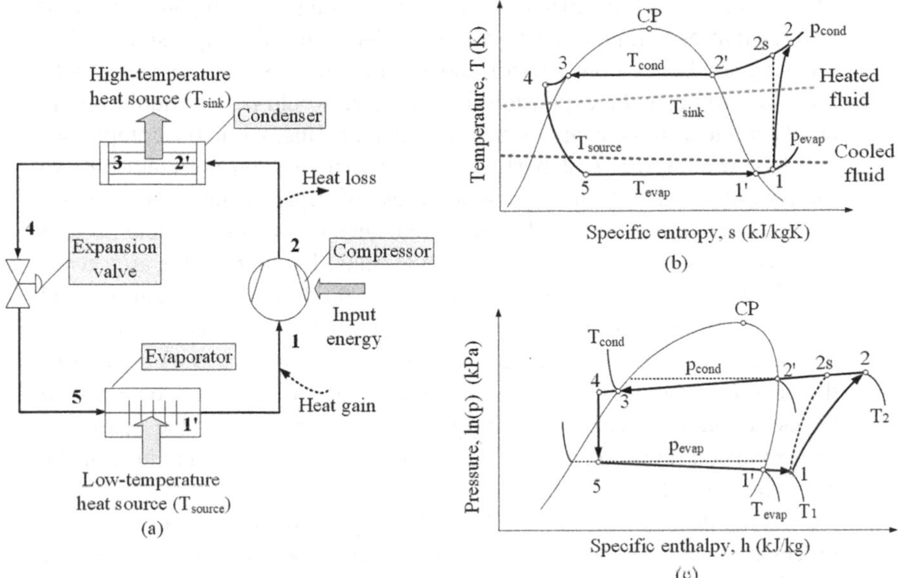

FIGURE 10.5 (a) Main mechanical components of a real, nonreversible subcritical heat pump operating in the heating mode only with a pure refrigerant; (b) thermodynamic cycle in T–s diagram; and (c) thermodynamic cycle in ln(p)–h diagram. cond., condensing; CP, critical point; evap., evaporating; EXV, expansion valve; h, specific enthalpy, p, pressure; s, specific entropy; T, temperature.

Figures 10.5b and c represent in the diagrams *T–s* and ln(*p*)–*h*, respectively, the real heat pump thermodynamic process using a pure refrigerant. Such a process consists in the following main processes:

i. The refrigerant leaving the evaporator as superheated vapor (state 1) is polytropically compressed up to the superheated state 2 (irreversible process 1-2); the electrical energy input is converted to shaft work to raise the pressure and temperature of the refrigerant; by increasing the vapor pressure, the condensing temperature is increased to a level higher than that of the heat source (T_{source}); at the end of the ideal (ideal, adiabatic) compression process (1-2*s*), the specific enthalpy (h_{2s}) of the superheated vapor is at pressure p_{cond}, but with the same entropy as the vapor before compression ($s_{2s} = s_1$); on the other hand, the final state 2 of the real (polytropic) compression process lies anywhere on the constant pressure line p_{cond}, on the right side of theoretical state point 2*s*.
ii. In the condenser, the superheated vapor is first desuperheated from the state 2 to saturated state 2′, and then undergoes a two-phase condensation at constant temperature (T_{cond}) and pressure (p_{cond}) (process 2′-3).
iii. Before leaving the condenser, the saturated liquid is subcooled at constant pressure to a lower temperature (process 3-4) in order to reduce the risks

of flashing within the expansion valve; subcooling of the liquid refrigerant ensures that only liquid and no vapor bubbles enter the expansion device, allowing for better flow control and smaller devices; also, if vapor bubbles were to enter the expansion device, there would be less liquid refrigerant available in the evaporator to produce cooling, while the compressor still has to compress that portion of the vapor that did not contribute to the cooling capacity; this represents a loss of capacity and efficiency; on the other hand, too much subcooling indicates that a considerable portion of the condenser volume is filled by single-phase liquid; thus, the area available for heat rejection from the condensing fluid is smaller than it could be; as a consequence, the saturation temperature is increased along with the saturation pressure and, as a result, the compressor work increases; the degree of subcooling at the condenser outlet is primarily determined by the amount of refrigerant charged to a heat pump and the setting of the expansion device primarily determines the degree of superheat at the evaporator outlet.

iv. During the desuperheating (2-2′), condensing (2′-3), and sub-cooling (3-4) processes, heat is rejected by the condenser to the heat sink medium (gas or liquid). Within all these processes, the refrigerant exhibits pressure drops (see Figure 10.4c).

v. After the condenser, the liquefied subcooled refrigerant enters the expansion valve where an expansion process at constant enthalpy (i.e., drop in pressure accompanied by a drop in temperature) takes place in order to reduce the refrigerant pressure at a level corresponding to an evaporating temperature (T_{evap}) below to the heat source temperature ($T_{sources}$) (process 4-5); the expansion valve controls the refrigerant flow into the evaporator in order to ensure its complete evaporation, maintain an optimum superheat in order to avoid the liquid refrigerant to enter the compressor, and also avoid excessive superheat that may lead to overheating of the compressor.

vi. The refrigerant then enters the evaporator in a two-phase (state 5), absorbs (recovers) heat from the heat source thermal carrier, and undergoes change from liquid–vapor mixture to saturated vapor at constant pressure (p_{evap}) and temperature (T_{evap}); inside the evaporator, the saturated vapor (state 1′) is slightly superheated up to state 1 (process 1′-1) before entering the compressor; the superheat is desirable for two reasons. First, it assures that all the refrigerant is evaporated where it should contribute to the cooling capacity. Second, superheating usually occurs within the evaporator for the purpose of insuring that the compressor receives only vapor without any liquid entrained that could damage the compressor; however, excessive superheat may lead to overheating of the compressor; at this point, the heat pump real thermodynamic cycle restarts.

The compression efficiency, which represents the ratio of the amount of work required to compress the refrigerant superheated vapor from p_{evap} to p_{cond} through the ideal (isentropic) compression process (1-2*s*) to the amount of work required to compress the same refrigerant superheated vapor from p_{evap} to p_{cond} via the actual (polytropic) compression process (1-2), can be expressed by the following expression (see Figure 10.3b and c).

$$\eta_{isentropic} = \frac{w_{ideal}}{w_{real}} = \frac{h_{2s} - h_1}{h_2 - h_1} \tag{10.46}$$

where

h_1 is the refrigerant specific enthalpy at the beginning of the real, polytropic compression process 1-2 (J/kg)

h_2 is the refrigerant specific enthalpy at the end of the real, polytropic compression process 1-2 (J/kg)

h_{2s} is the refrigerant specific enthalpy at the end of the ideal, adiabatic (isentropic) compression process 1-2*s* (J/kg)

It can be seen that the entire real, subcritical thermodynamic process represented in the *T–s* (Figure 10.4b) and ln(*p*)–*h* (Figure 10.4b) diagrams occurs below the critical point of the refrigerant being used. Heat absorption occurs by evaporation of the refrigerant at low temperature and pressure, and heat rejection takes place by condensing the refrigerant at a higher pressure and temperature, but always below that of the refrigerant critical point.

As can be seen in Figure 10.4b and c, there are some deviations of the heat pump real cycle from the theoretical cycle (Figure 10.5) due to several realistic effects, such as pressure drop or compressor inefficiencies (Radermacher and Hwang 2000). They can be summarized as follows:

i. The refrigerant enters the compressor not as a saturated vapor, but as a superheated vapor (state 1).
ii. During the polytropic compression, the compressor's work increases once again, as the entering suction vapor is more superheated; superheat horn presents a performance penalty because it increases the superheat horn of the heat pump Carnot cycle.
iii. Because the compression process 1-2 is polytropic and irreversible [involving heat transfer, pressure drops (due to internal friction of the mechanical parts), and increased specific volume that reduces the capacity for a given displacement], the actual state of the refrigerant at the compressor outlet is the point 2 located to the right of the endpoint 2*s* that would be achieved with isentropic compression; moreover, because of inefficiencies of the compressor, the actual discharge temperatures would be higher than the adiabatic discharge temperatures for any refrigerant (Stoecker 1998); for some refrigerants, such as ammonia, more intensive cooling of compressors is needed in addition to that provided by using ambient air as a cooling agent; for that, ammonia compressors must be equipped with water-cooled heads, thereby keeping valves cooler to prolong their life and preventing the breakdown of oil at high temperatures, with the flow rate of water regulated by a control valve that maintains an outlet cooling water temperature of approximately 45°C, and with an assumed cooling water temperature rise

of 15°C; the compressor discharge temperatures could not exceed 135°C, that is, approximately 16°C lower than if there had been no cooling of the head; the same precautions must however be applied to avoid condensation of refrigerant in the head of the cylinder, in order to (a) ensure that the temperature of the cooling water never falls below the heat pump condensing temperature, and (b) stop the supply of cooling water when the compressor is not in operation; the maximum pressure difference varies between 1,000 and 2,000 kPa depending upon the bore, stroke, and other construction features of the compressor.

iv. There are pressure losses in the evaporator and condenser as well as in all connecting tubing that increase the power requirement for the cycle, and the heat transfer to and from the various components; these pressure losses are indicated by the slopes of lines representing the evaporation (5-1) desuperheating (2-2′), condensation (2′-3), and subcooling (3-4) processes in the diagram $\ln(p)$–h (Figure 10.5c); due to the pressure drop in the evaporator, the compression process (1-2) starts at a lower pressure and, further, into the superheated vapor region than in Figure 10.4c, which causes a double penalty for the compressor because a higher pressure difference needs to be overcome by the compressor and the work increases due to the higher temperature of the superheated vapor inside the compressor suction line; the pressure drop of refrigerant in the suction line between the evaporator and compressor may also affect the compressor capacity because the pressure drop in the suction line increases the specific volume of the suction superheated vapor, so the mass rate of flow drops; however, this phenomenon is less penalizing at high evaporating temperatures (Radermacher and Hwang 2000); for example, the percentage reduction in refrigerating capacity for ammonia at two different evaporating temperatures and at drops in saturation temperature of 0.5°C and 1.0°C; at the lower saturated suction temperature caused by pressure drop in the suction line, the power requirement also decreases. However, if the compressor is already attempting to deliver full capacity, it cannot take advantage of this reduction in power to increase the refrigerating capacity (Stoecker 1998).

v. Heat losses from the high pressure side of the heat pump system improve cycle performance by decreasing the compressor load (McQuiston et al. 2005).

vi. Exposed surfaces on the low pressure side of the heat pump system generally are at a lower temperature than the ambient environment, and thus any heat gains generally degrade cycle performance by increasing the compressor load (McQuiston et al. 2005).

vii. As can be seen in the T–s diagram (Figure 10.5b), the isenthalpic expansion process provides simultaneously two penalties depending on the specific heat capacity of the refrigerant, that is, a small loss of work and loss of cooling capacity, resulting, when not extracted from the fluid, in an increased vapor quality; consequently, less liquid refrigerant is available to provide cooling, that is, heat recovery in the evaporator; in other words, during the expansion process of refrigerants with a large specific heat capacity, a relatively large amount of refrigerant has to evaporate to cool the remaining liquid to the evaporator saturation temperature and less liquid is left to provide cooling capacity (Radermacher and Hwang 2000).

10.7.2 Heat Pump Performance Indicators

In the heating mode, the mechanical vapor compression heat pumps based on the ideal Carnot cycle operates between two heat reservoirs having absolute temperatures $T_1 = T_{source} \approx T_{min} \approx T_{evap}$ (K) (heat source) and $T_2 = T_{sink} \approx T_{\max} \approx T_{cond}$ (K) (heat sink), respectively (see Figure 10.2b).

As shown by equation 10.40, the instantaneous heating (maximum) COP for such a Carnot (ideal) cycle is

$$COP_{heat\ pump}^{Carnot} = \frac{T_{source}}{T_{source} - T_{sink}} \approx \frac{T_{cond}}{T_{cond} - T_{evap}} = \frac{1}{1 - \dfrac{T_{evap}}{T_{cond}}} \tag{10.47}$$

where

T_{cond} is the condensing temperature (K)
T_{evap} is the evaporating temperature (K)

It can be concluded that (i) ideal reversible heat pumps operating under the same conditions have the same COP and (ii) real (practical) and, hence, irreversible heat pumps have lower coefficients of performance than an ideal reversible heat pump operating under the same conditions.

The actual (real) heat pumps used for heating may be evaluated in terms of an instantaneous heating COP that is the useful heating effect divided by the net energy input (McQuiston et al. 2005).

The heating $COP_{heat\ pump}^{real}$ of a real subcritical mechanical vapor compression heat pump is defined as the ratio between the condenser useful (supplied) thermal power output ($\dot{Q}_{cond}$) and the electrical power input at both compressor and blower ($\dot{W}_{compr+blower}$):

$$COP_{heat\ pump}^{real} = \frac{Usefull\ effect}{Net\ power\ input} = \frac{\dot{Q}_{cond}}{\dot{W}_{compr+blower}} \approx 1 + \frac{\dot{Q}_{evap}}{\dot{W}_{compr+blower}} < COP_{heat\ pump}^{Carnot} \tag{10.48}$$

where

$\dot{Q}_{evap}$, $\dot{W}_{compr+blower}$, and $\dot{Q}_{cond}$ are defined by Equations 10.42, 10.43, and 10.44, respectively.

It can be seen that $COP_{heat\ pump}^{real}$ of a heat pump is greater than 1, that is, a heat pump always supplies more heat than electrical energy consumed.

The heat pump energy efficiency is expressed by the COP defined as the total heat supplied by the condenser (kWh) divided by the compressor and blower electrical energy consumption (kWh) (Kiang and Jon 2007):

$$COP_{heat\ pump}^{real} = \frac{Q_{cond}}{E_{compr+blower}} = \frac{Q_{cond}}{\dot{W}_{compr+blower} * t} \tag{10.49}$$

where

$\dot{W}_{compr+blower}$ is the electrical power input (compressors and blowers)
t is the heat pump running time

In practice (i.e., real heat transfer systems with finite temperature differences across the heat exchangers, real working fluids, flow losses, and compressor efficiency), $COP_{heat\ pump}^{real}$ of mechanical vapor compression heat pumps varies between 40% and 60% of the maximum $COP_{heat\ pump}^{Carnot}$ and depends on the difference (temperature lift) between the condensation (or heat sink) and evaporation (or heat source) temperatures; the smaller the difference, the higher the $COP_{heat\ pump}^{real}$. For example, air-to-air (or air-to-water) heat pumps operating in mild climates may achieve $COP_{heat\ pump}^{real}$ up to 4.0, but at ambient temperatures below approximately –8°C their $COP_{heat\ pump}^{real}$ may drastically drop up to near 1.0.

To indicate how the real drying heat pump operation is close to the ideal optimum, the cycle efficiency can be expressed as the ratio between the ideal ($COP_{heat\ pump}^{Carnot}$) and real ($COP_{heat\ pump}^{real}$) coefficients of performance:

$$\eta = \frac{COP_{heat\ pump}^{Carnot}}{COP_{heat\ pump}^{real}} < 1 \tag{10.50}$$

Exercise E10.6

A single-stage mechanical vapor compression air-to-air heat pump coupled with a low-temperature hardwood dryer operates according to the cycle shown in Figure E10.6 using the HFO-1234yf as a refrigerant. The evaporating and condensing pressures are 1,018 and 1,641 kPa, respectively. If it is assumed that the vapor superheating at the evaporator outlet is 10°C, the compression process is isentropic ($ds = 0$) and the liquid subcooling at the condenser outlet is 0°C, determine,

a. The evaporating and condensing temperatures;
b. The theoretical COP;
c. The cycle efficiency compared to the Carnot COP.

Solution

a. Based on given evaporating (1,018 kPa) and condensing (1,641 kPa) pressures, the evaporating and condensing temperatures are determined from the thermodynamic properties tables or from the ln(p)–h diagram of refrigerant HFO-1234yf, or from an appropriate refrigerant properties software:

$$T_{evap} = 40°C$$

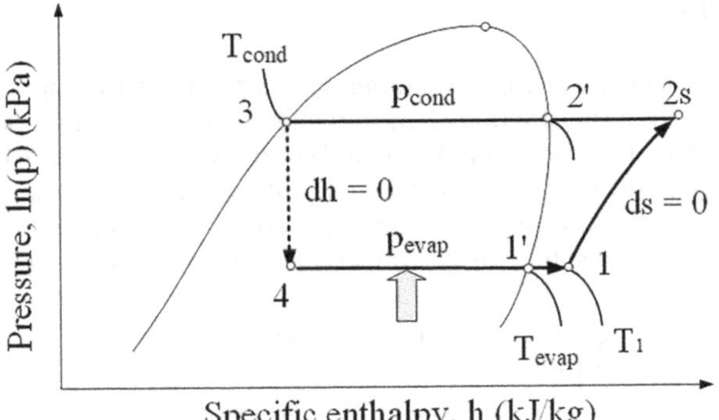

FIGURE E10.6 Theoretical cycle of the single-stage mechanical vapor compression air-to-air heat pump represented in diagram ln(p)–h.

$$T_{cond} = 60°C$$

b. Based on both evaporating (1,018 kPa) and condensing (1,641 kPa) pressures, the following specific enthalpies are determined from the thermodynamic properties tables or from the ln(*p*)–*h* diagram of refrigerant HFO-1234yf (see Figure E10.6):

$$h_1 \cong 400 \text{ kJ/kg}$$

$$h_{2s} \cong 412 \text{ kJ/kg}$$

$$h_3 = h_4 = 285.5 \text{ kJ/kg}$$

The theoretical COP is as follows

$$COP_{heat\ pump}^{theoretical} = \frac{h_{2s} - h_3}{h_{2s} - h_1} = \frac{412 - 285.5}{412 - 400} = 10.5$$

c. The Carnot heat pump COP (Equation 10.40) is as follows:

$$COP_{heat\ pump}^{Carnot} = \frac{1}{1 - \frac{T_{min}}{T_{max}}} \approx \frac{1}{1 - \frac{T_{evap}}{T_{cond}}} = \frac{1}{1 - \frac{313}{333}} = 16.6$$

where

$$T_{evap} = 40°C + 273°C = 313 \text{ K}$$

$$T_{cond} = 60°C + 273°C = 333 \text{ K}$$

The cycle efficiency compared to the Carnot COP is as follows:

$$\eta_{eff} = \frac{COP_{heat\ pump}^{theoretical}}{COP_{heat\ pump}^{ideal,\,Carnot}} = \frac{10.5}{16.5} 100 = 63.6\%$$

Exercise E10.7

A single-stage mechanical vapor compression air-to-air heat pump coupled with a medium-temperature softwood dryer operates according to a theoretical cycle with an evaporator pressure of (463 kPa) and condenser pressure of (913 kPa) using HFC-245fa as a refrigerant (see Figure E10.6). If it is assumed that the vapor superheating at the evaporator outlet is 20°C, the compression process is isentropic ($ds = 0$), and the liquid subcooling at the condenser outlet is 0°C, determine,

a. The evaporating and condensing saturated temperatures;
b. The theoretical COP;
c. The cycle efficiency compared to the Carnot COP.

Solution

a. Based on given evaporating (1,018 kPa) and condensing (1,641 kPa) pressures, as determined from the thermodynamic properties tables or from the ln(p)–h diagram of refrigerant HFC-245fa, or from an appropriate refrigerant properties software, the evaporating and condensing temperatures:

$$T_{evap} = 60°C$$

$$T_{cond} = 86°C$$

b. Based on both evaporating (463 kPa) and condensing (913 kPa) pressures, as determined from the thermodynamic properties tables and ln(p)–h diagram of refrigerant HFC-245fa (ASHRAE 2009) the following specific enthalpies (see Figure E10.6):

$$h_1 = 472 \text{ kJ/kg}$$

$$h_{2s} = 484 \text{ kJ/kg}$$

$$h_3 = 319.56 \text{ kJ/kg}$$

The theoretical heat pump COP is as follows:

$$COP_{heat\ pump}^{theoretical} = \frac{h_{2s} - h_3}{h_{2s} - h_1} = \frac{482 - 319.56}{500 - 474} = 6.2$$

c. The Carnot heat pump COP is (Equation 10.40) is as follows:

$$COP_{heat\ pump}^{Carnot} = \frac{1}{1 - \frac{T_{min}}{T_{max}}} = \frac{1}{1 - \frac{T_{evap}}{T_{cond}}} = \frac{1}{1 - \frac{333}{359}} = 13.8$$

where

$$T_{evap} = 60°C + 273 = 333 \text{ k}$$

$$T_{cond} = 86°C + 273 = 359 \text{ k}$$

The cycle efficiency compared to the Carnot COP is as follows:

$$\eta_{eff} = \frac{COP_{heat\ pump}^{theoretical}}{COP_{heat\ pump}^{ideal,\ Carnot}} = \frac{6.2}{13.8}100 = 45\%$$

Exercise E10.8

In a mechanical vapor compression high-temperature heat pump used for drying softwood, the refrigerant HFC-245fa enters the compressor at p_1 = 400 kPa, *abs* and T_1 = 70°C. The refrigerant mass flow rate is $\dot{m} = 4$ kg/s, and the compressor power input is $\dot{W} = 60$ kW. After compression, the refrigerant enters an air-cooled condenser at p_2 = 1,564 kPa, *abs* and T_2 = 115°C, and leaves it as a subcooled liquid at p_3 = 1,500 kPa, *abs* and T_3 = 95°C. The air enters the condenser at $T_{cond}^{air,\ in} = 66°C$ and leaves it at $T_{cond}^{air,\ out} = 82°C$

Based on the first law of thermodynamics applied to the compressor and condenser boundaries, respectively, determine the airflow rate required through the condenser.

Solution

Figure E10.8 shows the boundaries of the compressor (Figure E10.8a) and condenser (Figure E10.8b) with given parameters, as well as the compression process represented in the ln(*p*)–*h* diagram (Figure E10.8c).

From the pressure–enthalpy diagram of the refrigerant HFC-245fa, or from superheated vapor-constant pressure tables, or from a refrigerant properties software, the specific enthalpy at the compressor inlet and outlet, respectively, can be determined:

h_1 = 461 kJ/kg (specific enthalpy at the compressor inlet)
$h_2 = h_{refr,\ in}$ = 476 kJ/kg (specific enthalpy at the condenser inlet)
$h_3 = h_{refr,\ out}$ = 332 kJ/kg (specific enthalpy at the condenser outlet)

If it is assumed that there is not heat transfer across the condenser boundary, the changes of kinetic and potential energies inside the condenser are neglected, and, because $\dot{W}_{12} = 0$, the first law of thermodynamics can be written as follows:

$$\sum_{inlet}(\dot{m}_{in}h_{in}) = \sum_{outlet}(\dot{m}_{out}h_{out})$$

or

$$\dot{m}_{refr} * h_{refr,\ in} + \dot{m}_{air} * h_{air,\ in} = \dot{m}_{refr} * h_{refr,\ out} + \dot{m}_{air} * h_{air,\ out}$$

From air thermodynamic properties tables at atmospheric pressure, determine the following:

$$h_{air,\ in} = c_{p,\ in} * T_{cond}^{air,\ in} = 1.009\ \text{kJ/kg°C} * 66°\text{C} = 66.6\ \text{kJ/kg}$$

$$h_{air,\ out} = c_{p,\ out} * T_{cond}^{air,\ out} = 1.01\ \text{kJ/kg°C} * 82°\text{C} = 82.82\ \text{kJ/kg}$$

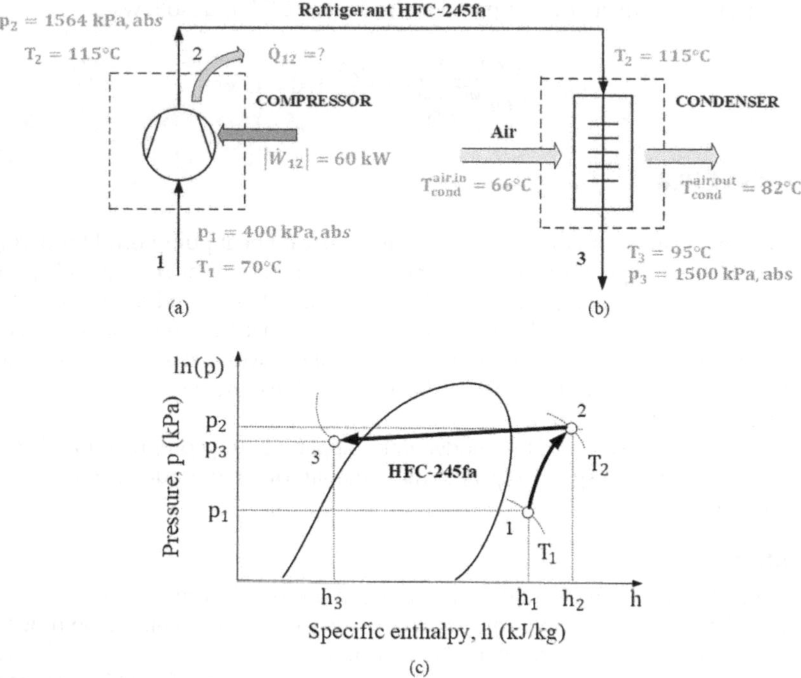

FIGURE E10.8 (a) Main data for the compressor and condenser thermodynamic processes; (b) compressor and condenser actual thermodynamic processes represented in diagram ln(p)–h.

Based on the energy balance equation, the airflow rate through the condenser will be as follows:

$$\dot{m}_{air} = \frac{\dot{m}_{refr}\left(h_{refr,\,in} - h_{refr,\,out}\right)}{h_{air,\,out} - h_{air,\,in}} = \frac{4*(476-332)}{82.82-66.6} = 35.5\ \text{kg/s} = 35.5*1.29\ \text{m}^3/\text{kg}$$

$$= 45.8\ \text{m}^3/s * 2118.8 = 97042\ CFM$$

Exercise E10.9

Design a single-stage mechanical vapor compression air-to-air heat pump coupled with a low-temperature lumber kiln that operates with HFC-245fa as a refrigerant according to the real cycle schematically represented in Figure 10.5a–c, in order to satisfy the following specifications:

CONDENSER

Heating capacity: 195 kW
Subcooling of refrigerant at the condenser exit: 5°C
Temperature ΔT of air across the evaporator: 30°C
Heat transfer effectiveness: 80%

COMPRESSOR

Isentropic compression efficiency: 70%

EVAPORATOR

Heat source temperature (drying air): 60°C
Temperature ΔT of air across the evaporator: 15°C
Superheat of refrigerant at the evaporator exit: 5°C
Heat transfer effectiveness: 80%

Determine

1. The thermodynamic properties of the refrigerant (i.e., pressures, temperatures, quality, and enthalpy) for all key points of the heat pumping cycle;
2. The refrigerant mass flow rate;
3. The evaporator thermal capacity;
4. The heat pump COP.

Solution

Figure 10.5c schematically shows in the ln(p)–h diagram the actual (real) heat pump cycle.

Given data allow schematically representing in temperature—area (T–A) diagrams the refrigerant and drying air temperature variations (Figure E9.9a and b) and determine the following parameters:

Condensing temperature is as follows:

$$T_{cond} = 90°C + 5°C = 95°C$$

Based on the condensing temperature (T_{cond} = 95°C), it can be found from a thermos-physical saturated vapor and liquid properties table for the refrigerant HFC-245fa (ASHRAE 2009), or from refrigerant properties software as NIST REFPROP (version 9.1) (www.nist.gov/srd/refprop. Accessed December 18, 2917), the corresponding condensing saturated pressure is as follows:

$$p_{cond} = 1\ 120\,\text{kPa}, abs$$

$$h_3 \approx 328\ \text{kJ/kgK}$$

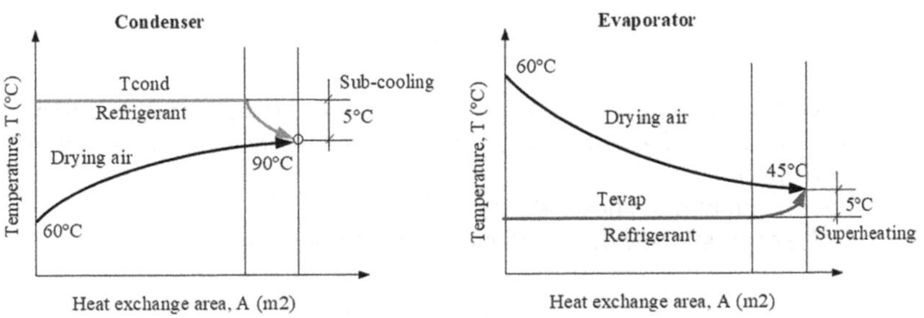

FIGURE E10.9 Temperature variations of refrigerant and drying air in (a) heat pump condenser; (b) heat pump evaporator.

The temperature of the subcooled liquid at the condenser outlet will be as follows:

$$T_4 = T_{cond} - 5°C = 95°C - 5°C = 90°C$$

The specific enthalpy of the subcooled liquid leaving the condenser can be computed with NIST REFPROP (version 9.1) software or easily found in the pressure–enthalpy diagram of refrigerant HFC-245fa.

Evaporating temperature is as follows:

$$T_{evap} = 45°C - 5°C = 40°C$$

Based on the evaporating saturated temperature (40°C), it can be found from the thermos-physical saturated vapor and liquid properties table for the refrigerant HFC-245fa (ASHRAE 2009), or calculated with refrigerant properties software as NIST REFPROP (version 9.1), the evaporating saturated pressure:

$$p_{evap} = 251.79 \text{ kPa}, abs$$

Temperature of the superheated vapor at the evaporator exit and compressor inlet is as follows:

$$T_1 = T_{evap} + 20°C = 40°C + 20°C = 60°C$$

At temperature T_1 and pressure p_{evap} the specific enthalpy of the superheated vapor leaving the evaporator at state 1 is as follows:

$$h_1 = 453 \text{ kJ/kg}$$

The compressor isentropic compression efficiency (70%) is defined as follows:

$$\eta_s = \frac{h_{2s} - h_1}{h_2 - h_1}$$

$$0.7 = \frac{h_{2s} - h_1}{h_2 - h_1} = \frac{486 - 453}{h_2 - 453} = \frac{33}{h_2 - 453}$$

$$0.7 * (h_2 - 453) = 33$$

$$0.7 * h_2 - 0.7 * 453 = 33$$

$$h_2 = 500.14 \text{ kJ/kg}$$

1. Table E10.9 summarizes the thermodynamic properties of the refrigerant HFC-245fa for all key points of the heat pumping cycle represented in Figure 10.5.
2. The refrigerant mass flow rate is as follows:

   The expression of the condenser thermal power is:

$$\dot{Q}_{cond} = \eta_{cond} * \dot{m}_r * (h_2 - h_4)$$

$$195 = 0.8 * \dot{m}_r * (500.14 - 328)$$

TABLE E10.9
Thermodynamic Properties of the Refrigerant HFC-245fa for All Key Points of the Heat Pumping Cycle (see Figure 10.4)

State	Pressure (kPa)	Temperature (°C)	Quality	Specific Enthalpy (kJ/kg)
1	251.79	60	–	453
2*s*	1,120	105	–	486
2	1,120	116	–	500.14
2′	1,120	95	1.0	573.5
3	1,120	95	0.0	333
4	1,120	90	–	328
5	251.79	40	0.4	328
1′	251.79	40	1.0	434.97

Note: Pressure drops inside evaporator and condenser are neglected.

The refrigerant mass flow rate can be thus determined as follows:

$$\dot{m}_r = \frac{195}{0.8 * 172.14} = 1.415\ \text{kg/s}$$

3. The evaporator thermal capacity is as follows:

$$\dot{Q}_{evap} = \eta_{evap} * \dot{m}_r * (h_1 - h_5) = 0.8 * 1.415 * (453 - 328) = 141.5\ \text{kW}$$

4. The compressor power requirement is as follows:

$$\dot{W}_{compr} = \dot{Q}_{cond} - \dot{Q}_{evap} = 195 - 141.5 = 53.5\ \text{kW}$$

5. The coefficient of performance is as follows:

$$COP = \frac{\dot{Q}_{cond}}{\dot{W}_{compr}} = \frac{195}{53.5} = 3.64$$

Exercise E10.10

A single-stage mechanical vapor compression air-to-air heat pump operates according to the real cycle schematically represented in Figure 10.5b and c. The compression process 1-2 is polytropic ($\delta q \neq 0$), and the expansion process 4-5 is isenthalpic ($dh = 0$). The refrigerant is the HFO-1234yf. The heat pump is coupled with a lumber kiln and operates with an average evaporating temperature of 20°C, average condensing temperature of 70°C, and average discharge temperature of 80°C. The superheating at the evaporator outlet is of 5°C and the subcooling amount at the condenser outlet, 8°C. The average refrigerant flow rate is 0.4 kg/s.

Neglecting the pressure drops inside the evaporator and condenser, determine

1. The total cooling and dehumidifying effect (i.e., the total thermal power extracted from the drying air flowing through the evaporator);
2. The total compressor power requirement;
3. The total heating effect (i.e., the total thermal power supplied to the lumber dryer by the evaporator);
4. The compression ratio;
5. The heat pump COP.

Solution

Based on given evaporating (20°C) and condensing (70°C) temperatures, can be determined from a thermos-physical saturated vapor and liquid properties table for the refrigerant HFO-1234yf, or from refrigerant properties software, the corresponding evaporating and condensing saturated pressures:

$$p_{evap} = 591.72 \text{ kPa}, abs$$

$$p_{cond} = 2044.53 \text{ kPa}, abs$$

The compressor suction temperature will be:

$$T_1 = T_{evap} + (\text{superheating}) = 20°\text{C} + 5°\text{C} = 25°\text{C}$$

Based on the evaporating saturated pressure (p_{evap} = 591.72 kPa, *abs*) and the compressor suction temperature (T_1 = 25°C), the specific enthalpy at state 1 (compressor suction) can be found from a thermos-physical superheating vapor properties table for the refrigerant HFO-1234yf, or from a refrigerant properties software:

$$h_1 = 378.2 \text{ kJ/kg}$$

Similarly, based on the condensing saturated pressure (p_{cond} = 2,044.53 kPa, *abs*) and the compressor discharge temperature (T_2 = 80°C), the specific enthalpy at state 2 (compressor discharge) can be found from a thermos-physical superheating vapor properties table for the refrigerant HFO-1234yf, or from refrigerant properties software:

$$h_2 = 415.0 \text{ kJ/kg}$$

The temperature of the subcooled refrigerant leaving the condenser is as follows:

$$T_4 = T_{cond} - (\text{sub-cooling}) = 80°\text{C} - 8°\text{C} = 72°\text{C}$$

The specific enthalpy at states 4 and 5 can be easily found from a pressure–enthalpy diagram or (with approximation) from a thermos-physical saturated liquid and vapor properties table for the refrigerant HFO-1234yf, or from refrigerant properties software:

$$h_4 = h_5 \approx 306 \text{ kJ/kg}$$

1. The total cooling and dehumidifying effect is as follows:

$$\dot{Q}_{5-1} = \dot{m}_r \left(h_1 - h_5\right) = 0.4 * (378.2 - 306) = 28.88 \text{ kW}$$

2. The total compressor power requirement is as follows:

$$\dot{W}_{2-4} = \dot{m}_r \left(h_2 - h_1\right) = 0.4 * (415 - 378.2) = 14.72 \text{ kW}$$

3. The total heating effect is as follows:

$$\dot{Q}_{2-4} = \dot{m}_r \left(h_2 - h_4\right) = 0.4 * (415.0 - 306) = 43.6 \text{ kW}$$

4. The compression ratio is as follows:

$$\pi = \frac{p_{cond}}{p_{evap}} = \frac{2044.53}{591.72} = 3.45$$

5. The heat pump real COP is as follows:

$$COP_{real} = \frac{\dot{Q}_{2-4}}{\dot{W}_{2-4}} = \frac{43.6}{14.72} = 2.96$$

Exercise E10.11

A single-stage mechanical vapor compression air-to-air heat pump operates according to the real cycle schematically represented in Figure 10.5b and c. The compression process 1-2 is polytropic ($\delta q \neq 0$), and the expansion process 4-5 is isenthalpic ($dh = 0$). The refrigerant is ammonia (R-717). The heat pump is coupled with a lumber kiln and operates with an average evaporating temperature of 30°C, average condensing temperature of 80°C, and average discharge temperature of 110°C. The superheating at the evaporator outlet is of 5°C and the subcooling amount at the condenser outlet is of 10°C. The average refrigerant flow rate is 0.4 kg/s.

Neglecting the pressure drops inside the evaporator and condenser, determine

1. The total cooling and dehumidifying effect (i.e., the total thermal power extracted from the drying air flowing through the evaporator);
2. The total compressor power requirement;
3. The total heating effect (i.e., the total thermal power supplied to the lumber dryer by the evaporator);
4. The compressor compression ratio;
5. The heat pump COP.

Solution

Based on given evaporating (30°C) and condensing (80°C) temperatures, which can be determined from a thermos-physical saturated vapor and liquid properties table for ammonia (R-717), or from refrigerant properties software, the corresponding evaporating and condensing saturated pressures:

$$p_{evap} = 1167 \text{ kPa}, abs$$

$$p_{cond} = 4142 \text{ kPa}, abs$$

The compressor suction temperature will be as follows:

$$T_1 = T_{evap} + (\text{superheating}) = 30°\text{C} + 5°\text{C} = 35°\text{C}$$

Based on the evaporating saturated pressure (p_{evap} = 1,167 kPa, *abs*) and the compressor suction temperature (T_1 = 35°C), the specific enthalpy at state 1 (compressor suction) can be found from a thermos-physical superheating vapor properties table for ammonia (R-717), or from refrigerant properties software:

$$h_1 \approx 1520 \text{ kJ/kg}$$

Similarly, based on the condensing saturated pressure (p_{cond} = 4,142 kPa, *abs*) and the compressor discharge temperature (T_2 = 110°C), the specific enthalpy at state 2 (compressor discharge) can be found from a thermos-physical superheating vapor properties table for the refrigerant ammonia (R-717), or from refrigerant properties software:

$$h_2 \approx 1630 \text{ kJ/kg}$$

The temperature of the subcooled refrigerant leaving the condenser is as follows:

$$T_4 = T_{cond} - (\text{sub-cooling}) = 80°\text{C} - 10°\text{C} = 70°\text{C}$$

The specific enthalpy at states 4 and 5 can be easily found from a pressure–enthalpy diagram or (with approximation) from a thermos-physical saturated liquid and vapor properties table for the refrigerant ammonia (R-717), or from refrigerant properties software:

$$h_4 = h_5 \approx 550 \text{ kJ/kg}$$

1. The total cooling and dehumidifying effect is as follows:

$$\dot{Q}_{5-1} = \dot{m}_r (h_1 - h_5) = 0.4 * (1,520 - 550 \text{ kJ/kg}) = 388 \text{ kW}$$

2. The total compressor power requirement is as follows:

$$\dot{W}_{2-4} = \dot{m}_r (h_2 - h_1) = 0.4 * (1,630 - 1,520) = 44 \text{ kW}$$

3. The total heating effect is as follows:

$$\dot{Q}_{2-4} = \dot{m}_r (h_2 - h_4) = 0.4 * (1,630 - 550) = 432 \text{ kW}$$

4. The compression ratio is as follows:

$$\pi = \frac{p_{cond}}{p_{evap}} = \frac{4{,}142}{1{,}167} = 3.55$$

5. The heat pump real COP is as follows:

$$COP_{real} = \frac{\dot{Q}_{2-4}}{\dot{W}_{2-4}} = \frac{432}{44} = 9.81$$

Exercise E10.12

A single-stage mechanical vapor compression air-to-air heat pump operates according to the real cycle schematically represented in Figure 10.5b and c. The compression process 1-2 is polytropic ($\delta q \neq 0$) and the expansion process 4-5 is isenthalpic ($dh = 0$). The refrigerant is the propane (R-290). The heat pump is coupled with a lumber kiln and operates with an average evaporating temperature of 28°C, average condensing temperature of 65°C, and average discharge temperature of 87°C. The superheating at the evaporator outlet is of 5°C and the subcooling amount at the condenser outlet is of 10°C. The average refrigerant flow rate is 0.4 kg/s.

Neglecting the pressure drops inside the evaporator and condenser, determine

1. The total cooling and dehumidifying effect (i.e., the total thermal power extracted from the drying air flowing through the evaporator);
2. The total compressor power requirement;
3. The total heating effect (i.e., the total thermal power supplied to the lumber dryer by the evaporator);
4. The compressor compression ratio;
5. The heat pump COP.

Solution

Based on given evaporating (28°C) and condensing (65°C) temperatures, which can be determined from a thermos-physical saturated vapor and liquid properties table for propane (R-290), or from refrigerant properties software, the corresponding evaporating and condensing saturated pressures:

$$p_{evap} = 1037 \text{ kPa}, abs$$

$$p_{cond} = 2365.4 \text{ kPa}, abs$$

The compressor suction temperature is as follows:

$$T_1 = T_{evap} + (\text{superheating}) = 28°\text{C} + 5°\text{C} = 33°\text{C}$$

Based on the evaporating saturated pressure (p_{evap} = 1,037 kPa, *abs*) and the compressor suction temperature (T_1 = 33°C), the specific enthalpy at state 1 (compressor suction) can be found from a thermos-physical superheating vapor properties table for propane (R-290), or from a refrigerant properties software:

$$h_1 \approx 950 \text{ kJ/kg}.$$

Similarly, based on the condensing saturated pressure (p_{cond} = 2,365.4 kPa, *abs*) and the compressor discharge temperature (T_2 = 87°C), the specific enthalpy at state 2 (compressor discharge) can be found from a thermos-physical superheating vapor properties table for the refrigerant propane (R-290), or from refrigerant properties software:

$$h_2 \approx 1025 \text{ kJ/kg}$$

The temperature of the subcooled refrigerant leaving the condenser is as follows:

$$T_4 = T_{cond} - (\text{sub-cooling}) = 65 - 10 = 55°\text{C}$$

The specific enthalpy at states 4 and 5 can be easily found from a pressure–enthalpy diagram or (with approximation) from a thermos-physical saturated liquid and vapor properties table for the refrigerant propane (R-290), or from a refrigerant properties software:

$$h_4 = h_5 \approx 660 \text{ kJ/kg}$$

1. The total cooling and dehumidifying effect is as follows:

$$\dot{Q}_{5-1} = \dot{m}_r (h_1 - h_5) = 0.4 * (950 - 660 \text{ kJ/kg}) = 116 \text{ kW}$$

2. The total compressor power requirement is as follows:

$$\dot{W}_{2-4} = \dot{m}_r (h_2 - h_1) = 0.4 * (1025 - 950) = 30 \text{ kW}$$

3. The total heating effect is as follows:

$$\dot{Q}_{2-4} = \dot{m}_r (h_2 - h_4) = 0.4 * (1025 - 660) = 146 \text{ kW}$$

4. The compression ratio is as follows:

$$\pi = \frac{p_{cond}}{p_{evap}} = \frac{2365.4}{1037} = 2.28$$

5. The heat pump real COP is as follows:

$$COP_{real} = \frac{\dot{Q}_{2-4}}{\dot{W}_{2-4}} = \frac{365}{30} = 4.86$$

10.7.3 Refrigerant Mixtures

During the past years, some refrigerant mixtures were introduced as working fluids for heat pumps as, for example, R-407C, a zeotropic refrigerant blend of difluoromethane (HFC-32, 32%) (that serves to provide the heat capacity), pentafluoroethane (HFC-125, 25%) (that contributes to decrease flammability), and 1,1,1,2-tetrafluoroethane (HFC-134a, 52%) (that contributes to reduce pressure).

The ozone depletion potential (ODP) of R-407C is zero (compared to 0.055 of HCFC-22) and its global warming potential (GWP) is 1770 (vs. 1810 of HCFC-22) based on integrated time of 100 years.

R-407C, with levels of pressure and moisture solubility approximately the same as those of HCFC-22, was intended as a replacement for HCFC-22 of which production will be phased out by 2020. R-407C is highly compatible with polyester and ether oils, but not compatible with mineral oils normally used with HCFC-22.

Exercise E10.13

A single-stage mechanical vapor compression air-to-air heat pump coupled with a wood dryer operates according to the cycle schematically represented in Figure E10.13. The compression process 1-2*s* is adiabatic ($\delta q = 0$), and the expansion process 4-5 is isenthalpic ($dh = 0$). The working fluid is the refrigerant zeotropic R-407C. The refrigerant dew point (i.e., the temperature on the saturated vapor curve) at the evaporating pressure is $T_{1''} = 20°C$ and the bubble point (i.e., the temperature on the saturated liquid curve) at the condensing temperature is $T_{3'} = 50°C$. The pressure losses in evaporator and condenser are neglected.

The superheating at the evaporator outlet is of 10°C, and the subcooling amount at the condenser outlet is of 10°C.

The compressor power input is 35 kW.

Neglecting the pressure drops inside the evaporator and condenser, determine

1. The refrigerant flow rate;
2. The total cooling and dehumidifying effect (i.e., the total thermal power extracted from the drying air flowing through the evaporator);
3. The total heating effect (i.e., the total thermal power supplied to the lumber dryer by the evaporator);
4. The compressor compression ratio;
5. The heat pump COP.

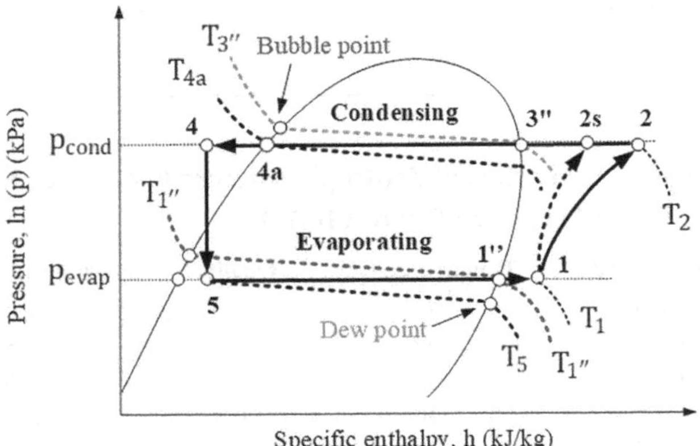

FIGURE E10.13 Theoretical cycle of the mechanical vapor compression heat pump with zeotropic mixture R-407C as refrigerant represented in ln(p)–h diagram (note: pressure drops inside evaporator and condenser not shown).

Solution

Based on given data, as determined from properties table and ln(*p*)–*h* diagram of R-407C the thermodynamic properties of the heat pump cycle (see Tables E10.13.1 and E10.13.2). It can be seen that the temperature glide in the evaporator is $T_{1'}$–$T_{5'}$ and the temperature glide in the condenser is $T_{3''}$–T_{4a}.

1. The refrigerant flow rate is as follows:
   From the compressor power input,

$$\dot{W}_{comp} = \dot{m}_r \left(h_2 - h_1\right)$$

results

$$\dot{m}_r = \frac{\dot{W}}{\left(h_2 - h_1\right)} = \frac{35}{465 - 430} = 1\ \text{kg/s}$$

2. The total cooling and dehumidifying effect is as follows:

$$\dot{Q}_{evap} = \dot{m}_r \left(h_1 - h_5\right) = 1 * (430 - 263) = 167\ \text{kW}$$

TABLE E10.13.1
Saturation Properties of Zeotropic Mixture R-407C (see Figure E10.13)

			Pressure		
Process	**Temperature (°C)**	**State**	**kPa**	**kPa**	**Specific Enthalpy (kJ/kg)**
Evaporation	20	1′	Liquid	1,023.4	230.1
	20	1″	Vapor	864.4	424.1
Condensation	50	3′	Liquid	2,185.9	281.9
	50	3″	Vapor	1,959.0	433.9

TABLE E10.13.2
Thermodynamic Properties of Zeotropic Mixture R-407C for Key Points of the Heat Pump Cycle (see Figure E10.13)

State	Pressure (kPa)	Temperature (°C)	Quality	Specific Enthalpy (kJ/kg)
1	864.4	30	–	430
2*s*	1,959.0	70	–	455
2	1,959.0	80	–	465
3″	1,959.0	50	1	281.9
3′	2,185.9	50	1	281.9
4	1,959.0	40	–	263
5	864.4	20	0.2	263
1″	864.4	20	1	424.1

3. The total heating effect is as follows:

$$\dot{Q}_{cond} = \dot{m}_r \left(h_2 - h_4\right) = 1*(465-263) = \dot{W}_{comp} + \dot{Q}_{evap} = 35+167 = 202 \text{ kW}$$

4. The compressor compression ratio is as follows:

$$\pi = \frac{p_{cond}}{p_{evap}} = \frac{1959.0}{864.4} = 2.266$$

5. The heat pump COP is as follows:

$$COP = \frac{\dot{Q}_{cond}}{\dot{W}_{comp}} = \frac{202}{35} = 5.77$$

10.8 IMPROVEMENTS OF HEAT PUMP CYCLES

Improvements of mechanical vapor compression heat pump technology, among many others, aim at (i) reducing the cycle thermodynamic losses; (ii) providing optimum amount of vapor superheat at the evaporator outlet and liquid subcooling at the condenser outlet (e.g., by using suction line vapor-to-liquid heat exchangers or mechanical subcoolers); (iii) recovering expansion energy losses (e.g., by using expanders or ejectors); (iv) achieving multistage cycles (e.g., with vapor, liquid, or two-phase refrigerant injection) (Park et al. 2015); and (v) optimizing the pressure drops in the compressor suction and discharge valves (if used), heat exchangers (leading to high heat transfer coefficients), and compressor suction pipes (in order to ensure sufficiently high vapor velocity for proper oil return to the compressor) (Radermacher and Hwang 2000; Park et al. 2015).

Because any advanced mechanical vapor compression more or less complex heat pump cycle involves additional costs, economic analysis on each available option is recommended.

10.8.1 Optimum Evaporator Superheating

When the refrigerant vapor at the heat pump evaporator outlet is saturated, the entire evaporator heat transfer surface area is used for the purpose of vaporization process with a maximum heat transfer rate. However, there is a risk to supply some moisture or refrigerant liquid (droplets) to the compressor suction line.

In order to ensure that the heat pump compressor is not exposed to any liquid that can be entrained in the suction line, some amount of superheat is required at the evaporator outlet, even if the evaporator saturation temperature is slightly lowered. Because of the specific heat of superheated vapor, the overall heat transfer coefficient and the average temperature difference during the superheating process are lower than for the refrigerant (pure or mixed) convective forced vaporization. These conditions result in much less heat absorbed from the moist drying air during the superheating process for a given evaporator heat transfer area.

In addition, the vapor superheat increases the compressor specific work (per unit mass of refrigerant) and reduces the compressor volumetric capacity; since the vapor density is also lower, the refrigerant mass flow rate is smaller, which reduces the heat pump's cooling and dehumidification capacities of heat pumps (Radermacher and Hwang 2000).

The refrigerant superheating inside the heat pump evaporators is controlled by expansion devices as thermostatic or electric expansion valves. The evaporator superheating amount must be carefully determined for each particular refrigerant used in heat pump-assisted dryers.

10.8.2 Condenser Subcooling

Typically, the state of the refrigerant leaving the heat pump condenser and entering the expansion device is subcooled liquid. Subcooling is usually achieved within the heat pump condenser once liquid refrigerant starts accumulating within the area close to the condenser exit. In practice, the refrigerant subcooling can be realized without internal heat exchanger or other similar devices.

Even the refrigerant (pure or zeotropic mixture) subcooling at the heat pump condenser outlet reduces the condenser total heat transfer area available for condensation, and increases the condenser saturation temperature, such a process reduces the enthalpy at the evaporator inlet and the throttling losses during the isenthalpic expansion (Pottker and Hrnjak 2015); consequently, it enhances the evaporator cooling and dehumidification capacity (by adding the heat quantity $\Delta q = h_4 - h_6$ to the evaporating heat $q_{evap} = h_1 - h_4$ provided during the heat pump cycle without subcooling) (see Figure 10.6) as well as the heat pump's heating COP (Linton et al. 1992; Primal and Lundqvist 2005; Corberan et al. 2008).

As a mass of refrigerant accumulates in the form of subcooled liquid at condenser exit, the two-phase heat exchange internal area of the condenser is reduced. This yields an increase in saturation temperature (not shown in Figure 10.6) while the liquid refrigerant exiting the condenser is cooled below saturation temperature ($T_5 < T_3$). The increase in subcooling is a result of both reduction of condenser exit temperature and increase of saturation temperature.

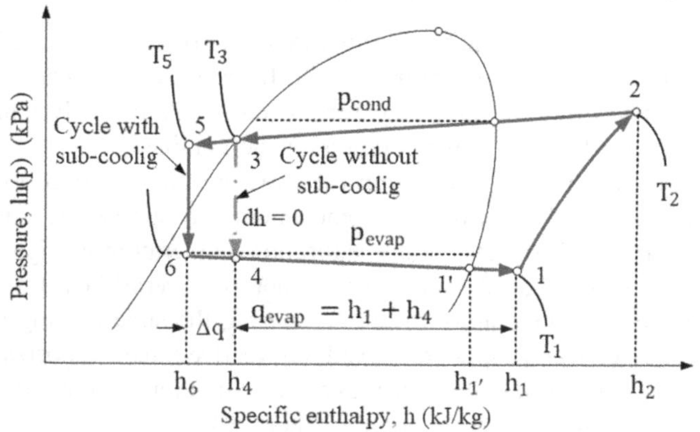

FIGURE 10.6 Comparison between heat pump cycles with and without condenser subcooling.

The enthalpy at the evaporator inlet is reduced from h_4 to h_6 while the refrigerant exiting the condenser is subcooled, thus enlarging the enthalpy difference across the evaporator by $\Delta q = h_4 - h_6$ (Figure 10.6).

The specific isentropic work of compression, however, first decreases at lower values of subcooling due to a reduction in the pressure ratio and an increase in the compressor inlet pressure even though the condensing pressure increases. This means that within lower values of subcooling, the effect of the suction pressure increase on the pressure ratio is dominant over that of the condensing pressure increase. At higher values of subcooling, however, as the condensing pressure rises sharply, its effect becomes dominant over that of the compressor inlet pressure increase, thus elevating the pressure ratio and, consequently, the isentropic specific work of compression. The increase in the isentropic efficiency is caused by the decrease in pressure ratio and the reduction of the compressor speed while matching the cooling capacity at different values of subcooling. Beyond its maximum value, the isentropic efficiency slightly drops mainly due to increasing pressure ratio (Pottker and Hrnjak 2015).

10.8.3 Internal Heat Exchanger

The internal (suction) heat exchanger, as the condenser is subcooling, serves to further subcool the refrigerant vapor prior entering the heat pump expansion valve and to increase the heat pump's evaporator cooling and dehumidifying capacity. The suction line (internal) heat exchanger is located between the condenser outlet and expansion valve inlet and between the evaporator outlet and compressor inlet. With internal heat exchangers, subcooling is internally achieved due the heat transfer between the cold refrigerant vapor exiting the evaporator and the hotter refrigerant liquid leaving the condenser (Figure 10.7).

As can be seen in Figure 10.7, when saturated liquid leaves the condenser as saturated liquid at state 3, the evaporator capacity is determined by the enthalpy difference $h_1 - h_4$. If the saturated liquid refrigerant is further subcooled from state 3 to state 5, the lower the enthalpy of the liquid becomes (h_5), and therefore, the enthalpy at the inlet of the evaporator becomes $h_5 < h_3$.

In the internal heat exchanger, the suction refrigerant vapor at state 1′ is superheated, and thus its enthalpy increased by $h_1 - h_{1''}$, while the inlet enthalpy to the evaporator (state 6) is lowered by $h_4 - h_6 = h_1 - h_{1'}$. This heat transfer enables the transfer of cooling capacity from the evaporator saturation temperature level, via the suction heat exchange, to the temperature level of the (warmer) suction gas as it is being superheated (Radermacher and Hwang 2000).

Since the refrigerant vapor entering the internal heat exchanger is superheated at state 1 (i.e., dry), its specific heat is less than the specific heat of the subcooled liquid at state 3. In this case, the thermal efficiency of the internal heat exchanger is defined as follows (Krakow and Lin 1988):

$$\eta_{IHEX} = \frac{T_3 - T_5}{T_3 - T_{1'}} \tag{10.51}$$

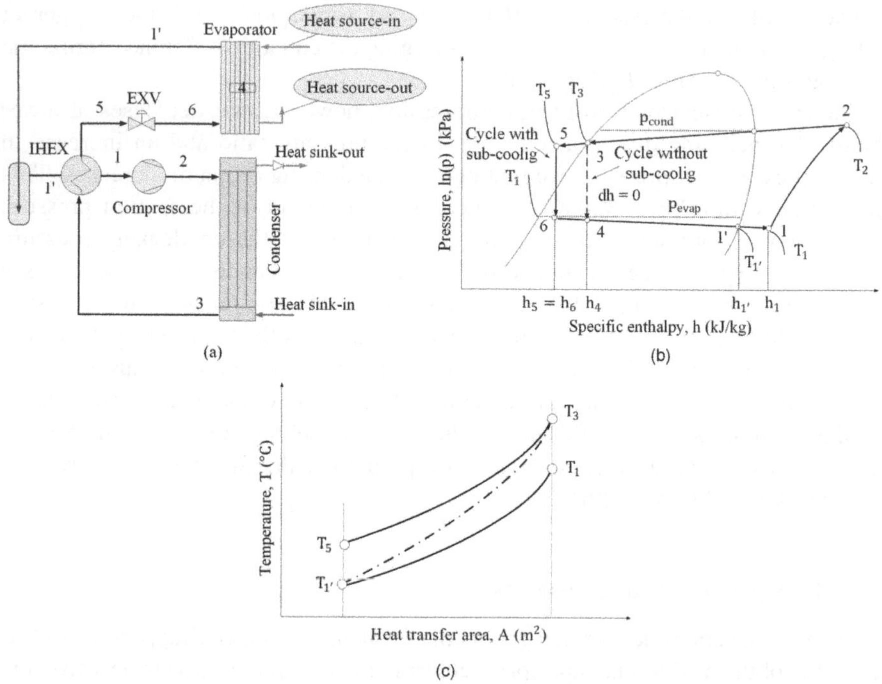

FIGURE 10.7 (a) Schematic of a heat pump with internal heat exchanger; (b) heat pump cycle with internal heat exchanger in ln(p)–h diagram; and (c) T–A diagram of the internal heat exchanger. EXV, expansion valve; IHEX, internal heat exchanger.

The heat transfer rate of the internal heat exchanger can be calculated as follows (Figure 10.7):

$$\dot{Q}_{IHEX} = \eta_{IHEX} * \dot{m}_{refr}\left(h_4 - h_6\right) = \dot{m}_{refr}\left(h_1 - h_{1'}\right) \tag{10.52}$$

where

$\dot{m}_{refr}$ is the refrigerant mass flow rate (kg/s)

The effects of internal heat exchanger on the heat pump's COP and compressor volumetric capacity depend on both operating conditions and refrigerants' thermos-physical properties (Domanski et al. 1994).

The benefit of the suction to liquid line internal heat exchange is higher when the refrigerant specific heat is high, because the refrigerants with high specific heats provide (i) lower increase in temperature during the compression process and (ii) less increase of the compressor work due to lack of vapor superheat. On the other hand, during the expansion process, a relatively large amount of refrigerant has to evaporate to cool the remaining liquid to the evaporator saturation temperature (Radermacher and Hwang 2000).

It can be seen that a reduced refrigerant temperature at the condenser outlet decreases the enthalpy of the evaporator inlet and this increases the evaporator cooling and dehumidification capacity. However, the increased temperature of the compressor inlet increases the vapor specific volume that lowers the refrigerant mass flow rate and decreases the compressor specific work and efficiency, and finally degrades the system performance. Therefore, these two aspects (i.e., reduced refrigerant temperature prior to expansion valve and increased temperature at the compressor inlet) should be carefully analyzed in heat pump applications with the internal heat exchanger (Park et al. 2015).

Thus, the benefit of the suction to liquid line internal heat exchanger depends on whether the benefit outweighs the penalty of increased compressor work or not, as well as on the working fluid used (Radermacher and Hwang 2000; Kim et al. 2004). All the previous observations show that the use of internal heat exchangers is controversial mainly because the reduction of refrigerant mass flow rate would typically lead to reduced cooling and dehumidification capacities. However, on the other hand, the relatively constant compressor work input increases the discharge temperature and enthalpy which leads to an increase in the driving temperature difference through the evaporator that can offset the penalty due to the reduced mass flow rate (Kim et al. 2004).

Since both subcooling and internal heat exchanger methods compete toward reducing the throttling losses, they significantly reduce the increase of the heat pump COP. However, besides the interference between condenser subcooling and internal heat exchanger, the use of both simultaneously seems provide more efficient air conditioning systems, especially with HFO-1234yf as a refrigerant (Pottker and Hrnjak 2015).

Due to the reduced pressure ratio (by 7.4%) of a heat pump with suction line heat exchanger at the same throttling opening, the cooling and dehumidifying capacity, and COP of the heat pump increased by 7.5% and 3.2%, respectively, compared to a baseline system (Kang et al. 2007).

The effectiveness of a tube-in-tube type suction line internal heat exchanger in a vapor compression cycle using HFO-1234yf as a refrigerant ranged from 17% to 25%, while the cooling capacity increased by 2%–9% (Navarro-Esbri et al. 2013).

Other experimental works (e.g., Cho et al. 2013) showed that the cooling capacity and COP of the HFO-1234yf system without the suction line heat exchanger decreased by up to 7% and 4.5%, respectively, compared to those of HFC-134a system, while those of the HFO-1234yf system with the suction line heat exchanger decreased by 1.8% and 2.9%, implying that similar performance could be achieved by the suction line heat exchanger.

An experimental study about the effect of condenser subcooling on the performance of an air conditioning system operating with R-134a and R-1234yf showed that the use of both the condensing subcooling and the suction line heat exchanger simultaneously yielded more efficient performance improvement, especially for R-1234yf (Pottker and Hrnjak 2015b).

By applying the internal heat exchanger in supercritical CO_2 heat pump cycles to reduce throttling losses, improvements of COPs have been reported (Lorentzen and Pettersen 1993) in comparison with those of other conventional refrigerant systems.

On the other hand, the suction line heat exchanger might be detrimental to system performance in systems using, for example, HCFC-22, and useful for systems using HFC-134a, HFC-404A, HFC-410A, R-290, and R-600 (Klein et al. 2000).

It can be concluded that by using suction line heat exchangers (i) COP improvement have been obtained with refrigerants as R-507A, HFC-134a, CFC-12, HFC-404A, R-290, R-407C, R-600, R-744, HFO-1234yf, and HFC-410A); (ii) no COP improvements have been reported with refrigerants as HCFC-22, HFC-32, and R-117); and (iii) the heat pump COP might either increase or decrease when the suction line heat exchanger is used, depending on the refrigerant working pressures, heat exchanger effectiveness, specific heat ratio, and the enthalpy (latent heat) of evaporation.

Exercise E10.14

A heat pump used as a dehumidifier device in wood drying operates according to a theoretical mechanical vapor compression cycle with internal heat exchanger (see Figure 10.7a). The working fluid is the refrigerant HFO-1234yf with saturated evaporating and condensing temperatures T_{evap} = 40°C and T_{cond} = 60°C, respectively. The evaporator superheating and liquid subcooling are both 10°C. Neglecting the pressure losses in evaporator and condenser, considering the compression process 1-2*s* as adiabatic ($ds = 0$), and assuming that the thermal efficiency of internal heat exchanger is 100%, determine

a. The specific cooling and dehumidifying effect;
b. The specific work consumed;
c. The heat pump theoretical COP $\left(COP_{heat\ pump}^{theoretical}\right)$.

Solution

a. Based on given data, can be found from table properties and ln(*p*)–*h* diagram of the refrigerant HFO-1234yf the following parameters (Figure E10.14):

$h_{1'}$ = 387.2 kJ/kg (specific enthalpy of saturated vapor at T_{evap} = 40°C)
$h_1 \cong 400$ kJ/kg (specific enthalpy of superheated vapor at the compressor inlet)
$h_{2s} \cong 412$ kJ/kg (specific enthalpy of superheated vapor at the compressor outlet)
$h_{3'}$ = 285.5 kJ/kg (specific enthalpy of saturated liquid at T_{cond} = 60°C)

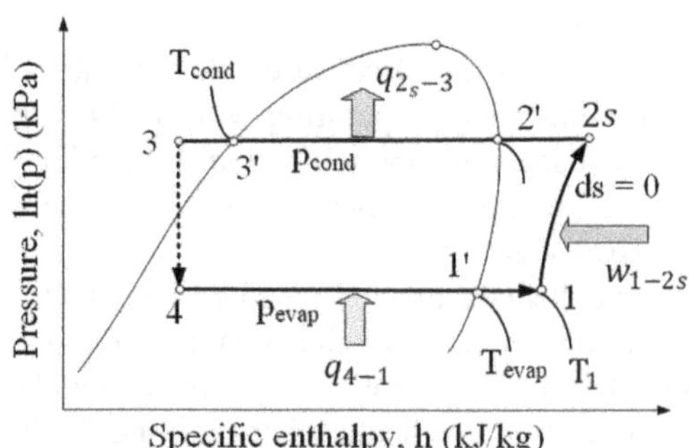

FIGURE E10.14 Theoretical mechanical vapor compression cycle with internal heat exchanger.

From the energy balance of the internal heat exchanger, can be determined the specific enthalpy of the subcooled liquid at the outlet of the internal heat exchanger:

$$h_1 - h_{1'} = h_{3'} - h_3$$

$$h_3 = h_{3'} - (h_1 - h_{1'}) = 285.5 - (400 - 387.2) = 272.7 \text{ kJ/kg} = h_4$$

The specific cooling and dehumidifying effect is as follows:

$$q_{4-1} = h_1 - h_3 = h_1 - h_4 = 400 - 272.7 = 127.3 \text{ kJ/kg}$$

b. The specific work consumed is as follows:

$$w_{1-2s} = h_{2s} - h_1 = 412 - 400 = 12 \text{ kJ/kg}$$

c. The heat pump theoretical COP is as follows:

$$COP_{heat\ pump}^{theoretical} \cong \frac{q_{4-1}}{w_{1-2s}} = \frac{127.3}{12} \cong 10.6$$

10.8.4 Vapor Injection

Subcritical mechanical vapor compression air-to-air and air-to-water heat pumps are widely used in the residential and institutional/commercial buildings for both cooling and heating applications. Such heat pumps work very well in climates with moderate ambient temperatures; but in colder climates, the degradation of coefficients of performance becomes significant when the outdoor temperature decreases under −8°C in the heating mode, or increases over 35°C in the cooling mode. Effectively, in the winter heating mode, low ambient temperatures result in low refrigerant density and mass flow rate at the compressor suction inlet port, reduced volumetric efficiency of the compressor, low isentropic efficiency of the compression process, and thus excessively high compressor power input. Therefore, the heat pump heating capacity and $COP_{heating}$ are reduced compared to those provided at ambient temperatures higher than −8°C. Contrarily, in the summer cooling mode, the refrigerant is compressed at higher pressure and temperature, which excessively increases the compressor discharge temperature and power input that result in lower cooling capacity and $COP_{cooling}$. High compressor discharge temperatures may also degrade the lubricating oil and thus reduce the reliability of the system. To partially fix these problems, refrigerant vapor injection techniques, where a part of the vapor after the expansion valve is extracted and injected to the compressor, may be applied in practice.

There are two vapor injection concepts (i) with phase separator (flash tank or economizer) (Figure 10.8a) and (ii) with internal heat exchanger (Figure 10.8b) (Aikins et al. 2013).

The vapor injection cycle can be separated into the flash tank and subcooler vapor injection cycle. For the flash tank cycle, refrigerant vapor for injection was provided

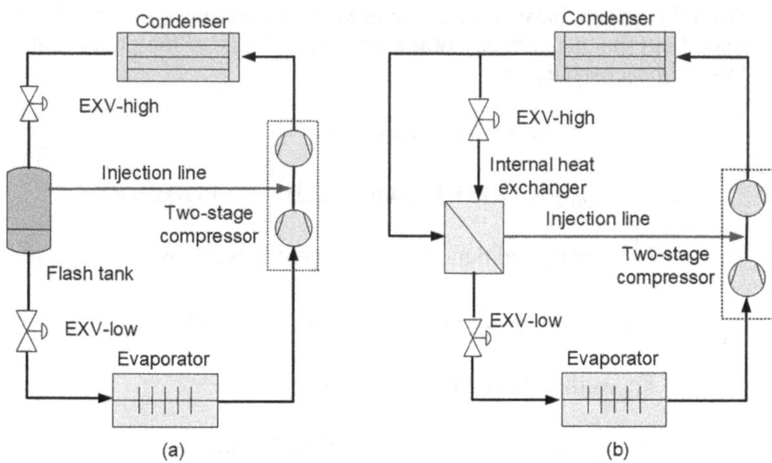

FIGURE 10.8 Heat pump concepts with refrigerant injection: (a) via a flash tank (phase separator); (b) via an internal heat exchanger (economizer). *EXV*, expansion valve.

by phase separation inside the flash tank. The flash tank cycle shows unexpected flooding in the compressor at high speeds due to the difficulty of accurately controlling the amount of injection. On the other hand, injected vapor in the subcooler cycle was generated by heat exchange using the temperature difference between before and after the high-stage expansion device in the subcooler. The subcooler cycle yields a lower heating performance and the flash tank cycle even though it allows for more stable and precise cycle control through the variation of the injected amount.

In the flash tank (phase separator) vapor injection cycle (Figures 10.8a), the flash tank receives partially expanded refrigerant diverted from the upper-stage expansion valve and injects it into the two-stage compressor at an intermediate pressure. Refrigerant has to be injected into the compressor in a superheated state in order to avoid injecting wet or liquid into the two-stage compressor. It can be seen that, due to the two-phase separation in the flash tank, the liquid entering the evaporator has lower enthalpy compared to that of a single-stage cycle. Thus, the enthalpy difference across the evaporator is greater than that of a single stage cycle and, moreover, the vapor injection reduces the refrigerant mass flow rate through the evaporator. However, the increased enthalpy difference increases the two-phase heat transfer area in the evaporator. Therefore, the overall effect is that the system capacity is increased, as well as the system *COP*. The saturated vapor from the flash tank also has lower temperature than that of the vapor in the compressor, which helps to reduce the compressor discharge temperature.

Contrary to flash tank injection cycle, internal heat exchanger vapor injection cycle (Figure 10.8b) may employ the evaporator outlet superheating amount as the control parameter to regulate the openings of both expansion valves by means of sensing bulbs and pressure balancing lines placed to the evaporator outlet.

In this cycle, after the condenser, the refrigerant liquid is separated into two paths: (i) the first flows through the upper-stage expansion valve *EXV—high* and (ii) the second enters the internal heat exchanger, where it provides subcooling to the refrigerant coming from the first circuit.

The two-phase refrigerant absorbs heat within the internal heat exchanger and becomes vapor, which is then injected to the compressor. At the same time, the subcooled liquid is further subcooled by the two-phase refrigerant and enters the lower-stage expansion valve *EXV—low*, the evaporator, and finally the compressor. Because of the (total) liquid subcooling process, after the lower-stage expansion valve *EXV–low*, the enthalpy difference across the evaporator becomes higher than that of the single-stage cycle.

Compared to liquid injection strategy that mainly aims at reducing the compressor discharge temperature, the vapor injection technique provides some other benefits such as (i) evaporator cooling capacity and *COPs* improvement because of reduced quality of the inlet vapor (Ma et al. 2003; Ma and Chai 2004; Hwang et al. 2004; Bertsch and Groll 2008; Cho et al. 2009); (ii) condenser heating capacity improvement in severe climates (i.e., heat pumping at ambient temperature lower than −8°C and air conditioning at ambient temperatures higher than 35°C) because of the added refrigerant from the injection line; (iii) system capacity can be varied by controlling the injected refrigerant mass flow rate, which allows some energy savings by avoiding intermittent operation of the compressor; (iv) lower compressor discharge temperature than that of conventional cycles because the injected vapor temperature is lower than that of the vapor at the compressor inlet; therefore, the compression process is closer to isentropic behavior; and (v) although the vapor injection reduces the refrigerant mass flow rate through the evaporator, the increased enthalpy difference increases the two-phase heat transfer area in this heat exchanger, and therefore the system capacity increases.

From a thermodynamic point of view, the performance of flash tank and internal heat exchanger cycles should have similar performance. Their working principle is to decrease the evaporator inlet enthalpy by two-stage expansion. The only difference is to achieve it by subcooling through the additional heat exchanger or by two-phase separation in the flash tank. However, the actual performance of the flash tank cycle is superior to that of the internal heat exchanger cycle.

The heating capacity and COP of the flash tank cycle are 10.5% and 4.3% higher than those of the internal heat exchanger cycle, respectively, mainly because the internal heat exchanger introduces additional pressure drops to the injected vapor, contrary to the flash tank which leads to a saturated state of refrigerant, and therefore there is no additional pressure drop in the vapor injection line. On the other hand, the internal heat exchanger cycle has a much wider operating range than that of the flash tank cycle (Ma and Zhao 2008).

The intermediate pressures of the refrigerant injection are selected to result in equal pressure ratios across the compressor stages, which minimize the compressor power.

For the saturation cycle, two-phase refrigerant is injected to compressor, and the amount is controlled to maintain the degree of superheat.

The main effect of the liquid injection in a heat pump working with HCFC-22 was to decrease the compressor discharge temperature by about 1.2°C (Winandy and Lebrun 2002).

The COP and heating capacity of the flash tank cycle using R-410A as a refrigerant were enhanced by up to 10% and 25%, respectively, at the ambient temperature of –15°C compared to those of the non-injection cycle (Heo et al. 2010; Heo et al. 2011; Heo et al. 2012).

The flash tank and sub-cooler cycles using R-410A as a refrigerant showed performance improvement over non-injection systems. The cooling capacity and COP were improved by 15% and 2%, respectively, at an ambient temperature of 46.1°C. The heating COP was improved up to 23% by the flash tank cycle at an ambient temperature of –17.8°C (Wang et al. 2009).

A heat pump system with scroll compressor with a vapor injection improved the cooling and heating COP up to 4% and 20%, respectively, compared to the non-injected case (Wang et al. 2009).

In the vapor injection cycle, the flash tank cycle has the possibility of unexpected flooding in the compressor despite its superior performance. When considering the stable and precise cycle control, the flash tank cycle would be the best selection as an alternative cycle.

10.8.5 Expansion Work Recovery

Expansion devices (capillary tubes or orifices and thermostatic or electric valves) provide a pressure difference between the condenser (or gas cooler, in the case of supercritical heat pump cycles) and the evaporator, and maintain proper refrigerant distribution through the evaporator tubes. The expansion process reduces the pressure of the refrigerant from the condenser level to the evaporator level in a process of (theoretically) constant enthalpy, and it is highly irreversible causing thermodynamic losses. According to thermodynamic second law analysis, using such devices results in cycle irreversibility and energy losses because any useful work is done during the expansion process. During the expansion process, a significant portion of the kinetic energy due to the passage from high pressure to the low pressure is dissipated in the fluid. The process is then not entirely isenthalpic, and these losses reduce the system heating and cooling efficiency.

An expander, a standard device in cryogenics (e.g., in air liquefaction) (Quack 1999, 2000), is a kind of reverse compressor that increases the heat pump heating capacity through performing a near-isentropic expansion, thus reducing the enthalpy of the refrigerant at the evaporator inlet, or by recovering the expansion energy, thus reducing the externally electrical power requirement of the compressor (Nickl et al. 2005).

The expander, which can be seen as a compressor operating in reverse, could be employed instead as an expansion device to generate isentropic condition in the throttling process, and thus recover the thermodynamic losses. The power output of such a device then can be supplied to the compressor to reduce compressor power input (Figure 10.9). An expander can be seen as a compressor operating in reverse. The expander cycle has a great potential to improve the vapor compression cycle performance. There are several options to extract the work by the expander such as (i) shaft of the expander can be combined with that of the compressor; (ii) the generated electricity can be used for the compressor; and (iii) multiple expanders can be utilized to recover the expansion loss more efficiently.

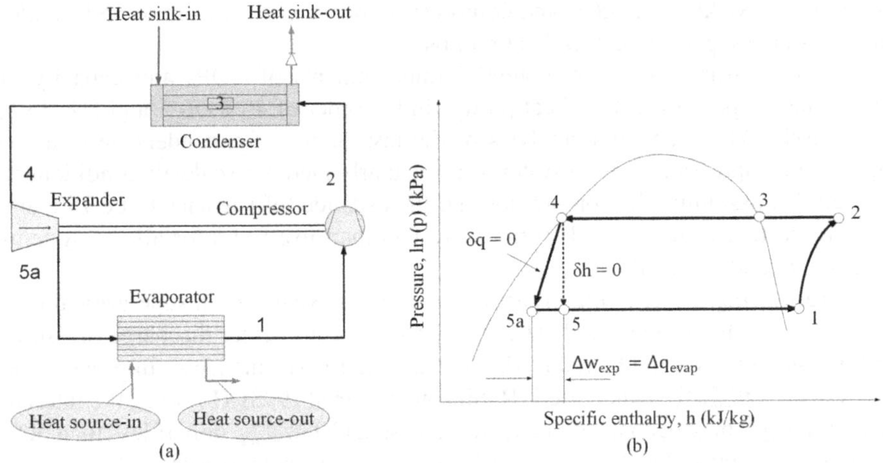

FIGURE 10.9 (a) Typical single-stage subcritical mechanical vapor compression heat pump with expander to recover expansion power; (b) thermodynamic process represented in ln(p)–h diagram.

The expander isentropic efficiency is around 50%, which usually provides overall efficiency gain of the heat pump on the order of 5% by (i) increasing the cooling and dehumidifying capacity of the evaporator through an isentropic process and (ii) by using the recovered expansion losses for assisting the compressor, which results in a reduced compressor power consumption.

Among the limitation on the expander cycle can be mentioned (i) the relatively low expander efficiency and (ii) internal leakage due to the large pressure differences inside such a device.

A vane-type expander with internal two-stage expansion process for HFC-410A refrigeration systems showed that the expander provided volumetric ratios up to 7.6 with isentropic efficiency of 55.9% at 2,000 rpm and, theoretically, improved the COP from 4.0 to 4.56, that is, by 14.2%, under design operating condition (condensation temperature of 54.4°C and evaporation temperature of 7.2°C) (Wang et al. 2012a, b).

The expander cycle has a great potential to improve the vapor compression cycle performance.

The expanders could be used for the special case of CO_2 supercritical heat pumps because of the following features of these thermodynamic systems: (i) very high pressure differences at relatively small volumetric flow rates; (ii) expansion into the two-phase region, mainly on the left side of the critical point; and (iii) amount of efficient work recovery, which is even more important for $COP_{supercritic}^{heating}$ improvement than supplying additional electrical power to CO_2 compressor.

With CO_2 as a refrigerant in supercritical heat pumps, the greater pressure difference between compressor discharge and suction pressures results in greater expansion losses thus making work recovery more feasible and more beneficial.

In the case of single-stage supercritical CO_2 heat pumps, the expansion work recovered with expanders may provide advantages such as (i) improved heating and cooling capacities, (ii) improved $COP_{supercritic}^{heating}$, and (iii) lower discharge pressures (up to 20%) (Nickl et al. 2003) and compression works compressors of both single- and two-stage supercritical CO_2 heat pumps.

First law estimations show potential improvements of COPs and capacity of single-stage supercritical CO_2 heat pumps in the order of 40%–70% and 5%–15%, respectively. Also, economic analysis of the installation of expanders on to existing vapor compression cooling systems, particularly medium-scale air conditioners, showed that assuming the compressor and the expander efficiencies to be 75% and 50%, respectively, the payback periods are less than five years for all the systems (Subiantoro and Ooi 2013).

By recovering a part of the expansion work with expanders in supercritical CO_2 heat pumps with intermediate heat exchangers, the compressor discharge pressures and pressure ratios could be reduced, the evaporator performances—improved, and the system $COP^{heating}_{supercritic}$—increased (Robinson and Groll 1998). However, the internal heat exchanger may become effective at increasing $COP^{heating}_{supercritic}$ only if less than 30% of the work recovered by the expander is used to reduce compressor work.

Using expanders in place of conventional expansion devices may reduce up to 50% of the exergy losses, resulting in 33% higher $COP^{heating}_{supercritic}$. However, although replacing the expansion valve with an expander can significantly improve the performance of supercritical CO_2 cycles, such additional devices may not be economically feasible for many practical applications, especially for small capacity supercritical CO_2 heat pump water heaters. The question is where the expanders should be integrated into the heat pump systems, and how the liquid must be distributed to the evaporators. In the case of two-stage supercritical CO_2 heat pumps, turbine-type expanders can transfer the expansion work recovered to directly drive, via gear boxes, the low-pressure, high-pressure compressor, or the low-pressure compressor provided with intermediate intercooler (Yang et al. 2007). Compared to single-stage compression heat pumps with expansion valves, the best performance is achieved by the expander directly driving the high-pressure compressor, that is, not via a gear box.

A three-stage reciprocating free piston expander directly powering the second stage compressor in a supercritical CO_2 refrigeration system (with liquid–vapor separator between the second and third stage expanders) with isentropic efficiency varying between 65% and 70% increased the $COP_{cooling}$ by 40% compared to a similar cycle using a conventional throttling valve (Nickl et al. 2005). When the shaft of a scroll-type expander is directly coupled to the first-stage compressor of a two-stage compression supercritical CO_2 system with intercooling, the compressor efficiency decreases as inlet pressure increases. The expander reduces the required external compressor work by about 12%, which results in an increase in the cooling capacity and $COP_{cooling}$ by 8.6% and 23.5%, respectively (Kim et al. 2008). An experimental two-cylinder reciprocating piston expander for work recovery in a supercritical CO_2 cooling system, with an isentropic efficiency of 10.27%, improved the $COP_{cooling}$ by 6.6% compared to the system using an expansion valve (Baek et al. 2005). A rolling piston expander in a supercritical CO_2 cycle experimentally recovered 14.5% of expansion work and showed that there is an optimal rotational speed of the expander which maximizes its efficiency, the amount of work recovered and the system $COP_{cooling}$. A swing piston expander experimentally achieved isentropic efficiency between 28% and 44% depending upon operating conditions. The expander efficiency increased as expander inlet temperature (gas cooler outlet temperature) increased, but the system heating capacity decreased (Haiqing et al. 2006).

In a hybrid heat pump system, the output power of the expander can be employed to drive the compressor of an auxiliary subcooling cycle, while the refrigerant at the outlet of condenser is subcooled by an evaporative cooler (She et al. 2014). In such a system, the main cycle includes an evaporator, a compressor, a water-cooled condenser, an evaporative cooler, and an expander, whereas the auxiliary subcooling cycle consisted of an evaporative cooler, a compressor, a water-cooled condenser, and a throttle valve. The expansion power recovered from the main refrigeration system is employed to supply for the compressor in the auxiliary subcooling cycle, which provided the cooling capacity of evaporative cooler to further subcool the CO_2 in the main cycle. Such a system that provides a significantly higher COPs, which indicated that the subcooling method using expansion power recovery (expander efficiency of 0.35), can become an efficient way to improve the performance of mechanical vapor compression heat pumps.

The limitation on the expander cycle is the low expander efficiency. Moreover, due to the large pressure differences inside the machine, one has to minimize the possibilities of internal leakage.

In Figure 10.9b, the state 5 represents the refrigerant state after the conventional isenthalpic expansion process ($dh = 0$) inside the thermostatic valve, and the point 5a represents the refrigerant state after the isentropic expansion process ($\delta q = 0$) in the expander.

During the isentropic expansion, a portion of the energy is removed in the form of work, and the produced power from the expander is 100% used for the compressor (Radermacher and Hwang 2000).

The power produced by the expander ($\dot{W}_{expander}$), that is extracted from the refrigerant stream and contributes at reducing the power requirement of the compressor, can be calculated as follows (see Figure 10.9b):

$$\dot{W}_{expander} = \dot{m}_{refr}\left(h_5 - h_{5a}\right) \tag{10.53}$$

where

$\dot{m}_{refr}$ is the refrigerant mass flow rate (kg/s)

The power produced by the expander ($\dot{W}_{expander}$) also represents an increase of evaporator cooling and dehumidifying capacity that, in this case, becomes

$$\dot{Q}_{evap} = \dot{m}_{refr}\left(h_1 - h_{5a}\right) > \dot{m}_{refr}\left(h_1 - h_5\right) \tag{10.54}$$

10.8.6 Reduction of Compressor Work

Air-to-air and air-to-water heat pumps are attractive technologies for either domestic hot water or space heating applications (Hepbasli and Kalinci 2009; Bourke and Bansal 2010), but their heating capacity and coefficients of performance degrade at low or very low ambient temperature conditions (Bertsch and Groll 2008).

In order to partially remediate this issue, a lot of methods have been proposed, such as using ejectors (also known as jets, injectors, or jet pumps) and refrigerant (liquid or vapor) injection (Heo et al. 2011; Chua et al. 2010), as well as hybrid systems such as dual-source and compression–absorption heat pumps (Lazzarin 2012).

The ejector is mainly composed of a nozzle, a mixing chamber, and a diffuser (Figure 10.10a). Basically, the ejector is designed to mix the high pressure fluid with low-pressure fluid. The refrigerant from the condenser flows through the ejector and exits the nozzle, which creates a low pressure at the nozzle outlet. The low pressure draws a refrigerant from the low-temperature evaporator outlet. High-pressure fluid from the condenser and low-pressure fluid from the low-temperature evaporator could be mixed in the mixing chamber. The pressure of mixed refrigerant recovers at the diffuser section. The ejector is an energy converter that transforms expansion losses into kinetic energy and then back to an increase pressure so that the compression work is reduced.

In such cycles, the high-pressure refrigerant from the condenser or gas cooler (in the case of supercritical heat pumps) enters the nozzle of the ejector where its velocity is increased and pressure is decreased. This lower pressure draws refrigerant vapor from the evaporator into the ejector mixing chamber where the pressure increases. A diffuser is utilized to increase the refrigerant pressure while also lowering the velocity. Compressed refrigerant then enters a liquid–vapor separator from which vapor is drawn into the compressor via the suction accumulator, while the separated liquid reenters the evaporator. In other words, inside the ejector, the refrigerant from the high pressure side of the system (condenser or gas cooler) expands in the motive nozzle and the high-speed two-phase flow at the motive nozzle outlet mixes with the evaporator flow (Sarkar 2008).

The ejector applied vapor compression cycle has a large potential to improve the performance but there are some limitations such as a narrow range of operation due to the fixed geometry.

Figure 10.10b shows the thermodynamic process of a single evaporator ejector cycle in the ln(p)–h diagram.

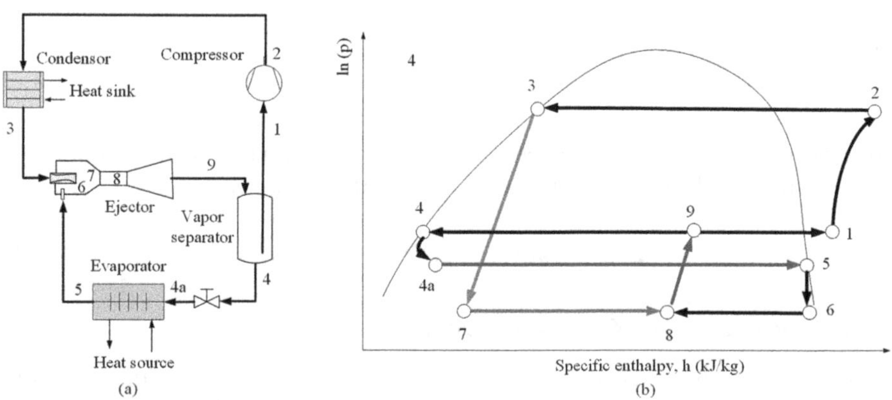

FIGURE 10.10 (a) Schematic representation of an ejector-based heat pump system; (b) thermodynamic process of the single evaporator ejector cycle in the ln(p)–h diagram.

Generally, the isentropic efficiency of both the nozzle and the diffuser is assumed to be 70%.

The ejector expansion devices have advantages such as (Kornhauser 1990; Yari 2009) (i) low cost; (ii) no moving parts; (iii) the ability to handle two-phase flow, making it attractive for the development of high-performance heat pump systems; and (iv) reduce the compressor work by raising the suction pressure to a level higher than that in the evaporator, leading to the improvement of system overall COPs.

The performance of the ejector cycle can be increased at the low heat sink temperature. The application of the adjustable ejector enables the potential of the energy savings in multiple evaporator refrigeration system.

The thermodynamic performance of the ejector expansion refrigeration cycle using CFC-12 as a refrigerant based on a constant mixing pressure model provided a $COP_{cooling}$ improvements of up to 21% over the standard cycle under standard operating conditions (i.e., evaporator temperature of –15°C and condenser temperature of 30°C) (Kornhauser 1990).

However, improvements in the order of 3.9%–7.6% have been achieved in the subsequent experiments with HFC-134a (Harrell and Kornhauser 1995).

Compared to conventional systems with expansion valves, supercritical CO_2 ejectors may improve the $COP_{cooling}$ and the cooling capacity by 7% and 8%, respectively (Elbel and Hrnjak 2008).

A CO_2–NH_3 ejector expansion cascade cycle achieved maximum COPs approximately 7% higher compared to the conventional cycle. Also, the exergy destruction rate decreased by about 8% because of lower exergy losses through expansion valves and the system second law efficiency increased by 5% compared to the equivalent conventional system (Dokandari et al. 2014).

A solar-assisted ejector cooling system integrated an ejector refrigeration system with a mechanical vapor compression air conditioning heat pump. The ejector was driven by the refrigerant vapor evaporated by a solar collector. The intercooler heat exchanger acted as the evaporator of the ejector cooling system and subcools the liquid refrigerant leaving the condenser of an air conditioner with inverter-type compressor (Huang et al. 2014). The use of such an ejector cooling concept in the first stage of the solar-assisted ejector cooling and heating system reduced the condensing temperature of the air conditioner by 7.3°C–12.6°C and the compressor input power by 34.5%–81.2%. Also, the overall system COPs, including auxiliary energy consumptions, increased by 33%–43%.

An ejector expansion transcritical CO_2 cycle based on a constant pressure mixing model improved the COP by more than 16% compared to the basic cycle for typical air conditioning applications (Li and Groll 2005).

For a refrigeration cycle using the HFC-134a as a refrigerant and ejector as an expansion device, heat source and heat sink with temperatures varying from 6°C to 18°C and from 25°C to 40°C, respectively, and operating with compressor pressure ratios and discharge temperatures lower than those of the typical vapor compression cycles, the amounts of COP improvements decrease as the heat sink temperatures increase (Wongwises and Disawas 2005).

With an ejector expansion transcritical CO_2 cycle based on the constant pressure mixing model, the COP improvement may be higher (more than 16% compared to the basic vapor compression cycle) especially at low inlet cold water temperature. However, if the separator does not properly separate the liquid and vapor from each other in the single-phase ejector cycle (see Figure 10.10), the performance of the ejector cycle would be deteriorated mainly because the liquid that leaves the separator at the vapor port could be supplied to the compressor instead to the evaporator. In this case, the cooling capacity decreases and the compressor work increases, both contributing at lowering the heat pump COPs (Li and Groll 2005).

REFERENCES

Aikins, K.A., S.H. Lee, J.M. Choi. 2013. Technology review of two stage vapor compression heat pump system. *International Journal of Air-Conditioning and Refrigeration* 21(3) (14 pages) (https://doi.org/10.1142/S2010132513300024, Accessed August 15, 2015).

Alefeld, G. 1987. Efficiency of compressor heat pumps and refrigerators derived from the second law of thermodynamics. *International Journal of Refrigeration* 10(6):331–341.

ASHRAE. 2009. *Fundamentals Handbook*, SI edition. American Society of Heating, Refrigerating and Air-Conditioning Engineers, Inc., Atlanta, GA.

Baek, J.S., E.A. Groll, P.B. Lawless. 2005. Piston-cylinder work producing expansion device in a transcritical cabbon dioxide cycle. Part I: Experimental investigation. *International Journal of Refrigeration* 28:141–151.

Bertsch, S.S., Groll, E.A. 2008. Two-stage air-source heat pump for residential heating and cooling applications in northern U.S. climates. *International Journal of Refrigeration* 31:1282–1292.

Boltzmann, L. 1877a. Bemerkeungen ¨uber einige Probleme der mechanische W¨armtheorie, Sitzungberichte der Akademie der Wissenschaften zu Wien, mathematisch-naturwissenschaftliche Klasse, 75, 62–100. Reprinted in *Boltzmann* (1909) Vol. 2, pp. 116–122. Partial English translation in Brush (2003), pp. 362–376.

Boltzmann, L. 1877b. Uber die Beziehung eines allgemeinen mechanischen ¨ Satzes zum zweiten Satze der W¨armtheorie, Sitzungberichte der Akademie der Wissenschaften zu Wien, mathematisch-naturwissenschaftliche Klasse, 76, 373–435. Reprinted in *Boltzmann* (1909) Vol. 2, pp. 164–223.

Bourke, G., P. Bansal. 2010. Energy consumption modeling of air source electric heat pump water heaters. *Applied Thermal Engineering* 30:1769–1774.

Carnot, S. 1824. *Réflexion sur la puissance motrice du feu sur les machines propres à développer cette puissance*, Paris, chez Bachelier, Libraire.

Cho, H., C. Baek, C. Park, Y. Kim. 2009. Performance evaluation of a two-stage CO_2 cycle with gas injection in the cooling mode operation. *International Journal of Refrigeration* 32:40–46.

Cho, H., H. Lee, C. Park. 2013. Performance characteristics of an automobile air conditioning system with internal heat exchanger using refrigerant R-1234yf. *Applied Thermal Engineering* 61(2):563–569.

Chua, K.J., S.K. Chou, W.M. Wang. 2010. Advances in heat pump systems: A review. *Applied Energy* 87(12):3611–3624.

Ciconkov, R. 2001. *Refrigeration Solved Examples*. Faculty of Mechanical Engineering, University "SV. Kiril I Metodij", Skopje, Macedonia.

Clausius, Rudolf. 1867. *The Mechanical Theory of Heat*. Taylor and Francis, London, eBook.

Corberan, J.M., I.O. Martinez, J. Gonzalves. 2008. Charge optimisation study of a reversible water-to-water propane pump. *International Journal of Refrigeration* 31:716–726.

Dokandari, D.A., A.S. Hagh, S.M.S. Mahmoudi. 2014. Thermodynamic investigation and optimization of novel ejector-expansion CO_2/NH_3 cascade refrigeration cycles (novel CO_2/NH_3 cycle). *International Journal of Refrigeration* 46:26–36.

Domanski, P.A., D.A. Didion, J.P. Doyle, C. Ryu, Y. Kim. 1994. Evaluation of suction line liquid line heat exchange in the refrigeration cycle. *International Journal of Refrigeration* 17(7):487–493.

Elbel, S., P. Hrnjak. 2008. Experimental validation of a prototype ejector designed to reduce throttling losses encountered in transcritical R744 system operation. *International Journal of Refrigeration* 31(3):411–422.

Glassley, W.E. 2015. Geothermal energy. In *Renewable Energy and the Environment*, 2nd edition, edited by Abbas Ghassemi (series), CRC Press – Taylor & Francis Group, Boca Raton, FL.

Haiqing, G., M. Yitaai, L. Minxia. 2006. Some design features of a CO_2 swing piston expander. *Applied Thermal Engineering* 26:237–243.

Harrell, G.S., A.A. Kornhauser. 1995. Performance tests of a two-phase ejector. In *Proceedings of the 30th Intersociety Energy Conversion Engineering Conference*, Orlando, FL, July 30–August 4, pp. 49–53.

Heo, J., M.W., Jeong, Y. Kim. 2010. Effects of flash tank vapor injection on the heating performance of an inverter-driven heat pump for cold regions. *International Journal of Refrigeration* 33:848–855.

Heo, J., M.W. Jeong, C. Baek, Y. Kim. 2011. Comparison of the heating performance of air-source heat pumps using various types of refrigerant injection. *International Journal of Refrigeration* 34:444–453.

Heo, J., H. Kang, Y. Kim. 2012. Optimum cycle control of a two-stage injection heat pump with a double expansion sub-cooler. *International Journal of Refrigeration* 35:58–67.

Hepbasli, A., Y. Kalinci. 2009. A review of heat pump water heating systems. *Renewable and Sustainable Energy Reviews* 13:1211–1229.

Herold, K.E. 1989. Performance limits of thermodynamic cycles. In *Proceedings of the ASME Winter Annual Meeting,* AES-Vol. 8, December 10–15, pp. 41–45.

Huang, B.J., W.Z. Ton, C.C. Wu, H.W. Ko, H.S. Chang, H.Y. Hsu, J.H. Liu, J.H. Wu, R.H. Yen. 2014. Performance test of solar-assisted ejector cooling system. *International Journal of Refrigeration* 39:172–185.

Hwang, Y., A. Celik, R. Radermacher. 2004. Performance of CO_2 cycles with a two-stage compressor. In *Proceedings of the International Refrigeration and Air Conditioning Conference at Purdue*, July 12–15.

Joule, J.P. 1850. On the mechanical equivalent of heat. In *Philosophical Transactions of the Royal Society of London,* Royal Society, Vol. 140, pp. 61–82 (www.jstor.org/stable/108427, Accessed June 8, 2016).

Kang, H., K., Choi, C. Park, Y. Kim. 2007. Effects of accumulator heat exchangers on the performance of a refrigeration system. *International Journal of Refrigeration* 30:282–289.

Kiang, C.S., C.K. Jon. 2007. Heat pump drying systems. In *Handbook of Industrial Drying*, 3rd edition, edited by A.S. Mujumdar, CRC Press - Taylor & Francis, Boca Raton FL.

Kim, M.H., J. Pettersen, C.W. Bullard. 2004. Fundamental process and system design issues in CO_2 vapor compression systems. *Progress in Energy and Combustion Science* 30:119–174.

Kim, H.J., J.M. Ahn, S.O. Cho, K.R. Cho. 2008. Numerical simulation on scroll expander-compressor unit for CO_2 transcritical cycles. *Applied Thermal Engineering* 28:1654–1661.

Klein, S.A., D.T. Reindl, K. Brownell. 2000. Refrigeration system performance using liquid-suction heat exchangers. *International Journal of Refrigeration* 23(8):588–596.

Kornhauser, A.A. 1990. The use of an ejector as a refrigerant expander. In *Proceedings of the* 1990 *USNC/IIR-Purdue Refrigeration Conference*, West Lafayette, IN, July 17–20, pp. 10–19.

Krakow, K.I., S. Lin. 1988. A numerical model of industrial dehumidifying heat pumps, Concordia University, *Report submitted to Supply and Services Canada in fulfillment of Contract Serial No. 23284-7-7040/01-ST,* June 29.

Lazzarin, R.M. 2012. Dual source heat pump systems: Operation and performance. *Energy in Buildings* 52:77–85.

Li, D., E.A. Groll. 2005. Transcritical CO_2 refrigeration cycle with ejector-expansion device. *International Journal of Refrigeration* 28:766–773.

Linton, J.W., W.K. Snelson, P.V. Hearty. 1992. Effect of condenser liquid sub-cooling on system for refrigerants CFC-12, HFC-134a and HFC-152a. *ASHRAE Transactions* 98:146–160.

Lorentzen, G., J. Pettersen. 1993. A new, efficient and environmentally benign system for car air-conditioning. *International Journal of Refrigeration* 16(1):4–12.

Ma, G., Q. Chai, Y. Jiang. 2003. Experimental investigation of air-source heat pump for cold regions. *International Journal of Refrigeration* 26:12–18.

Ma, G.Y., Q.H. Chai. 2004. Characteristics of an improved heat-pump cycle for cold regions. *Applied Energy* 77:235–247.

Ma, G.Y., H.X. Zhao. 2008. Experimental study of a heat pump system with flash-tank coupled with scroll compressor. *Energy and Buildings* 40(5):697–701.

McQuiston, F.C., J.D. Parker, J.D. Spitler. 2005. *Heating, Ventilating, and Air Conditioning Analysis and Design*, 6th edition. John Wiley & Sons, Inc., New York.

Minea, V. 2016. Advances in heat pump-assisted drying technology. In *Advances in Drying Technology*, edited by V. Minea, CRC Press - Taylor & Francis, Boca Raton FL, pp. 1–116.

Navarro-Esbri, J., F. Moles, A. Barragan-Cervera. 2013. Experimental analysis of the internal heat exchanger influence on a vapor compression system performance working with R1234yf as a drop-in replacement for R134a. *Applied Thermal Engineering* 59:153–161.

Nickl, J., G. Will, W.E. Kraus, H. Quack. 2003. Third generation CO_2 expander. In *Proceedings of the 21th International Congress of Refrigeration*, Washington, DC, 17–22 August.

Nickl, J., G. Will, H. Quack, W.E. Kraus. 2005. Integration of a three-stage expander into a CO_2 refrigeration system. *International Journal of Refrigeration* 28(8):1219.

Park, C., H. Lee, Y. Hwang, R. Radermacher. 2015. Recent advances in vapor compression cycle technologies. *International Journal of Refrigeration* 60:118–134.

Pottker, G., P. Hrnjak. 2015. Effect of the condenser sub-cooling on the performance of vapor compression systems. *International Journal of Refrigeration* 50:155–164.

Primal, F., P. Lundqvist. 2005. Refrigeration systems with minimized refrigerant charge: System design and performance. In *Proceedings of the Institution of Mechanical Engineers, Part E: Journal of Process Mechanical Engineering*, pp. 127–138.

Quack, H. 1999. Arbeitsleistende Expansion in Kaltdampf - Kältekreisläufen. *DKV-Tagungsbericht* 26:109–123.

Quack, H. 2000. Cryogenic expanders. In *Proceedings of the 18th International Cryogenic Engineering Conference* (ICEC 18), Mumbai, February 21–25, pp. 33–40.

Radermacher, R., Y. Hwang. 2000. *Vapor Compression Heat Pumps with Refrigerant Mixtures*. Center for Environmental Energy Engineering Department of Mechanical Engineering, University of Maryland, College Park, MD.

Rice, C.K. 1993. Influence of HX size and augmentation on performance potential of mixtures in air-to-air heat pumps. *ASHRAE Transactions* 99(Part 2):665–679.

Robinson, D.M., E.A. Groll, 1998. Efficiency of transcritical CO_2 cycles with and without en expansion turbine. *International Journal of Refrigeration* 21(7):577–589.

Sarkar, J. 2008. Optimization of ejector-expansion transcritical CO_2 heat pump cycle. *Energy* 33(9):1399–1406.

Sauer, H.J. Jr., R.H. Howell. 1983. *Heat Pump Systems*. A Wiley – Inter Science Publication, John Wiley & Sons, New York.

She, X., Y. Yin, X. Zhang. 2014. A proposed sub-cooling method for vapor compression refrigeration cycle based on expansion power recovery. *International Journal of Refrigeration* 43:50–61.

Stoecker, W.F. 1998. *Industrial Refrigeration Handbook*, McGraw-Hill, New York.
Subiantoro, A., K.T. Ooi. 2013. Economic analysis of the application of expanders in medium scale air-conditioners with conventional refrigerants, R1234yf and CO_2. *International Journal of Refrigeration* 36:1472–1482.
Tsilingiris, P.T. 2008. Thermophysical and transport properties of humid air at temperature range between 0 and 100°C. *Energy Conversion and Management* 49(5):1098–1110.
Wang, X., Y. Hwang, R. Radermacher. 2009. Two-stage heat pump system with vapor-injected scroll compressor using R-410A as a refrigerant. *International Journal of Refrigeration* 32:1442–1451.
Wang, M., Y., Zhao, F. Cao, G. Bu, Z. Wang. 2012a Simulation study on a novel vane-type expander with internal two-stage expansion process for R-410A refrigeration system. *International Journal of Refrigeration* 35:757–771.
Wang, X., K. Amrane, P. Johnson. 2012b. AHRI Low Global Warming Potential (GWP) Alternative Refrigerants Evaluation Program (Low-GWP AREP). Purdue e-Pubs.
Winandy, E.L., J. Lebrun. 2002. Scroll compressors using gas and liquid injection: Experimental analysis and modelling. *International Journal of Refrigeration* 25(8):1143–1156.
Wongwises, S., S. Disawas. 2005. Performance of the two-phase ejector expansion refrigeration cycle. *International Journal of Heat and Mass Transfer* 48(19&20):4282–4286.
Yang, J.L., Y.T. Ma, S.C. Liu. 2007. Performance investigation of transcritical carbon dioxide two-stage compression cycle with expander. *Energy* 32:237–245.
Yari, M. 2009. Performance analysis and optimization of a new two-stage ejector-expansion transcritical CO_2 refrigeration cycle. *International Journal of Thermal Science* 48(10):1997–2005.

11 Refrigerants

11.1 CLASSIFICATION

The heat pump industry currently uses both pure and mixture (blend) refrigerants. Selection of such refrigerants as working fluids for mechanical vapor compression heat pumps depends on their thermodynamic properties (which determines in a large degree the thermal performances and economic efficiencies of real heat pumps) and environmental impacts (mainly due to leakages in the atmosphere and greenhouse gas emissions associated with energy use) (Radermacher and Hwang 2000).

Refrigerants are classified as follows (http://ozone.unep.org/new_site/fr/montreal_protocol.php. Accessed March 4, 2016):

- **Chlorofluorocarbons (CFCs)** such as CFC-11 (trichlorofluorométhane) and CFC-12 (dichlorodifluoromethane) have a strong destructive impact on the ozone layer if released into the atmosphere and important influence upon the global warming; the manufacture of these refrigerants was already discontinued in January 1996.
- **Hydrochlorofluorocarbons (HCFCs)**, such as HCFC-22, with reduced impact on the ozone layer and on the greenhouse effect, have been used during the last two decades as short-term alternatives to CFCs.
- **Hydrofluorocarbons (HFCs)**, such as HFC-134a (1,1,1,2-tetrafluoroethane) and HFC-410A, harmless to the ozone layer and with less than 1% of the global warming effect of all greenhouse gases, are used today as long-term alternatives to CFCs and HCFCs; although HFCs do not contain chlorine atoms and are very stable fluids, their global warming potential (GWP) values are relatively high (see Table 11.1); HFCs are continuously subjected to progressive phase-down in order to eliminate all adverse environmental effects;
- **Natural refrigerants** as carbon dioxide (CO_2, R744) and ammonia (NH_3, R717), and flammable hydrocarbons (HCs) (R-600-propane, R-600A-isobutane, ethylene, propylene); all these fluids are harmless to the ozone layer, and have less or no influence upon global warming;
- **Low-GWP refrigerants** as HFO-1234yf, HFO-1234ze(E), and HFO-365mfc for low- and high-temperature heat pumps.

ANSI Standard 34 (2010) also classifies the refrigerants according to their toxicity and flammability in six safety groups coming from the least to the most hazardous.

Fluorinated gases (as CFCs), had a significant impact on climate change. It was estimated that they have accounted for 12% to the increased levels of greenhouse gases since the beginning of the industrial revolution.

Consequently, in 1987, Montreal Protocol regulated the production and trade of ozone depleting substances. As a result, CFCs were completely phased out in the developed countries, and extensive R&D activities have been initiated to find new, more environmental friendly refrigerants. Thanks to the phase-out of CFCs under the Montreal Protocol, atmospheric concentrations of these gases are declining.

Due to their excellent thermodynamic characteristics, HCFCs (as HCFC-22) and HFCs (as HFC-134a) have been extensively used during the past two decades in heat pump applications as replacements for CFCs. However, a disadvantage of these synthetic refrigerants is their strong contribution to the greenhouse effect in case of massive leakage.

Today, even they do not deplete the ozone layer, most of HFCs used as replacements are potential global warming gases (Kauffeld 2012).

11.2 GREENHOUSE GAS EMISSIONS

The phenomenon of greenhouse gas emission, due to the usage of energy and leak of refrigerants, contributes to changing the global climate (Radermacher and Hwang 2000).

Emissions of HFCs currently contribute with around 1% (≈0.4 Gigaton of CO_2 equivalent per year) (1 Gigaton = 10^{12} tons) to global greenhouse gas emissions combined. They are growing by 8%–9% annually and have the potential to increase significantly in the future. Global emissions of HFCs in 2050 (estimated at 5.5–8.8 Gigaton of CO_2 equivalent per year) will be equivalent to 9%–19% of projected global CO_2 emissions in business-as-usual scenarios. This is due to their use as replacements for CFCs and HCFCs and because of rapidly increasing demand for refrigeration, air-conditioning and heat pumping in emerging economies.

The impact of a refrigerant containing chlorine compounds on ozone destruction is estimated by using the ozone depletion potential (ODP) index which is a measure of its destructive potential on stratospheric ozone layer relative to depletion caused by an equal amount of a reference substance, generally CFC-11 of which ODP is 1.0. Ozone depletion harms living creatures on earth, increases the incidence of skin cancer and cataracts, and poses risks to the human immune system (Radermacher and Hwang 2000).

The GWP index quantifies the capability of refrigerants to absorb infrared radiation relative to carbon dioxide. It compares the amount of heat trapped by a given mass of gas over a certain time period, usually 20,100 or 500 years, compared to that of CO_2, a natural gas assigned as a reference with GWP = 1. For example, the GWP of methane (a natural gas) over 100 years is 25, which means that if the same mass of methane and CO_2 were released to the atmosphere, then methane will trap 25 times more heat than CO_2 over the next 100 years (Kauffeld 2012). The GWPs of synthetic refrigerants are, generally, 1,300–2,100 times higher compared to that of the natural refrigerant CO_2 taken as a reference.

Even though HFCs have lower-GWP indexes than many of CFCs and HCFCs they are replacing, their GWPs are still very high, that is, in the range of hundreds to thousands of times greater than that of CO_2, mainly because of the presence of fluorine atoms. Some HFCs break down relatively quickly in the atmosphere, namely those with a double carbon bond, resulting in a short atmospheric lifetime as short as 11 days, while others have an extremely long atmospheric life (e.g., HFC-23) with a lifetime of 270 years resulting in a very high GWP of 14,800 (Kauffeld 2012).

The total equivalent warming impact (TEWI) index, used as an indicator for environmental impact of heat pumps and refrigeration systems over their entire lifetime, is the sum of the direct refrigerant emissions (due to the leaks of refrigerants as HFCs in the atmosphere and their interaction with heat radiation), expressed in terms of equivalent CO_2 (direct effect) and the indirect emissions of CO_2 (due to the emission of CO_2 by consuming energy generated through the combustion of fossil fuels) over its service life (indirect effect).

Generally, direct effects of common refrigerants are negligible, but indirect effects are significant because of energy consumption. Depending on the application, refrigerant charge, leakage rate, and annual energy consumption, the CO_2 emissions associated with the energy consumption of heat pump systems can contribute more to greenhouse gas emissions than the refrigerant emissions.

The TEWI index can be calculated with the following equation (ASHRAE 2013):

$$\mathrm{TEWI} = n * l * \mathrm{GWP} + m_{refr}(1-\alpha_{rec})\mathrm{GWP} + n * \beta * E \tag{11.1}$$

where

n is the heat pump technical life (years)
l is the annual leakage rate of the heat pump (3%–8%)
GWP is the global warming potential of the refrigerant for a period of 100 years (–)
m_{refr} is the refrigerant charge (kg)
α_{rec} is the refrigerant recovery and/or recycling factor (0.70%–0.85%)
β is the CO_2 emission factor
$E = \dot{W}_{compr} * t$ is the annual energy consumption (kWh)
$\dot{W}_{compr}$ is the compressor shaft input power (kW)
t is the heat pump running time (hours/year)

11.3 PURE REFRIGERANTS

As pure refrigerants, the heat pump industry uses today pure low-, medium- and high-temperature HCFs, as well as natural fluids. Their thermodynamic properties (e.g., temperature, pressure, enthalpy, specific volume, and entropy) and processes are commonly represented in two-dimensional thermodynamic diagrams, such as temperature–entropy and pressure–enthalpy.

11.3.1 Temperature–Entropy Diagram

Figure 11.1 schematically represents the temperature–entropy (T–s) diagram for pure refrigerants.

The left part of the dome curve represents the thermodynamic states of refrigerant saturated liquid and those of saturated vapor on the right side.

The subcooled liquid region is located to the lower entropy, left side of the saturated liquid curve, while the superheated vapor region is located to the right side of

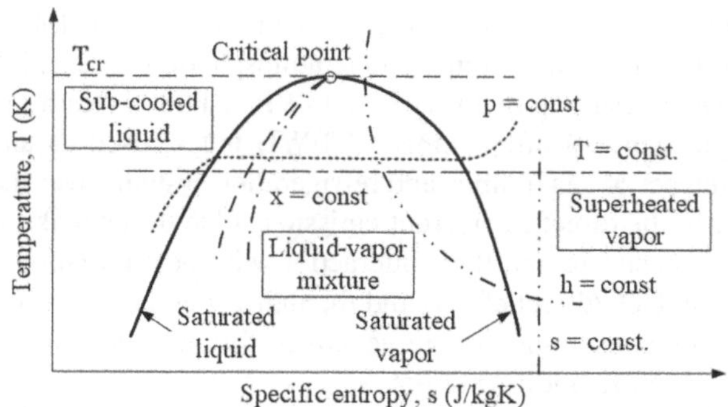

FIGURE 11.1 Schematic of the temperature–entropy diagram for pure refrigerants.

the saturated vapor curve. Both saturated liquid and saturated vapor curves culminate in the critical point on top of the saturated dome in an inflection point (critical point), where the curvature of isobars changes from positive to negative and where the saturation liquid and vapor states become the same as the pressure increases. Above the critical pressure and temperature, no liquefaction will take place.

In the two-phase region, that is, under the saturation dome, the isobar lines (p = *const.*) are horizontal. From the saturated vapor line, the slope of constant pressure curves increases sharply into the superheated vapor region. On the other hand, the isobar curves drop quickly from the saturated liquid line into the subcooled liquid region (Radermacher and Hwang 2000).

11.3.2 Pressure–Enthalpy Diagram

In the pressure–enthalpy ln(p)–h diagram of a pure refrigerant (Figure 11.2), highly utilized in the heat pump technology, the ordinate shows the logarithm of the pressure, and the abscissa shows the specific enthalpy. The boundary line between the subcooled liquid range and the two-phase region is the saturated liquid curve, and the boundary between the superheated vapor and the two-phase range is the saturated vapor curve.

The two-phase region is found under the saturation dome, the area where the isotherms are horizontal. To the left of the two-phase region is the subcooled liquid region and to the right is the superheated vapor region.

The curves of saturated liquid and saturated vapor converge at the critical point. Above this point there is no distinction between the liquid and vapor. Because both pressure and temperature are constant during the phase change processes, the isotherms are straight lines in the two-phase region located below the saturated dome.

In the superheated vapor region, the isotherms (T = *const.*) curve very quickly to the vertical and approach ideal gas behavior for which the enthalpy is independent of the pressure. The isentropic (constant entropy, s = *const.*) lines, very important to determine compressor work, have positive slopes in the superheated vapor region. At the contact with the vapor saturation line, the isentropic lines have no abrupt changes in the slope. The isochores (constant specific volume, v = *const.*) curves, important in determining the volumetric capacity of heat pump compressors, show a much flatter slope.

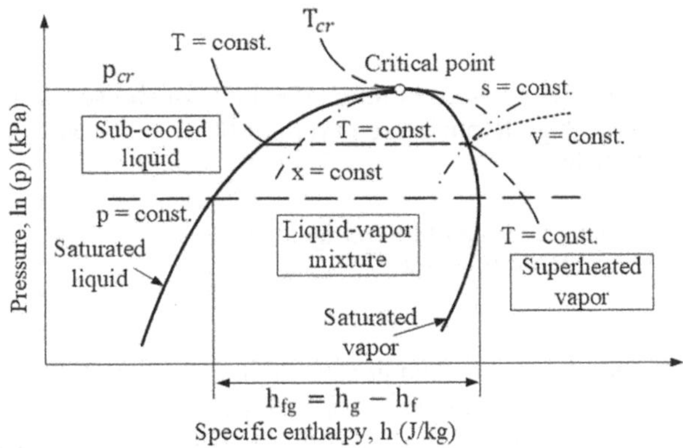

FIGURE 11.2 Schematic of the pressure–enthalpy diagram for pure refrigerants.

The thermodynamic properties of saturated liquid such as specific enthalpy, specific entropy, density, and specific volume are designed by h_f, s_f, ρ_f, and v_f, respectively, and those of saturated vapor by h_g, s_g, ρ_g, and v_g.

In the saturated region, situated under the saturation dome, the specific enthalpy, specific entropy, and specific volume are expressed as follows:

$$h = xh_g + (1 - x)h_f \tag{11.2}$$

$$s = xs_g + (1 - x)s_f \tag{11.3}$$

$$v = xv_g + (1 - x)v_f \tag{11.4}$$

where

x is the quality of saturated vapor (–)

For isobar/isotherm processes, for example, vaporization and condensation of pure refrigerants, ln(p)–h diagram allows calculating the enthalpy (latent heat) of vaporization or condensation for a given saturation pressure as the enthalpy difference between saturated vapor and saturated liquid:

$$h_{fg} = h_g - h_f \tag{11.5}$$

where

h_g is the specific enthalpy of saturated vapor (J/kg)
h_f is the specific enthalpy of saturated liquid (J/kg)

The thermodynamic properties (e.g., enthalpy, entropy, and specific volume) of sub-cooled (or compressed) liquid are strongly dependent on temperature (rather than pressure), and thus may be approximated by the corresponding values for saturated liquid (h_f, s_f, and v_f) at the existing temperature.

Exercise E11.1

The temperature and the pressure of refrigerant HFC-245fa used working fluid in a heat pump-assisted high-temperature dryer are 60°C and 1 MPa, respectively. Determine the following:

a. The thermodynamic state of the refrigerant,
b. The enthalpy of the refrigerant.

Solution

a. From ASHRAE's table (Refrigerant 245fa Properties of Saturated Liquid and Saturated Vapor) or ASHRAE's Pressure-Enthalpy Diagram for Refrigerant 245fa (ASHRAE 2009), it can found that at a pressure of 1 MPa, the saturation temperature is 90°C, higher than the actual refrigerant temperature of 60°C. Consequently, the refrigerant is a subcooled liquid.
b. In ASHRAE's Pressure-Enthalpy Diagram for Refrigerant 245fa, the thermodynamic state (point) of the refrigerant can be found at the intersection of the isobar line 1 MPa with the isotherm curve 60°C. From this point, moving down to the horizontal *x*-axis, the refrigerant specific enthalpy is 281 kJ/kg.

Exercise E11.2

The temperature and the pressure of the low-GWP refrigerant HFO-1234yf used as working fluid in a heat pump-assisted low-temperature dryer are 100°C and 0.1 MPa, respectively. Determine the following:

a. The thermodynamic state of the refrigerant,
b. The enthalpy of the refrigerant.

Solution

a. From a manufacturer's thermodynamic properties Table of refrigerant HFO-1234yf (e.g., www.chemours.com/Refrigerants/en_US/assets/downloads/k24239_Opteon_yf_thermo_properties_SI.pdf. Accessed October 3, 2017), it can be found that at the given pressure 0.1 MPa, the saturation temperature is 40°C. Because the actual temperature (100°C) is higher than the saturation pressure, the refrigerant is in a superheated state.
b. The refrigerant's specific enthalpy, determined from the manufacturer's thermodynamic Tables (i.e., superheated vapor properties at constant pressure), is 463.1 kJ/kg.

Exercise E11.3

The saturated temperature of the refrigerant HFC-134a used as working fluid in a heat pump-assisted low-temperature dryer is of 30°C. Determine the following:

a. The saturated pressure,
b. The specific enthalpy of the saturated liquid,
c. The specific enthalpy of the saturated vapour,
d. The specific enthalpy (latent heat) of vaporization.

Solution

a. From ASHRAE's table (Refrigerant 134a Properties of Saturated Liquid and Saturated Vapor) or ASHRAE's Pressure-Enthalpy Diagram for Refrigerant 134a (ASHRAE 2009), it can found that at the saturated temperature 30°C, the saturated pressure is 0.77 MPa.
b. From ASHRAE's table (Refrigerant 134a Properties of Saturated Liquid and Saturated Vapor) (ASHRAE 2009), at the saturated temperature 30°C, the specific enthalpy of the saturated liquid is $h_f = 241.72$ kJ/kg.
c. Also from ASHRAE's table (Refrigerant 134a Properties of Saturated Liquid and Saturated Vapor), at the saturated temperature 30°C, the specific enthalpy of the saturated vapor is $h_g = 414.82$ kJ/kg.
d. The specific enthalpy (latent heat) of vaporization is as follows:

$$h_{fg} = h_g - h_f = 414.82 - 241.72 = 173.1 \text{ kJ/kg}$$

11.3.3 Alternatives for HFC Pure Refrigerants

As previously noted, the heat pump industry faces a complex process aiming at complete phase-out of HCFCs and HFCs in both old and new equipment by replacing them with new safe, affordable, and energy efficient alternatives in order to contribute to the global reduction of greenhouse gas emissions (Kauffeld 2012).

There is no single alternative that will replace HFCs in all applications. Moreover, some barriers to the adoption of existing alternative refrigerants and technologies exist, as national regulations and standards that inhibit the use of flammable and/or toxic alternatives, insufficient supply of components, increased investment costs, and lack of relevant skills among technicians. Such barriers can be overcome through revised legislation and technical standards, training and technical assistance, infrastructure developments, financial subsidies, as well as taxes on HCFC and HFC refrigerants.

To be acceptable for heat pump industry, future refrigerants are expected to meet all, or the major part of, criteria as (Radermacher and Hwang 2000) follows: (i) they should have properties as close as possible to those of the original HFCs, such as zero ozone depleting potential (ODP) and low GWP; (ii) they should be chemically inert (i.e., noncorrosive) with construction materials and fluids, such as compressor lubricants; (iii) they should be chemically stable in order to prevent decomposition, for example, during high temperature and pressure at the end of compression process; (iv) they should be nontoxic for manufacturing and service personnel of refrigerants, (v) they should be nonflammable and without danger of explosion; (vi) they should provide high energetic efficiency; (vii) they should be inexpensive and easily available; (viii) they should have pressure–temperature properties that must fit with those of particular applications; (ix) they should have relatively high enthalpy (latent heat) of vaporization because it directly impacts on the compressor size and the system's overall cost and performances.

For replacing HCFC-22 refrigerant, widely used in the past in heat pump industry, fluids as the following have been considered: HFC-134a (with lower vapor pressure than that of HCFC-22 resulting in a lower volumetric capacity), HFC-407C (with vapor pressure similar to that of HCFC-22 and performance within a ±6% variation), and HFC-410A (with 50% higher vapor pressure and better performance within ±10% variation) (Radermacher and Hwang 2000).

Currently, with the need to replace fluorinated refrigerants, new refrigerants are being evaluated, with quite different conditions of pressure and temperature, which is a challenge for the design of new heat pump components like evaporators, condensers, and compressors. Generally, original heat pump systems must be readjusted to fit the specific characteristics of new refrigerants, which create the need for new criteria of design and optimization (Copetti et al. 2013).

11.3.4 Low- and Medium-Temperature Refrigerants

As alternatives for replacing commonly used HFC low- and medium-temperature refrigerants (fluids with relatively low critical temperature, normally allowing the industrial heat pumps working in the subcritical regime) that have GWPs higher than 1,000, for example, HFC-134a and HFC-410a (see Table 11.1), new fluids, such as low-GWP (HFO) compounds, and others occurring naturally in nature, such as ammonia, CO_2, and hydrocarbons (see Table 11.3), can be successfully used.

Hydrofluoroolefins (HFOs) are synthetic refrigerants that promise to be a part of solution to the environmental problems. Among them, there are pure compound refrigerants such as HFO-1234yf (2,3,3,3-tetrafluoropropene) and HFO-1234ze(E) (trans 1,3,3,3-tetrafluoropropene) (Calm 2008; Brown et al. 2009) as promising low-GWP substitutes to HFC-134a with minimal modifications of existing heat pump systems.

The GWPs of HFO-1234yf (~4) and HFO-1234ze(E) (~6) are very low, but these new refrigerants have more complex molecule structures than most HFCs and are

TABLE 11.1
Some Properties of Low- and Medium-Temperature Pure Refrigerants and of Their Potential Replacements

Refrigerant	ODP[a]	GWP[b]	Critical Temperature (°C)	Critical Pression (MPa)
Present Refrigerants				
HFC-134a	0	1,300	101.08	4.0603
HFC-410A	0	1,890	72.13	4.9261
Potential Replacements				
HFO-1234yf	0	4	124.69	3.382
HFO-1234ze(E)	0	6	109.35	3.632

[a] Based on ODP = 1 of refrigerant CFC-11.
[b] Based on GWP = 1 of refrigerant CFC-11 over 100 years.

more costly (McLinden et al. 2014), slightly flammable, and relatively not stable (Brian and Sumathy 2011; Longo and Zilio 2013).

It is very difficult to find low-GWP substitutes for traditional HFC refrigerants with no flammability, as a weak chemical stability and/or a large chemical reactivity are necessary to obtain low GWP (Longo et al. 2016).

Extensive studies for HFO-1234yf and HFO-1234ze(E) refrigerants showed that they are less toxic and flammable than hydrocarbons (Brown et al. 2009). Hence, their characteristics make them attractive not only for vehicle air conditioning, but also for air-conditioning systems, refrigeration systems, and heat pumps (Brown et al. 2009).

HFO refrigerants are classified as "A2L" in the ASHRAE standard 34 safety group (ASHRAE 2009; ANSI 2010), because of their mild flammability. In addition, because of their relatively low operation pressure, the pressure drops that occur in the tubes and heat exchangers have a significant influence on system performance.

Specifically, HFO-1234yf has emerged as a HFC-134a replacement for automotive applications (Del Col et al. 2013).

HFO-1234yf has good thermos-physical properties, very similar to HFC-134a, and can be directly charged into conventional refrigeration and heat pump systems with only minor modifications. However, as an alternative to HFC-134a in mobile air conditioning systems, HFO-1234yf achieves cooling capacities and coefficients of performance 4% and 8% lower than those of HFC-134a, respectively. However, at the same operating conditions, HFO-1234yf provides lower compression ratios and lower discharge temperatures. The cooling capacity and COPs obtained with HFO-1234yf can be about 9% and between 5% and 30% lower, respectively, than those obtained with HFC-134a, and that these performances further decrease when the condensing temperature increases.

On the other hand, HFO-1234ze(E) is considered as the best medium-pressure, low-GWP refrigerant available on the market. It is an energy-efficient alternative to traditional refrigerants (such as HFC-134a) in air-cooled and water-cooled chillers for institutional and commercial buildings, as well as in other applications like heat pumps, refrigerators, vending machines, dryers, supermarket cascade systems, etc. Field tests with air-cooled chillers using HFO-1234ze(E) instead, for example, propane (R-290) showed significantly lower energy consumption.

In some jurisdictions, as those of the European Union, low-GWP alternatives as HFO-1234yf and HFO-1234ze(E) have already won significant market share in some sectors, with over 90% of new domestic refrigerators/freezers and approximately 25% of new industrial air conditioners (Kauffeld 2012).

11.3.5 High-Temperature Refrigerants

In heat pump applications, the present most common refrigerants, in particular HFCs, are limited to heat delivery temperatures of maximum 80°C. To provide heat at higher temperatures (i.e., at 90°C–120°C) by recovering industrial waste heat at temperatures up to around 50°C–60°C, and thus to reduce the specific energy consumption (kWh/product unit), several synthetic and natural fluids are available, or will be developed, for industrial heat pumps (Table 11.2).

TABLE 11.2
Some Properties of Available High-Temperature Refrigerants for Heat Pumps

Refrigerant	ODP[a]	GWP[b]	Critical Temperature (°C)	Critical Pressure (MPa)
Water (H_2O, R-718)	0	0	374.14	22.09
Air (not a pure substance)	0	0	−140.7	3.72
HFC-236fa	0	8,060	124.9	3.12
HFC-245fa	0	858	154.01	3.651
HFO-1234ze(E)	0	6	109.35	3.632
HFO-1234ze(Z)	0	<1	150.1	3.97
HFO-365mfc	0	804	189.85	3.266

[a] Based on ODP =1 of CFC-11.
[b] Based on GWP = 1 of CFC-11 over 100 years.

Environmentally friendly refrigerants such as water (R-718), HFC-236fa, HFC-245fa, HFO-1234ze(E), HFO-1234ze(Z), and HFO-365mfc can be used in cascade industrial high-temperature heat pump systems in order to recover waste heat at temperatures as high as 80°C and supply heat up to 140°C (and even 160°C) with relatively high COPs (Boblin and Bourin 2012; Kondou and Koyama 2015).

HFC-245fa (1,1,1,3,3-pentafluoropropane) is a hydrofluorocarbon without chlorine, nearly non-toxic, with no ODP and practically nonbiodegradable with a lifetime of 7.2 years when it eventually does escape into the atmosphere. Compared with HFC-134a, it has lower pressures at the same temperature levels (Wang et al. 2000). However, the GWP value of HFC-245fa is relatively high (858) (i.e., 858 times the global warming effect of CO_2). As a refrigerant, HFC-245fa provides comparable and, in some cases, improved performance in high-temperature heat pumps developed, for example, for industrial wood drying (Minea 2004), to reheat circulating water or to generate steam from waste heat available at 50°C–60°C.

HFC-245fa has a relative high critical temperature (152°C) and has been used as a substitute for CFC-123 and CFC-11 in high-temperature heat pumps and heat recovery ORC power cycles (Angelino and Invernizzi 2003; Saleh et al. 2007; Eames et al. 2007; Wang et al. 2010; Minea 2014).

Unsaturated HFO composed of hydrogen, fluorine, and carbon, are rather unstable due to the double carbon binding and have high reactivity and, therefore, shorter lifetimes in the troposphere resulting in very low GWPs.

The HFO refrigerants are either single substances (e.g., HFC-1234yf) or mixtures with other fluids (Kauffeld 2012).

Because of their high critical temperatures, low-GWP fluids such as HFO-1234ze(E) and HFO-365mfc can be also used as high-temperature refrigerants for heat pump-assisted dryers.

However, there are two main safety issues: (i) flammability and (ii) potential to form dangerous acids. A number of HFOs, including HFO-1234yf and

HFO-1234ze(E), are mild flammable, while others are not, for example, HFO-1336mzz(Z) and HFO-1233zd(E).

On the other hand, flammable refrigerants have greater probability to bring greater risk of formation of dangerous substances under high temperatures. In other words, under the influence of high operating or ambient temperatures, or fire, unsaturated halogenated hydrocarbons, HFOs are easily decomposable to form hydrogen fluoride, a gas that easily form acids that would cause skin, eye, and throat irritation, and could lead to human's death (Kauffeld 2012).

The low-GWP refrigerant HFO-1234ze(Z) is a low-GWP alternative for HFC-134a (GWP_{100} = 1,300) and to HFC-245fa (GWP_{100} = 858) because of its very similar thermodynamic properties and extremely low GWP ($GWP_{100} \leq 6$) (Brown et al. 2009; Fukuda et al. 2014; Cavallini et al. 2014).

HFO-1234ze(E), with a relatively high critical temperature (109.35°C), could be considered a moderate high-temperature refrigerant, and the isomer HFO-1234ze(Z) are today attractive candidates substitute for HFC-236fa and HFC-245fa in high temperature heat pumps because of their low GWP (Brown at al. 2009; Fukuda et al. 2014).

HFO-1234ze(Z) exhibits a relatively high critical temperature (150.1°C) that allows operating subcritical cycles at high temperatures required by industrial heat pumps. Therefore, it seems to be the most promising low-GWP refrigerant for high-temperature and very high-temperature heat pumps. During condensation, HFO-1234ze(Z) shows heat transfer coefficients and frictional pressure drop higher than HFC-236fa, as well as weak sensitivity to saturation temperature and great sensitivity to refrigerant mass rate. For example, at mass flux higher than 15 kg/m^2·s, the forced convection condensation heat transfer coefficients increase by 30%. By doubling the refrigerant mass flux, HFO-1234ze(Z) exhibits higher heat transfer coefficients (between 48% and 82%) and frictional pressure drop (73%–82%) than those of HFC-236fa under the same operating conditions.

R-365mfc, another hydrofluorocarbon (1,1,1,3,3-pentafluorobutane), is suitable as a refrigerant of high-temperature heat pumps aiming at generating steam using waste heat because of its very high critical temperature (189.85°C), but its GWP value is relatively high (i.e., 804 at 9.9 years atmospheric lifetime).

11.3.6 Natural Refrigerants

The ozone layer protection was initially addressed by the advent of HFCs, but global warming would not accept long-term use of HFCs (Calm and Domanski 2004). For application fields such as refrigeration and heat pumps, another generation of refrigerants with zero ODP and zero (or negligible) GWP, short atmospheric lifetime, and high energy efficiency are required.

As long-term options for eliminating the influence of synthetic refrigerants on global climate change, natural refrigerants, containing neither chlorine nor fluorine molecules, such as ammonia (NH_3, R-717), hydrocarbons (e.g., propane R-290, propylene R-1270, and isobutane R-600a), water, and carbon dioxide (CO_2, R-744) (Artemenko and Mazur 2007; Calm 2008; Cavallini et al 2014) (see Table 11.3), must be more and more used in industrial heat pump systems as environmental-friendly

TABLE 11.3
Properties of Most Promising Natural Refrigerants for Heat Pumps

Refrigerant	ODP[a]	GWP[b]	Critical Point	
			Temperature (°C)	Pressure (MPa)
Ammonia (NH_3, R-717)	0	0	132.25	11.483
Propane (R-290)	0	3	96.7	4.30
Isobutane (R-600a)	0	3	134.7	3.677
Propylene (R-1270)	0	3	91.43	4.58
Water (H_2O, R-718)	0	0	373.946	22.06
Carbon dioxide (CO_2, R-744)	0	1	30.98	7.474

[a] Based on ODP = 1 of CFC-11.
[b] Based on GWP = 1 of CFC-11 over 100 years.

alternatives. The ODPs of these natural refrigerants are zero and most of them have close to zero GWPs in comparison to HFCs.

From an environmental point of view, water (H_2O, R-718) is the perfect refrigerant. However, some of its thermos-physical properties, such as very low vapor pressure and freezing point at 0°C, limit its use as refrigerant to applications at very high temperature (above 100°C), because at such temperatures, the working pressures are much lower compared to those of other refrigerants. A disadvantage of the use of water as a refrigerant is its low density of the gaseous phase, which requires relatively high compressor capacities.

11.3.6.1 Ammonia

Ammonia (NH_3, R-717), an alkaline fluid with ODP = 0 and GWP = 0 (Table 11.3), is not miscible with mineral oil and has very good thermodynamic properties. It is suitable for usage in large industrial, low-temperature refrigeration systems, but also in medium- and high-temperature heat pump-assisted dryers (generally by using two-stage cycles with intermediate cooling between the compression stages) because it can provide temperatures as high as 80°C, and even 90°C with higher energy efficiency compared to that of HFC systems.

Today, over 90% of the large industrial refrigeration installations use ammonia (R-717), whereas the market share of ammonia is only 5% (India and China) to 25% (Europe and Russia) for smaller industrial refrigeration systems (Kauffeld 2012).

Industrial ammonia systems are in general 15% more energy efficient than their HFC-counterparts. Applying improvements such as reduced condensing temperature, increased evaporation temperature, variable speed compressors, and multistage systems, the energy consumption of the ammonia plants can be further reduced.

When compared with synthetic CFCs, HCFCs, and HFCs refrigerants, ammonia offers advantages as follows: (i) excellent heat transfer and thermos-physical properties; (ii) self-alarming refrigerant in case of accidental leakage; industrial refrigeration systems using ammonia are very tight due to the pungent smell of ammonia, while systems using HFCs still show current leakage rates of 8%–10% annually; (iii) high latent heat (e.g., five to six times that of the halocarbons); (iv) lower mass and volume flow rates for a given heat transfer rate; and (v) higher theoretical COP than that of refrigerants as HCFC-22 and HFC-134a.

The limitations of ammonia as a refrigerant can be described as follows: (i) has a pungent odor and thus a high warning effect; (ii) special safety measures must be provided because of its toxicity at high concentrations in air; (iii) is flammable within a narrow temperature band; (iv) at concentrations in air between 15% and 30% (by volume), ammonia-air mixtures can be ignited; (v) gaseous ammonia is susceptible to strongly react with nitrogen oxides and strong acids; (vi) in contact with water and carbon dioxide, ammonia may form ammonium hydroxide and ammonium carbonate, respectively; (vii) since ammonia is corrosive to copper, brass, and aluminum, steel is the most commonly used material in refrigeration and heat pump installations; (viii) provides higher compression pressure ratios than those of HFC-134a, but comparable with those of HCFC-22 at similar operating conditions; and (ix) achieves higher compressor discharge temperatures than those of HFC-134a and HCFC-22 at the same operating conditions.

Most of these technical limitations can be resolved with proper design and technological development of industrial ammonia refrigeration and heat pump systems.

11.3.6.2 Hydrocarbons

Hydrocarbons, such as propane (R-290) and isobutene (R-600a) (see Table 11.3), can be natural alternative refrigerants to halogenated substances in refrigeration and heat pump systems (Domanski 1998; Granryd 2001; Thome et al. 2008; Mohanraj et al. 2009).

As refrigerants, the hydrocarbons present advantages as follows: (i) are available in nature at relatively low costs; (ii) have small molecular weights; (iii) have excellent thermodynamic and transport properties, similar to those of most of HCFCs and HFCs; some thermophysical properties (e.g., latent heat and liquid thermal conductivity) are usually higher than those of the fluorocarbons; therefore, higher heat transfer coefficients and COPs are expected with hydrocarbons as refrigerants in the same range of operating temperatures and pressures; (iv) may be used as drop-in substitutes to HFCs in heat pumps using the same technology as long as the regulations are respected in terms of refrigerant charge limit and certification of equipment; (v) are suitable for use in heat pumps providing heat at temperatures higher than 80°C; however, hydrocarbons have not yet entered the market of high-temperature heat pumps because of safety concerns (Granryd 2001); (vi) offer excellent efficiency; for example, by using propane (R-290), up to 15% higher refrigeration efficiency could be achieved compared to conventional household refrigerators using traditional HFCs refrigerants; (vii) have zero ODPs and insignificant GWPs (about 3); (viii) are non-toxic and nearly odorless; (ix) are compatible with most of common lubricants; mineral oils, widely used for HCFCs and HFCs, are miscible with hydrocarbons and, thus, the hygroscopic synthetic can be avoided; (x) are compatible with most of the common materials; (xi) heat pump systems with large hydrocarbon charges initially designed for HFCs would need about 40%–50% of that charge when operating on hydrocarbons; by using mini-channel heat exchangers, the refrigerant charge inside the heat pumps can be further reduced by up to 80%; (xii) for large heat pump systems, generally using reciprocating, rotary, and centrifugal compressors, regular monitoring and maintenance measures (as gas sensors and air removal devices) are required, and, thus, safety measures are less restrictive (Palm 2008); (xiii) with some refrigeration systems with scroll compressors, originally designed for HFC-407C and HFC-404A, using hydrocarbons such as propane (R-290) and propylene (R-1270), any problems concerning flammability have been reported (Arnemann et al. 2012).

Among the limitations of hydrocarbons that reduce their widespread in refrigeration and heat pump industries, the following can be mentioned: (i) high flammability with relatively low limits; however, for small- and medium-scale applications such as domestic refrigerators, vending machines, plug-in bottle coolers, chest freezers, food service cabinets, and small exhaust air heat pumps, where the refrigerant charge is very low ($\leq 150\, g/kW_{cooling}$), this is usually not a concern (Granryd 2001; Palm 2008); if part of the refrigerant circuit is in an occupied space (direct systems), the charge limit is 1.5 kg per system; for a room air conditioner, for example, not more than about 500 g of hydrocarbon is allowed; if a heat pump is placed in an unoccupied space (indirect system), the charge of hydrocarbons is limited to 5 kg (EN 378 2008); (ii) significantly more expensive precautions are required to prevent ignition (that can happen, for example, through defects of the winding insulation inside the compressors) than those required for ammonia systems; (iii) explosion proof technologies are required for the development of refrigeration and heat pumps with hydrocarbons; (iv) although hydrocarbons have similar thermos-physical properties with HFCs, it would be required to blend some of them for special applications, thus it would be needed to handle such mixtures and design the refrigeration systems (Miyara 2008); (v) today, heat pumps with hydrocarbons are available for capacities <20 kW, but unit prices are about 5% higher compared to those of common heat pumps using HFCs due to safety requirements.

11.3.6.3 Carbon Dioxide

Because of its negligible impact on climate change (no ODP and negligible direct GWP when used as a refrigerant in closed cycles), carbon dioxide (CO_2, R-744) is another promising natural refrigerant for usage in direct or indirect refrigeration systems, as well as in sub- and supercritical heat pumps (especially domestic water heating systems), often in combination with ammonia in cascade-type installations (Lorentzen and Pettersen 1993; Lorentzen 1994, 1995; Kim and Kim 2002).

Carbon dioxide is a fluid non-toxic at low concentrations (but can be harmful in higher concentrations), nonflammable, colorless, odorless, noncorrosive, heavier than air, relatively inexpensive and readily available (actually recovered from other industrial processes), compatible with most of common lubricants, and with excellent thermos-physical and transport properties.

The maximum allowable concentration of CO_2 for a workplace is 5,000 ppm or 0.5%. Immediate danger to health and life exists for CO_2 concentration over 4% (by volume) in air (40,000 ppm). Above 10% (by volume) in breathing air, CO_2 has a numbing effect and is immediately lethal above 30% (by volume).

As a refrigerant, CO_2 operates with significantly higher pressures (up to 15 MPa on the high pressure side) than other refrigerants, which require stronger materials and/or larger wall thicknesses.

The critical temperature of a refrigerant is an important parameter to consider. Generally, in conventional mechanical vapor compression heat pumps, the condensing temperatures are kept well below the critical temperature of the common refrigerants. As the critical temperature of CO_2 is low (31.1°C at a critical pressure of 7.474 MPa), the CO_2 heat pumps will operate in supercritical modes at high ambient temperatures where heat rejection takes place by cooling the compressed superheated vapor at supercritical high-side pressures (in gas coolers) while the low-side conditions remain subcritical. Usually, the energy efficiency of supercritical heat pump systems is lower than that of conventional subcritical systems. This characteristic can be partially compensated by application of an internal heat exchanger, which has a greater positive impact on energy efficiency in the supercritical CO_2 process than with other refrigerants. There is an optimal high side pressure for every CO_2 gas cooler exit temperature, that is, the high-side pressure has to be adjusted depending on the temperature of the cooling air or water in order to achieve optimum performance. The control of a CO_2 refrigeration system must account for this characteristic and constantly adjust the high-side pressure in order to ensure low energy consumption.

The high working pressures required and the low critical temperature limit the usage of CO_2 as a working fluid in certain refrigeration and heat pump applications. For example, if used in conventional (subcritical) heat pumps, the low critical temperature of CO_2 the heat cannot be delivered at temperatures greater than this value. On the other hand, the volumetric refrigeration capacity of CO_2 is much higher than that of traditional refrigerants, allowing system designs with smaller volumes.

However, the following two main factors restrict the widespread use of the CO_2 system (Miyara et al. 2012): (i) the system operates at a significantly high pressure, which leads to higher initial cost because of the requirement of high-pressure bearings; moreover, the safety during operation should also be carefully considered; (ii) large throttling losses and irreversibility occur during the expansion process, which decrease the whole system performance.

Some mixtures of CO_2 with low-GWP working fluids (e.g., CO_2/R-41 and CO_2/R-32) can reduce the high-side pressures in heat pump water heaters (Dai et al. 2015).

With zeotropic mixtures as CO_2/HFC-134a and CO_2/R-290, used in auto-cascade refrigeration systems, when the CO_2 concentration decreases, the COP tends to increase while the high-side pressure declines (Kim and Kim 2002). The discharge pressure of CO_2/R-290 mixture decreases with an increase of the R-290 fraction. Furthermore, adding R-290 to CO_2 improves the system energy efficiency but reduces the cooling capacity (Kim et al. 2008). CO_2 blends with other HFCs as R-125, R-32, and R-23 are preferable as low-temperature stage working fluid used in cascade refrigeration systems (DiNicola et al. 2005). Blends of CO_2/R-600 and CO_2/R-600a as working fluids for medium- and high-temperature heat pumps for simultaneous heating and cooling applications showed higher COPs and lower high-side pressure (Sarkar and Bhattacharyya 2009) as well as higher cooling capacity (Niu and Zhang 2007) in comparison with similar systems using pure CO_2.

An zeotropic mixture of CO_2/RE170 for a hot water heat pump, with the CO_2 mass concentration ranging from 0 to 1, provided higher $COPs_{heat\ pump}$ than those working with pure CO_2 (Onaka et al. 2010).

11.4 CONVENTIONAL REFRIGERANT MIXTURES

During the past two decades increasing effort was made to use refrigerant mixtures formed by combining pure substances such as CFC-12, HCFC-22, and HFC-134a for heat pump applications primarily to meet environmental concerns, even their thermo-physical properties were more complex than those of pure fluids.

Azeotropic and zeotropic refrigerant mixtures can be (i) hydrocarbon based (e.g., R-1270/R-290/HFC-152a) as alternatives to R-502 in low-temperature refrigeration applications; (ii) HFC based-mixtures (e.g., HFC-404A/HFC-407C/HFC-410A) as potential alternatives to HCFC-22 in refrigeration and heat pump systems; the refrigerant HCFC-22, historically widely used in heat pump industry, was among the first refrigerants replaced with fluid mixtures environmentally benign and equivalent performance for minimum hardware changes (Radermacher and Hwang 2000); the selection of refrigerant mixtures as candidates to replace HCFC-22 has been based on criteria such as toxicity, stability, flammability, boiling point and ODP; (iii) carbon dioxide based (e.g., R-744/HFC-41, R-744/R-32, R-744/R-23, and R-744/HFC-125) as working fluids in low-temperature applications; and (iv) ammonia-based mixtures (e.g., R-717/R-170) having lower compressor discharge temperature compared to R-717 and good miscibility with mineral oil, thereby reducing the usage of highly hygroscopic synthetic oils.

11.4.1 Azeotropic Refrigerant Mixtures

The word *azeotropic* denotes a mixture of at least two substances whose vapor phase has the same composition as its liquid phase. Thus, such a mixture behaves like a pure substance and, therefore, exhibits no temperature glide (Fronk and Garimella 2013).

In other words, azeotropic (or non-zeotropic) refrigerants are mixtures in which the concentrations of the species in the vapor and liquid phases are equal, and thus, the dew and bubble point curves are touching each other at least at one point. Azeotropic mixtures evaporate and condense at constant temperatures acting as a single substances in both liquid and vapor phases although they contain two or more pure refrigerants.

The thermodynamic properties of azeotropic refrigerants can be represented in similar T–s (see Figure 11.1) and ln(p)–h (see Figure 11.2) diagrams as those of pure refrigerants. Both these diagrams show that the bubble point and dew point are on the same isotherm, that is, within the two-phase range, the isotherms are parallel to the isobars.

11.4.2 Zeotropic Mixtures

Zeotropic (also called non-azeotropic) refrigerants (e.g., HFC-245fa/HCF-134a, which is used as the working fluid for heat pumps generating steam at >120°C) are chemical mixtures (blends) of two (or more) fluids of which the composition of the liquid phase within the two-phase range is always different from that of the vapor phase, because the very different boiling points of the individual components.

They undergo phase changes at varying temperatures during constant-pressure evaporation/boiling and condensation process. In other words, these processes achieve temperature glides, contrary to pure single-component refrigerants that evaporate and condense at constant temperatures.

Although the individual components all evaporate at the same time, the component with the lowest boiling point evaporates much more than the other components with higher normal boiling points. As a result, the component with the lowest normal boiling point will be completely evaporated first, while the remaining components will continue to boil. But since their boiling points are now higher, the mean evaporating temperature will rise steadily over the two-phase range. The temperature at the end of the two-phase range of an evaporator, that is, dew-point temperature, is therefore always higher than the inlet temperature of the evaporator.

A similar situation arises with the condenser, only that the two-phase range is passed through in the opposite direction. The temperature of the refrigerant at the outlet of a condenser (bubble point temperature) is therefore always lower than at its inlet (dew-point temperature). This phenomenon is called temperature glide and must be given special consideration when designing evaporators and condensers of heat pump dryers.

The evaporation (convective boiling) and condensation processes are thus non-isothermal, and dew and bubble point curves do not touch each other over the entire composition range. If the temperature glide fits with the temperature change of the source/sink heat, it will contribute to higher heat pump COPs. A drawback of zeotropic mixtures is the preferential leakage of more volatile component(s) leading to change in the mixture composition over the time.

The first difference between the diagrams of pure fluids and mixtures is that the zeotropic diagram is valid only for a given mass fraction. For a different mass fraction, all the lines on this diagram shift. Thus, it is only possible to show one diagram at a time with a full set of lines of constant properties. If one attempted to superimpose a second diagram for a different mass fraction, the graphical representation would become incomprehensible. Typically, only one mass fraction is circulating in a vapour compression cycle.

Second, due to the temperature glide in the two-phase region, the isotherms in the two-phase region have a negative slope. Depending on the nature of the pure fluids that constitute the mixture, the isotherms in the two-phase region may be concave, convex or straight lines. When one considers a constant pressure evaporation process that commences with saturated liquid and ends with saturated vapor, it becomes apparent that the beginning of the evaporation process starts at a given isotherm and ends at an isotherm of higher temperature. Thus, the temperature glide is correctly represented (Radermacher and Hwang 2000).

11.4.2.1 Temperature–Entropy Diagram

The mechanical vapor compression cycle of a heat pump working with a refrigerant mixture with a given mixture mass fraction can be represented in a *T–s* diagram where the critical point is defined as the point at which the dew and bubble lines meet (Figure 11.3).

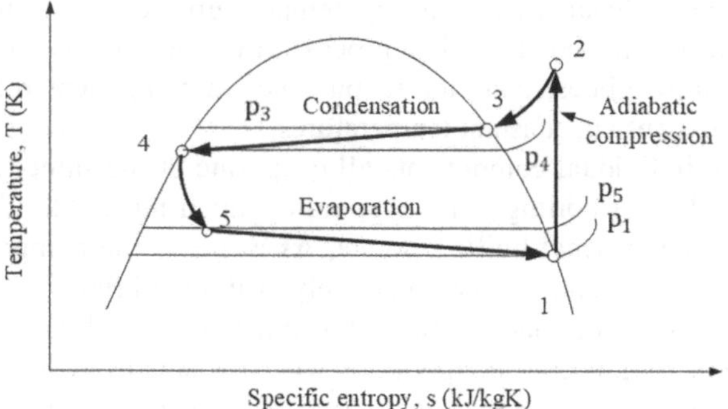

FIGURE 11.3 Theoretical cycle of the mechanical vapor compression heat pump with zeotropic mixtures as refrigerants represented in the temperature–entropy diagram.

It can be seen that in the two-phase region, the isobars have a positive slope. When evaporating, the most volatile component will boil off first and the least volatile component will boil off last.

Figure 11.3 shows that for a given pressure, the temperature will change in the liquid–vapor mixture region. This results in a gliding evaporation and condensing temperatures along the heat transfer surface.

In practice, the saturation temperature at the inlet of the heat pump evaporator will be lower than at the outlet. The evaporating line 5–1 indicates a pressure drop (*p*5–*p*1) that causes temperature drop (or temperature glide). The amount of this temperature glide influences the mixture selection for any heat pump application.

Contrary to the evaporator, in the heat pump condenser, the temperature glide of the refrigerant mixture caused by the pressure drop becomes larger. In the condenser, the saturation temperature at the inlet will be higher than at the outlet.

Because the condensing pressures are higher than evaporating temperatures, the temperature glides in evaporator are always smaller than those in the condenser (Radermacher and Hwang 2000).

11.4.2.2 Pressure–Enthalpy Diagram for Refrigerant Mixtures

The real mechanical vapor compression cycles of heat pumps working with refrigerant zeotropic mixtures can be also represented in ln(p)–h diagrams (Figure 11.4) where the state points are the same as those of thermodynamic cycles for pure refrigerants. However, in the two-phase region, the isotherms show temperature glides. For zeotropic mixtures, the pressure–enthalpy diagrams seem to have the same general topology as those for pure fluids, however, there are two important differences, that is, in the evaporator the temperature glide is $T_{1'}–T_{1''}$ and in the condenser, the temperature glide is $T_{3'}–T_{3''}$ (Figure 11.4)

At the heat pump condenser, the temperature glide of zeotropic refrigerant mixtures results in smaller temperature differences and the condenser performance decreases. Thus, the heat pump sizing based on the dew-point temperature, the condensers must be larger compared to refrigerants without temperature glide. This ultimately leads, however, to lower condensing pressures, which can have a positive impact on the efficiency of the overall heat pump installation.

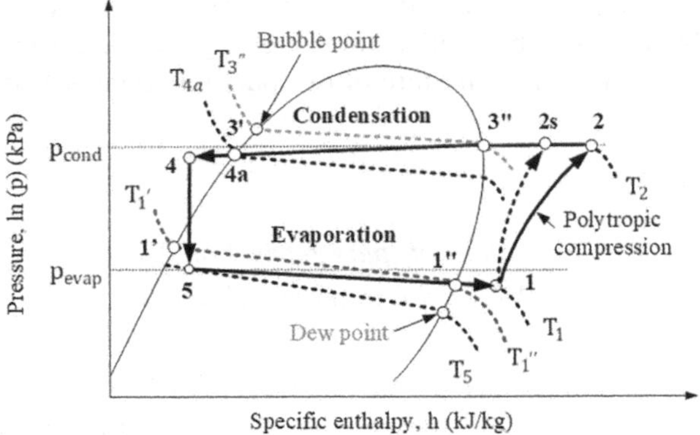

FIGURE 11.4 Theoretical cycle of the mechanical vapor compression heat pump with zeotropic mixtures as refrigerants represented in ln(p)–h diagram.

11.4.2.3 Concentration–Temperature Diagram

The thermodynamic variables of conventional refrigerant mixtures (e.g., temperature, pressure, enthalpy, specific volume, entropy, and mass fraction/concentration) and processes can be represented in diagrams such as temperature–concentration (mass fraction), temperature–entropy, temperature–enthalpy, pressure–enthalpy, pressure–mass fraction, and enthalpy–mass fraction.

In practice, the most useful diagram is the temperature–concentration (mass fraction) diagram (Figure 11.5).

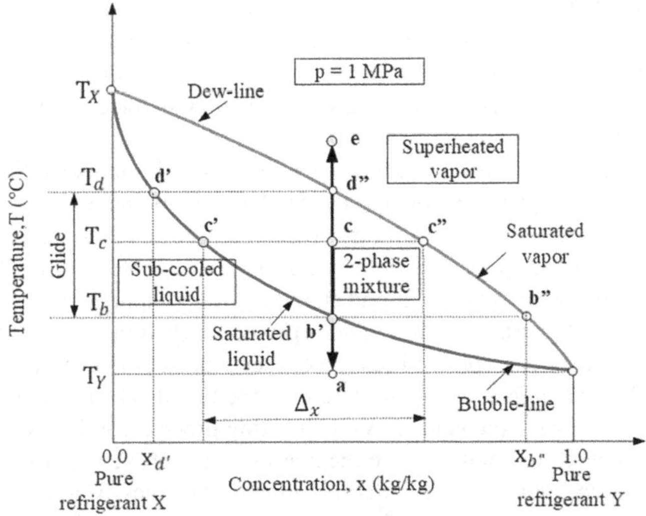

FIGURE 11.5 Evaporation and condensation processes for zeotropic refrigerant mixtures represented in the temperature–concentration (mass fraction) diagram.

A refrigerant mixture can have many different concentrations (or mass fractions) ($0.0 \le x \le 1.0$) that describe the relative amount of a given component. The concentration of the refrigerant *X* in a mixture of two pure refrigerants *X* and *Y* is defined as follows $\left(\frac{kg_X}{kg_{X+Y}}\right)$:

$$x = \frac{Mass\ of\ pure\ refrigerant\ X}{Mass\ of\ refrigerants\ X\ and\ Y} \tag{11.6}$$

The temperature–mass fraction (*T*–*x*) diagram (Figure 11.5) shows that when the liquid and vapor phases of a mixture coexist in equilibrium, the saturation temperature (T_{sat}) varies with the mass fraction at constant pressure, in contrast to a pure fluid where it remains constant.

The mass fraction axis ranges from 0 (pure component *X*) to 1.0 (pure component *Y*). The area below the bubble line represents subcooled liquid. The area above the dew line represents superheated vapor. The area enclosed by the bubble and dew lines is the two-phase region. The boiling point for a mixture of mass fraction *x* is located on the bubble line at that mass fraction.

The bubble line indicates the saturation temperature at which the first vapor bubble is formed for the specified pressure and mass fraction when the temperature is raised from the subcooled liquid region to the saturation value.

The dew line indicates the temperature at which the first liquid droplet is formed when a superheated gas mixture of a given mass fraction is cooled at constant pressure.

As can be seen in Figure 11.5, in the case of zeotropic mixtures at equilibrium (e.g., *X* + *Y*), at a given temperature between the dew and bubble temperature, the concentrations of the vapor and liquid phases are indeed different. It can also be inferred that as the mixture begins to condense, the concentration of the more volatile component in the vapor phase (in this case, *X*) increases, and the dew-point temperature decreases. Thus, unlike a single-component fluid, the condensation process is not isothermal.

An evaporation process at constant pressure begins with subcooled liquid at point a. Points with a single prime denote the liquid phase and points with a double prime denote the vapor phase. As the mixture is heated, the temperature increases and the bubble line reached (point b′). Here, the first vapor bubble is formed. The mass fraction of the first vapor bubble is found at point b″. This vapor is in thermal equilibrium with the liquid phase at point b′. The vapor is enriched in *Y* as compared to the liquid, and has a mass fraction $x_{2''}$. This is a consequence of the fact that, at the same temperature, *Y* has a higher vapor pressure than the liquid. Thus, the vapor phase is bound to contain more *Y* molecules. As the heating process continues, the evaporation process proceeds to point c where the mass fraction of the vapor in equilibrium with the remaining liquid is represented by point c″. The mass fraction of the liquid is now indicated by c′. At this point, the amount of *Y* in the remaining liquid has been reduced as compared to point b′ while the vapor is enriched in *Y*. However, the vapor contains a lower fraction of *Y* and more of *X* than at point b′. The same is hue for the remaining liquid. As the evaporation process proceeds, the state points of the liquid and vapor phases continue to follow the bubble and dew point lines. When point d″ is reached, the evaporation process is completed. The vapor has the same mass fraction as the original subcooled liquid and the mass fraction of the last liquid droplet that evaporated is indicated by point d′. Further heating produces superheated vapor at point e.

In heat pumps, the condensing process follows the refrigerant compression process that does not change the mass fraction (x) of a zeotropic refrigerant mixture $X + Y$, but the pressure and temperature are increased up to point e representing superheated vapor (Figure 11.5). Within the heat pump condenser, the superheated vapor at state e is cooled at constant mass fraction up to state d″ (saturated vapor), and the first liquid droplet appears with the mass fraction of $x_{d''}$. In other words, heat is removed until the dew-point temperature is reached at state d″, where the first drop of condensate formed has a composition of refrigerant X corresponding to state d′ situated on the bubble point curve. At the end of the condensation process (state b') is reached with a mass fraction of the last bubble of $x_{b''}$. If the liquid is subcooled, it reaches state a.

In practice, the entire system does not remain at thermodynamic equilibrium, and the local concentrations and temperature at the interface are different from those of the liquid and vapor bulk. (Fronk and Garimella 2013).

The $T - x$ diagram demonstrates very well how the mass fractions change in both evaporator and condenser heat exchangers that influence transport properties and heat transfer coefficients.

Since during the evaporation and condensation processes, the mass fractions of vapor and liquid are different, a system leak can cause a shift in the mass fraction of the remaining refrigerant mixture. Vapor leaks are more effective in changing the mass fraction.

11.5 UNCONVENTIONAL REFRIGERANT MIXTURES

The heat transfer performance of various thermal devices may be augmented by active and passive techniques. One of the passive techniques is the addition of ultrafine particles (called nanoparticles) to the common heat transfer fluids so that the thermal transport properties of the prepared suspension (called nanofluid) will be enhanced as compared to the base fluid.

A nanofluid is a special class of refrigerant in which the nanoparticles are suspended and well dispersed in the base fluid (as a refrigerant) in order to improve its thermos-physical properties and heat transfer performances (Maxwell 1873; Choi 1995a, 1995b).

The superior thermos-physical properties of nanorefrigerants in comparison to the base refrigerants is as follows: (i) the use of nanorefrigerants will lead to smaller and lighter refrigeration systems and (ii) the refrigeration systems functioning on nano-refrigerants will consume less compressor power, that is, they will be more energy efficient (Wang et al. 1999).

Nanorefrigerants are a special type of nanofluids which are mixtures of nanoparticles and refrigerants and have a broad range of applications in diverse fields as heat pumps (Celen et al. 2014). Nanorefrigerants are unconventional mixtures of base (host) fluids and nanoparticles. The base fluid is a conventional heat transfer fluid, such as water (the most commonly used, having the greatest aptitude to suspend non-coated nanoparticles, in comparison with other base fluids such as ammonia, hydrocarbons, and HFCs), light oils, ethylene glycol or refrigerants, while nanoparticles are colloidal suspensions, such as metallic (Cu, Au) or metal oxides (Al_2O_3, TiO_2, ZrO_2), with very small dimensions (1–100 nm) allowing limiting fouling, sedimentation, erosion, and high pressure drops (Henderson et al. 2010).

Such mixtures provide much higher thermal conductivity and boiling heat transfer coefficients due to the nature of base fluid (e.g., highly wetting), particles' concentration, surface properties, and higher heat transfer areas, as well as better dispersion stability with predominant Brownian motion that prevents gravity settling and agglomeration of particles, reduced particle clogging and pumping power as compared to the base fluid (Choi 1995a, 1995b; Barber et al. 2011).

In mechanical vapor compression heat pumps, nanoparticles are also added to lubricants in order to enhance the solubility of mineral oils and, at the same time, improve the convective boiling heat transfer in evaporators up to 30%, depending on the type of refrigerant and nanoparticles used (Peng et al. 2009). Such heat transfer improvement is normally achieved by varying the particle concentration in the base fluid in order to increase its thermal conductivity and surface wettability, while changing the fluid viscosity and density (Barber et al. 2011). Most researches also confirmed an enhancement in the critical heat flux during nanofluid convective boiling (by up to 53%), which would allow for more compact and effective industrial heat pumps. However, this enhancement of the critical heat flux could be related to the nanoparticles (as alumina—Al_2O_3, zinc oxide, and diamond water-based nanofluids) deposition and cluster on the heat transfer surfaces, especially in microchannel heat exchangers (Lee and Mudawar 2007; Kim et al. 2010; Barber et al. 2011).

Adding nanoparticles to refrigerants may also lead to increase of pressure drops. However, if the penalty in pressure drops is not considerable in comparison with heat transfer enhancement, the use of nanorefrigerant can be helpful in industrial heat pumps.

In the case of a domestic refrigerator using TiO_2/CFC-12 as a nanorefrigerant working fluid, the compression work has been reduced by 11% and the COP increased by 17% (Sabareesh et al. 2012). A refrigeration system using Al_2O_3/R-600a as nanorefrigerant reduced the compressor power input by about 11.5% compared to the conventional POE oil, while the system COP was enhanced by about 19.6% (Kumar et al. 2013). With TiO_2-R600a as the nano-refrigerant working fluid, domestic refrigerators reduced the energy consumption by 9.6%, while the COP increased by 19.6% (Bi et al. 2008). However, it can be noted that in heat pump applications, where the main heat transfer processes are convective vaporization (flow boiling) and condensation, increasing the internal forced convective heat transfer coefficients will have little or any impact on the heat exchangers' overall heat transfer coefficients. As a consequence, this may not improve the heat pumps' overall performance because the actual limits for the global heat transfer are the air-side and/or water-side heat transfer coefficients that are much small compared to internal convective vaporization/flow boiling and condensing heat transfer coefficients.

The thermal conductivity, viscosity, specific heat, latent heal, density, and surface tension are some of the most important thermos-physical properties of nanorefrigerants.

The thermal conductivity of nanorefrigerants, which can be measured by techniques such as transient hot wire and plane source, temperature oscillation, laser flash, photo-acoustic method (Bianco et al. 2015), and micron-scale beam deflection (Putnam et al. 2006; Das et al. 2007), depends on both thermal conductivity of nanoparticles and base fluid, and on the nanoparticle volume fraction, size, base fluid material, pH level, and sonication (Hwang et al 2006; Jiang et al. 2009).

The effective thermal conductivity of the base fluid increases due to the addition of nanoparticles (Choi 1995a; Wang et al. 1999; Yoo et al. 2007). The addition of nanoparticles to refrigerants generally leads to an increase of the fluid kinematic viscosity which decreases when the temperature increases (Alawi and Sidik 2015) and flow pressure drop depending on temperature, nanoparticle mass fractions, and diameter.

The specific heat of nanofluids mainly depends on the temperature, type and volume concentration of nanoparticles, and the type of base fluid. Generally, if the specific heat of the nanoparticles is lesser than the specific heat of base fluid, the specific heat of nanorefrigerants deteriorates.

The pressure drop in nanorefrigerants is bound to increase because of the addition of nanoparticles. The increase in pressure drop is mainly due to the increase in the effective viscosity of the nanorefrigerant. A higher pressure drop in the evaporator will marginally increase the pressure lift between compressor suction and compressor discharge that will consequently lead to slightly higher power consumption for compressing the same amount of refrigerant vapor.

REFERENCES

Alawi, O.A., N.A.C. Sidik. 2015. The effect of temperature and particles concentration on the determination of thermos- and physical properties of SWCNT nanorefrigerant. *International Communications on Heat Mass Transfer* 67: 8–13.

Angelino, G., C. Invernizzi. 2003. Experimental investigation on the thermal stability of some new zero-ODP refrigerants. *International Journal of Refrigeration*, 26:51–58.

ANSI. 2010. *Standard 34–2010. Designation and Safety Classification of Refrigerants.* American Society of Heating, Refrigerating, and Air-Conditioning Engineers, Atlanta.

Arnemann, H., B. Bella, N. Kaemmer, J. Nohales. 2012. Scroll compressor assessment with R-290 and R-1270. In *Proceedings of the 10th IIR Gustav Lorentzen Conference on Natural Refrigerants*, Delft, the Netherlands, paper no. 308.

Artemenko, S., V. Mazur. 2007. Azeotropy in the natural and synthetic refrigerant mixtures. *International Journal of Refrigeration* 30(5), 831–839.

ASHRAE. 2009. *Handbook Fundamentals*, SI edition, ASHRAE Inc., Atlanta, Georgia.

Barber, J., D. Brutin, L. Tadrist. 2011. A review on boiling heat transfer enhancement with nanofluids. *Nanoscale Research Letters* 6:280, (http://www.nanoscalereslett.com/content/6/1/280, accessed December 15, 2017).

Bi S., L. Shi, L. Zhang. 2008. Application of nanoparticles in domestic refrigerators. *Applied Thermal Engineering* 28:1834–1843.

Bianco, V., O. Manca, S. Nardini, K. Vafai. 2015. *Heat Transfer Enhancement with Nanofluids*, 1st ed. CRC Press, Florida.

Brian, T.A., K. Sumathy. 2011. Transcritical carbon dioxide heat pump systems: a review. *Renewable and Sustainable Energy Reviews* 15:4013–4029.

Brown, J.S., C. Zilio, A. Cavallini. 2009. The fluorinated olefin R-1234ze(Z) as a high-temperature heat pumping refrigerant. *International Journal of Refrigeration* 32:1412–1422.

Calm, J.M. 2008. The next generation of refrigerants - historical review, considerations, and outlook. *International Journal of Refrigeration* 31(1):1123–1133.

Calm, J.M., P.A. Domanski. 2004. R-22 replacement status. *ASHRAE Journal* 46(8):29–39.

Cavallini, A., C. Zilio, J.S. Brown. 2014. Sustainability with prospective refrigerants. *Energy Research* 38:285–298.

Celen, A., A. Cebi, M. Aktas, O. Mahian, A.S. Dalkilik, S. Wongwises. 2014. A review of nanorefrigerants: flow characteristics and applications. *International Journal of Refrigeration* 44:125–140.

Choi, S.U.S. 1995a. Enhancing thermal conductivity of fluids with nanoparticles. In *Proceedings of the 1995 ASME International Mechanical Engineering Congress and Exposition*, November 12–47, San Francisco, CA, USA.

Choi, S.U.S. 1995b. Enhancing thermal conductivity of fluids with nanoparticles. In: Singer, D.A., Wang, H.P. (Eds.), *Developments and Applications of Non-Newtonian Flows.* ASME, New York, pp. 99–105. FED-231/MD–66.

Copetti, J.B., M.H. Macagnan, Z. Zinani. 2013. Experimental study on R-600a boiling in 2.6 mm tube. *International Journal of Refrigeration* 36:325–334.

Dai, B., C. Dang, M. Li, H. Tian, Y. Ma. 2015. Thermodynamic performance assessment of carbon dioxide blends with low-global warming potential (GWP) working fluids for a heat pump water heater. *International Journal of Refrigeration* 56: 1–14.

Das, S.K., S.U.S. Choi, W. Yu, T. Pradeep. 2007. *Nanofluids: Science and Technology*, 1st ed. John Wiley & Sons, New Jersey.

Del Col, D., S. Bortolin, D. Torresin, A. Cavallini. 2013. Flow boiling of R1234yf in a 1 mm diameter channel. *International Journal of Refrigeration* 36:353–362.

DiNicola, G., G. Giuliani, F. Polonara, R. Stryjek. 2005. Blends of carbon dioxide and HFCs as working fluids for the low-temperature circuit m cascade refrigerating systems. *International Journal of Refrigeration* 28:130–140.

Domanski, P.A. 1998. Refrigerants for the 21st century. *Journal of Research (NIST JRES)* 103(5):1–5.

Eames, I.W., A.E. Ablwaifa, V. Petrenko. 2007. Results of an experimental study of an advanced jet-pump refrigerator operating with R245fa. *Applied Thermal Engineering* 27:2833–2840.

EN 378. 2008. European Standard—Refrigerating systems and heat pumps—Safety and environmental requirements.

Fronk, B.M., S. Garimella. 2013. Water-coupled carbon dioxide microchannel gas cooler for heat pump water heaters: Part I-Experiments. *International Journal of Refrigeration* 34(1):7–16.

Fukuda, S., C. Kondou, N. Takata, S. Koyama. 2014. Low GWP refrigerants R-1234ze (E) and R-1234ze (Z) for high temperature heat pumps. *International Journal of Refrigeration* 40:161–173.

Granryd, E. 2001. Hydrocarbons as refrigerants—an overview. *International Journal of Refrigeration* 24:15–24.

Henderson, K, Y.G. Park, L. Liu, A.M. Jacobi. 2010. Flow-boiling heat transfer of R-134a-based nanofluids in a horizontal tube. *International Journal of Heat and Mass Transfer* 53(5–6):944–951.

Hwang, Y.J., Y.C. Ahn, H.S. Shin, C.G. Lee, G.T. Kim, H.S. Park. 2006. Investigation on characteristics of thermal conductivity enhancement of nanofluids. *Current Applied Physics* 6:1068–1071.

Jiang, W., G. Ding, H. Peng. 2009. Measurement and model on thermal conductivities of carbon nanotube nanorefrigerants. *International Journal of Thermal Sciences* 48: 1108–1115.
Kauffeld, M. 2012. Availability of low GWP alternatives to HFCs—Feasibility of an early phase-out of HFCs by 2020. Environmental Investigation Agency (EIA) 62/63 (www.eia-international.org). Produced and published by Environmental Investigation Agency (www.eia-global.org).
Kim, S., M. Kim. 2002. Experiment and simulation on the performance of an auto-cascade refrigeration system using carbon dioxide as a refrigerant. *International Journal of Refrigeration* 25:1093–1101.
Kim, J.H., J.M. Cho, M.S. Kim. 2008. Cooling performance of several CO_2/propane mixtures and glide matching with secondary heat transfer fluid. *International Journal of Refrigeration* 31:800–806.
Kim, S.J., T. McKrell, J. Buongiorno, L.W. Hu. 2010. Subcooled flow boiling heat transfer of dilute alumina, zinc oxide, and diamond nanofluids at atmospheric pressure. *Nuclear Engineering and Design* 240(5):1186–1194.
Kondou, C., S. Koyama. 2015. Thermodynamic assessment of high-temperature heat pumps using low-GWP HFO refrigerants for heat recovery. *International Journal of Refrigeration* 53:126–141.
Kumar, R. R., K. Sridhar, M. Narasimha. 2013. Heat transfer enhancement in domestic refrigerator using R600a/mineral oil/nano-Al_2O_3 as working fluid. *International Journal of Computational Engineering Research* 3(4):42–51.
Lee, J., I. Mudawar. 2007. Assessment of the effectiveness of nanofluids for single-phase and two-phase heat transfer in micro-channels. *International Journal of Heat Mass Transfer* 50(3–4):452–463.
Longo, G.A., C. Zilio. 2013. Condensation of the low GWP refrigerant HFC1234yf inside a brazed plate heat exchanger. *International Journal of Refrigeration* 36:612–621.
Longo, G.A., S. Mancin, G. Righetti, C. Zilio. 2016. Saturated flow boiling of HFC-134a and its low GWP substitute HFO-1234ze(E) inside a 4 mm horizontal smooth tube. *International Journal of Refrigeration* 64:32–39.
Lorentzen, G. 1994. Revival of carbon dioxide as a refrigerant. *International Journal of Refrigeration* 17:292–301.
Lorentzen, G. 1995. The use of natural refrigerants: a complete solution to the CFC/HCFC predicament. *International Journal of Refrigeration* 18:190–197.
Lorentzen, G., Pettersen, J. 1993. A new, efficient and environmentally benign system for car air-conditioning. *International Journal of Refrigeration* 16:4–12.
Maxwell, J.C. 1873. *Treatise on Electricity and Magnetism*. Clarendon Press, Oxford.
McLinden, M.O., A.F. Kazakov, J.S. Brown, P.A. Domanski. 2014. A thermodynamic analysis of refrigerants: possibilities and trade-off for low-GWP refrigerants. *International Journal of Refrigeration* 38:80–92.
Minea, V. 2004. Heat pumps for wood drying—new developments and preliminary results, *Proceedings of the 14th International Drying Symposium*, Sao Paulo, Brazil, August 22–25, Volume B:892–899.
Miyara, A. 2008. Condensation of hydrocarbons—A review. *International Journal of Refrigeration* 31:621–632.
Minea, V. 2014. Power generation with ORC machines using low-grade waste heat or geothermal renewable energy. *Applied Thermal Engineering* 69:143–154.
Miyara, A., Y. Onaka, S. Koyama. 2012. Ways of next generation refrigerants and heat pump/refrigeration systems. *International Journal of Air Conditioning and Refrigeration* 20(1):11. DOI: 10.1142/S2010132511300023.
Mohanraj, M., S. Jayaraj, S. Muraleedharan. 2009. Environment friendly alternatives to halogenated refrigerants—A review. *International Journal of Greenhouse Gas Control* 3:108–119.

Niu, B., Y. Zhang. 2007. Experimental study of the refrigeration cycle performance for the R744/R290 mixtures. *International Journal of Refrigeration* 30:37–42.

Onaka, Y., A. Miyara, K. Tsubaki. 2010. Experimental study on evaporation heat transfer of CO_2/DME mixture refrigerant in a horizontal smooth tube. *International Journal of Refrigeration* 33:1277–1291.

Palm, B. 2008. Hydrocarbons as refrigerants in small heat pump and refrigeration systems—A review. *International Journal of Refrigeration* 31:552–563.

Peng, H., G. Ding, W. Jiang, H. Hu, Y. Gao. 2009. Heat transfer characteristics of refrigerant-based nanofluid flow boiling inside a horizontal smooth tube. *International Journal of Refrigeration* 32(6):1259–1270.

Putnam, S.A., D.G. Cahill, P.V. Braun, Z. Ge, R.G. Shimmin. 2006. Thermal conductivity of nanoparticle suspensions. *Applied Physics* 99:284–308.

Radermacher, R., Y. Hwang. 2000. *Vapor Compression Heat Pumps with Refrigerant Mixtures.* Center for Environmental Energy Engineering Department of Mechanical Engineering University of Maryland, College Park, MD.

Sabareesh, K., N. Gobinath, V. Sajithb, S. Das, C.B. Sobhan. 2012. Application of TiO_2 nanoparticles as a lubricant-additive for vapor compression refrigeration systems - an experimental investigation. *International Journal of Refrigeration* 35(7):1989–1996.

Saleh, B., G. Koglbauer, M. Wendland, J. Fischer. 2007. Working fluids for low-temperature Organic Rankine Cycles. *Energy* 32:1210–1221.

Sarkar, J., S. Bhattacharyya. 2009. Assessment of blends of CO_2 with butane and isobutane as working fluids for heat pump applications. *International Journal of Thermal Sciences* 48:1460–1465.

Thome, J.R., L. Cheng, G. Ribatski, L.F. Vales. 2008. Flow boiling of ammonia and hydrocarbons: a state-of-the-art review. *International Journal of Refrigeration* 31:603–620.

Wang, X., X. Xu, S.U.S. Choi. 1999. Thermal conductivity of nanoparticle-fluid mixture. *Journal of Thermo-Physics and Heat Transfer* 13(4):474–480.

Wang, X.D., L. Zhao, J.L. Wang. 2000. Experimental investigation on the low-temperature solar Rankine Cycle System using R-245fa. *Energy Conversion and Management* 52:946–952.

Wang, X.D., L. Zhao, J.L. Wang, W.Z. Zhang, X.Z. Zhao, W. Wu. 2010. Performance evaluation of a low-temperature solar Rankine Cycle System utilizing R-245fa. *Solar Energy* 84:353–364.

Yoo, D. H., K.S. Hong, H.S. Yang. 2007. Study of thermal conductivity of nanofluids for the application of heat transfer fluids. *Thermochimica Acta* 455:66–69.

Zyhowski, G. J., M. M. Spatz, S.Y. Motta. 2002. An overview of properties applications of HFC-245fa, Purdue University, Purdue e-Pubs, International Refrigeration and Air Conditioning, Conference School of Mechanical Engineering.

12 Heat Exchangers without Phase Changes

12.1 CLASSIFICATION

Heat exchangers without phase change(s) are sometimes employed as refrigerant desuperheaters, internal heat exchangers (IHEs) and air- or water-cooled refrigerant sub-coolers in mechanical vapor compression heat pump-assisted wood dryers.

In industrial processes, heat exchangers without phase change(s) used for cooling and/or heating are usually classified according to their flow arrangement and type of construction.

In the most simple concentric (double) tube heat exchangers without phase changes, the hot and cold fluids move in the opposite (counter-flow) directions (Figure 12.1a) or in the same (parallel-flow) direction (Figure 12.2a).

The heat capacity rates of hot and cold fluids are, respectively (Incropera and DeWitt 2002),

$$\dot{C}_{p,hot} = \dot{m}_{hot}\overline{c}_{p,hot} \tag{12.1}$$

$$\dot{C}_{p,cold} = \dot{m}_{cold}\overline{c}_{p,cold} \tag{12.2}$$

where

$\dot{C}_{p,hot}$ is the heat capacity rate of the hot fluid (W/K)
$\dot{C}_{p,cold}$ is the heat capacity rate of the cold fluid (W/K)
$\dot{m}_{hot}$ and $\dot{m}_{cold}$ are the mass flow rates of hot and cold fluids, respectively (kg/s)
$\overline{c}_{p,hot}$ and $\overline{c}_{p,cold}$ are the average specific heats of hot and cold fluids, respectively (J/kg·K)

In the case of counter-flow heat exchangers without phase change where the heat capacity rates of both fluids are equal $\left(\dot{C}_{p,hot} = \dot{C}_{p,cold}\right)$, the temperature difference ΔT is constant throughout the exchanger, i.e., $\Delta T_1 = \Delta T_2$ (Figure 12.1b).

12.2 DESIGN METHODS

Two basic methods are used to size, analyze, or select heat exchangers: (i) the logarithmic mean temperature difference (*LMTD*) method, (ii) the effectiveness-number of transfer units (effectiveness-*NTU* or *ε-NTU*) method (Radermacher and Hwang 2000; Incropera and DeWitt 2002; McQuiston et al. 2005).

Their scope is to predict the total rate of heat transfer between two fluids, generally based on the following equation that relates the total heat transfer rate to quantities such as the inlet and outlet fluid temperatures, the overall heat transfer coefficient,

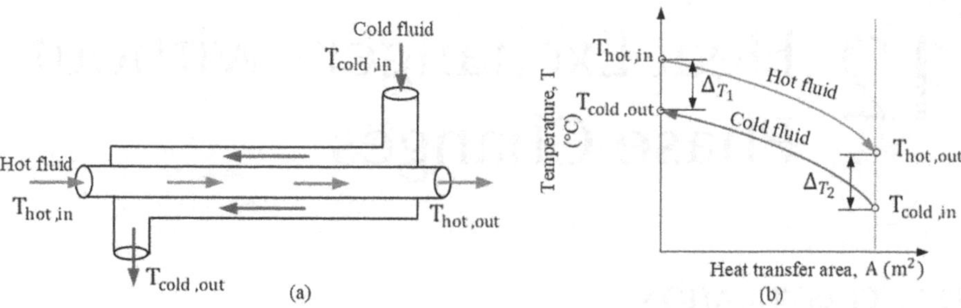

FIGURE 12.1 Concentric (double-tube) counter-flow heat exchanger without phase change(s): (a) schematic; (b) temperature–heat exchange area diagram. *T*, temperature; in, inlet; out, outlet.

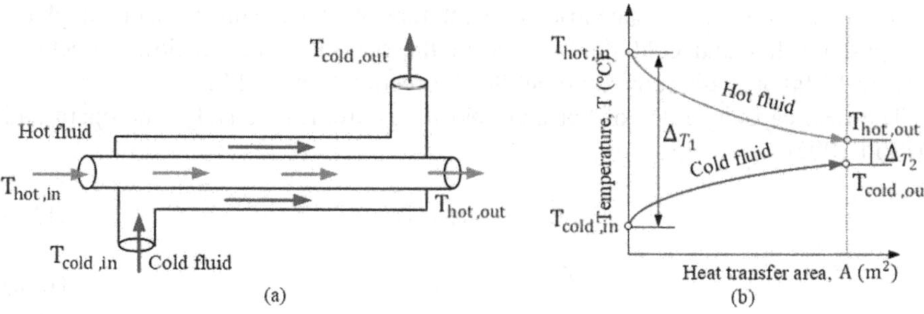

FIGURE 12.2 Concentric (double-tube) parallel-flow heat exchanger without phase change(s): (a) schematic; (b) temperature–heat exchange area diagram (for legend, see Figure 12.1).

and the total heat transfer surface area (McQuiston and Parker 1988; Radermacher and Hwang 2000; Incropera and DeWitt 2002; McQuiston et al. 2005):

$$\dot{Q} = \bar{U} * A * \Delta T_{mean} \tag{12.3}$$

where

$\bar{U}$ is the mean overall heat transfer coefficient (W/m²·K)
A is the surface area associated with $\bar{U}$ (m²)
ΔT_{mean} is the mean temperature difference between the fluids (°C)

If the flow paths in the heat exchanger are not simply counterflow or parallel flow but are quite complex, the concept of a correction factor F is used (McQuiston et al. 2005):

$$\dot{Q} = \bar{U} * A * F * \Delta T_{mean} \tag{12.4}$$

where

ΔT_{mean} is calculated in the same manner as the *LMTD* for an "equivalent" counter-flow heat exchanger (°C)

When one fluid undergoes phase change, or the two fluids have a very similar heat capacity, then $F = 1$.

The mean temperature difference between the two fluids (ΔT_{mean}) is used because the temperature difference is variable from one place to another in the exchanger. The prediction of average values of average overall heat transfer coefficient $(\bar{U})$ is based on the knowledge of internal and external convective heat transfer coefficients, mostly experimentally determined (McQuiston et al. 2005).

12.2.1 *LMTD* Method

For both counter and parallel flows, an appropriate expression for *LMTD* can be derived with assumption as follows (McQuiston et al. 2005): (i) the mean overall heat transfer coefficient $\bar{U}$, the fluid mass flow rates ($\dot{m}_{hot}$ and $\dot{m}_{hot}$), and the fluid heat capacity rates ($\dot{C}_{p,hot}$ and $\dot{C}_{p,cold}$) are constants; (ii) there is no heat loss or gain external to the heat exchanger, and there is no axial conduction in the heat exchanger; (iii) single bulk temperature applies to each stream at a given cross section (see Figure 12.1b and Figure 12.2b):

$$LMTD = \frac{\Delta T_1 - \Delta T_2}{\ln \dfrac{\Delta T_1}{\Delta T_2}} \tag{12.5}$$

where

ΔT_1 and ΔT_2 are temperature differences as shown (defined) in Figure 21.1b and b

The calculation of *LMTD* is very simple when the inlet and outlet temperatures of both fluids are known. If three of these temperatures are known, together with the fluid heat capacity rates, the fourth temperature may be calculated an energy balance. However, in practice, only two of temperatures represented in Figure 12.1a and b are known, and thus trial-and-error (iteration) procedures are required (McQuiston et al. 2005).

If the heat exchanger heat transfer area (*A*) and the operational conditions ($\bar{U}$ and *LMTD*) are given, the total heat transfer rate $(\dot{Q})$ can be calculated.

On the other hand, if the target total heat transfer rate $(\dot{Q})$ is given, only one parameter among the heat exchanger heat transfer area (*A*) and the operational conditions ($\bar{U}$ and *LMTD*) is calculated, while the other two parameters are given (Radermacher and Hwang 2000).

For moisture condensers (heat pump evaporators) (see Chapter 13) the *LMTD* should be separately calculated for the two-phase (vaporization) and vapor superheating regions. Similarly, for heat pump refrigerant-to-air condensers (see Chapter 14), the *LMTD* should be separately calculated for superheated, two-phase (condensation), and subcooled regions.

The total heat transfer rates will be the sum of heat transfer rates at each region:

$$\dot{Q} = \sum \dot{Q}_i = \sum \bar{U}_i * A_i * LMTD_i \quad (12.6)$$

where

$\bar{U}_i$ is the average overall heat transfer coefficient for each region (W/m²·K)
A_i is the total heat transfer area for each region (m²)
$LMTD_i$ is the logarithmic mean temperature difference for each region (K)

Sometimes, direct knowledge of the *LMTD* is not available, and thus the *NTU* method is used.

Exercise E12.1

An air-to-air mechanical vapor compression heat pump used in a heat pump-assisted wood dryer is equipped with a desuperheating countercurrent heat exchanger aiming at heating $\dot{m}_{water} = 5\,\text{kg/s}$ of process water from $T_{water,\ in}$ = 30°C to $T_{water,\ out}$ = 50°C. The superheated HFC-134a vapor enters the desuperheater at 85°C with a specific enthalpy of $h_{refr,\ in}$ = 453 kJ/kg and leaves it at 70°C with a specific enthalpy of $h_{refr,\ out}$ = 430 kJ/kg. Knowing that the thermal efficiency of the desuperheater is $\eta_{desuper}$ = 85%, determine

1. The desuperheater heat transfer rate
2. The refrigerant mass flow rate

Solution

The average isobaric specific heat of water between 30°C and 50°C is $\bar{c}_{p,\ water} = 4.189\ \text{kJ/kg}\cdot\text{K}$.

1. Desuperheater heat transfer rate can be calculated as follows:

$$\dot{Q}_{desuper,water} = \dot{m}_{water} * \bar{c}_{p,water} * \left(T_{water,out} - T_{water,in}\right) = 5 * 4.189 * (50 - 30) = 418.9\ \text{kW}$$

2. Refrigerant mass flow rate results from the following energy balance equation:

$$\eta_{desuper} * m_{refr} * \left(h_{refr,in} - h_{refr,out}\right) = \dot{Q}_{desuper,water}$$

$$\dot{m}_{refr} = \frac{\dot{Q}_{desuper,water}}{\eta_{desuper} * \left(h_{refr,in} - h_{refr,out}\right)} = \frac{418.9}{0.85 * (453 - 430)} = 21.4\ \text{kg/s}$$

Exercise E12.2

An air-to-air mechanical vapor compression heat pump using HFC-134a as a refrigerant and installed in a heat pump-assisted wood dryer is equipped with an IHE (see Figure 10.7). Knowing that (i) the refrigerant mass flow rate is $\dot{m}_{refr} = 21.4\,\text{kg/s}$, (ii) the evaporating temperature is T_{evap} = 30°C, (iii) the condensing temperature is T_{cond} = 70°C, (iv) vapor superheating is 10°C, (v) liquid subcooling is 10°C, and (vi) thermal efficiency of IHE is η_{IHX} = 90%, determine the heat transfer rate of the IHE.

Solution

From the ln(*p*)–*h* diagram of the refrigerant HFC-134a, at T_{cond} = 70°C and for 10°C liquid subcooling, results $h_{liquid,\,in} \approx 304$ kJ/kg and $h_{liquid,\,out} \approx 287$ kJ/kg.

The thermal capacity of the IHE is as follows:

$$Q_{IHE} = \eta_{IHX} * \dot{m}_{refr} * \left(h_{liquid,in} - h_{liquid,out}\right) = 0.9 * 21.4 * (304 - 287) = 327.42\ \text{kW}$$

12.2.2 *NTU* Method

For compact finned-tube heat exchangers (i.e., with surface areas per unit volume greater than 656 m²/m³) where at least one fluid is a vapor, the *ε-NTU* method can be applied in order to eliminate the trial-and-error (iteration) procedure of the *LMTD* method when only the inlet fluid temperatures are known (McQuiston and Parker 1988; Incropera and DeWitt 2002; McQuiston et al. 2005). This method involves the use of an average value of the overall heat transfer coefficient $\left(\bar{U}\right)$.

The heat exchanger effectiveness is defined as follows:

$$\varepsilon = \frac{Actual\ heat\ transfer\ rate}{Maximum\ possible\ heat\ transfer\ rate} \tag{12.7}$$

The actual heat transfer rate $\left(\dot{Q}\right)$ (in W) is given by (see Figures 12.1 and 12.2)

$$\dot{Q} = \dot{C}_{p,hot}\left(T_{hot,in} - T_{hot,out}\right) = \dot{C}_{p,cold}\left(T_{cold,out} - T_{cold,in}\right) \tag{12.8}$$

The maximum possible heat transfer rate could, in principle, be achieved in a counterflow heat exchanger of infinite length where one of the fluids would experience the maximum possible temperature difference (i.e., $T_{hot,\,in} - T_{cold,\,in}$), as represented in Figure 12.3.

To satisfy the energy balance, the fluid experiencing the maximum temperature change must be the one with the minimum value of heat capacity rate $\dot{C}$.

Consequently, the maximum possible heat transfer rate is expressed by

$$\dot{Q}_{max} = \dot{C}_{p,min}\left(T_{hot,in} - T_{cold,in}\right) \tag{12.9}$$

where

$\dot{C}_{min} = \min\left(\dot{C}_{p,hot}, \dot{C}_{p,cold}\right)$ is the minimum heat capacity rate between hot and cold fluids (W/K)

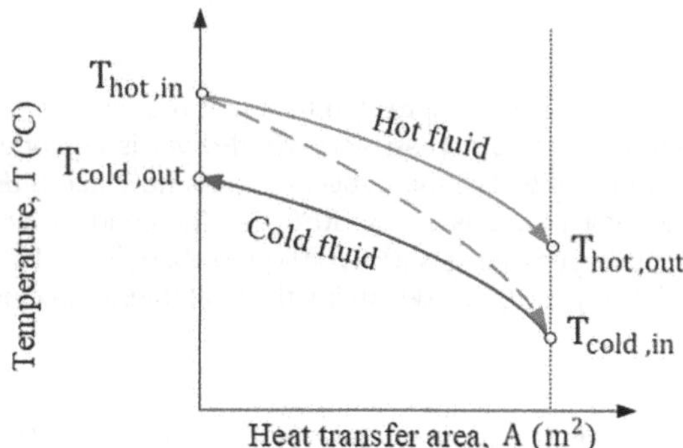

FIGURE 12.3 Maximum temperature change in a counter-flow heat exchanger of infinite length.

For $\dot{C}_{hot} = \dot{C}_{\min}$, using Equations 12.8 and 12.9, the heat exchange effectiveness is as follows:

$$\varepsilon = \frac{\dot{q}}{\dot{q}_{\max}} = \frac{\dot{C}_{hot}\left(T_{hot,in} - T_{hot,out}\right)}{\dot{C}_{\min}\left(T_{hot,in} - T_{cold,in}\right)} = \frac{\left(T_{hot,in} - T_{hot,out}\right)}{\left(T_{hot,in} - T_{cold,in}\right)} \tag{12.10}$$

For $\dot{C}_{cold} = \dot{C}_{\min}$, the heat exchange effectiveness becomes

$$\varepsilon = \frac{\dot{q}}{\dot{q}_{\max}} = \frac{\dot{C}_{cold}\left(T_{cold,out} - T_{cold,in}\right)}{\dot{C}_{\min}\left(T_{hot,in} - T_{cold,in}\right)} = \frac{\left(T_{cold,out} - T_{cold,in}\right)}{\left(T_{hot,in} - T_{cold,in}\right)} \tag{12.11}$$

It is therefore necessary to have two expressions for the effectiveness (Equations 12.10 and 12.11). When effectiveness is known, the outlet temperature may be easily computed.

For example, when $\dot{C}_{hot} < \dot{C}_{cold}$,

$$T_{hot,out} = \varepsilon * \left(T_{cold,in} - T_{hot,in}\right) + T_{hot,in} \tag{12.12}$$

also,

$$T_{cold,out} = \frac{\dot{q}}{\dot{C}_{cold}} + T_{cold,in} = \frac{\dot{C}_{hot}}{\dot{C}_{cold}}\left(T_{hot,in} - T_{hot,out}\right) + T_{cold,in} \tag{12.13}$$

or,

$$T_{cold,out} = \frac{\dot{C}_{hot}}{\dot{C}_{cold}}\varepsilon * \left(T_{hot,in} - T_{cold,in}\right) + T_{cold,in} \tag{12.14}$$

It may also be observed that flow configuration is unimportant when $\frac{C_{\min}}{C_{\max}} = 0$. This corresponds to the situation of one fluid undergoing a phase change (e.g., refrigerant evaporating or condensing) where the isobaric specific heats $\left(\bar{c}_p\right)$ are infinite.

For parallel-flow heat exchangers, irrespective of whether the minimum heat capacity rate is associated with the hot or cold fluid, the heat exchanger effectiveness is expressed as follows (Incropera and DeWitt 2002):

$$\varepsilon = \frac{1 - \exp\left\{-NTU\left[1 + \frac{\dot{C}_{\min}}{\dot{C}_{\max}}\right]\right\}}{1 + \frac{\dot{C}_{\min}}{\dot{C}_{\max}}} \tag{12.15}$$

Similarly, for counter-flow heat exchangers, the heat exchanger effectiveness has been found as follows:

$$\varepsilon = \frac{1 - \exp\left\{-NTU\left[1 - \frac{\dot{C}_{\min}}{\dot{C}_{\max}}\right]\right\}}{1 - \frac{\dot{C}_{\min}}{\dot{C}_{\max}} \exp\left[-NTU\left(1 - \frac{\dot{C}_{\min}}{\dot{C}_{\max}}\right)\right]} \quad \text{for} \quad \frac{\dot{C}_{\min}}{\dot{C}_{\max}} < 1 \tag{12.16}$$

and

$$\varepsilon = \frac{NTU}{1 + NTU} \quad \text{for} \quad \frac{\dot{C}_{\min}}{\dot{C}_{\max}} = 1 \tag{12.17}$$

The NTU parameter is defined as $\frac{U * A}{\dot{C}_{\min}}$. When the NTU method is utilized, the heat transfer rate is generally computed from the temperature change for either fluid. For example,

$$\dot{Q} = \dot{C}_{hot}\left(T_{hot,in} - T_{hot,out}\right) \tag{12.18}$$

However, in heat exchanger design calculations, it is more convenient to work with relations of the form

$$NTU = f\left(\varepsilon, \frac{\dot{C}_{\min}}{\dot{C}_{\max}}\right) \tag{12.19}$$

For parallel-flow heat exchangers, the relation for NTU is as follows:

$$NTU = -\frac{\ln\left[1 - \varepsilon\left(1 + \frac{\dot{C}_{\min}}{\dot{C}_{\max}}\right)\right]}{1 + \frac{\dot{C}_{\min}}{\dot{C}_{\max}}} \tag{12.20}$$

For counter-flow heat exchangers, it is:

$$NTU = \frac{1}{\frac{\dot{C}_{\min}}{\dot{C}_{\max}} - 1} \ln\left(\frac{\varepsilon - 1}{\varepsilon * \frac{\dot{C}_{\min}}{\dot{C}_{\max}} - 1}\right) \quad \text{for} \quad \frac{\dot{C}_{\min}}{\dot{C}_{\max}} < 1 \tag{12.21}$$

and

$$NTU = \frac{\varepsilon}{1 - \varepsilon} \quad \text{for} \quad \frac{\dot{C}_{\min}}{\dot{C}_{\max}} = 1 \tag{12.22}$$

For both *LMTD* and *NTU* design methods, an average value of the overall coefficient $(\bar{U})$ must be known.

For a simple heat exchanger without fins, the average overall coefficient $\bar{U}$ is given by (McQuiston and Parker 1988):

$$\frac{1}{\bar{U}A} = \frac{1}{h_{out}A_{out}} + \frac{\delta}{k_{wall}A_{mean}} + \frac{1}{h_{in}A_{in}} + \frac{R_{f,in}}{A_{in}} + \frac{R_{f,out}}{A_{out}} \tag{12.23}$$

where

h_{out} is the heat transfer coefficient on the outside ($W/m^2 \cdot K$)
h_{in} is the heat transfer coefficient on the inside ($W/m^2 \cdot K$)
δ is the thickness of the separating wall (m)
k_{wall} is the thermal conductivity of the separating wall (W/m K)
A_{in} is the interior heat transfer area (m^2)
A_{out} is the exterior heat transfer area (m^2)
A_{mean} is the mean heat transfer area (m^2)
$R_{f,in}$ is the interior fouling factor (m^2K/W)
$R_{f,out}$ is the exterior fouling factor (m^2K/W)

In general, the areas A_{out}, A_{mean}, and A_{in} are not equal and $\bar{U}$ may be referenced to any one of these three.

If we get $A = A_{out}$, then (McQuiston and Parker 1988)

$$\frac{1}{\bar{U}_{out}} = \frac{1}{h_{out}} + \frac{\delta}{k\frac{A_{mean}}{A_{out}}} + \frac{1}{h_{in}\frac{A_{in}}{A_{out}}} + \frac{R_{f,in}}{\frac{A_{in}}{A_{out}}} + R_{f,out} \tag{12.24}$$

REFERENCES

Incropera, F.P., D.P. DeWitt. 2002. *Fundamentals of Heat and Mass Transfer*, 5th edition. John Willey & Sons, Inc., New York.

McQuiston, F.C., J.D. Parker. 1988. *Heating, Ventilating, and Air Conditioning – Analysis and Design*, 3rd edition. John Wiley & Sons, Inc., New York, 746 pp.

McQuiston, F.C., J.D. Parker, J.D. Spitler. 2005. *Heating, Ventilating, and Air Conditioning Analysis and Design*, 6th edition, John Wiley & Sons, Inc., New York, 615 pp.

Radermacher, R., Y. Hwang. 2000. *Vapor Compression Heat Pumps with Refrigerant Mixtures*, edited by Center for Environmental Energy Engineering, Department of Mechanical Engineering, University of Maryland, College Park, MD.

13 Moisture Condensers

13.1 INTRODUCTION

Heat pumps employed for wood drying are heating-only devices that use two main types of heat exchangers: (i) moist air-to-refrigerant evaporator(s) for sensible and latent heat recovery and (ii) refrigerant-to-air condenser(s) for sensible heat delivery to the wood drying enclosure.

Moist air-to-refrigerant evaporators (also called moisture condensers) are finned-tube heat exchangers where unmixed fluids (moist air and refrigerants) move roughly perpendicular (cross flow) to each other (Figure 13.1a).

Because a refrigerant convective flow boiling (vaporization) process occurs with heat capacity rate $\dot{C}_{p,refr} \rightarrow \infty$, it remains at a nearly uniform saturation temperature during vaporization ($T_{refr,sat} = const.$), while the temperature of the hot moist air decreases from $T_{moist\ air,in}$ to $T_{moist\ air,out}$ (Figure 13.1b).

As can be seen in the Mollier diagram (Figure 13.1c), the moist air is first cooled (process 1-*a*), and then the water vapor is condensed (process *a*-2) on the external finned tubes of moisture condensers.

13.2 CONSTRUCTION

The augmentation of heat transfer from condensing moisture (water vapor) to vaporizing (cooling) refrigerant is a challenging task because it can reduce the size of moisture condensers in a significant manner (Incropera and DeWitt 2002). The optimum performance is achieved by using heat pump evaporators with large face area, minimum depth and fin spacing, low airflow rates, as high evaporating temperature as possible, and optimum superheat across the coil.

Moisture condensers are compact, extended surface, direct expansion heat pump evaporators. They are provided with dense arrays of finned tubes (Figure 13.2a, b) in order to achieve very large (e.g., 2,700 m^2/m^3) heat transfer surface area per unit volume on the moist air-side.

The major components of air-to-refrigerant evaporators are tubes and tube sheets, fins, and drain pans. Fins are used on the air-side to increase the heat transfer area because the thermal resistance associated with the airflow is too high compared with that of the refrigerant. The refrigerant circuitry also plays a major role in the design, construction, and operation of moisture condensers.

The performances of air-to-refrigerant heat exchangers (moisture condensers) are affected by factors such as the type of refrigerant- and air-side heat transfer surfaces, fin spacing, tube pitch, depth row pitch, refrigerant circuitry, and air velocity distribution over the frontal surface of the heat exchanger (Stoecker 1998; Domanski et al. 2004).

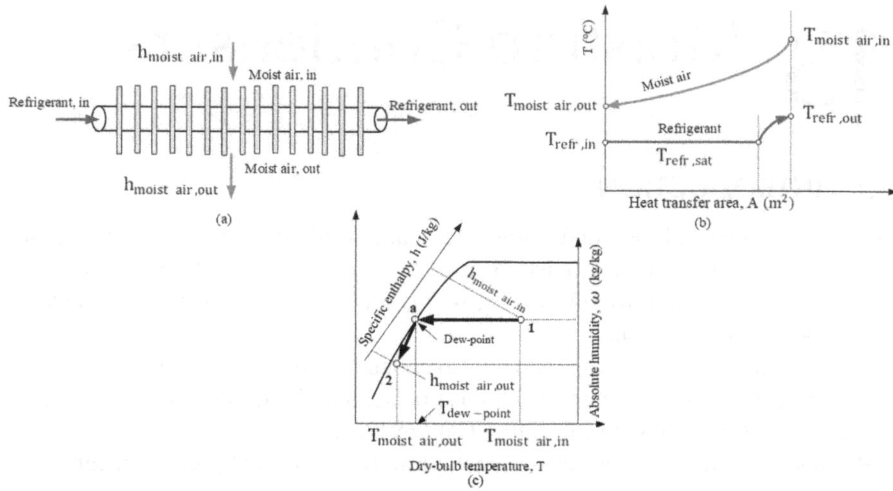

FIGURE 13.1 Cross-flow air-to-refrigerant finned-tube heat exchanger with phase change (heat pump evaporator or moisture condenser): (a) schematic; (b) temperature variations represented in a temperature–heat exchange area diagram; (c) moist air process represented in the Mollier diagram. h, specific enthalpy; in, inlet; out, outlet; T, temperature; refr, refrigerant; sat, saturation.

13.2.1 Tubes

Under dehumidifying conditions, the geometry of fin-and-tube moisture condensers of heat pump-assisted dryers (e.g., coil configuration, number of tube rows, columns, modules and circuits, tube outer diameters, and longitudinal tube pitch) drastically affects the performance of heat transfer capability, pressure drops on both refrigerant and air-sides, the refrigerant and air mass flow rates, drying time, and system overall energy consumption (Wang et al. 1999a, b; Erdem 2015).

The tubes of moisture condensers are generally circular and horizontally oriented. At each end of coils, heavy plates support the tubes by having holes through which the tubes pass (Figure 13.2a,b). The pattern of these holes defines whether the tubes are in-line or staggered. A coil with a staggered-fin pattern provides improved heat transfer but with a slight increase in air pressure drop (Stoecker 1998).

For moisture condensers with fewer number of tube rows, the surface heat transfer is more sensitive to wetting conditions at low airflow rates. Increasing in the number of rows did not always increase the heat transfer coefficients that depend on the Reynolds number. However, even with a small number of tube rows, relatively high heat transfer coefficients have been obtained because of turbulent vortex pattern associated with the number of tube rows (Rich 1975). As the number of rows in the moisture condenser increases, the heat transfer area also increases. In heat pump used for wood drying, increasing the number of rows from 2 to 4 reduces the energy consumption and drying time by 12% and 14%, respectively. However, further increasing the number of rows does not significantly change the energy consumption or drying time (Erdem 2015).

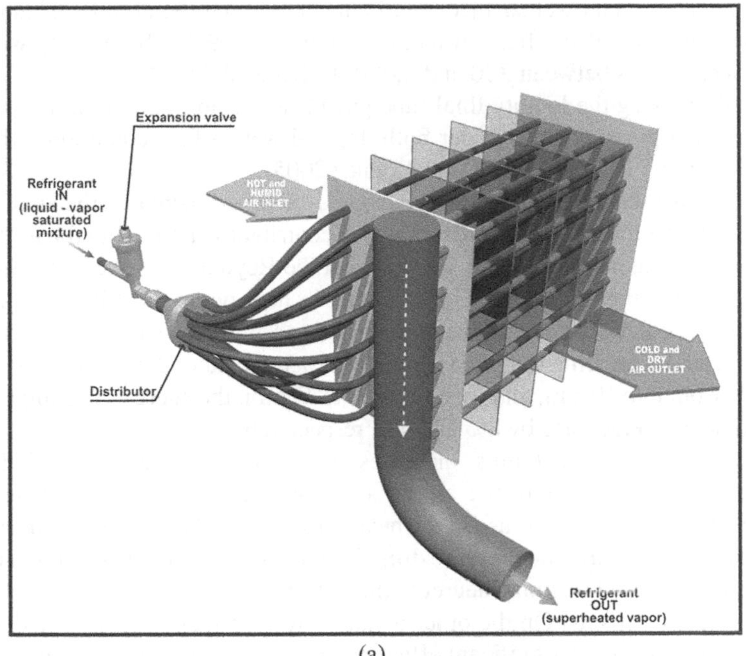

(a)

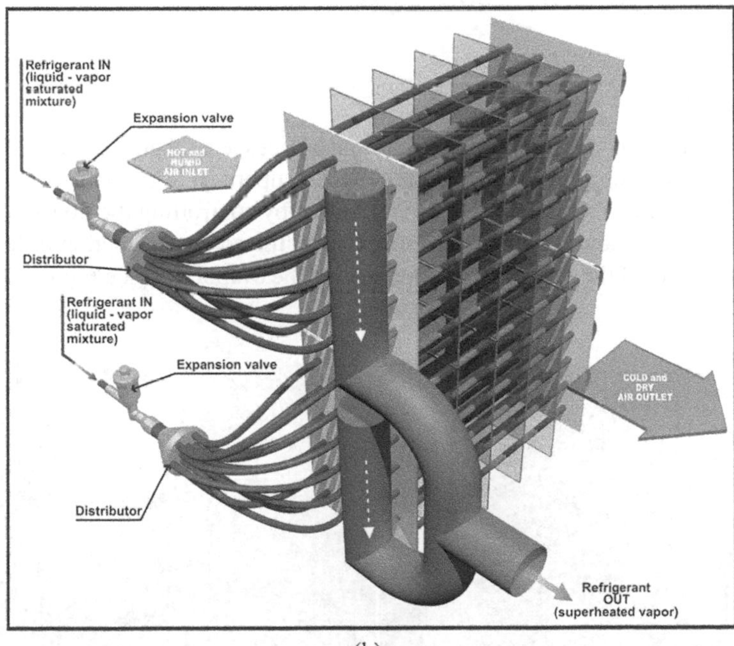

(b)

FIGURE 13.2 Schematic representations of typical finned-tube moisture condensers (heat pump evaporators): (a) with one refrigerant circuit; (b) with two refrigerant circuits (note: schematics not to scale).

Under both dry and wet surface conditions, increasing the number of tube rows decreases the value of the friction factor (Wang et al. 1999a, b), notably when the Reynolds number is between 300 and 2,000 (Halici et al. 2001).

Also, decreasing the longitudinal tube pitch leads to an increase in heat transfer coefficient and pressure drop under both dry and wet surface conditions when the Reynolds number is >500 (Halici and Taymaz 2005).

On the other hand, as the outer diameters of tubes increase, the total cross-sectional air-side finned area decreases which contributes at reducing the heat transfer in the evaporator (even the air velocity and the Reynolds number increase), and the total air-side pressure drop. For example, in heat pump used in wood drying, increasing the outer diameter of the tube from 6 to 8 mm may cause the energy consumption and drying time to increase by 9%. When the outer diameter of the tube is increased from 8 to 10 mm, and then from 10 to 12 mm, the energy consumption and drying time could increase by 8% and 4%, respectively.

As the number of columns increases, the cross-sectional area of the airflow increases, and therefore the air velocity and the Reynolds number decrease. However, because of the increase in the heat transfer area, the evaporator' heat transfer increases and the air-side pressure drop. In heat pump dryers, increasing the number of columns from 2 to 4 may decrease the energy consumption and drying time by 8% and 4%, respectively. On the other hand, using evaporators with more than four columns does not have a significant effect on the energy consumption and/or drying time (Erdem 2015).

13.2.2 Refrigerant Distribution

Moisture condensers use refrigerant expansion devices (e.g., thermostatic, electric or electronic valves) in order to regulate the overall superheat at the evaporators' exit, commonly between 4°C and 7°C. This is achieved by controlling the flow rate of saturated refrigerant (liquid–vapor mixture) entering the coil as a function of temperature and pressure of the refrigerant vapor leaving the evaporator (see Figures 13.2a, b and 13.3a).

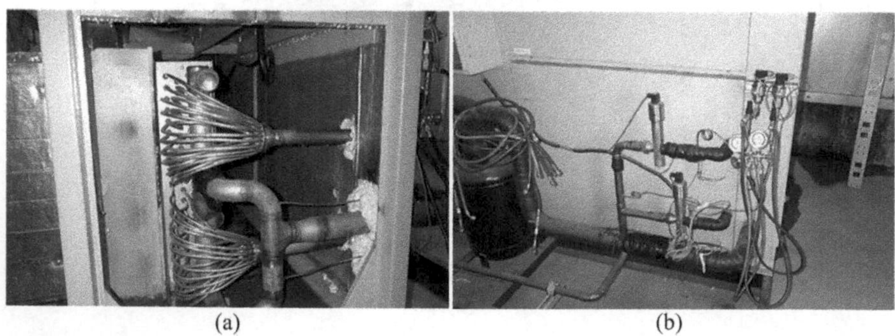

(a) (b)

FIGURE 13.3 (a) View of refrigerant distribution to a heat pump dryer evaporator with two distinct circuits; (b) view of electric expansion valves and compressor suction accumulator (Pictures reproduced with permission from NYLE Systems LLC).

After the expansion valves, refrigerant distributors are usually provided in order to blend the vapor/liquid mixture and evenly distribute it to each individual row of tubes, thus avoiding the degradation of the evaporator's heat transfer capacity.

Refrigerant distributors are provided inside with properly sized nozzles (or orifices) where the refrigerant velocity increases, providing a pressure drop and a "flash" temperature in the evaporator. The refrigerant tubes after the distributors (Figure 13.3b) are of the same lengths and diameters, ensuring a nearly identical pressure drop between the distributor and each coil circuit. Without such distributors, the vapor–liquid stream may not be properly mixed and distributed, and then some tube circuits may receive mostly vapor and others mostly liquid.

Inside moisture condensers, the refrigerant flows through parallel refrigerant circuits, which are designed to optimize between the benefit of improved refrigerant heat transfer and the penalty of refrigerant pressure drop. Ideally, the tube circuits inside the evaporator coils are designed to be identical to each other for consistent pressure drops and flash temperatures. There are an extremely large number of refrigerant circuitry possibilities aiming at providing maximum heat transfer rates and evaporator capacities. Optimized refrigerant circuit arrangements may reduce the needed heat transfer area by about 5% for the same evaporating capacity. The moisture condensers perform optimally when the superheat at individual circuit exits matches the desired overall superheat in the exit manifold (Lee and Domanski 1997; Choi et al. 2003).

Refrigerant-side maldistribution may lead to refrigerant dry-out in certain sections of the evaporator with significant reduction in cooling and dehumidifying capacity.

However, the evaporator heat transfer rates and capacities are much more sensitive to moist air maldistribution than to refrigerant maldistribution, causing further degradation of the coil cooling and dehumidifying capacity. Consequently, designing optimized refrigerant circuitries is particularly difficult if the airflow is not uniformly distributed over the coil surface (Chwalowski et al. 1989; Domanski et al. 2004).

In heat pump dryers that, generally, operate at medium and high vaporization temperatures, the issue of providing minimum refrigerant velocities inside each circuit of the evaporators in order to push through and out the refrigerant is not a concern, as it is for heat pump or refrigeration systems running at lower vaporizing temperatures (e.g., <0°C).

13.2.3 Fins

The use of (extended surface) fins is a way of increasing the effective heat transfer surface area without increasing the actual size of the heat pump evaporating (moisture condensing) coils. The main role of plate (or circular) fins is thus to maximize the external, moist air-side heat transfer area. This is because, in the case of moisture condensers, the major thermal resistance is on the air-side (e.g., 0.01305 m^2K/W), while on the refrigerant side, the thermal resistance is much lower (e.g., 0.000833 $m^2 \cdot K/W$, i.e., up to 20 times lower). This comparison suggests that, to increase the overall heat transfer rate attention should be directed toward increasing the moist air-side heat transfer surface, and not on that of the refrigerant side. That is because, if the heat transfer coefficient on the refrigerant side is, for example, doubled, the overall heat transfer coefficient would increase by only 3%.

Other methods to increase the overall heat transfer coefficient, thus the heat transfer rate, would be to increase the moist air-side average heat transfer coefficient ($\bar{h}_{moist\ air}$) by increasing the air velocity. However, by increasing the air velocity, additional electrical blower power is required that ultimately appears as a parasite cooling load; generally, up to 10%–20% of the heat removed by the evaporator come from the blower and its motor. That means that the air velocity which provides optimum air-side heat transfer coefficients requires a reasonably sized blower and motor (Stoecker 1998). According to one of fans' laws, the blower power requirement increases exponentially as airflow rate increases. For example, the blower input electrical power increases more than eight times as airflow rate increases from 0.4 to 1 kg/s.

The moisture condenser coils, as shown in Figure 13.2a, b, are equipped with fins formed from flat metal plates that are then punched on the tubes inserted in the holes. The fins must form good bonds to the tubes because, otherwise, there will be additional heat transfer resistance through air gaps. For that when the tubes are in position, they are expanded against the collar of the fins either hydraulically or mechanically to provide good thermal contact with the fins (Rich 1973).

As the fin pitch of heat pump evaporators increases, the air-side heat transfer coefficient, for a given velocity, increases only slightly. For four-row coils having 7 mm-diameter tubes, the effect of varying fin pitch on the air-side heat transfer performance and friction characteristics is negligible, that is, independent of fin pitch.

For plain plate fin configurations ranging from 8 to 14 fins per inch (1 in. = 25.4 mm), the effect of longitudinal tube pitch on the air-side is negligible for both the air-side heat transfer and pressure drop. However, the heat transfer performance increases with reduced fin pitch (Rich 1973; Wang et al. 1999). For air-side Reynolds numbers >1,000, the effect of fin pitch on the air-side friction pressure drop is negligibly small.

As both fin depth and tube spacing increase, with all other variables constant, the air-side heat transfer coefficient decreases. However, by reducing the tube spacing and the tube diameter, an increase in the air-side heat transfer coefficient occurs (Shepherd 1956).

13.2.4 Air Distribution

Airflow rate, one of the most important parameters of heat pump dryers, should be matched with every component of the drying system, and thus must be carefully determined.

Evaporator air-side velocities may vary due to the geometry of the heat exchanger installation, nearness of the blower, blockage of the air filter, and other factors. Nonuniform airflow can cause some circuits to have excessive superheat, while others may remain two-phase at the evaporator exit. In such situations, some circuits inefficiently use coil area when transferring heat with superheated vapor instead of two-phase refrigerant (Choi et al. 2003).

Airflow maldistribution (together with the refrigerant inside tubes) could lead to significant loss in the heat transfer capacity of the compact evaporators (Fagan 1980; Kandlikar 1990a, b; Kirby et al. 1998). The evaporator capacity degradation due to both nonuniform refrigerant and airflow distribution can be as much as 30%.

13.2.5 Water Draining

Water condensed from moist air either forms a thin layer, or appears as discrete droplets on the finned-heat exchanger surfaces. The water film or droplets are then swept from the surface due to gravity and/or drag forces. The condensed water is sometimes entrained in the air stream because of carryover resulting in undesirable moisture blowout, a situation highly undesirable that must be avoided. The effect of the retained water on the heat transfer rate is found to be small, while its effect on the pressure drop may be significant.

Sometimes, the air passages may become partly blocked resulting in considerable increase in pressure drop. With the gradual reduction in the free flow area, the increased pressure drop causes a cyclic but sudden outburst of water from the blocked passages, a cyclic phenomenon. To avoid the problem, proper drainage path for condensed water must be provided (Kandlikar 1990a, b). For that, the channels on the moist air-side are specially designed to promote the drainage of the condensed water from the moisture condenser, thereby reducing the blowout of water droplets in the outlet air stream (Nakasawa et al. 1984).

Recent developments aiming at improving the water film drainage include coating the fin surface with a thick coating of a hydrophilic film (~1 μ). The primary effect of the coating is to reduce or even avoid the water holdup in the evaporator, especially in compact heat exchangers with narrow passages. It may result in up to 30% reduction in the air-side pressure drop and fan power requirements, and up to 3% increase in the heat transfer rate (Ito et al. 1977).

Finally, all moisture condensers must be equipped with drain pans where the condensate is collected and drained to convenient destinations.

13.2.6 Materials

In the past, copper was most commonly used for both fins' and tubes' construction of heat pump evaporators with halocarbon refrigerants, mainly because it had a high thermal conductivity and was relatively inexpensive.

With rising copper costs, the industry widely adopted aluminum for both tubes and plain fins. Aluminum tubes can be extruded with a wide range of passage sizes and tube expansion methods that provide good fin–tube thermal contact for a wide range of fin thicknesses. The degradation in thermal performance, due to the lower thermal conductivity of aluminum, was offset by using more of a less-expensive material.

Typical combinations of tube/fin materials are (Stoecker 1998) as follows: (i) copper tube/aluminum fin or aluminum tube/aluminum fin for moisture condensers operating with halocarbon refrigerants; (ii) carbon steel tube/carbon steel fin for evaporator coils using ammonia inside the tubes; (iv) stainless steel tube/stainless steel fin when special cleaning provisions are required on the air-side.

The advances in fin-making technologies allow the use of very thin fins, but in spite of this advance, most heat pump evaporators currently rely on round-tube designs with plain fins, similar to flat plates (see Figure 13.2a, b). In residential and commercial heat pump evaporating coils with thin aluminum fins, the fin spacing may be 470 per meter, while in industrial heat pumps, they are usually built with 118 or 158 fins per meter (Stoecker 1998).

13.3 FROSTING AND OTHER UNDESIRABLE ISSUES

For many applications, as in heat pump drying processes, if the drying temperature drops, let us say, below approximately 15°C, the heat pump evaporators can be exposed to moisture freezing conditions. That means that the surfaces of evaporator coils may operate at temperatures below the dew-point temperature of the air. In such a situation, the moisture from the air first condenses to liquid water, then freezes to ice. In this case, frost or ice buildup may form on the evaporator coils blocking airflow through the coil while the heat pump may continue to run, and thus affecting the heat pump operating parameters and overall drying (cooling and dehumidification) performances.

If ice formation is extreme, nearly all of the airflow across the coil is blocked and the heat pump system does not cool and dehumidify the drying air.

The frost has detrimental effects such as (Stoecker 1998) (i) increased resistance to heat transfer, (ii) restriction of airflow, and (iii) irreversible damage to the heat pump's compressor.

As frost accumulates, the airflow rate decreases and the pressure drop through the evaporator increases if the fan input power is not progressively increased. The reductions of airflow rate and air velocity reduce the mean overall heat transfer coefficient, one the most serious effect.

As a basic rule, any heat pump used as a dehumidifier in drying systems must be designed with non-frosting evaporator coil(s), that is, they must never form frost or build ice. In other words, when heat pump-assisted dryers are used, one of the fundamental design and operational aspects is the minimum allowed evaporating temperature and airflow rate of the drying air at the heat pump's evaporator inlet.

In industrial heat pump-assisted drying processes, in addition to inappropriate drying temperature setting (i.e., too low), frost or ice buildup may also occur on the evaporator coils because of the following: (i) improper refrigerant charge, (ii) uncontrolled refrigerant leakage, (iii) inadequate (maldistribution) airflow across the moisture condenser coil, (iv) incorrect sizing or inappropriate operation of thermostatic expansion valve(s), (v) other heat pump control (e.g., suction and differential pressure switches) defects, (vi) dirt air filters, and (vii) coil damaged fins.

The potential causes of frost or ice buildup on the heat pumps' finned-refrigerant evaporators (moisture condensers) can be summarized as follows:

a. Too low temperature and/or flow rate of drying air through the evaporator coil.
b. Defective operation of expansion valves.
c. Dirty or residual debris (such as soot, organic particles from the wood dried products, mold spores, and unusual materials or substances, such as

fiberglass insulation, and large trash fragments like paper or leaves) blocking airflow over the evaporator coil by forming a gray mat on the fins of the evaporator coil; debris stick particularly quickly to this surface because of the combination of close spacing of the cooling fins and the fact that condensate forming on the coil keeps the surface damp; in the case of dirty heat transfer surfaces, the evaporator coils must be cleaned in-place, an operation that often requires cutting refrigerant lines, removal of the coil and other components for cleaning, and reinstallation, pulling a vacuum on the refrigerant lines, and recharge with refrigerant; such service and repair may involve significant expense, although there are some simpler infield cleaning methods using foams and sprays.

d. Dirty, clogged, or blocked air filters that can lead to lost cooling and dehumidifying capacity of moisture condensers, because they reduce the airflow rate through the heat pump refrigerant evaporator.
e. Damaged evaporator coil fins over more than 10% of the coil surface; small areas of damaged cooling fins can be straightened and cleaned up using a cooling coil comb, but the evaporator coils with extensive physical damage need to be replaced.
f. If an evaporator coil is damaged or leaky due to corrosion, it must be replaced, a hard and costly work.
g. Heat pump's blower defects or dirty squirrel cage, or malfunctioning.

13.4 MOISTURE CONDENSATION

Condensation of moisture (water vapor) on the outside heat transfer surfaces of heat pump evaporators (moisture condensers) takes place when the temperature of pure water vapor is reduced below its saturation temperature at the same pressure by contact with the heat exchanger cool surfaces. In this way, the saturated water vapor is converted into water liquid (condense) by releasing its sensible heat and enthalpy (latent heat) of condensation. The total (sensible and latent) condensing heat is thus transferred to the evaporator external surface. This cooling and dehumidifying processes is complex because of the simultaneous heat and mass transfer particular characteristics. Subsequently, the heat from the evaporator external surface is transferred through the pipe wall to the refrigerant flowing inside the tube (that vaporizes) and carried away (Incropera and DeWitt 2002). Thus, the purpose of heat pump evaporators is to recover and transfer heat from the hot moist air, making it cooler and less humid.

13.4.1 MOISTURE CONDENSATION ON HORIZONTAL PLAIN TUBES

Condensation heat transfer of moisture on single or multiple (bank) circular tubes is influenced by several factors as the tube layout (e.g., horizontal, vertical, plain finned, in-line staggered, number of tube rows, pitch-to-diameter ratio, etc.), vapor (nature, flow direction, velocity) and condensate (flow mode, rate of inundation, freezing, etc.).

Horizontal orientation of plain tubes is more common when pure vapor of fluids with high thermal conductivity and surface tension higher than 0.05 Pa m (as water vapor) is to be condensed (Hiroshi 2017) (Figure 13.4).

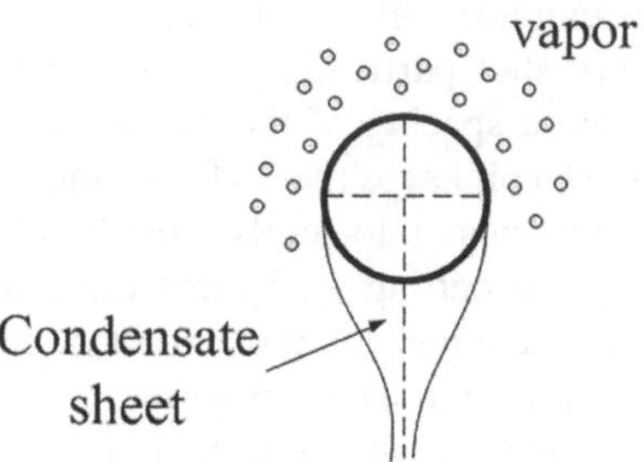

FIGURE 13.4 Moisture condensation on a single horizontal plain tube.

The laminar film condensation of vapor on horizontal tubes (as well as on vertical plane surfaces) (see Section 13.4.4) can be investigated with the following assumptions (Rose 1988a, b): (i) pure saturated vapor at uniform saturation temperature; (ii) uniform wall temperature, or constant condensing heat flux; (iii) motion of condensate controlled solely by gravity and viscosity; (iv) negligible effects of vapor drag and acceleration of condensate; (v) negligible shear stress on the condensate surface due to the vapor; (vii) heat transfer across the condensate film occurs by pure conduction only; (vii) constant properties of film condensate; (vii) condensate film thickness is much less than the tube radius.

Despite the involvement of many factors in determining the pure vapor-to-surface heat transfer coefficient during condensation and the complexity associated with the film condensation, even for simpler geometries, complete solutions to the problem of laminar film condensation based on computer simulations have shown that the Nusselt assumptions lead to accurate results for the practical range of operating conditions (Rose 1988).

For example, the condensation of pure vapor on integral finned-tube surfaces with the condensate evacuated by gravity without subcooling, the solution consisted of the superposition of the condensation on the horizontal tube surface and the condensation on a vertical plate, both using the Nusselt solution for laminar film condensation for these geometries (Beatty and Katz 1948).

For horizontal integral tubes with fins having rectangular cross sections, the prediction of film condensation on the flooded and unflooded surfaces, based on Nusselt assumptions and including the effect of fin efficiency, provide pertinent results (Adamek and Webb 1990).

In the case of water vapor (moisture) condensation over a single horizontal tube (Figure 13.4), the mean Nusselt number ($\overline{Nu}$) (equation 13.1) and the mean specific heat transfer coefficient ($\bar{h}_d$, W/m$^2\cdot$K) (equation 13.2), respectively, can be expressed as follows (Dhir et al. 1971; Rose 1988a, b):

$$\overline{Nu}_d = \frac{\bar{h}_d * d}{k} = 0.729\left[\frac{\rho * \Delta\rho * g * h_{fg} * d^3}{\eta * k * (T_{sat} - T_s)}\right]^{1/4} \tag{13.1}$$

$$\overline{h}_d = 0.729\left[\frac{\rho_f * (\rho_f - \rho_g) * g * h_{fg} * k_f^3}{\mu_f * d * (T_{sat} - T_s)}\right]^{1/4} \tag{13.2}$$

where

ρ_f is the liquid density (kg/m^3)
ρ_g is the vapor density (kg/m^3)
g is the gravitational acceleration (m/s^2)
h_{fg} is the vapor enthalpy (latent heat) of condensation evaluated at the vapor saturation temperature (J/kg)
k_f is the liquid thermal conductivity (W/m · K)
μ_f is the liquid dynamic viscosity (Pa · s = kg/s · m)
d is the tube external diameter (m)
T_{sat} is the vapor saturation temperature (K)
T_s is the outside cool surface temperature (K)

In Equations 13.1 and 13.2, the characteristic length is the tube external diameter (d), and the reference temperature (T_{ref}) for the evaluation of properties of condensate (liquid) film is (Rose 1988a, b) as follows:

$$T_{ref} = \frac{1}{3}T_{sat} + \frac{2}{3}T_{tube} \tag{13.3}$$

where

T_{sat} is the saturated temperature of the bulk vapor (K)
T_{tube} is the temperature of the tube outside surface (K)

By including the advection effects, Rohsenow (1956) proposed to improve the previous correlations (Equations 13.1 and 13.2) by replacing h_{fg} by:

$$h_{fg}^* = h_{fg} + 0.68 * \overline{c}_p (T_{sat} - T_{tube}) \tag{13.4}$$

where

$\overline{c}_p$ is the isobaric specific heat capacity of the condensate (J/kg · K)

With this correction, for the horizontal tube case, the coefficient in equations 13.1 and 13.2 becomes 0.695:

$$\overline{h}_d = 0.695\left[\frac{\rho_f * (\rho_f - \rho_g) * g * h_{fg}^* * k_f^3}{\mu_f * d * (T_{sat} - T_{tube})}\right]^{1/4} \tag{13.5}$$

The average condensation rate per unit length of a horizontal single tube (kg/s · m) may be determined from the following relation:

$$\dot{m}_1 = \frac{\dot{q}}{h_{fg}} = \frac{\overline{h}_d (\pi d) T_{sat} - T_{tube}}{h_{fg}} \tag{13.6}$$

where

$\dot{q}$ is the heat transfer rate to the unit length tube surface (W/m)

h_{fg} is the water vapor enthalpy (latent heat) of condensation (J/kg)

$\bar{h}_d$ is the average heat transfer coefficient given by equations 13.2 or 13.5 ($W/m^2 \cdot K$)

d is the tube external diameter (m)

T_{sat} is the water vapor saturation temperature (K)

T_{tube} is the tube surface temperature (K)

In the case of water vapor condensation on a column of N horizontal circular tubes (Figure 13.5a), condensation is the subject of the combined effects of vapor shear and falling (inundation) condensate from upper tubes. The saturated vapor may be stationary or flowing downward, horizontal, upward, or in any other direction. The mode of condensate inundation depends on the inundation rate and vapor velocity. At low vapor velocities, condensate drains in discrete drops, then in condensate columns, and then in a condensate sheet as the condensate rate increases. At high vapor velocity, the condensate leaving the tube is disintegrated into small drops and impinges on the other tubes. For downward flow of vapor, the condensate inundation rate depends only on the condensation rate at the upper rows.

Both the falling of condensate under gravity from higher tubes and the mode of condensate drainage influence the performance of multi-tube condensers. This is due to the fact that condensate dropping off the first (top) row tube joins the condensate of the tube, which lies in the second row, thereby increasing the thickness of the condensate layer on the second row tube. This also changes the velocity profile of

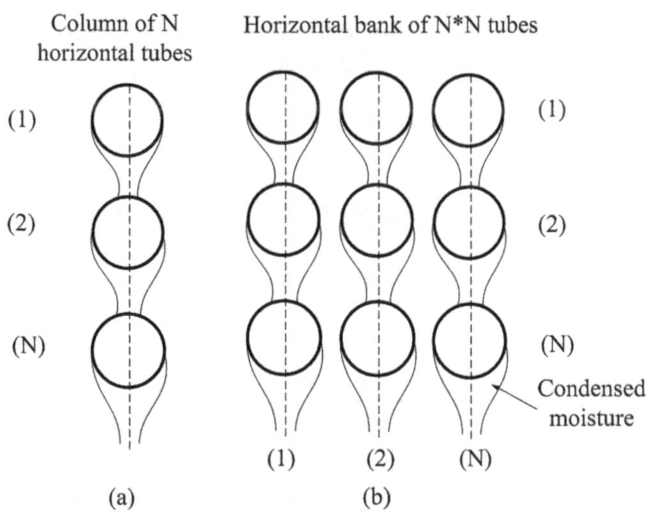

FIGURE 13.5 Schematic of moisture condensation with continuous condensate sheet: (a) on a column of N horizontal circular tubes; (b) on a vertical tier (N * N) of horizontal circular tubes.

the condensate layer thereon. The combined condensate of the second row tube, after dropping off, joins that of the third row tube, thereby changing its thickness and velocity profile. This trend continues further from the third row tube to the fourth row tube and so on (Nusselt 1916).

For a vertical tier of $N * N$ horizontal tubes with negligible vapor velocity (Figure 13.5b), and by assuming that the condensate drains (falls) in a continuous sheet, and that heat transfer to the condensate sheet from a tube onto the tube below between the tubes and the momentum gain (as the sheet falls freely under gravity), are neglected, the average convection heat transfer coefficient may be expressed as follows (Rose 1988a, b; Incropera and DeWitt 2002):

$$\bar{h}_d = 0.729\left[\frac{\rho_f * (\rho_f - \rho_g) * g * h_{fg} * k_f^3}{N * \mu_f * d * (T_{sat} - T_s)}\right]^{1/4} \tag{13.7}$$

or

$$\bar{h}_{d,N} = h_{d,1} * N^{-\frac{1}{4}} \tag{13.8}$$

where $h_{d,1}$ is the heat transfer coefficient for the top (upper) tube given by equations 13.2 or 13.5 ($W/m^2 \cdot K$).

Reduction of the average convection coefficient with increasing number of tubes (N) can be attributed to an increase in the average film thickness for each successive tube.

Under the same assumptions, for a complete array of $N * N$ horizontal circular tubes, the condensate mass flow rate per unit length will be as follows:

$$\dot{m}_N = N^2 * \dot{m}_1 \tag{13.9}$$

where $\dot{m}_1$ is the condensate mass flow rate per unit length ($kg/s \cdot m$) of the first row of tubes (given by equation 13.6).

13.4.2 Moisture Condensation on Vertical Surfaces

If a pure (i.e., without non-condensable gases), single-component, saturated (at the saturated temperature T_{sat}) passive vapor condenses on a clean, uncontaminated vertical plate surface at uniform temperature T_s, a film condensation will develop without ripples from the leading edge (Figure 13.6).

The film condensate wets the cool surface of the vertical plate providing a thermal resistance to heat transfer between the saturated vapor and the surface of the vertical plate. This thermal resistance increases with the condensate thickness in the vertical flow direction.

In film-wise condensation process, the drops initially formed quickly coalesce to produce a continuous film of liquid on the surface through which heat is transferred to condense more liquid. This means that the liquid (condensate) film originating

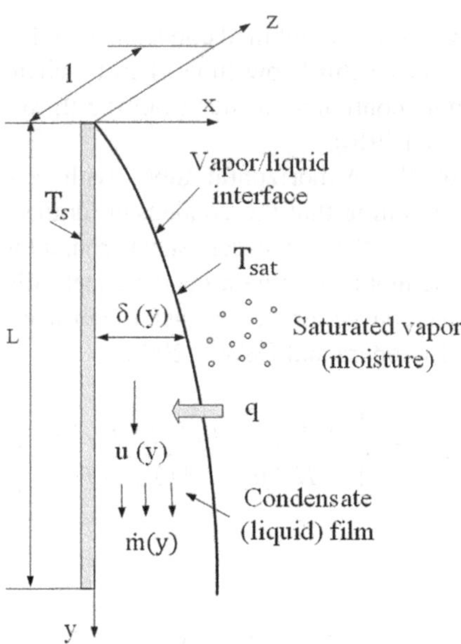

FIGURE 13.6 Schematic of film condensation of a quiescent saturated vapor (moisture) on a vertical plate surface. $\dot{m}$, condensate flow rate; $\dot{q}$, heat flux density; T_{sat}, temperature of saturated vapor; T_s, temperature of vertical plate surface; u, condensate velocity.

at the top of the vertical plate, flows continuously downward under the influence of gravity, adding additional, new condensate (Incropera and DeWitt 2002).

In addition to the assumptions made for condensation over horizontal tubes (Nusselt 1916), more assumptions are made for film condensation on vertical plates (Incropera and DeWitt 2002) (Figure 13.6) as follows: (i) flow of liquid film is laminar with constant properties; (ii) gravity is the only external force acting on the film; (iii) the adjoining vapor is stagnant and does not exert drag on the film; (iv) the sensible cooling of the film is negligible with respect to the latent heat; (v) the saturation temperature of the interface is that of the vapor pressure curve of the fluid; (vi) heat transfer from vapor to the liquid occurs only by condensation at the interface; (vii) negligible shear stress at the liquid–vapor interface, so there is no need to consider the vapor velocity or thermal boundary layers; (viii) negligible momentum and energy transfer by advection in the condensate film because of low velocities associated with the film; (ix) the heat transfer across the laminar condensation film occurs only by conduction with linear temperature distribution and conductive thermal resistance from the interface to the wall.

As can be seen in Figure 13.6, the thickness $\delta(y)$ and the condensate mass flow rate $\dot{m}(y)$ increase with increasing y because of continuous condensation at the liquid–vapor interface, which is at the saturation temperature (T_{sat}). There is then heat transfer from this interface through the film to the surface, which is maintained at $T_s < T_{sat}$. In the most general case, the vapor may be superheated and may be part of a mixture containing one or more non-condensable gase (Kandlikar 1990; Incropera and DeWitt 2002).

The velocity component at any location y in the condensate film is as follows:

$$u(y)=\frac{g*(\rho_f-\rho_g)*\delta^2}{\mu_f}\left[\frac{y}{\delta}-\frac{1}{2}\left(\frac{y}{\delta}\right)^2\right] \tag{13.10}$$

where $\delta(y)$ is the local thickness of the film at any distance y from the top of the vertical plate surface which can be expressed as follows (Incropera and DeWitt 2002):

$$\delta(y)=\left[\frac{4k_f*\mu_f(T_{sat}-T_s)*y}{g*\rho_f(\rho_f-\rho_g)*h_{fg}}\right]^{1/4} \tag{13.11}$$

The film thickness is similar to the conduction length through a solid of the same thickness, directly proportional to the temperature difference ($T_{sat}-T_s$), such that a larger temperature difference results in a higher condensation rate.

With the condensate mass flow given by $\dot{m}(y)=\rho_f*\bar{u}(y)*l*\delta$, the Reynolds number may be expressed as follows:

$$Re_\delta=\frac{4*\dot{m}(y)}{\mu_f*l}=\frac{4*\rho_f*\bar{u}(y)*\delta}{\mu_f} \tag{13.12}$$

where

$\bar{u}$ is the film average velocity (m/s)
δ is the film thickness, that is the characteristic length (m)
L is the width of the vertical plate (m)

For $Re_\delta \leq 30$, the film is wave-free laminar, for which the Reynolds number can be expressed as follows:

$$Re_\delta=\frac{4g*\rho_f*(\rho_f-\rho_g)\delta^3}{3\mu_f^2} \tag{13.13}$$

For higher Re_δ, ripples or waves form on the condensate film. At $Re_\delta \approx 1800$, the transition from laminar to turbulent flow is complete (Incropera and DeWitt 2002).

The mean heat transfer coefficient for the condensation of vapor on the entire cold vertical plate is (Collier and Thome 1994; Incropera and DeWitt 2002) as follows:

$$\bar{h}_L=0.943\left[\frac{g*\rho_f*(\rho_f-\rho_g)*k_f^3*h_{fg}}{\mu_f(T_{sat}-T_s)*L}\right]^{1/4} \tag{13.14}$$

where

μ_f is liquid dynamic viscosity (Pa·s)
k is the liquid thermal conductivity (W/m·K)
ρ_f is the liquid density (kg/m^3)

ρ_g is the vapor density (kg/m^3)
$\Delta\rho$ is the difference between liquid and vapor density (kg/m^3)
$T_{sat}-T_s$ is the vapor-to-condensing surface temperature difference (°C)
L is the plate height (m)
g is the gravitational acceleration (m/s^2)

Using this equation, all liquid properties should be evaluated at the film temperature $T_f = \frac{T_{sat}+T_s}{2}$, while h_{fg} should be evaluated at T_{sat}.

Consequently, the average Nusselt number then has the form

$$\overline{Nu_L} = \frac{\bar{h}_L * L}{k_f} = 0.943\left[\frac{g*\rho_f*(\rho_f-\rho_g)*h_{fg}*L^3}{\mu_f*k_f*(T_{sat}-T_s)}\right]^{1/4} \tag{13.15}$$

Condensation at vertical plate surfaces introduces a negative velocity that causes the velocity gradient at the wall to increase, thereby reducing the boundary layer thickness and increasing the wall shear stress. Similar effects are seen for heat transfer resulting in an increase in the wet surface heat transfer coefficient (h_{wet}) during dehumidification over the dry surface heat transfer coefficient (h_{dry}) (Kandlikar 1990a, b).

The condensate mass flow rate per unit width (kg/sm) is expressed as follows (Incropera and DeWitt 2002):

$$\dot{m}(y) = \frac{g*\rho_f*(\rho_f-\rho_g)*\delta^3}{3\mu_f} \tag{13.16}$$

The total heat transfer rate ($\dot{Q}$) to the vertical surface may be obtained in the following form of Newton's law of heat transfer by convection:

$$\dot{Q} = \bar{h}_L * A*(T_{sat}-T_s) \tag{13.17}$$

where

Q is the total heat transfer rate (W)
$\bar{h}_L$ is the mean heat transfer coefficient given by equation 13.14 (W/m$^2\cdot$K)
A is the heat transfer surface (m^2)

The total condensate mass rate (kg/s) may then be determined from the following relation:

$$\dot{m} = \frac{\dot{Q}}{h_{fg}} = \frac{\bar{h}_L * A*(T_{sat}-T_s)}{h_{fg}} \tag{13.18}$$

where

$\dot{m}$ is the total condensate mass rate (kg/s)
h_{fg} is the enthalpy (latent heat) of condensation (J/kg)

13.4.3 Moisture Condensation on Integral Finned Tubes

The augmentation in heat transfer rate from the condensing moisture to the in-tube evaporating refrigerants can reduce the size and costs of heat pumps' evaporators in a significant manner.

The heat transfer rate from the condensing moisture (water vapor) to the heat exchange surface (wall) of moisture condensers can be improved by the following (Kandlikar 1990a, b; Rathod Pravin et al. 2011): (i) lowering the temperature of the heat transfer surface; (ii) enhancing the condensing side heat transfer coefficient at constant vapor to tube surface temperature difference; (iii) providing double enhancement methods, that is on both air (by using tubes with external rectangular or circular integral fins) and refrigerant (by enhancing the wall roughness or using internal fins) sides.

13.4.3.1 Schematic of a Moisture Condenser

In heat pump drying industry, enhanced finned tubes with continuous rectangular (or circular) fins (similar to vertical plates, see Figure 13.6) are extensively used for air-to-refrigerant evaporators (Incropera and DeWitt 2002) (also called moisture condensers or dehumidifying coils). Such devices operate under conditions where only sensible (as in refrigerant-to-air condensers) and both sensible and latent heat transfer (as in air-to-refrigerant evaporators) occur simultaneously (Thome 2004; Rathod Pravin et al. 2011).

Extended (or finned) surfaces may take on many forms ranging from the simple rectangular plate (fin) of uniform cross section (e.g., height l, thickness y, width s; see Figure 13.7) to complex patterns attached to tubes. The heat exchanger surface on the moist air-side consists of *prime* surface and *finned* surface (Figure 13.7). The tubes and fins are spaced quite apart so that the moisture condensed from the air on the coil external surface readily drains off without obstructing the flow of air, and thus the air-side pressure drop is relatively small. As shown in Section 13.2.3, for optimum design of refrigerant finned-tube evaporators, fin spacing is one of the most important geometric variable.

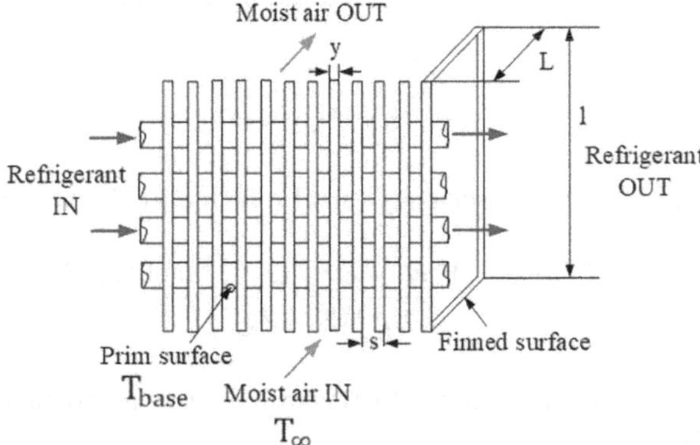

FIGURE 13.7 Schematic of a continuous plate fin–tube moisture condenser (note: drawing not to scale). l, fins' height; L, fins' length; s, fins' thickness; T_∞, air bulk temperature; T_{base}, base tube's temperature.

The moisture removed (vaporized) from the lumber stack(s) inside the wood drying enclosures condenses under both laminar and turbulent flow regimes on vertical, continuous fins of heat pump evaporators, while the drying air flows parallel to the fins and normal to the horizontal tubes (with typical diameters of around 20 mm). Initially, the liquid drops formed quickly coalesce to produce a continuous film-wise of liquid on the surfaces of vertical fins through which heat is transferred from the condensate to the heat pump evaporator. Heat transfer resistances associated with the tube and fin are (i) the resistance due to tube itself and (ii) the resistance due to the contact between the fin and the tube.

The flow of the condensate on finned tubes is three-dimensional and involves effects of gravity, shear stress (surface tension) at the condensate surface (due to vapor velocity), temperature difference at the liquid–vapor interface (essentially confined to a region in the vapor having a thickness of a few mean free paths of the vapor molecules and where equilibrium conditions cannot prevail), and inundation (i.e., condensate from higher or upstream surfaces impinging on lower or downstream surfaces) (Niknejab and Rose 1981).

13.4.3.2 Fin Efficiency

The finned-tube moisture condensers generally operate under dry and wet (or frosted) conditions with huge impacts on their overall thermal performances (Oskarsson et al. 1990a, b). Fin thermal efficiency must, thus, be separately calculated for both dry and wet surfaces that simultaneously occur with heat pump moisture condensers.

13.4.3.2.1 Dry Surfaces

Because fins do not have uniform temperatures, their fin efficiency, used to characterize the heat transfer rate, can be defined as follows (McQuiston et al. 2005):

$$\eta_{fin} = \frac{\textit{Actual heat transfer}}{\textit{Heat transfer with fins all at the base temperature } T_{base}} \tag{13.19}$$

where

T_{base} is the temperature of the tube base (°C) (see Figure 13.7).

The following definition of fin efficiency can also be used (McQuiston et al. 2005):

$$\eta_{fin} = \frac{(T_\infty - T_{base}) * m * k_{fin} * A_{fin,cross} * \tanh(m * l)}{\bar{h} * A * (T_\infty - T_{base})} \tag{13.20}$$

Or, if a straight fin of height l (Figure 13.7) is used and an adiabatic tip is assumed, (Incropera and DeWitt 2002):

$$\eta_{fin} = \frac{\tanh(ml)}{ml} \tag{13.21}$$

where

T_∞ is the average bulk air temperature (°C)

T_{base} is the base temperature (°C)

$\bar{h}$ is the air-side average heat transfer coefficient (W/m$^2\cdot$K) (assumed constant to define the parameter m)

$$m = \left(\frac{\bar{h} * P}{k_{fin} * A_{fin,cross}} \right)^{0.5}$$

l is the fin height (m)

P is the fin perimeter (m)

k_{fin} is the fin thermal conductivity (W/m$\cdot$K)

$A_{fin,\,cross}$ is the fin cross-sectional area (m^2)

Since the base on which the fins are mounted also transfers heat, another parameter, similar to fin efficiency (called the surface effectiveness), is defined.

The overall surface efficiency ($\bar{\eta}_{surface}$), a parameter indicating how effectively the entire surface is utilized for the heat transfer, is defined as the actual total heat transfer rate to the fin and base ($\dot{Q}_{total}$) divided by the maximum possible heat transfer rate to the fin and base ($\dot{Q}_{max}$) that would result if the entire fin surface, as well as the exposed base, are maintained at the base temperature (T_{base}) (McQuiston and Parker 1994; Incropera and DeWitt 2002):

$$\bar{\eta}_{surface} = \frac{\textit{Actual heat transfer for fin and base}}{\textit{Heat transfer for fin and base when the fin is at the base temperature } T_{base}}$$

or

$$\bar{\eta}_{surface} = \frac{\dot{Q}_{total}}{\dot{Q}_{max}} = \frac{\dot{Q}_{total}}{\bar{h} * A_{total} * (T_\infty - T_{base})}$$

$$= \frac{\bar{h} * A_{base} * (T_\infty - T_{base}) + \bar{h} * A_{fin} * \eta_{fin} * (T_\infty - T_{base})}{\bar{h} * A_{total} * (T_\infty - T_{base})} \quad (13.22)$$

where

$\bar{\eta}_{surface}$ is the overall surface efficiency (–)

$\dot{Q}_{total}$ is the total heat flow rate from the total surface area A_{total} associated with both the fins and the exposed portion of the base (often termed the prime surface) (W)

$\dot{Q}_{max}$ is the maximum possible heat transfer rate to the fin and base that would result if the entire fin surface, as well as the exposed base, are maintained at the base temperature T_{base} (W)

$\bar{h}$ is the air-side, average convective, heat transfer coefficient (W/m$^2\cdot$K)

A_{base} is the base heat transfer area (m^2)

A_{fin} is the fin heat transfer area (m^2)

$A_{total} = A_{base} + A_{fin}$ is the total (base plus fins) heat transfer surface (m^2)

T_∞ is the air average bulk temperature (°C)
T_{base} is the base temperature (°C)

The fin base temperature may be different than the prime surface temperature due to the contact resistance at the fin base. In contrast to the fin efficiency (η_{fin}), which characterizes the performance of a single fin, the overall surface efficiency ($\overline{\eta}_{surface}$) characterizes an array of fins and the base surface to which they are attached (Incropera and DeWitt 2002).

Assuming the air-side, average convective, heat transfer coefficient ($\overline{h}$) constant over the fin and base, the overall surface efficiency can be expressed as follows:

$$\overline{\eta}_{surface} = \frac{A_{base} + \eta_{fin} * A_{fin}}{A_{base} + A_{fin}} = \frac{1 + \dfrac{A_{fin}}{A_{base}}\eta_{fin}}{1 + \dfrac{A_{fin}}{A_{base}}} \quad (13.23)$$

or

$$\overline{\eta}_{surface} = 1 - \frac{A_{fin}}{A_{base} + A_{fin}}\left(1 - \eta_{fin}\right) \quad (13.24)$$

where
A_{fin} is the surface area of a single fin (m^2)
$\dfrac{A_{fin}}{A_{base} + A_{fin}}$ is the ratio of fin surface area to the total surface area (–)
$A_{tot} = A_{base} + A_{fin}$ is the total (base plus fin) surface area (m^2)
η_{fin} is the fin efficiency (–)

The rate of heat transfer becomes (W):

$$\dot{Q} = \overline{h} * \left(A_{base} + A_{fin}\right) * \overline{\eta}_{surface} * \left(T_\infty - T_{base}\right) \quad (13.25)$$

If there are N fins in the array, the total surface area is (Incropera and DeWitt 2002) as follows:

$$A_{total} = N * A_{fin} + A_{base} \quad (13.26)$$

A_{fin} is the surface area of a single fin (m^2)

In this case, the total convective heat transfer rate from the fins and the prime (un-finned) surface may be expressed as follows:

$$\dot{Q}_{total} = N * \eta_{fin} * \overline{h} * A_{fin} * \left(T_\infty - T_{base}\right) + \overline{h} * A_{base} * \left(T_\infty - T_{base}\right) \quad (13.27)$$

where
$\overline{h}$ is the average convection coefficient assumed to be equivalent for the finned and base (prime) surfaces (J/kg · K)
η_{fin} is the efficiency of a single fin (–)

Hence,

$$\dot{Q}_{total} = \bar{h} * \left[N * \eta_{fin} * A_{fin} + \left(A_{total} - N * A_{fin}\right)\right](T_\infty - T_{base})$$
$$= \bar{h} * A_{total}\left[1 - \frac{N * A_{fin}}{A_{total}}\left(1 - \eta_{fin}\right)\right](T_\infty - T_{base}) \quad (13.28)$$

By using equation 13.24, it follows that

$$\dot{Q}_{total} = \bar{h} * A_{total} * \bar{\eta}_{surface} \quad (13.29)$$

If fins are machined as an integral part of the wall from which they extend, there is no contact resistance at their base. However, more commonly, fins are manufactured separately and are attached to the wall.

13.4.3.2.2 Wet Surfaces

The heat transfer at any location over a wet fin surface consists of a sensible heat transfer component that depends on the temperature difference between the moist air and the fin surface, and a latent heat transfer component that depends on the absolute humidity difference between the moist air and the saturated air at the fin temperature (Kandlikar 1990a, b).

Different methods are available for calculating the surface efficiency fins by considering the presence of condensed moisture (water). The efficiency of wet fins (i.e., fins with combined heat and mass transfer) can be calculated as follows (Threlkeld 1970):

$$\eta_{fin,\,wet} = \frac{\tanh\left(\dot{m}_{wet} * l\right)}{\dot{m}_{wet} * l} \quad (13.30)$$

where

$\dot{m}_{wet}$ is the moisture condensation flow rate (kg/s)
l is the fin height (m)

If the enthalpy potential of saturated air is used, the efficiency of wet fins can be calculated as follows (Kandlikar 1990a, b):

$$\eta_{fin,\,wet} = \frac{h_{wet} - \bar{h}_{sat,\,fin}}{h_{wet} - \bar{h}_{sat,\,base}} \quad (13.31)$$

where

h_{wet} is the air specific enthalpy under wet surface condition ($W/m^2 \cdot K$)
$\bar{h}_{sat,\,fin}$ is the saturated air specific enthalpy at the fin mean temperature T_{fin} (°C)
$\bar{h}_{sat,\,base}$ is the saturated air specific enthalpy at the base mean temperature T_{base} (°C)

Generally, the wet fin efficiency is lower than that of dry fins with only sensible heat transfer.

13.4.3.3 Air-Side Heat Transfer Coefficients

The total enthalpy change of the drying air flowing over the evaporator finned-tube heat transfer area is the sum of the specific enthalpy change due to temperature drop, or sensible heat transfer (Δh_{sens}), and the specific enthalpy change due to condensation, or latent heat transfer (Δh_{lat}):

$$\Delta h_{tot} = \Delta h_{sens} + \Delta h_{lat} \tag{13.32}$$

In addition to dry or wet operating conditions, the air-side heat transfer and the technical life of heat pump evaporators are influenced by the fin-tube thermal contact resistance; refrigerant and airflow maldistributions; and surface fouling, corrosion, and other damages.

Fouling may occur when deposits form on the surfaces of the heat pump evaporators reducing the efficiency of the global heat transfer. Thus, when choosing a heat pump moisture condenser (refrigerant evaporator), the qualities of the drying air need to be taken into consideration.

Several analytical models of different degree of complexity (e.g., Beatty and Katz 1948; Webb and Rudy 1982, 1985; Adamek and Webb 1990) have been developed to calculate the air (fin)-side heat transfer coefficients during moisture condensation of pure water vapor on horizontal integral finned tubes (Rose 1988; Rathod Pravin et al. 2011).

One of the first condensation models (Beatty and Katz 1948) was based on Nusselt's equation and assumptions for film condensation on vertical plain surfaces.

The average moisture condensation heat transfer coefficient $\bar{h}$ is obtained as the area weighted average of the condensation coefficients on the base area between fins $\left(\bar{h}_{base}\right)$ and that of the fins $\left(\bar{h}_{fin}\right)$:

$$\bar{h} * \bar{\eta}_{surface} = \bar{h}_{base} * \frac{A_{base}}{A} + \eta_{fin} * \bar{h}_{fin} \frac{A_{fin}}{A} \tag{13.33}$$

where

$\bar{h}$ is the average moisture condensation heat transfer coefficient (W/m$^2\cdot$K)
$\bar{\eta}_{surface}$ is the surface average efficiency (–)
$A = A_{base} + A_{fin}$ is the total surface area (m^2)
A_{base} is the base area between the fins (m^2)
A_{fin} is the surface area of the fins (m^2)

For calculating heat transfer coefficient corresponding to the tube base (i.e., area between fins) ($\bar{h}_{base}$) is used Nusselt correlation (Equation 13.2):

$$\bar{h}_{base} = 0.729 \left[\frac{\rho_f * (\rho_f - \rho_g) * g * h_{fg} * k_f^3}{\mu_f * d * (T_{sat} - T_s)} \right]^{1/4}$$

The corresponding equation for fins to that of vertical plates (equation 13.14) is as follows:

$$h_L = 0.943\left[\frac{g * \rho_f * (\rho_f - \rho_g) * k_f^3 * h_{fg}}{\mu_f (T_{sat} - T_s) * L}\right]^{1/4}$$

For circular fins, the characteristic length of the fins for condensation L (m) is the surface area of one side of the fin divided by the fin diameter d_{fin} (m):

$$L = \frac{\pi\left(d_{fin}{}^2 - d^2\right)}{4 * d_{fin}} \tag{13.34}$$

By replacing the coefficient 0.729 by 0.689 to better match their experimental data, the following equation has been obtained (Beatty and Katz 1948):

$$\bar{h} * \bar{\eta}_{surface} = 0.689\left[\frac{\rho_f * (\rho_f - \rho_g) * g * h_{fg} * k_f^3}{\mu_f * d * (T_{sat} - T_s)}\right]^{1/4}\left[\frac{A_{fin}}{A * d^{0.25}} + 1.3\frac{\eta_{fin} * A_{fin}}{A * L^{0.25}}\right] \tag{13.35}$$

13.5 REFRIGERANT-SIDE VAPORIZATION

The refrigerant side of mechanical vapor compression air-to-air heat pumps concerns at least a compressor (that move the refrigerant), an evaporator (that vaporize the refrigerant), a condenser (that condense the refrigerant), the expansion device (that reduce the refrigerant pressures from high to low values), and sometimes a suction line (liquid subcooling) internal heat exchanger.

The knowledge of refrigerant-side flow patterns, pressure drops, and convective heat transfer coefficients and rates, much more complex (Collier 1981) and uncertain parameters than those of single-phase flows, is essential to design phase-changing heat exchangers such as the heat pump evaporators and condensers (Westwater 1983; Kandlikar 1990a, b).

Boiling and convective (forced) evaporation are thermodynamic processes consisting in heat addition that converts a liquid into its vapor. They are the most common processes in heat pump evaporators (McQuiston et al. 2005). Boiling process occurs when vapor bubbles are formed at a heated solid surface without fluid flow, while the convective (forced) evaporation occurs when a liquid of a pure or mixed refrigerant is superheated at the surface and vaporizes at liquid–vapor interfaces within a flowing fluid. If boiling and convective evaporation occur simultaneously in a flowing fluid, the coexisting heat transfer mechanisms are described as forced convective vaporization (or flow boiling) where the flow is due to a bulk motion of the fluid, as well as to buoyancy effects characterized by rapid changes from liquid to vapor in the flow direction (Incropera and DeWitt 2002).

13.5.1 Two-Phase Flow Patterns

The flow pattern of the liquid and vapor phases relative to the tube wall during the two-phase flow of refrigerants in smooth horizontal tubes (see Figure 13.8) depends

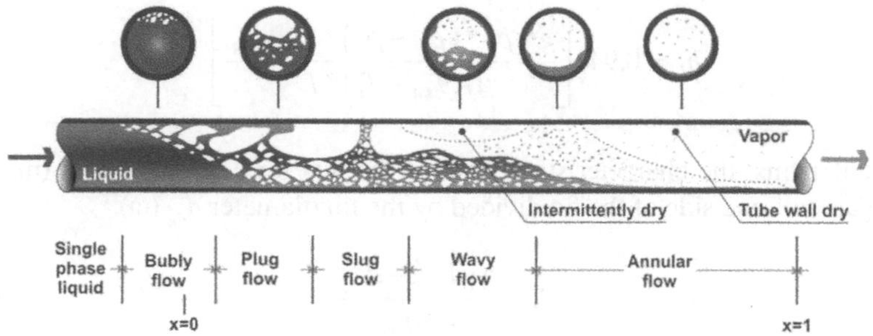

FIGURE 13.8 Vaporization flow regime map for an internally non-enhanced horizontal tube.

on the fluid properties, characteristics of the heat transfer surface and, in some cases, on gravity effects (Backer 1960; Kandlikar 1990a, b; Griffith 1998; Incropera and DeWitt 2002).

13.5.1.1 Internally Non-Enhanced Tubes

The flow patterns of refrigerants in horizontal non-enhanced (plain, smooth) tubes, which can be recognized by simple visual inspection or spectral analysis, are usually displayed in the form of flow regime maps, often applied to specific tubes sizes and refrigerants. They depend on refrigerant mass (volume) rates, vapor title, and thermos-physical properties such as density, viscosity, and surface tension. Because of several transitions and instabilities governed by different fluid properties, the attempts to find generalized maps allowing covering different fluids and tubes of different sizes had limited success (Chiriac and Dumitrescu 2006).

Flow pattern maps for many refrigerants flowing in horizontal tubes have been determined for both low and high mass velocities, as well as different tube diameters, heat flux densities, and evaporating temperatures (e.g., Kattan et al. 1998a, b, c). Since a flow regime becomes unstable as transition to another flow regime approaches, the multiphase transitions are rather unpredictable since they may depend on the wall's roughness or the initial (inlet) conditions. Hence, the flow pattern boundaries are not distinctive lines (or curves) but more or less poorly defined transition zones in almost all flow pattern maps.

One of the first flow regime map proposed by Collier and Thome (1994) (Figure 13.8) shows that if a single-phase, subcooled refrigerant liquid (title $x = 0$) enters an internally non-enhanced horizontal tube, an entrance length of 30–50 diameters is necessary to establish fully developed turbulent flow. In this region, the heat transfer initially occurs by forced convection, and once nucleate boiling is initiated, vapor bubbles that appear at the tube surface grow are carried into the mainstream of the liquid and are strongly influenced by the flow velocity.

In the bubbly flow zone, the vapor bubbles are dispersed in the liquid with a high concentration of bubbles in the upper half of the tube due to their buoyancy. When shear forces are dominant, the bubbles tend to disperse uniformly in the tube. Later, these large vapor bubbles coalesce and form a continuous core of vapor with the liquid flowing in an annular film along the wall. At high values of the mass flux, the annular flow pattern occurs over most of the tube length, whereas at low values, the flow may become stratified with liquid moving along the bottom of the tube and vapor along the top. As the flow progresses down the tube, the liquid film on the walls becomes thinner and eventually disappears as all of the liquid evaporates.

As the vapor title (volume fraction) increases, individual bubbles coalesce to form plugs and slugs of vapor. In the plug flow regime, the liquid phase is continuous along the bottom of the tube, below the elongated vapor bubbles, smaller than the tube diameter. Slug flow regime occurs at higher vapor velocities, the diameters of elongated bubbles becoming similar in size to the tube diameter (Incropera and DeWitt 2002). The plug and slug flow regimes are followed by stratified and annular mist (not explicitly shown in Figure 13.8) regimes in which the liquid forms a film that moves along the inner surface, while vapor moves at a larger velocity through the core of the tube.

In the stratified flow, at low liquid and vapor velocities, complete separation of the two phases occurs. The vapor goes to the top and the liquid to the bottom of the tube, separated by an undisturbed horizontal interface.

By increasing the vapor velocity in the stratified flow, waves are formed on the interface and travel in the direction of flow. The amplitude of the waves is notable and depends on the relative velocity of the two phases. However, their crests do not reach the top of the tube. The waves climb up the sides of the tube, leaving thin films of liquid on the wall after the passage of the wave. This is a stratified wavy flow. At high vapor fractions (titles), the top of the tube with its thinner film becomes dry first, so that the annular film covers only part of the tube perimeter, and thus this is then classified as stratified wavy flow.

Further increasing the gas velocity, these interfacial waves become large enough to wash the top of the tube. This is an intermittent flow regime characterized by large amplitude waves intermittently washing the top of the tube with smaller amplitude waves in between. Large amplitude waves often contain entrained bubbles. The top wall is nearly continuously wetted by the large amplitude waves and the thin liquid films left behind. Intermittent flow is also a composite of the plug and slug flow regimes.

At larger vapor flow rates, the liquid forms a continuous annular flow (film) around the perimeter of the tube, the liquid film being thicker at the bottom than the top. The interface between the liquid annulus and the vapor core is disturbed by small amplitude waves and droplets may be dispersed in the gas core.

Finally, the wall dry-out occurs at the top of the horizontal tube, first where the annular liquid film is thinner, and then dry-out proceeds around the perimeter of the tube along its length until reaching the bottom where the liquid film disappears. Thus, dry-out in the horizontal tube takes place over a range of vapor qualities, beginning as an annular flow and ending when the fully developed mist flow regime is reached. With increasing vapor quality, the heat transfer coefficient decreases.

13.5.1.2 Internally Enhanced Tubes

Internally enhancements of horizontal tubes, such as micro-fins, twisted tapes, or helical inserts, can have a significant effect on refrigerant flow pattern, pressure, and heat transfer coefficients.

Internally micro-fin tubes, for example, characterized by numerous fins (e.g., 48–70 per inch) (1 in. = 25.4 mm), short in height (e.g., 0.12–0.25 mm), show the most promise devices for heat pump applications because they (i) increase the vaporization (and condensation) heat transfer significantly, while producing only minor increases in pressure drops; (ii) decrease the annular to stratified wavy flow threshold at lower mass velocities by achieving complete wetting of the tube perimeter; (iii) displaces the annular stratified wavy transition to lower mass velocities; and (iv) create more turbulence in the liquid film and increases liquid entrainment in the central vapor core.

Depending on geometrical parameters, such as peak and valley shape, fin height, number of fins, and spiral angle, as well as on configurations, such as helical wire and herringbone (Thome 2004), the average vaporization heat transfer may increase by 45% to 238% over the reference smooth tube values, while the pressure drop could increase by only 4% to 102% over a range of Reynolds number from 5,000 to 11,000 (Huang and Pate 1988).

13.5.2 Pressure Drops

Most of the theoretical and experimental researches on pressure drops during two-phase forced convective evaporation of refrigerants have been conducted with pure and zeotropic refrigerant mixtures flowing in smooth tubes, are equipment- and fluid-dependent, and thus should not be extrapolated outside the specified ranges or to other refrigerants. On the refrigerant side, the available pressure drop and heat transfer correlations for micro-finned tubes are still far from satisfactory.

13.5.2.1 Pure Refrigerants

Inside horizontal non-heated (adiabatic) straight tubes, the total pressure drop in single-phase flow is the result of the action of three factors: (i) wall friction (due to the fluid viscosity), (ii) gravity (due to elevation) forces, and (iii) fluid momentum (due to the fluid acceleration) change (Griffith 1998; McQuiston et al. 2005):

$$\Delta p = \Delta p_f + \Delta p_m + \Delta p_g \qquad (13.36)$$

where

Δp is the total pressure drop, equal to final pressure less initial pressure (Pa)
Δp_f is the frictional pressure drop, mainly depending on the liquid viscosity and mass flux (Radermacher and Hwang 2005) (Pa)
Δp_m is the momentum (acceleration) pressure drop (Pa)
Δp_g is the gravitational pressure drop (Pa)

For horizontal tubes with adiabatic flow conditions, the gravitational pressure drop can be neglected ($\Delta p_g = 0$) as well as the acceleration pressure drop that is generally much smaller than the frictional pressure drop.

Therefore, in a circular tube, only the frictional pressure loss may be considered and can be calculated with the Darcy–Weisbach equation (Weisbach 1845; Incopera and DeWitt 2002; Field and Hrnjak 2007):

$$\Delta p_f = f * \frac{L}{2} * \rho * \frac{\overline{u}^2}{d} \tag{13.37}$$

where

f is Darcy friction factor, which is a function of the Reynolds number and the tube roughness (–)
L is the tube length (m)
ρ is the density of the fluid (kg/m^3)
$\overline{u}$ is the fluid average velocity (m/s)
d is the tube internal diameter (m)

For laminar flow inside circular tubes, the friction (or Moody) factor is only dependent on Reynolds, in the form:

$$f = \frac{64}{Re} \tag{13.38}$$

where

Re is the Reynolds number $\left(Re = \frac{\overline{u} * d}{\nu} \right)$
$\overline{u}$ is the fluid average velocity (m/s)
$\nu = \mu/\rho$ is the kinematic viscosity (m^2/s)
μ is the fluid absolute (dynamic) viscosity (Pa·s)
ρ is the fluid density (kg/m^3)

For turbulent in-tube flow of subcooled liquid and superheated vapor, Blasius friction factor varies in the range $0.006 < f < 0.06$ and can be calculated as follows (Blasius 1913; Radermacher and Hwang 2005):

$$f = \frac{0.0791}{Re^{1/4}} \quad \text{for } 2.1 * 10^3 < Re < 10^5 \tag{13.39}$$

The friction factor (f) can also be determined from the Moody diagram by assuming that the pure liquid is flowing in the tube at the mixture mass velocity. At qualities below 70%, it is generally low, but higher at qualities above 70%.

Two-phase (vapor–liquid) pressure drops are usually much higher than would occur for either phase flowing along at the same mass rate and can be determined according to both homogeneous (where the two-phase mixture is treated as a single-phase flow and it is assumed that the liquid and vapor phases are flowing at the same velocity with averaged properties for density and viscosity) and separated flow models (where the two phases are distinct) (Field and Hrnjak 2007).

For homogeneous flows, the two-phase average Reynolds number ($\overline{Re_{tp}}$) is calculated based on average fluid properties:

$$\overline{Re_{tp}} = \frac{G * d}{\overline{\mu}} \tag{13.40}$$

where

G is the fluid mass (velocity) flux (kg/m$^2 \cdot$ s)

$\bar{\mu}$ is the fluid average absolute (dynamic) viscosity (Pa $\cdot$ s)

For nonadiabatic, two-phase, convective (vaporization) flows in horizontal tubes, as for the refrigerant flowing through smooth tubes in condensation processes, the velocities of the two phases differ, and the acceleration pressure drop may be very important (Radermacher and Hwang 2005; McQuiston et al. 2005).

The continuity equation for each phase allows determining a relation between the liquid (u_f, m/s) and vapor (u_g, m/s) phase velocities, the void fraction (α), and the flow quality (x) (Griffith 1998).

The liquid- and vapor-phase velocities, respectively, are defined as follows:

$$u_f = \frac{\dot{V}_f}{A(1-\alpha)} \tag{13.41}$$

$$u_g = \frac{\dot{V}_g}{A * \alpha} \tag{13.42}$$

where

$\dot{V}_f$ is the liquid volume flow rate (m^3/s)

$\dot{V}_g$ is the vapor (gas) phase volume flow rate (m^3/s)

A is the tube cross-sectional area (m^2)

α is the void fraction (dimensionless)

The relationship between the void fraction, velocity ratio, and the vapor quality is as follows (Griffith 1998):

$$\left(\frac{1-\alpha}{\alpha}\right) = \left(\frac{\dot{V}_g}{\dot{V}_f}\right)\left(\frac{\rho_g}{\rho_f}\right)\left(\frac{1-x}{x}\right) \tag{13.43}$$

where

ρ_g is the vapor density (kg/m^3)

ρ_f is the liquid density (kg/m^3)

x is the vapor quality (dimensionless)

Frictional pressure drop in two-phase flow is due to the combination of interactions between the refrigerant and the walls of the tube, as well as between the liquid and vapor phases. Many empirical correlations to predict two-phase frictional pressure drop exist, but they have been developed for specific conditions (Adams et al. 2006) and no general theory exists to predict the frictional losses.

To get the two-phase pressure gradient, the single-phase flow pressure gradient is first calculated and then multiplied by the corresponding two-phase multiplier Φ^2.

The two-phase multiplier (Φ^2) relates the two-phase frictional pressure drop in terms of either single-phase liquid or vapor pressure drop, based on two assumptions

(Martinelli and Nelson 1948; Lockhart and Martinelli 1949; Adams et al. 2006): (i) the static pressure drop for both liquid and vapor phases are the same regardless of the flow pattern as long as the changes of static pressure in the radial direction are not significant; (ii) the sum of the volumes occupied by vapor and liquid at any instant is equal to the total volume of the pipe, which is the mass continuity equation. For example, if $\left(\frac{\Delta p}{\Delta L}\right)_f$ is the single-phase pressure gradient based on the liquid flow rate in the tube based on the Reynolds number $Re_{fo} = \frac{G*d}{\mu_{fo}}$, the pressure gradient of the two-phase mixture, $\left(\frac{\Delta p}{\Delta L}\right)_{tp}$, is determined as follows:

$$\left(\frac{\Delta p}{\Delta L}\right)_{tp} = \Phi_f^2 \left(\frac{\Delta p}{\Delta L}\right)_f \tag{13.44}$$

where

G is the mass flow velocity (kg/m$^2\cdot$s)
d is the tube interior diameter (m)
μ_{fo} is the dynamic viscosity of the liquid flowing alone in the tube (Pa$\cdot$s)

Likewise, if it is assumed that the entire flow is vapor, the pressure gradient can be calculated by using the Reynolds number $Re_{go} = \frac{G*d}{\mu_{go}}$ (where μ_{go} is the dynamic viscosity of the vapor flowing alone in the tube, in Pa$\cdot$s) and calculating the pressure gradient $\left(\frac{\Delta p}{\Delta L}\right)_g$ and the two-phase multiplier Φ_g^2.

To characterize separated flows, the following parameter has been developed (Lockhart and Martinelli 1949):

$$X_{tt} = \left(\frac{1-x}{x}\right)^{0.875} \left(\frac{\rho_g}{\rho_f}\right)^{0.5} \left(\frac{\mu_f}{\mu_g}\right)^{0.125} \tag{13.45}$$

where

x is the vapor title (–)
ρ_f and μ_f are the liquid density (kg/m^3) and dynamic viscosity (Pa$\cdot$s), respectively
ρ_g and μ_g are the vapor density (kg/m^3) and dynamic viscosity (Pa$\cdot$s), respectively

For single-phase flow, the total pressure drop in two-phase flow (i.e., the gradient of pressure in the direction of flow) also comprised three contributing effects (Adams et al. 2006): (i) frictional dissipation ("*f*"), (ii) momentum (acceleration) ("*m*"), and (iii) gravitational ("g"):

$$\frac{dp}{dx} = \left(\frac{dp}{dx}\right)_f + \left(\frac{dp}{dx}\right)_m + \left(\frac{dp}{dx}\right)_g \tag{13.46}$$

For the pressure drop during convective vaporization in horizontal tubes, the Martinelli–Nelson correlation (1948) is as follows:

$$\Delta p_{MN} = \frac{2 * f_{fo} * G^2 * L}{d * \rho_f}\left[\frac{1}{x}\int_0^x \phi_{fo}^2 dx\right] + \frac{G^2}{\rho_f}\left[\frac{x^2}{\alpha}\left(\frac{\rho_f}{\rho_g}\right) + \frac{(1-x)^2}{(1-\alpha)} - 1\right] \quad (13.47)$$

where

Δp_{MN} is the Martinelli–Nelson total two-phase pressure drop (Pa)
f_{fo} is the frictional coefficient considering the two-phase flow as a liquid flow (–)
G is the fluid mass flux ($kg/m^2 \cdot s$)
x is the vapor quality (–)
L is the tube length (m)
d is the tube inner diameter (m)
ρ_f is the liquid density (kg/m^3)
ρ_g is the vapor density (kg/m^3)
ϕ_{fo}^2 is the local two-phase frictional multiplier (–)
α is the void fraction

As can be seen in equation 13.47, to obtain total pressure drop during the two-phase vaporization process, the Martinelli–Nelson correlation requires the evaluation of the local two-phase frictional multiplier $\left(\phi_{f0}^2\right)$ and vapor void fraction (α). These two parameters can be calculated with the following relations, respectively equations 8.20 and 8.21 (Radermacher and Hwang 2005).

$$\phi_{fo}^2 = \phi_f^2 (1-x)^{1.75} \quad (13.48)$$

where

$$\phi_f^2 = \left(1 + \frac{1}{x^{1/2}}\right)^2$$

$$\alpha = \left(1 + x^{0.8}\right)^{-0.378}$$

For some common refrigerants (such as HCFC-22 and HFC-152a), the following modified version of Martinelli–Nelson correlation matches data for both pure and mixtures fluids within 8% deviation (Jung and Radermacher 1989; Radermacher and Hwang 2005):

$$\Delta p_{tp} = \frac{2 * f_{f0} * G^2 * L}{d * \rho_f}\left[\frac{1}{\Delta x}\int_{x1}^{x2} \phi_{tp}^2 dx\right] \quad (13.49)$$

where

$f_{f0} = 0.046 Re^{-0.2}$ (–)
G is the specific mas flux ($kg/s \cdot m^2$)

D is the tube diameter (m)
ρ_f is the liquid density (kg/m^3)

$$\phi_{tp}^2 = 30.78 * x^{1.323} * (1-x)^{0.477} * p_r^{-0.7232}$$

where
$p_r = \dfrac{p}{p_{critical}}$ is the reduced pressure (–)
p is the actual refrigerant pressure (Pa)
$p_{critical}$ is the refrigerant critical pressure (Pa)

13.5.2.2 Refrigerant Mixtures

The pressure drops of refrigerant mixtures flowing in horizontal tubes depend on the nature of pure fluid components and their respective mixing percentages (Jung and Radermacher 1989, 1993). For example, by modifying the Martinelli–Nelson correlation for refrigerant mixtures as CFC-13/CFC-12 (5/95, 10/90, and 15/85 wt.%), the total pressure drop (Δp_{tp}) could match the available experimental data within 30% by using the following equation (Singal et al. 1983a, b):

$$\Delta p_{tp} = 0.87(1+C)^{2.66}\,\Delta p_{MN} \tag{13.50}$$

where
C is the concentration of CFC-13 in the binary mixture CFC-13/CFC-12 (decimals)
Δp_{MN} is the total pressure drop given by the Martinelli–Nelson correlation 13.47 (Pa).

For the refrigerant mixture R-32/R-125, the frictional pressure drop correlation for two-phase flow matches experimental data within 8% deviation is as follows (Souza and Oimenta 1995):

$$\Delta p_{tp} = \frac{2 * f_{f0} * G^2 * L}{d * \rho_f}\left[\frac{1}{\Delta x}\int_{x_1}^{x_2} \phi_{tp,o}^2 dx\right] \tag{13.51}$$

where
$f_{f0} = 0.079Re^{-0.25}$

$$\phi_{tp,o}^2 = 1 + \left(\Gamma^2 - 1\right) * x^{1.75}\left(1 + 0.9524 * \Gamma * X_{tt}^{0.4126}\right)$$

$$\Gamma = \left(\frac{\rho_f}{\rho_g}\right)^{0.5}\left(\frac{\mu_g}{\mu_f}\right)^{0.125}$$

ρ_f is the liquid phase density (kg/m^3)
ρ_g is the vapor phase density (kg/m^3)

μ_f is the liquid phase dynamic viscosity (Pa · s)
μ_g is the vapor phase dynamic viscosity (Pa · s)
X_{tt} is the Lockhart–Martinelli parameter (see equation 13.45)

The pressure drops of refrigerant mixtures flowing in horizontal tubes could be higher or lower than that of pure fluids they are supposed to replace. For example, the pressure drop of R-32/R-125 (50/50 wt.%) mixture during forced convective vaporization is approximately 24%–38% lower pressure drop than that of pure HCFC-22 (Wijaya and Spatz 1994). Also, the pressure drop of HCFC-22/CFC-114 mixture in forced convective evaporation inside an internally finned horizontal tube, correlated by the Lockhart–Martinelli parameter, is about 40% higher than that in a smooth tube, independently of molar concentration (Miyama et al. 1988).

13.5.3 In-Tube Vaporization

Most heat pump heat exchangers involve two-phase convective vaporization of pure synthetic (fluorinated, chlorinated, and brominated) halocarbons (such as CFCs, HCFCs, and HFCs) or mixed zeotropic and azeotropic refrigerants, as well as inorganic refrigerants like carbon dioxide, ammonia, water, and air inside smooth (plain, non-enhanced) tubes.

Inside horizontal tubes, there are two interacting mechanisms for refrigerants' boiling: (i) forced convection that dominates when the refrigerant flow rates are high; heat is transferred from the wall to the flowing liquid and evaporation occurs at the liquid–vapor interface; (ii) nucleate boiling at the wall surface that dominates when the heat transfer rate is high or the refrigerant flow rate is low.

The relations for heat transfer over the entire boiling range are based on a combination of these two basic mechanisms. The two-phase convective (vaporization) heat transfer coefficients of such processes depend on fluid properties, flow pattern and regime, heat flux density, mass velocity, vapor quality, tube geometry, and one or more empirical constants (Kandlikar 1990; Kattan et al. 1998a, b, c; Zürcher et al. 2008).

13.5.3.1 Pure Halocarbon Refrigerants

Halocarbons describe organic compounds based on carbon chains attached to hydrogens in which a number of hydrogens have been replaced by a combination of the halogens (e.g., fluorine, chlorine, bromine, and iodine). A number of halocarbons, particularly those based on chains of only a few carbons, are used as refrigerants. Halocarbons based on chlorine and fluorine, such as CFCs, when released, rise, and are broken apart by ultraviolet radiation from the sun in the upper atmosphere where they react to destroy the protective ozone layer. As a result, the use of such ozone-depleting substances are banned today.

The knowledge of heat transfer coefficients contributes to avoid some operational problems due to under- or overdesign of heat pump equipment and reduce their costs (Kandlikar 1990a, b). The mechanism of two-phase evaporation of pure refrigerants inside the horizontal tubes of heat pump evaporators is complex. As heat is added to the refrigerant, it progressively evaporates, and the velocity increases until the refrigerant leaves the evaporator saturated or superheated. The variations in the heat transfer coefficient along the evaporator tube are associated with different patterns of flow as the fraction of vapor and the velocity change along the tube (Stoecker 1998) (see Section 13.5.1) (Figure 13.9).

Correlations for the refrigerant-side heat transfer coefficient are classified into correlations for single-phase flows (e.g., subcooled liquid and superheated vapor) and correlations for two-phase (convective boiling) flows (Gnielinski 1976; Kandlikar 1987; Huang and Pate 1988).

For single-phase flow, the Dittus–Boelter correlation is (Khanpara 1986; Khanpara et al. 1986; Levenspiel, O. 1998; Jin and Spitler 2002):

$$Nu = 0.023Re^{0.8}Pr^{n} \tag{13.52}$$

where

n is 0.4 for the refrigerant being heated (e.g., inside a heat pump evaporator) and 0.3 for the refrigerant being cooled (e.g., inside a heat pump condenser).

Most of the available phenomenological correlations (Kandlikar 1990; Webb and Gupte 1992) for two-phase convective evaporation heat transfer coefficients during in-tube flow of pure refrigerants combine nucleate boiling and convective evaporation, most of them being restricted to one fluid (Kandlikar 1987, 1990a, b), but not limited to specific flow geometry.

Generally, two-phase convective evaporation models are classified as follows: (i) superposition models where the total heat transfer coefficient is the sum of nucleate boiling and bulk convection contributions (e.g., Chen 1966; Gungor-Winterton 1986, 1987); (ii) enhancement models where the two-phase heat transfer coefficient enhances the single-phase heat transfer coefficient of a flowing liquid by a two-phase

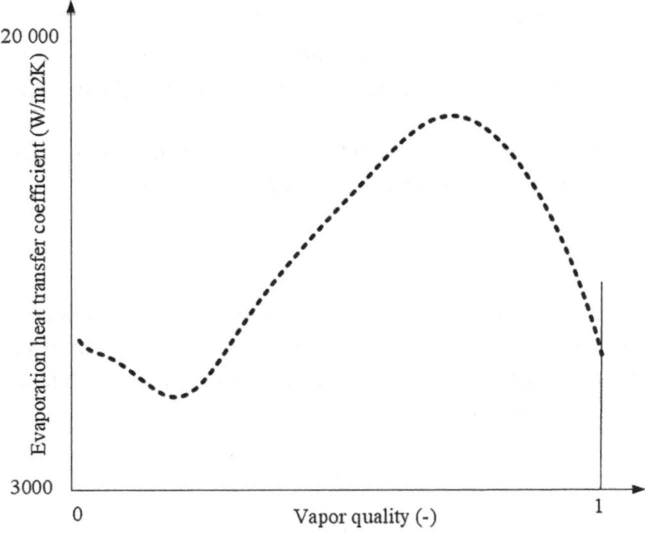

FIGURE 13.9 Relative variation of two-phase convective evaporation heat transfer inside a horizontal tube.

enhancement factor) (e.g., Kandlikar 1883, 1987; Shah 1982); and (iii) asymptotic models (e.g., Liu and Winterton 1988, 1999).

Using the Forster and Zuber (1955) equation for pool boiling to predict the nucleate (pool) boiling heat transfer coefficient ($h_{nb} = h_{pool}$), Chen (1966a, b) divided the heat transfer during vaporization in vertical tubes into (i) nucleate (pool) boiling contribution based on Foster and Zuber equation (1955) and (ii) forced convection contribution based on the single-phase (liquid only) Dittus–Boelter (1930) equation.

The relatively simple superposition (additive) model assumes that the total heat flux density $(\dot{q})$ is the sum of nucleate (pool) boiling $(\dot{q}_{nb})$ and convective evaporation $(\dot{q}_{cv})$ components:

$$\dot{q} = \dot{q}_{nb} + \dot{q}_{cv} \tag{13.53}$$

The terms $\dot{q}_{nb}$ and $\dot{q}_{cv}$ are calculated at the existing wall superheat value:

$$\Delta T_{wall} = T_{wall} - T_{sat} \tag{13.54}$$

where

T_{wall} is the wall temperature (°C)
T_s is the refrigerant saturated temperature (°C)

By dividing equation 13.53 by the wall superheat, the superposition model may be written in terms of the component heat transfer coefficients:

$$h = h_{nb} + h_{cv} \tag{13.55}$$

where

h is the superposed total heat transfer coefficient (W/m$^2 \cdot$K)
h_{nb} is the nucleate (pool) boiling heat transfer coefficient (W/m$^2 \cdot$K)
h_{cv} is the convective evaporation heat transfer coefficient (W/m$^2 \cdot$K)

The nucleate boiling and convective contributions are superposed to give the following overall two-phase heat transfer coefficient (Chen 1966a, b):

$$\overline{h}_{tp} = F * h_f + S * h_{nb} \tag{13.56}$$

where $F < 1$ is the Chen's empirical enhancement factor (that reflects the influence of much higher velocities, and hence forced convection heat transfer in the two-phase flow compared to the single-phase, liquid-only flow) correlated against the Lockhart–Martinelli parameter as follows:

$$F = 1 + 1.8 * X_{tt}^{-0.79} \tag{13.57}$$

h_f is the liquid heat transfer coefficient (W/m$^2 \cdot$K)
h_{nb} is the nucleate boiling (or pool) heat transfer coefficient (W/m$^2 \cdot$K)

$S < 1$ is the Chen's in-tube flow suppression factor that shows the lower effective superheat available in forced convection as opposed to pool boiling, due to the thinner boundary layer (Bogles et al. 1981); it assumes that the liquid velocity suppresses nucleate boiling, empirically correlated as a function of the two-phase Reynolds number (Re_{tp}):

$$S = \frac{1}{1 + 2.53 * 10^{-6} Re_{tp}^{1.17}} \tag{13.58}$$

In equation 13.56, the bulk convection term is enhanced by the two-phase correction factor [$F(X_{tt})$]accounting for the higher vapor velocities at the higher vapor quality, while the suppression factor [$S(Re_{tp})$]accounts for the thinner boundary layer at higher vapor qualities. The enhancement factor F and the suppression factor S in the Chen correlation have been correlated to a number of operating and system variables (Cooper 1984; Gungor and Winterton 1987).

For annular flow, the two-phase convection multiplier F can be calculated as follows (Collier 1981):

$$F = \frac{h_{cv}}{h_f} = \left(\frac{Re_{tp}}{Re_f} \right)^{0.8} \tag{13.59}$$

where

h_{cv} is the convection heat transfer coefficient (J/kg · K)
h_f is the convection heat transfer coefficient of the liquid phase flowing alone (J/kg · K)
Re_{tp} is the two-phase Reynolds number ($F^{1.25} * Re_f$) (dimensionless)
Re_f is the liquid phase Reynolds number $\left[\frac{d * G(1-x)}{\mu_f} \right]$ (dimensionless)
x is the vapor title (–)
d is the tube inner diameter (m)
G is the mass flux velocity (kg/m^2 · s)
μ_f is the fluid dynamic viscosity (Pa · s)

Based on the Chen superposition correlation, Gungor and Winterton (1986) developed the following correlation for flow boiling in horizontal tubes:

$$h = E * h_f + S * h_{nb} \tag{13.60}$$

where E is the forced convection enhancement parameter (–) defined as follows:

$$E = 1 + 24000 * Bo^{1.16} + 1.37 * X_{tt}^{-0.86} \tag{13.61}$$

Bo is the boiling number defined as the ratio of mass flux perpendicular to the wall due to boiling to the total (axial) mass flux $\left(\frac{\dot{q}}{h_{fg} * G} \right)$ (Shah 1976, 1978, 1982, 1982)

$\dot{q}$ is the heat flux density (W/m²)
h_{fg} is the enthalpy (latent heat) (J/kg)
G is the specific mass flux (kg/m²·s)
h_f is the heat transfer coefficient for the liquid phase flowing alone (W/m²·K)
S is the forced convection suppression parameter (–)
h_{nb} is the nucleate (pool) boiling heat transfer coefficient (W/m²·K)

For laminar flow of refrigerant liquids inside tubes, the recommended correlation for predicting the average heat transfer coefficient ($\bar{h}$, W/m²·K) can be predicted by the following correlation (McQuiston et al. 2005):

$$\frac{\bar{h} * D}{k_f} = 1.86\left[Re_D * Pr\frac{d}{L}\right]^{1/3}\left(\frac{\mu}{\mu_{wall}}\right)^{0.14} \quad \text{for} \left[Re_D Pr\frac{D}{L}\right]^{1/3} > 20 \quad (13.62)$$

where
d is the tube inner diameter (m)
k_f is the liquid thermal conductivity evaluated at arithmetic mean bulk temperature (W/m·K)
μ is the fluid dynamic viscosity evaluated at arithmetic mean bulk temperature (Pa·s)
μ_{wall} is the fluid dynamic viscosity evaluated at the wall temperature (Pa·s)

In equation 13.62, properties should be evaluated at the arithmetic mean bulk temperature, except for μ_{wall}.

Even though laminar flow may sometimes occur, the forced convection turbulent flow is the most important regime. For turbulent flow of refrigerants flowing inside tubes, dominant regime in evaporators of industrial heat pumps, the most widely used correlation for the single-phase (liquid) heat transfer coefficient (h_f) applied under conditions of $Re_D > 10{,}000$, $0.7 < Pr < 100$, and $L/D > 60$ (where fluid properties should be evaluated at the arithmetic mean bulk temperature of the fluid) is the Dittus–Boelter equation (McQuiston et al. 2005):

$$h_f = 0.023\left[\frac{G(1-x)d}{\mu_f}\right]^{0.8} Pr_f^n\left(\frac{k_f}{d}\right) \quad (13.63)$$

where
G is the mean mass flux on the cross section of the tube (kg/m²·s)
d is the internal diameter of the tube
μ_f is the dynamic viscosity of the liquid
x is the vapor quality (–)
k_f is the thermal conductivity of the liquid
Pr_f is the Prandtl number of the liquid
$n = 0.4$ for $T_{wall} > T_{bulk}$
$n = 0.3$ for $T_{wall} < T_{bulk}$

For a tube roughness of 1 μm, the nucleate boiling heat transfer coefficient is (Cooper 1984) as follows:

$$h_{nb} = 55 * p_r^{0.12}\left(-log_{10} p_r\right)^{-0.55} M^{-0.5} \dot{q}^{0.67} \quad (13.64)$$

where

p_r is the reduced pressure (–)
M is the molecular mass (kg/kmol)
$\dot{q}$ is the density of heat flux (W/m^2)

$$S = \frac{1}{1 + 1.15 * 10^{-6} * E^2 * Re_f^{1.17}} \quad (13.65)$$

Shah (1976, 1978, 1982, 1982) proposed a widely accepted (Collier 1981) empirical correlation in both graphic and analytic forms for flow boiling in vertical and horizontal tubes. Instead of the suppression factor, this empirical correlation uses the boiling (*Bo*) and convection (*Co*) numbers. The convection (*Co*) number replaces the Lockhart–Martinelli parameter X_{tt}.

Shah's correlation also considers two distinct mechanisms, that is, nucleate boiling and forced convection, but, instead of adding the two contributions together, the larger of the two calculated heat transfer coefficients is chosen.

Bo and *Co* numbers are defined as follows:

$$Bo = \frac{\dot{q}}{h_{fg} * G} \quad (13.66)$$

and

$$Co = \left(\frac{1-x}{x}\right)^{0.8}\left(\frac{\rho_g}{\rho_f}\right)^{0.5} \quad (13.67)$$

where

$\dot{q}$ is the heat flux density (W/m^2)
h_{fg} is the fluid enthalpy (latent heat) of vaporization (kJ/kg)
G is the mass flux velocity (kg/m$^2 \cdot$ s)
x is the vapor quality (–)
ρ_g is the vapor density (m^3/kg)
ρ_f is the liquid density (m^3/kg)

For horizontal tubes, the Shah's enhancement model defines the two-phase heat transfer coefficient as follows:

$$h_{tp} = E * h_f \quad (13.68)$$

where h_f is the single-phase heat transfer coefficient for the liquid phase flowing alone (W/m$^2 \cdot$ K).

E is the enhancement factor of the single-phase heat transfer coefficient which is a function of the boiling number (Bo), the convection number (Co),and the Froude number (Fr_f) accounting for partial wall wetting occurring in horizontal tubes:

$$E = f\left(Bo, Co, Fr_f\right) \tag{13.69}$$

where

$$Fr_f = \frac{G^2}{g * d * \rho_f^2} \tag{13.70}$$

Shah's correlation predicts the two-phase heat transfer coefficient as being dependent on mass flux velocity, as follows:

$$\frac{h_{tp}}{h_f} = 230 * Bo^{0.5} = 230 * \left(\frac{\dot{q}}{G * h_{fg}}\right)^{0.5} \tag{13.71}$$

where

h_f is the single-phase heat transfer coefficient for the liquid-phase flowing alone ($W/m^2 \cdot K$)
Bo is the boiling number defined by equation 13.66
$\dot{q}$ is the heat flux density (W/m^2)
G is the mean mass flux on the cross section of the tube ($kg/m^2 \cdot s$)
h_{fg} is the fluid enthalpy (latent heat) of vaporization (kJ/kg)

Based on Kutaeladze's (1961) theory for subcooled convective evaporation of water in plain tubes and on theory for natural convection, the following asymptotic model has been proposed for flow boiling (vaporization) inside horizontal tubes (Liu and Winterton 1988, 1999):

$$h = \sqrt{\left(E * h_f\right)^2 + \left(S * h_{pool}\right)^2} \tag{13.72}$$

where the two-phase correction (E and S) factors, obtained for water, refrigerants, and other fluids, are empirically defined as follows:

$$E = \left[1 + x * Pr_f\left(\frac{\rho_f}{\rho_g} - 1\right)\right]^{0.35} \tag{13.73}$$

$$S = \frac{1}{1 + 0.055 * E^{0.1} * Re_f^{0.16}} \tag{13.74}$$

Among pure fluorinated fluid used as refrigerants in high-temperature heat pumps, there is HFC-245fa (1,1,1,3,3-pentafluoropropane). It is a hydrofluorocarbon (Table 13.1) that has a relatively high GWP value (950) (i.e., 950 times the global warming effect of CO_2), no ozone depletion potential, is nearly nontoxic, nonreactive, nonflammable,

TABLE 13.1
Some Physical Properties of Refrigerant HFC-245fa

–	Unit	Value
Molecular formula	–	$CF_3CH_3CHF_2$
Molecular weight	–	134
Boiling point	°C	15.3
Liquid density at 20°C	kg/m^3	1.32
Freezing point	°C	< −107
Vapor pressure at 20°C	kPa	123
Vapor thermal conductivity	mW/mK	14

non-mutagenic, noncorrosive toward all commonly used metals including carbon steel, stainless steel, copper, and brass (but there is a concern with the use of aluminum due to the reactive nature of this metal), and, practically, nonbiodegradable with a lifetime of 7.2 years when it eventually does escape into the atmosphere. R-245fa has a high degree of thermal and hydrolytic stability, at temperatures ranging from 75°C to 200°C, and exhibits acceptable miscibility in a wide range of polyols.

This refrigerant provides favorable properties, as low saturated pressures at relatively high saturated temperatures (Figure 13.10) and, in some cases, improved performances in high-temperature heat pumps developed, for example, for wood drying industry (Minea 2004), to reheat circulating water and to generate steam from waste heat available at 50°C–60°C (Zyhowski et al. 2002).

Flow patterns, pressure drop, and heat transfer characteristics of HFC-245fa flowing in horizontal circular tubes (Ong and Thome 2009; Pike-Wilson and Karayiannis 2014), mini-channels with diameters between 0.96 and 2.95 mm (Kandlikar and Grande 2003; Charnay et al. 2014), and circular micro-channels with diameters around 1 mm (Bortolin et al. 2009) have been extensively studied with saturation temperatures varying from 60°C to 80°C, mass velocity from 100 to 400 $kg/m^2 \cdot s$ (La Yssac et al. 2015), mass heat flux from 5 to 85 kW/m^2, and vapor quality from 0.05 to 0.8.

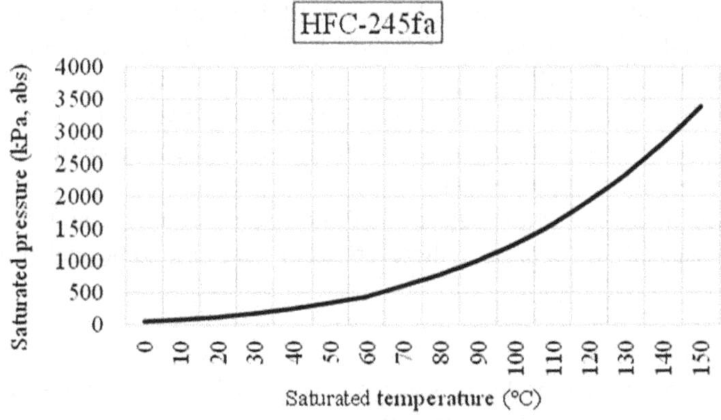

FIGURE 13.10 Saturated pressures as functions of saturated temperatures of refrigerant HFC-245fa.

Two-phase frictional pressure drops of refrigerant HFC-245fa, a fluid with high liquid and low vapor density, are dependent upon the average kinetic energy of the flow, hydraulic diameter, and friction factor. As mass flux increases, pressure drop increases related to the square of mass flux. Also, the pressure drop increases with quality, with the exception of some very high-quality conditions. Starting with a quality of zero, pressure drop can be seen to increase nearly linearly for most of the quality range. Moreover, at high qualities, the pressure drop begins to depart from the linear relationship, causing a flattening out and even a downward bend of the two-phase pressure drop curve. This departure from a linear relationship is seen most dramatically at high mass fluxes and with refrigerants that have a low vapor density, such as R-245fa (Coleman and Garimella 1999).

Saturation temperature also plays an important role in two-phase pressure drop primarily due to its effect on vapor density. Increasing the saturation temperature decreases the two-phase pressure drop because as saturation temperature increases, vapor density also increases. As higher vapor density corresponds to lower pressure drop due to lower relative vapor velocities, increasing the saturation temperature decreases the two-phase pressure drop for that fluid. Vapor density is inversely related to the two-phase pressure drop of a refrigerant. Additionally, liquid density plays a little role in determining the magnitude pressure drop as the liquid to vapor density ratio is greater for R-245fa than for ammonia, yet ammonia creates a greater pressure drop due to its lower vapor density (Adams et al. 2006).

The vaporization heat transfer coefficient of HFC-245fa at mass velocity ranging between 200 and 400 kg/s m^2 saturation temperature around 31°C shows to be highly dependent on the heat flux, while the effect of mass velocity is less important in round 0.96 mm mini-channels. With regard to the effect of vapor quality, the heat transfer coefficient decreases when vapor quality increases, and this has been verified for a large range of heat flux values (between 30 and 78 kW/m^2) (Bortolin et al. 2009) and increases with rise in the saturated temperature (Tibiriça and Ribatski 2010; Vakili-Farahani et al. 2013).

13.5.3.2 Natural Refrigerants

Because of relatively high ozone depletion and global warming potential indexes of commonly used refrigerants (e.g., HCFC-22 and HFC-134a), a considerable attention has been focused on the application of natural refrigerants (Cavallini 1996), as ammonia (R-717) (Lorentzen 1988), hydrocarbons (Granryd 2001), such as propane (R-290), butane (R-600), and isobutene (R-600a), as well as carbon dioxide (R-744) (Lorentzen 1994) in heat pump systems.

Among the most susceptible natural fluids to be used as refrigerants for industrial heat pump-assisted wood dryers, there are ammonia and some hydrocarbons (as propane).

13.5.3.2.1 Ammonia

Ammonia has been used for decades as a refrigerant of choice for large-scale refrigeration and heat pump industrial applications. As for other refrigerants, many experiments on two-phase flow patterns and pressure drops, as well as pool and flow (evaporation) boiling heat transfer of ammonia inside horizontal tubes, have been undertaken in the past. However, no formal database is available on two-phase flow heat transfer of ammonia (Ohadi et al. 1996).

For a plain horizontal tube (21.6 mm inner diameter and 10 m length) with liquid entering at saturation conditions with flow rates between 60 and 1,200 *L*/*h*, the overall average heat transfer coefficient for ammonia two-phase convective flow has been correlated (over a temperature range of 30°C–75°C, with deviations of about ±15%) by using the following empirical relation (Malek and Colin 1983):

$$\bar{h}_{tp} = \kappa * G^a * q^b * p_{red}^c \tag{13.75}$$

where

$\bar{h}_{tp}$ is the overall mean two-phase heat transfer coefficient (W/m$^2\cdot$K)
$\kappa = 1.59$, $a = -0.08$, $b = 0.53$, and $c = 0.18$ are experimental fitted constants (–)
G is the ammonia mass flux (2–20 kg/m$^2\cdot$s)
$\bar{q}$ is the ammonia average density of heat flux (W/m^2)
$p_{red} = \dfrac{p}{p_{critical}}$ is the reduced pressure (0.1–0.33) (–).

For the annular flow region, by coupling energy and mass equations, the two-phase heat transfer coefficient ($\bar{h}_{tp}$) can be calculated by using the following equation (Ohadi et al. 1996):

$$\bar{h}_{tp} = \frac{k_f}{\delta} Nu_f \tag{13.76}$$

where

$\bar{h}_{tp}$ is the two-phase heat transfer coefficient (W/m$^2\cdot$K)
k_f is the thermal conductivity of the liquid phase (W/m$\cdot$K)
δ is the thickness of the thermal boundary layer (m)
Nu_f is the film Nusselt number calculated utilizing the tabulated values of Collier (1981).

Few experimental test data are available for evaporation of ammonia inside horizontal tubes, excepting those of Shah (1975), Chaddock and Buzzard (1986) (with pure ammonia evaporated in one 13.4-mm internal diameter tube at saturation temperatures from –21°C to –35°C and mass velocities up to 130 kg/m$^2\cdot$s), and Zürcher et al. (1999) (with ammonia evaporating at 4°C, mass velocities from 20 to 140 kg/m$^2\cdot$s, vapor qualities from 1% to 99%, and heat fluxes from 5 to 58 kW/m^2). In several tests, it was observed (Zürcher et al. 1999) that the Gungor–Winterton correlation (1987) provides very poor accuracy for ammonia.

Most of the available refrigerant vaporization heat transfer coefficients inside smooth tubes are only applicable to fully turbulent flow regimes under ammonia oil-free conditions in direct expansion heat pumps (Chaddock and Buzzard 1986). However, even ammonia is not miscible with mineral oil in most cases, the lubrication oil leaking from the compressor forms a layer of film with poor conductivity on the tube wall and leads to an increase in the heat transfer resistance at the wall, thus causing a strong reduction of the heat transfer coefficient (Gross 1994; Boyman et al. 2004).

At high ammonia vapor quality, the oil tends to climb up the walls of the pipe as high as two-third of the tube diameter as a thin annular flow film (Shah 1975) acting as an additional layer of thermal resistance (Chaddock and Buzzard 1986). For example, ammonia mixed with an immiscible synthetic oil (with concentration varying from 0% to 3% by weight) and flowing in a horizontal plain tube with an inner diameter of 14 mm caused a considerable reduction of the heat transfer coefficient, especially when the oil concentration exceeded 1% (Boyman et al. 2004).

The significant decrease of the heat transfer coefficient of ammonia in the presence of immiscible oils led to the development of synthetic oils miscible with ammonia. For example, by using miscible poly-glycol oil of up to 3% (by mass) with mass velocities varying between 30 and 65 kg/m$^2\cdot$s (Boyman et al. 1997) and, also, polyalkylene glycols (Zürcher et al. 1998) with mass velocities from 50 to 80 kg/m$^2\cdot$s, no major negative effects on the heat transfer coefficients occur. However, in order to use such new oils, the heat pump systems have to be totally exempt of impurities and without any hygroscopic traces.

The presence of water in ammonia causes a reduction of the heat transfer coefficient of the evaporator (Panchal et al. 1981). Experimental works showed that at water concentrations below 0.1%, the reduction in the overall heat transfer coefficient is less than 1%. At 1% water concentration, the heat transfer coefficient could decrease by up to 20%–50% (Ohadi et al. 1996).

On other hand, at constant heat flux varying from 60 to 1,400 kW/m^2 and reduced pressures between 0.3 and 0.95, the ammonia heat transfer coefficient increases when the working pressure increases (Barthau 1976).

Finally, substantial heat transfer augmentation was observed for micro-fin tubes at mass velocities below 80 kg/m$^2\cdot$s, while little or no augmentation was observed at higher mass velocities (Zurcher et al. 1999).

13.5.3.2.2 Hydrocarbons

Hydrocarbons (HC) are already used as refrigerants in many applications such as domestic refrigerators and small capacity air-conditioning and residential heat pumps (Thonon 2008). Hydrocarbons as propane (R-290) (a nontoxic natural refrigerant compatible with most conventional materials used in heat pump equipment and miscible with commonly used lubricants), butane (HC-600), and isobutane (HC-600a) exhibit GWPs as low as 3 (for HC-600a), 3.3 (for R-290), and 4 (for HC-600).

The most salient features (flow patterns, pressure drops, and heat transfer) of hydrocarbons used as refrigerants are (note: not in order of importance and/or relevance) as follows:

a. Hydrocarbon refrigerants are highly flammable, being classified as class A3; leakage of hydrocarbons in heat pumps is thus a real challenge because most of the commercial and industrial systems present annual leakages up to 15%–20% of the total charges (Palm 2007).
b. Charges of hydrocarbons replacing HCFCs and/or HFCs in heat pump systems must be reduced as much as possible to avoid explosion caused by accidental leakage (Palm 2007; Oh et al. 2015); one of the best solutions is to use reduced-size geometry mini- or micro-channel heat exchangers

that, in addition, have more efficient heat transfer performances than conventional types, due to their higher heat transfer area per refrigerant volume and minimize the amount of material required for their manufacture (Vlasie et al. 2004; Wojtan et al. 2005; Copetti et al. 2013).

c. Hydrocarbons have thermodynamic and transport properties, such as thermal conductivity and saturated pressures very similar to those, for example, of HFC-134a.
d. Hydrocarbons have latent heats higher compared to those (almost twice) of banned HCFC refrigerants.
e. Most of the existing heat pumps today use HFC-134a as a refrigerant, can use hydrocarbons (e.g., propane) without major mechanical modifications.
f. However, in heat pumps with conventional and mini- or micro-channel heat exchangers (Fernando et al. 2004; Blanco Castro et al. 2005; Dupont and Thome 2005; Bertsch et al. 2008; Fernando et al. 2008; Choi et al. 2009; Mastrullo et al. 2014) using hydrocarbons, the refrigerant flow rates, pressure drops, and heat transfer coefficients are significantly different compared to those obtained with artificial refrigerants; therefore, the system designing and sizing rules have to be established specifically for hydrocarbons.
g. Even the physical properties of HCs and operating parameters have a great influence on flow patterns, some of the available flow pattern maps are used today to identify flow patterns of hydrocarbons (Thome et al. 2008).
h. Few studies exist about the lubricant oil effects on flow boiling of hydrocarbons (Thome et al. 2008).
i. The average two-phase pressure drops of hydrocarbons are much higher than those of the refrigerant HCFC-22, for a constant heat flux of 44 kW/m^2, saturation temperature of 22°C, and two mass velocities (240 and 440 kW/m$^2\cdot$s), the magnitude of the pressure drop is higher for R-600a than for HFC-134a; as the mass flux increases at a given heat flux, the pressure drop increases, mainly due to the lower density of R-600a, both in liquid and vapor phases.
j. To predict the heat transfer coefficients for forced flow boiling of hydrocarbons, some known correlations, as those of Chen (1966a, 1966b), Kandlikar (1990), and Liu and Winterton (1991) can be used (Wojtan et al. 2005); the heat transfer correlations of Shah (1976) and Gungor-Winterton (1986) predict well the experimental data, while the Kandlikar correlation (1987) is the best one (Lee et al. 2006); heat transfer coefficient correlations proposed by Shah (1978) and Tran et al. (1996) also show accurate predictions compared to most of experimental data.
k. Heat transfer coefficients of propane and butane measured in a horizontal plain tube with inner diameter of 2.46 mm are much higher than those of HCFC-134a, while the corresponding two-phase frictional pressure drops are lower at equal heat fluxes and mass fluxes (Wen and Ho 2005).
l. At low vapor quality, the local heat transfer coefficients of hydrocarbons, such as propane, isobutene, and propylene depend strongly on heat flux, are higher than those of HCFC-22 and HFC-134a, and they became independent with increasing vapor quality (Shin et al. 1997; Watel and Thonon 2002; Copetti et al. 2011).

Other experimental works about convective boiling pressure drops and heat transfer coefficients of propane in horizontal mini-channels showed the following (Oh et al. 2015): (i) pressure drop of propane is well predicted by the models proposed by Mishima and Hibiki (1996), Friedel (1979), and Chang et al. (2000); (ii) pressure drop (a function of the mass flux, tube diameter, surface tension, density, and viscosity) increases with the increase of mass and heat fluxes, as well as with the decrease in the inner tube diameter and saturation temperature; (iii) a new pressure drop correlation was developed on the basis of the Lockhart–Martinelli (1949) model, defined as a function of the two-phase Reynolds number and the two-phase Weber number; (iv) heat transfer coefficient of propane is also affected by the mass flux, heat flux, inner tube diameter, and saturation temperature; it increases with the decrease in the inner tube diameter and the increase of the saturation temperature.

Experimental results for the flow boiling (vaporization) of R-600a in horizontal mini-channels under variations in the mass velocity, heat flux, and vapor quality showed the following (Copetti et al. 2013): (i) at low vapor quality, there is a significant influence of heat flux on the heat transfer coefficient. At high vapor quality, for high mass velocities, this influence tends to disappear and the heat transfer coefficients decrease; (ii) for the highest heat flux value, the heat transfer coefficient is almost independent of mass velocity; (iii) the frictional pressure drops depend significantly on the mass flux and vapor quality; (iv) compared to R-134a, the R-600a refrigerant has a similar behavior with respect to the effect of different parameters; however, R-134a achieved higher heat transfer coefficients and higher frictional pressure drops for similar operation conditions; for the heat transfer coefficient, the correlation of Kandlikar and Balasubramanian (2004) and Bertsch et al. (2009b) best fitted the experimental results.

13.5.3.3 Low-GWP Refrigerants

The hydrofluoroolefines (HFO) fluids, such as HFO-1234yf and HFO-1234ze(E), are promising alternatives to the present HFC refrigerants.

13.5.3.3.1 HFO-1234yf

Among low-GWP fluids, the refrigerant HFO-1234yf has zero ODP and a very low GWP (4), meeting the today's environmental requirements (Minor and Spatz 2008).

Some results about the convective boiling heat transfer experiments performed with HFO-1234yf flowing in horizontal round tubes can be summarized as follows:

a. No major difference was observed between the flow patterns of HFO-1234yf and HCF-134a refrigerants in the same operating conditions, that is, mass velocities from 300 to 500 kg/s m^2, saturation temperature of 10°C, and flow inside a 6.70 mm-circular tube (Padilla et al. 2011).
b. In general, pressure drops of HFO-1234yf are lower than those of HCF-134a under the same flow conditions (Padilla et al. 2011).
c. The thermos-physical properties of HFO-1234yf are similar to those of HFC-134a.

d. Several theoretical and experimental studies with HFO-1234yf looked into the feasibility of direct substitution of HFC-134a with minimum modifications in mobile air conditioning (Zilio et al. 2011; Jarall 2012), and refrigeration and heat pump systems.
e. A reduction in the coefficient of performance and cooling capacity of automotive air conditioning systems has been reported when HFO-1234yf is used as a drop-in alternative (Minor and Spatz 2008a, b).
f. Refrigerant mass flux and vapor quality have insignificant effects on the heat transfer coefficients at low vapor quality regions (Longo et al. 2016);
g. At same operating conditions (i.e., same heat flux and vapor quality), HFO-1234yf and HFC-134a display similar values of the heat transfer coefficients, but HFO-1234yf shows lower pressure drops by 10%–12% compared to HFC-134a (Del Col et al. 2013).
h. In micro-fin tubes using HFO-1234yf and HFC-134a, simulations showed that the overall heat transfer coefficient of HFO-1234yf is about 10% lower compared to that of HFC134a (Mendoza-Miranda et al. 2015);
i. For a given cooling capacity, HFO-1234yf achieves lower energy performances than those using HFC-134a as "drop in" automotive air conditioning systems (Zilio et al. 2011; Del Col et al. 2013).

13.5.3.3.2 HFO-1234ze(E)

HFO-1234ze(E) is another promising substitute for HFC-134a, showing a GWP lower than 1 together with pressure and volumetric properties close to those of HFC-134a (Grauso et al. 2013).

Recent experimental studies showed the following:

a. The flow patterns observed during the experiments for HFO-1234ze(E), starting from low vapor qualities, are, in order, slug, intermittent, annular, and dry-out flow regimes, at all the mass fluxes investigated.
b. The frictional pressure gradients increase strongly with vapor quality reaching a maximum, and then decreasing for both HFO-1234ze(E) and HFC-134a refrigerants during all operating conditions.
c. During flow boiling of HFO-1234ze(E) in a horizontal smooth stainless steel tube (6 mm of inner diameter) with mass fluxes between 146 and 520 kg/s m^2, the evaporating temperature varying from –2.9°C to +12.1°C and heat fluxes from 5 and 20.4 kg/m^2, and a wide range of the vapor quality, HFO-1234ze(E) provided heat transfer coefficients very similar to HFC-134a and HFO-1234yf.
d. During vaporization inside small, commercial, brazed plate heat exchangers (Longo et al. 2014), the frictional pressure drops of HFO-1234ze(E) is slightly higher than those of HFC-134a and HFO-1234yf.
e. The local heat transfer coefficients of HFO-1234ze(E) increase slowly with vapor quality until the occurrence of the onset of dry-out at high vapor qualities, are close to those of HFC-134a, and present similar trends with vapor quality, frictional pressure drops, mass rates, and heat fluxes (Grauso et al. 2013; Mendoza-Miranda et al. 2015; Longo et al. 2016).

f. The increase of the mass flux of HFO-1234ze(E) leads to a significant increase of local heat transfer coefficients for vapor qualities over 20%–30% (Grauso et al. 2013).
g. The local heat transfer coefficients are very similar at all the investigated operating conditions, showing the same trends with vapor quality, with the only differences for the earlier dry-out inception for HFO-1234ze(E) and the heat transfer coefficients of HCF-134a being about 15% higher than those of HFO-1234ze(E) at low vapor qualities.
h. The increase of the saturation temperature of HFO-1234ze(E) leads to negligible variations of the local heat transfer coefficients in the whole range of vapor quality (Grauso et al. 2013; Hossain et al. 2013).
i. Boiling heat transfer characteristics of HFO-1234yf/oil mixture flowing inside a 7 mm micro-fin tube at mass velocities ranging from 100 to 400 $kg/s \cdot m^2$ increase strongly with the mass fluxes and are more sensitive to the mass fluxes in the high vapor quality region. At moderate and high heat fluxes, the average heat transfer coefficients decrease strongly with the oil concentration (Han et al. 2013).
j. The comparison of the experimental local heat transfer coefficients of HFO-1234ze(E) as a function of vapor quality obtained at T_{sat} = 7°C for several mass fluxes (from 146 to 520 $kg/m^2 \cdot s$) and for two heat fluxes (5 and 20 kg/m^2) with those of HFC-134a at the same operating conditions showed that (i) the local heat transfer coefficients of the two refrigerants are quite similar at all the investigated operating conditions and the same trend with vapor quality and (ii) at low mass velocities, the heat transfer coefficients of HFO-1234ze(E) are up to 33% less than those of HFC-134a, but they become similar at higher mass fluxes.
k. Very similar variations were found with each operating parameter between HFO-1234ze(E) and HFC-134a refrigerants, that is, strong increase with mass flux, slightly increase with heat flux (mostly at low-medium vapor qualities), and negligible effect of saturation temperature.

13.5.3.4 Refrigerant Mixtures

The interest for the use of zeotropic and azeotropic mixtures (blends) of refrigerants as working fluids in heat pump applications is increasing, due to potential advantages in energy efficiency and capacity (Schulz 1985), even the prediction of their performances is complicated by uncertainty in thermos-physical properties (Ross et al. 1987).

Even refrigerant mixtures offer some advantages such as capacity control and lower pressure ratios across the heat pump compressor (Arora 1967; Mulroy and Didion 1986), their use has not yet been popular due, in part, to the lack of information on actual pressure drops and heat transfer coefficients (Jung et al. 1989).

Several questions about the flow and heat transfer mechanisms are still of the actuality, for example (i) what are the main reasons for the degradation of heat transfer coefficients during forced convective evaporation of refrigerant mixtures; (ii) is the full suppression of nucleate boiling easier to achieve with mixtures than with pure components; and (iii) what are the impacts of heat and mass fluxes, vapor quality, and pressure on forced convective evaporation of refrigerant mixtures.

The reasons for the transport properties and heat transfer coefficient degradation of zeotropic mixtures, up to 50% as compared to pure fluids (Ross et al. 1987), can be summarized as follows (Radermacher and Hwang 2005): (i) increase of the local saturation temperature because the more volatile components evaporates before other component(s) during the bubble growth process and (ii) large changes of mixture physical properties with vapor mass fraction (quality).

Several experimental works showed that

a. The heat transfer (nucleate boiling, vaporization, and condensation) characteristics of two-phase (zeotropic and azeotropic) refrigerant mixtures are quite different from that of pure refrigerants (Wang and Chato 1995).
b. Generally, the measured pool boiling heat transfer coefficients of forced convective evaporation inside horizontal tubes with zeotropic mixtures are lower than that of the ideal pool boiling heat transfer coefficient predicted from that of pure fluid components, especially in internally enhanced tubes, for example, in the case of CFC-11/R-113 refrigerant mixture (Fink et al. 1982).
c. The boiling heat transfer coefficients degrade as compared to the ideal heat transfer coefficients; the main and most complex contribution to the heat transfer coefficient degradation is the reduction of the effective superheating (Radermacher and Hwang 2005), for example, the degradation of the boiling heat transfer coefficient compared to the ideal boiling heat transfer coefficient of R-32/HFC-134a (30/70 wt.%) (Torikoshi and Ebisu 1993) and of HCFC-22/R-142b (60/40 wt.%) (Rohlin 1994) mixtures were 30% and 40%, respectively, for the mass fluxes of 91 $kg/m^2 \cdot s$ and 182 $kg/m^2 \cdot s$, respectively, at a saturation temperature of 0°C.
d. The critical heat flux for mixtures can be either higher or lower than that of pure fluids (Radermacher and Hwang 2005).
e. The heat transfer coefficients of zeotropic mixtures does not change linearly with mass fraction.
f. For near-azeotropic mixtures with small temperature differences between saturated liquid and vapor at constant pressures (e.g., less than 3°C for CFC-12/ HCFC-22), the evaporation heat transfer coefficient lies between those of its component (Hihara et al. 1987).
g. During forced convective evaporation of CFC-12/R-13 mixture in a horizontal tube at low heat fluxes, the heat transfer coefficients are only affected by changes in thermos-physical properties, and, in the nucleate boiling region, a significant reduction of heat transfer coefficients occurs (Jain and Dher 1983).
h. In other works, the average heat transfer coefficients for binary mixtures of CFC-13 (5%–20%) and CFC-12 have been found slightly greater than those for pure CFC-12 (Singal et al. 1983b, 1984).
i. The nucleate boiling contribution could be significant at high vapor qualities (Wattelet 1994; Wattelet et al. 1994).

j. At low mass velocities (<100 kW/m^{2}s) and medium densities of heat fluxes (≈10 kW/m^2), the heat transfer coefficients during forced convective evaporation of HCFC-22/R-114 mixture at the top of a horizontal tube are very low due, probably, to the stratified flow in this region.
k. The average heat transfer coefficients of the refrigerant mixtures are always lower than those of pure components, no matter what the quality is.
l. At high mass flux (e.g., G = 300 kg/m$^2 \cdot$ s), high heat flux (e.g., 20 kW/m^2), and low-quality ($x < 0.4$) regions, the average heat transfer coefficients for the mixtures are lower than those for pure components. However, in the high-quality region ($x > 0.4$), the average heat transfer coefficients of the 50% mass fraction mixture are between those of HCFC-22 and R-114. Thus, the higher the quality, the lower were the heat transfer coefficients of R-114, while the heat transfer coefficients of both the mixture and HCFC-22 increase slightly, a phenomenon attributed to the smaller latent heat of R-114, thus easy to dry out at the top of the tube in the high-quality region (Yoshida et al. 1990).
m. The heat transfer coefficients of mixtures such as HCFC-22/R-114 in the annular flow region at reduced pressure of 0.08, and heat flux and mass flow rates of 10–45 kW/m^2 and 16–46 g/s, respectively, are up to 36% lower than the ideal values under the same flow condition, the variations in physical properties account for 80% of the heat transfer degradation and the other 20% may be caused by mass transfer resistance (Jung et al. 1989).
n. For the convective boiling heat transfer coefficients of HCFC-22/R-114 and CFC-12/R152a mixtures, it has been concluded that (Jung et al. 1989) (i) the flow boiling characteristics of zeotropic mixtures are different from those of azeotropic mixtures that behave similar to the pure fluids; (ii) the actual flow boiling heat transfer coefficients of mixtures are lower than the ideal heat transfer coefficients of pure fluids, for example, at 65% vapor quality, the actual heat transfer coefficient of mixture was lower by 19%, 36%, and 35% than that of ideal values for the HCFC-22 compositions of 0.23, 0.47, and 0.77, respectively; up to 80% of this heat transfer coefficient degradation is estimated to be due to the physical property variation; the mass transfer resistance is estimated to contribute less than 20% of the heat transfer coefficient degradation; (iii) flow boiling heat transfer coefficients of mixtures are functions of the mixture mass fraction vapor quality, and heat flux; and (iv) convective boiling is dominant when the vapor quality is high; the heat transfer coefficient is not affected by the heat flux since the nucleate boiling is suppressed.
o. Based on the experimental study of horizontal flow boiling heat transfer with non-azeotropic mixtures of HCFC-22 and R-114, the following conclusions can be drawn (Jung et al. 1989): (i) a full suppression of nucleate boiling has been observed for both pure and mixed refrigerants at qualities above 10%–30% depending upon the test conditions; flow boiling heat transfer coefficients for HCFC-22/R-114 mixtures in the convective evaporation region are always lower than the values calculated by an ideal mixing rule with two pure component coefficients; the degradation of heat transfer coefficients for

mixtures from the ideal values was up to 36% depending on the overall composition; (ii) under the suppression of nucleate boiling, the physical property variation associated with mixtures is responsible for the degradation of heat coefficients from the ideal values up to 27%, which accounts for up to 80% of the observed heat transfer degradation seen with mixtures; (iii) the effect due to the additional mass transfer resistance on heat transfer coefficients for HCFC-22/R-114 mixtures seems small, <10% of the total heat transfer coefficient under the suppression of nucleate boiling.

p. The mixtures of R-152a/R-13B1 achieve significantly lower heat transfer coefficients in horizontal forced convective evaporation than either pure refrigerant. In addition (Ross et al. 1987) (i) mixtures can reach full suppression of nucleate boiling easier than the pure fluids due to the mass transfer resistance; (ii) the heat transfer coefficients of mixtures are severely degraded (up to 50%) compared to pure fluids; (iii) for the evaporation-dominated heat transfer regime, Chen's correlation (1966a, b) applied well for both pure fluids and mixtures with a Prandtl number correction for pure fluids and without correction for mixtures; this means there is some nucleation for pure fluids while nucleate boiling is suppressed for mixtures; (iv) for the mixtures, there is a circumferential variation of the heat transfer coefficient opposite to that found for pure fluids; this phenomenon means that there was a circumferential gradient in mass fraction and interfacial temperature.

q. With R-13B1/R-152a mixture flowing in a stainless steel tube (2.7 m long, 0.9 cm ID) heated via a direct current, the local and average heat transfer coefficients for mixtures were significantly lower than those for either pure component. The results were compared with the available correlations, and it was concluded that the local heat transfer correlations for pure fluids did not adequately predict the heat transfer for zeotropic mixtures.

r. Experimental tests conducted with R-13B1/R-152a mixtures showed a substantial degradation of heat transfer coefficients with an R-13B1/R-152a mixture (Radermacher et al. 1983; Ross et al. 1987; Jung et al. 1989).

s. The non-azeotropic refrigerant mixtures with variable temperature during phase change at constant pressure increase the coefficient of performance of heat pumps up to 32% compared to HCFC-22 (Mulroy et al. 1988).

t. The average heat transfer coefficients for a binary mixtures such as CFC-13/CFC-12 (with a composition of CFC-13 in CFC-12 between 0% and 20% by mass in intervals of 5%) flowing in horizontal stainless steel tubes (2.35 m long, 0.95 cm inner diameter) are slightly greater than those for pure CFC-12 (Singal et al. 1983b).

u. A large reduction of the local heat transfer coefficients was observed in the region where nucleate boiling was dominant during forced convective evaporation of mixtures of refrigerants CFC-11 and R-114 in a horizontal smooth tube, with a difference of about 20 K in boiling points, in the mixtures compared with those of pure components; however, in the regions where convective evaporation was dominant, the reduction of heat transfer coefficients could be explained mainly by the change of fluid properties (Murata et al. 1991);

v. Other experimental studies achieved with CFC-123/HFC-134a with a difference of about 54 K in boiling points showed that even in the region where convective evaporation was dominant, there was some reduction in heat transfer coefficients compared with those of an equivalent pure fluid, although not as significant as that in the region where nucleate boiling dominated.

13.6 DESIGN OF MOISTURE CONDENSERS

Heat exchangers operating with humid air, such as a heat source and refrigerants as heat sink, are called moist air condensers, refrigerant evaporators, or dehumidifiers. In such finned-tube heat exchangers, some of the moisture present in the air may condense (or even freeze), over the outside fins and tube surfaces modifying the flow field, while inside the tubes, partial or total vaporization of the refrigerant may cause inhomogeneous distribution of the flow. For proper design of moisture condensers, information of the heat transfer rates as functions of several variables of the system must be determined, usually from empiric (experimental) correlations (Kandlikar 1990; Pacheco-Vega et al. 2001).

Moisture condensers are complex, critical components of heat pumps due to their various geometrical configurations, large number of variables, and operating requirements involved, especially in heat pump drying applications. Appropriate design and practical sizing and selection of finned-tube moisture condensers, aiming at optimizing the thermal and hydraulic processes, are challenging tasks. Their type and size depend on the nature of drying process, material cost, manufacturability, space constraints, thermal requirements for wet surfaces, and refrigerant properties (i.e., temperature, density, viscosity, pressures, chemical composition, etc.). Moisture condensers are most often selected via computer programs, even if hand calculations are possible via several iterations. The computer design and sizing programs generally assume steady-state conditions with nonhomogeneous, two-phase flow inside the moisture condensers' tubes and outside air in cross-flow, and allow specifying the moisture condenser operating conditions, fin-and-tube heat exchanger parameters, refrigerant and airflows, and the type of fins for enhanced tubes (McQuiston et al. 2005).

Researches on the cooling and dehumidifying coils are limited due to the complex simultaneous heat and mass transport of moist air to cooled-fin surfaces (Oskarsson et al. 1990a, b; Wang and Hihara 2003).

13.6.1 Overall Heat Transfer Coefficients

The overall heat transfer coefficients (U) (also known as U-values) depend on sensible and latent heat transfers on both air-side (from hot moist air to the external surface of the evaporator finned tubes) and refrigerant side (from the tube wall to the vaporizing refrigerant) of finned-tube moisture condensers (Figure 13.11).

In heat pump dryers, the moisture condensers operate partially under external dry-surface conditions (i.e., during sensible heat recovery and refrigerant superheating) and, partially, under external wet surface conditions (i.e., during the moisture condensation resulting in condensing enthalpy recovery).

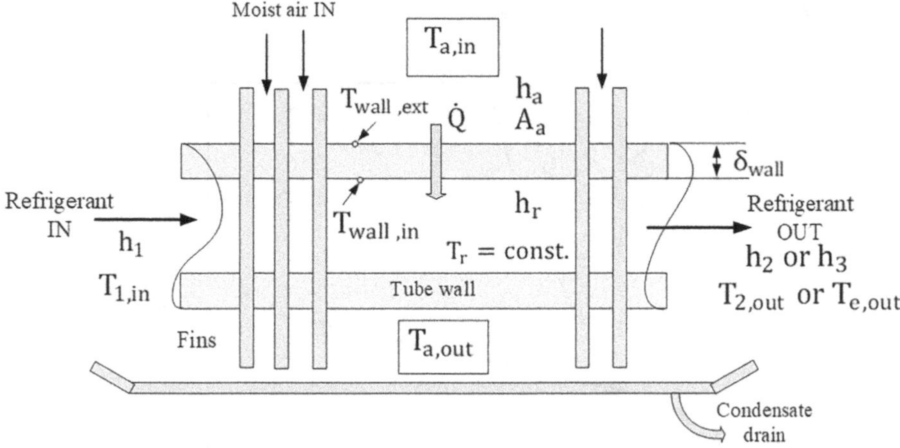

FIGURE 13.11 Schematic representation of a finned-tube moisture condenser tube (heat pump evaporator) and its main heat transfer parameters (notes: internal and external fouling layers are not shown; schematic not to scale). a, air; in, inlet; h, specific enthalpy; out, outlet; i, heat transferred; r, refrigerant; T, temperature.

The refrigerant coming from the expansion valve as a low-quality mixture of saturated liquid and vapor enters the moisture condenser tubes, vaporizes during its passage through the coil, and finally leaves it at a slightly superheated (2°C–5°C) state. The temperature of the refrigerant is constant during the phase change process when the heat capacity of the saturated refrigerant mixture is infinite.

The process by which the refrigerant changes from liquid to vapor is a combination of boiling and evaporation, although a clear distinction between the two is usually not made. In boiling, vapor is produced at the solid surface and bubbles up through the liquid to the surrounding vapor, while in evaporation, the liquid changes phase at the interface between the liquid and the vapor.

No really universal methods are available to correlate and predict the combined boiling and evaporating (i.e., vaporization) heat transfer coefficients. The actual fluid temperature has little effect on the heat transfer, and the proposed equations can all be put in a form that relates the heat transfer coefficient to the difference between the surface temperature and saturation temperature of the fluid at that pressure.

On the air-side of moisture condensers, both cooling and dehumidification processes consisting in heat and mass (water) transfers occur simultaneous. The state of the moist air (state 1 in Figure 13.12) can move toward the saturation curve (100% relative humidity, RH) of which average temperature is that of the wetted (tube and fins) surface. Typically, the air dry-bulb temperature and the humidity ratio both decrease as the air flows through the exchanger. Therefore, sensible and latent heat transfer occurs simultaneously (McQuiston et al. 2005).

When moist air at state 1 is cooled over a finned-tube surface, several situations can occur (Wang and Hihara 2003): (i) if the air dew-point temperature (*DEW*) is

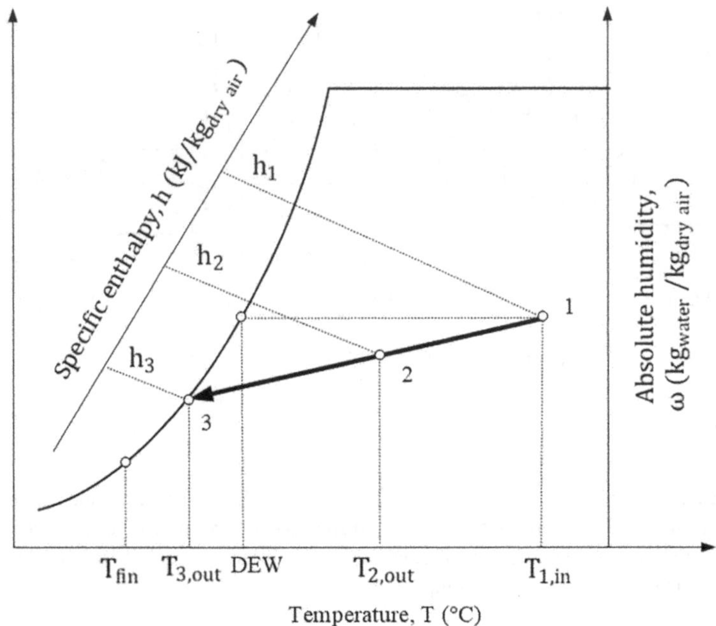

FIGURE 13.12 Possible cooling and dehumidifying processes of the moist air over the finned-tube surface of a moisture condenser.

less than the tube outer surface temperature ($DEW \le T_{2,out}$), the process is totally dry; in the dry region, standard design methods, such as *LMTD* or $\varepsilon - NTU$, can be employed for designing the moisture condensers; (ii) if the DEW is equal to, or greater than the air temperature leaving the evaporator ($DEW \ge T_{3,out}$), the condensing process is totally wet; in other words, when the heat exchanger surface in contact with moist air is at a temperature below the air dew point, condensation of vapor will occur; (iii) if the dew-point temperature is between the final air temperature and the surface (*DEW*) temperature ($T_{2,out} > DEW > T_{3,out}$), the condensing process is partially wet.

By neglecting the fouling and thermal efficiency of both smooth internal tube surface and finned-tube external surface, there are three series thermal resistances.

On the air-side, the thermal resistance is defined as follows:

$$R_a = \frac{1}{h_a * A_a} \tag{13.77}$$

where

- h_a is the heat transfer coefficient on the air-side (W/m$^3 \cdot$K) determined, for example, by using Webb and Gupte correlation (1992)
- A_a is the total heat transfer area (tube plus fins) on the dry air-side (m^2)

The dominant thermal resistance is on the air-side (Wang et al. 1999a, b). To improve the overall heat transfer and reduction the resistance in air-side, the surface of the tube is finned.

On the air-side of the heat exchanger, the heat, mass, and friction coefficients should be obtained from correlations based on test data, since the analogy method is unreliable (McQuiston et al. 2005).

The thermal resistance of the circular (cylindrical) tube wall is as follows:

$$R_{wall} = \frac{\ln\left(\frac{d_{ext}}{d_{int}}\right)}{2\pi * L * k_{wall}} \tag{13.78}$$

where

d_{ext} is the external diameter of circular tubes (m)
d_{int} is the internal diameter of circular tubes (m)
L is the condensing tube length (m)
k_{wall} is the tube wall thermal conductivity (W/m · K).

Finally, the thermal resistance on the refrigerant-side is as:

$$R_{refr} = \frac{1}{h_{refr} * A_{refr}} \tag{13.79}$$

where h_{refr} is the heat transfer coefficient on the refrigerant-side $\left(\frac{W}{m^2k}\right)$ determined, for example, by using Liu and Winterton (1991) asymptotic correlation; in particular, for the refrigerant HCFC-22 vaporizing at temperatures from 4.4 to 26.7°C in a tube diameter of 8.7 mm and 2.4 m long, the refrigerant side heat transfer coefficient can be determined with the following correlation (Altman et al. 1960):

$$Nu = \frac{h_{refr} * d_{int}}{k_{refr}} = 0.0225\left(Re^2 * K_f\right)^{0.375} \tag{13.80}$$

where

h_{refr} is the refrigerant-side heat transfer coefficient $\left(\frac{W}{m^2k}\right)$
d_{int} is the tube interior diameter (m)
k_{refr} is the refrigerant thermal conductivity $\left(\frac{W}{mk}\right)$
Re is the Reynolds number (–)
$K_f = \frac{\Delta x * h_{fg}}{L * g}$ is the boiling number (–)
Δx is the difference of refrigerant quality (generally, between 0.2 and 1) (–)
h_{fg} is the water enthalpy (latent heat) of condensation $\left(\frac{J}{kg}\right)$

L is the tube section length (m)

g is the gravitational acceleration $\left(\frac{m}{s^2}\right)$

By applying the electrical analogy, the total thermal resistance (R_{tot}) $\left(\frac{K}{W}\right)$ can be expressed as follows:

$$R_{tot} = R_{air} + R_{wall} + R_{refr} = \frac{1}{h_{air} * A_{air}} + \frac{\ln\left(\frac{d_{ext}}{d_{int}}\right)}{2\pi * L * k_{wall}} + \frac{1}{h_{refr} * A_{refr}} \tag{13.81}$$

The overall heat transfer coefficient ($\bar{U}$) $\left(\frac{W}{m^2k}\right)$ can be derived by using the total thermal resistance as follows:

$$\bar{U} = \frac{1}{R_{tot}} = \frac{1}{\frac{1}{h_{air} * A_{air}} + \frac{\ln\left(\frac{d_{ext}}{d_{int}}\right)}{2\pi * L * k_{wall}} + \frac{1}{h_{refr} * A_{refr}}} \tag{13.82}$$

Depending on operating conditions, the mean overall heat transfer coefficients can be determined by using more or less exact correlations for both h_{air} and h_{refr}.

The simple design method of plain surface (un-finned) cooling coils (Goodman 1936), performed by using the enthalpy potential, neglecting the liquid film resistance, and assuming the refrigerant-side temperature and heat transfer coefficient constant, and assuming that the wet surface heat transfer coefficient to be the same as that of the dry surface, cannot be extended to evaporators with large fin surface areas, and with significant variations in the refrigerant-side temperatures and heat transfer coefficients along the refrigerant flow direction (Kandlikar 1990).

A more accurate analysis of the heat and mass transfer processes during dehumidification on the coil finned surfaces (Threlkeld 1970), considers the fin efficiency under wet surface conditions by using the enthalpy potential, includes the moisture film thermal resistance, considers the refrigerant-side heat transfer coefficient constant, and employs a logarithmic mean enthalpy potential between the heat exchanger inlet and outlet which permits variations in the refrigerant temperature.

The finned-tube evaporators employing continuous plate fins can be designed by dividing them into two-phase and superheated regions on the refrigerant side (where the refrigerant temperature and the heat transfer coefficient are assumed to be constant) (Goldstein 1983), and into dry and wet region on the air-side. Further, the moisture condenser can be divided into small steps on the two sides forming a number of elements which are individually analyzed. The average heat transfer coefficients for refrigerant superheated process are estimated by Dittus–Bolter correlation applicable to turbulent flows in circular tubes.

During normal heat exchanger operation of moisture condensers, surfaces are often subject to fouling by fluid impurities, rust formation, or other reactions between the fluid and the wall material. The subsequent deposition of a film or scale on the surface can greatly increase the resistance to heat transfer between the fluids. This effect can be treated by introducing an additional thermal resistance, termed the fouling factors of which values depend on the operating temperature, fluid velocity, and length of service of the heat exchanger (Incropera and DeWitt 2002). Therefore, in practice, the total thermal resistances comprise the air-side, air-side fouling, tube wall, fin-to-tube contact, refrigerant-side fouling, and refrigerant-side thermal resistances. The fouling factors are variable during heat exchanger operation (increasing from zero for a clean surface, as deposits accumulate on the surface) (Incropera and DeWitt 2002).

Accordingly, with inclusion of surface fouling and fin (extended surface) effects, the overall heat transfer coefficient must be expressed as follows (Incropera and DeWitt 2002):

$$\frac{1}{U}=\frac{1}{\eta_{air}h_{air}A_{air}}+\frac{1}{\eta_{air}h_{a,f}A_{air}}+\frac{1}{R_cA_c}+\frac{1}{2\pi k_{wall}L}ln\frac{d_{ext}}{d_{int}}+\frac{1}{\eta_{refr}h_{r,f}A_{refr}}$$

$$+\frac{1}{\eta_{refr}*h_{refr}A_{refr}} \tag{13.83}$$

where

U is the overall heat transfer coefficient (W/m$^2\cdot$K)

η_{air} is the air-side overall surface efficiency of enhanced finned tubes (–); for industrial in-use coils, η_{air} generally ranges between 0.3 and 0.7;

h_{air} is the air-side heat transfer coefficient (W/m$^2\cdot$K)

A_{air} is the air-side heat transfer area (m^2)

$h_{a,f}$ is the air-side fouling factor (heat conductance) (W/m$^2\cdot$K); for industrial moisture condensers in operation for more than 5 years, the air-side fouling factor may be of 2.84 kW/m$^2\cdot$K

R_c is the thermal contact conductance between the tube and fin surfaces in contact where a temperature drop exists; it is defined as the ratio between this temperature drop and the average heat transfer density across the interface (W/m$^2\cdot$K); the contact resistance between moisture condenser's tubes and fins (that can range between 10 and 16 kW/m$^2\cdot$K, with mean values of around 13 kW/m$^2\cdot$K) appears because the heat is exchanged across an interface where two surfaces are in imperfect contact (Wang and Chi 2000; Wang et al. 2000)

A_c is the contact area between fins and tubes (m^2)

k_{wall} is tube wall thermal conductivity (W/m$\cdot$K)

L is the total length of circular tubes (m)

d_{ext} is the external diameter of circular tubes (m)

d_{int} is the internal diameter of circular tubes (m)

η_{refr} is the refrigerant-side overall surface efficiency (–)

$h_{r,f}$ is the refrigerant-side fouling factor (conductance) (W/m$^2 \cdot$K)
h_{refr} is the refrigerant-side heat transfer coefficient (W/m$^2 \cdot$K)
A_{refr} is the refrigerant-side tube surface area (m^2)

The calculation of overall heat transfer coefficients depends on whether they are based on the air-side or refrigerant-side surface area; if $A_{refr} \neq A_{air}$, then $U_{refr} \neq U_{air}$.

Based on the air-side heat transfer area, equation 13.83 becomes

$$\frac{1}{U_{air}} = \frac{1}{\eta_{air} h_{air}} + \frac{1}{\eta_{air} h_{a,f}} + \frac{A_{air}}{R_c A_c} + \frac{A_{air} * \ln\left(\frac{d_{ext}}{d_{int}}\right)}{2\pi * L * k_{wall}} + \frac{1}{\eta_{refr} h_{r,f} \frac{A_{refr}}{A_{air}}} + \frac{1}{\eta_{refr} * h_{refr} \frac{A_{refr}}{A_{air}}} \tag{13.84}$$

In equations 13.81, 13.83, and 13.84 the wall thermal conduction resistance term $\left(\frac{1}{2\pi k_{wall} L} \ln \frac{d_{ext}}{d_{int}}\right)$ is often neglected, since a thin wall of large thermal conductivity is generally used. Also, one of the convection coefficients is often much smaller than the other and hence dominates determination of the overall coefficient. For example, if one of the fluids is air and the other is a liquid–vapor mixture experiencing boiling (or condensation), the air-side convection coefficient is much smaller, which justifies the usage of fins to enhance air-side convection heat transfer.

13.6.2 Heat Transfer Rate

In heat pump-assisted wood drying, the actual evaporating temperatures of the heat pump depend on the lumber moisture content and the dry- and wet-bulb temperatures of the drying process. They are usually imposed by the kiln designer, or by the kiln operators, via specific schedules. In such systems, the heat pump's refrigerant evaporating temperature must always be lower than the actual dry-bulb temperature of the drying air.

The heat pump cooling and dehumidifying capacity always decreases as the evaporating temperature drops (Figure 13.13a), and also as the condensing temperature increases (Figure 13.13b).

Compared to the influence of the evaporating temperature, each degree change in the condensing temperature affects the refrigerating capacity, however, to a lesser extent than a degree change in evaporating temperature (Figure 13.13b). The reason for this difference is that changes in the evaporating temperature exert a considerable effect on the specific volume entering the compressor, while the condensing temperature does not (Stoecker 1998). At high evaporating temperatures, the increase in cooling and dehumidifying capacity is approximately 4%/°C and at low evaporating temperatures, near the maximum pressure ratios of reciprocating compressors, the decrease in refrigerating capacity is about 9%/°C.

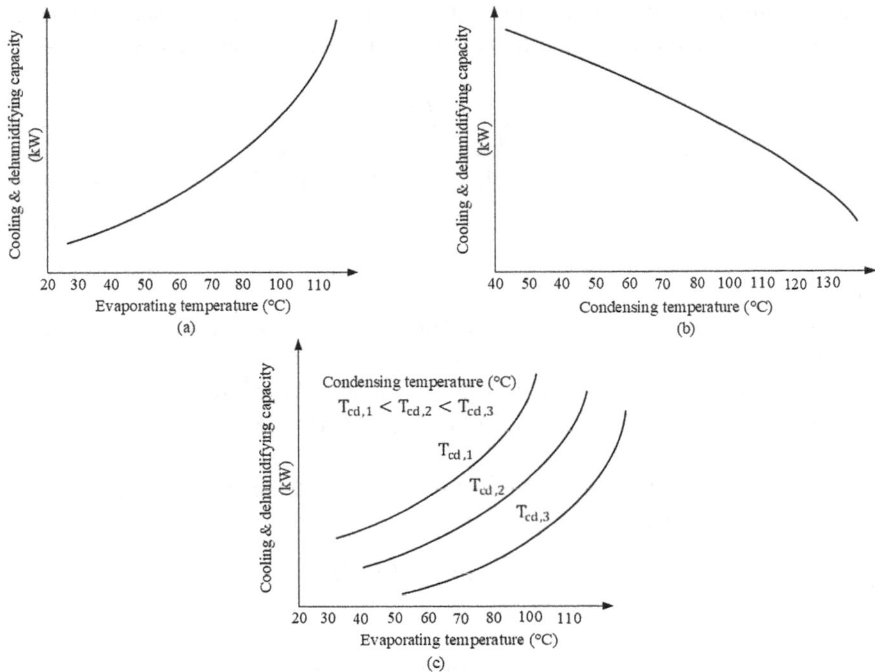

FIGURE 13.13 Cooling and dehumidifying capacity variation with evaporating and condensing temperatures.

The comparison of the influences of evaporating and condensing temperatures on the refrigerating capacity also appears on the complete map of refrigeration capacity, as shown in Figure 13.13c, as controlled by the evaporating and condensing temperatures.

The design of heat pump evaporators requires the following data: (i) the air-side parameters, (ii) the air mass flow rate through the evaporator, and (iii) the average evaporation temperature.

The evaporator heat transfer rate can be determined based on the following conservation of energy equation (see Figure 13.11):

$$\dot{Q}_{evap} = \eta_{air}\dot{m}_{air}\left(h_{air,in} - h_{air,out}\right) = \dot{m}_{refr}\left(h_{refr,out} - h_{refr,in}\right) \tag{13.85}$$

where

$\dot{Q}_{evap}$ is the evaporator heat transfer rate (W)
η_{air} is the air-side heat transfer efficiency (–)
$\dot{m}_{air}$ is the air mass flow rate (kg/s)
$\dot{m}_{refr}$ is the refrigerant mass flow rate (kg/s)
h is the air and refrigerant enthalpy at the evaporator inlets and outlets (J/kg) (see Figure 13.11).

By accounting both sides fouling and using equation 13.83, the moisture condenser (heat pump evaporator) total heat transfer rate (*W*) can be calculated as follows:

$$\dot{Q}_{evap} = \frac{LMTD}{\frac{1}{h_{air}A_{air}\eta_{air}} + \frac{1}{\eta_{air}h_{a,f}A_{air}} + \frac{1}{R_c A_c} + \frac{1}{2\pi k_{all}L} ln\frac{d_{ex}}{d_{int}} + \frac{1}{\eta_{refr}h_{r,f}A_{refr}} + \frac{1}{\eta_{refr}h_{refr}A_{refr}}} \tag{13.86}$$

where

$h_{air} = h_{sens} + h_{lat}$ is the air-side total heat transfer coefficient accounting for composed for sensible and latent heat transfer coefficients (W/m^2·K)
h_{sens} is the air-side sensible heat transfer coefficient (W/m^2·K)
h_{lat} is the air-side latent heat transfer coefficient (W/m^2·K)

The heat transfer rate between the moist drying air and the evaporating refrigerant can be also expressed as follows:

$$\dot{Q}_{evap} \approx \bar{U} * A * LMTD \tag{13.87}$$

where

$\dot{Q}_{evap}$ is the total heat transfer rate of the refrigerant evaporator (moisture condenser) (W)
A is the average heat transfer area (m^2)
$LMTD = \frac{T_{a,\,in} - T_{a,\,out}}{ln\frac{T_{a,\,in} - T_{refr}}{T_{a,\,out} - T_{refr}}}$ is the logarithmic mean temperature difference (°C) (see Figure 13.11)
$T_{a,in}$ is the drying air inlet temperature (°C)
$T_{a,out}$ is the drying air outlet temperature (°C)
T_{refr} is the refrigerant evaporating (constant) temperature (°C)

The heat exchanger design consists of selecting an appropriate heat exchanger type and determining the size, that is, the heat transfer surface area *A* required to achieve the desired outlet temperature.

For performing a heat exchanger design, both *LMTD* method and the *NTU* approach can be used to obtain equivalent results. However, depending on the nature of the problem, the *NTU* approach may be easier to implement (Incropera and DeWitt 2002).

The use of the *LMTD* method is facilitated by knowledge of the air and refrigerant inlet and outlet temperatures. Typically, the fluid inlet temperatures and flow rates, as well as a desired air or refrigerant outlet temperatures, are prescribed.

Based on Equation 13.87, the required heat transfer area can be calculated as follows:

$$A \approx \frac{\dot{Q}_{evap}}{\bar{U} * LMTD} \tag{13.88}$$

Alternatively, the heat exchanger type and size may be known, and the objective is to determine the heat transfer rate and the fluid outlet temperatures for prescribed fluid flow rates and inlet temperatures.

Exercise E13.1

In a heat pump-assisted wood dryer, the drying air enters the heat pump evaporator (moisture condenser) with a constant volumetric flow rate of $\dot{V} = 3.25\,\text{m}^3/\text{s}$, *DBT* of 40°C, and *RH* = 50%. The drying air leaves the heat pump evaporator at *DBT* = 23.9°C and *RH* = 100%.

Determine,

1. The cooling and dehumidification capacity of the heat pump evaporator,
2. The rate of moisture removal from the drying moist air.

Solution

Based on given *DBT* = 40°C and relative humidity *RH* = 50%, the specific volume of the air entering the moisture condenser results from a Mollier diagram or a moist air properties software: $v = 0.921$ m3/kg.

The mass flow rate of dry air is as follows:

$$\dot{m}_{air} = \frac{\dot{V}}{v} = \frac{3}{0.921} = 3.25\,\text{kg/s}$$

The specific enthalpy of air entering the moisture condenser results from a Mollier diagram or a moist air properties software:

$$h_{air,\,in} = 102.4\,\text{kJ/kg}$$

The specific enthalpy of leaving the moisture condenser also results from a Mollier diagram or a moist air properties software:

$$h_{air,\,out} = 73.1\,\text{kJ/kg}$$

1. The moisture condenser's cooling and dehumidification capacity is as follows:

$$Q = \dot{m}_{air}\left(h_{air,\,in} - h_{air,\,out}\right) = 3.25(102.4 - 73.1) = 95.224\,\text{kW}$$

2. The absolute humidity of air entering the moisture condenser results from a Mollier diagram or a moist air properties software:

$$\omega_{air,\,in} = 0.02395\ \text{kg}_{water}/\text{kg}_{dry\,air}$$

The absolute humidity of leaving the moisture condenser also results from a Mollier diagram or a moist air properties software:

$$\omega_{air,\,out} = 0.019\,\text{kg}_{water}/\text{kg}_{dry\,air}$$

The rate of moisture removal from the drying moist air is as follows:

$$m_{water} = \dot{m}_{air} * \left(\omega_{air,\,in} - \omega_{air,\,out}\right) = 3.25(0.02395 - 0.019) = 0.01608\,\text{kg/s}$$

REFERENCES

Adamek, T., R.L. Webb. 1990. Prediction of film condensation on horizontal integral fin tubes. *International Journal of Heat and Mass Transfer* 33(8):1721–1735.

Adams, D.C., J. Burr, P. Hrnjak, T. Newell. 2006. Two-phase pressure drop of CO2, ammonia and R-245fa in multiport aluminum microchannel tubes. *Purdue University Purdue e-Pubs, International Refrigeration and Air Conditioning Conference, School of Mechanical Engineering.*

Altman, M., F.W. Staub, R.H. Norris. 1960. Local heat transfer and pressure drop for refrigerant-22 condensing to horizontal tubes. *Chemical Engineering Progress Symposium Series* 56(30):151–160.

Arora, C.P. 1967. Power savings in refrigerating machines using mixed refrigerants. In *Proceedings of the 12th International Congress of Refrigeration,* Madrid, Vol. 2:397–409.

Backer, O. 1960. Designing pipelines for simultaneous flow of oil and gas. *Section Pipeline Engineer*, PH 69.

Barthau, G. 1976. Experimental investigation of convective boiling of ammonia at high pressure. *Heat and Mass Transfer Source Book, 5th All-Union Conference,* Scripta Publishing, Minsk, Ukraine, pp. 106–110.

Beatty, K.O., D.L. Katz. 1948. Condensation of vapors on outside of finned tubes. *Chemical Engineering Progress* 44:55–70.

Bertsch, S.S., E.A. Groll, S.V. Garimella. 2008. Review and comparative analysis of studies on saturated flow boiling in small channels. *Nanoscale Microscale Thermo-Physics Engineering* 12(3):187–227.

Bertsch, S.S., E.A. Groll, S.V. Garimella. 2009. Effects of heat flux, mass flux, vapor quality, and saturation temperature on flow boiling heat transfer in micro-channels. *International Journal of Multiphase Flow* 35:142–154.

Blanco Castro, J., J. Urchueguía, J.M. Corberán, J. Gonzálvez. 2005. Optimized design of a heat exchanger for an air-to-water reversible heat pump working with propane (R290) as refrigerant: Modelling analysis and experimental observations. *Applied Thermal Engineering* 25(14&15):2450–2462.

Blasius, P.R.H. 1913. Das Aehnlichkeitsgesetz bei Reibungsvorgangen in Flüssigkeiten. *Forschungsheft* 131:1–41.

Bogles, E., J.G. Collier, J.M. Delhaye, G.F. Hewittand, F. Mayinger. 1981. *Two Phase Flow and Heat Transfer in the Power and Process Industries*. McGraw-Hill, New York, 250 pp.

Bortolin, S., D. Del Col, L. Rosseto. 2009. Flow boiling of R-245fa in a single circular micro-channel. In *Proceedings of the 2nd Micro and Nano Flows Conference*, West London, UK, 1–2 September.

Boyman, T. 1997. Experiments and experiences with small vapor compression refrigeration plants with ammonia and miscible oil, heat transfer issues in natural refrigerants. In *Proceedings of IIR Conference in Nov. 1997*, College Park, USA, pp. 203–220.

Boyman, T., P. Aecherli, A.S.W. Wettestein. 2004. Flow boiling of ammonia in smooth horizontal tubes in the presence of immiscible oil. *International Refrigeration and Air Conditioning Conference at Purdue*, July 12–15, Paper 656, (http://docs.lib.purdue.edu/iracc/656, accessed August 19, 2016).

Cavallini, A. 1995. Working fluids for mechanical refrigeration. In *Proceedings of the 19th International Congress of Refrigeration*, The Hague, August 20–25.

Chaddock, J., G. Buzzard. 1986. Film coefficients for in-tube percentages of mineral oil. *ASHRAE Transactions* 92(1A):22–40.

Chang, Y.J., S.K. Chiang, T.W. Chung, C.C. Wang. 2000. Two-phase frictional characteristics of R-410A and air-water in a 5 mm smooth tube. *ASHRAE Transactions* DA-00-11-3, Vol. 109, pp. 792–797.

Charnay, R., R. Revellin, J. Bonjour. 2014. Flow boiling characteristics of R-245fa in a minichannel at medium saturation temperatures. *Experimental Thermal and Fluid Science* 59:184–194.

Chen, J.C. 1966a. Correlation for boiling heat transfer to saturated fluids in convective flow. *Industrial & Engineering Chemistry Process Design and Development* 5:322–329.

Chen, J.C. 1966b. A correlation for boiling heat transfer to saturated fluids in convective flow. *Industrial and Engineering Chemistry, Process Design and Development* 5 (3):322–329.

Chiriac, F., R. Dumitrescu. 2006. *Compléments of Heat and Mass Transfer (in Rumanian)*, AGIR Publisher, Bucharest, 239 pp.

Choi, J.M., W.V. Payne, P.A. Domanski. 2003. Effects of non-uniform refrigerant and air flow distributions on finned-tube evaporator performance. In *Proceedings of the 21st International Congress of Refrigeration*, Washington, DC.

Choi, K.-I., A.S. Pamitran, J.-T. Oh, K. Saito. 2009. Pressure drop and heat transfer during two-phase flow vaporization of propane in horizontal smooth mini-channels. *International Journal of Refrigeration* 32(5):837–845.

Chwalowski, M., D.A. Didion, P.A. Domanski. 1989. Verification of evaporator computer models and analysis of performance of an evaporator coil. *ASHRAE Transactions* 95(1):1229–1236.

Coleman, J.W., S. Garimella. 1999. Characterization of two-phase flow patterns in small diameter round and rectangular tubes. *International Journal of Heat and Mass Transfer* 42(15):2869–2881.

Collier, J.G. 1981. *Convective Boiling and Condensation*. 2nd edition, McGraw-Hill, New York.

Collier, J.G., J.R. Thome. 1994. *Convective Boiling and Condensation*. Clarendon Press, Oxford, England, UK, 640 pages.

Cooper, M.G. 1984. Saturation nucleate pool boiling. In *Proceedings of the 1st U.K. National Conference on Heat Transfer* 2:785–793, University of Leeds, 3–5 July.

Copetti J.B., M.H. Macaganan, F. Zinani. 2013. Experimental study on R-600a boiling in 2.6 mm tube. *International Journal of Refrigeration* 33:325–334.

Del Col, D., S. Bortolin, D. Torresin, A. Cavallini. 2013. Flow boiling of R1234yf in a 1 mm diameter channel. *International Journal of Refrigeration* 36:353–362.

Dhir, V.K., J.H. Lienhard. 1971. Laminar film condensation on plane and axisymmetric bodies in non-uniform gravity. *Journal of Heat Transfer* 93(1):97–100 (DOI: 10.1115/1.3449773; https://www.researchgate.net/publication/23898319. Accessed March 5, 2017).

Dittus, F.W., L.M.K. Boelter. 1930. Heat transfer in automobile radiator of the tubular type. *University of California at Berkley Publications in Engineering* 2:443–461.

Domanski, P.A., D. Yashar, K.A. Kaufman, R.S. Michalski. 2004. An optimized design of finned-tube evaporators using the learnable evolution model. *HVAC&R Research* 10(2):201–211.

Dupont, V., J.R. Thome. 2005. Evaporation in micro-channels: Influence of the channel diameter on heat transfer. *Microfluid Nanofluid* 1:119–127.

Erdem, S. 2015. The effects of fin-and-tube evaporator geometry on heat pump performance under dehumidifying conditions. *International Journal of Refrigeration* 57:35–45.

Fagan, T.M. 1980. The effects of air flow maldistributions on air-to-refrigerant heat exchanger performance. *ASHRAE Transactions* 86(2):699–713.

Fernando, P., B. Palm, P. Lundqvist, E. Granryd. 2004. Propane heat pump with low refrigerant charge: Design and laboratory tests. *International Journal of Refrigeration* 27(7):761–773.

Fernando, P., B. Palm, T. Ameel, P. Lundqvist, E. Granryd. 2008. A mini-channel aluminium tube heat exchanger – Part II: Evaporator performance with propane. *International Journal of Refrigeration* 31(4):681–695.

Field, B.S., P.S. Hrnjak. 2007. *Two-phase Pressure Drop and Flow Regime of Refrigerants and Refrigerant-oil Mixtures in Small Channels.* Air Conditioning and Refrigeration Center University of Illinois (https://pdfs.semanticscholar.org/.pdf; Accessed December 31, 2017).

Fink, J. 1982. Forced convection boiling of a mixture of Freon-11 and Freon-113 flowing normal to a cylinder. In *Proceedings 7th Int. Heat Transfer Conference,* Munich, Vol. 4:207–212.

Foster, H.K., N. Zuber. 1955. Dynamics of vapor bubbles and boiling heat transfer. *AIChE Journal* 1:531–535.

Friedel, L. 1979. Improved friction pressure drop correlations for horizontal and vertical two-phase pipe flow. In *the European Two-phase Flow Group Meeting, Paper E2*, June, Ispra, Italy.

Gnielinski, V. 1976. New equations for heat and mass transfer in turbulent pipe and channel flow. *International Chemical Engineering* 16(2):359–368.

Goldstein, S.D. 1983. A mathematically complete analysis of plate-fin heat exchangers. *ASHRAE Transactions* 89(2): 447–469.

Goodman, W. 1936. Dehumidification of air with coils. *Refrigerating Engineering* 82:225–275.

Granryd, E. 2001. Hydrocarbons as refrigerants - an overview. *International Journal of Refrigeration* 24:190–197.

Grauso, S., R. Mastrullo, A.W. Mauro, J.R. Thome, G.P. Vanoli. 2013. Flow pattern map, heat transfer and pressure drops during evaporation of R-1234ze(E) and R134a in a horizontal, circular smooth tube: Experiments and assessment of predictive methods. *International Journal of Refrigeration* 36:478–491.

Griffith, O. 1998. Two-phase flow, Section 14, In *Handbook of Heat Transfer*, edited by W.M. Rohsenow, J.P. Hartnett, Y.I. Cho, 3rd Edition, McGraw-Hill, New York.

Gross, U. 1994. Einfluss von Oel auf die Zwangskonvektion im Verdampferrohr. *Ki Luft- und Kaeltetechnik.* 6:269–274.

Gungor, K.E., R.H.S. Winterton. 1986. A general correlation for flow boiling in tubes and annuli. *International Journal of Mass and Heat Transfer* 29(3):351–358.

Gungor, K.E., R.H.S. Winterton. 1987. Simplified general correlation for saturated flow boiling and comparisons of correlations with data. *The Canadian Journal of Chemical Engineering* 65(1):148–156.

Halici, F., I. Taymaz. 2005. Experimental study of the air-side performance of tube row spacing in finned tube heat exchangers. *Heat and Mass Transfer* 42(9):817–822.

Halici, F., I. Taymaz, M. Gunduz. 2001. The effect of the number of tube rows on heat, mass and momentum transfer in flat-plate finned tube heat exchangers. *Energy* 26(11):963–972.

Han, X., P. Li, X. Yuan, Q. Wang, G. Chen. 2013. The boiling heat transfer characteristics of the mixture HFG-1234yf/oil inside a micro-fin tube. *International Journal of Heat and Mass Transfer* 67:1122–1130.

Hermes, C.J.L. 2012. Conflation of ε – NTU and EGM design methods for heat exchangers with uniform wall temperature. *International Journal of Heat and Mass Transfer* 55:3812–3817.

Hermes, C.J.L. 2013. Thermodynamic design of condensers and evaporators: Formulation and applications. *International Journal of Refrigeration* 36:633–640.

Hihara, E., T. Saito. 1991. A binary mixture operated heat pump. In *Proceeding of ASME/JSME Thermal Engineer Conference,* Vol. 3, pp. 297–304, Reno, NV (United States), 17–22 March.

Hiroshi, H. 2017. *Tube Banks, Condensation Heat Transfer.* (www.thermopedia.com/fr/content/1210; Accessed January 22, 2017).

Hossain, Md. A., Y. Onaka, H.M.M. Afroz, A. Miyara, 2013. Heat transfer during evaporation of R-1234ze(E), R-32, R-410A and a mixture of R-1234ze(E) and R-32 inside a horizontal smooth tube. *International Journal of Refrigeration* 36:465–477.

Huang, K.H., M.B. Pate. 1988. A model of an air-conditioning condenser and evaporator with emphasis on in-tube enhancement. *International Refrigeration and Air Conditioning Conference*. Paper 79 (http://docs.lib.purdue.edu/iracc/79, accessed March 7, 2017).

Incropera, F.P., D.P. DeWitt. 2002. *Fundamentals of Heat and Mass Transfer*, 5th edition. John Willey & Sons, New York, 470 pp.

Ito, M., H. Kimura, T. Senshu. 1977. Development of high efficiency air-cooled heat exchanger. *Hitachi Review* 28(10):323–326.

Jain, V.K., P.L. Dher. 1983. Studies on flow boiling of mixtures of refrigerants R-12 and R-13 inside a horizontal tube. In *Proceedings of the 16th International Congress of Refrigeration*, August 23–30, Paris.

Jarall, S. 2012. Study of refrigeration system with HFO-1234yf as a working fluid. *International Journal of Refrigeration* 35:1668–1677.

Jin, H., J.D. Spitler. 2002. A parameter estimation based model of water-to-water heat pumps for use in energy calculation programs. *ASHRAE Transactions* 108(1):3–17.

Jung, D.S., M. McLinden, R. Radermacher, D. Didion. 1989a. A study of flow boiling heat transfer with refrigerant mixtures. *International Journal of Heat and Mass Transfer* 32(9):1751–1764.

Jung, D.S., M. McLinden, R. Rademacher, D. Didion. 1989b. Horizontal flow boiling heat transfer experiments with a mixture of R22/R114. *International Journal of Heat Mass Transfer* 32:131:145.

Jung, D.S., R. Radermacher. 1989. Prediction of pressure drop during horizontal annular flow boiling of pure and mixed refrigerants. *International Journal of Heat Mass Transfer* 32(12):2435–2446.

Jung, D.S., R. Radermacher. 1993. Prediction of evaporation heat transfer coefficient and pressure drop of refrigerant mixtures in horizontal tubes. *International Journal of Refrigeration* 16(3):201–209.

Kakaç, S., H. Liu. 2002. *Heat Exchangers: Selection, Rating, and Thermal Design*. CRC Press, Boca Raton, FL.

Kandlikar, S.G. 1987. A general correlation for saturated two-phase flow boiling heat transfer inside horizontal and vertical tubes. In *Proceedings of ASME Winter Annual Meeting*, Boston, MA, December 14–18.

Kandlikar, S.G. 1990a. A general correlation for saturated two-phase flow boiling heat transfer inside horizontal and vertical tubes. *Journal of Heat Transfer Journal of Heat Transfer* 112:219–228.

Kandlikar, S.G. 1990b. Thermal design theory for compact evaporators, In *Compact Heat Exchangers*, edited by R.K. Shah, A.D. Kraus, D. Metzger. Hemisphere, New York, pp. 245–286.

Kandlikar, S.G. 1991. A model for correlating flow boiling heat-transfer in augmented tubes and compact evaporators. *ASME Journal of Heat Transfer* 113(4):966–972.

Kandlikar, S.G., P. Balasubramanian. 2004. An extension of the flow boiling correlation to transition, laminar and deep laminar flows in mini-channels and micro-channels. *Heat Transfer Engineering* 25:86–93.

Kandlikar, S.G., W.J. Grande. 2003. Evolution of microchannel flow passages –thermo-hydraulic performance and fabrication technology. *Heat Transfer Engineering* 24(1):3–17.

Kattan, N., J.R. Thome, D. Favrat. 1998a. Flow boiling in horizontal tubes. Part 1: Development of adiabatic two-phase flow pattern map. *ASME Journal of Heat Transfer* 120:140–147.

Kattan, N., J.R. Thome, D. Favrat. 1998b. Flow boiling in horizontal tubes. Part 2: New heat transfer data for five refrigerants. *ASME Journal of Heat Transfer* 120:148–155.

Kattan, N., J.R. Thome, D. Favrat. 1998c. Flow boiling in horizontal tubes. Part 3: Development of a new heat transfer model based on flow patterns. *ASME Journal of Heat Transfer* 120:156–165.

Khanpara, J.C. 1986. *Augmentation of in-tube evaporation and condensation with micro-fin tubes using R-113 and R-22*. Ph.D. dissertation, Iowa State University, Ames, IA.

Khanpara, J.C., A.E. Bergles, M.R. Pate. 1986. Augmentation of R-113 in-tube evaporation with micro-fin tubes. In *Proceeding of ASME Winter Annual Meeting*, Anaheim, California, December 7–12, pp. 21–32; *ASHRAE Transactions* 92(2):506–524.

Kirby, E.S., C.W. Bullard, W.E. Dunn. 1998. Effect of airflow non-uniformity on evaporator performance. *Transactions of the American Society of Heating, Refrigeration, and Air Conditioning Engineers* 104(2):755–762.

Kutateladze, S.S. 1961. Boiling heat transfer. *International Journal of Heat and Mass Transfer* 4:31–45.

La Yssac, T., S. Lips, R. Revellin. 2015. Assessment of stratification during horizontal two-phase flow of R-245fa: Intermittent and annular flows, In *Proceedings of the 22nd Mechanical French Congress*, Lyon, France, 24–28 August.

Lee, J., P.A. Domanski. 1997. Impact of air and refrigerant maldistribution on the performance of finned-tube evaporators with R-22 and R-407C. *Building Environment Division National Institute of Standards & Technology*, ARTI MCLR Project Number 665–54500, pp. 1–31.

Lee, H.S., J.J. Yoon, J.D. Kim, P.K. Bansal. 2006. Heat transfer and pressure drop characteristics of hydrocarbon refrigerants. *International Journal of Heat and Mass Transfer* 49:1922–1927.

Liu, Z., R.H.S. Winterton. 1988. Wet wall flow boiling correlation with explicit nucleate boiling term, In *Multi-Phase Transport and Particulate Phenomena*, edited by T. Nejat Veziroglu, Hemisphere, Washington, DC, Vol. 1, pp. 419–432.

Liu, Z., R.H.S. Winterton. 1991. A general correlation for saturated and subcooled flow boiling in tubes and annuli based on a nucleate pool boiling equation. *International Journal of Heat and Mass Transfer* 34(11):2759–2766.

Liu, Z., R.H.S. Winterton. 1999. A general correlation for saturated and subcooled flow boiling in tubes and annuli, based on a nucleate pool boiling equation. *International Journal of Heat and Mass Transfer* 34(11):2759–2766.

Lockhart, R.W., R.C. Martinelli. 1949. Proposed correlation of data for isothermal two-phase, two-component flow in pipes. *Chemical Engineering Progress* 45(1):39–48.

Longo, G.A., S. Mancin, G. Righetti, C. Zilio. 2016. Saturated flow boiling of HFC-134a and its low GWP substitute HFO-1234ze(E) inside a 4 mm horizontal smooth tube. *International Journal of Refrigeration* 64:32–39.

Longo, G.A., C. Zilio, G. Righetti, J.S. Brown. 2014. Experimental assessment of the low GWP refrigerant HFO-1234ze(Z) for high temperature heat pumps. *Experimental Thermal and Fluid Science* 57:293–300.

Lorentzen, G. 1988. Ammonia: An excellent alternative. *International Journal of Refrigeration* 11:248–252.

Lorentzen, G. 1994. Revival of carbon dioxide as a refrigerant. *International Journal of Refrigeration* 17:292–301.

Malek, A., R. Colin. 1983. Ébullition de l'ammoniac en tube long. Transfert de chaleur et pertes de charges en tube vertical et horizontal. *Centre Technique des Industries Mécaniques*, Senlis, France, CETIM-14-011, pp. 1–65.

Martinelli, R.C., B. Nelson. 1948. Prediction of pressure drop during forced-circulation boiling of water. *Transactions ASME* 70(6):695–702.

Mastrullo, R., A.W. Mauro, L. Menna, G.P. Vanoli. 2014. Replacement of R404A with propane in a light commercial vertical freezer: A parametric study of performances for different system architectures. *Energy Conversion and Management* 82:54–60.

McQuiston, F.C. 1975. Fin efficiency with combined heat and mass transfer. *ASHRAE Transactions* 81(1):350–384.

McQuiston, F.C. 1976. Heat, mass and momentum transfer in a parallel plate dehumidifying exchanger. *ASHRAE Transactions* 82(2):87–105.

McQuiston, F.C., J.P. Parker. 1994. *Heating, Ventilating and Air Conditioning – Analysis and Design*, John Wiley & Sons, New York.

McQuiston, F.C., J.D. Parker, J.D. Spitler. 2005. *Heating, Ventilating, and Air Conditioning Analysis and Design*, 6th edition. John Wiley & Sons, Inc, Hoboken, NJ.

Mendoza-Miranda, J.M., J.J. Ramirez-Minguela, V.D. Munoz-Carpio, J. Nauarro-Esbri. 2015. Development and validation of a micro-fin tubes evaporator model using R134a and R1234yf as working fluids. *International Journal of Refrigeration* 50:32–43.

Minea, V. 2004. Heat pumps for wood drying – new developments and preliminary results, In *Proceedings of the 14th International Drying Symposium*, Sao Paulo, Brazil, August 22–25, Volume B, pp. 892–899.

Minor, B., M. Spatz. 2008a. HFO-1234yf: low GWP refrigerant update. *International Refrigeration and Air Conditioning Conference, Purdue University* (http://docs.lib.purdue. edu/iracc/937, accessed January 12, 2017).

Minor, S., M.A. Spatz. 2008b. HFO-1234yf: a low GWP refrigerant for MAC. *VDA Alternative Refrigerant Winter Meeting*, Saalfelden, Salzburg.

Mishima, K., T. Hibiki. 1996. Some characteristics of air–water two-phase flow in small diameter vertical tubes. *International Journal of Multiphase Flow* 22 703–712.

Miyama, A. 1988. Pressure drop of condensation and boiling inside a horizontal tube with refrigerant mixtures R22/114. In *Proceedings of the 25th Japan National Heat Transfer Symposium*, pp. 85–87.

Mulroy, W., D. Didion. 1986. The performance of a conventional residential sized heat pump operating with a non-azeotropic binary refrigerant mixture. *NBSIR 863422, NBS*, Gaithersburg, MD 20899.

Mulroy, W., M. Kauffeld, M. McLinden, D. Didion. 1988. An evaluation of two refrigerant mixtures in a breadboard air conditioner. *International Institute of Refrigeration Conference, Purdue University*, July.

Murata, K., et al. 1991. An experimental investigation on forced convection boiling of refrigerant mixtures. In *Proc. 18th Int. Congress of Refrigeration,* Montreal, Vol. 2:494–498.

Nakasawa, T., K. Tanabe, K. Ushikubo, M. Hiraga. 1984. Performance evaluation of serpentine evaporator for automotive air conditioning system. *SAE Technical Paper Series*, Paper No. 840384.

Niknejab, J., J.W. Rose. 1981. Interphase matter transfer – an experimental study of condensation of mercury. *Proceedings of the Royal Society A, Mathematical Physical and Engineering Sciences* 378:305–327, London.

Nusselt, W. 1916. Die Oberflachenkondensation des Wasserdampfes (in German). *Z. Vereines Deutsch. Ing.* 60: 541–546, 569–575.

Oh, J.T., K.I. Choi, N.B. Chien. 2015. Pressure drop and heat transfer during a two-phase flow vaporization of propane in horizontal smooth mini-channels. DOI: 10.5772/60813 (http://www.intechopen.com/books/heat-transfer-studies-and-applications/pressure-drop-and-heat-transfer-during-a-two-phase-flow-vaporization-of-propane-in-horizontal-smooth, accessed March 14, 2017).

Ohadi, M.M., S.S. Li, R. Radermacher, S. Dessiatoun. 1996. Critical review of available correlations for two-phase flow heat transfer of ammonia. *International Journal of Refrigeration* 19(4):272–284.

Ong, C.L., J.R. Thome. 2009. Flow boiling heat transfer of R 134a, R236fa and R245fa in a horizontal 1.030 mm circular channel. *Experimental Thermal and Fluid Science* 33: 651–663.

Oskarsson, S.P., K.I. Krakow, S. Lin. 1990a. Evaporator models for operation with dry, wet, and frosted finned surfaces - Part I: Heat transfer and fluid flow theory. *ASHRAE Transactions* 96(1):373–381.

Oskarsson, S.P., K.I. Krakow, S. Lin. 1990b. Evaporator models for operation with dry, wet, and frosted finned surfaces - Part II: Evaporator models and verification. *ASHRAE Transactions* 96(1):381–392.

Oskarsson, S.P., K.I. Krakow, S. Lin. 1990c. Evaporator models for operation with dry, wet, and frosted finned surfaces. Part I: Heat transfer and fluid flow theory. *ASHRAE Transactions* 96(1):373–381.

Pacheco-Vega, A., M. Sen, K.T. Yang, R.L. McClain. 2001. Neural network analysis of fin-tube refrigerating heat exchanger with limited experimental data. *International Journal of Heat and Mass Transfer* 44(2001):763–770.

Padilla, M., R. Revellin, P. Haberschill, A. Bensafi, J. Bonjour. 2011. Flow regimes and two-phase pressure gradient in horizontal straight tubes: Experimental results for HFO-1234yf, R-134a and R-410A. *Experimental Thermal and Fluid Science* 35(6):1113–1126.

Palm, B. 2007. Refrigeration systems with minimum charge of refrigerant. *Applied Thermal Engineering* 27(10):1693–1701.

Panchal, C.B., J.J. Lorenz, D.L. Hillis. 1981. The effects of ammonia contamination by water on OTEC power system performance. Technical Report, ANL/OTEC-PS-8 Argonne National Laboratory.

Parker, J.D., J.H. Boggs, E.F. Blick. 1969. *Introduction to Fluid Mechanics and Heat Transfer.* Addison-Wesley, Reading, MA.

Pike-Wilson, E.A., T.G. Karayiannis. 2014. Flow boiling of R245fa in 1.1 mm diameter stainless steel, brass and copper tubes. *Experimental Thermal and Fluid Science* 59:166–183.

Pirompugd, W., C.-C. Wang, S. Wongwises. 2008. Finite circular fin method for wavy fin-and-tube heat exchangers under fully and partially wet surface conditions. *International Journal of Heat and Mass Transfer* 51(15–16):4002–4017.

Radermacher, R., Y. Hwang. 2005. *Vapor Compression Heat Pumps with Refrigerant Mixtures.* CRC Pres – Taylor & Francis Group, Boca Raton, FL.

Radermacher, R., H. Ross, D. Didion. 1983. Experimental determination of forced convective evaporation heat transfer coefficients for non-azeotropic refrigerant mixtures. *ASME National Heat Transfer Conference*, ASME 83-WA/HT54.

Rathod Pravin, P., R. Kumar, A. Gupta. 2011. Enhancement of condensation heat transfer over horizontal integral fin tubes – a review study. *Journal of Engineering Research and Studies*, E-ISSN 0976-7916, Article 17, Vol. II(2):76–81, April–June (www.technicaljournalsonline.com/jers/, accessed March 13, 2016).

Rich, D.G. 1973. The effect of fin spacing on the heat transfer and friction performance of multi-row, smooth plate fin tube heat exchangers. *ASHRAE Transactions* 79:135–145.

Rich, D.G. 1975. The effect of the number of tubes rows on heat transfer performance of smooth plate fin-and-tube heat exchangers. *ASHRAE Transactions* 81:307–317.

Rohlin, P. 1997. Heat transfer coefficient of zeotropic mixtures and their pure components in horizontal flow boiling -an experimental study. In *Proceedings of ASME International Mechanical Engineering Congress and Exposition*, Dallas, USA, November 16–21.

Rohsenow, W.M. 1956. Heat transfer and temperature distribution in laminar film condensation. *ASME Transactions* 78:16–45.

Rohsenow, W.M., J.R. Hartnett, Y.I. Cho. 1985. *Handbook of Heat Transfer.* 3rd edition, McGraw-Hill, New York.

Rohsenow, W., J.P. Hartnette, E.N. Ganic. 1985. *Handbook of Heat Transfer Fundamentals.* McGraw-Hill, New York, 1413 pages.

Rose, J.W. 1988a. Some aspects of condensation heat transfer theory. *International Communications in Heat and Mass Transfer* 15(4):449–473.

Rose, J.W. 1988b. Fundamentals of condensation heat transfer: Laminar film condensation. *ASME International Journal (Series II)* 31(3):357–375.

Ross, H., R. Radermacher, M. Di Marzo, D. Didion. 1987. Horizontal flow boiling of pure and mixed refrigerants. *International Journal of Heat Mass Transfer* 30(5):979–992.

Shah, M.M. 1975. Visual observations in an ammonia evaporator. *ASHRAE Transactions* 81(1):295–306.

Shah, M.M. 1976. A new correlation for heat transfer during boiling flow through pipes. *ASHRAE Transactions* 82(2):66–86.

Shah, M.M. 1978. Chart correlation for saturated boiling heat transfer: Equations and further study. *ASHRAE Transactions* 2673:185–196.

Shah, M.M. 1982. Chart correlation for saturated boiling heat transfer: Equations and further study. *ASHRAE Transactions* 88(1):185–196.

Shah, R.K., D. Sekulic. 2003. *Fundamentals of Heat Exchanger Design*. Wiley, New York.

Shepherd, D.G. 1956. Performance of one-row tube coils with thin-plate fins, low velocity forced convection. *Heating, Piping, and Air Conditioning* 28:137–144.

Shin, J.Y., M.S. Kim, S.T. Ro. 1997. Experimental study on forced convective boiling heat transfer of pure refrigerants and refrigerant mixtures in a horizontal tube. *International Journal of Refrigeration* 20(4):267–275.

Singal, L.C., C.P. Sharma, H.K. Varma. 1983a. Pressure drop during forced convection boiling of binary refrigerant mixtures. *International Journal of Multiphase Flow* 9(3):309–323.

Singal, L.C., C.P. Sharma, H.K. Vanna. 1983b. Experimental heat transfer coefficient for binary refrigerant mixtures of R-13 and R-12. *ASHRAE Transactions* 2747:175–188.

Singal, L.C., C.P. Sharma, H.K. Varma. 1984. Heat transfer correlations for the forced convection boiling of R12-R13 mixtures. *International Journal of Refrigeration* 7(5): 278–284.

Souza, A.L., M. Oimenta. 1995. Prediction of pressure drop during horizontal two-phase flow of pure and mixed refrigerants. *Journal of Heat Transfer ASME, Cavitation and Multiphase Flow* 210:161–171.

Srinivasan, J.V., R.K. Shah. 1997. Condensation in compact heat exchangers. *Journal of Enhanced Heat Transfer* 4(4):237–256.

Stoecker, W.F. 1998. *Industrial Refrigeration Handbook*. McGraw-Hill, New York.

Thome, J.R. 2004. *Engineering Data Book III*, www.thermalfluidscentral.org, e-Books Home, accessed September 28, 2016.

Thonon, B. 2008. A review of hydrocarbon two-phase heat transfer in compact heat exchangers and enhanced geometries – review. *International Journal of Refrigeration* 31:633–642.

Thonon, B. 2008. A review of hydrocarbon two-phase heat transfer in compact heat exchangers and enhanced geometries – review. *International Journal of Refrigeration* 31:633–642.

Threlkeld, J.L. 1970. *Thermal Environmental Engineering*. Prentice-Hall, Inc., Englewood Cliffs, NJ.

Tibiriça, C.B., G. Ribatski. 2010. Flow boiling heat transfer of R134a and R245fa in a 2.3 mm tube. *International Journal of Heat and Mass Transfer* 53:2459–2468.

Tran, T.N., M.W. Wambsganss, D.M. France. 1996. Small circular and rectangular-channel boiling with two refrigerants. *International Journal of Multiphase Flow* 22: 485–498.

Vakili-Farahani, F., B. Agostini, J.R. Thome. 2013. Experimental study on flow boiling heat transfer of multiport tubes with R0245fa and R-1234ze(E). *International Journal of Refrigeration* 36:335–352.

Vlasie, C., H. Macchi, J. Guilpart, B. Agostini. 2004. Flow boiling in small diameter channels. *International Journal of Refrigeration* 27:191–201.

Wang, S.P, J.C. Chato. 1995. On heat transfer with mixtures: Part 2: Boiling and evaporation. *ASHRAE Transactions Symposia*, paper CH-95-23-3:1387–1401.

Wang, C.C., K.Y. Chi. 2000. Heat transfer and friction characteristics of plain fin-and-tube exchangers. Part I: New experimental data. *International Journal of Heat and Mass Transfer* 43:2681–2691.

Wang, C.C., K.Y. Chi, C.J. Chang. 2000. Heat transfer and friction characteristics of plain fin-and-tube exchangers. Part II: Correlation. *International Journal of Heat and Mass Transfer* 43:2693–2700.

Wang, C.C., Y.J. Du, Y.J. Chang, W.H. Tao. 1999a. Airside performance of herringbone fin-and-tube heat exchangers in wet conditions. *Canadian Journal of Chemical Engineering* 77(6):1225–1230.

Wang, J., E. Hihara. 2003. Prediction of air coil performance under partially wet and totally wet cooling conditions using equivalent dry-bulb temperature method. *International Journal of Refrigeration* 26:293–301.

Wang, C.C., J.Y. Jang, N.F. Chiou. 1999b. A heat transfer and friction correlation for wavy fin-and-tube heat exchangers. *International Journal of Heat and Mass Transfer* 42:1919–1924.

Watel, B., Thonon, B. 2002. An experimental study of flow boiling of propane in a plate fin heat exchanger. *Journal of Enhanced Heat Transfer* 9:1–15.

Wattelet, J.P. 1994. Heat transfer flow regimes of refrigerants in a horizontal-tube evaporator. *Ph.D. Thesis*. University of Illinois, Urbana-Champaign.

Wattelet, J.P., I.C. Chato, A.L. Souza, B.R. Christoffersen. 1994. Evaporative characteristics of R-12, R-134a, and a mixture at low mass fluxes. *ASHRAE Transactions* 100(1):603–615.

Webb, R.L., N.S. Gupte. 1992. A critical review of correlations for convective vaporization in tubes and tube banks. *Heat Transfer Engineering* 13(3):58–81.

Webb, R.L., T.M. Rudy. 1982. Theoretical model for condensation of horizontal integral-fin tubes. *American Institute of Chemical Engineers (AIChE) Symposium Series* 225 (79):11–18.

Webb, R.L., T.M. Rudy. 1985. An analytical model to predict condensate retention on horizontal integral-fin tubes. *Journal of Heat Transfer* 107:361–368.

Weisbach, J. 1845. *Lehrbuch der Ingenieur- und Maschinen-Mechanik*, Braunschwieg. Librairie Philosophique J. Vrin, (Paris, Frankreich).

Wen, M.Y., C.Y. Ho 2005. Evaporation heat transfer and pressure drop characteristics of R-290 (propane), R-600 (butane), and a mixture of R-290/R-600 in the three-lines serpentine small-tube bank. *Applied Thermal Engineering* 25(17&18):2921–2936.

Westwater, J.W. 1983. Boiling heat transfer in compact and finned heat exchangers. *NATO Advanced Series E: Applied Science* 64:827–857, Martinus Nijhoff Publications, Boston, MA.

Wijaya, H., M. Spatz. 1994. *Two-phase Flow Heat Transfer and Pressure Drop Characteristics of R-22 and AZ-20.* Herrick Laboratories, Purdue University.

Wojtan, L., T. Ursenharcher, J.R. Thome. 2005. Investigation of flow boiling in horizontal tubes. *International Journal of Heat and Mass Transfer* 48(1&2):2955–2985.

Yan, Y.Y., T.F. Lin. 1999. Evaporation heat transfer and pressure drop of refrigerant R-134a in a plate heat exchanger. *ASME Journal of Heat Transfer* 118(1):118–127.

Yoshida, S., T. Matsunaga, H. Mori, K. Ohishi. 1990. Heat Transfer to non-azeotropic mixtures of refrigerants flowing in a horizontal evaporator tube. *Transactions of the Japan Society of Mechanical Engineers, Part B 56*(524):1084–1089.

Zilio, C., J.S. Brown, G. Schiochet, A. Cavallini. 2011. The refrigerant R-1234yf in air conditioning systems. *Energy* 36: 6110–6120.

Zürcher, O., D. Favrat, J.R. Thome. 1998. Évaporation de mélanges d'ammoniac et d'huile dans des tubes. *Rapport final*, Office Fédéral de l'Énergie, Berne, Avril.

Zürcher, O., J.R. Thome, D. Favrat. 1999. Evaporation of ammonia in a smooth horizontal tube: Heat transfer measurements and predictions. *Journal of Heat Transfer* 121:89–101.

Zyhowski, G.J., M.M. Spatz, S.Y. Motta. 2002. *An Overview of Properties Applications of HFC-245fa.* Purdue University, Purdue e-Pubs, International Refrigeration and Air Conditioning, Conference School of Mechanical Engineering.

14 Refrigerant-to-Air Condensers

14.1 INTRODUCTION

The term *air-side* of heat pumps concern evaporators (also called moisture condensers that cool and dehumidify the moist drying air) (see Figure 13.1), condensers (that reheat the return drying air), and blowers (that circulate the air through evaporators and condensers) (Jin and Spitler 2002a). In heat pump dryers, the air-cooled condensers are finned-tube coils (Stoecker 1998), where unmixed (e.g., refrigerants and air) fluids move roughly perpendicular (cross-flow) to each other (Figure 14.1a).

The refrigerant-to-air condensers, most common in heat pump-assisted drying applications, are fin-and-tube heat exchangers that reject heat from the refrigerant circulating inside horizontal tubes to the outside air. One (or several) fan(s) forces the air to circulate across the coil between the vertical fins and over the horizontal tubes thus intensifying the heat transfer.

When the refrigerant leaves the compressor, it enters the condenser as a superheated vapor and leaves as a subcooled liquid (see Figure 14.1). The condenser can be separated into three sections according to the refrigerant thermodynamic states, that is, superheated, saturated, and subcooled. The specific heat rejected can be found by evaluating the refrigerant enthalpies or temperatures at the inlets and outlets of each section.

In more detailed words, the refrigerant enters the heat pump condenser coming from the compressor as superheated vapor at temperature $T_{refr,in}$. The superheated vapor is first cooled (desuperheated) to saturation ($T_{refr,sat}$) (process 2-*a*), condensed (with the specific heat $c_p \rightarrow \infty$) at constant temperature ($T_{refr,sat} = const.$), and pressure, and then slightly subcooled up to temperature $T_{refr,in}$ transferring heat to the cooler air stream flowing over the outside finned tubes, while the temperature of the colder air increases from $T_{air,in}$ to $T_{air,out}$ (Incropera and DeWitt 2002) (Figure 14.1b). It can be seen that in the desuperheating and subcooling sections, both the refrigerant and the heated air change in temperature, while during the phase change the temperature of the refrigerant is constant as it condenses. Although the majority of the heat transfer from the refrigerant to air occurs during the phase change section, up to 15%–10% does occur during desuperheating and subcooling processes.

14.2 TYPICAL CONSTRUCTION

The refrigerant-to-air condensers can be located outside (see Figure 9.7) or inside (see Figure 9.9) the drying chambers, and oriented vertically (Figure 14.2a) or horizontally (Figure 14.2b). They are fin-and-tube heat exchangers generally with circular (round, cylindrical) internally smooth (or enhanced) tubes and plate fins on the air-side (Cavallini et al. 2003; Doretti et al. 2013).

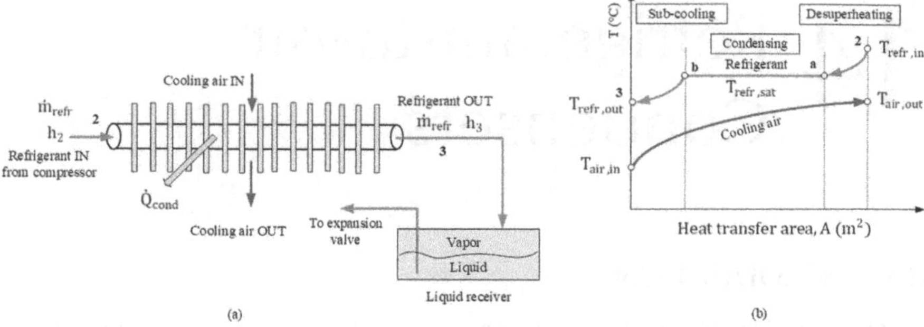

FIGURE 14.1 Cross-flow finned-tube condenser providing refrigerant vapor desuperheating and two-phase condensation, and liquid subcooling: (a) schematic; (b) temperature–heat exchange area diagram. h, specific enthalpy; $\dot{m}_r$, refrigerant mass flow rate; $\dot{Q}_{cond}$, condenser heat transfer rate; T, temperature.

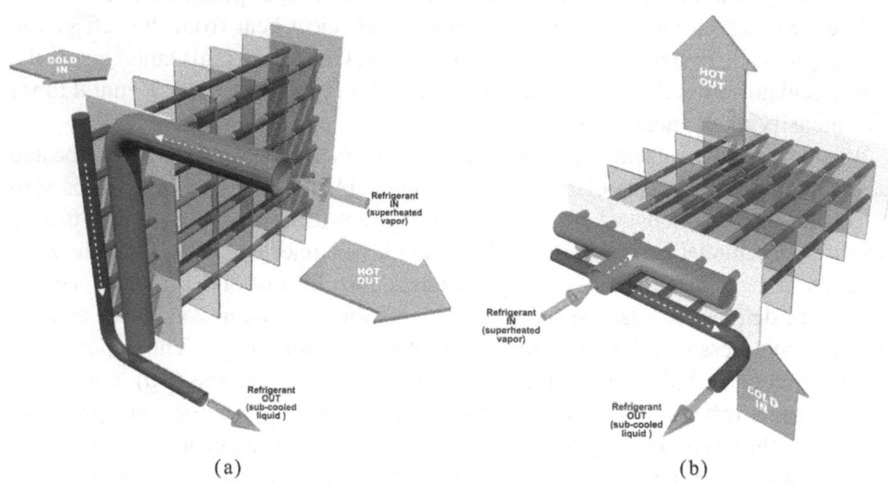

FIGURE 14.2 Air-cooled refrigerant-to-air condensers: (a) vertical arrangement; (b) horizontal arrangement (note: schematics not to scale).

Currently, fin-and-tube condensers are made of copper tubing with aluminum fins, both tubes and fins of aluminum, or combination of other special materials, such as copper–nickel, carbon steel, or stainless steel. Standard fin thicknesses vary from 0.0065 to 0.016 in. (1 in. = 25.4 mm) depending on the materials. Fin spacing may include 4–20 fins per inch (1 in. = 25.4 mm) depending on tubes' outside diameters. Usually, the core tubes penetrate directly into the headers to a depth no greater than 0.12 inches without the use of intermediate adapter tubes. The return bends must be of the same material as the core tubing, with wall thickness no less than that of the core tube. In operation, the finned tubes and fans should be kept as clean as possible (i.e., free from dirt so that maximum airflow can take place) to maintain a maximum heat transfer efficiency. If the flow of air is blocked, the condenser thermal efficiency will be negatively affected and the heat pump compressor could fail in the medium or long term.

The refrigerant-to-air condenser coils are factory pressurized. For drying applications, they shall be designed to withstand at a pressure of over 2758 kPa and a maximum 150°C temperature, and, then, completely submerged in warm water containing special wetting and final cleaning agents for leak testing. At the end, the standard copper tube coils, for example, are air tested at a minimum of 2172 kPa pressure, then filled with nitrogen assuring that they remain leak free and clear of internal contamination prior factory- or in-field installation.

The large-scale heat pump condensers drain the liquid refrigerant into high-pressure liquid receivers (see Figure 14.1a) which provide sufficient volume for storage all refrigerant charge. The piping arrangement allows the outlet pipe to not touch the bottom of the liquid receiver, thus avoiding passing solid contaminants into the system. In industrial heat pump-assisted drying systems, the liquid level in the refrigerant receivers is almost constantly rising or falling because the liquid flow rate from the condenser is not precisely the same as the rate demanded of the low-side pressure (evaporator) of the system. When an excess flow rate enters the receiver, the vapor is compressed building up the pressure in the receiver and, temporarily, restricting the flow of condensate from the condenser. If the flow rate demanded by the heat pump exceeds temporarily the rate provided by the condenser, the pressure in the receiver drops and some of the liquid vaporizes (Stoecker 1998).

14.3 REFRIGERANT-SIDE CONDENSATION

Air-cooled heat exchangers with in-tube condensation of refrigerants are widely used in industrial heat pump-assisted drying systems. They transfer sensible heat and enthalpy (latent heat) of condensation from the desuperheated, and then partially or totally condensed refrigerant vapor, and finally from the subcooled liquid, to the heated drying air. The heat pump condensers working with pure halogenated [hydrochloroflurocarbons (HCFC), hydrofluorocarbon (HFC)], low GWP (global warming potential) (HFO), natural fluids, and refrigerant mixtures are designed and optimized in order to achieve minimal pressure drops and high-condensation heat transfer rates at specific operating conditions (Fronk and Garimella 2013).

14.3.1 Pure Halogenated Refrigerants

14.3.1.1 Flow Pattern

The flow pattern of two-phase (liquid and vapor) refrigerants condensing in horizontal tubes depends on the following parameters: (i) refrigerant flow rate and quality, (ii) physical properties of the two phases, and (iii) tube geometry (e.g., round, mini- or micro-channel, smooth, internally enhanced, etc.). Flow patterns of pure halogenated refrigerants during two-phase condensation inside horizontal plain tubes are complex phenomenon affected by simultaneous mass, momentum, and heat transfer rates (Rayleigh 1880).

The variety of the flow patterns reflects the different ways that the vapor and liquid phases are distributed in horizontal tubes, which causes the pressure drop and condensing heat transfer mechanisms to be different in the different flow patterns.

Several maps for in tube-side condensation of pure halogenated (HFC or similar) refrigerants in horizontal smooth tubes exist (Taitel and Dukler 1976; Breber et al. 1980; Sardesai et al. 1981; Tandon et al. 1982, Cavallini et al. 2002; Jassim et al. 2008; van Rooyen et al. 2010). Figure 14.3a shows one of them, illustrating the main (simplified) flow patterns that typically occur.

It can be seen that, during condensation of pure refrigerants inside horizontal tubes, the main flow regimes (annular and stratified) are similar to those for two-phase evaporation (Palen et al. 1979; Thome et al. 2003; Wolverine Tube 2006). However, compared to in-tube vaporization (see Section 13.5), in-tube condensation of refrigerants presents the following differences: (i) depending on the heat pump thermodynamic cycle, the refrigerant inlet vapor may be superheated, saturated ($x = 1$), or wet ($x < 1$) (where x is the vapor title); (ii) the top of the tube is wetted (coated) by a thin condensate (liquid) film in both annular and stratified regimes, and thus, no dry-out may occur; and (iii) the two-phase flow is usually dominated by vapor shear (as in the annular flow) or gravity forces (as in stratified, wavy, and slug flow regimes).

The principal regime in heat pump condensers is the annular flow where the interfacial shear stress dominates the gravitational force and results in a nearly symmetric annular film with a high-velocity vapor core. If the flow is fully developed, the condensate flows almost symmetrically distributed as a film on the inside tube wall with a film thickness (δ) (Figure 14.3b). The annular flow converts to stratified regime with the liquid flowing along the bottom of the tube to the end, mainly under the action of a hydraulic gradient (Figure 14.3c). The very complex flow structure of annual flow is usually assumed to apply to the intermittent and mist flow regimes, too.

The fully stratifying flow regime (where fresh condensate is continually formed at the top of the tube and, then, stratifies under the influence of gravity) with all the liquid in the lower portion of the tube has a thin layer of condensate around the upper perimeter (Figure 14.3c). Slug flow may appear when large interfacial waves grow sufficiently in amplitude to block the entire cross section at some transversal sections. As condensation continues, the slugs coalesce into a predominantly liquid flow with large bubbles, a flow within a flow referred as plug regime that may occur at the end of the condensation process (Cavallini et al. 2003).

The type of flow regime could be determined in function of the dimensionless vapor velocity (Breber et al. 1979):

$$j_g^* = \frac{G * x}{\sqrt{g * d * \rho_g \left(\rho_f - \rho_g\right)}} \tag{14.1}$$

where

j_g^* is the dimensionless vapor velocity (–)

$G = \dfrac{\dot{m}}{A}$ is the mass flux (velocity) (kg/m$^2 \cdot$ s)

x is the vapor title (–)

$\dot{m}$ is the refrigerant mass flow rate (kg/s)

A is the cross-sectional area (m^2)

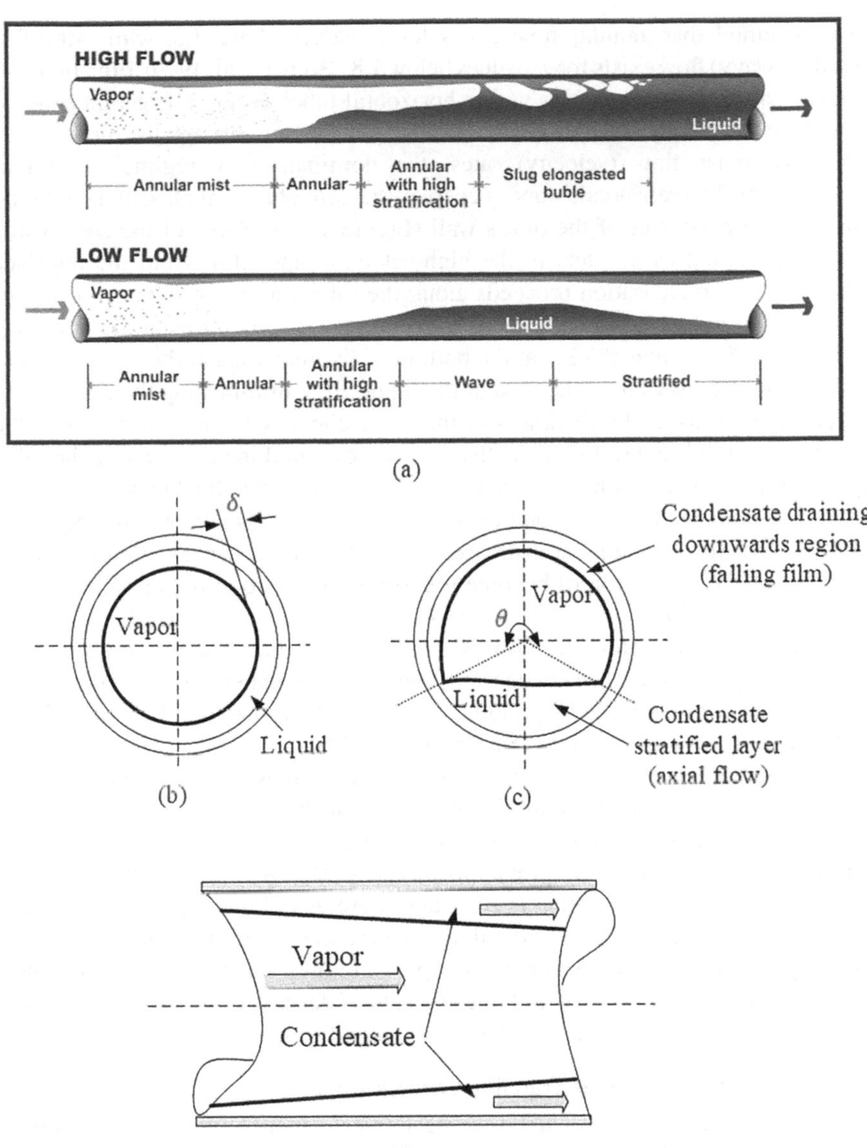

FIGURE 14.3 (a) Simplified two-phase flow patterns during the condensation of pure refrigerants inside horizontal tubes at high and low flow rates; (b) cross section of idealized annular flow regime; (c) cross section of idealized stratified flow regime for low vapor velocities; (d) longitudinal section of condensate flow for large vapor velocities.

g is the gravitational acceleration (m/s^2)
d is the inner tube diameter (m)
ρ_g is the vapor density (kg/m^3)
ρ_f is the liquid density (kg/m^3)

It was assumed that annular flow exists for j_g^* values above 1.8, while stratified (including wavy) flow exists for j_g^* values below 1.8 (Breber et al. 1979). Equation 14.1 suggests that the flow conditions within horizontal tubes during refrigerant condensation depend strongly on the velocity of the vapor flowing through the tube.

At high mass flow (velocity) rates, the dominant flow regime is annular (Figure 14.3a). The refrigerant vapor occupies the core of the annulus, and the liquid film is on the perimeter of the tube's wall (Figure 14.3b). Some of the condensing liquid is entrained as droplets in the high-velocity vapor flow from the interface of the film. As condensation proceeds along the tube, the vapor velocity decreases due to condensation and, thus, there is a decrease in vapor shear on the interface. The liquid film becomes thicker at the bottom of the tube than at the top. New condensate formed adds to the thickness of the liquid film, diminishing the diameter of the core of annulus as the thickness of the outer condensate layer increases in the flow direction (Rohsenow 1973). As the quantity of liquid increases along the tube, slug flow is encountered which gives way to plug or elongated bubble flow. The tube runs full of liquid beyond the point at which all the liquid is condensed, which has two effects: (i) the entrainment decreases, since there is less shear to break off the droplets and (ii) gravity forces become relatively more important, causing marked asymmetry in the film.

At low mass flow rates, the flow is stratified and the condensation occurs as depicted by Figure 14.3c. The condensate flows from the upper portion of the tube to the bottom, and, then, in the longitudinal direction (Incropera and DeWitt 2002) in laminar and/or turbulent regimes (Kattan et al. 1998a).

The pressure loss that occurs with a vapor–liquid flow is usually much higher than would occur for either phase flowing along at the same mass rate. The total pressure loss along a tube depends on the following factors (McQuiston et al. 2005): (i) friction, due to viscosity; (ii) gravitational, due to change of elevation; in horizontal flows, the change in elevation is zero, and there would be no pressure drop due to this factor; (iii) acceleration of the fluid—where there is a small change in vapor density or little evaporation occurring, the pressure drop due to acceleration is usually small. However, in flows with large changes of density, or where evaporation is present, the acceleration pressure drop may be significant.

14.3.1.2 Condensing Heat Transfer Correlations

In the refrigerant condensers of air-to-air mechanical vapor compression heat pumps, the refrigerant superheated vapor is first cooled (desuperheated), then condensed, and finally, the resulted liquid, subcooled. In the vapor desuperheating and liquid subcooling processes, the pressure drops and heat transfer coefficients are predicted based on the correlations of single-phase superheated vapor and subcooled liquids, respectively.

In-tube condensation of refrigerants is more efficient inside small tube diameters ($0 < 25$ mm) and at relatively high mass fluxes, where the liquid/vapor shear interactions become important. In such tubes, the flow regimes as well as the transitions from annular to stratified flow may significantly influence the heat and mass transfer processes (Traviss and Rohsenow 1973; Taitel and Dukler 1976; Tandon et al. 1982; Coleman and Garimella 1999, 2003).

As noted, during the relatively complex process of condensation, the refrigerant flow regimes change continuously as the refrigerant passes through the tube from the superheated vapor state at the inlet up to subcooled liquid state at the outlet. As a consequence, the condensing heat transfer coefficients vary throughout the horizontal condensing tubes (Stoecker 1998), as shown in Figure 14.4.

At the entrance to the condenser tube of the superheated vapor, the forced convection heat transfer coefficient is low. Then, it increases and will be at its highest value during annular flow. As more and more condensed liquid flows with the vapor, the surface available for condensation decreases. Near the end of the condenser tube, the coefficient drops because the process approaches that of convection heat transfer to a liquid. Thus, the lower condensing heat transfer coefficient occurs near the end of the condenser tube when all or most of the vapor has condensed (Figure 14.4). The reason is that backing liquid into an air-cooled condenser shifts some heat transfer area into the liquid subcooling mode, which achieves low heat transfer coefficients.

Because, during two-phase condensation processes, the saturated mixtures of vapor and liquids vary considerably in composition and hydrodynamic behavior, it is generally not possible to describe all conditions with one general correlation. In addition, the condensing heat transfer coefficients are affected by various factors and parameters as (Kew and Cornwell 1997; Mehendale et al. 2000; Cavallini et al. 2003; Kandlikar and Grande 2003; Thome et al. 2003; Cheng et al. 2007; Harirchian and Garimella 2010; Li et al. 2010; Doretti et al. 2013; Fang et al. 2013): (i) the flow regime (particularly when the inertial and gravitational forces are relatively important), (ii) refrigerant properties, (iii) tube inner diameter, (iv) condensing (saturation) temperature, (v) heat flux density, (vi) refrigerant mass velocity, and (vii) refrigerant vapor quality. However, in-tube condensation process is independent of the wall temperature difference ($T_{sat} - T_{wall}$) for most operating conditions, except at low-refrigerant mass flow rates (velocities).

Many experimental data are available for the heat transfer during condensation of pure halogenated refrigerants (HCFC-22, R-32, R-125, HFC-410A, HFO-236ea, and

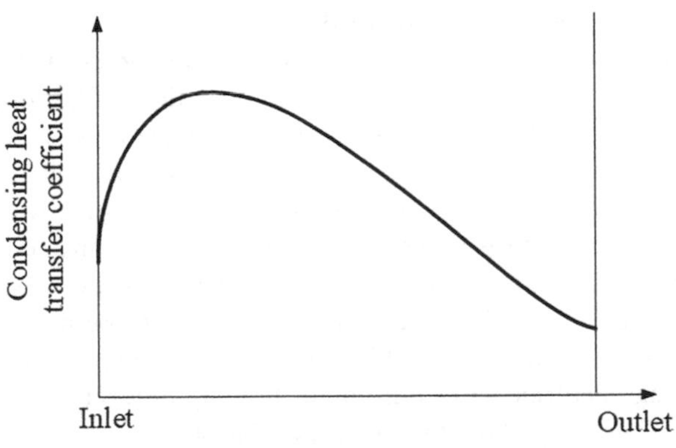

FIGURE 14.4 Relative variation of refrigerants condensing heat transfer coefficients with the distance along a horizontal tube from the inlet to the outlet.

HFC-134a) inside horizontal plain tubes with internal diameters $d > 3$ mm, reduced pressures $p_{red} = \frac{p}{p_{cr}} < 0.75$ (where p_{cr} is the refrigerant critical pressure), and density ratios $\frac{\rho_f}{\rho_g} > 4$ (where ρ_f is the liquid density and ρ_g is the vapor density, both expressed in kg/m^3), wide ranges of mass velocities, saturation temperatures, vapor qualities, and temperature differences $T_{sat} - T_{wall}$ (Cavallini et al. 2001, 2003).

Based on such data, many theoretical and semiempirical correlations for both annular and stratified flow regimes (e.g., Dalkilic and Wongwises 2009) exist for predicting the condensation heat transfer coefficients. However, as previously noted, there is not a single correlation that can be universally applied to various tube geometries, flow regimes, and boundary conditions. To select an appropriate correlation, it is necessary to identify what flow regime is expected to be dominant in each particular application (Zhang et al. 2015). In the fully developed annular flow regime, where the phase change process is dominated by vapor shear or gravity forces, the condensing heat transfer coefficients of pure refrigerants vary with refrigerant's mass velocity (G), vapor quality (x), and saturation temperature (T_{sat}). In this regime, where there is a thin condensate film on the entire tube wall, while the vapor phase flows in the central core, the heat transfer is governed by vapor and liquid turbulence (Boyko and Kruzhilin 1967; Cavallini and Zecchin 1971, 1974; Shah 1979; Tang 1997).

In the shear-dominated annular flow (Kosky and Staub 1971; Traviss et al. 1972), two different approaches were used for the calculation of condensing heat transfer coefficient: (i) interfacial shear model and (ii) two-phase multiplier model (Shah 1979; Haraguchi et al. 1994; Tang 1997).

Some condensing heat transfer coefficients have been determined as functions of the interfacial vapor–liquid shear stress and of the liquid film thickness requiring prediction of the frictional pressure gradient (Cavallini and Zecchin 1971; Kosky and Staub 1971; Traviss et al. 1972; Moser et al. 1998). At high mass velocity and low quality, where stratified, including wavy flow, could prevail with substantial convective heat transfer in the bottom part of the tube, the heat transfer in the liquid pool is not negligible (Dobson and Chato 1998).

In the case of stratified flow dominated by gravity forces, a thick condensate layer flows along the bottom of the tube, while a thin liquid film forms on the wall in the upper portion of the tube. The condensing heat transfer occurring under gravity-driven flow regimes (including the stratified, wavy, and slug flow regions) is affected by temperature difference between saturation and tube wall temperatures ($T_{sat} - T_{wall}$), and is referred to as film condensation. The heat transfer through the film is treated by a classical Nusselt (1916) analysis, while at very low mass velocities, the heat transfer through the thick condensate layer can either be neglected (Cavallini and Zecchin 1974; Jaster and Kosky 1976) or treated as a convective process.

For condensing heat transfer of pure fluids flowing in turbulent annular regime, Akers et al. (1959) developed the following correlation:

$$h_{cond} = 0.0265 \frac{k_f}{d} Re_{eq}^{0.8} Pr_f^{0.33} \quad \text{for} \quad Re_{eq} > 5 * 10^4 \tag{14.2a}$$

$$h_{cond} = 5.03\frac{k_f}{d}Re_{eq}^{0.8}Pr_f^{0.33} \quad \text{for} \quad Re_{eq} < 5*10^4 \tag{14.2b}$$

where

h_{cond} is the condensing heat transfer coefficient (W/m²·K)
k_f is the liquid thermal conductivity (W/mk)
d is the tube inner diameter (*m*)
Re_{eq} is the Reynolds number at equivalent mass flux (–), defined as follows:

$$Re_{eq} = Re_f + Re_g\left(\frac{\mu_g}{\mu_f}\right)\left(\frac{\rho_f}{\rho_g}\right)^{0.5}$$

$$Re_f = \frac{G*d*(1-x)}{\mu_f}$$

$$Re_g = \frac{G*d*x}{\mu_g}$$

Pr_f is the liquid Prandtl number (–)
Re_f is the liquid Reynolds number (–)
Re_g is the vapor Reynolds number (–)
G is the mass velocity (kg/m²·s)
x is the vapor title (–)
μ_f is the liquid dynamic viscosity (Pa·s)
μ_g is the vapor dynamic viscosity (Pa·s)

Based on experimental data of CFC-12 (chloroflurocarbon) and HCFC-22 refrigerants condensing in horizontal smooth tubes, Traviss et al. (1973) proposed the following semiempirical correlation:

$$h_{cond} = \frac{k_f}{d}F(X_{tt})\frac{Re_f^{0.9}Pr_f}{F_2} \tag{14.3}$$

where

h_{cond} is the condensing heat transfer coefficient (W/m²·K)
k_f is the liquid thermal conductivity (W/mk)
d is the tube inner diameter (m)

$$F(X_{tt}) = 0.15\left(X_{tt}^{-1} + 2.85X_{tt}^{-0.476}\right)$$

$$X_{tt} = \left(\frac{\mu_f}{\mu_g}\right)^{0.1}\left(\frac{1-x}{x}\right)^{0.9}\left(\frac{\rho_g}{\rho_f}\right)^{0.5}$$

F_2 is a constant depending on Re_f (–)

For condensation of pure halogenated fluids such as CFC-11, CFC-12, HCFC-22, and HFC-134a $(d = 7–40\text{ mm}; 0 < x < 100\%; 21°\text{C} < T_{cond} < 310°\text{C};$

$10.8 < G < 210\ \text{kg/m}^2\cdot\text{s}; 0.158 < \dot{q} < 1893\ \text{kW/m}^2$), the condensing heat transfer coefficient (h_{cond}) ($\text{W/m}^2\cdot\text{K}$) can be calculated with Shah (1976) correlation:

$$Nu_{cond}^{Shah} = 0.023 Re_f^{0.8} Pr_f^{0.4} \frac{k_f}{d}\left[1 + \frac{3.8}{p_{red}^{0.38}}\left(\frac{x_g}{1-x_g}\right)^{0.76}\right] \quad (14.4)$$

where

Nu_{cond}^{Shah} is Nusselt number (–) (Shah 1976)
Re_f is the liquid phase Reynolds number (–)
Pr_f is the liquid phase Prandtl number (–)
k_f is the liquid thermal conductivity (W/mK)
d is the tube inner diameter (m)
x_g is the vapor fraction (–)

$p_{red} = \dfrac{p}{p_{cr}}$ is the refrigerant reduced pressure (–)
p is the refrigerant actual pressure (Pa)
p_{cr} is the refrigerant critical pressure (Pa)

In order to fit the experimental data with a wide range of mass flows, where different types of flow patterns may occur, Bivens and Yokozedi (1979) have modified the Shah correlation as follows (for $d = 10\ \text{mm}; 10 < x < 99\%; 2°\text{C} < T_{cond} < 40°\text{C}; 175 < G < 650\ \text{kg/m}^2\cdot\text{s}; 0.35 < \dot{q} < 30.9\ \text{kW/m}^2$) for synthetic refrigerants as HCFC-22 and R-502:

$$Nu = Nu_{cond}^{Shah}\left(0.78738 + \frac{6187.89}{G^2}\right) \quad (14.5)$$

where

Nu_{cond}^{Shah} is Shah's correlation (Shah 1976) (equation 14.4)
G is the refrigerant mass (velocity) flux ($\text{kg/m}^2\cdot\text{s}$)

For film condensation inside horizontal tubes, the following new correlation was developed for the condensing heat transfer coefficient (Shah 1979a, 1979b; Eckels and Pate 1990):

$$\bar{h}_{cond} = h_{fo}\left(\frac{\mu_f}{14\mu_g}\right)^n \left\{(1-x)^{0.8} + \left[\frac{3.8 * x^{0.76}(1-x)^{0.04}}{Pr_f^{0.38}}\right]\right\} \quad (14.6)$$

where

$\bar{h}_{cond}$ is the mean condensing heat transfer coefficient ($\text{W/m}^2\cdot\text{K}$)
h_{fo} is the heat transfer coefficient assuming all refrigerant mass flowing as liquid phase ($\text{W/m}^2\cdot\text{K}$) calculated as follows:

$$h_{fo} = 0.023 Re_{fo}^{0.8} Pr_f^{0.4} \quad (14.7)$$

$Re_{fo} = \dfrac{G(1-x)*d}{\mu_f}$ is Reynolds number assuming liquid phase flowing alone (–)

G is the refrigerant mass (velocity) flow (kg/m$^2\cdot$s)
x is the refrigerant vapor quality (title) (–)
d is the tube interior diameter (m)
μ_f is the vapor dynamic viscosity (Pa·s)

$$n = 0.0058 + 0.557 * p_{red}$$

$p_{red} = \dfrac{p}{p_{cr}}$ is the reduced pressure (–)

p is the refrigerant actual pressure (Pa)
p_{cr} is the refrigerant critical pressure (Pa)

$Pr_f = \dfrac{c_{p,f} * \mu_f}{k_f}$ is liquid Prandtl number (–)

The presence of refrigerant lubricating oil in circulation inside the condensers tends to significantly decrease the condensing heat transfer coefficient mainly because the lubricants coat the tube interior surfaces and increase the resistance to heat transfer, thereby raising the surface temperature of the refrigerant. However, when the oil separators are operational, there is at least 25% increase in the condensing heat transfer coefficients for refrigerant as HFC-134a and CFC-12 (Eckels and Pate 1990).

The following shear-based correlation, originally developed for vertical flow condensation (Chen et al. 1987), would be appropriate for horizontal tubes:

$$Nu = 0.018 \left(\frac{\rho_f}{\rho_g}\right)^{0.39} \left(\frac{\mu_g}{\mu_f}\right)^{0.078} Re_f^{0.2} \left(Re_{fo} - Re_f\right)^{0.7} Pr_f^{0.65} \tag{14.8}$$

where
ρ_f is the liquid density (kg/m^3)
ρ_g is the vapor density (kg/m^3)
μ_g is the vapor dynamic density (Pa·s)
μ_g is the liquid dynamic density (Pa·s)

$Re_{fo} = \dfrac{G*d}{\mu_f}$ is the Reynolds number corresponding to the liquid flowing alone inside the tube (–)

$Re_f = \dfrac{G*(1-x)*d}{\mu_f}$ is the liquid Reynolds number based on liquid phase parameters (–)

Pr_f is the Prandtl number based on liquid phase parameters (–)
G is the refrigerant mass (velocity) flow rate (kg/m$^2\cdot$s)
d is the tube inner diameter (m)

For convective condensation in turbulent annular regime, the heat transfer coefficient (h_{cond}) can be correlated by the following expression, also applied to refrigerants' convective evaporation (Kattan et al. 1998b), albeit with different empirical constants c and n:

$$h_{cond} = c * Re_f^n * Pr_f^{0.5} \frac{k_f}{\delta} \tag{14.9}$$

where

h_{cond} is the condensing heat transfer coefficient (W/m$^2\cdot$K)
d is the internal tube diameter (m)
δ is the liquid film thickness for annular flow (m); assuming $\delta \ll d, \delta = \frac{d(1-\varepsilon)}{4}$
ε is the void fraction (–)
Re_f is the liquid film Reynolds number based on the mean velocity of the liquid, defined as follows:

$$Re_f = \frac{4G(1-x)\delta}{(1-\varepsilon)\mu_f} = \frac{Gd(1-x)}{\mu_f}$$

G is the mass (flux) velocity (kg/m$^2\cdot$s)
x is the vapor title (–)
μ_f is the liquid dynamic viscosity (Pa$\cdot$s)
Pr_f is the liquid Prandtl number (–)

By assuming that the falling film on the internal perimeter from top to bottom of the horizontal tube is laminar in fully stratified flows, the average film condensation heat transfer coefficient can be calculated by using the Nusselt (1916) theory:

$$\bar{h}_{film} = 0.728\left[\frac{\rho_f(\rho_f - \rho_g) * g * h_{fg} * k_f^3}{\mu_f * d * (T_{sat} - T_{wall})}\right]^{1/4} \tag{14.10}$$

where

$\bar{h}_{film}$ is the average film condensation heat transfer coefficient (W/m$^2\cdot$K)
ρ_f is the liquid density (kg/m^3)
ρ_g is the vapor density (kg/m^3)
g is the gravitational acceleration (m^2/s)
h_{fg} is the refrigerant enthalpy of condensation (J/kg)
k_f is the liquid thermal conductivity (W/mk)
μ_f is the liquid dynamic viscosity (Pa$\cdot$s)
d is the tube inner diameter (m)
T_{sat} is the refrigerant saturation temperature (°C)
T_{wall} is the wall surface temperature (°C)

The Nusselt number calculations corresponding to both annular and stratified (including wavy, slug and plug) flows can be expressed as follows (Dobson et al. 1993):

$$Nu = \alpha(X_{tt}) * \left[\frac{g * \rho_f * (\rho_f - \rho_g) * d^3 * h_{fg}}{\mu * \Delta T * k_f} \right]^{0.25} \tag{14.11}$$

where

$$\alpha(X_{tt}) = \frac{0.375}{X_{tt}^{0.23}}$$

$$Nu = 0.023\, Re_f^{0.8} Pr_f^{0.3} * \beta(X_{tt}) \tag{14.12}$$

where

$$\beta(X_{tt}) = \frac{2.61}{X_{tt}^{0.805}}$$

$X_{tt} = \left(\frac{\rho_g}{\rho_f}\right)^{0.5} \left(\frac{\mu_f}{\mu_g}\right)^{0.1} \left(\frac{1-x}{x}\right)^{0.9}$ is Lockhart–Martinelli parameter (–)

For condensing saturated vapor, ASHRAE (2009) recommends calculating the condensation average heat transfer coefficient with the following correlation:

$$Nu_f = 13.8 * Pr_f^{1/3} \left(\frac{h_{fg}}{c_{p,f} * \Delta T} \right)^{1/6} \left[\frac{d * G_g}{\mu_f} \left(\frac{\rho_f}{\rho_g} \right)^{1/2} \right]^{0.2} \tag{14.13}$$

where

$Nu_f = \dfrac{\bar{h} * d}{k_f}$ is the Nusselt number (–)

$\bar{h}$ is the average film condensation heat transfer coefficient ($W/m^2 \cdot K$)
d is the tube interior diameter (m)
G_g is the vapor mass velocity ($kg/m^2 \cdot s$)
k_f is the liquid thermal conductivity (W/mK)
Pr_f is the liquid Prandtl number (–)
h_{fg} is the refrigerant enthalpy (latent heat) of condensation (J/kg)
$c_{p,f}$ is the liquid isobaric specific heat ($J/kg \cdot K$)
ΔT is the difference between the fluid saturation temperature (T_{sat}) and the wall surface temperature (T_{wall})
G_g is the vapor mass velocity ($kg/m^2 \cdot s$)
μ_f is the liquid dynamic viscosity ($Pa \cdot s$)
ρ_f is the liquid density (kg/m^3)
ρ_g is the vapor density (kg/m^3)

Even equation 14.13 has been developed for condensing saturated vapor, little error is introduced for superheated vapor when the wall temperature is below the saturation temperature (ASHRAE 2009).

In laminar flow of single-phase subcooled refrigerants inside tubes, the recommended correlation for predicting the condensing average heat transfer coefficient is (McQuiston et al. 2005):

$$Nu = \frac{\bar{h} * d}{k_f} = 1.86 * \left[Re_f * Pr_f * \frac{d}{L} \right]^{0.33} \left(\frac{\mu_f}{\mu_{wall}} \right)^{0.14} \tag{14.14}$$

where

$\bar{h}$ is the condensing average heat transfer coefficient (W/m$^2 \cdot$K)
d is the tube interior diameter (m)
k_f is the liquid thermal conductivity (W/mK)

$Re_d = \dfrac{u * d}{\nu_f}$ is the liquid Reynolds number (–)
u is the liquid velocity (m/s)
ν_f is the liquid kinematic viscosity (m^2/s)

$Pr_f = \dfrac{\nu_f}{\alpha}$ is the liquid Prandtl number (–)
α is the liquid thermal diffusivity (m^2/s)

In equation 14.14, the refrigerant properties should be evaluated at the arithmetic mean bulk temperature, except for μ_{wall} which is evaluated at the wall temperature. In the transition from laminar to turbulent flow, defined approximately by 2,000 < Re_d < 10,000, the prediction of heat transfer and friction coefficients is uncertain during transition. The usual practice is to avoid the region by proper selection of tube size and flow rate (McQuiston et al. 2005).

Under turbulent flow conditions (Re_d < 10,000, 0.7 < Pr < 100 and $\dfrac{L}{d}$ > 60) of the single-phase subcooled liquid, the Dittus and Boelter (1930) correlation is generally used (Thome et al. 2003; McQuiston et al. 2005; ASHRAE 2009):

$$Nu_f = \frac{\bar{h} * d}{k_f} = 0.023 * Re_d^{0.8}\ Pr^n \tag{14.15}$$

or

$$\bar{h} = 0.023 \frac{k_f}{d} \left(\frac{G_f * d}{\mu_f} \right)^{0.8} \left(\frac{\mu_f * c_{p,f}}{k_f} \right)^n \tag{14.16}$$

where

$n = 0.4$ for $T_{wall} > T_{bulk}$
$n = 0.3$ for $T_{wall} > T_{bulk}$
T_{wall} is the temperature of the tube interior wall (°C)
T_{bulk} is the refrigerant bulk temperature (°C)

In equations 14.13–14.16, all fluid properties should be evaluated at the arithmetic mean bulk temperature of the fluid (McQuiston et al. 2005; ASHRAE 2009).

Some experimental studies on condensing heat transfer coefficients in horizontal round tubes of pure halogenated refrigerant provided the following conclusions (note: not in order of importance and/or relevance):

a. The condensation heat transfer coefficients increase with the refrigerant mass (velocity) flux.
b. At high mass fluxes, where the flow regime is generally annular and the liquid and vapor velocities increase, the condensation heat transfer coefficients rise significantly with the vapor quality. This is because, when the vapor quality increases, the liquid film thickness becomes thinner, thus the thermal resistance decreases (Zhuang et al. 2016).
c. At low mass fluxes and vapor qualities, the flow pattern is gravity-driven with a thick liquid layer at the bottom of the tube that affects the condensing heat transfer. Therefore, the condensation heat transfer coefficients are less sensitive to mass flux and low vapor quality.
d. The condensing heat transfer coefficient of HFC-410A is about 3%–6% higher than that of HCFC-22 at low refrigerant vapor quality since the liquid thermal conductivity and viscosity of HFC-410A are lower than those of HCFC-22 (Wijaya and Spatz 1995).
e. During flow condensation in circular tubes of refrigerant R-152a, at saturation temperatures between 40°C and 50°C, mass flux varying from 200 to 800 $kg/m^2 \cdot s$ and vapor quality from 0.1 to 0.9, both condensing heat transfer coefficients and pressure drops increase with mass (velocity) flux and vapor quality but decrease with saturation temperature (Liu et al. 2013).
f. The heat transfer coefficient of high-temperature refrigerant HFO-245fa condensing in a horizontal tube of 8 mm inner diameter increases with the increase in velocity, condensation temperature, and superheat of inlet vapor (Liu et al. 2015).

14.3.2 Natural Refrigerants

14.3.2.1 Ammonia

Applied to ammonia condensing in horizontal tubes, all previous heat transfer correlations (Section 14.3.1.2) give relatively close values to each other depending on the mass (velocity) flow rates. Among these equations, Chen correlation (1987) (equation 14.8) seems the most conservative, while Traviss correlation (1972) (equation 14.3) gives the higher condensation heat transfer coefficients for the same mass (velocity) flow rates (Vollrath et al. 2003) (Figure 14.5).

The heat transfer coefficient of ammonia for the saturated vapor condensation in a steel tube with copper plate fins has been calculated by using the following empirical equation (Chepurnenko et al. 1993; Ohadi et al. 1996):

$$h_{cond,\,NH_3} = 711,\,780\left(\eta_{fin} * \dot{q}\right)^{-0.105}\left(\frac{L}{d}\right)^{-0.69} \tag{14.17}$$

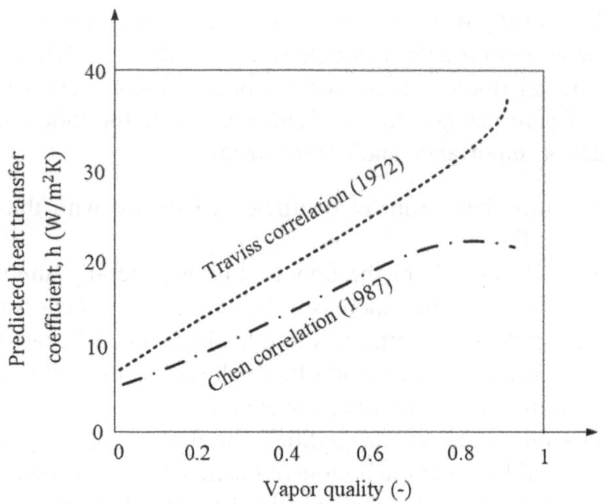

FIGURE 14.5 Comparison of condensation heat transfer coefficients for single flow regime of ammonia at mass flow rate $G = 160$ kg/m²·s.

where

$h_{cond,\,NH_3}$ is the heat transfer coefficient of ammonia (W/m²·K)
η_{fin} is the fin efficiency (–)
$\dot{q}$ is the heat transfer density (W/m²)
L is the length of the horizontal tube (m)
d is the tube inner diameter (m)

14.3.2.2 Hydrocarbons

With zero ozone depleting potential (ODP), low GWP, and high thermodynamic performance, most of hydrocarbons are candidates as refrigerants for heat pump and refrigeration systems. Generally, hydrocarbons have smaller liquid densities than most of the fluorocarbons, thus, the amount of heat pump charges may decrease significantly by using hydrocarbons. This may help further reduce the direct greenhouse emissions of refrigerants. Although, compared to fluorocarbon refrigerants, the hydrocarbons have superior thermophysical properties and are widely used in domestic refrigerators, for example, R-170 (ethane) and its mixtures (Zhuang et al. 2016), their flammability prevents the wide application to larger systems, such as industrial heat pumps.

A considerable number of experimental and theoretical investigations for pressure drop and heat transfer during the condensation of hydrocarbons in horizontal tubes have been realized. Some results can be summarized as follows (Chang et al. 2000; Lee et al. 2006; Vera-Garcia et al. 2007):

a. The condensation heat transfer coefficients of pure n-pentane, R-290 (propane), and R-600 (*n*-butane) in compact plate heat exchangers at the operating pressures ranging from 150 to 1800 kPa decrease with increasing the Reynolds number (Thonon and Bontemps 2002).

b. The condensation heat transfer coefficients and pressure drops of hydrocarbons increase as the vapor quality and mass (velocity) flux increase (Wen et al. 2006).
c. During condensation of R-290 (propane) and R-600a (isobutane) in horizontal tubes with inner diameters between 6 and 10 mm, mass velocities of 35.5–210.4 kg/m$^2 \cdot$s, and condensing temperature of 40°C, the pressure drops and local condensation heat transfer coefficients increase as the inner tube diameter decreases (Lee and Son 2010).
d. In-tube condensation heat transfer coefficients of smooth (plain) tubes are well correlated with equations obtained from experimental studies with pure fluorocarbons (see Section 14.3.2.2). In other words, because the thermodynamic and transport properties of hydrocarbons are similar to those of the fluorocarbons, or at least there are no large differences, it might be considered that characteristics of heat transfer and pressure drop during condensation of hydrocarbons are also similar to those of the fluorocarbons.
e. The condensing heat transfer coefficients experimentally determined have been compared with some of previously proposed correlations (Traviss et al. 1972; Cavallini and Zecchin 1974; Shah 1979); the Cavallini–Zecchin correlation (Cavallini and Zecchin 1974) got the values the most close to the experimental results (within about ∓20%).
f. Calculated average condensing coefficients of R-290 (propane) and R-600a (isobutane) using Shah correlation (1976) are greater by 220%–223% and 177%–187% in comparison to CFC-12 and HFC-134a, respectively (Mathur 2000).
g. For similar operating conditions, hydrocarbons provide higher heat transfer coefficients compared to those obtained with HCFC-22 (Thonon 2005; Lee et al. 2006; Miyara 2008).
h. Experimental $COP_{heat\ pump}$ values of R-290 (propane) are higher than those of HFC-404A and HFC-410A by 5%–12% and 0%–9%, respectively, while the Life Cycle Climate Performance (LCCP) of HFC-410A is lower than that of R-290 (propane) as long as the annual emission is kept below 10% (Hwang et al. 2007).
i. Drop-in experiments with R-290 (propane) into a heat pump with plate condenser used as a substitute of R-407C (a zeotropic mixture) showed that R-290 (propane) provides higher $COPs_{heat\ pump}$, that is, 27% in the cooling mode and 9%–15% in the heating mode (Martinez et al. 2006).
j. Analytical evaluations including the optimization process of finned-tube evaporators and condensers showed that R-290 (propane) achieves the highest system $COPs_{heat\ pump}$ compared to HFC-134a, HCFC-22, HFC-410A, and R-32 (Domanski and Yashar 2006).
k. At condensing temperatures of R-600a (isobutane) in the range of 35°C–45°C and mass fluxes between 150 and 300 kg/m^{2}s, the condensing heat transfer coefficients are higher than those of HCFC-22 by at least 31%, even the pressure drops are about 50% higher (Lee et al. 2006).

l. Theoretical simulations showed that the heat transfer coefficient of R-290 (propane) is about 40% higher than that of HFC-410A, and the pressure drop is about twice higher (Spatz and Motta 2004).
m. Pressure gradients, estimated from physical properties of R-290 and R-600a (isobutane) for single-phase flow and two-phase flow during condensation, are significantly higher than that for CFC-12 and HFC-134a (Mathur 2000).
n. The condensing heat transfer coefficients inside an 8 mm tube of pure hydrocarbons R-290 (propane), R-600a (isobutane), R-600 (*n*-butane), and R-1270 (propylene) are higher than those of HCFC-22 at the same mass velocities (Chang et al. 1997, 2000).

14.3.3 Low-GWP Refrigerants

Among the new low-GWP refrigerants, HFO-1234yf and HFO-1234ze(E) seem the most promising for use in industrial heat pumps.

a. HFO-1234yf

   Some of experimental and theoretical results concerning the condensation of the low-GWP refrigerant HFO-1234yf in horizontal smooth tubes can be summarized as follows (note: not in order of importance and/or relevance):

   i. The condensing heat transfer coefficients show weak sensitivity to saturation temperature and great sensitivity to refrigerant mass flux (Longo and Zilio 2013); for example, at low refrigerant mass flux (<20 kg/m$^2\cdot$s), the heat transfer coefficients are not dependent on mass flux, and the condensation is controlled by gravity; for higher refrigerant mass flux (>20 kg/m$^2\cdot$s), the heat transfer coefficients depend on mass flux and forced convection condensation occurs.
   ii. HFO-1234yf exhibits heat transfer coefficients lower (10%–12%) and frictional pressure drop lower (10%–20%) than those of HFC-134a under the same operating conditions (Longo and Zilio 2013).
   iii. Condensation heat transfer coefficients of HFO-1234yf are lower as compared to those of HFC-134a at mass fluxes ranging between 200 and 1000 kg/s$\cdot$m^2 and 40°C saturation temperature (Pottker and Hrnjak 2015).
   iv. During condensation of HFO-1234yf inside a 1 mm diameter tube, the condensing heat transfer coefficients are comparable to those of HFC-134a.
   v. $COP_{heat\ pump}$ obtained with HFC-1234yf is up to 19% higher compared to that achieved with HFC-134a, while the condenser subcooling can be more beneficial for heat pumps operating with HFC-1234yf than those using HFC-134a (Pottker and Hrnjak 2015).
   vi. The experimental analysis for HFC-134a and HFC-1234yf confirmed that the presence of an internal heat exchanger may significantly reduce the $COP_{heat\ pump}$ increase due to condenser subcooling, since both improvements (i.e., simultaneous liquid subcooling and vapor superheating via an internal heat exchanger) compete toward reducing the throttling losses. Without internal heat exchanger, the maximum $COP_{heat\ pump}$ improvement

due to condenser subcooling is equal to 18% against 9% for the system with internal heat exchanger (Pottker and Hrnjak 2015).

b. HFO-1234ze(E)
The condensation process of the potential low-GWP refrigerant HFO-1234ze(E) presents features as the following:
   i. Due to its similar thermos-physical properties and heat transfer characteristics, HFO-1234ze(E) is currently considered to be a potential replacement for HFC-134a, especially for car air-conditioning apparatus, and for HFO-245fa in high-temperature heat pumps and organic Rankine cycles.
   ii. At saturation temperatures of 30°C, 40°C, and 50°C, the effect of saturation temperature on heat transfer coefficient of HFO-1234ze(E) is not as significant as mass flux because the thermos-physical properties of this refrigerant do not vary drastically over these operating conditions; the vapor density of refrigerant increases with pressure, which results in lower vapor velocity in order to maintain the same mass flux; also, the decrease in liquid thermal conductivity, latent heat, and increase in pressure results in higher resistance on the refrigerant side; therefore, with increase in pressure, the overall resistance on the refrigerant side increases which reduces the condensing heat transfer coefficient (Agarwal and Hrnjak 2015).
   iii. The pressure drops of the refrigerant HFO-1234ze(E) are significantly higher than HFC-134a, mainly because, at the same mass flux, the lower vapor density of HFO-1234ze(E) results in higher velocity compared to those of HFC-134a which results in higher slips at the liquid vapor interface and thus, increased frictional pressure drops; this aspect should be considered while using the refrigerant HFO-1234ze(E) as a drop-in replacement of HFC-134a.
   iv. For mass fluxes of 100–300 $kg/m^2 \cdot s$, saturation temperatures of 30°C–50°C, and vapor titles from $x = 0.05$ to superheat ($x = 1$), the refrigerant HFO-1234ze(E) condensing inside a horizontal smooth tube with 6.1 mm inner diameter achieves condensation heat transfer coefficients approximately 10% lower compared to those of HFC-134a (Agarwal and Hrnjak 2015).
   v. With HFO-1234ze(E) condensing in a horizontal tube (mass flux of 100–450 $kg/m^2 \cdot s$ and saturation temperature of 35°C–45°C), a reduction by 20%–45% of the condensing heat transfer performance as compared to R-32 has been achieved (Hossain et al. 2012).

14.3.4 Refrigerant Mixtures

In the heat pump industry, the restrictions on the use of CFC and HCFC due to their high ODP and/or GWP has increased the interest for refrigerant mixtures, which may mitigate negative environmental impacts while maintaining desirable thermodynamic performance. The use of mixtures as working fluids has been shown to have great promise in utilizing low-grade waste heat due to the ability to match source temperature glide with the corresponding zeotropic mixtures' saturation temperature glides.

Mixtures generally consist of condensable and miscible refrigerants (i.e., water vapor and ammonia) or condensable refrigerants and non-condensable gases (i.e., water and air). Mixtures of condensable fluids can be further classified based on the relation of the concentrations of each component in each phase at equilibrium.

The coupled heat and mass transfer phenomena during condensation of refrigerant mixtures in horizontal smooth and enhanced tubes are of great interest to numerous industries, as mechanical vapor compression and absorption heat pumps (Fronk and Garimella 2013).

The mechanisms of coupled heat and mass transfer during condensation in smooth and enhanced tubes of binary or multicomponent refrigerants are complex due to continuously variation of vapor and liquid compositions along the condenser length, the increased importance of vapor/liquid shear interactions, and the effect of two-phase flow patterns. Compared to their pure refrigerant components, the heat and mass transfer resistances in the vapor and liquid phases of zeotropic mixtures increase and, thus, the overall heat transfer resistance also increases (Cavallini et al. 2003; Garimella 2006; Dalkilic and Wongwises 2009; Lips and Meyer 2011; Fronk and Garimella 2013).

Considerable experimental and modeling researches have been conducted on the condensation of binary and multicomponent mixtures of synthetic (CFC, HCFC, and HFC), natural, and hydrocarbon refrigerants in both smooth and enhanced horizontal tubes with interior diameters varying from 2.45 to 25.4 mm. The experimental heat transfer coefficients were reported either as averages for the entire condensation process (i.e., 100% quality change) or for some smaller vapor quality change.

For modeling the heat and mass transfer during condensation process of zeotropic mixtures inside horizontal tubes (Cavallini et al. 2002a, 2002b, 2002c) without non-condensable gases, the following assumptions are generally made (Silver 1947; Bell and Ghaly 1973): (i) the liquid and vapor compositions are in equilibrium at the vapor bulk temperature; (ii) liquid and vapor enthalpies are those of the equilibrium phases at the vapor bulk temperature; (iii) the mass transfer in the vapor phase is neglected; (iv) sensible heat is transferred from the bulk vapor to the interface by convective heat transfer, where the heat transfer coefficient is calculated for the given geometry assuming only vapor is present, using vapor bulk properties and vapor mass flux; (v) the total latent heat of condensation, and sensible heat of cooling of the condensate and vapor are transferred through the entire thickness of the condensate; and (vi) the calculation method is often referred to as the equilibrium (Silver–Bell–Ghaly) model because complete equilibrium is assumed in both phases. For the refrigerant zeotropic HFC-407C, the Silver–Bell–Ghaly corrected model had an absolute average deviation of 21%, whereas for mixtures of R-125/HFO-236ea, the average absolute deviation was 22%.

Several existing empirical and/or semiempirical correlations to predict heat transfer during condensation process of pure refrigerants inside smooth horizontal tubes could be inaccurate if applied without modifications to refrigerant mixtures (Cavallini et al. 2003).

For predicting the condensation heat transfer coefficient of HCFC-22/CFC-12 mixture flowing inside a horizontal smooth tube, with a mean deviation of $\mp$15%

from the experimental data, the following empirical correlation has been successfully used (Tandon et al. 1986):

$$h_{cond} = 2.82\frac{k_f}{d} Ph^{-0.365} Re_g^{0.146} Pr_f^{0.33} \quad (14.18)$$

where

h_{cond} is the condensation heat transfer coefficient (W/m$^2 \cdot$ s)

$$Ph = \frac{h_{fg}}{c_{p,f}\left(T_{sat} - T_{wall}\right)}$$

h_{fg} is the refrigerant enthalpy (latent heat) of condensation (J/kg)
$c_{p,f}$ is the liquid isobaric specific heat (J/kg · K)
T_{sat} is the refrigerant condensing (saturation) temperature (°C)
T_{sat} is the tube wall temperature (°C)

Different equilibrium and nonequilibrium models, as well as empirical- and energy conservation-based approaches, have been used to determine the condensing heat transfer coefficients of refrigerant mixtures and the condenser heat exchange area (Fronk and Garimella 2013).

Most of the experimental studies on refrigerant mixtures condensation inside horizontal smooth finned tubes focused on (Wang and Chato 1995) the following: (i) thermal resistance of the vapor diffusion layer affecting the condensation, (ii) influence of the flow direction of vapor during the condensation (horizontal or vertical tube), (iii) turbulence in the vapor generated by fins, and (iv) enhancement of the condensation performance.

Some experimental studies on condensing heat transfer coefficients in horizontal round tubes of mixed refrigerant provided the following conclusions (note: not in order of importance and/or relevance):

a. The condensing heat transfer coefficient of R-407C (a zeotropic blend of R-32, R-125, and HFC-134a) is between 30% and 50% of that of HCFC-22, and is not dependent on the changes of condensing pressure.
b. Comparing the condensing heat transfer coefficients of pure halogenated refrigerants HCFC-22, HFC-134a, HFC-410A with those of zeotropic mixture R-407C in smooth and micro-fin tubes, it was observed that (Eckels and Tesene 1999) (i) the condensing heat transfer coefficients for the micro-fin tube are significantly higher than those for the smooth tube for all refrigerants tested; (ii) HFC-134a has the highest heat transfer coefficients in both types of tubes, while R-407C has the lowest heat transfer coefficients; and (iii) the heat transfer coefficient enhancement of the micro-fin tube over the smooth tube varies from 2.2 to 2.5 at the lowest mass flux and from 1.2 to 1.6 at the highest mass flux.
c. Experiments with a R-114/CFC-12 mixture condensing in a horizontal tube of 12.7 mm interior diameter and 1.5 m length, at a nominal mass flux of

180 kg/m^2·s, and average densities of heat flux varying between 4.22 and 4.5 kg/m^2 showed no discernible flow regime transitions from annular to stratified with varying mixture composition (Stoecker and Kornota 1985).

d. For a heat pump operating with a CFC-12/CFC-114 zeotropic mixture, the average measured condensation heat transfer coefficients ranged from 800 to 2800 W/m^2·K. As the concentration of CFC-114 increased from 10% to 70% at varying mass fluxes ($45 < G < 50$ kg/m^2·s) to obtain a constant cooling load, the average heat transfer coefficient decreased, reaching a minimum at approximately 50%–60% mass concentration of CFC-114; these results imply that the mass transfer resistance in the zeotropic mixture degrades the heat transfer. The expected increase in efficiency of heat pump systems working with this zeotropic mixture was less than that predicted by theoretical simulations by approximately 6% (Launay 1981; Stoecker and McCarthy 1984).

e. The condensation heat transfer coefficients of CFC-12/CFC-114 mixture showed that (Stoecker and McCarthy 1984; Stoecker and Kornota 1985) (i) in the high-temperature (desuperheating) region of condensers, no change of mixture mass fraction occurs; (ii) in the last region where the phase changes, the condensing heat transfer coefficients are highest; and (iii) as the title of CFC-114 increases, the heat transfer coefficients decrease, a degradation probably due to the slip between the liquid and vapor.

f. Almost all experimental studies showed degradation in condensation heat transfer coefficients of refrigerant mixtures.

g. Additionally, most of the mixtures exhibit temperature glides ranging from near 0 K for the near-azeotropic mixtures to over 80 K for mixtures of ammonia and water, which led to additional sensible heat transfer resistance in the vapor phase, and further reduction of heat transfer rates.

h. The condensation heat transfer coefficients of the mixture HCFC-22/CFC-12 are lower than those of HCFC-22 and higher than those of CFC-12 (Tandon et al. 1986).

i. The condensing heat transfer coefficients of HCFC-22/CFC-114 mixture in horizontal tubes (Koyama et al. 1990) (i) are smaller than those of pure refrigerant components; (ii) are higher in a micro-fin tube by about 30% as compared to a plain tube; this enhancement is only half as compared to that of the pure refrigerant enhancement; (iii) the maximum degradation of the condensing heat transfer coefficient of HCFC-22/CFC-114 is about 20% at 34–43 wt.% of HCFC-22 mass fraction; and (iv) for HCFC-22/R-114 mixture, the circumferential wall temperature is uniform, which might be due to the opposite changes of liquid–vapor interfacial temperature distribution.

j. The degradation of the condensing heat transfer coefficient of R-32/HFC-134a mixture (30/70 wt.%), compared to the condensing heat transfer coefficients of the pure components, is 36% and 29%, respectively, for mass fluxes varying between 91 and 182 kg/m^2·s at the saturation temperature of 50°C.

k. Compared to that of HCFC-22, the condensing heat transfer coefficient of refrigerant mixture R-32/HFC-134a (30/70 wt.%) increases at lower vapor qualities, but decreases at higher vapor quality; in annular flow regime, the condensing heat transfer coefficients match better with existing correlations at the higher vapor quality with mean deviation of about 30%.

l. The heat transfer coefficients and pressure drops during condensation of non-azeotropic mixture dimethyl ether (DME)/CO_2 inside a horizontal smooth tube have been measured experimentally with mass fractions of 39/61% and 21/79%, and mass fluxes varying from 200 to 500 kg/m²s. It was observed that (i) the increase in mass fraction of CO_2 in the mixture decreases the heat transfer coefficient and the pressure drop; (ii) at higher refrigerant mass fluxes, the effect of mass transfer resistance on the heat transfer decreases; and (iii) a heat pump operating with a binary refrigerant mixture CO_2/DME (90/10 wt.%) achieves almost the same maximum heating $COP_{heat\,pump}$ value as a heat running with pure CO_2 but at considerably lower discharge pressure.
m. The hydrocarbon mixture R-170 (ethane)/R-1270 (propylene) is recommended to replace HCFC-22 in residential heat pumps, mainly because (Park et al. 2010) (i) the heat pump heating capacities of R-170/R-1270 mixture (with R-170 concentrations of 2%–10% (by mass) were between 4% and 20% higher than those of HCFC-22; (ii) the heat pump heating capacities increase as the concentration of R-170 (ethane) in the mixture increases; and (iii) the heat pump's compressor discharge temperatures of R-170/R-1270 mixture are lower than those of HCFC-22, which may lead to the system improved reliability and lifetime.
n. The hydrocarbon mixture R-170 (ethane)/R-290 (propane) would be an attractive replacement for HCFC-22 in industrial refrigeration and heat pump systems because the energy efficiency may be 7%–9% higher compared to that of pure HCFC-22, mainly because of improved thermodynamic and transport properties of the mixture (Cleland et al. 2009).
o. Both hydrocarbon mixtures R-170 (ethane)/R-116 (hexafluoroethane) and R-170 (ethane)/R-23/R-116 show good performance as refrigerants used in the low-stage loop in the two-stage cascade system instead of the refrigerant R-503. The coefficients of performance of the R-170 blends are higher and the compression ratios lower, probably due to the vapor increased pressures which contribute at increasing the compressor efficiency by reducing the compression ratio (Gong et al. 2009).

14.4 AIR-SIDE PRESSURE DROP

The air-side pressure drop can be separated into two components: (i) the pressure drop due to the tubes and (ii) the pressure drop due to the fins (Rich 1973):

$$\Delta p_{tot} = \Delta p_{fin} + \Delta p_{tube} \tag{14.19}$$

where

Δp_{tube} is the pressure drop due to tubes (Pa)
Δp_{fin} is the pressure drop due to fins (Pa)

The pressure drop (Pa) due to the fins can be expressed as follows:

$$\Delta p_{fin} = f_{fin} * \overline{v} * \frac{G_{max}^2}{2} \frac{A_{fin}}{A_c} \tag{14.20}$$

where

f_{fin} is the fins' friction factor (–)
v is the mean specific volume (m³/kg)
G_{max} is the mass velocity (flux) through minimum free-flow cross-sectional area (kg/m²·s)
A_{fin} is the fins' surface area (m²)
A_c is the minimum free-flow cross-sectional area (m²)

The friction factor depends on Reynolds number but is independent of fin spacing. For fin spacing between 3 and 14 fins per inch (1 in. = 25.4 mm), the fin friction factor is (Rich 1973) as follows:

$$f_{fin} = 1.70 Re_l^{-0.5} \tag{14.21}$$

where

Reynolds number is based on the transverse (in the direction of airflow) tube spacing:

$$Re_l = \frac{G * \delta_l}{m} \tag{14.22}$$

G is the refrigerant mass flux (kg/m²·s)
δ_l is the longitudinal tube spacing (parallel to airflow) (m)
m is the fin parameter (–)

The pressure drop over the refrigerant tubes can be calculated with the following relationships developed for over the banks of plain tubes (Zukauskas and Ulinskas 1988):

$$\Delta p_{tube} = Eu_{corr} \frac{G^2}{2r_{out}} z \tag{14.23}$$

where

$Eu_{corr} = C_g * C_z * Eu$ is the corrected Euler number (that is related to the tube friction factor and depends on the Reynolds number and the tube geometry) (–)
r_{out} is the tubes' outside radius (m)
z is the number of rows (–)
C_g is the staggered array geometry factor (–)
C_z is the average row correction factor (–). For tube banks with a small number of transverse rows, the average row correction factor (C_z) is applied because the pressure drop over the first few rows will be different than the pressure

drop over the rest of the rows; C_z is thus the average of the individual row correction factors

Eu is Euler number (–)

The Euler number (Eu) is used to characterize energy losses in the refrigerant flow. It expresses ratio between the flow pressure drop caused by a restriction (e.g., friction) and the inertia forces:

$$Eu = \frac{\Delta p}{\rho u^2} \tag{14.24}$$

where

$\Delta p = p_2 - p_1$ is the pressure differential (Pa)

p_2 is the upstream pressure (Pa)

p_1 is the downstream pressure (Pa)

ρ is the density of the fluid (kg/m^3)

u is the refrigerant velocity (m/s)

14.5 AIR-SIDE HEAT TRANSFER COEFFICIENTS

The calculation of air-side heat transfer coefficients of refrigerant-to-air dry condensers is based on the Colburn j-factor, which can be defined as (Rich 1973; Sadler 2000) follows:

$$j = St * Pr^{2/3} \tag{14.25}$$

where

j is the Colburn j-factor (–)

$St = \frac{h_{air}}{\bar{\rho} * \bar{u} * c_p}$ is the Stanton number (–)

$Pr = \frac{\nu}{a}$ is the Prandtl number (–)

Substituting the appropriate values for the Stanton number gives the following relationship for the convective heat transfer coefficient as a function of *j*-factor (Kim and Bullard 2002):

$$h_{air} = \frac{j * c_p * G_{max}}{Pr^{2/3}} = \frac{j * \bar{\rho} * \bar{u} * c_p}{Pr^{2/3}} \tag{14.26}$$

where

h_{air} is the air-side heat transfer coefficient (W/m$^2 \cdot$ K)

G_{max} is the mass flux through minimum flow area (kg/m$^2 \cdot$ s)

$\bar{\rho}$ is the air mean density (kg/m^3)

u is the air average velocity (m/s)

$\bar{c}_p$ is the air average isobaric specific heat (J/kg $\cdot$ K)

For a four-row finned-tube heat exchanger, the j-factor is (McQuiston and Parker 1994) defined as follows:

$$j_4 = 0.2675 * JP + 1.325 * 10^{-6}$$

where

$$JP = Re_d^{-0.4}\left(\frac{A_{tot}}{A_{fin}}\right)^{-0.15}$$

Re_d is Reynolds number based on the outside diameter of the tubes (d_{out}) and the maximum mass flux G_{max} (–)
A_{fin} is the air-side fin heat transfer area (m^2)
A_{tot} is the total air-side heat transfer area (fins and tubes) (m^2)

14.6 DESIGN OF REFRIGERANT-TO-AIR CONDENSERS

Refrigerant-to-air condensers are heat exchangers with fairly uniform wall temperature employed in heat pump-assisted drying systems. There are two well-established methods available for the thermal heat exchanger design (Kakaç and Liu 2002; Shah and Sekulic 2003; McQuidton 2005): (i) the log-mean temperature difference (LMTD) method (see Section 12.2.1) and (ii) the effectiveness number of transfer units ($\varepsilon - NTU$) approach (see Section 12.2.2), where ε is defined as the ratio between the actual heat transfer rate and the maximum amount that can be transferred and the number of transfer units (NTU) compares the thermal size of compact heat exchangers with their capacity of heating the fluid. Furthermore, the $\varepsilon - NTU$ approach avoids the number of iterations required by the $LMTD$ for outlet temperature calculations (Hermes 2013).

14.6.1 Overall Heat Transfer Coefficient

After calculating the refrigeration-side and air -side heat transfer coefficients, the overall heat transfer coefficient may be calculated based on the following relation (Kim and Bullard 2002):

$$\frac{1}{\bar{U}} = \frac{1}{h_{refr} * A_{refr}} + \frac{\ln\left(\frac{d_{ext}}{d_{int}}\right)}{2\pi * L * k_{wall}} + \frac{1}{\eta_{air} * h_{air} * A_{air}} \quad (14.27)$$

where
$\bar{U}$ is the overall heat transfer coefficient (W/m$^2\cdot$K)
h_{refr} is the refrigerant heat transfer coefficient (W/m$^2\cdot$K)
A_{refr} is the refrigerant-side heat transfer area (m^2)
d_{ext} is the external diameter of circular tubes (m)
d_{int} is the internal diameter of circular tubes (m)
L is the total length of circular tubes (m)

k_{wall} is the tube wall thermal conductivity of the tube wall material (copper or aluminum) (W/mK)
η_{air} is the air-side overall surface heat transfer efficiency (–)
h_{air} is the air-side heat transfer coefficient (W/m$^2\cdot$K)
A_{air} the air-side heat transfer area (m^2)

It can be seen that the overall heat transfer coefficient (equation 14.27) accounts for three thermal resistances: (i) forced boiling (vaporizing) thermal resistance on the refrigerant-side, (ii) conductive thermal resistance through the tube cylindrical wall, and (iii) convective thermal resistance on the air-side.

The air-side heat transfer efficiency is defined as follows:

$$\eta_{air} = 1 - \frac{A_{fin}}{A_{air}}\left(1 - \eta_{fin}\right) \tag{14.28}$$

where

η_{air} is the air-side overall surface heat transfer efficiency (–)
A_{fin} is the fin heat transfer area (m^2)
A_{air} is the overall air-side surface heat transfer area (m^2)
$\eta_{fin} = \dfrac{\tanh(ml)}{ml}$ is the fin efficiency (–)

$$m = \sqrt{\frac{2 * h_{air}}{k_{fin} * F_\delta}\left(1 + \frac{F_\delta}{T_d}\right)}$$

$$l = \frac{F_h}{2} - F_\delta$$

F_δ is the fin thickness (m)
F_h is the fin height (m)
T_d is the tube row depth (m)

14.6.2 Heat Transfer Rate

Based on the energy balance of the refrigerant condenser, its steady-state thermal capacity (also called heat transfer rate or total heat of rejection) can be calculated from the refrigerant-side enthalpy change and flow rate, or from the air-side mass flow rate, specific heat and temperature change as follows:

$$\dot{Q}_{cond} = \overline{\eta}_{cond} * \dot{m}_{refr} * (h_2 - h_3) = \dot{m}_{air} * \overline{c}_{p,air} * (T_I - T_M) \tag{14.29}$$

where

$\dot{Q}_{cond}$ is the refrigerant condenser thermal capacity (or heat transfer rate) (kW)
$\overline{\eta}_{cond}$ is the overall thermal efficiency of the refrigerant-to-air condenser (–)
$\dot{m}_{refr}$ is the refrigerant mass flow rate (kg/s)
h_2 is the specific enthalpy of refrigerant superheated vapor entering the condenser (kJ/kg)
h_4 is the specific enthalpy of refrigerant subcooled liquid leaving the condenser (kJ/kg)

$\dot{m}_{air}$ is the air mass flow rate (kg/s)
$\bar{c}_{p,air}$ is the average isobaric specific heat of air (kJ/kg·K)
T_M is the dry-bulb temperature of air entering the condenser (C)
T_I is the dry-bulb temperature of air leaving the condenser (C)

The same steady-state thermal capacity (heat transfer rate) of refrigerant condensers can be written as follows:

$$\dot{Q}_{cond} = \bar{U} * A * (LMTD) = U * A * \left(\frac{T_{air,\,out} - T_{air,\,in}}{ln \frac{T_{cond} - T_{air,\,in}}{T_{cond} - T_{air,\,out}}} \right) \quad (14.30)$$

where
$\dot{Q}_{cond}$ is the thermal capacity (heat transfer rate) of the refrigerant-to-air condenser (kW)
$\bar{U}$ is the overall heat transfer coefficient expressed based on Equation 14.27 (W/m^2·K)
A is the condenser heat transfer area (m^2)
$LMTD = \frac{\Delta T_{max} - \Delta T_{min}}{ln \frac{\Delta T_{max}}{\Delta T_{min}}} = \frac{T_{air,\,out} - T_{air,\,in}}{ln \frac{T_{cond} - T_{air,\,in}}{T_{cond} - T_{air,\,out}}}$ is the logarithmic mean temperature difference (C)

where
$\Delta T_{max} = T_{cond} - T_{air,in}$
$\Delta T_{min} = T_{cond} - T_{air,out}$
T_{cond} is the condensing temperature (C)

Some design methods assume that the refrigerant-to-air condenser operates with the constant temperature equal to the refrigerant saturation temperature. If the single-phase zones of the condenser (i.e., vapor desuperheating and liquid subcooling) are neglected, this model reduces to a one-zone heat exchanger. In this case, an average condensing temperature ($\bar{T}_{cond}$) is defined as the weighted average of the actual temperatures occurring in the three zones (single-phase desuperheating, two-phase condensation, and single-phase undercooling) (Lemort and Bertagnolio 2010).

In practice, manufacturers of condensers provide performance data allowing selecting the appropriate heat exchanger for given applications at specific design parameters. By applying some fundamentals of heat transfer, it is possible to translate catalog data to non-design conditions. The strategy in extending catalog data to non-design conditions is usually to compute the *UA* value (the product of the overall heat transfer coefficient and the heat transfer area), and, for situations where the *UA* remains essentially constant, apply the *UA* value to the new set of operating conditions. The temperature profiles are somewhat complex because of desuperheating and subcooling, but to approximate, assume the condensing temperature prevails throughout the condenser (Stoecker 1998).

Exercise E14.1

The refrigerant-to-air condenser of a mechanical vapor compression heat pump heats $\dot{m}_{air} = 3\,\mathrm{m}^3/\mathrm{s}$ of air from $T_{air,in}$ = 50°C to $T_{air,out}$ = 65°C by condensing the refrigerant HFO-1234yf at saturated pressure and temperature of p_{sat} = 2044 kPa,abs and T_{sat} = 70°C, respectively. The condensate (liquid) leaves the condenser at saturation state. Knowing the enthalpy (latent heat) of condensation h_{fg} = 96.3 kJ/kg, the condenser thermal efficiency η_{cond} = 96%, the air average isobaric specific heat $\bar{c}_{p,air} = 1.0075\,\mathrm{kJ/kg \cdot K}$, and the condenser overall heat transfer coefficient $\bar{U} = 600\,\mathrm{W/m^2K}$, calculate the following:

1. Refrigerant mass flow rate,
2. Condenser internal heat transfer surface area.

Solution

Based on the given data,

$$\Delta T_{max} = T_{cond} - T_{air,\,in} = 70 - 50 = 20°\mathrm{C}$$

$$\Delta T_{min} = T_{cond} - T_{air,\,out} = 70°\mathrm{C} - 65°\mathrm{C} = 5°\mathrm{C}$$

The LMTD is expressed as follows:

$$LMTD = \frac{\Delta T_{max} - \Delta T_{min}}{\ln \dfrac{\Delta T_{max}}{\Delta T_{min}}} = \frac{20-5}{\ln \dfrac{20}{5}} = \frac{15}{1.386} = 10.8°\mathrm{C}$$

The air-side heat transfer rate is expressed as follows:

$$\dot{Q}_{air} = \dot{m}_{air} * \bar{c}_{p,\,air} * \left(T_{air,\,out} - T_{air,\,in}\right) = 3 * 1.0075 * (65 - 50) = 45.33\ \mathrm{kW}$$

1. The refrigerant mass flow rate can be calculated as follows:

$$\dot{m}_{refr} = \frac{\dot{Q}_{air}}{\eta_{cond} * h_{fg}} = \frac{45.33}{0.96 * 96.3} = 0.49\ \mathrm{kg/s}$$

2. Using equation 14.32, the refrigerant-side required heat transfer area can be calculated as follows:

$$A = \frac{\dot{Q}_{cond}}{\bar{U} * (LMTD)} = \frac{45.33}{0.6 * 10.8} = 7\ \mathrm{m}^2$$

14.6.3 Refrigerant-Side Pressure Drop

The refrigerant-side's two-phase pressure drop includes an acceleration component (De Souza et al. 1995):

$$\Delta p = G^2\left[\frac{x_{out}^2}{\rho_g * \varepsilon_{out}} + \frac{(1-x_{out})^2}{\rho_f(1-\varepsilon_{out})}\right] - \left[\frac{x_{in}^2}{\rho_g * \varepsilon_{in}} + \frac{(1-x_{in})^2}{\rho_f(1-\varepsilon_{in})}\right] \quad (14.31)$$

where

G is the mass velocity (kg/m^{2}s)
x_{in} is the inlet vapor title (–)
x_{out} is the outlet vapor title (–)
ρ_f is the liquid density (kg/m^3)
ρ_g is the vapor density (kg/m^3)
ε is the void fraction (Zivi 1964)

The pressure drop correlation also includes a frictional component, which can be expressed as follows:

$$\Delta p_{frictional} = \Delta p_{fo} \frac{1}{\Delta x} \int \Phi_{fo}^2 dx \tag{14.32}$$

In this correlation, the frictional pressure drop for the whole mixture flowing as a liquid (Δp_{fo}) is given by the following:

$$\Delta p_{fo} = \frac{2 * \varepsilon_{fo} * G^2 * L}{\rho_f * d} \tag{14.33}$$

where

$\varepsilon_{fo} = \dfrac{0.079}{Re_{fo}^{1/4}}$ is the liquid Fanning friction coefficient for the whole mixture flowing as a liquid (-)
L is tube length (m)
d is the tube inner diameter (m)

The two-phase friction multiplier (i.e., the ratio of the two-phase pressure gradient to the friction pressure gradient if the total mixture flows as a liquid) is given by the following:

$$\Phi_{fo}^2 = 1 + \left(\chi^2 - 1\right) * x^{1.75} \left(1 + 0.9524 * \chi * X_{tt}^{0.4126}\right) \tag{14.34}$$

where

$\chi = \left(\dfrac{\rho_f}{\rho_g}\right)^{0.5} \left(\dfrac{\mu_f}{\mu_g}\right)^{0.125}$ is the physical property index.

REFERENCES

Akers, W.W., H.A., Deans, O.K. Crosser. 1959. *Condensing Heat Transfer Within Horizontal Tubes.* Chemical Engineering Progress Symposium Series 55:171–176.

ASHRAE. 2009. *Fundamentals Handbook*, SI Edition. American Society of Heating, Refrigerating and Air-Conditioning Engineers, Inc., Atlanta, GA.

Agarwal, R., P. Hrnjak. 2015. Condensation in two phase and desuperheating zone for R-1234ze(E), -134a and R-32 in horizontal smooth tubes. *International Journal of Refrigeration* 50:171–183.

Bell, K.J., M.A. Ghaly. 1973. An approximate generalized design method for multi-component/partial condensers. *American Institute of Chemical Engineers Symposium Series* 69:72–79.

Bivens, D. B., A. Yokozeki. 1994. Heat transfer coefficient and transport properties for alternative refrigerants. In *Proceedings of the International Refrigeration Conference*, Purdue, Indiana, Paper 263, pp. 299–304.

Boyko, L.D., G.N. Kruzhilin. 1967. Heat transfer and hydraulic resistance during condensation of steam in a horizontal tube and in a bundle of tubes. *International Journal of Heat and Mass Transfer* 10:361–73.

Breber, G., J.W. Palen, J. Taborek. 1979. Prediction of horizontal tube-side condensation of pure components using flow regime criteria. *Journal of Heat Transfer* 102(3): 471–76.

Breber, G., J. Palen, J. Taborek. 1980. Prediction of horizontal tube-side condensation of pure components using flow regime criteria. *Journal of Heat Transfer* 102:471–476.

Cavallini, A., R. Zecchin. 1971. High velocity condensation of organic refrigerants inside tubes. In *Proceedings of the 13th International Congress of Refrigeration*, Washington, DC, Vol. 2, pp. 193–200.

Cavallini, A., R. Zecchin. 1974. A dimensionless correlation for heal transfer in forced convection condensation. In *Proceedings of the 6th International Heat Transfer Congress*, Tokyo, Vol. 3, pp. 309–313.

Cavallini, A., D. Del Col, L. Doretti, G.A. Longo, L. Rossetto. 2000. Condensation heat transfer of new refrigerants: Advantages of high pressure fluids. In *Proceedings of the 8th International Refrigeration Conference at Purdue*, 25–28 July, pp. 177–184.

Cavallini, A., G. Censi, D. Del Col, L. Dorctti, G.A. Longo, L. Rossello. 2001. Experimental invcstigation on condensation heat transfer and pressure drop of new HFC refrigerants (R-134a, R-125, R-32, R-410A, R-236ea) in a horizontal tube. *International Journal of Refrigeration* 24:73–87.

Cavallini, A, G. Censi, D. Del Col, L. Doretti, G.A. Longo, L. Rossetto. 2002a. In-tube condensation of halogenated refrigerants. *ASHRAE Transactions* 108(1):146–161.

Cavallini, A., G. Censi, D. Del Col, L. Doretti, G.A. Longo, L. Rossetto. 2002b. Condensation of halogenated refrigerants inside smooth tubes. *HVAC&R Research* 8:429–451.

Cavallini, A., G. Censi, D. Del Col, L. Doretti, G.A. Longo, L. Rossetto. 2002c. Intube condensation of halogenated refrigerants. *ASHRAE Transactions* 108(1), Paper 4507.

Cavallini, A., G. Censi, D. Del Cola, L. Doretti, G.A. Longo, L. Rossetto, C. Zilio. 2003. Condensation inside and outside smooth and enhanced tubes – A review of recent research. *International Journal of Refrigeration* 26:373–392.

Cavallini, A., D. Del Col, L. Doretti, M. Matkovic, L. Rossetto, C. Zilio, G. Censi. 2006. Condensation in horizontal smooth tubes: A new heat transfer model for heat exchanger design. *Heat Transfer Engineering* 27 (8):31–38.

Chang Y.J., C.C. Wang. 1997. A generalized heat transfer correlation for louvered fin geometry. *International Journal of Refrigeration* 40(3):533–544.

Chang, Y.S., M.S. Kim, S.T. Ro. 2000. Performance and heat transfer characteristics of hydrocarbon refrigerants in a heat pump system. *International Journal of Refrigeration* 23:232–242.

Chen, S.L., F.M. Gerner, C.L. Tien. 1987. General film condensation correlations. *Experimental Heat Transfer* 1:93–107.

Cheng, P., H.Y. Wu, F.J. Hong. 2007. Phase-change heat transfer in microsystems. *Heat Transfer* 129(2):101–108.

Chepurnenko, V. P., A. E. Lagutin, N. I. Gogol. 1993. An investigation of heat exchange during condensation of ammonia inside a pipe at low heat flow densities. In *Energy Efficiency in Refrigeration and Global Warming Impact*. In *Proceedings of Refrigeration Science and Technology*, Belgium, May. *International Institute of Refrigeration, Commissions B1/B2*, pp. 243–249.

Cleland, D.J., R.W. Keedwell, S.R. Adams. 2009. Use of hydrocarbons as drop-in replacements for HCFC-22 in on-farm milk cooling equipment. *International Journal of Refrigeration* 32(6):1403–1411.

Coleman, J.W., S. Garimella. 1999. Characterization of two-phase flow patterns in small diameter round and rectangular tubes. *International Journal of Heat Mass Transfer* 42:2869–2881.

Coleman, J. W., S. Garimella. 2003. Two-phase flow regimes in round, square and rectangular tubes during condensation of refrigerant R134a. *International Journal of Refrigeration* 26:117–128.

Dalkilic, A.S., S. Wongwises. 2009. Intensive literature review of condensation inside smooth and enhanced tubes. *International Journal of Heat Mass Transfer* 52:3409–3426.

Dittus, F.W., M.L.K. Boelter. 1930. Heat transfer in automobile radiators of the tubular type. *University of California Publications on Engineering*, Berkeley, CA, Vol. 2(13), p. 443.

Dobson, M.K., J.P. Wattelet, J.C. Chato. 1993. *Optimal Sizing of Two-Phase Heat Exchangers*. ACRC Technical Report 42, University of Illinois at Urbana-Champaign.

Dobson, M.K., J.C. Chato. 1998. Condensation in smooth horizontal tubes. *ASME Journal of Heat Transfer* 120:193–213.

Domanski, P.A., D. Yashar. 2006. Comparable performance evaluation of HC and HFC refrigerants in an optimized system. In *Proceedings of 7th IIR Gustav Lorentzen Conference on Natural Working Fluids*, Trondheim, Norway.

Doretti, L., C. Zilio, S. Mancin, A. Cavallini. 2013. Condensation flow patterns inside plain and micro-fin tubes: A review. *International Journal of Refrigeration* 36:567–587.

Eckels, S.J., M. B. Pate. 1990. Evaporation and condensation heat transfer coefficients for a HCFC-124/HCFC-22/ HFC-152a blend and CFC-12. *International Refrigeration and Air Conditioning Conference at Purdue*. Paper 106 (http://docs.lib.purdue.edu/iracc/106, accessed December 12, 2017).

Eckels, S.J., B. Tesene. 1999. A comparison of R-22, R-134a, R-410a, and R-407C condensation performance in smooth and enhanced tubes: Part I - heat transfer. *ASHRAE Transactions* 105(2):428–451.

Fang, X., Z. Zhou, D. Li. 2013. Review of correlations of flow boiling heat transfer coefficients for carbon dioxide. *International Journal of Refrigeration* 36:2017–2039.

Fronk, B.M, S. Garimella. 2013. In-tube condensation of zeotropic fluid mixtures: A review. *International Journal of Refrigeration* 36:534–561.

Garimella, S. 2006. Condensation in mini-channeis and micro-channels, in *Heat Transfer and Fluid Flow in Mini-Channels and Micro-Channels*, edited by S. Kandlikar. Elsevier, Amsterdam, pp. 295–494.

Gong, M., Z. Sun, J. Wu, Y. Zhang, C. Meng, Y. Zhou. 2009. Performance of R-170 mixtures as refrigerants for refrigeration at −80°C temperature range. *International Journal of Refrigeration* 32(5):892–900.

Haraguchi, H., S. Koyama, T. Fujii. 1994. Condensation of refrigerants HCFC22, HFC134a and HCFC123 in a horizontal smooth tube (2nd report, proposal of empirical expressions for local heat transfer coefficient). *Transactions of the JSME* 60 (574):245–252.

Hermes, C. J. L., C., Melo, C. F.T. Knabben. 2013. Alternative test method to assess the energy performance of frost-free refrigerating appliances. *Applied Thermal Engineering* 50(1):1029–1034.

Hwang, Y. Jr., R. Radermacher. 2007. Comparison of R-290 and two HFC blends for walk-in refrigeration systems. *International Journal of Refrigeration* 30(4):633–641.

Harirchian, T., S.V. Garimella. 2010. A comprehensive flow regime map for microchannel flow boiling with quantitative transition criteria. *International Journal of Heat and Mass Transfer* 53(13–14):2694–2702.

Hossain, M.A., Y. Onaka, A. Miyara. 2012. Experimental study on condensation heat transfer and pressure drop in horizontal smooth tube for R1234ze(E), R32 and R410A. *International Journal of Refrigeration* 35(4):927–938.

Incropera, F.P., D.P. DeWitt. 2002. *Fundamentals of Heat and Mass Transfer*, 5th ed. John Willey & Sons, Hoboken, NJ.

Jassim, L.W., T.A. Newell, J.C. Chato. 2008. Prediction of two-phase condensation in horizontal tubes using probabilistic flow regime maps. *International Journal of Heat and Mass Transfer* 51:485–496.

Jaster, H., P.G. Kosky. 1976. Condensation heat transfer in a mixed flow regime. *International Journal of Heat and Mass Transfer* 19:95–99.

Jin, H., J.D. Spitler. 2002. A parameter estimation based model of water-to-water heat pumps for use in energy calculation programs. *ASHRAE Transactions* 108(1):3–17.

Kakaĉ, S., H. Liu. 1998. *Heat Exchangers Selection, Rating and Thermal Design*. CRC Press, Boca Raton, FL, p. 323.

Kandlikar, S.G., W.J. Grande. 2003. Evolution of micro-channel flow passages – Thermohydraulic performance and fabrication technology. *Heat Transfer Engineering* 24(1):3–17.

Kattan, N., J.R. Thome, D. Favrat. 1998a. Flow boiling in horizontal tubes: Part 1—Development of adiabatic two-phase flow pattern map. *Journal of Heat Transfer* 120:140–147.

Kattan, N., J.R. Thome, D. Favrat. 1998b. Flow boiling in horizontal tubes: Part 3—Development of a new heat transfer model based on flow pattern. *Journal of Heat Transfer* 120:156–165.

Kew, P.A., K. Cornwell. 1997. Correlations for the prediction of boiling heat transfer small-diameter channels. *Applied Thermal Engineering* 17:705–715.

Kim, M., C.W. Bullard. 2002. Air-side thermal hydraulic performance of multi-louvered fin aluminum heat exchangers. *International Journal of Refrigeration* 25:390–400.

Kosky, P.G., F.W. Staub. 1971. Local condensing heat transfer coefficients in the annular flow regime. *ATChE Journal* 17:1037–1043.

Koyama, S., A. Miyara, H. Takamatsu, T. Fujii. 1990. Condensation heat transfer of binary refrigerant mixtures of R22 and R114 inside a horizontal tube with internal spiral grooves. *International Journal of Refrigeration* 13(4):256–263.

Koyama, S., J. Yu, A. Ishibashi. 1998. Condensation of binary refrigerant mixtures in a horizontal smooth tube. *Thermal Science Engineering* 6(1):123–129.

Jaster, H., P.G. Kosky. 1976. Condensation heat transfer in a mixed flow regime. *International Journal of Heat and Mass Transfer* 19(1):95–99.

Launay, P.F. 1981. *Improving the Efficiency of Refrigerators and Heat Pumps by Using a Nonazeotropic Mixture of Refrigerants*. Oak Ridge National Laboratory.

Lee, H.S., J. Yoon, J.D. Kim, P.K. Bansal. 2006. Condensing heat transfer and pressure drop characteristics of hydrocarbon refrigerants. *International Journal of Heat and Mass Transfer* 49:1922–1927.

Lee, H., C. Son. 2010. Condensation heat transfer and pressure drop characteristics of R-290, R-600a, R-134a, and R-22 in horizontal tubes. *Heat and Mass Transfer* 46:571–284.

Lemort, V., S. Bertagnolio, 2010. A generalized simulation model of chillers and heat pumps to be calibrated on published manufacturer's data. In *Proceedings of International Symposium on Refrigeration Technology*, August 6–7, Zhuha, China, 11p.

Li, W., Z. Wu. 2010. A general criterion for evaporative heat transfer in micro/mini-channels. *International Journal of Heat and Mass Transfer* 53(9–10):1967–1976.

Lips, S., J.P. Meyer. 2011. Two-phase flow in inclined tubes with specific reference to condensation: A review. *International Journal of Multiphase Flow* 37:845–859.

Liu, N., J.M. Li, J. Sun, H.S. Wang. 2013 Heat transfer and pressure drop during condensation of R-152a in circular and square micro-channels. *Experimental Thermal and Fluid Science* 47:60–67.

Liu, S., V. Huo, Z. Liu, L. Li, J. Ning. 2015. Theoretical research on R-245fa condensation heat transfer inside a horizontal tube. *Engineering* 7:261–271.

Longo, G.A., C. Zilio. 2013. Condensation of the low GWP refrigerant HFC1234yf inside a brazed plate heat exchanger. *International Journal of Refrigeration* 36:612–621.

Martinez, I.O., J. Gonzalvez, J.M. Corberan. 2006. Charge and COP optimization of a reversible water to water unit using propane as alternative refrigerant to R-407C. In *Proceedings of 7th IIR Gustav Lorentzen Conference on Natural Working Fluids*, Trondheim, Norway.

Mathur, G.D. 2000. Hydrodynamic characteristics of propane (R-290), isobutane (R-600a), and 50/50 mixture of propane and isobutene. *ASHRAE Transactions* 106:571–582.

McQuiston, F. C., J. P. Parker. 1994. *Heating Ventilating and Air-Conditioning Analysis and Design*, John Wiley & Sons, New York.

McQuiston, F.C., J. D. Parker, J. D. Spitler. 2005. *Heating, Ventilating, and Air Conditioning Analysis and Design*, 6th ed. John Wiley & Sons, Inc., New York.

Mehendale, S.S., A.M. Jacobi, R.K. Shah. 2000. Fluid flow and heat transfer at micro- and meso-scales with application to heat exchanger design. *Applied Mechanics Reviews* 53:175–193.

Miyara, A. 2008. Condensation of hydrocarbons – A review. *International Journal of Refrigeration* 31:621–632.

Moser, K.W., R.L. Webb, B. Na. 1998. A new equivalent Reynolds number model for condensation in smooth tubes. *ASME Journal of Heat Transfer* 20:410–417.

Mathur, G.D. 2000. Hydrodynamic characteristics of propane (R-290), isobutane (R-600a), and 50/50 mixture of propane and isobutene. *ASHRAE Transactions* 106:571–582.

Nusselt, W. 1916. Die Obertlachenkondensation des Wasser dampfes. *Zeitschrift des Vereines deutscher Ingenieure* 60(541–546):569–75.

Ohadi, M. M., S. S. Li, R. Radermacher, S. Dessiatoun. 1996. Critical review of available correlations for two-phase flow heat transfer of ammonia. *International Journal of Refrigeration* 19(4):272–284.

Palen, J.W., G. Breber, J. Taborek. 1979. Prediction of flow regimes in horizontal tube-side condensation. *Journal of Heat Transfer Engineering* 1(2):47–57 (http://dx.doi.org/, accessed March 12, 2017).

Park, K., D.G. Kang, D. Jung. 2010. Condensation heat transfer coefficients of HFC-245fa on a horizontal plain tube. *Journal of Mechanical Science and Technology* 24(9):1911–1917.

Pottker, G., P. Hrnjak. 2015. Effect of the condenser sub-cooling on the performance of vapor compression systems. *International Journal of Refrigeration* 50:155–164.

Rayleigh, L. 1880. On the stability, or instability, of certain fluid motions. In *Proceedings of the London Mathematical Society* 11, 57-70 (also in: *Scientific Papers*, Vol. 1, Cambridge University Press, Cambridge, 1889, 474–487).

Rich, D. G. 1973. The effect of fin spacing on the heat transfer and friction performance of multi-row, smooth plate fin-and-tube heat exchangers. *ASHRAE Transactions* 79(2):137–145.

Rohsenow, W.M. 1973. Condensation, in *Handbook of Heat Transfer*, edited by W.M. Rohsenow, J.P. Hartnett and Young I. Cho, Chapter 14, McGraw Hill, New York, pp. 14.1–14.63.

Sadler, E.M. 2000. Design analysis of a finned-tube condenser for residential air conditioner using R-22. In *Partial Fulfillment of the Requirements for the Degree Master of Science in Mechanical Engineering*, Georgia Institute of Technology.

Shah, M.M. 1976. New correlation for heat transfer during boiling flow through pipes. In *Proceedings of Annual Meeting*, June 27–July 1, Vol. 82(part 2): 66–86.

Shah, M.M. 1979a. A general correlation for heat transfer during film condensation inside pipes. *International Journal of Heat and Mass Transfer* 22:547–556.

Shah, M.M. 1979b. An improved and extended general correlation for heat transfer during condensation in plain tubes. *HVAC&R Research* 15(5):889–913.

Shah, R.K., T.E. Sekulic. 2003. *Fundamentals of Heat Exchanger Design*. John Wiley & Sons, New York, Chapters 10 and 11.

Sardesai, R.G., R.G. Owen, D.G. Pulling. 1981. Flow regimes for condensation of a vapour inside a horizontal tube. *Chemical Engineering Science* 36:1173–1180.

Silver, L. 1947. Gas cooling with aqueous condensation. *Transactions of the Institution of Chemical Engineers* 25:30–42.

Spatz, M.W., S.F.Y. Motta. 2004. An evaluation of options for replacing R22 in medium temperature refrigeration systems. *International Journal of Refrigeration* 27:475–483.

Stoecker, W.F., C.J. McCarthy. 1984. *The Simulation and Performance of a System Using and R-12/R-114 Refrigerant Mixture*. Oak Ridge National Laboratory.

Stoecker, W.F., E. Kornota. 1985. Condensing coefficients when using refrigerant mixtures. *ASHRAE Transactions* 91:1351–1367.

Stoecker, W.F. 1998. *Industrial Refrigeration Handbook*. McGraw-Hill, New York.

Taitel, Y., A.E. Dukler. 1976. A model for predicting flow regime transitions in horizontal and near horizontal gas–liquid flow. *AIChE Journal* 22(2):47–55.

Tandon, T.N., H.K. Varma, C.P. Gupta. 1982. A new flow regime map for condensation inside horizontal tubes. *ASME Journal of Heat Transfer* 104: 763–768.

Tandon, T.N., H.K. Varma, C.P. Gupta. 1986. Generalized correlation for condensation of binary mixtures inside a horizontal tube. *International Journal of Refrigeration* 9:134–136.

Tang, L. 1997. Empirical study of new refrigerant flow condensation inside horizontal smooth and micro-fin tubes. *Ph.D. thesis*, University of Maryland at College Park.

Thome, J.R., J. El Hajal, A. Cavallini. 2003. Condensation in horizontal tubes, part 2: New heat transfer model based on flow regimes. *International Journal of Heat and Mass Transfer* 46:3365 3387.

Thonon B., A. Bontemps. 2002. Condensation of pure and mixture of hydrocarbons in compact heat exchangers; experiments and modelling. *Heat Transfer Engineering* 23 (6):3–17.

Thonon, B. 2005. A review of hydrocarbon two-phase heat transfer in compact heat exchangers and enhanced geometries. In *Proceedings of the IIR Conferences: Thermo-Physical Properties and Transfer Processes of Refrigerants*, Vicenza, Italy.

Traviss, D.P., W.M. Rohsenow, A.B. Baron. 1972. Forced convection inside tubes: A heat transfer equation for condenser design. *ASHRAE Transactions* 79:157–165.

Traviss, D.P., W.M. Rohsenow. 1973. Flow regimes in horizontal two-phase flow with condensation. *ASHRAE Transactions* 79:31–39.

van Rooyen, E., Christians, M., Liebenberg, L., Meyer, J.P., 2010. Probabilistic flow pattern-based heat transfer correlation for condensing intermittent flow of refrigerants in smooth horizontal tubes. *International Journal of Heat and Mass Transfer* 53:1446–1460.

Vera-García, F., J.R. García-Cascales, J.M. Corberán-Salvador, J. Gonzálvez-Maciá, D. Fuentes-Díaz. 2007. Assessment of condensation heat transfer correlations in the modelling of fin and tube heat exchangers. *International Journal of Refrigeration* 30(6):1018–1028.

Vollrath, J. E., P. S. Hrnjak, T. A. Newell. 2003. *An Experimental Investigation of Pressure Drop and Heat Transfer in an In-tube Condensation System of Pure Ammonia*. Air Conditioning and Refrigeration Center (ACRC), University of Illinois at Urbana-Champaign, 97 pp.

Wang, S., J.E. Chato. 1995. Review of recent research on heat transfer with mixtures, Part 1: Condensation. *ASHRAE Transactions* 101(1):1387–1401.

Wen, M.Y., C.Y. Ho, J.M. Hsieh. 2006. Condensation heat transfer and pressure drop characteristics of R-290 (propane), R-600 (butane), and a mixture R290/R600 in the serpentine small-tube bank. *Applied Thermal Engineering* 26(16):2045–2053.

Wijaya, H., M.W. Spatz. 1995. Two-phase flow heat transfer and pressure drop characteristics of R-22 and R-32/R125. *ASHRAE Transactions* 118:1020–1027.

Wolverine Tube. 2006. *Engineering Data Book III.* Condensation inside tubes. Chapter 8 (www.prnewswire.com/news-releases/wolverine-tube-inc, accessed November 1, 2017).

Zukauskas, A.A., R.V. Ulinskas, V. Katinas. 1988. *Fluid Dynamics and Flow-Induced Vibrations of Tube Banks*, edited by J. Karni. Hemisphere, New York, p. 290.

Zhang, H., X. Fang, H. Shang, W. Chen. 2015. Review – Flow condensation heat transfer correlations in horizontal channels. *International Journal of Refrigeration* 59:102–111.

Zhuang, X.R., M.Q. Gong, X. Zou, G.F. Chen, J.F. Wu. 2016. Experimental investigation on flow condensation heat transfer and pressure drop of R-170 in a horizontal tube. *International Journal of Refrigeration* 66:105–120.

Zivi, S.M. 1964. Estimation of steady-state steam void-fraction by means of the principle of minimum entropy production. *Journal of Heat Transfer* 86:247–252.

15 Heat Pump Modeling

15.1 GENERALITIES

Accurate modeling and simulation of each component of heat pumps allow for predicting the behavior of the entire system (Afjei et al. 1992). By assembling the models of the basic components (compressors, evaporators, condensers, and expansion devices), the behavior and the efficiency of actual heat pumps can be determined, and the results extrapolated for the whole heat pump cycle at different operating conditions (Bourdouxhe et al. 1994; Koury et al. 2001; Jin and Spitler 2002; Ding 2007; Hermes and Melo 2008; Porkhial et al. 2002; Lemort and Bertagnolio 2010).

Real air-to-air heat pumps used for industrial wood drying (see Figure 10.5a–c) can be modeled with acceptable accuracy over relative large ranges of thermal and hydraulic operating conditions encountered in practice. It can be done by using conventional governing equations (Krakow and Lin 1988) (e.g., conservation of mass and energy generally solved by means of iterations) of the main mechanical/electrical (compressors, blowers) and thermal (evaporators, condensers, internal heat exchangers, expansion devices) components.

For modeling the heat pumps coupled with batch lumber dryers, the thermos-physical properties of both heat pump refrigerant and dryer drying air, as well as some basic characteristics, such as type of kiln and control sequences, initial properties and load of the dried product, drying rate and drying time, moisture exfiltration, should be available and/or determined empirically, analytically, or provided by the manufacturers.

The objective is to determine the steady-state cooling/dehumidifying and heating capacities and energy (power) requirements, and thus, evaluate the efficiency of both heat pumping and drying processes (Krakow and Lin 1988; Byrne et al. 2013).

15.2 TYPES OF HEAT PUMP MODELS

Simulation models of mechanical vapor compression heat pumps are classified according to various criteria. For example, the steady-state equation-fit models treat the heat pump as a black box and fit its performance to one or more equations, while the deterministic models are based on applying thermodynamic laws and fundamental heat and mass transfer relations to individual components (Hamilton and Miller 1990).

The equation-fitting modeling approach, most suitable when the product catalog data (e.g., compressor) is available, alleviates the need for measured data and requires less computational time (Stoecker and Jones 1982; Allen and Hamilton 1983; Hamilton and Miller 1990; Shelton and Weber 1991; Stefanuk et al. 1992; Gordon and Ng 1994).

Jin and Spitler (2002) developed such a steady-state simulation model for a water-to-water reciprocating vapor compression heat pump, which requires only input data available from manufacturers' published catalogs.

Calibration of heat pump models can be achieved by analyzing the information provided by the compressor manufacturer (e.g., model, performance points and number of compressors, type of condenser and evaporator, etc.), and, then, by sensitive analysis and manual and/or automatic validation (Lemort and Bertagnolio 2010).

The deterministic approaches require, among other (relatively few) input parameters (such as compressor speed, displacement volume, and clearance ratio), the saturated condensing and evaporating temperatures (and corresponding pressures), and the amounts of subcooling and superheating, data usually available from manufacturers' catalogs in order to predict the overall performances of mechanical vapor compression heat pumps (Parise 1986; Cecchini and Marchal 1991; Bourdouxhe et al. 1994; Jin and Spitler 2002).

On the other hand, empirical and semiempirical (Bourdouxhe 1994) models, based on physical descriptions of heat pumps, can predict the performances of main components (compressors, evaporators, condensers, etc.) at both full and part loads and compare the cycle performances at different heat source and heat sink temperatures (McLinden and Radermacher 1985; Domanski et al. 1992, 1994).

15.3 GENERAL PROCEDURE

Modeling procedures usually are based on both theoretical and real (practical) heat pump cycles. The modeled heat pumps consist of four main components: evaporator (see Chapter 13), condenser (see Chapter 14), compressor, and expansion device. Other components are neglected due to the comparatively small contribution to the thermodynamic analysis for the entire system.

Heat pumps used in wood drying operate only in the heating mode for which the main data required are, among many others, the following: (i) initial mass of lumber, (ii) lumber initial and final moisture contents, (iii) total mass of moisture expected to remove, (iv) average dry- and wet-bulb drying temperatures, and (v) expected running time of heat pump.

Some simplified modeling approaches for heat pump dryers may not require experimental and component specification data from heat pump manufacturers' catalogs and, thus, may eliminate the requirement for any internal measurements, such as the refrigerant mass flow rate, temperature, and pressure. In these cases, all the calculations are made based on the external measurements of air flow rates and temperatures for both source and load sides. These data are generally available in the heat pump manufacturers' catalogs (Jin and Spitler 2002).

For modeling the heat pumps operating in the heating mode only, as those for wood drying, the following simplified procedure may be used (Jin and Spitler 2002):

a. Calculate the evaporator and condenser thermal efficiencies, saturated temperatures and pressures, as well as the corresponding specific enthalpies;

b. Determine the refrigerant state at the compressor suction port by adding the superheat to the evaporating temperature and its specific enthalpy;
c. Determine the compressor suction and discharge thermodynamic states by adding or subtracting the pressure drop, as well as the corresponding specific volumes;
d. Calculate the refrigerant mass flow rate and the compressor total power input required;
e. Calculate the cooling and dehumidifying capacity of the heat pump running in the heating mode.

15.4 TYPES OF COMPRESSORS FOR HEAT PUMPS

Heat pump compressors have a number of design parameters and characteristics (e.g., type of refrigerant, capacity, performance, cost, sound, etc.) that must be evaluated in order to choose the best model for a given drying application.

Generally, there are two large categories of compressors: positive displacement and dynamic compressors (ASHRAE 2008).

Positive displacement compressors achieve compression through the reduction of the compression chamber volume through work applied to the compressor's mechanism typically driven by an electric motor. Low pressure refrigerant vapor enters the compression chamber through the suction port, and mechanical work decreases the volume of the chamber causing the vapor pressure to rise. The high pressure refrigerant is then allowed to escape the chamber through the discharge port.

For industrial heat pump dryers, the majority of reciprocating compressors is of the positive displacement (most of them being limited to compression ratios on the order of 8:1) and scroll type.

Most positive displacement compressors (hermetic, semi-open, or open) can be single- or multistage devices.

The single-stage positive displacement compressors are directly driven through a pin and connecting rod from the crankshaft (typically used to translate the motion of a rotating shaft into linear motion of a piston) (McQuiston et al. 2005).

During the last two decades, novel single- and two-stage reciprocating compressors have been adapted and/or developed in response to high-temperature/pressure mechanical vapor compression heat pump cycles using natural refrigerants as ammonia (NH_3), carbon dioxide (CO_2), and also, HFCs as HFC-410A, and to associated lubricant issues.

Two-stage compressors are normally combined with intercooling, that is, the process of desuperheating the discharge superheated vapor from the first-stage compressor, for example, by direct contact with liquid refrigerant maintained at the intercooling (high-stage suction) pressure (Minea 2016).

For two-stage compression heat pumps, the optimum intermediate pressure ($p_{opt,int}$) is given by the following equation:

$$p_{opt,\,int} = \sqrt{p_{discharge,\,sat} - p_{suction,\,sat}} \tag{15.1}$$

where

$p_{opt,int}$ is the optimum intermediate compression pressure (Pa)

$p_{discharge,sat}$ is the saturation pressure corresponding to the refrigerant discharge state (Pa)

$p_{suction,sat}$ is the saturation pressure corresponding to the refrigerant suction state (Pa)

Two-stage compression heat pump systems using ammonia as a refrigerant have been developed because the combination of compression ratio limits of reciprocating or rotary vane compressors and refrigerant discharge superheat limit the ability to provide useful heat pumping in a single-stage compression cycle. Effectively, reciprocating and rotary vane compressors have physical compression ratio (i.e., ratio of discharge to absolute suction pressure) and oil temperature limits depending on the refrigerant discharge temperature. During the compression process, the ammonia pressure increases and, because of its low heat capacity, the temperature increases dramatically. High discharge temperatures tend to increase the rate of compressor lubricating oil breakdown as well as increasing the likelihood of compressor material fatigue. To avoid this, an external source of cooling for the compressor (water or refrigerant-cooled heads) must be used. To reduce the compressors' energy consumption for a required compression ratio, one approach is to keep the compressor temperature low during running by using suction cold gas or external means, and undergo isothermal compression process by transferring heat from the compression chamber (Chua et al. 2010).

On the other hand, dynamic (e.g., centrifugal) compressors have mechanical elements which are rotating at a high speed to increase the pressure of large refrigerant vapor volumes. This is achieved by a continuous transfer of angular momentum to the vapor from the rotating member (kinetic energy) followed by a conversion of this momentum into a pressure rise (McQuiston et al. 2005). Some centrifugal compressors present the following limitations (ASHRAE 2008): (i) inadequate physical size, (ii) noise, and (iii) low efficiency levels.

The compressors can be further classified by type of drive enclosure (hermetic, semi-hermetic, and open), motor drive (electrical or thermal), and capacity control (single or variable speed).

Hermetic compressors, generally of relatively small capacities, contain the motor (as a part of the compressor crankshaft) and compressor assembly mounted within the same pressure hermetic steel housing (casing, vessel) permanently sealed with the refrigerant circulating through both, thus avoiding rotating seals through which refrigerant can leak. Such a configuration does not provide access for servicing internal parts in the field, but the heat dissipated by the motor operation is allowed to be transferred to the compressed refrigerant vapor.

Semi-hermetic compressors are devices with the motor within an accessible enclosure assembled with bolts, flanges, and gaskets allowing many of the mechanical parts (i.e., piston, connecting rod, crankshaft, valves, etc.) to be in-field accessed, replaced, or repair (McQuiston et al. 2005; Willingham 2009).

In open-drive compressors, extensively used in large-scale industrial wood heat pump-assisted drying installations, the shaft extends through a seal in the crankcase for an external drive. In this way, the crankshaft is exposed and allows for the motor to be decoupled from the unit, providing an easy and accurate method for measuring shaft speed and torque. In such compressors, a two-piece shaft seal assembly eliminates seal leakage. A carbon ring, in combination with a neoprene bellows, seals tightly against a highly polished seat. The seal assembly is completely surrounded by an oil bath for maximum reliability over a wide temperature range. Positive pressure lubrication is provided by a large capacity manually reversible oil pump, an automatic pressure regulator and an oil filtering system. An efficient crankcase heater prevents liquid refrigerant accumulation in the crankcase during shutdown. In wood drying applications, as the dehumidifying load becomes lower, the capacity control automatically adjusts compressor capacity to as low as 25% of full design load, thus reducing the electrical power requirements. Another advantage to this configuration is that it dramatically reduces the amount of waste heat generated from the motor (which has an adverse effect on the performance of the compressor) that ends up being absorbed by the refrigerant. Typical examples are the ammonia compressors that are manufactured only as open type because of the incompatibility of the refrigerant and hermetic motor materials (McQuiston et al. 2005; ASHRAE 2008; Willingham 2009). Open-belt or direct-drive reciprocating compressors used for industrial heat pump drying applications are generally compatible with medium- and high-temperature refrigerants as HFC-134a, HFC-245fa, and HFC-236fa.

Generally, the heat pump performance is expressed by the coefficient of performance ($COP_{heat\,pump}$). For heat pumps with hermetic or semi-hermetic compressors, the compressor input power used to define $COP_{heat\,pump}$ includes the combined operating efficiencies of both the motor and compressor. However, for open-type compressors, it does not include the motor efficiency being equal to the input power to shaft.

Depending on the heat output amounts, the mechanical vapor compression heat pumps use several types of compressors (McQuiston et al. 2005; Willingham 2009): (i) scroll (in small and medium heat pumps up to 100 kW heat rates output), (ii) reciprocating (up to approximately 500 kW), (iii) screw (up to around 5 MW), and (iv) turbo compressors in large systems above 2 MW, as well as oil-free turbo compressors (above 250 kW).

Generally, the heat pumps' compressors run at partial loads (McQuiston et al. 2005). In relatively small heat pump dryers, the compressor can be started and stopped when the demand for cooling/dehumidifying is satisfied. However, it is not desirable to start and stop the compressors at frequent intervals in industrial-scale heat pump dryers with large compressors having two or more cylinders. In this case, the compressor capacity can be varied by unloading each cylinder independent of the others, a control action initiated by sensing the suction pressure and carried out by an electric solenoid. For example, if one of two cylinders is unloaded, the capacity of the compressor would be decreased by about 50%.

Other methods consist of using variable-speed compressors, holding the intake valves open (it is the most widely used method), or varying the clearance volume (McQuiston et al. 2005). The inverter-based variable speed compressor technology offers significant potential of energy savings, especially for reciprocating and scroll compressors.

The vast majority of reciprocating compressors are available with variable-speed drives to run the drive motor at speeds that match the heat pump load heating at any given time and thus, reduce energy costs by as much as 30% compared with those of traditional fixed-speed compressors. For reciprocating compressors, the usual range of speed is between 800 and 1800 rpm. There is no precise value of the maximum speed, but the life expectancy of the compressor is influenced by the speed. The minimum speed is generally dictated by the oil pump, which requires minimum speed in order to develop an acceptable pressure rise of the oil. On the other hand, the impact on suction and discharge valves increases with speed, and the amount of frictions usually require piston speeds at approximately 4 m/s, providing piston translation of 1200 rpm (Stoecker 1998).

For large heat pump-assisted wood drying systems, two or more compressors are used, each with capacity control, for further flexibility in matching the cooling and dehumidification capacity with the lumber thermal dewatering (moisture vaporization) rate.

Some compressor manufacturers (e.g., Danfoss-Turbocor) developed variable-speed two-stage compact centrifugal, totally oil-free compressors with HFC-134a and HFO-1234ze(E) as refrigerants, with capacities ranging from 210 to 700 kW focusing air-cooled water chillers for heating, ventilation, and air conditioning applications. These compressors have no metal-to-metal contact of rotating components, and include magnetic bearings, variable speed permanent magnet motors, and intelligent digital electronic solid-state controls. Semi-hermetic piston compressors for supercritical CO_2 heat pumps with maximum allowable pressure on the high pressure side of the compressor of 163 bars are today available on the market (e.g., DORIN products; http://www.dorin.com/en/News-46/. Accessed November 21, 2015). To overcome the large efficiency losses associated with the throttling processes of supercritical CO_2 heat pumps, due to very high pressure differences required across the compressors, a number of developments have been achieved for various types of positive displacement compressors, mainly of the vane type, which combine compression with some recovery of work from the expansion process (Stošić et al. 2002).

In the field of industrial heat pumps and refrigeration, the use of reciprocating and vane compressors is decreasing, while the number of screw compressors is expected to increase. The main advantage of screw compressors over scroll compressors is their fairly large pressure ratio and their excellent part load characteristic. Also, there are no net radial or axial forces exerted on the main screw or drive shaft components due to the work of compression. Since compression occurs symmetrically and simultaneously on opposite sides of the screw, the compression forces are canceled out. The only vertical loads exerted on the main screw bearings are due to gravity. Since the discharge end of the screw is vented to suction, the suction gas pressure is exerted on both ends of the screw resulting in balanced axial loads (www.vilter.com, Accessed October 12, 2014).

Screw compressors are simple, reliable, and compact positive displacement rotary volumetric machines, in which the moving parts all operate with pure rotational motion enabling them to operate at higher speeds with less wear than most other types of compressor. They essentially consist of a pair of meshing helical lobed rotors contained in a casing. Together, these form a series of working chambers by means of views from opposite ends and sides of the machine (www.carlylecompressor.com. Accessed September 15, 2014). Screw compressors are up to five times lighter than equivalent (same capacity) reciprocating compressors and have a nearly ten times longer operating life between overhauls. Furthermore, their internal geometry is such that they have a negligible clearance volume, and leakage paths within them decrease in size as compression proceeds. Thus, provided that the running clearances, between the rotors and between the rotors and their housing, are small, they can maintain high volumetric and adiabatic efficiencies over a wide range of operating pressures and flows (Stošić 2002).

Disadvantages of screw compressors include critical lubrication, incompatibility with solid contaminants, and poor performance at low suction pressures.

Significant R&D advances and great improvements in performance have been achieved during the last years in the screw compressors' mathematical modeling and computer simulation of the heat and fluid flow processes, as well as in rolling element bearings and rotor profiles (e.g., with very small tolerances of the order of 5 μm or less at affordable costs allowing reducing the internal leakages) and manufacturing. However, pressure differences across screw compressor rotors impose heavy loads on them and create rotor deformation, which is of the same order of magnitude as the clearances between the rotors and the casing. Consequently, the working pressure differences between entry and exit at which twin screw compressors can operate reliably and economically are limited. These pressure differences create very large radial and axial forces on the rotors whose magnitude and direction is independent of the direction of rotation. Current practice is for a maximum discharge pressure of 85 bars and a maximum difference between suction and discharge of 35 bars (Stošić 2002). Even though the development of screw compressors is now advanced, significant improvement through better rotor profiling, housing ports, seals, lubrication systems, and the entire compressor design is still required.

High-pressure single-stage ammonia screw compressors have been developed for industrial heat pumps (www.emersonclimate.com/en-us/brands/vilter/Pages/Vilter.aspx, accessed October 1, 2015). This technology allows for sustained operation with ammonia at extremely high saturated condensing temperatures than conventional compressors. Integrated, for example, into existing ammonia refrigeration systems, the high-pressure ammonia heat pumps provide a cost-effective solution to harnessing and converting waste heat of rejection to high-grade hot water, up to 90°C. A new high-pressure double screw ammonia compressor for high-temperature heat pumps was developed in Germany in the range from 165 to 2,838 kW drive power. It allows achieving discharge pressures up to 63 bars and maximum condensing temperature up to 90°C with high $COPs_{heat\ pump}$. For example, with a heat source at 35°C and a heat sink at 80°C, a single-stage ammonia heat pump using such a compressor can reach a $COP_{heat\ pump}$ of 5.0 at a heating capacity of 14 MW (source: GEA Refrigeration Technologies; www.gea.com/global/en/products/gea-refrigeration.jsp, Accessed June 2015. Accessed September 1, 2016).

New water vapor (steam) high-speed radial turbo-compressors, dedicated to industrial heat pumps using waste heat at temperature varying between 50°C and

200°C, are under development (Todd et al. 2008). As working fluid, steam has excellent thermodynamic properties at high temperatures to meet high $COP_{heat\ pump}$ values, is nontoxic, and has zero greenhouse potential. The new compressor unit is designed for a pressure ratio up to 3, which is equivalent to a temperature lift about 30°C. To achieve a higher temperature lift, two-stage serial coupling compressors can deliver temperature lifts between 30°C and 60°C. Among focused applications are the high-temperature heat pump dryers where the new turbo-compressors can be configured in series or parallel to match the operational specification of actual drying systems. Preliminary analyses have shown that the new turbo-compressor-based heat pump drying concepts can achieve $COPs_{heat\ pump}$ between 4 and 7 (IEA 2015a, b).

15.4.1 Thermodynamic Compression Cycles

The performance of the most used compressors (reciprocating, scroll, or screw) strongly impacts on that of the whole heat pump system (Lemort and Bertagnolio 2010). These compressors can be modeled using the curve (data)-fits and compressor maps provided by the manufacturers, while for the other components (as evaporators and condensers), fundamental correlations and detailed design data are required (Fischer and Rice 1983; Radermacher and Hwang 2000).

15.4.1.1 Idealized Compression Cycle

The idealized, reversible single-stage cycle of a reciprocating compressor is similar to the refrigerant vapor evaluating adiabatically inside a cylinder with piston. In such cycles, pressure losses in the valves, and intake and exhaust manifolds are neglected (Figure 15.1) (McQuiston et al. 2005).

The pressures of the refrigerant at the inlet and outlet of the cylinder are $p_{suction}$ and $p_{discharge}$, respectively. Similarly, the specific volumes of the refrigerant at the inlet and outlet of the cylinder are $v_{suction}$ and $v_{discharge}$, respectively.

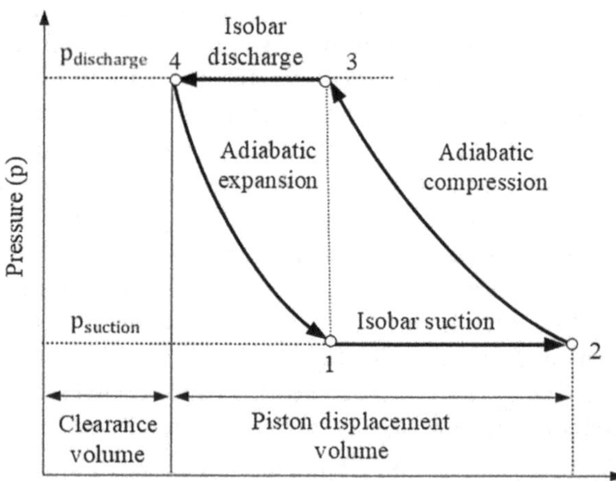

FIGURE 15.1 Schematic of the compressor ideal compression cycle similar to the refrigerant vapor evaluating inside a cylinder with piston.

As shown in Figures 15.1, during the compressor operation, the refrigerant pressure and the specific volume in the compression chamber vary according to the following four main thermodynamic sequential processes (Threlkeld 1962; ASHRAE 2008; Willingham 2009): (i) the vapor is drawn into the compression chamber at low pressure $p_{suction}$ during the suction isobar process 1–2; (ii) the vapor is adiabatically compressed from pressure $p_{suction}$ to pressure $p_{discharge}$ (process 2–3); (iii) the higher-pressure vapor is pushed out during the discharge isobar process 3–4; and (iv) finally, the vapor is adiabatically expanded from the high pressure $p_{discharge}$ to low pressure $p_{suction}$ (4–1), followed by the next compressor cycle.

The compressor suction state (point 2) is defined as the state of the refrigerant with the piston at the bottom dead center position, that is, the point of maximum cylinder volume (Figure 15.1). The discharge condition (state 4) is defined as the state of the refrigerant with the piston at the top dead center position, that is, the point of minimum cylinder volume. The pressures and temperatures at the compressor inlet and outlet states can typically be measured directly with the use of thermocouples and pressure transducers installed directly in the pipes on the suction and discharge lines. However, obtaining properties at the suction and discharge states is typically not possible.

15.4.1.2 Real Compression Cycle

In practice, ideal compression conditions of single-stage reciprocating compressors never occur because of various factors (McQuiston et al. 2005; ASHRAE 2008): (i) pressure drops through the compressor suction and discharge ports, internal muffler, and lubricant separator; (ii) suction and discharge valves' inefficiencies caused by imperfect mechanical action; (iii) compressor motor losses; (iv) deviation from isentropic compression primarily because of refrigerant vapor and mechanical friction, as well as heat transfer in the compression chamber; (v) over-compression that may occur when pressure in the compression chamber reaches discharge pressure before finishing the compression process; (vi) under-compression that may occur when the compression chamber reaches the discharge pressure after finishing the compression process; (vii) heat gain by refrigerant from cooling the motor; (viii) internal heat exchange between the compressor and the suction vapor; (ix) internal vapor leakage; (x) oil circulation; and (xi) clearance losses because the vapor remaining in the compression chamber after discharge re-expands during the suction cycle and thus limits the mass of fresh vapor that can be brought into the compression chamber. However, the work required to compress the clearance volume vapor is just balanced by the work done in the expansion of the clearance volume vapor.

Combined to other factors, specific to the heat pump real thermodynamic cycle (e.g., pressure drops through suction accumulator, across liquid strainer and check valves, suction and discharge manifolds, shutoff valves, etc.), all these factors contribute to heat pump decreased cooling and dehumidifying capacity and increased power input requirement for real reciprocating compressors.

Figure 15.2 shows the pressure-specific volume diagram for a real compressor cycle where it is assumed that (i) the same polytropic exponent n applies to the compression process 2–3 and the expansion process 4–1; (ii) heat transfer during the exhaust process is negligible. The refrigerant vapor enters the compressor at state a'.

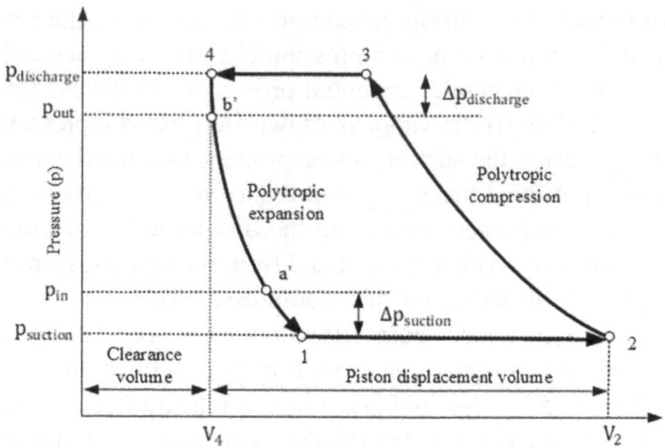

FIGURE 15.2 Schematic representation of the real compression cycle.

As it passes through the suction valve and into the cylinder, there is a drop in pressure ($\Delta p_{suction}$), then an isobaric heating-up in the suction pipe is due to the heat transfer, bringing the refrigerant vapor to state 1.

Even the pressure in the compressor's cylinder is assumed constant during the intake process 1–2 ($p_1 = p_2$) and the exhaust process (3–4) ($p_3 = p_4$), the pressure losses in the valves must be considered. The refrigerant vapor is compressed from state 2 to state 3 in a polytropic process where heat transfer may occur. As the compressed vapor passes through the discharge valve, there is a drop in pressure ($\Delta p_{discharge}$) leaving the refrigerant vapor at state b' when it exits the compressor. The refrigerant vapor remaining in the clearance volume expands in a polytropic process from state b' to state a'. State 1 is different from that at a' because of the mixing of the expanded clearance volume vapor and the intake vapor (Hannay et al. 1999; Grodent et al. 1999; Winandy et al. 2002a, b; Duprez et al. 2007).

The polytropic expansion process 4–1 is represented by the following equation (Willingham 2009):

$$V_1 = V_4 \left(\frac{p_{discharge}}{p_{suction}} \right)^{1/n} \tag{15.2}$$

where

V_1 is the volume of refrigerant in the cylinder after re-expansion (m^3)
V_4 is the volume of refrigerant in the cylinder before re-expansion (m^3)
n is the polytropic exponent (–)
$p_{discharge}$ is the compressor discharge pressure (Pa)
$p_{suction}$ is the compressor suction pressure (Pa)

In most theoretical studies, the pressure drops across the suction and discharge valves are assumed to occur via adiabatic processes. This means that, while the pressure has been reduced, the enthalpy of the refrigerant vapor at the inlet state, just upstream of the suction valve, is equal to the enthalpy of the vapor just downstream of the suction valve. Similarly, the enthalpy at the outlet state, just downstream of the discharge valve, is equal to the enthalpy of the refrigerant vapor just upstream of the discharge valve. This is important because the enthalpy of refrigerant vapor in the cylinder cannot be easily measured but its inlet and outlet states can.

15.5 MODELING RECIPROCATING COMPRESSORS

The purpose of the compressor of mechanical vapor compression heat pump cycles is to circulate the refrigerant, basically to increase its pressure and thus, create a pressure differential between the condenser and evaporator. Mechanically, it is the most complex and, often, the most expensive single component of heat pumps (McQuiston et al. 2005).

The design of reciprocating compressors consisting in pistons moving inside cylinders, aiming at determining the refrigerant mass flow rate, power requirement, and compression efficiency, is highly specialized and their modeling can be achieved at several levels of complexity (Radermacher and Hwang 2000; Byrne et al. 2013).

The variables needed to simulate the unsteady behavior of the heat pump reciprocating compressors by using empirical or semiempirical mathematical equations (Negrao et al. 2010) are the input power, the discharge temperature, and the mass flow rate. The last two variables are used in the other components of the system, such as evaporator(s) (see Chapter 13), condenser(s) (see Chapter 14), and expansion device to simulate the complete heat pump system. They can be obtained with sufficient accuracy using semiempirical models with constant polytropic coefficients.

The mathematical models for modeling reciprocating compressors can be divided into the following categories: (i) empirical polynomial fit (where the correlations are fitted to calorimeter data) (AHRI 2015; ASHRAE 2010); (ii) semiempirical (based on simple thermodynamic correlations, also fitted to experimental data); and (iii) detailed (requiring large amount of data only available from manufacturers), as Computational Fluid Dynamics (CFD) model (Yasar and Koças 2007; Pereira et al. 2008) in which the refrigerant flow and heat transfer within the cylinder and through the valves are considered three dimensional and turbulent (Dufour et al. 1995; Popovic and Shapiro 1995; Jähnig et al. 2000; Kim and Bullard 2002; Winandy et al. 2002a; Srinivas et al. 2002; Longo and Gasparella 2003; Hermes and Melo 2006; Navarro et al. 2007a, b; Duprez et al. 2007; Elhaj et al. 2008; Negrao et al. 2010).

Even less accurate over some ranges of operating conditions, empirical or semiempirical models are simple and convenient for the design and analysis, and good approach for parametric studies. Such methods calculate the refrigerant mass flow rate and the power input requirement from the knowledge of up to five or six operating parameters (Popovic and Shapiro 1995; Jähnig et al. 2000; Choi and Kim 2002; Duprez et al. 2007; Cuevas and Lebrun 2009) that may be found in the technical datasheets of compressors.

With the map-based models, the compressor performance is modeled by curve fitting steady-state experimental data. One curve fit method is to define the mass flaw rate, discharge state, and compressor power as functions of the suction conditions and pressure ratio (Murphy and Galdschmidt 1985). Although the map-based models are very accurate for a given fluid at steady state, they are not generally applicable to refrigerants and compressors other than the ones it was based upon.

More detailed, thus, more complex models of reciprocating compressors (AHRI 2015; ASHRAE 2008; Armstrong et al. 2009a, b) imply the calculation of a set of differential equations for transient mass, momentum, and energy balances with appropriate initial and boundary conditions. They require a large number of input data (parameters) concerning the compressor internal geometry, which generally are difficult to obtain or known only by the constructor and nonavailable in the datasheets (Todescat and Fagotti 1992; Corberan et al. 2000; Rigola et al. 2000; Pérez-Segarra et al. 2003; Rigola et al. 2003; Rigola et al. 2005).

The most complex compressor models are the distributed parameter models that provide better accuracy but needs larger computing times. In such models, the complete set of energy, momentum, and continuity equations is somewhat simplified (MacArthur 1984) and then solved. The compression process is assumed to be polytropic. The heat transfer between different compressor components and thermal storage is accounted for by utilizing constant heat transfer coefficients (Radermacher and Hwang 2000).

In the compressor polytropic model, it is assumed that the compressor rotates at a constant speed and does not account for volumetric efficiency. In other words, the volume flow rate though the compressor is constant, and the compressor power and discharge pressure and temperature are determined by assuming the compression with a constant polytropic exponent (Welsby et al. 1988; Domanski and McLinden 1992). The goal of polytropic models is to determine (i) the refrigerant flow rate by multiplying the volumetric efficiency with the theoretical maximum mass flow rate and (ii) the outlet state.

15.5.1 Basic Assumptions

According to the degree of complexity, several assumptions can be made when modeling a compressor (Haberschill et al. 1994): (i) the compression process is isentropic or polytropic with constant exponents depending on the refrigerant type; (ii) the lubricant has negligible effects on refrigerant properties and compressor operation; (iii) there are isenthalpic pressure drops at the suction and discharge valves; (iv) there are constant suction and discharge pressures and temperatures; (v) the maximum internal pressure is equal to the discharge pressure; and (vi) there is no heat loss at the compressor.

15.5.2 Refrigerant Mass Flow Rate

The capacity of a heat pump compressor at a given operating condition is function of the mass of refrigerant vapor compressed per unit time.

In order to determine the refrigerant mass flow rate through reciprocating compressors, which generally is not given in heat pump manufacturers' catalogs, up to ten parameters depending solely on the evaporating and condensing temperatures can be used. The suction and discharge pressures play important roles in varying the magnitude of the theoretical mass flow rate. These two pressures are different from the evaporating and condensing pressures due to the pressure drop across suction and discharge valves. The inclusion of pressure drops across the suction and discharge valves led to a more accurate prediction for reciprocating compressor model (Popovic and Shapiro 1995). In other words, the mass flow rate depends on the refrigerant properties at the compressor suction and discharge states. However, the actual suction pressure is lower than the inlet pressure and the discharge pressure is actually higher than the outlet pressure. Determination of these pressure drops is crucial to obtaining accurate mass flow models.

The suction-side pressure drop is more critical than the discharge-side pressure drop having effects on the change in the mass flow rate. For the same polytropic exponent, a 10% drop in suction pressure results in the mass flow rate being reduced by approximately 15%, while a 10% rise in discharge pressure results in the mass flow rate being reduced by approximately 3%. The mass flow rate is thus more sensitive to the suction-side pressure drop (Willingham 2009).

For a steady state, steady-flow isentropic process 2–3 (see Figure 15.2), the refrigerant mass flow rate that enters the reciprocating compressor ($\dot{m}_1$, in kg/s) must also leave it ($\dot{m}_2$, in kg/s), according to the following mass balance relation (Radermacher and Hwang 2000):

$$\dot{m}_1 = \dot{m}_2 = \dot{m} \tag{15.3}$$

For an ideal compressor with clearance, due to the re-expansion of the refrigerant vapor in the clearance volume, the ideal refrigerant mass flow rate of the compressor is a decreasing function of the pressure ratio (Lemort and Bertagnolio 2010) and can be expressed as follows (Jin and Spitler 2002; McQuiston et al. 2005):

$$\dot{m}_{refr,ideal} = \frac{\dot{V}_{swept}}{v_{suction}} \eta_{compr,ideal} \tag{15.4}$$

where

$\dot{m}_{refr,ideal}$ is the ideal refrigerant mass flow rate (kg/s)

$\dot{V}_{swept}$ is the compressor swept volume rate (or the geometric displacement of compressor piston) (m^3/s)

$v_{suction}$ is the specific volume at the suction port calculated as a function of pressure and temperature by employing the refrigerant superheating properties (m^3/kg)

$\eta_{compr,ideal}$ is the volumetric efficiency of the ideal compressor (–), given as follows:

$$\eta_{compr,ideal} = 1 - C\left[\left(\frac{p_{discharge}}{p_{suction}}\right)^{1/k} - 1\right] \tag{15.5}$$

where

C is the clearance factor (–)
$p_{discharge}$ is the compressor discharge pressure (Pa)
$p_{suction}$ is the compressor suction pressure (Pa)
k is the refrigerant isentropic coefficient (–)

The clearance factor (C) is defined as the ratio of the clearance volume (V_4) (i.e., the volume of refrigerant left in the cylinder just after the discharge process has been completed) to the piston displacement volume (V_2–V_4), where V_2 is the total compressor cylinder volume (see Figure 15.2):

$$C = \frac{Clearance\ volume}{Piston\ displacement} = \frac{V_4}{V_2 - V_4} \quad (15.6)$$

Due to the irreversibility of the compression process, piston–cylinder leakages, suction and discharge valve throttling, suction gas heating, and refrigerant vapor-to-cylinder wall heat transfer, the volumetric efficiency of real compressors ($\eta_{compr,\ real}$) will be smaller than the volumetric efficiency of ideal compressors ($\eta_{compr,\ ideal}$) as well as the actual mass flow rate ($\dot{m}_{refr,real}$).

By analogy, the actual mass flow rate of a real reciprocating compressor can be defined as follows:

$$\dot{m}_{refr,real} = \frac{\dot{V}_{swept}}{v_{suction}} \eta_{compr,real} \quad (15.7)$$

where

$\dot{m}_{refr,real}$ is the compressor (real) actual mass flow rate (kg/s)
$\eta_{compr,\ real}$ is the volumetric efficiency of the real compressor (–)

Although for real compression process the polytropic exponent must be determined experimentally, it may be approximated by the isentropic exponent k when other data are not available. Typical values are $k = 1.3$ for HFC-134a and $k = 1.6$ for HCFC-22 (McQuiston et al. 2005).

The compressor real volumetric flow rate can be expressed as follows:

$$\dot{V}_{refr,real} = \frac{\dot{V}_{suction}}{\eta_{compr,real}/100} \quad (15.8)$$

where

$\dot{V}_{refr,real}$ is the compressor real volumetric flow rate (m^3/s)
$\dot{V}_{suction}$ is the refrigerant volumetric rate at the compressor inlet port (m^3/s)
$\eta_{compr,\ real}$ is the volumetric efficiency of the real compressor (%)

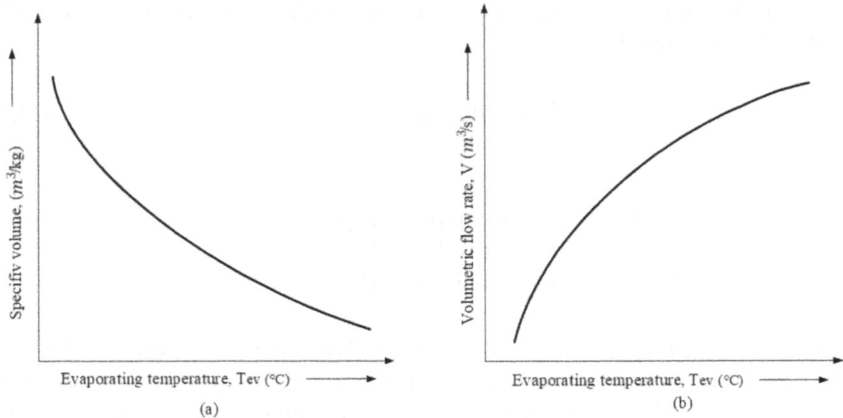

FIGURE 15.3 (a) Variation of refrigerant specific volume at the compressor suction port with the heat pump evaporating temperature; (b) variation of refrigerant volumetric flow rate at the compressor suction with the heat pump evaporating temperature.

Because the refrigerant specific volume at the compressor inlet decreases with the evaporating temperature (Figure 15.3a), the refrigerant volumetric flow rate at the compressor inlet drops off when the heat pump evaporating temperature drops (Figure 15.3b) (Stoecker 1998).

15.5.3 Power Input Requirement and Heat Generation

For the isentropic (adiabatic) compression process represented in Figure 15.2 (process 2–3), we have

$$p_2 v_2^k = p_3 v_3^k \tag{15.9}$$

where

p is the refrigerant pressure at states 2 and 3, respectively (kPa)
v is the refrigerant specific volume at states 2 and 3, respectively (m³/kg)
k is the refrigerant adiabatic exponent (–)

Similarly, for the re-expansion process 4–1 (Figure 15.2), we have

$$p_4 v_4^k = p_1 v_1^k \tag{15.10}$$

where

p is the refrigerant pressure at states 4 and 1, respectively (kPa)
v is the refrigerant specific volume at states 4 and 1, respectively (m³/kg)

If the compression is ideal (i.e., adiabatic and frictionless), the power requirement is (Stoecker 1998) as follows:

$$\dot{W} = \dot{m}_{refr,\,ideal} * w_{compr} \tag{15.11}$$

where
$\dot{W}$ is the total compressor power requirement (W)
$\dot{m}_{refr,\,ideal}$ is the refrigerant ideal mass flow rate (kg/s)
w_{compr} is the ideal compressor specific work (J/kg)

According to the energy balance based on the first law of thermodynamics and applied to the steady-state isentropic compressor process 2–3, the thermal and electrical powers supplied to the compressor have to leave it without considering compressor heat losses and internal frictions (Allen and Hamilton 1983) (see Figure 15.2):

$$\dot{m}_{refr} * h_2 + \dot{W}_{compr,\,input} = \dot{m}_{refr} * h_3 \tag{15.12}$$

or

$$\dot{W}_{compr,\,input} = \dot{m}_{refr}\left(h_3 - h_2\right) \tag{15.13}$$

where
$\dot{W}_{compr,\,input}$ is the compressor total ideal power input (W)
$\dot{m}_{refr}$ is the refrigerant mass flow rate (kg/s)
h_2 is the refrigerant specific enthalpy at the compressor inlet (state 2) (J/kg)
h_3 is the refrigerant specific enthalpy at the compressor outlet (state 3) (J/kg)

The compressor specific work, which is always higher than its isentropic counterpart because of thermodynamic and mechanical losses, can thus be calculated as follows:

$$w_{compr} = h_3 - h_2 = \frac{\dot{W}_{compr,\,input}}{\dot{m}_{refr}} \tag{15.14}$$

The isentropic compression specific work of an ideal reciprocating compressor with clearance can be also expressed as (Jin and Spitler 2002) follows:

$$w_{compr,\,ideal} = \frac{k-1}{k} p_{suction} * v_{suction}\left[\left(\frac{p_{discharge}}{p_{suction}}\right)^{\frac{k-1}{k}} - 1\right] \tag{15.15}$$

where
k is the refrigerant adiabatic exponent (–)

An expression for the compressor polytropic work may be derived subject to the same assumptions used to obtain volumetric efficiency (McQuiston et al. 2005):

$$w_{polytropic} = \frac{n-1}{n} p_{suction} * v_{suction} \left[\left(\frac{p_{discharge}}{p_{suction}} \right)^{\frac{n-1}{n}} \right] \quad (15.16)$$

where

n is the refrigerant polytropic exponent (–)

If it is assumed that the temperature change from state 3 to state 4 is negligible, the specific volume at state 3 is equal to the specific volume at state 4. In this case, the compressor ideal specific work required is represented by the enclosed area (1–2–3–4) of the diagram (Figure 15.2):

$$w_{compr} = \int_2^3 v dp - \int_1^4 v dp \quad (15.17)$$

where

w_{compr} is the compressor ideal specific work (J/kg)

The actual power input for a real reciprocating compressor can also be calculated by the following equation (Jin and Spitler 2002; McQuiston et al. 2005):

$$\dot{W}_{compr,\,real} = \eta * \dot{W}_{compr,\,input} + \dot{W}_{compr,\,loss} \quad (15.18)$$

where

$\dot{W}_{compr,\,real}$ is the real compressor power input (W)

η is the loss factor used to define the electromechanical loss that is supposed to be proportional to the theoretical power (–)

$\dot{W}_{compr,\,input} = \frac{\dot{m}_{refr} * w_{compr,\,ideal}}{\eta_{compr,\,mech}}$ is the compressor theoretical power input (W)

$\dot{W}_{compr,\,loss}$ is the constant part of the electromechanical power losses (W)

$\dot{m}_{refr}$ is the refrigerant mass flow rate (kg/s)

$w_{compr,\,ideal}$ is the isentropic compression specific work (–)

$\eta_{compr,\,mech}$ is the compressor mechanical efficiency (–)

At a given condensing temperature, the ideal compression work decreases as the evaporating temperature (and the refrigerant mass flow rate, as well) increases (Figure 15.4a). When the evaporating temperature reaches the condensing temperature, the ideal compression work becomes zero.

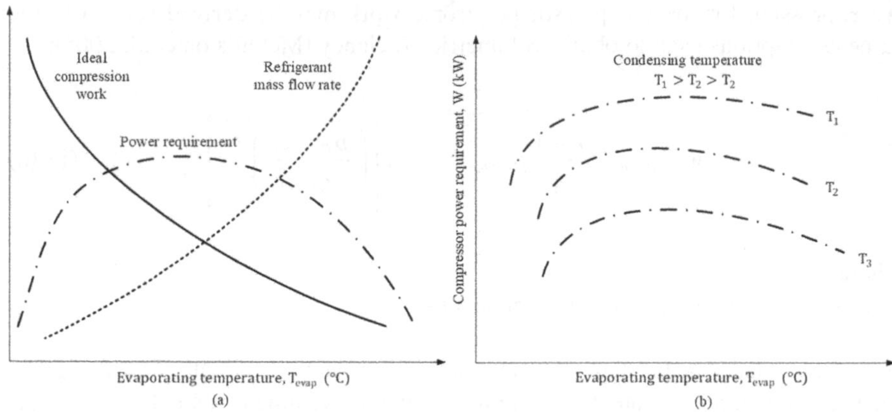

FIGURE 15.4 (a) Trends of ideal compression work and refrigerant mass flow rate as the evaporating temperature increases, and the condensing temperature remains constant; (b) Trends of actual compressor power requirements as functions of evaporating and condensing temperatures (note: graphics not to scale).

Figure 15.4a shows another possible situation that can occur with heat pump-assisted dryers. If the dryer has been out of service and then is brought into service, the heat pump evaporating temperature starts high and progressively drops. During such a pull-down of evaporating temperature, the compressor power requirement passes through the peak, and unless it does so quickly, the compressor motor might overload. In other words, the compressor power would be expected to be low at both very low and very high evaporating temperatures; between those extremes the power requirement reaches a peak (Stoecker 1998).

The heat rejected by the compressor is a function of the electrical efficiency of the compressor motor and is determined by the conservation of energy represented by the following:

$$\dot{Q}_{comp} = \dot{W}_{comp}\left(\frac{1}{e_{cm}} - 1\right) \tag{15.19}$$

where

e_{cm} is the compressor motor efficiency (decimals)

On the other hand, the actual compressor power requirement always increases with an increase in condensing temperature, at least within the normal range of operation (Figure 15.4b).

To properly select and operate the heat pump compressors, it is useful to understand how the evaporating and condensing temperatures affect the heat pump cooling and dehumidifying capacity, as well as the power requirements of the compressor. Each component of heat pump affects the performance of the total system, and the two operating variables to which the compressor is most sensitive are the evaporating and condensing temperatures.

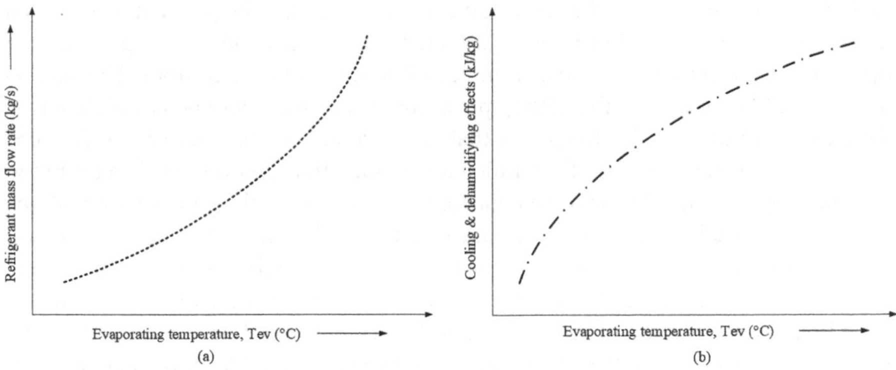

FIGURE 15.5 Trends of (a) refrigerant mass flow rate and (b) cooling and dehumidifying capacity of heat pumps as functions of evaporating temperatures.

Figure 15.5a shows the influence of evaporating temperature on the refrigerant mass flow rate. Figure 15.5b shows the influence of evaporating temperature on the cooling and dehumidification effect (Stoecker 1998).

Typically, the reciprocating compressors operate with pressure ratios between about 2.5 and 9. In this range of pressure ratios, the power required by the compressor increases as the suction pressure and temperature increase; thus, the evaporating temperature raises, for example, when the cooling and dehumidification loads of the heat pump evaporator increase. As a result, the power requirement of the compressor increases, often resulting in the overload of the motor that drives the compressor.

As previously noted, in the case of small heat pump-assisted dryers, the match of the heat pump cooling and dehumidification capacity with the lumber thermal dewatering (moisture vaporizing) rate can be provided by simply cycling the compressor on/off, that is, by an intermittent running approach.

In drying, if the capacity (size) of the heat pump compressor is too small, drying times will be prolonged and the risk of staining will be high. Contrary, if the capacity (size) of the heat pump compressor is too large, it will cycle on and off frequently, which will shorten its service life, but no lumber damage will be incurred if the controls are working properly.

Because almost all industrial-scale heat pump dryers are subject to varying cooling and dehumidification loads and thus operate at part load, most multicylinder reciprocating compressors are equipped with cylinder unloaders that must be activated by special intervention during the peak power requirement and when the evaporating temperature drops below the normal operating setting. In other words, the compressor capacity must be reduced, otherwise the evaporating temperature would drop until the heat pump cooling and dehumidification capacity matches the lumber thermal dewatering (moisture vaporizing) rate. The cylinder unloading is the standard technique for adjusting the heat pumping capacity of reciprocating compressors. The percent of full capacity closely matches the percent of total cylinders pumping, but the power requirement runs several percent higher than the linear relationship. The multicylinder compressors generally are unloaded by holding the suction valve open on one cylinder. During its intake stroke, the piston draws suction gas into the compressor, but then on its return stroke, instead of compressing the vapor, the piston pushes the refrigerant back into the suction manifold. The suction valve can be held open by valve-lifting pins, which in turn are actuated by oil from the lubricating oil pump or discharge gas that is controlled by solenoid valves. In some designs, the normal position of the unloaders is such that oil pressure is required to activate compression. This arrangement automatically starts the compressor with the cylinders unloaded, because oil pressure is not available until the compressor is at least partially up to speed. The two variables most frequently sensed to regulate the operation of the cylinder unloaders are the suction pressure or the outlet temperature of the liquid being chilled in the evaporator (Stoecker 1998). The modulation of heat pump capacity to at least 25% of full capacity by using variable-frequency drives is one of the most advanced techniques.

As a rule of thumb, for heat pump-assisted softwood dryers, between approximately 560 W (0.75 hp) and 735 W (1.0 hp) of compressor nominal power input would be required for every MBF (1 MBF = Thousand Board Feet = 2.359 m^3) of lumber to be dried (Minea 2004).

On average, heat pump-assisted wood drying kilns consume approximately 300–500 kWh/MBF (or 1,100–1,800 MJ/MBF) (1 MBF = 1 Thousand Board Feet = 2.36 m^3) of electrical energy (for heat pump compressors, blowers, circulating fans, controls, etc.).

15.5.4 Compressor Efficiency

Because the compressor piston cannot entirely displace all of the working volume of the cylinder, a clearance volume is necessary. In addition, because of inefficiencies of valves and ports, the actual amount of the circulated refrigerant vapor through the compressor is less than the theoretical volume. Factors such as the friction due to the mechanical rubbing of metal parts and the friction of the flow of refrigerant are losses that reduce the compression efficiency. In irreversible compression processes, the entropy of the refrigerant increases and it cannot be returned to its initial state without violating the second law of thermodynamics by requiring a reduction in entropy (Willingham 2009). The ratio of the actual amount of refrigerant taken in to the theoretical compressor displacement is called the efficiency of the compressor (Radermacher and Hwang 2000).

For the irreversible compression process, the compressor actual efficiency is defined as follows:

$$\eta_{compr} = \frac{\Delta h_{ideal}}{\Delta h_{actual}} = \frac{w_{ideal}}{w_{actual}} \tag{15.20}$$

where

η_{compr} is the compressor actual efficiency (–)
$\Delta h_{ideal} = w_{ideal}$ is the ideal (adiabatic, isentropic) specific compressor work (J/kg)
$\Delta h_{actual} = w_{actual}$ is the actual (irreversible) compressor work (J/kg)

Δh_{ideal} is the specific enthalpy variation during the ideal (isentropic) compression proves (J/kg)
Δh_{actual} is the specific enthalpy variation during the real (polytropic) compression proves (J/kg)

As can be seen in Figure 15.6, the value of compressor efficiency (η_{compr}) drops at higher compression ratios because of increased forces of the rubbing parts, such as shafts on bearing and piston rings on cylinders. There is also a drop-off of η_{compr} at low compression ratios. This reduced efficiency is probably due to flow friction. At a compression ratio of 1.0, the value of Δh_{ideal} is zero, thus any actual work, even though small, drives η_{compr} to zero. For most advanced reciprocating compressors, η_{compr} is about 70% at high compression ratios and around 80% at low compression ratios (Stoecker 1998).

The knowledge of the theoretical compression efficiency is important because η_{compr} allows to (i) compare the effectiveness of two different compressors and (ii) quickly estimate the compression work for an operating compressor by determining Δh_{ideal} and dividing by η_{compr}.

In practice, several compressor efficiencies are defined (Shaffer and Lee 1976; Stoecker 1998; Radermacher and Hwang 2000):

a. Volumetric efficiency (η_{vol}, %) accounts for the geometric compressor displacement and is defined as the ratio of actual refrigerant mass (or volumetric) flow rate to ideal refrigerant mass (or volumetric) flow rate:

$$\eta_{vol} = \frac{Actual\ refrigerant\ mass\ (or\ volumetric)\ flow\ rate}{Ideal\ refrigerant\ mass\ (or\ volumetric)\ flow\ rate} * 100 \qquad (15.21)$$

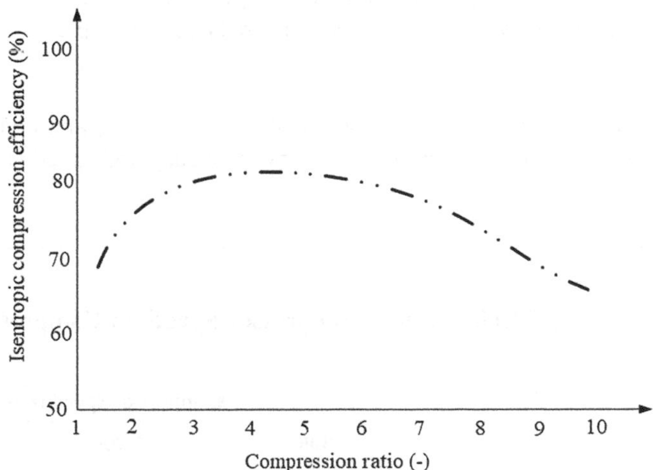

FIGURE 15.6 Trend of the isentropic compression efficiency variation with the compressor compression ratio (note: graphic not to scale).

or

$$\eta_{vol} = \frac{\textit{Refrigerant volumetric flow rate entering the compressor}\left(\mathrm{m^3/s}\right)}{\textit{Compressor displacement rate}\left(\mathrm{m^3/s}\right)} * 100 \tag{15.22}$$

In other words, the compressor volumetric efficiency is the ratio of the actual mass of the refrigerant vapor compressed to the mass of refrigerant vapor that could be compressed if the intake volume equaled the piston displacement (McQuiston et al. 2005).

The volumetric efficiency, a key term in explaining trends in the refrigerating capacity and power requirement of reciprocating compressors, accounts for re-expansion of the refrigerant in the clearance volume and the density change of the refrigerant prior entering the compressor (Radermacher and Hwang 2000).

The compressor displacement rate is the volume rate swept through by the pistons during the suction strokes. The volumetric efficiency is less than 100% because of such factors as leakage past the piston rings, pressure drop through the suction and discharge valves, heating of the suction gas when it enters the cylinder by the warm cylinder walls, and the re-expansion of refrigerant vapor remaining in the cylinder following discharge.

Volumetric efficiency can also be expressed as follows:

$$\eta_{vol} = \frac{\dot{m}_{refr} * v_3}{PD} \tag{15.23}$$

where

$\dot{m}_{refr}$ is the refrigerant mass flow rate (kg/s)
PD is the piston displacement in volume per unit of time ($\mathrm{m^3/s}$)
v_3 is the specific volume at the end of the polytropic compression process 2–3 ($\mathrm{m^3/kg}$)

As can be seen in Table 15.1, for specified evaporating and condensing temperatures, both compressor volumetric efficiency and speeds influence

TABLE 15.1
Effect of Volumetric Efficiency and Compressor Speed on Compressor Efficiency

	Compressor Speed (rpm)		
	800	**1,200**	**1,600**
Volumetric efficiency (%)	80	80	80
Compressor overall efficiency (%)	80	76	72

the compressor overall efficiency (Stoecker 1998). It can be seen that the overall compression efficiency drops as the compressor speed increases.

The overall heat pump cooling and dehumidifying capacity (rate) can be calculated as follows:

$$\dot{Q}_{evap} = \dot{DR} * \eta_{vol} * \frac{\Delta h_{evap}}{v_2} \tag{15.24}$$

where

$\dot{Q}_{evap}$ is the overall heat pump cooling and dehumidifying rate (kW)
$\dot{DR} = \dot{V}_2 - \dot{V}_4$ is the displacement rate (m³/s)
η_{vol} is the actual volumetric efficiency (%)
Δh_{evap} is the variation of refrigerant specific enthalpy during the vaporization process (J/kg)
v_2 is the refrigerant specific volume at the compressor inlet (m³/kg)

b. Overall isentropic efficiency ($\bar{\eta}_{isen}$, %) considers only what occurs within the compression volume and is a measure of the deviation of actual compression from ideal isentropic compression. It is defined as the ratio of work required for isentropic compression of the refrigerant vapor to work delivered to the refrigerant vapor within the compression volume:

$$\bar{\eta}_{isen} = \frac{\textit{Ideal isentropic compression work (from shell inlet to outlet)}}{\textit{Actual motor power}} \tag{15.25}$$

For a multicylinder or multistage compressor, this equation applies for each individual cylinder or stage. The isentropic model is the simplest model, and independent of the physical characteristics of compressors and various types of refrigerants.

Typically, both volumetric and isentropic efficiencies can be directly obtained from experimental data or varied as a design parameter (Radermacher and Hwang 2000).

Sometimes, the polytropic efficiency is used instead of the isentropic efficiency, defined as follows (Radermacher and Hwang 2000):

$$\eta_{pol} = \frac{\textit{Polytropic compression specific work}}{(\textit{Refrigerant specific enthalpy change}) * (\textit{refrigerant mass flow rate})} \tag{15.26}$$

where the polytropic compression specific work is given by equation 15.16.

Together with the power requirement, the heat pump cooling and dehumidifying capacity is a key characteristic of a compressor (Stoecker 1998).

c. Mechanical efficiency (η_{mec}) is the ratio of power delivered to the refrigerant vapor (also called indicated power) to power input to the compressor shaft:

$$\eta_{mec} = \frac{Indicated\ power}{Compressor\ shaft\ power} \tag{15.27}$$

Both volumetric efficiencies and mechanical can be measured, or taken from manufacturer's data, and they depend on the actual detail of the equipment used.

d. Isentropic (reversible, adiabatic) efficiency (η_{isen}) of the actual compression process 1–2 versus the ideal (adiabatic) process (1–2s, where $s_1 = s_{2s}$) is the ratio of work required for isentropic compression of the refrigerant vapor to work input to the compressor shaft:

$$\eta_{isen} = \frac{Ideal\ isentropic\ compression\ work}{Compressor\ shaft\ power} = \frac{h_{2s} - h_1}{h_2 - h_1} < 1 \tag{15.28}$$

e. Motor efficiency (η_{motor}) is the ratio of work input to the compressor shaft to actual work input to the motor:

$$\eta_{motor} = \frac{Shaft\ power}{Actual\ motor\ power} \tag{15.29}$$

f. Superheat compression efficiency is defined as follows:

$$\eta_{superheat} = \frac{Ideal\ isentropic\ compression\ work\ (from\ shell\ inlet\ to\ outlet)}{Ideal\ isentropic\ compression\ work\ (from\ suction\ port\ to\ outlet)} \tag{15.30}$$

g. Finally, the total compressor efficiency ($\eta_{total,\ compr}$) can be expressed as follows:

$$\eta_{total,compr} = \frac{Power\ required\ for\ isentropic\ compression}{Power\ input\ to\ compressor\ motor} \tag{15.31}$$

15.6 MODELING SCROLL COMPRESSORS

Scroll compressors are hermetical-, orbital-, rotary-, or positive displacement-type machines that use two (a fixed and an orbiting) fitted together, spiral-shaped scroll members geometrically identical, closely machined, and assembled 180° out of phase (Willingham 2009) to compress the refrigerant vapor (McQuiston et al. 2005). The orbiting scroll rotates around the fixed scroll during the compression process. The pockets formed between the scrolls are reduced in size as the gas is moved inward toward the discharge port (ASHRAE 2009).

Scroll compressors present extended contact surfaces between the orbiting scroll and the fixed scrolls than between moving and static parts of other types of compressors.

The surface roughness of moving parts can today be minimized and controlled by advanced manufacturing technologies (Jiang et al. 2003). Capacities are controlled either by opening and closing porting holes between the suction side and the compression chambers (McQuiston et al. 2005) or using variable-speed drive, or by range from approximately 3.5–53 kW with current use in residential, commercial, and industrial heat pump applications (as for single- or multiple-rack heat pump dryers).

For small heat pumps (e.g., with thermal output powers ranging from 1.5 to 6.5 kW), the inverter losses may decrease the $COPs_{heat\ pump}$ by 2%–5% (Cuevas and Lebrun 2009).

The isentropic efficiency of inverter-driven scroll compressors may decrease substantially with decreasing suction quality in heat pump systems using the liquid injection technology (Afjei et al. 1992). Same drawback may occur in heat pump applications with low temperature lifts (i.e., low compression pressure ratios) (Cuevas and Lebrun 2009; Winandy et al. 2002a).

Scroll technology became very popular in the latter part of the 20th century mainly because of several performance advantages as low noise and vibration (because of fewer moving parts compared to other compressor technologies), low torque variations, reliability and tolerance to refrigerant droplets, large suction and discharge ports (that reduce pressure losses and heat transfer to suction vapor), absence of valves and re-expansion volumes and continuous flow process (resulting in high volumetric efficiency over wide range of operating conditions), high efficiency, and virtually no maintenance required because most scroll compressors are hermetical (Winandy 2002a; ASHRAE 2008).

Scroll compressors are approximately 10% more efficient than the standard reciprocating compressors because (Qureshi and Tassou 1996; Wang et al. 2009) (i) the suction and discharge processes are separate, meaning that no heat is added to the suction gas as it enters the compressor; (ii) the compression process is performed slowly and therefore, fluctuations in driven torque are only 10% of those of reciprocating compressors; and (iii) the scroll compression mechanism enables the elimination of the suction and discharge valves, which are a source of pressure losses in reciprocating compressors.

In addition, scroll compressors have better reliability because they have fewer moving parts and can operate better under liquid slugging conditions (Winandy et al. 2002a, 2002b). Compact and hermetic scroll CO_2 compressors allowing for reducing the potential for leaks have been designed specifically for use in supercritical CO_2 heat pump systems. They are 30%–60% lighter than reciprocating piston compressors of similar capacity. A specific limitation in the use of CO_2 in air conditioning, refrigeration, and heat pump systems is that the range of operating pressures and temperatures required is close to the critical point of CO_2. Hence, the losses associated with throttling are much larger than those associated with conventional refrigerants. It follows that some recovery of power is required from the expansion process in order to achieve an acceptable *COP* from a CO_2 cycle.

The geometrical, semiempirical, or empirical techniques developed for hermetic scroll compressors modeling are based on several assumptions and require a minimum number of parameters (e.g., polytropic coefficient, leakage flow rate or effective clearance factor, and heat losses to the environment) (Byrne et al. 2012).

The scroll compressor models require the knowledge of at least seven parameters generally found in the technical datasheets (as the global heat transfer coefficient at suction, the ratio between the swept volume and the exhaust volume, the swept volume and the compressor rotation speed) of the compressors or fitted in such a way that the calculated mass flow rate and electrical power match those given in these datasheets (Winandy 2002a; Byrne et al. 2009, 2012; Lemort and Bertagnolio 2010).

Geometrical models usually focus on the suction, compression and discharge chambers, refrigerant flow in the suction and discharge ports, radial and flank leakage, and heat transfer between the refrigerant and the scroll wraps (Chen et al. 2002a, b; Ooi and Zhu 2004; Tseng and Chang 2006; Blunier et al. 2009; Sun et al. 2010), or on tangential and axial leakage and their impact on performance (Wang et al. 2005; Rong and Wen 2009). Pressure distribution on the scrolls surfaces and shaft for different types of scroll wraps (Qiang et al. 2013a, b) and the bypass behavior used to prevent over-compression and frictional losses in bearings (Liu et al. 2009, 2010) are also included in some geometrical models.

Semiempirical models assess the performances of heat pumps under various operating conditions, the output variables generally being the mass flow rate, compressor input power, and discharge temperature (Byrne et al. 2012).

The refrigerant mass flow rates, compressor power inputs, discharge temperatures, and suction and discharge thermal capacities have been modeled based on both geometric data and semiempirical equations of several variable speed scroll compressors with or without refrigerant injection (Winandy et al. 2002a; Winandy and Lebrun 2002; Cuevas and Lebrun 2009; Cuevas et al. 2010; Guo et al. 2011; Li 2013). Average deviations less than 3% of refrigerant mass flow rates and power input requirements for scroll compressors have been reported (Duprez et al. 2007, 2010; Cuevas et al. 2012).

Modeling works carried out with empirical formulas for isentropic efficiency lead to relatively good agreements for the mass flow rate (±7%) and the electric power requirements (±5%) (Kinab et al. 2010).

Many scroll compressor models neglect the impact of lubricant in the compression process while giving results with high accuracy (Li 2012, 2013; Duprez et al. 2007; Duprez et al. 2010). Neglecting oil behavior seems reasonable because with a variation of oil sump temperature, the solubility of refrigerant in oil and the heating-up of refrigerant vary with compensatory effects.

In the case of scroll compressors, the suction heating-up of refrigerant before entering the scrolls is less influenced by the oil temperature and thus, the oil behavior in scroll compressor modeling can be neglected since the oil sump temperature is about 10°C lower than that of reciprocating compressors (Navarro et al. 2012).

15.7 EXPANSION DEVICES

In heat pump systems, expansion devices are required to regulate the flow of the refrigerant and to provide safe operating limits for all of the components. From the standpoint of system control, the expansion device, along with other devices required to control the evaporator temperature or compressor load, is the most important component in the mechanical vapor compression heat pump cycles. Proper operation of the expansion device, either manually or automatically, can substantially optimize the heat pump's drying performances (McQuiston et al. 2005; ASHRAE 2008).

The simplest expansion device model is a thermodynamic expansion model that assumes an isenthalpic expansion with negligible heat loss, that is, the inlet enthalpy equals the outlet enthalpy.

The basic functions of an expansion device used in refrigeration systems are as follows: (i) reduce the refrigerant pressure from condensing to evaporating pressure, and maintain a pressure difference between the condenser and evaporator; the temperature of the refrigerant during expansion is also reduced so that the refrigerant can absorb heat from the heat source in the evaporator; (ii) regulate the refrigerant flow to the evaporator at, ideally, the same rate as it is drawn off by the compressor and also proportional to the evaporator actual cooling and dehumidifying load; and (iii) provide an optimum amount of vapor superheating (normally between 5 and 10 K) at the evaporator outlet in order to prevent saturated vapor or liquid refrigerant to not enter the compressor in order to protect it.

The expansion devices used in heat pump systems can be divided into fixed (where the flow area remains fixed) and variable opening types (where the flow area changes with changing refrigerant mass flow rates). Their types, which depend on the capacity and operating mode of the heat pump application, can be (Radermacher and Hwang 2000) as follows: (i) hand (manual) expansion valves, not used when an automatic control is required; (ii) capillary tubes, used for small capacity heat pump applications; (iii) short tube orifices, used only in some special applications mainly because, even the geometry of such devices is very simple and at lower cost, it is difficult to accurately control the heat pumping process with different refrigerants and boundary conditions; (iv) constant pressure or automatic expansion valves; (v) thermostatic expansion valve, used for higher capacity heat pump applications; (vi) float-type expansion valves; (vii) high- or low-side float valves; and (viii) electric and electronic expansion valves.

Of the above types of expansion devices, capillary tubes and orifices belong to the fixed opening type, while the rest of them belong to the variable opening type.

Capillary tubes, long thin tubes placed between the condenser and the evaporator, are not valves but replace the expansion valve in many small applications. The small diameter and long length of tubes produce large pressure drops. Effective control of the system results because the tube allows the flow of liquid more readily than vapor. Although the capillary tube operates most efficiently at one particular set of conditions, there is only a slight loss in efficiency at off-design conditions in small refrigeration and heat pump systems. The main advantages of the capillary tube are their simplicity and low cost, being not subject to wear. However, the very small bore of the tube is subject to plugging if precautions are not taken to maintain a clean system or to avoid ice formation (McQuiston et al. 2005).

The most efficient mechanical control of refrigerant is the thermostatic expansion valve. The thermostatic expansion valve works by measuring and controlling superheat in the evaporator. Superheat is an increase in temperature of the gaseous refrigerant above the temperature at which the refrigerant vaporizes. The expansion valve is designed so that the temperature of the refrigerant at the evaporator outlet must have 4°C–7°C of superheat before more refrigerant is allowed to enter the evaporator.

Superheat is a direct measure of the work done, or heat absorbed, by the evaporator. Controlling superheat allows the thermostatic expansion valve to meter the proper amount of refrigerant into the evaporator under all load conditions and still prevent flood-back from damaging the compressor.

An ideal refrigerant control device would be non-refrigerant specific, have a very wide load range, be able to be set remotely, and control temperature directly. Electronically controlled valves meet these requirements.

A thermal expansion valve in heat pump systems uses a temperature sensing bulb to open a valve, allowing liquefied refrigerant to move from the high pressure side of the system to the low pressure side before entering the evaporator. As the evaporator temperature decreases, pressure on the bulb decreases, allowing a spring to close the valve.

Two types of expansion valves are used in practice: (i) externally equalized valve and (ii) internally equalized valve.

In the externally equalized expansion valve, a thermal bulb with a small line filled, for example, with CO_2 is attached to the evaporator tailpipe. If the temperature on the tail pipe raises, the CO_2 vapor will expand and cause pressure against the diaphragm. This expansion will then move the seat away from the orifice, allowing an increased refrigerant mass flow rate. As the tailpipe temperature drops, the pressure in the thermal bulb also drops, allowing the valve to restrict the refrigerant flow as required by the evaporator. In this way, the pressure of the refrigerant entering the evaporator is fed back to the underside of the diaphragm through the internal equalizing passage. Expansion of the CO_2 vapor in the thermal bulb must overcome the internal balancing pressure before the valve will open to increase refrigerant flow. A spring is installed against the valve and adjusted to a predetermined setting at the time of manufacture. This is the superheat spring, which prevents slugging of the evaporator with excessive liquid. The adjusted tension of the spring is the determining factor in the opening and closing of the expansion valve. During opening or closing, the spring tension retards or assists valve operation as required. Normally, the spring is never adjusted in the field. Tension is adjusted from 4° to 15° as required for the unit on which it is to be installed. This original setting is sufficient for the life of the valve, and special equipment is required in most cases to accurately calibrate this adjustment

In the externally equalized expansion valve (Figure 15.7a), the refrigerant enters as a high-pressure liquid, flows through a metered orifice, and changes from a high-pressure liquid to a low-pressure liquid.

In practice, thermostatic expansion valves are normally selected from manufacturers' catalogs based on the refrigeration capacity, type of refrigerant, and operating temperature range.

Certain practical problems are encountered with expansion devices if either the selection or its operation is not proper. An oversized expansion device will overfeed the refrigerant or hunt (too frequent closing and opening) and not achieve the balance point. It may allow more refrigerant to flow to the evaporator and cause flooding and consequent slugging of the compressor with disastrous results.

A small valve, on the other hand, passes insufficient quantity of the refrigerant so that balance point may occur at a lower temperature. The mass flow rate

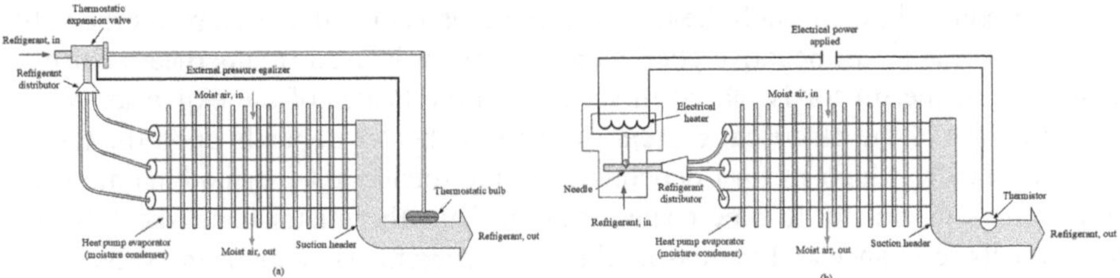

FIGURE 15.7 Heat pump evaporator (moisture condenser) with (a) thermostatic expansion valve with external pressure equalizer; (b) electronic expansion valve.

through the expansion valve depends upon the pressure difference between condenser and evaporator.

Since the area available for refrigerant flow in the expansion device is normally very small, there is a danger of valve blockage due to some impurities present in the system. Hence, it is essential to use a filter/strainer before the expansion device, so that only refrigerant flows through the valve and solid particles, if any, are blocked by the filter/strainer. Normally, the automatic expansion valve and thermostatic expansion valves consist of in-built filter/strainers. However, when a capillary tube is used, it is essential to use a filter/dryer ahead of the capillary to prevent entry of any solid impurities and/or unbound water vapor into the capillary tube.

Any number of evaporators may be operated in parallel on the same compressor, but each of them must be individually selected and controlled to match the heat pump cooling and dehumidifying loads.

Some of the advantages of the thermostatic expansion valve compared to other types of expansion devices are as follows: (i) excellent control of cooling and dehumidifying capacity of heat pumps as the supply of refrigerant to the evaporator matches the load; (ii) ensures that the evaporator operates efficiently by preventing starving under high load conditions; and (iii) protects the compressor from slugging by ensuring a minimum degree of superheat under all conditions of load, if properly selected.

Compared to capillary tubes, thermostatic expansion valves are more expensive and proper precautions should be taken at the installation. For example, the feeler bulb must always be in good thermal contact with the refrigerant tube. The feeler bulb should preferably be insulated to reduce the influence of the ambient air. The bulb should be mounted such that the liquid is always in contact with the refrigerant tubing for proper control.

The electronically controlled electric valves (Figure 15.7b) have the same role as the thermostatic expansion valve, but electricity is utilized to assist in part of the control process. These valves may be heat motor operated, magnetically modulated, pulse-width modulated, or step motor driven. The control may be by either digital or analog electronic circuits, which gives flexibility not possible with thermostatic valves (McQuiston et al. 2005). The electronically controlled electric valves consist of an orifice and a needle in front of it. The needle moves up and down in response to magnitude of current in the heating element. A small resistance allows more current to flow through the heater of the expansion valve, as a result the valve opens wider. A small negative coefficient thermistor is used if superheat control is desired. The thermistor is placed in series with the heater of the expansion valve. The intensity of the electrical current through the heater depends upon the thermistor resistance, which in turn depends upon the refrigerant thermodynamic state. Exposure of thermistor to superheated vapor permits thermistor to self-heat thereby lowering its resistance and increasing the heater current. This opens the valve wider and increases the mass flow rate of refrigerant. This process continues until the vapor becomes saturated, and some liquid refrigerant droplets appear. The liquid refrigerant will cool the thermistor and increase its resistance. Hence in presence of liquid droplets, the thermistor offers a large resistance, which allows a small current to flow through the heater making the valve opening narrower. The control of this valve is independent of refrigerant and refrigerant pressure; hence it works in reverse flow direction also.

Some of electronically controlled electric valves available are as follows: (i) solenoid (or pulse) and (ii) step motor.

Solenoid (or pulse) valves are commonly used as shutoff valves, that is, no ability to modulate the refrigerant flow. In such valves, when the magnetic coil surrounding the plunger is energized, the magnetic field lifts the plunger. To be successful used as modulation valves, the solenoids must be opened and shut rapidly in response to a signal generated by a controller; but because of many sudden starts and stops of the refrigerant flow, "water hammer" or vibration may occur and cause valve or system damage.

In the step motor valves, more sophisticated and more expensive, a small motor does not rotate continuously, but, instead, rotates a fraction of a revolution for each signal sent by the controller. Some step valves are designed for up to 6,000 steps, so good resolution or control of refrigerant flow is possible.

There are two general types of step motors: (i) unipolar (where the electrical current flows in only one direction) and (ii) bipolar (powered by signals that change polarity).

Among other types of expansion valves can be mentioned: (i) float expansion valves, normally used with flooded evaporators in large capacity ammonia heat pumps; a float-type expansion valve opens or closes depending upon the liquid level as sensed by a float in order to maintain a constant liquid level in the float chamber; depending upon the location of the float chamber, a float-type expansion valve can be either a low-side float valve or a high-side float valve; (ii) low-side float valves maintain a constant liquid level in a flooded evaporator or a float chamber attached to the evaporator; when the load on the system increases, more amount of refrigerant evaporates from the evaporator; as a result, the refrigerant liquid level in the evaporator or the low-side float chamber drops momentarily; the float then moves in such a way that the valve opening is increased and more amount of refrigerant flows into the evaporator to take care of the increased load and the liquid level is restored; the reverse process occurs when the load falls, that is, the float reduces the opening of the valve and less amount of refrigerant flows into the evaporator to match the reduced load; (iii) high-side float valves that maintain the liquid level constant in a float chamber that is connected to the condenser on the high pressure side; when the load increases, more amount of refrigerant evaporates and condenses; as a result, the liquid level in the float chamber rises momentarily; the float then opens the valve more to allow a higher amount of refrigerant flow to cater to the increased load, and as a result the liquid level drops back to the original level; the reverse happens when the load drops; since a high-side float valve allows only a fixed amount of refrigerant on the high pressure side, the bulk of the refrigerant is stored in the low pressure side (evaporator).

Comparative studies between the performance of thermostatic and electronic expansion valve when replacing a heat pump or refrigeration system's refrigerant under steady-state operation have been conducted in order to determine if the two types of expansion valves produce similar or different results regardless of which refrigerant is used (Aprea and Mastrullo 2002). Generalized steady-state and transient mathematical models for thermostatic expansion valves have been also developed (Eames et al. 2014).

Generally, the thermostatic expansion valves are modeled in both steady-state (Chi and Didion 1982; Conde and Sutera 1992) and transient (James and James 1987) modes of operation as parts of mechanical vapor compression heat pumps as a whole, and theoretical results are validated by experimental measurements (Mithraratne and Wijeysundera 2002).

REFERENCES

Afjei, T., P. Sutz, D. Favra. 1992. Experimental analysis of an inverter-driven scroll compressor with liquid injection. In *Proceedings of the International Compressor Engineering Conference*, Paper 845, Purdue (http://docs.lib.purdue.edu/icec/845, Accessed March 23, 2017).

AHRI. 2015. *Standard 540 for Performance Rating of Positive Displacement Refrigerant Compressors and Compressor Units*. Air-Conditioning and Refrigeration Institute.

Allen, J.J., J.F. Hamilton. 1983. Steady-state reciprocating water chiller models. *ASHRAE Transactions* 89(2A):398–407.

Aprea, C., R. Mastrullo. 2002. Experimental evaluation of electronic and thermostatic expansion valves performances using R22 and R407C. *Applied Thermal Engineering* 22:205–218.

Armstrong, P.R., W. Jiang, D. Winiarski, S. Katipamula, L.K. Norford, R.A. Willingham. 2009a. Efficient low-lift cooling with radiant distribution, thermal storage and variable-speed chiller controls Part I: Component and subsystem models. *HVAC&R Research* 15(2):366–401.

Armstrong, P., W. Jiang, D. Winiarski, S. Katipamula, L.K. Norford. 2009b. Efficient low-lift cooling with radiant distribution, thermal storage and variable-speed chiller controls. Part II: Annual energy use and savings estimates. *HVAC&R Research* 15(2):402–433.

ASHRAE. 2008. *ASHRAE Handbook, HVAC Systems and Equipment*, Chapter 37, Compressors.

ASHRAE. 2009. *Handbook Fundamentals*, SI edition, ASHRAE Inc., Atlanta, Georgia.

ASHRAE. 2010. ASHRAE/ANSI Standard 23.1–2010. *Methods of Testing for Rating Positive Displacement Refrigerant Compressors and Condensing Units.* American Society of Heating, Refrigerating and Air-Conditioning Engineers, Inc., Atlanta.

Blunier, B., G. Cirrincione, Y. Herve, A. Miraouia. 2009. A new analytical and dynamical model of a scroll compressor with experimental validation. *International Journal of Refrigeration* 32:874–891.

Bourdouxhe, J.-P.H., M. Grodent, J.J. Lebrun, C. Saavedra, K.L. Silva. 1994. A toolkit for primary HVAC system energy calculation—Part 2: Reciprocating chiller models. *ASHRAE Transactions* 100(2):774–786.

Byrne, P., J. Miriel, Y. Lenat. 2009. Design and simulation of a heat pump for simultaneous heating and cooling using HFC or CO_2 as a working fluid. *International Journal of Refrigeration* 32:1711–1723.

Byrne, P., R. Ghoubali, J. Miriel. 2012. Scroll compressor modelling for heat pumps using hydrocarbons as refrigerants. *International Journal of Refrigeration* 41:1–13.

Byrne, P., J. Miriel, Y. Lénat. 2013. Modelling and simulation of a heat pump for simultaneous heating and cooling. *Building Simulation* 593:219–232.

Cecchini, C., D. Marchal. 1991. A simulation model of refrigerant and air-conditioning equipment based on experimental data. *ASHRAE Transactions* 97(2):388–393.

Chen, Y., N.P. Halm, E.A. Groll, J.E. Braun. 2002a. Mathematical modeling of scroll compressors – Part I: Compression process modelling. *International Journal of Refrigeration* 25:731–750.

Chen, Y., N.P. Halm, J.E. Braun, E.A. Groll. 2002b. Mathematical modeling of scroll compressors – Part II: Overall scroll compressor modeling. *International Journal of Refrigeration* 25:751–764.

Chi, J., D. Didion. 1982. A simulation model of the transient performance of a heat pump. *International Journal of Refrigeration* 5:176–184.

Choi, J.M., Y.C. Kim. 2002. The effects of improper refrigerant charge on the performance of a heat pump with an electronic expansion valve and capillary tube. *Energy* 27(4):391–404.

Chua, K.J., S.K. Chou, W.M. Wang. 2010. Advances in heat pump systems: A review. *Applied Energy* 87(12): 3611–3624.

Conde, M.R., P. Sutera. 1992. A mathematical simulation model for thermostatic expansion valves. *Heat Recovery Systems and CHP* 12(3):271–282.

Corberan, J.M., J. Gonzalvez, J. Urchueguia, A. Calas. 2000. Modelling of refrigeration compressors. In *Proceedings of the 15th International Compressor Engineering Conference at Purdue University*, West Lafayette, IN, July 25–28, pp. 571–578.

Cuevas, C., J. Lebrun. 2009. Testing and modelling of a variable speed scroll compressor. *Applied Thermal Engineering* 29(2):469–478.

Cuevas, C., N. Fonseca, Y. Lemort. 2012. Automotive electric scroll compressor: Testing and modeling. *International Journal of Refrigeration* 35:841–849.

Cuevas, C., J. Lebrun, Y. Lemort, E. Winandy. 2010. Characterization of a scroll compressor under extended operating conditions. *Applied Thermal Engineering* 30:605–615.

Ding, G. 2007. Recent developments in simulation techniques for vapour-compression refrigeration systems. *International Journal of Refrigeration* 30:1119–1133.

Domanski. P.A., McLinden, M.O. 1992. A simplified cycle simulation model for the performance rating of refrigerants and refrigerant mixtures. *International Journal of Refrigeration* 15(2):81–88.

Domanski, P.A., M.O. McLinden. 1992. A simplified cycle simulation model for the performance rating of refrigerants and refrigerant mixtures. *International Journal of Refrigeration* 15(2):81–88.

Domanski, P.A., D.A. Didion, W.J. Mulroy, J. Parise. 1994. *A Simulation Model and Study of Hydrocarbon Refrigerants for Residential Heat Pump Systems*. (https://www.nist.gov/node/593616, Accessed September 26, 2016).

Dufour, R., J. Der Hagopian, J. Lalanne. 1995. Transient and steady state dynamic behaviour of single cylinder compressors: Prediction and experiments. *Journal of Sound and Vibration* 181(1):23–41.

Duprez, M.E., E. Dumont, M. Frère. 2007, Modelling of reciprocating and scroll compressors. *International Journal of Refrigeration* 30:873–886.

Duprez, M.E., E. Dumont, M. Frere. 2010. Modeling of scroll compressors –Improvements. *International Journal of Refrigeration* 33:721–728.

Eames, I.W., A. Milazzo, G. G. Maidment. 2014. Modelling thermostatic expansion valves. *International Journal of Refrigeration* 38:189–197.

Elhaj, M., F. Gu, A.D. Ball, A. Albarbar, M. Al-Qattan, A. Naid. 2008. Numerical simulation and experimental study of a two-stage reciprocating compressor for condition monitoring. *Mechanical Systems and Signal Processing* 22:374–389.

Fischer, S.K., C.K. Rice. 1983. *The Oak Ridge Heat Pump Models: I. A Steady-State Computer Design Model for Air-to-Air Heat Pumps*, ORNUCON-80/R1, August.

Fischer, S.K., C.K. Rice, W.L. Jackson. 1988. *The Oak Ridge Heat Pump Design Model: Mark III Version Program Documentation*. Report No. ORNL/TM-10192. Oak Ridge National Laboratory, Oak Ridge, TN.

Gordon, J.M., K.C. Ng. 1994. Thermodynamic modeling of reciprocating chillers. *Journal of Applied Physics* 75(6):2769–2774.

Grodent, M., J. Lebrun, E. Winandy. 1999. Simplified modelling of an open-type reciprocating compressor using refrigerants R22 and R410A. 2nd part: Model. In *Proceedings of the 20th International Congress of Refrigeration*, Sidney, September 19–24, Vol. 3 (Paper 800).

Guo, J.L., J.Y. Wu, R.Z. Wang, S. Li. 2011. Experimental research and operation optimization of an airsource heat pump water heater. *Applied Energy* 88:4128–4138.

Haberschill, P., M. Laflemand, S. Borg, M. Mondot. 1994. Hermetic compressor models determination of parameters from a minimum number of tests. In *Proceedings of International Compressor Engineering Conference,* Purdue University, July 19–22 (Paper 969).

Hamilton, J.F., J.L. Miller. 1990. A simulation program for modeling an air-conditioning system. *ASHRAE Transactions* 96(1):213–221.

Hannay, C., J. Lebrun, D. Negoiu, E. Winandy. 1999. Simplified modelling of an open-type reciprocating compressor using refrigerants R22 and R410A. 1st part: Experimental analysis. In *Proceedings of the 20th International Congress of Refrigeration*, Sidney, September 19–24, Vol. 3 (paper 657).

Hermes, C.J.L., C. Melo. 2006. A first-principles simulation model for the start-up and cycling transients of household refrigerators. *International Journal of Refrigeration* 31:1341–1357.

Hydeman, M., N. Webb, P. Sreeharan, S. Blanc. 2010. *Development and Testing of a Reformulated Regression-based Electric Chiller Model.* International Symposium on Refrigeration Technology, Zhuhai, China.

IEA. 2015a. Industrial energy-related systems and technologies Annex 13, IEA Heat Pump Program Annex 35, Application of Industrial Heat Pumps, final report, Part 1.

IEA. 2015b. Industrial energy-related systems and technologies Annex 13, IEA Heat Pump Program Annex 35, Application of Industrial Heat Pumps, final report, Part 2.

Jähnig, D.I., D.T. Reindl, S.A. Klein. 2000. A semi-empirical method for representing domestic refrigerator/freezer compressor calorimeter test data. *ASHRAE Transactions* 106(2).

James, K.A., R.W. James. 1987. Transient analysis of the thermostatic expansion valves for refrigeration system evaporators using mathematical models. *Transactions of the Institute of Measurement and Control* 9(4):198–205.

Jin, H., J.D. Spitler. 2002. A parameter estimation based model of water-to-water heat pumps for use in energy calculation programs. *ASHRAE Transactions* 108(1):3–17.

Jiang, Z., O.K. Harrison, K. Cheng. 2003. Computer-aided design and manufacturing of scroll compressors. *Journal of Materials Processing Technology* 138:145–151.

Kim, M. H., C. W. Bullard. 2002. Thermal performance analysis of small hermetic refrigeration and air conditioning compressors. *JSME International Journal, Series B* 45(4):857–864.

Kinab, E., D. Marchio, P. Riviere, A. Zoughaib. 2010. Reversible heat pump model for seasonal performance optimization. *Energy and Buildings* 42:2269–2280.

Koury, R.N.N, M. Machado, K.A.R. Ismail. 2001. Numerical simulation of a variable speed refrigeration system. *International Journal of Refrigeration* 24:192–200.

Krakow, K.I., S. Lin. 1988. A numerical model of industrial dehumidifying heat pumps, Concordia University, Report submitted to Supply and Services Canada in fulfillment of Contract Serial No. 23284-7-7040/01-ST, June 29.

Lemort, V., S. Bertagnolio. 2010. *A Generalized Simulation Model of Chillers and Heat Pumps to be Calibrated on Published Manufacturer's Data.* International Symposium on Refrigeration Technology, Zhuhai, China.

Li, W. 2012. Simplified steady-state modeling for hermetic compressors with focus on extrapolation. *International Journal of Refrigeration* 35:1722–1733.

Li, W. 2013. Simplified steady-state modeling for variable speed compressor. *Applied Thermal Engineering* 50:318–326.

Liu, Y., C. Hung, Y. Chang. 2009. Mathematical model of bypass behaviors used in scroll compressor. *Applied Thermal Engineering* 29:1058–1066.

Liu, Y., C. Hung, Y. Chang. 2010. Design optimization of scroll compressor applied for frictional. *International Journal of Refrigeration* 33:615–624.

Longo, G.A., A. Gasparella. 2003. Unsteady state analysis of the compression cycle of a hermetic reciprocating compressor. *International Journal of Refrigeration* 26:681–689.

MacArthur, J.W. 1984. Analytical representation of the transient energy interactions in vapor compression heat pumps. *ASHRAE Transactions* 90(1B):982–996.

McLinden, M., R. Radermacher. 1985. Methods for comparing the performance of pure and mixed refrigerants in the vapor compression cycle. *International Journal of Refrigeration* 10:318–325.

McQuiston, F.C., J.D. Parker, J.D. Spitler. 2005. *Heating, Ventilation, and Air Conditioning: Analysis and Design*, 6th edition, Wiley, Hoboken, NJ.

Minea, V. 2004. Heat pumps for wood drying—new developments and preliminary results. In *Proceedings of the 14th International Drying Symposium* (IDS 2004), Sao Paulo, Brazil, August 22–25,Volume B, pp. 892–899.

Minea, V. 2016. Advances in heat pump-assisted drying technology. In *Advances in Drying Technology*, edited by V. Minea, CRC Press, Boca Raton, FL, pp. 1–116.

Mithraratne, P., N.E. Wijeysundera. 2002. An experimental and numerical study of hunting in thermostatic-expansion-valve controlled evaporators. *International Journal of Refrigeration* 25:992–998.

Murphy, W.E., V.W. Galdschmidt. 1985. Cyclic characteristics of a typical residential air conditioner-modeling of start-up transients. *ASHRAE Transactions* 92(1A):186–202.

Negrao, C.O.R, R.H. Erthal, D. E.V. Andrade, L.W. Silva. 2010. An algebraic model for transient simulation of reciprocating compressors. In *International Compressor Engineering Conference at Purdue*, July 12–15.

Navarro, E., E. Granryd, J.F. Urchueguía, J.M. Corberán. 2007a, A phenomenological model for analyzing reciprocating compressors. *International Journal of Refrigeration* 30:1254–1265.

Navarro, E., J.F. Urchueguía, J.M. Corberán, E. Granryd. 2007b. Performance analysis of a series of hermetic reciprocating compressors working with R290 (propane) and R407C. *International Journal of Refrigeration* 30:1244–1253.

Navarro, E., J.M. Corberan, I.O. Martinez-Galvan, J. Gonzalvez. 2012. Oil sump temperature in hermetic compressors for heat pump applications. *International Journal of Refrigeration* 35:397–406.

Ooi, K.T., J. Zhu. 2004. Convective heat transfer in a scroll compressor chamber a 2-D simulation. *International Journal of Thermal Sciences* 43:677–688.

Parise, J.A.R. 1986. Simulation of vapor-compression heat pumps. *Simulation* 46(2):71–76.

Pereira, E.L.L., C.J. Deschamps, F.A. Ribas Jr. 2008. Performance analysis of reciprocating compressors through computational fluid dynamics. *Journal of Process Mechanical Engineering* 222(4):183–192.

Pérez-Segarra, C.D., J. Rigola, A. Oliva. 2003. Modelling and numerical simulation of the thermal and fluid dynamic behaviour of hermetic reciprocating compressors. Part 1: Theoretical basis. *HVAC & R Research* 9 (2):215–235.

Popovic, P., H.N. Shapiro. 1995. A semi-empirical method for modeling a reciprocating compressor in refrigeration system. *ASHRAE Transactions* 101(2):367–382.

Porkhial, S., B. Khastoo, M. R. Modarres Razavi. 2002. Transient characteristic of reciprocating compressors in household refrigerators. *Applied Thermal Engineering*, 22:1391–1402.

Qiang, J., B. Peng, Z. Liu. 2013a. Dynamic model for the orbiting scroll based on the pressures in scroll chambers-Part I: Analytical modeling. *International Journal of Refrigeration*, doi:10.10 16/j.ijrefrig.20 13.02.004.

Qiang, J., B. Peng, Z. Liu. 2013b. Dynamic model for the orbiting scroll based on the pressures in scroll chambers-Part II: Investigations on scroll compressors and model validation, *International Journal of Refrigeration*, doi: 10.10 16/j.ijrefrig.20 13.02.0.

Qureshi, T.Q., S.A. Tassou. 1996. Variable speed capacity control in refrigeration systems. *Applied Thermal Engineering* 16(2):103–113.

Radermacher, R., Y. Hwang. 2000. *Vapor Compression Heat Pumps with Refrigerant Mixtures*, edited by Center for Environmental Energy Engineering, Department of Mechanical Engineering, University of Maryland, College Park, MD.

Rigola, J., C.D. Pérez-Segarra, A. Oliva. 2003. Modelling and numerical simulation of the thermal and fluid dynamic behaviour of hermetic reciprocating compressors. Part 2: experimental investigation. *HVAC & R Research* 9(2):237–249.

Rigola, J., C.D. Peérez-Segarra, A. Oliva. 2005. Parametric studies on hermetic reciprocating compressors. *International Journal of Refrigeration* 28:253–266.

Rigola, J., C.D. Peérez-Segarra, A. Oliva, J.M. Serra, M. Escribà, J. Pons. 2000. Advanced numerical simulation model of hermetic reciprocating compressors, parametric study and detailed experimental validation. In *Proceedings of the 15th International Compressor Engineering Conference at Purdue University*, West Lafayette, IN, pp. 23–30.

Rong, C., A. Wen. 2009. Discussion on leaking characters in meso-scroll compressor. *International Journal of Refrigeration* 32:1433–1441.

Tseng, C.H., Y.C. Chang. 2006. Family design of scroll compressors with optimization. *Applied Thermal Engineering* 26:1074–1086.

Shaffer, R.W., W.D. Lee. 1976. Energy consumption in hermetic refrigerator compressors. In *Proceedings of the Purdue Compressor Technology Conference at Purdue*, West Lafayette, IN, July 6–9.

Shelton, S.V., E.D. Weber. 1991. Modeling and optimization of commercial building chiller/cooling tower systems. *ASHRAE Transactions* 97(2):1209–1216.

Srinivas, M.N., C. Padmanabhan. 2002. Computationally efficient model for refrigeration compressor gas dynamics. *International Journal of Refrigeration* 25:1083–1092.

Stefanuk, N.B.M., J.D. Aplevich, M. Renksizbulut. 1992. Modeling and simulation of a superheat-controlled water-to-water heat pump. *ASHRAE Transactions* 98(2):172–184.

Stoecker, W.F., J.W. Jones. 1982. *Refrigeration and Air Conditioning*, 2nd edition. McGraw-Hill, New York.

Stoecker, W.F. 1998. *Industrial Refrigeration Handbook*. McGraw-Hill, New York.

Stošić, N. 2002. *Screw Compressors in Refrigeration and Air Conditioning*, Centre for Positive Displacement Compressor Technology, City University, London, EC1V 0HB, England. (http://lms.iknow.com/pluginfile.php/28915.pdf, Accessed April 10, 2017)

Stosic, N., I.K. Smith, A. Kovacevic. 2002. Optimization of screw compressor design. In *Proceedings of XVI International Compressor Engineering Conference at Purdue*, July 2002.

Sun, S., Y. Zhao, L. Li, P. Shu. 2010. Simulation research on scroll refrigeration compressor with external cooling. *International Journal of Refrigeration* 33:897–906.

Threlkeld, J.L. 1962. *Thermal Environmental Engineering*. Prentice-Hall, Englewood Cliffs, NJ.

Todescat, M.L., F. Fagotti. 1992. Thermal energy analysis in reciprocating hermetic compressors. In *Proceedings of the International Compressor Engineering Conference at Purdue University*, West Lafayette, Indiana, July 14–17, pp. 1419–1428.

Todd, B. J., T. Douglas, T. Reindl. 2008. ROTREX turbo compressor for water vapor compression (IEA HPP Annex 13/35 2015).

Wang, B., X. Li, W. Shi. 2005. A general geometrical model of scroll compressors based on discretional initial angles of involute. *International Journal of Refrigeration* 28:958–966.

Wang, B., W. Shi, L. Han, X. Li. 2009. Optimization of refrigeration system with gas-injected scroll compressor. *International Journal of Refrigeration* 32(7):1544–1554.

Winandy, E., J. Lebrun. 2002b. Scroll compressors using gas and liquid injection: Experimental analysis and modelling. *International Journal of Refrigeration* 25:1143–1156.

Willingham, R.A. 2009. *Testing and Modeling of Compressors for Low-lift Cooling Applications*. Degree of Master of Science in Mechanical Engineering, Massachusetts Institute of Technology.

Welsby, P., S. Devotta, P.J. Diggory. 1988. Steady- and dynamic-state simulation of heat pumps. Part II: Modeling of a motor driven water-to-water heat pump. *Applied Energy* 32:239–262.

Willingham, R.A. 2009. *Testing and Modeling of Compressors for Low-lift Cooling Applications*. Submitted to the Department of Mechanical Engineering in partial fulfillment of the requirements for the degree of Master of Science in Mechanical Engineering at the Massachusetts Institute of Technology, 138 pp.

Winandy, E., J. Lebrun. 2002. Scroll compressors using gas and liquid injection: experimental analysis and modelling. *International Journal of Refrigeration* 25:1143–1156.

Winandy, E., C. Saavedra, J. Lebrun. 2002a. Experimental analysis and simplified modelling of a hermetic scroll refrigeration compressor. *Applied Thermal Engineering* 22:107–120.

Winandy, E., O.C. Saavedra, J. Lebrun. 2002b. Simplified modelling of an open-type reciprocating compressor. *International Journal of Thermal Science* 41:183–192.

Yasar, O., M. Koças. 2007. Computational modeling of hermetic reciprocating compressors. *International Journal of High Performance* 21(1):30–41.

16 Case Studies

16.1 LOW-TEMPERATURE HEAT PUMP-ASSISTED HARDWOOD DRYING

16.1.1 Introduction

Hardwoods such as American (Yellow, Gray, Black) and European (Silver, Downy) Birches, Aspen, Elm, Eucalyptus, Maple, Oak (Red, White), Poplar, and Walnut are usually dried in convective-type kilns at low temperatures (around 55°C) by using energy sources as fossil fuels (oil, natural gas, propane) or biomass (wood waste, bark). Electrically driven, low-temperature mechanical vapor compression air-to-air heat pumps are also used in combination with fossil fuels (or electricity) as supplementary and backup energy sources.

16.1.2 Experimental Site

A conventional, laboratory-scale, air-forced dryer, having a maximum capacity of 8,000 MFB (18.8 m^3) (1 MFB = 1,000 Thousand Feet Board = 2.36 m^3) lumber, equipped with variable speed, reverse rotation fans, and two types of heating coils (steam and electricity), has been coupled with a 5.6 kW (nominal compressor power input) single-stage mechanical vapor compression air-to-air heat pump using HFC-134a as a refrigerant (Figure 16.1a).

The heat pump main components (constant-speed compressor, blower, evaporator, condenser, liquid subcooler, refrigerating piping, and controls) are installed inside a mechanical room adjacent to the dryer chamber (Figure 16.1b and c).

The steam is supplied at variable flow rates via a precision modulating steam valve (SV) by a natural gas-fired boiler with annual average thermal efficiency of 80%.

The system's drying control software includes several programmable modes, such as wood preheating and conditioning, heat pump-assisted (hybrid, with fossil or electrical backup heating), and conventional (with steam or electricity as heat sources only). It displays data such as the compressor, fans, steam heating valves and air vents current status, elapsed operating time, dry- and wet-bulb temperatures, compressor suction and discharge pressures and temperatures, and alarms. The drying strategy is based on the intermittent operation of the heat pump's compressor according to the actual wet-bulb temperature measured inside the kiln.

About 30 electrical and thermodynamic parameters (temperatures, pressures, electrical powers, etc.) were measured at 15 second intervals, then averaged and saved every 2 minutes.

Among more than 100 laboratory experimental drying cycles, 5 representative performance tests performed with 2 hardwood species from eastern Canada (Yellow

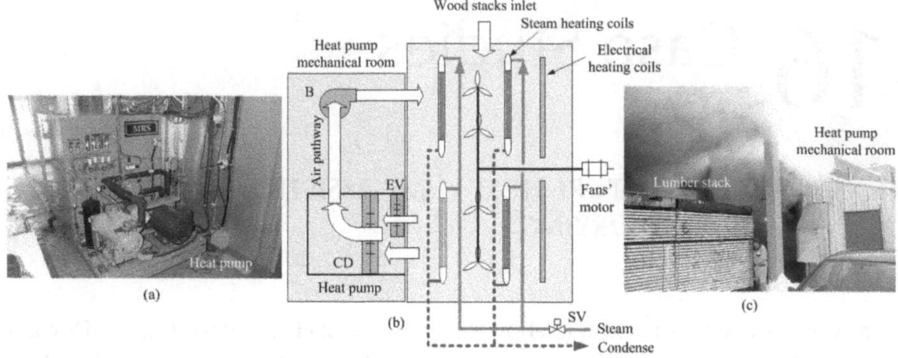

FIGURE 16.1 Experimental heat pump-assisted dryer for low-temperature hardwood drying: (a) front view of the low-temperature heat pump (picture reproduced with permission from MRS Dehumidification Inc.); (b) schematic representation of the dryer and low-temperature heat pump (not to scale); (c) view of a hardwood stack at the dryer inlet (picture reproduced with permission from FPInnovations). B, blower; CD, condenser; EV, evaporator; SV, steam solenoid valve.

TABLE 16.1
Representative Heat Pump-Assisted Performance Tests

Test Number	Drying Mode	Hardwood Species	Moisture Contents (%, Dry Basis) Initial	Final
#1	All-electrical: electrical heat pump with electricity for preheating and supplementary heating	Yellow birch	29.1	7.4
#2	All-electrical: electrical heat pump with electricity for preheating and supplementary	Hard maple	40.7	7.8
#3	Hybrid: electrical heat pump with fossil energy (natural gas) for preheating and supplementary heating	Hard maple	31.1	7.5
#4	Conventional: no heat pump with electricity for preheating and supplementary heating	Hard maple	36.4	7.5
#5	Hybrid: electrical heat pump with fossil energy (natural gas) for preheating and supplementary heating	Yellow birch	75.9	7.6

birch and Hard maple) have been selected in order to compare their drying performances (Table 16.1).

16.1.3 Typical Drying Schedule

The performance drying hybrid test #5, achieved with electrical heat pump and natural gas for preheating and supplementary heating, was selected as a typical, representative example for the heat pump operating parameters and dehumidification performance.

As for the rest of laboratory tests, the drying schedule for this representative performance test with heat pump has been established at the beginning of the drying process as a function of the wood species (Yellow birch) to be dried. The drying schedule was however modified during the drying cycle in order to better match the actual dryer's dry- and wet-bulb temperatures with their set values. In other words, the schedule modifications aimed at improving the correlation between the actual wood thermal dewatering (vaporization) rate and the heat pump dehumidification capacity.

Table 16.2 shows the real (actual) drying schedule of the representative test #5, including the steps' time and cumulative number of hours, set dry- and wet-bulb temperatures, and actual compressor hourly percentages of running times (%) for each drying step.

Dry- and wet-bulb set temperatures, as well as the compressor hourly percentages of running, have been determined based on periodical measurements of moisture content of lumber, and of the air wet-bulb temperature inside the kiln, as well as on the author's and operator's personal experience.

During the preheating step, the steam heating coils provided heat to increase the temperature of the kiln structure and equipment, and the wood stack. The heat pump started up when the actual wet-bulb temperature equaled its setting point.

As can be seen in Table 16.2, the preheating step (when the compressor was not running) lasted 4.75 hours, while the total drying time (preheating plus heat pump-assisted drying) was of 215 hours. The heat pump thus ran intermittently (see Figure 16.3b) in the hybrid (dehumidifying) mode during 210.25 hours.

The initial and the final average moisture contents of the lumber (Yellow birch) stack were of 75.9% (dry basis) and 7.6% (dry basis), respectively (Table 16.1), with actual (measured) frequency distribution of boards' final moisture contents shown in Figure 16.2 (Minea et al. 2005).

TABLE 16.2
Drying Schedule of Heat Pump-Assisted Drying Test #5 (with Yellow Birch)

Drying Step	Time (Hours)	Cumulative Time (Hours)	Set Dry-Bulb Temperature (°C (°F))	Set Wet-Bulb Temperature (°C (°F))	Percentage of Compressor Hourly Running Time (%)
Preheating	4.75	0.00	37.7	32.2	0
2	71.75	4.75	37.7	32.2	100
3	29.50	76.50	40.5	38.9	85
4	24.00	106.00	43.3	34.4	75
5	12.33	130.00	46.1	34.4	65
6	30.83	142.33	54.4	34.4	65
7	16.58	173.17	57.2	35.0	75
8	3.50	189.75	60	35.0	85
9	21.75	193.25	60	35.0	100
Total time	–	215.00	–	–	–

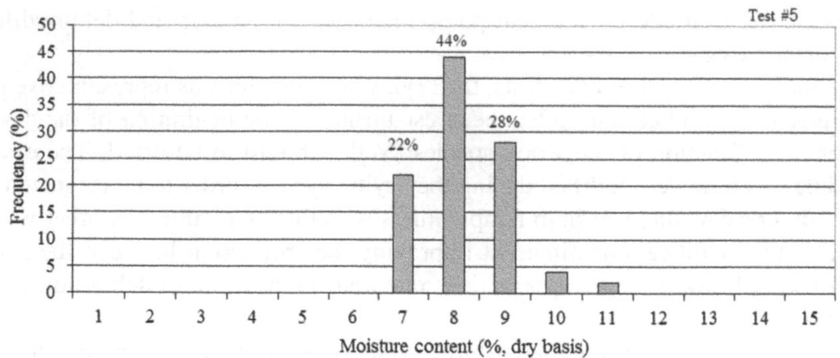

FIGURE 16.2 Frequency distribution of final moisture contents (dry basis) (heat pump-assisted performance test #5).

The heat pump's compressor (and blower) ran only when the actual dryer wet-bulb temperature was higher than the preset values. When the actual wet-bulb temperature became higher than its set value, the compressor hourly running time was increased. Contrarily, when the actual wet-bulb temperature approached its set value, the compressor hourly running time was decreased (see Figure 16.5).

To perform such a control, empirical correlations allowed setting the optimum running and stopping periods of time, that is, the intermittency factors defined as the fraction of time during which the heat pump is running (t_{on}) and the total time of the heat pump on/off running cycle (t_{total}):

$$i = \frac{t_{on}}{t_{total}} = \frac{t_{on}}{t_{on} + t_{off}} \tag{16.1}$$

For example, when the hourly percentage of operation was set at 60%, the heat pump ran for 30 minutes and shut down during the next 20 minutes.

16.1.4 Experimental Results

Heat pump-assisted performance test #5 has been performed with a 16.45 m^3 stack of Yellow birch composed of 101.6 × 101.6 mm and 2.44 m length boards (in English units: 4″ × 4″ and 8 feet length).

16.1.4.1 Operating Parameters

The last column of Table 16.2 shows the hourly percentages (%) of the heat pump's compressor running time during the heat pump-assisted performance test #5.

These hourly percentages (see also Figure 16.5a and b) have been progressively decreased from 100% to 65% during the first part of the drying cycle, then increased from 65% to 100% toward the end of the drying cycle, while the heat pump blower ran continuously.

Figure 16.3a represents the compressor intermittent running profile during the entire heat pump-assisted drying test #5, and Figure 16.3b shows the intermittent running profile of heat pump's compressor during three consecutive hours.

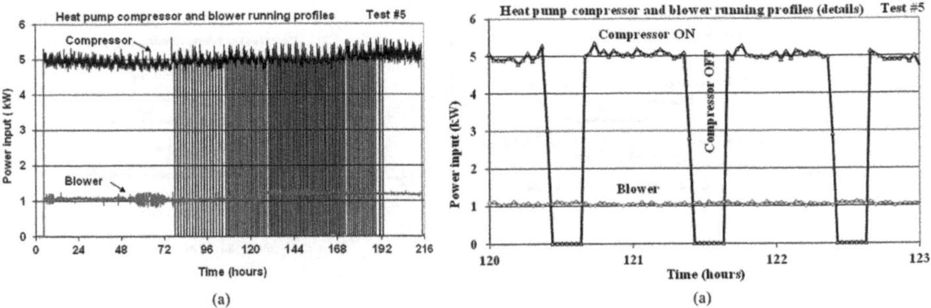

FIGURE 16.3 Running profile of heat pump compressor and blower: (a) during the entire drying cycle; (b) during 3 hours (detail) (heat pump-assisted performance test #5).

Figure 16.4 shows the profile of steam flow rates supplied for preheating the lumber stack and, then, as supplementary (backup) heating during the heat pump-assisted test #5.

Figure 16.5a and b represents the correlation between the heat pump's hourly percentages of operation, and the set (blue lines) and actual (red lines) wet-bulb temperatures during the representative test #5. It can be seen that both actual dry- and wet-bulb temperatures have quasi-perfectly fitted their respective set values during the entire duration of the drying test #5.

Finally, the compressor suction and discharge pressures and temperatures measured during the performance test #5 are shown in Figure 16.6a and b, respectively. It can be seen that both discharge pressures and temperatures have continuously increased toward the end of the drying cycle because the dryer dry-bulb temperature has also increased.

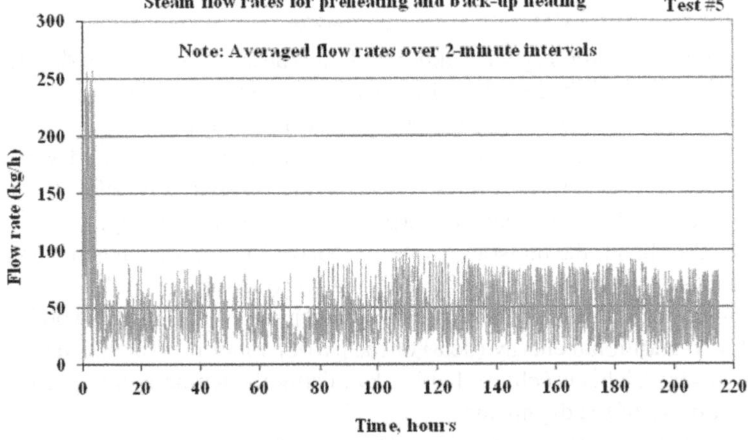

FIGURE 16.4 Profile of steam supplied for lumber preheating and supplementary (and backup) heating (heat pump-assisted performance test #5) (note: steam flow rates averaged at each 2-minute interval).

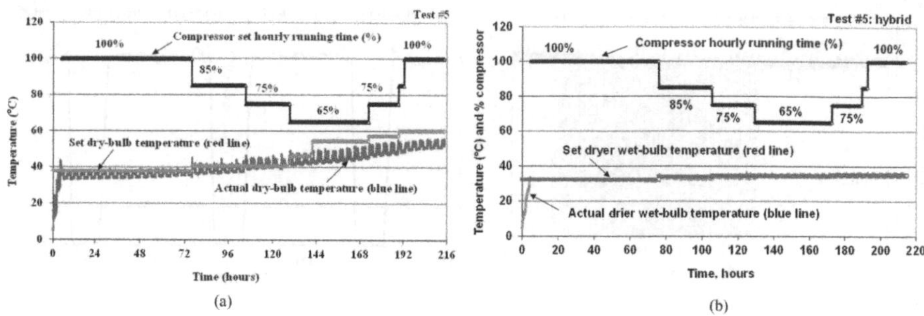

FIGURE 16.5 Compressor hourly running time (%) and (a) set and actual dry-bulb temperatures; (b) set and actual wet-bulb temperatures (heat pump-assisted performance test #5).

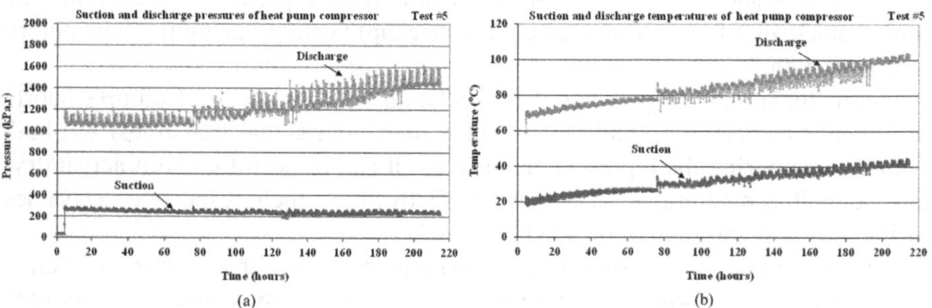

FIGURE 16.6 Suction and discharge parameters of heat pump compressor: (a) temperatures; (b) pressures (heat pump-assisted performance test #5).

16.1.4.2 Drying Performances

16.1.4.2.1 Moisture Extraction (Condensation)

Generally, for every wood species and board thickness, there is a safe drying rate at which the moisture can be removed. In other words, there is a rate at which the wood can be dried with little or no significant degradation or damage.

Exceeding a maximum safe vaporization rate increases the risks of drying defects (splits, cracks, or checks). However, drying at a rate substantially below the safe rate also causes a risk of drying defects (warp, stain, and uneven drying).

On the other hand, the moisture (water) vaporization rate depends on the amount of heat supplied and the capacity of the drying air to absorb it.

To maintain a constant drying rate, the water molecules in the wood must be supplied with additional energy (heat) or the partial vapor pressure of the drying air has to be lowered. This is achieved either by raising the temperature or reducing the relative humidity of the drying air.

The mass of moisture condensed strongly depends on the wood species and initial and final moisture contents, as well as on the drying method used.

With heat pumps as dehumidifiers, the drying rate can be expressed in terms of average mass of moisture extracted per hour.

During the heat pump-assisted performance test #5, the total mass of moisture condensed was of 1,935 kg (Figure 16.7), including about 5% of water losses by evaporation and other uncontrolled losses. This total amount of moisture condensation corresponds to an average moisture extraction rate of 9.2 kg/h (Minea et al. 2005).

As can be seen in Figure 16.8a, test #1 (all-electrical), performed with Hard maple at relatively low initial moisture content (29.1%), provided 2.5 times less water as compared to the heat pump-assisted (hybrid) test #5, where the initial average moisture content of Hard maple was much higher (75.9%).

The fiber saturation point (FSP) is the physical state where the cell cavities are completely devoid of free water and their walls are still completely saturated. For hardwood, the moisture content of the FSP is about 25% (dry basis). It can be seen that with an initial moisture contents of 29.1% (all-electrical test #1, with Yellow birch), the mass of water extracted below the FSP was much higher than the volume extracted above the FSP (Figure 16.8b). However, when the wood initial moisture content was 40.7% (test #2-hybrid, with Hard maple), the volume of water condensed above the FSP was practically equal to the volume removed below this point. Also, when the initial moisture content was significantly higher than 40.7% (75.9% in test #5-hybrid with Yellow birch), the mass of water extracted above the FSP was about three times higher than the quantity removed below the FSP.

In the case of heat pump-assisted test #5 performed with Yellow birch, the FSP has been reached after 99.57 hours after the heat pump started-up. Above the FSP (PSF > 25%), the mass of moisture extracted was of 1,447 kg, which is equivalent to an average moisture extraction rate of 14.53 kg/h. On the other hand, below the FSP (PSF < 25%), the mass of moisture extracted was of 488 kg, which is equivalent to an average moisture extraction rate of 4.4 kg/h during 110.7 hours. Consequently, the rate of moisture extraction above the FSP (PSF > 25%) was 3.3 times higher compared to that achieved below the FSP (PSF < 25%) (Table 16.3).

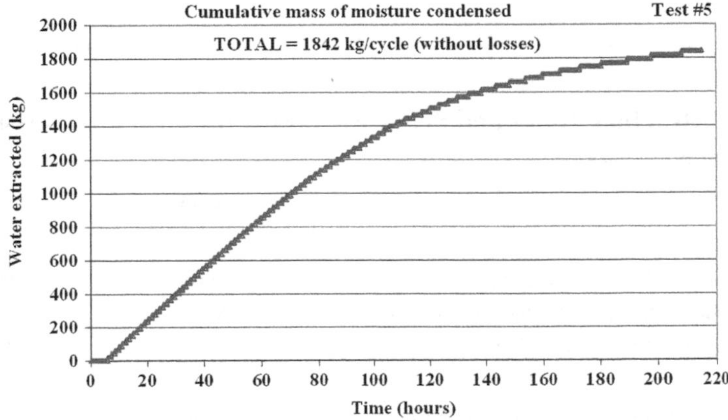

FIGURE 16.7 Cumulative mass of moisture condensed, the losses of condensed moisture (heat pump-assisted performance test #5)

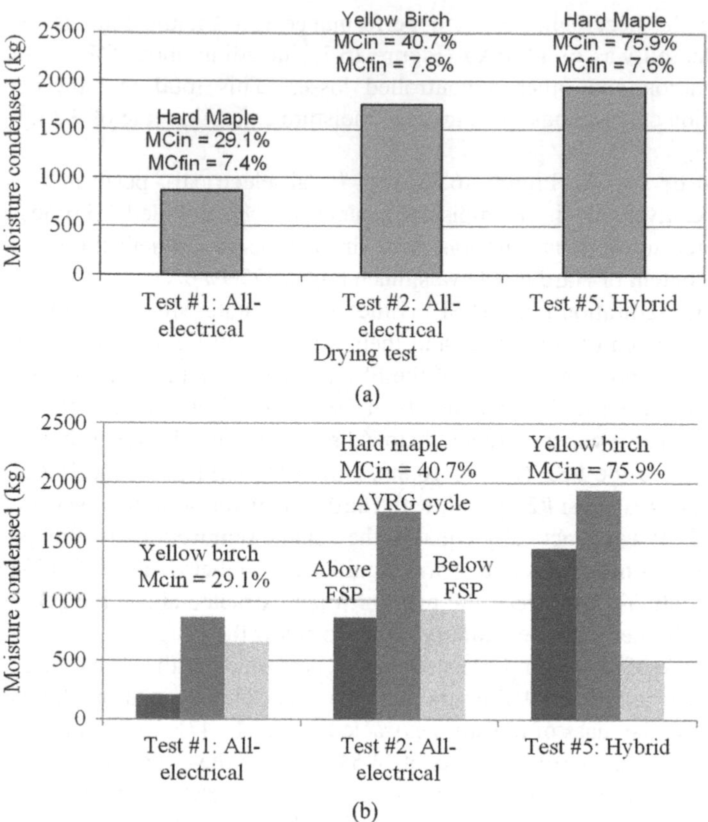

FIGURE 16.8 Masses of moisture extracted (condensed) (a) as functions of wood species, initial and final moisture contents, and drying methods; (b) versus fiber saturation point (FSP = 25%). AVRG, average per cycle.

TABLE 16.3
Moisture Extraction Above and Below FSP during the Heat Pump-Assisted Performance Test #5

Parameter	Unit	Value
Above FSP		
Heat pump running time above the FSP (>25%)	hours	99.57 hours
Mass of moisture extracted above the FSP (>25%)	kg	1,447 kg
Moisture extraction rate above the FSP (>25%)	kg/s	14.53 kg/s
Below FSP		
Heat pump running time below the FSP (<25%)	hours	110.73
Mass of moisture extracted below the FSP (<25%)	kg	488 kg
Moisture extraction rate below the FSP (<25%)	kg/s	4.4 kg/s

16.1.4.2.2 Energy Consumption

Energy consumptions of heat pump-assisted drying performance test #5 that accounted for seven steps after the preheating step according to the schedule shown in Table 16.2 are summarized in Table 16.4.

It can be seen in Table 16.4 and Figure 16.9 that the compressor (with 20%) and the natural gas (steam) (with 66%) were the most important energy consumers. The dryer central fans and then heat pump blower assumed 9% and 5% of the total energy consumption, respectively. The relatively high steam energy consumption can be attributed to the relatively poor thermal insulation and its relatively high air leakage rates of the experimental dryer located outdoors in a cold region (eastern Canada). However, this situation is representative for most of real, industrial-scale kilns found in such cold climates.

The total natural gas consumption of the heat pump-assisted performance test #5 for both preheating and heat pump-assisted drying steps was of 4,727 kg (or 382 m^3 of natural gas at the inlet of the natural gas-fired boiler operating with 80% average thermal efficiency), which represents a reduction by 47% versus the conventional drying test performed with natural gas only (test #4, see Table 16.1).

Compared to natural gas consumption of conventional test #4 (809 m^3), the natural gas consumption of the hybrid drying test #5 (350 m^3) was 56% lower. Consequently, by assuming 40 drying cycles per year, the natural gas saving would account for about 15,000 m^3/year.

TABLE 16.4
Electrical and Fossil Energy Consumptions during the Heat Pump-Assisted Performance Test #5

Component	Preheating Step (kWh)	Heat Pump-Assisted Drying Steps (kWh)	Total Drying Cycle (kWh)	Percentage versus Total (%)
Heat pump compressor	–	902	902	20
Heat pump blower	–	234	234	5
Dryer central fan(s)	8.8	391.2	400	9
Fossil energy (natural gas)	248	2,785	3,033	66
Total cycle energy consumption	–	–	4,569	100

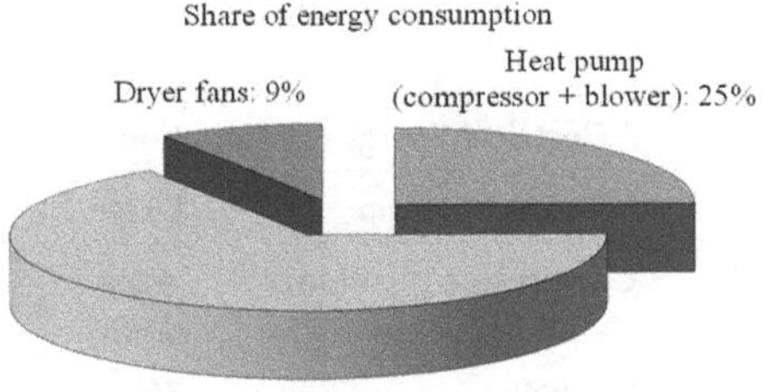

FIGURE 16.9 Share of energy consumption (heat pump-assisted performance test #5).

The equivalent emissions of CO_2 have consequently been reduced by 30.6 tons per year and per drier. This takes into consideration approximately 40kg of CO_2 emissions per year due to the heat pump additional electric energy consumption.

If 500 similar small-scale hardwood driers with similar low-temperature heat pumps would be installed, the annual reduction in greenhouse emission would be of 15,300 tons of CO_2. This estimation is done for the province of Québec (eastern Canada), where almost 100% of the electricity production is hydroelectric, and the regional conversion factor is 0.00122kg of CO_2 per kWh of electrical energy consumed.

The energy consumptions of all-electrical (test #1 and test #2, with Yellow birch and Hard maple, respectively) drying tests (Table 16.5), of the conventional (test #4, with Hard maple), and of heat pump-assisted (hybrid) drying test #3 (with Yellow birch) (Table 16.6) have been compared with those of the heat pump-assisted test #5.

It can be seen that the heat pump (compressor and blower) assumed 30% (test #1) and 21% (test #2) of the total electrical energy consumptions of all-electrical drying tests, respectively, while the energy consumption of electric backup coils accounted for 62% (test #1) and 61% (test #2), and that of the dryer fan for 8% (test #1) and 11% (test #2).

On the other hand, the equivalent energy consumption of the hybrid test #3 (with heat pump and natural gas) was more than 50% lower as compared to the

TABLE 16.5
Electrical Energy Consumption during All-Electrical Drying Tests

		Energy Consumption			
		Heat Pump		Dryer	Backup
Test Number	Hardwood Species	Compressor (kWh)	Blower (kWh)	Fan (kWh)	Electricity (kWh)
#1: all-electrical	Yellow birch	747	151	229	1,828
#2: all-electrical	Hard maple	872	258	451	2,441

TABLE 16.6
Electrical and Fossil Energy Consumptions during the Conventional and Hybrid Drying Tests

		Electrical Energy Consumption			Running Time versus Total Cycle Time	
		Heat Pump		Dryer	Heat Pump	Backup
Test Number	Hardwood Species	Compressor (kWh)	Blower (kWh)	Fans (kWh)	Compressor (%)	Steam (%)
#3: hybrid	Hard maple	650	129	360	65	39
#4: CONV	Hard maple	–	–	638	–	70
#5: hybrid	Yellow birch	902	234	391.2	84.7	40

energy consumption of the conventional test #4 that used natural gas as a unique heating energy source.

It can be also seen that when the initial moisture content (MC_{in}) was relatively high (75.9%—hybrid performance test #5), the compressor operated 84.7% of the total drying time. However, when the MC_{in} was significantly lower (31.1%—hybrid performance test #3), the compressor ran only 65% of the time.

16.1.4.2.3 Dehumidification Efficiency

The dehumidification efficiency of heat pump dryers is well expressed by the heat pump Specific Moisture Extraction Rate ($SMER_{heat\ pump}$), defined as the ratio between the mass of water extracted and the heat pump total electrical energy consumption (compressor and blower) $\left(\frac{kg_{water}}{kWh_{heat\ pump}}\right)$.

Table 16.7 compares the $SMER_{heat\ map}$ of test #5 with those of all-electrical heat pump-assisted tests #1 and #2. It can be seen that:

i. When the initial moisture content (MC_{in}) was relatively low (e.g., 29.1% in test #1, with Yellow birch), the FSP was reached after 25 hours from a total of 138.1 hours; when MC_{in} was higher (e.g., 75.9% in test #5, also with Yellow birch), it was achieved after 104 hours of operation from a total of 210.3 hours.
ii. In tests #1 and #5, both performed with Yellow birch, the water extraction rates above the FSP were practically similar (13 kg/h and 14.53 kg/h, respectively), even if the total water mass removed during the test #1 was more than five times lower than the volume extracted during the test #5 (1,447 kg).
iii. The average $SMER_{heat\ pump}$ above FSP increased by 10.3% (from 2.06 to 2.5 $\frac{kg_{water}}{kWh}$) when MC_{in} rose from 29.1% (all-electrical test #1) to 75.9% (hybrid test #5), while below FSP, they were practically equal (0.82 vs. 0.87 $\frac{kg_{water}}{kWh}$); the moisture extraction rate above the FSP is thus little sensitive to the mass of moisture extracted.

TABLE 16.7
Dehumidifying Performances of Heat Pump-Assisted Performance Test #5

		Specific Moisture Extraction Ratio ($SMER_{heat\ pump}$)		Comparison
Test Number	**MC_{in} (% [d.b.])**	**Above FSP (MC > 25%) (kg_{water}/kWh_{hp})**	**Below FSP (MC < 25%) (kg_{water}/kWh_{hp})**	**Above FSP versus Bellow FSP (Times)**
#1: all-electrical	29.1	2.06	0.82	2.25
#2: all-electrical	40.7	2.3	1.19	1.9
#5: hybrid	75.9	2.5	0.87	2.85

FSP, fiber saturation point; MC_{in}, initial moisture content.

iv. The $SMER_{heat\ pump}$ values below FSP decreased 2.4 times (test #1) and 2.95 times (test #5), respectively, as compared to their respective values above the FSP.
v. Below the FSP, the moisture extraction rates were proportional to the gap between the FSP and final moisture contents (about 17.5%) and practically equal (4–5 kg/h), regardless the dried species (Yellow birch and Hard maple, respectively), and the respective water masses removed;
vi. Above FSP, the average $SMER_{heat\ pump}$ of laboratory heat pump-assisted performance test #5 was $2.5\frac{kg_{water}}{kWh_{heat\ pump}}$ (Table 16.7), which represents a very good performance for a dehumidification process.

16.1.4.2.4 Energy Costs

The energy costs of experimental heat pump-assisted drying test #2 (all-electrical) and test #5 (hybrid) have been reduced by 20% and 23%, respectively, compared to the energy costs of the conventional heat pump-assisted drying test #4 performed with natural gas only (Figure 16.10).

16.1.5 Conclusions

A 5.6-kW low-temperature heat pump coupled to one 18.8 m^3 wood dryer was tested with electricity and natural gas (steam) as backup energy sources. The heat pump electrical energy consumption (compressor and blower) varied between 25% and 30% of the total equivalent energy consumption of all-electrical or hybrid drying cycles. The dryer fan and the electrical (or fossil) backup energy generally assumed 8%–9%, and 62%–66% of the total drying energy consumptions, respectively. For initial moisture contents higher than 41%, the total water quantities extracted above the FSP were up to 2.85 times higher than those removed below the FSP. Consequently, in these cases, the dehumidification efficiency of the low-temperature drying heat pump was up to approximately three times higher above than below the FSP. Finally, the hybrid drying cycles (performed with electrical heat pump and natural gas as supplementary energy source) have reduced the natural gas consumption by 56%

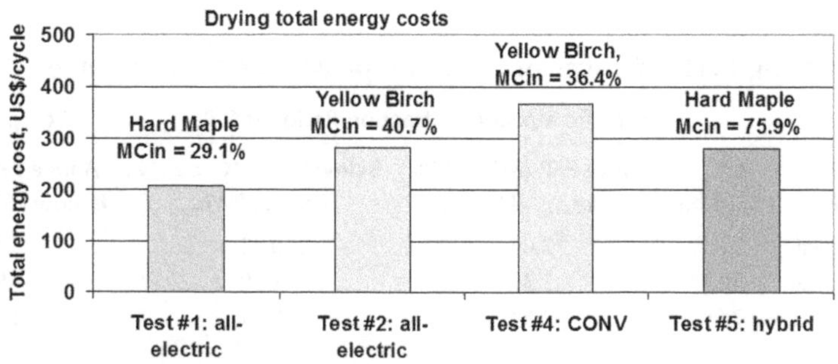

FIGURE 16.10 Comparison of total energy costs (US$ 2004). MC_{in}, initial moisture content.

and the equivalent energy costs by 21.5%, compared to the conventional drying cycle with natural gas as unique heating source.

16.2 HIGH-TEMPERATURE SOFTWOOD DRYING—LABORATORY SCALE

16.2.1 Introduction

For drying large quantities of softwood, commercial high-temperature dry kilns operating at temperatures of 94°C–115°C are generally used in most large wood producing countries, such as the United States, Canada, and New Zealand.

16.2.2 Experimental Site

A conventional, laboratory-scale, air-forced lumber dryer equipped with variable speed, reverse rotation fans, and two types of heating coils (steam and electrical) (Figure 16.11a) was coupled with a custom-made 7 kW (nominal compressor power input) electrically driven high-temperature, single-stage mechanical vapor compression heat pump (Figure 16.11d) using HFC-236fa as refrigerant (Minea et al. 2004). The critical temperature and pressure of the high-temperature refrigerant HFC-236fa (1,1,1,3,3,3-hexafluoropropane) are 124.9°C and 3.2 MPa (abs), respectively.

The experimental laboratory air-forced lumber dryer with maximum wood charge capacity of 18.8 m^3 (≈8 MBF) (MBF, Thousand Board Feet) (Figure 16.11c) is representative of the majority of industrial-scale wood dryers operating in the Canadian cold climate. Inside the dryer, the dry-bulb temperature may vary from 43°C (for low-temperature drying tests) to more than 115°C (for high-temperature drying tests).

The heat pump compressor, blower, evaporator, condenser, electronic expansion valve (for controlling refrigerant flow with a two-phase bipolar motor and a microprocessor driven by the refrigerant pressure and temperature), refrigerant piping, and air ducts are installed inside a mechanical room next to the drying chamber (Figure 16.11b and c).

At the experimental site, steam was generated by a natural gas-burned boiler with an average annual thermal efficiency of 80% or by electricity via electrical coils. Its flow rate and pressure (0.96 MPa, a) are controlled by high-precision proportional integral derivative pressure regulators that makes it possible to optimize energy consumption for both preheating and backup. The horizontally installed steam finned coils are equipped with condensate collectors, dilatation compensators, and vapor drains.

A customized control software, which integrates the dryer and the heat pump, includes several programmable modes, such as wood preheating and conditioning, hybrid (with heat pump and backup heating), and conventional (with electrical or steam heating). The control software displays data such as the compressor, fans, heating valves and air vents current status, elapsed operating time, dry- and wet-bulb temperatures, compressor suction and discharge pressures and temperatures, alarms, etc., and makes it possible to modify the dry- and wet-bulb temperatures and the

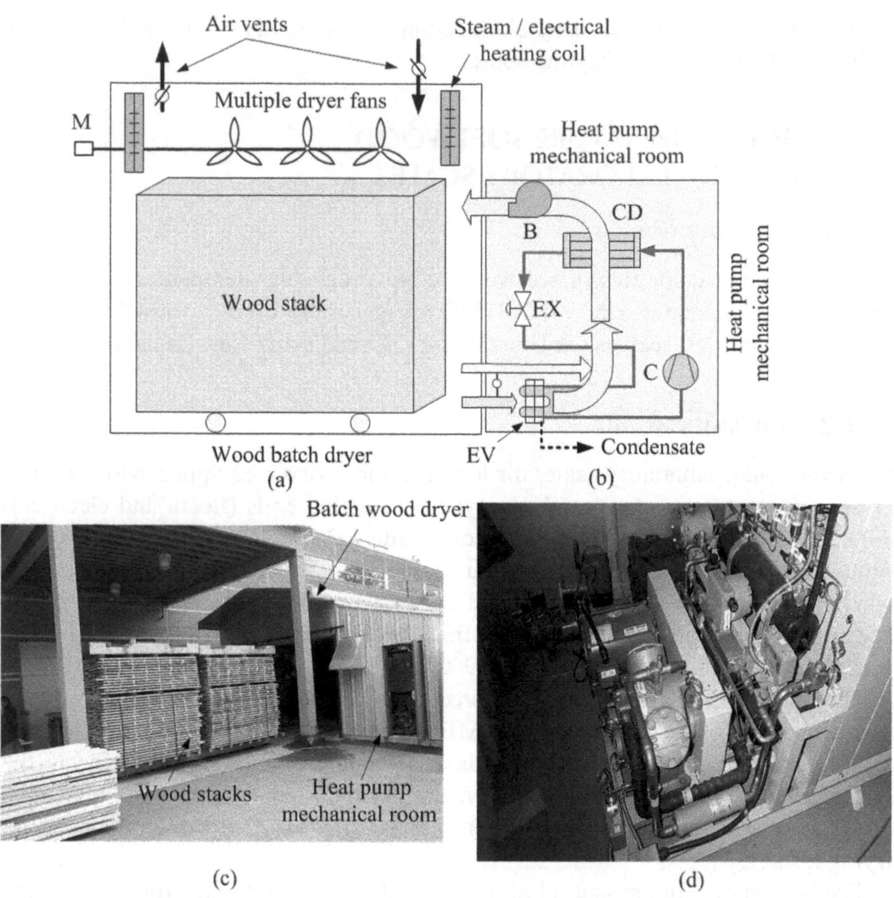

FIGURE 16.11 (a) Schematic representation of the laboratory-scale dryer (not to scale); (b) schematic diagram of the drying air and refrigerant circuits inside the heat pump mechanical room (not to scale); (c) view of softwood lumber stack entering the dryer (picture reproduced with permission from FPInnovation); (d) front view of the compact high-temperature heat pump (picture reproduced with permission from NYLE Systems LLC). B, blower; CD, condenser; M, fan electrical motor; EV, evaporator; EX, expansion valve.

number of central fan rotation reversals, as well as the duration of each step in the drying schedules.

About 30 electrical, thermodynamic, and hydraulic parameters, such as temperature (type T thermocouples), pressure (4–20 mA transducers), electrical current and power (watt transducers), flow rate (CORIOLIS force-based flowmeters), were measured at a 15 second interval, then averaged and saved every 2 minutes, and then transferred to a central data analysis software on a daily basis (Figure 16.12).

Among dozens of experimental performances tests, one conventional drying test (test #1-CONV) using only steam (produced with natural gas) and four heat pump-assisted performance drying tests (#2, #3, #5, and #6) using steam as supplementary (and backup) energy source have been selected as representative (see Tables 16.10

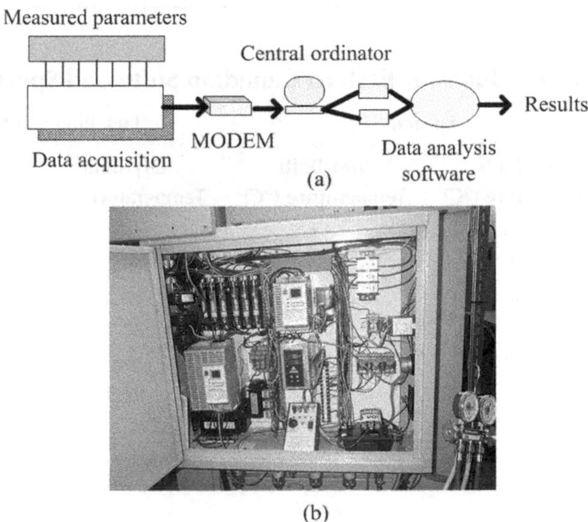

FIGURE 16.12 (a) Schematic diagram of the data acquisition system; (b) front view of the data acquisition panel (reproduced with permission from FP Innovation Canada).

TABLE 16.8
Information Data for Heat Pump-Assisted Performance Test #6

Parameter	Information
Softwood species	White spruce
Boards' dimensions	2″ × 4″ × 8 feet length (1 foot = 12″)
Initial volume of the lumber stack	16.45 m^3
Initial average moisture content	43.5% (dry basis)
Preheating and supplementary heat source	Steam from natural gas-fired boiler (80% thermal efficiency)

and 16.11). Moreover, to illustrate several operating parameters and performance indicators, heat pump-assisted performance test #6 (see Table 16.8) has been chosen as a base for comparisons.

16.2.3 Drying Schedules

The actual drying schedules of all experimental tests have been adopted according to drying schedule index for eastern Canadian softwood species.

16.2.3.1 Typical Schedule

For medium- and high-temperature kiln drying processes of softwood species, a schedule index (as that shown in Table 16.9) exists for eastern Canada (Cech and Pfaf 2000) (Table 16.9).

According to this index, for Spruce (a species that is easy to dry normally), for example, with moisture contents between 40% and 30% (dry basis), the set dry- and

TABLE 16.9
Index of Drying Schedules for Eastern Canadian Softwood Species

Species	Spruce		Jack Pine and Balsam Fir	
Moisture Content (%)	Dry-Bulb Temperature (°C)	Wet-Bulb Temperature (°C)	Dry-Bulb Temperature (°C)	Wet-Bulb Temperature (°C)
>40	71.0	66.0	n/a	n/a
40–30	76.5	65.5	n/a	n/a
>35	n/a	n/a	82.2	79.4
35–25	n/a	n/a	87.7	79.4
<30	82.2	54.4	n/a	n/a
<25	n/a	n/a	93.3	65.5

n/a, not available.

wet-bulb temperatures could be 76.5°C and 65.5°C, respectively. For moisture contents below 30% (dry basis), the set dry- and wet-bulb temperatures could be 82.2°C and 54.4°C, respectively.

On the other hand, for Jack pine and Balsam fir, for example, when moisture content is above 35% (dry basis), the set dry- and wet-bulb temperatures could be 82.2°C and 79.4°C, respectively. For moisture contents between 35% and 25% (dry basis), the set dry- and wet-bulb temperatures could be 87.7°C and 79.4°C, respectively. Finally, for moisture contents below 25% (dry basis), the set dry- and wet-bulb temperatures could be 93.3°C and 65.5°C, respectively.

16.2.3.2 Revised Drying Schedules

The typical drying schedules (as that shown in Table 16.9) have been revised (adapted) to the actual heat pump-assisted drying process according to the quality, board dimensions, initial temperatures, and moisture content of each dried species.

Table 16.10 shows the revised drying schedules used for all representative experimental tests. It can be seen that each heat pump-assisted test began with a preheating step (when the heat pump was not allowed to run), with set dry-bulb temperatures between 87.7°C and 93.3°C, and set wet-bulb temperatures between 68.3°C and 79.4°C, in order to destroy the stain-producing organisms. Each of the following five steps has been set for maximum 10 hours, and the sixth for maximum 20 hours. However, the dryer operator generally decided to pass from one step to anther before reaching the maximum set drying time of each drying step based on the actual measured moisture contents.

When the set wet-bulb temperature of the drying step #1 was reached at the end of each preheating step, the heat pump started. From then, the steam modulating valve allowed additional heat to be supplied to the dryer in order to keep the actual dry-bulb temperature close to the set points. Additional heat was required to compensate heat losses because the dryer is installed outdoor, is relatively poorly insulated and airtight, and operates in a very cold climate.

TABLE 16.10
Actual Drying Schedules Used for the Representative Heat Pump-Assisted Drying Tests

	Preheating			Step 1			Step 2			Step 3			Step 4			Step 5		
Test Number	Set DB (°C)	Set WB (°C)	Actual Time (hours)	Set DB (°C)	Set WB (°C)	Set Time (hours)	Set DB (°C)	Set WB (°C)	Set Time (hours)	Set DB (°C)	Set WB (°C)	Set Time (hours)	Set DB (°C)	Set WB (°C)	Set Time (hours)	Set DB (°C)	Set WB (°C)	Set Time (hours)
#1-CONV	93.3	62.7	–	79.4	68.3	10	79.4	65.5	10	79.4	64.4	10	82.2	64.4	10	82.2	62.7	20
#2	87.7	68.3	5.03	76.6	62.7	10	79.4	62.7	10	79.4	62.7	10	82.2	62.7	10	82.2	60.0	20
#3	87.7	68.3	5.56	76.6	62.7	10	79.4	62.7	10	79.4	62.7	10	82.2	62.7	10	82.2	60.0	20
#5	87.7	73.8	8.28	79.4	62.7	10	79.4	62.7	10	79.4	62.7	10	82.2	62.7	10	82.2	60.0	10
#6	93.3	79.4	8.03	85.0	155	10	85.0	68.3	10	87.7	71.1	10	90.5	71.1	10	93.3	71.1	20

CONV, conventional test; *DB*, dry-bulb temperature; *WB*, wet-bulb temperature.

The heat pump compressor ran continuously as long as the actual wet-bulb temperature of the drying air was higher than its set point. When the actual dryer wet-bulb temperature dropped below its set point, the heat pump shut down. When the actual dryer wet-bulb temperature exceeded again the set point, the heat pump started again, but after a 5 minute delay.

During the actual drying schedules, which included up to five steps, the speed of the compressor was set to vary the frequency of the electrical current between 30 and 60 Hz, while the airflow rate of the heat pump blower varied between 10% and 100% in order to match the lumber thermal dewatering (vaporization) rate with the heat pump dehumidification capacity.

The heat pump stopped automatically according to the prescheduled time of the last drying step or manually after periodically measuring the moisture content of the lumber.

16.2.4 Experimental Results

Table 16.11 shows the dried species, real drying times, and initial and final average moisture contents of the conventional drying test #1 and of the four heat pump-assisted drying tests (#2, #3, #5 and #6). It can be seen that the initial average lumber moisture content, defined as the ratio between the total mass of moisture (water) present in a piece of lumber and its anhydrous (oven-dry) mass, varied from 34.4% (heat pump-assisted test #2) to 60% (conventional test #1) at the dryer inlet. Also, the final average lumber moisture content varied from 14.7% (heat pump-assisted test #5) to 16.9% (heat pump-assisted test #6) at the dryer outlet.

16.2.4.1 Dryer Operating Parameters

Table 16.12 shows, for each heat pump-assisted test (see Table 16.11), the average temperatures of the drying air entering the heat pump evaporator, as well as the air entering (after the mixing process, see Figure 16.11b) and leaving the condenser.

Figure 16.13 represents, for the drying performance test #6 chosen as a representative example, the measured dry-bulb temperatures of the drying air entering the heat pump's evaporator and leaving the heat pump condenser (i.e., entering the

TABLE 16.11
Measured Drying Times and Moisture Contents during the Representative Heat Pump-Assisted Drying Tests

Test Number	Species	Drying Time (Hours)		Moisture Contents (d.b., %)	
		Preheating	Heat Pump-Assisted	Initial	Final
#1-CONV	White spruce	–	54.75	60.2	16.0
#2	White spruce	5.03	32.47	34.4	16.8
#3	White spruce	5.56	30.80	47.0	16.6
#5	White spruce	8.20	43.77	41.1	14.7
#6	White spruce	8.03	41.00	43.5	16.9

TABLE 16.12
Average Temperatures of the Drying Air through the Heat Pump (see Figure 16.11)

Test Number	Evaporator Inlet (°C)	Condenser Inlet[a] (°C)	Condenser Outlet (Return to Dryer) (°C)	
			Dry-Bulb	Wet-Bulb
#2	74.79	65.48	81.38	64.14
#3	75.50	65.58	80.64	70.45
#5	75.31	66.43	82.17	64.54
#6	82.95	74.15	86.70	67.52

[a] Mixed drying air.

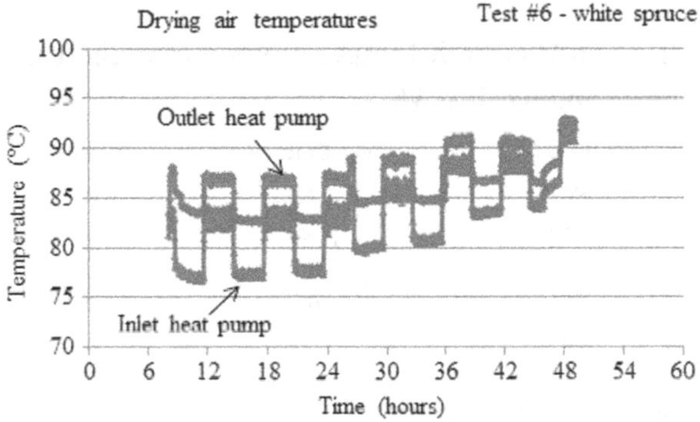

FIGURE 16.13 Actual dry-bulb temperatures of the drying air entering the heat pump evaporator and leaving the heat pump condenser (entering the dryer) (heat pump-assisted performance test #6).

dryer). Upward and downward temperature variations are due to the periodical reversals of the dryer central fan rotation.

For the same heat pump-assisted performance test #6, Figure 16.14 shows the set (see also Table 16.13) and actual dry- and wet-bulb temperatures of the drying air inside the dryer. On the actual dry-bulb temperature curve, only the maximum values are relevant because they were measured only on the heat pump (inlet) side, while the dryer was controlled using the dry-bulb temperatures measured alternately on the discharge side of the dryer central fan.

The wet-bulb temperatures were manually set during drying test #6 in order to avoid the intersection of the actual wet-bulb temperature curve. This means that the minimum difference between the actual and set wet-bulb temperatures was permanently kept at all times around 3°C. Thus, the heat pump compressor ran continuously, as shown, for example, in Figure 16.18).

The average moisture contents of the softwood stack, which helped decide when to stop the drying cycles, were measured manually using the ovendry standard method.

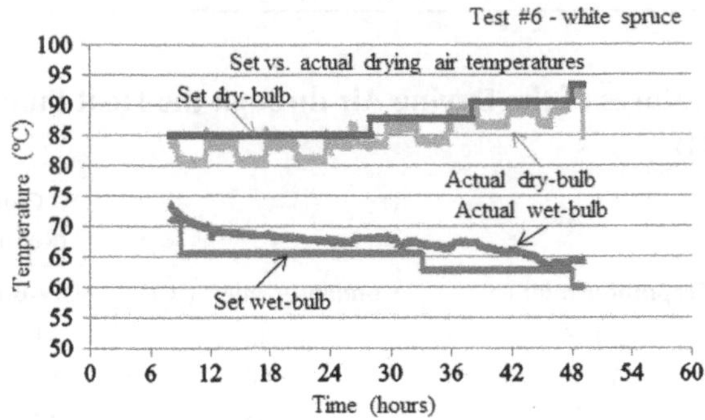

FIGURE 16.14 Set and actual dry- and wet-bulb temperatures of the drying air (heat pump-assisted performance test #6).

TABLE 16.13
Measured Set Dry- and Wet-Bulb Temperatures during the Heat Pump-Assisted Performance Test #6

Drying Step	Dry-Bulb (°C)	Wet-Bulb (°C)	Spray	Heat Pump
Preheating	94	72	ON	OFF
Step 1	85	65	OFF	ON
Step 2	87	65	OFF	ON
Step 3	90	63	OFF	ON
Step 4	94	60	OFF	ON

OFF, device not operating; *ON*, device operating.

According to this method, periodically (8–12 hours) six representative lumber board samples (including the extreme conditions and variability of the expected kiln drying behavior) were removed from the softwood stack and weighed manually with a precision electronic scale, and the moisture content was calculated at that time (see Table 16.11 for heat pump-assisted performance test #6). Ovendrying of lumber board samples was performed in a forced convection oven maintained at 103°C until all water was completely removed, usually after a period of 24 hours. The board samples were weighed using the same procedure immediately after removing them from the oven, and their moisture content was calculated by dividing the mass (weight) of the removed water by the ovendry mass of the board, and multiplying the result by 100. For heat pump-assisted performance drying test #6, all results are presented in Table 16.14.

16.2.4.2 Heat Pump Operating Parameters

In order to match the heat pump dehumidification capacity with the dryer thermal dewatering (vaporizing) rate, the speeds of the heat pump compressor and blower were periodically changed simply by varying the frequency of the electrical current supplied to these devices. Consequently, the heat pump compressor ran continuously with the main average parameters shown in Table 16.15.

TABLE 16.14
Measured Moisture Contents of Board Samples during the Heat Pump-Assisted Performance Test #6

	2,590.4		2,450.8		2,598.2		2,264.2	
	Sample #1		Sample #2		Sample #3		Sample #4	
Dry Mass (g)	Mass (g)	MC (%)	Mass (g)	MC (%)	Mass (g)	MC (%)	Mass (g)	MC (%)
–	5,028	94.1	3,985	62.6	3,518	35.4	3,034	34.0
–	4,746	83.2	3,758	53.3	3,363	29.4	2,883	27.3
–	3,858	48.9	3,375	35.3	3,179	22.9	2,696	19.1
–	3,482	34.4	3,221	31.4	3,111	19.7	2,625	15.9
–	3,237	25.0	3,008	22.7	2,959	13.9	2,467	9.0

	2,474.9		2,201.4		
	Sample #5		Sample #6		
Dry Weight (g)	Mass (g)	MC (%)	Mass (g)	MC (%)	MC (%, d.b.) Average (%)
–	3,346	35.2	3,225	46.5	51.3
–	3,179	28.5	3,064	39.2	43.5
–	3,043	23.0	2,895	28.3	29.5
–	2,971	20.0	2,748	24.8	24.3
–	2,796	13.0	2,592	17.1	16.9

MC, moisture content.

TABLE 16.15
Measured Average Thermodynamic Parameters of the Heat Pump

	Compressor Suction			
Test Number	Pressure (kPa,a)	Temperature (°C)	Evaporation Saturated Temperature (°C)	Evaporator Superheating (°C)
#2	494	60.5	45.5	14.9
#3	473	59.4	44.0	15.7
#5	473	61.1	43.9	17.6
#6	480	67.0	44.9	22.4
	Compressor Discharge			
Test Number	**Pressure (kPa,a)**	**Temperature (°C)**	**Condensing Saturated Temperature (°C)**	**Condenser Subcooling (°C)**
#2	1,439	94.4	85.8	3.5
#3	1,411	93.9	85.0	3.4
#5	1,453	97.8	86.2	3.2
#6	1,599	104.0	91.0	3.9

For the same heat pump-assisted test #6, Figures 16.15 and 16.16 show the actual variations of the compressor suction and discharge pressures and temperatures, respectively.

In the case of heat pump-assisted test #6, the average evaporating and condensing (saturation) temperature was 44.9°C (Figure 16.17), while the condensing temperature varied between 86°C (at the beginning) and 97.5°C (toward the end of the drying process). This made it possible to supply hot and dry air to the dryer at average temperatures up to 93°C, that is, in average, up to 10°C higher than the dryer average dry-bulb drying temperature.

Table 16.16 shows other measured parameters for all heat pump-assisted drying tests, such as compressor, blower, and dryer fans power inputs; refrigerant flow rates; compression ratios; and evaporator's and condenser's pressure drops. For example, in the case of heat pump-assisted test #6, the average compressor electrical power input was of 7.24 kW, while the blower electrical power input varied from 0.94 kW, when the heat pump started, to 1.49 kW, at the end of the drying cycle (Figure 16.18). The average refrigerant (HFC-236fa) mass flow rate was stabilized around 24 kg/min

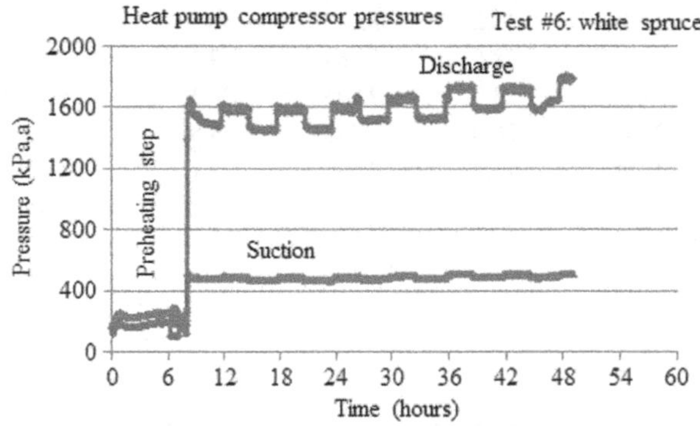

FIGURE 16.15 Average compressor suction and discharge pressures (heat pump-assisted performance test #6).

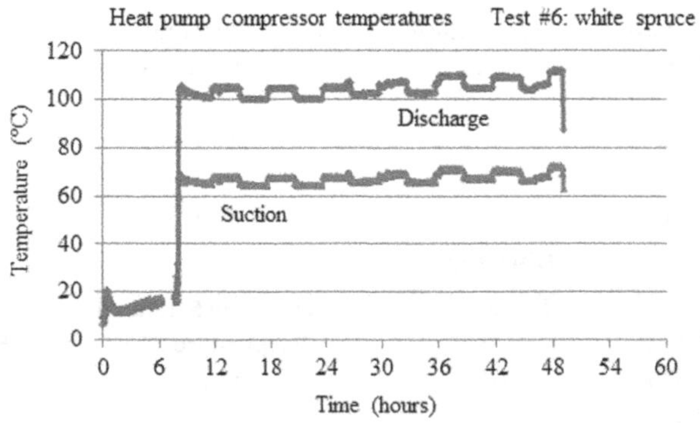

FIGURE 16.16 Average compressor suction and discharge temperatures (heat pump-assisted performance test #6).

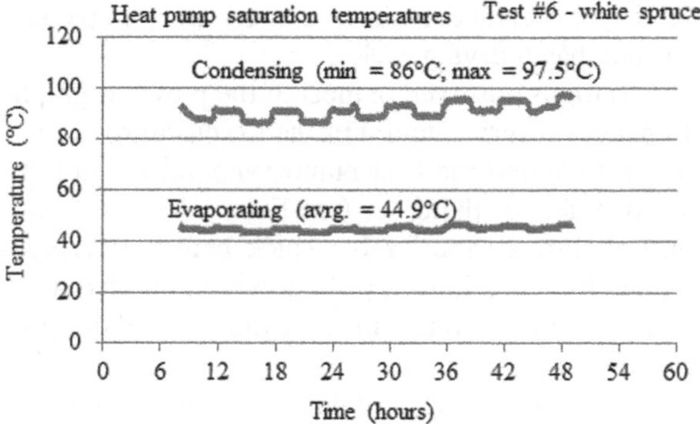

FIGURE 16.17 Average heat pump evaporating and condensing temperatures (heat pump-assisted performance test #6).

TABLE 16.16
Other Measured Average Thermodynamic Parameters of the Heat Pump

	Electrical Power Input					Pressure Drop	
Test Number	**Compressor (kW [average])**	**Blower (kW)**	**Dryer Fans (kW)**	**Refrigerant Flow Rate (kg/min)**	**Compression Ratio**	**EVAP (kPa,a)**	**COND (kPa,a)**
Test #2	6.93	0.90–1.60	2.2	23.30	3.3	6.1	3.6
Test #3	6.87	0.80–1.27	2.4	24.04	3.5	7.3	4.6
Test #5	6.90	0.82–1.37	2.3	23.84	3.6	7.6	4.2
Test #6	7.24	0.94–1.49	2.2	23.90	3.9	8.3	5.2

EVAP, evaporator; *COND*, condenser.

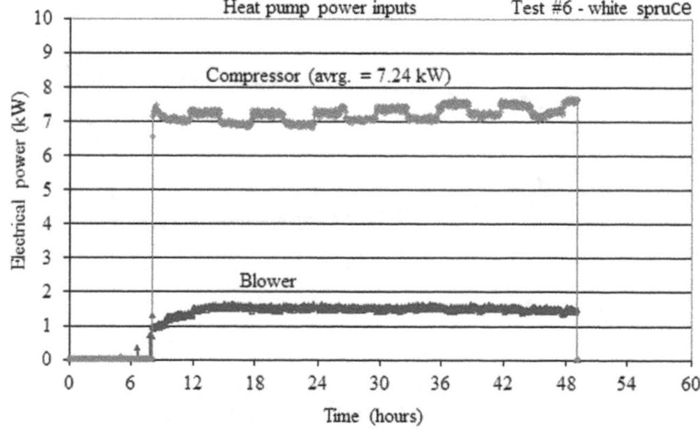

FIGURE 16.18 Average electrical power input of the heat pump compressor and blower (heat pump-assisted performance test #6).

(Figure 16.19). This experimental data shows that the heat pump operated adequately during all experimental batch drying cycles.

The parameter variations observed in most of the previous graphs are attributed to the periodical changes (every 2 hours) of the dryer fan rotation. This happened because the drying air entered the heat pump evaporator and left the heat pump condenser on the same side of the dryer (see Figure 16.11a). Consequently, when the drying air passed through the lumber stack before entering the heat pump evaporator, its temperature was lower by about 5°C compared to when the drying air entered the heat pump evaporator directly, that is, without going through the lumber stack.

For the representative performance test #6, Figure 16.20 shows an example of the actual average heat pump thermodynamic cycle.

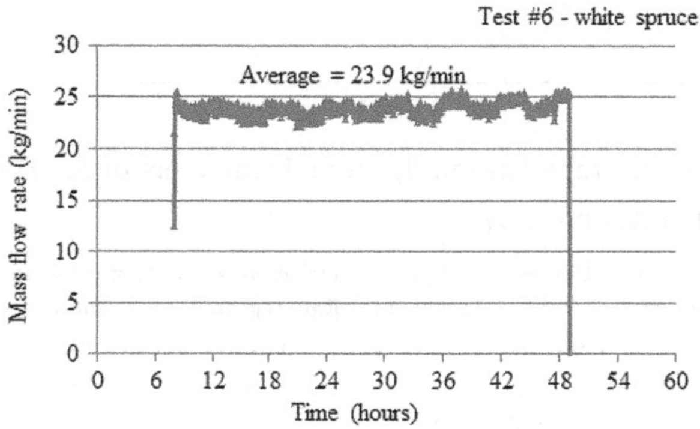

FIGURE 16.19 Refrigerant (HFC-236fa) mass flow rate (heat pump-assisted performance test #6).

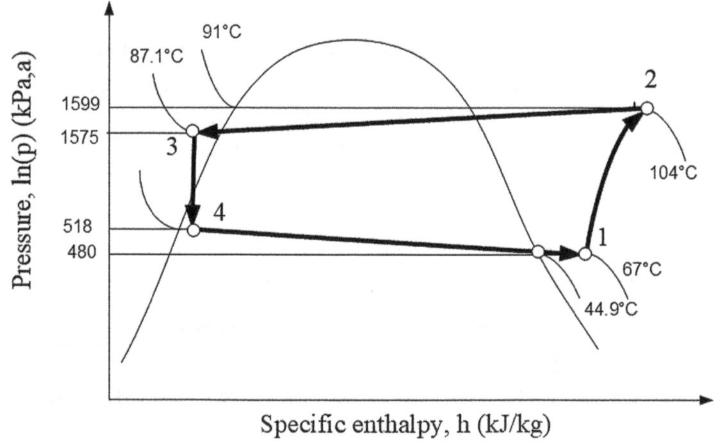

FIGURE 16.20 Example of the actual average heat pump thermodynamic cycle (heat pump-assisted performance test #6).

During heat pump-assisted test #6, the average thermal power recovered from the dryer was 28.2 kW, while the average thermal power supplied to the dryer reached 42.2 kW (Table 16.17).

Consequently, the average heat pump coefficient of performance, defined as the thermal power supplied to the dryer (42.2 kW) divided by the compressor average power input (7.24 kW), was 5.8.

16.2.4.3 Energy Consumption

Table 16.18 summarizes total energy consumption (i.e., during the preheating step and the following heat pump-assisted drying steps). The heat pump (compressor and blower) consumed about 11% of total energy. It can be seen that, compared to conventional drying test #1 performed with natural gas as unique energy source, the total energy consumption (electricity plus fossil) of heat pump-assisted drying tests has been reduced between 20.3% (test #6) and 44.6% (test #2).

Table 16.19 groups together the main data and assumptions used for calculating the energy performance of heat pump-assisted drying tests. They include some measured parameters and energy prices, such as electrical energy (CAN$0.0445/kWh) and natural gas (CAN$0.39/m³).

TABLE 16.17
Measured Heat Pump's Coefficients of Performance during the Heat Pump-Assisted Performance Test #6

Refrigerant Average Flow Rate (kg/min)	Compression Enthalpy (kJ/kg)	Compressor Average Power Input (kW)	Thermal Power Recovered (kW)	Thermal Power Supplied (kW)	Coefficient of Performance (COP)
24	17.5	7.24	28.2	42.2	5.8

TABLE 16.18
Measured Total Energy Consumptions

		Electricity			Fossil Energy		Total Energy Consumption
Test Number	Total Time (Hours)	Heat Pump (C + B) (kWh)	Dryer Fans (kWh)	Total (kWh)	Steam (kWh)	Natural Gas (m³)	Electricity + Fossil (kWh/cycle)
#1-CONV	54.75	0.00	121.55	121.55	5,607.40	495.40	6,102.8
#2	37.47	276.97	83.18	360.15	3,136.00	242.43	3,378.43
#3	36.36	246.40	87.26	333.66	3,141.00	242.81	3,383.81
#5	51.97	343.16	108.10	451.26	5,546.44	428.76	5,975.2
#6	48.00	348.80	106.08	454.88	4,514.50	348.99	4,863.49

B, heat pump blower; *C*, heat pump compressor; *CONV*, conventional test.

TABLE 16.19
Data Used for Energy Performance Calculations

Test Number	#2	#3	#5	#6
Measured parameters				
Softwood initial volume (m^3)	7	7	7	7
Average heat pump compressor input power (kW)	6.93	6.87	6.90	7.24
Average heat pump blower input power (kW)	1.60	1.13	0.94	1.49
Total heat pump input power (compressor + blower) (kW)	8.53	8.00	7.84	8.72
Average dryer fans input power (kW)	2.22	2.40	2.08	2.21
Total mass of condensed moisture (water) (including 5% losses) (kg_{water})	534	451	n/a	472
Additional data and assumptions				
Electricity price (C$/kWh) (institutional rate, 2003)	0.0445	0.0445	0.0445	0.0445
Natural gas price (C$/$m^3$) (2003)	0.39	0.39	0.39	0.39
Higher calorific value of natural gas (kWh/m^3)	9.94	9.94	9.94	9.94
Lower calorific value of natural gas (kWh/m^3)	11.07	11.07	11.07	11.07
Average thermal efficiency of natural gas-burned boiler (%)	80	80	80	80
Moisture (water) evaporation enthalpy (kJ/kg)	2,213	2,213	2,213	2,213
Average flow rate of moisture exfiltration (kg_{water}/h)	3.2	3.2	3.2	3.2
Total moisture loss (kg_{water})	119.90	116.35	166.30	153.60

C$, Canadian dollar (2003); *n/a*, nonavailable.

Moisture losses, before condensation on the heat pump evaporator, were estimated on average at $3.2\frac{kg_{water}}{h}$, and were mainly attributed to the dryer ventilation system combined with water vapor leakage through the doors and air vents, and with vapor condensation on the outside dryer walls.

16.2.4.4 Heat Pump Drying Performances

Figure 16.21 shows the cumulative mass of moisture (water) extracted by the heat pump evaporator during heat pump-assisted performance drying test #6. The total mass of moisture extracted (472 kg) includes 5% of (estimated) uncontrolled losses (e.g., natural evaporation, leakage, etc.) from the actual moisture condensed.

Heat pump dehumidification performance is usually determined by the Specific Moisture Extraction Ratio ($SMER_{heat\ pump}$) defined as the ratio of the moisture extracted from the dried material and total heat pump energy consumption (mainly, compressor and blower) $\left(\frac{kg_{water}}{kWh_{heat\ pump}}\right)$.

Other dehumidification performance indicators include Specific Energy Consumption ($SEC_{heat\ pump}$) defined as the *SMER* reciprocal number $\left(\frac{kWh_{heat\ pump}}{kg_{water}}\right)$. Table 16.20 gives both parameters, which were determined experimentally for all heat pump-assisted drying tests.

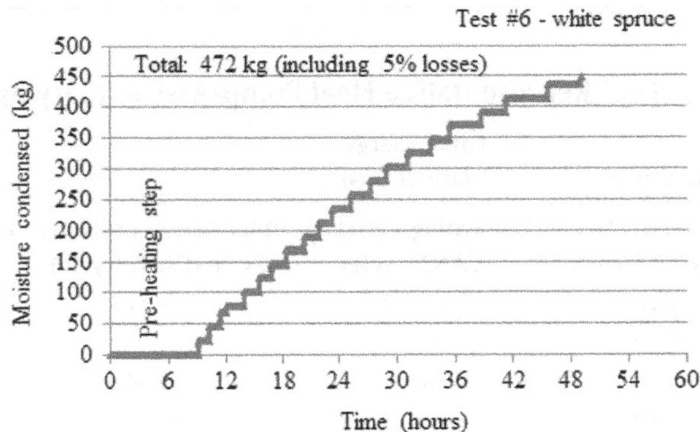

FIGURE 16.21 Cumulative mass of moisture extracted, including 5% of estimated condensation losses (heat pump-assisted performance test #6).

TABLE 16.20
Measured Dehumidification Performances of Representative Heat Pump-Assisted Drying Tests

Test Number	$SMER_{heat\ pump}\left(\frac{kg_{water}}{kWh_{heat\ pump}}\right)$	$SEC_{heat\ pump}\left(\frac{kWh_{heat\ pump}}{kg_{water}}\right)$
#2	1.93	0.52
#3	1.83	0.55
#5	n/a	n/a
#6	1.35	0.74

SMER, Specific Moisture Extraction Rate; *SEC*, Specific Energy Consumption.

16.2.4.5 Energy Costs

Table 16.21 compares the total energy costs per drying cycle of the conventional test #1 with those of the four heat pump-assisted drying tests. It can be seen that the total energy costs (expressed in CAN$/cycle) of heat pump-assisted drying tests have been reduced between 26% (test #5) and 56.7% (test #3) compared to conventional drying test.

16.3 HIGH-TEMPERATURE SOFTWOOD DRYING—INDUSTRIAL SCALE

16.3.1 Introduction

Traditionally, softwood species are used for construction lumber. For the production of such value-added commodity products, they are kiln dried at temperatures higher

TABLE 16.21
Total Energy Costs of Representative Heat Pump-Assisted Drying Tests

	Electricity	Fossil Energy (Natural Gas)		
Test Number	Heat Pump + Dryer Fans (CAN$/cycle)	Preheating + Backup (CAN$/cycle)	Total Energy Costs per Cycle (CAN$/cycle)	Reduction versus CONV. (%)
#1-CONV	5.41	247.70	253.11	–
#2	16.03	94.55	110.57	56.3
#3	14.85	94.70	109.54	56.7
#5	20.08	167.22	187.30	26.0
#6	20.24	136.11	156.35	38.2

CAN$, Canadian dollar (2003).

than 80°C because of shorter drying times, lower energy costs, and improved dimensional stability.

As a clean energy technology compared with traditional heat-and-vent dryers, high-temperature heat pump-assisted dryers offer additional, interesting benefits for drying resinous timber.

16.3.2 Experimental Site

16.3.2.1 Dryer

The sawmill site is located in eastern Canada (province of Québec). It consists of two air-forced conventional dryers, each including 1,500 kW steam heating coils. An oil-burned boiler of 4,900 kW output capacity and 82% thermal efficiency supplies both dryers with high-pressure saturated steam (Minea 2004) (Figure 16.22a).

Inside the heat pump-assisted kiln, a 56 kW multiple-blade dryer fan with outdoor electrical motor (*FM*) provides forced circulation of the drying air through the stacks of wood. Mural deflectors and inversion of the rotation of the dryer fan at every 3 hours at the beginning and at every 2 hours at the end of the drying cycles contribute to achieve uniform drying of lumber boards. To avoid air implosion risks, three air vents (located on the dryer roof) open when the central fan changes its rotation direction and also when the actual dry-bulb temperature exceeds the set point.

The average quantity of softwood entering the heat pump-assisted dryer is around 335 m^3 (142 MFB—*Thousand Feet Board*; 1 MFB ≈ 0.002359737 m^3). Figure 16.22c shows one of the two softwood lumber stacks at the kiln inlet door.

16.3.2.2 High-Temperature Heat Pumps

One of the existing conventional kilns (Figure 16.22a) has been equipped with two 65 kW (compressor nominal power input) high-temperature (HP-1 and HP-2) heat pumps (Figure 16.22b).

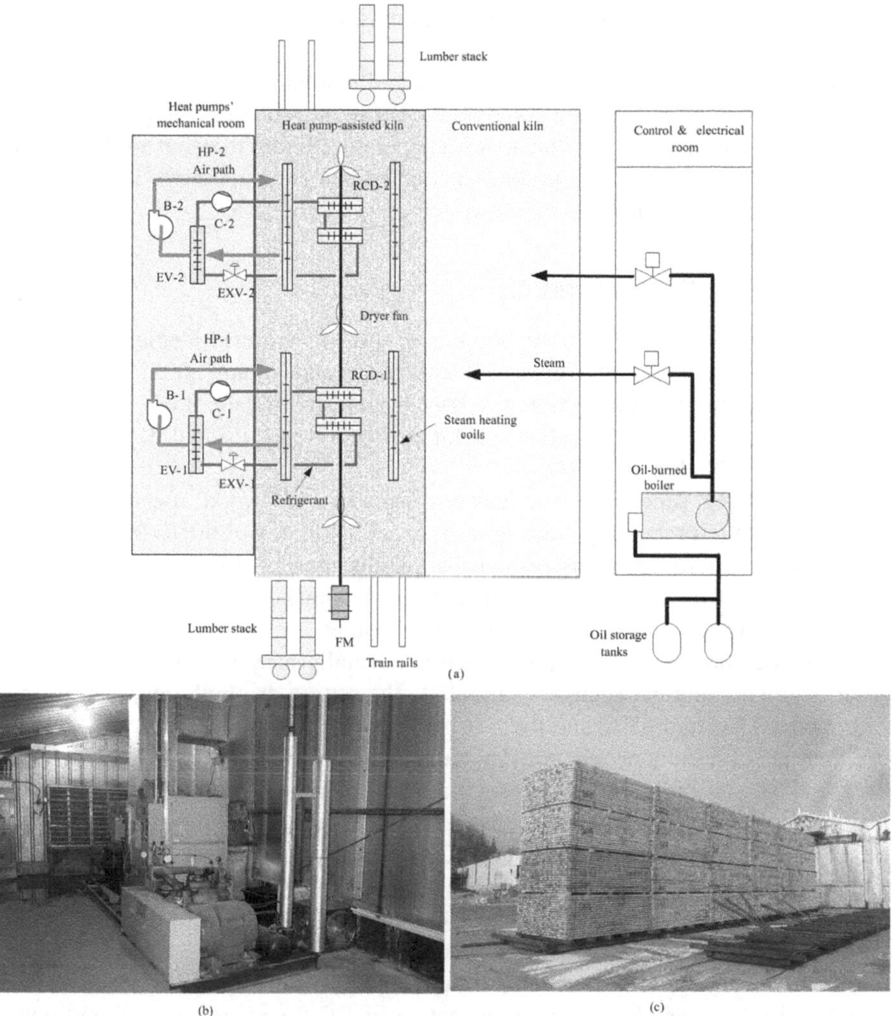

FIGURE 16.22 Industrial-scale high-temperature heat pump-assisted softwood drying system: (a) schematic of the experimental site (note: not to scale); (b) view of one of the two high-temperature heat pumps (reproduced with permission from NYLE Systems LLC); (c) view of one of the two softwood lumber stacks prior entering the heat pump-assisted kiln (reproduced with permission from GERARD CRÊTE & FILS Inc.). HP, heat pump; B, blower; C, compressor; RCD, remote condenser; EV, evaporator; EXV, expansion valve; FM, fan electrical motor.

Each high-temperature heat pump includes a single-stage reciprocating compressor, an evaporator, and a variable speed blower, as well as electronic expansion valves, all installed within an adjacent mechanical room (Figures 16.22a and b). Designed for industrial processes, both open- and belt-driven compressors are provided with oil pumps, external pressure relief valves, and crankcase heaters. Expansion valves are incorporated into microprocessor-based process controllers that display the set and actual process temperatures. The heat pumps' condensers are remote installed inside the drying enclosure (Figure 16.22a).

The refrigerant (HFC-245fa) is a nontoxic and nonflammable fluid, having a relatively high critical temperature compared to the highest process temperature and a normal boiling point less than the lowest temperature likely to occur in the system. Moreover, the saturation vapor pressure at highest design temperature is not so high as to impose design limitations on the system.

16.3.3 Typical Drying Schedule

The basis for determining the softwood kiln drying seasoning strategies depends on how moisture moves through softwood boards (Keey and Nijdam 2002). Historically, various drying programs and strategies have been tested in laboratory kiln in order to determine the risks of defects as shrinkage, warp, and residual drying stresses (Girard et al. 2005; Fortin 2009).

Typical drying schedules with eastern Canadian softwood species generally include 6–8 hours preheating steps at average temperatures of 90°C–95°C in order to destroy the microorganisms responsible for discoloring the sapwood. The drying conditions of the following drying steps are established based upon the type of softwood species and initial average moisture contents.

For White spruce, for example, which is normally easy to dry, at initial average moisture content between 40% and 25%, the setting dry-bulb temperatures are set between 83°C and 85°C and the wet-bulb temperatures at about 63°C. At moisture contents of less than 25%, the dry-bulb temperatures are generally set at about 80°C and the wet bulb-temperatures set at 63°C. On the other hand, with Balsam fir, which is harder to dry, when the initial average moisture content is above 30%, the dry-bulb temperatures are set at 82°C–83°C and the wet-bulb temperatures set at about 79°C–80°C. At average moisture contents lower than 25%, the setting dry-bulb temperatures may attain 93°C–94°C, and wet-bulb temperatures of around 71°C–72°C.

Changes in dry- and wet-bulb temperature settings are achieved on predetermined time-based schedules. The scope is to not exceed the average time of traditional drying cycles for the same species of dried wood. For White spruce, for example, steps 1–3 may generally last 10 hours; then step 4, 20 hours; and step 5, 10–20 hours, depending upon the lumber actual average moisture content. In the case of Balsam fir, each of the first four or five drying steps may last up to 30 hours, while the last step may last up to 15 hours.

16.3.4 Performance Drying Tests

More than 800 industrial-scale tests have been achieved. They allowed improving the design of the initial heat pump-assisted drying system and control strategies.

In this section, two representative performance tests with White spruce (#70 and #88) and one test with Balsam fir (#176) (see Table 16.22) have been selected.

Succinctly are presented the main parameters of the drying air (Section 16.3.4.1), the main operating parameters of one of the two high-temperature heat pumps

TABLE 16.22
Parameters of Moist Drying Air Entering and Leaving the Evaporator of Heat Pump HP-1 during the Heat Pump-Assisted Performance Test #88 (see Figure 16.25)

	Start (1s–2s)		End (1e–2e)	
Parameter	**Entering Evaporator (EE)**	**Leaving Evaporator (LE)**	**Entering Evaporator (EE)**	**Leaving Evaporator (LE)**
Relative humidity (%)	53.0	80.0	30	70
Dry-bulb temperature (°C)	82.2	60.0	82.2	54.4
Pressure (kPa)	101,325	101,325	101,325	101,325
Wet-bulb temperature (°C)	67.91	57.48	57.0	47.99
Absolute humidity $\left(\frac{kg_{water}}{kg_{dry-air}}\right)$	0.2386	0.13017	0.11255	0.07351
Specific enthalpy (kJ/kg)	694.75	400.13	381.02	245.81

EE, entering evaporator; *LE*, leaving evaporator.

(HP-1) (Section 16.3.4.2), and the heat pump's dehumidification (drying) performances (Section 16.3.4.3).

16.3.4.1 Parameters of Drying Air

As can be seen in Figure 16.23, during test # 88, performed with White spruce, the actual (measured) wet-bulb temperature closely followed the set values, which allowed the heat pump's compressor to run continuously. For that, the set wet-bulb temperature has been modified twice, that is, when the gap between the actual (measured) and the previous set values reached 1.5°C.

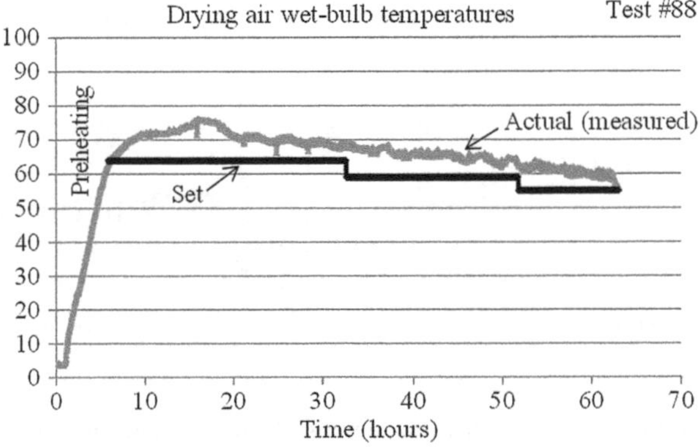

FIGURE 16.23 Set and actual (measured) wet-bulb temperatures of moist drying air (heat pump-assisted performance test #88).

During the test #88, at the evaporator inlet, the relative humidity of the drying air has been almost constant at around 88%, except at the end of the cycle when it dropped, in average, to 70% (Figure 16.24a).

Figure 16.24b shows that, immediately after the preheating step, the absolute humidity of the drying air increased up to $0.35\frac{kg_{water}}{kg_{dry\ air}}$. The average gradient of the air absolute humidity across the heat pump HP-1 evaporator varied from $0.214\frac{kg_{water}}{kg_{dry\ air}}$, immediately after the compressors started-up, to about $0.039\frac{kg_{water}}{kg_{dry\ air}}$ toward the end of the drying cycle. The up and down values of measured relative and absolute humidity shown in Figure 16.24a and b were caused by the periodical changes in the rotation direction of the dryer central fans.

Also, based on the measured (average) absolute humidity and dry- and wet-bulb temperatures of the heat pump-assisted performance test #88 (Table 16.23), the drying air entering (EE), and leaving (LE) the heat pump evaporator (i.e., moisture condenser), two dehumidification processes have been represented in Mollier diagram (Figure 16.25), respectively, at the beginning (1s–2s) (i.e., just before the heat pump started) and the end (1e–2e) (i.e., when the heat pump shut down).

16.3.4.2 Heat Pump Operating Parameters

Table 16.23 groups together the main parameters of the three heat pump-assisted performance drying tests (#70, #88, and #176).

For drying White spruce to obtain 17.2% and 20.6% (dry basis) average final moisture contents, respectively, heat pumps HP-1 and HP-2 have run 61 hours (or about 2.54 days). Similarly, for drying with Balsam fir to obtain 20.7% (dry basis) average final moisture content, heat pumps HP-1 and HP-2 have run 151.4 hours (or about 6.3 days). These numbers do not include the duration (approximately 6 hours) of preheating steps when both heat pumps didn't work.

The compressor of heat pump HP-1 ran with an average electrical power input of 65.12 kW (Figure 16.26a) and average compression ratios of 5.5. Figure 16.26b

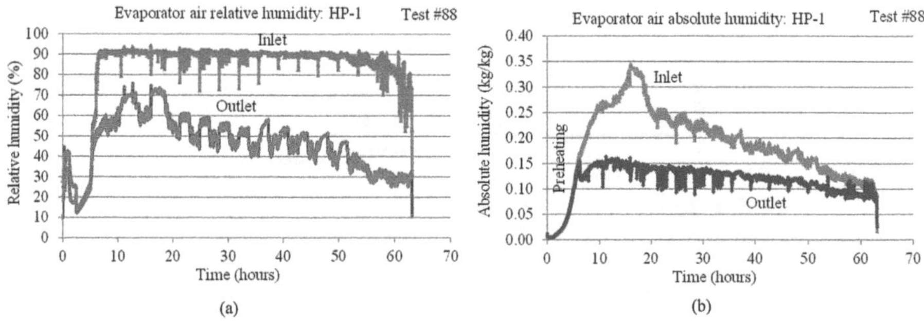

FIGURE 16.24 (a) Relative humidity of the moist drying air entering and leaving the evaporator of heat pump HP-1; (b) absolute humidity of the moist drying air entering and leaving the evaporator of heat pump HP-1 (preliminary heat pump-assisted performance test #88). HP-1, heat pump #1.

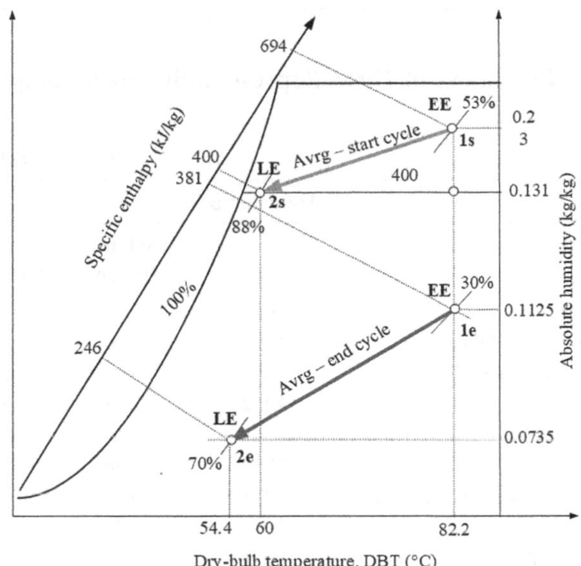

FIGURE 16.25 Average dehumidifying processes of drying air at the beginning and at the end of the heat pump-assisted test #88 represented in Mollier diagram. EE, entering evaporator; e, end; LE, leaving evaporator; s, start.

shows the operating profile of the blower of heat pump HP-1 that ran with an average power input of 0.62 kW. Moreover, the compressor of heat pump HP-1 ran with stable suction and discharge pressures (Figure 16.27a) and temperatures (Figure 16.27b), as well as with average refrigerant subcooling of 8°C. The evaporating temperatures ranged between 41°C and 45°C, and the condensing temperatures varied around 105°C, about 20°C higher than the kiln actual dry-bulb temperature.

The average coefficients of performance ($COP_{heat\,pump}$) of heat pump HP-1, defined as useful thermal energy delivered divided by the total electrical energy consumption (compressor plus blower), were 4.23 (test #70), 4.6 (test #88), and 3.46 (test #176) (Table 16.23).

16.3.4.3 Moisture Extraction

As can be seen in Table 16.23, the total mass of moisture extracted (condensed) by the heat pump HP-1 only (including 5% of uncontrollable condensate losses) was of 9,454 kg (test #70), 8,656 kg (test #88, see also Figure 16.28) and 13,550 kg (test #176).

Based on the measured masses of moisture extracted by heat pump HP-1 and the actual running times, the total moisture extraction rates were of $313\frac{kg_{water}}{h}$ (test #70), $263.2\frac{kg_{water}}{h}$ (test #88), and $178.8\frac{kg_{water}}{h}$ (test #176), respectively. These numbers do not include the venting moisture losses (on average, $90\frac{kg_{water}}{h}$) but account for 5% of condensate moisture losses.

TABLE 16.23
Main Energy Performances of Three Representative Heat Pump-Assisted Drying Tests

Test Number	–	#70	#88	#176
Parameter	**Heat Pump**	–	–	–
Softwood Species	–	White Spruce	White Spruce	Balsam Fir
Heat pump running time (excluding preheating steps) (hours)	HP-1	61.00	61.3	151.4
	HP-2	61.00	61.3	151.4
Average compressor power input (kW)	HP-1	65.12	63.36	61.0
	HP-2	62.78	58.50	57.14
Compressor electrical energy consumption (kWh)	HP-1	3,972	3,884	9,235.4
	HP-2	3,830	3,586	8,651.0
Blower electrical energy consumption (kWh)	HP-1	13.42	16.6	28.7
	HP-2	14.03	39.5	107.5
Total heat pump electrical energy consumption (kWh)	HP-1	3,985.42	3,900.6	9,263.7
	HP-2	3,986.03	3,625.5	8,679.7
Mass of moisture condensed (including 5% of condensate losses) (kg)	HP-1	9,454	8,656	13,550
	HP-2	9,655	8,478	13,531
Final moisture content (%, dry basis)	–	17.2	20.6	20.7
Heat pump average COP[a] (–)	HP-1	4.23	4.6	3.46
	HP-2	3.70	4.07	3.00
Heat pump average SMER[a] $\left(\frac{kg_{water}}{kWh}\right)$	HP-1	2.38	2.13	1.46
	HP-2	2.52	2.36	1.54
Heat pump average SEC[a] $\left(\frac{kWh}{kg_{water}}\right)$	HP-1	0.42	0.47	0.68
	HP-2	0.40	0.42	0.64

[a] Based on the heat pump's compressor and blower total energy consumptions.

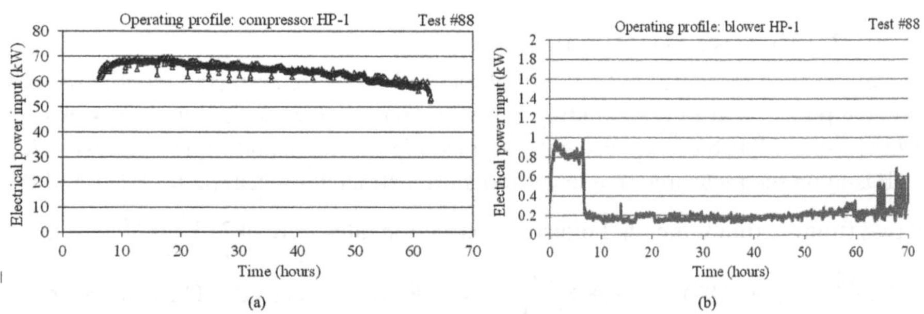

FIGURE 16.26 Operating profiles of heat pump compressor (a) and blower (b) (heat pump-assisted performance test #88). HP-1, heat pump #1.

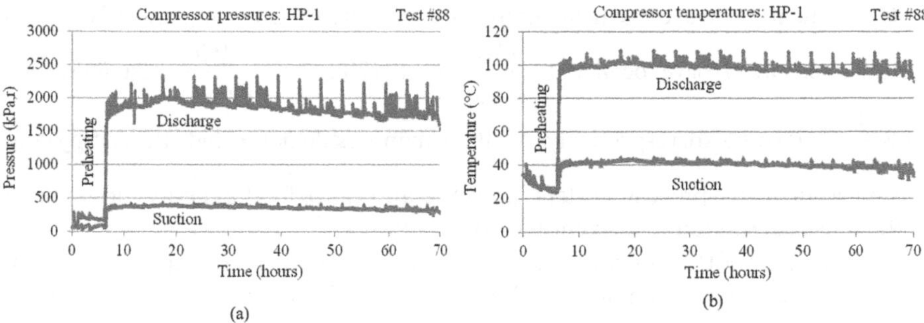

FIGURE 16.27 (a) Compressor suction and discharge pressures; (b) compressor suction and discharge temperatures (heat pump-assisted performance test #88). HP-1, heat pump #1.

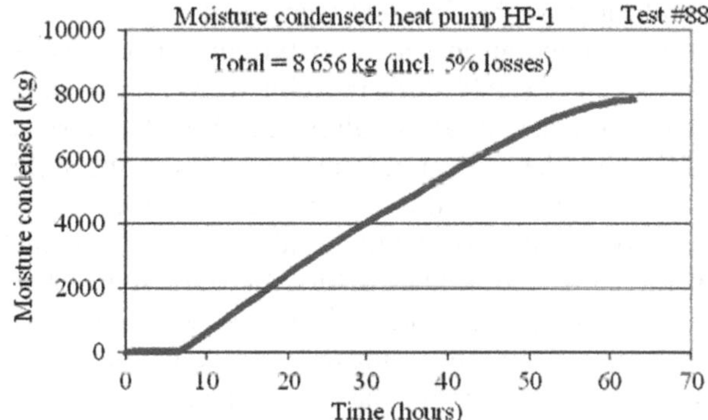

FIGURE 16.28 Cumulative mass of moisture extracted by heat pump HP-1 (heat pump-assisted performance test #88).

16.3.4.4 Energy Consumptions and Costs

Compared to the energy consumption of conventional drying cycles with oil as unique energy source, the energy consumptions of the high-temperature heat pump-assisted drying tests (Table 16.23) were between 27% and 57% lower. The average reduction in specific energy costs, compared to the costs of conventional softwood drying cycles, was estimated at 35%.

In average, the heat pumps (compressors plus blowers) and the dryer's central fans assumed 72% and 28%, respectively, of the total energy consumptions of most of heat pump-assisted experimental tests.

16.3.4.5 Dehumidification Performances

The average heat pumps' Specific Moisture Extraction Rate ($SMER_{heat\ pump}$) values (measured), defined as the amount of moisture extracted by the heat pump (kg) and the total energy consumption (compressor and blower, expressed in kWh, have been $2.38\frac{kg_{water}}{kWh}$ (test #70), $2.13\frac{kg_{water}}{kWh}$ (test #88), and $1.46\frac{kg_{water}}{kWh}$ (test #176), respectively.

Consequently, the average heat pumps' Specific Energy Consumption ($SEC_{heat\,pump}$) values (measured) have been $0.42\frac{kWh}{kg_{water}}$ (test #70), $0.47\ \frac{kWh}{kg_{water}}$ (test #88), and $0.68\frac{kWh}{kg_{water}}$ (test #176), respectively. All these numbers do not include the energy consumption during the preheating steps, as well as allowance for the energy consumption of the kiln's central fan or the venting moisture losses.

For about 39,600 m^3 of annually dried softwood lumber, the specific cost for drying has been estimated at 14.75 US\$/m^3 (2004 US\$) including kiln operation, electrical and fossil energy consumption, equipment depreciation, and insurance.

16.3.5 Operating Lessons Learned

Spread over several months, the first development step of high-temperature heat pump prototypes was aimed at ensuring a maximum operational stability of the thermodynamic parameters, establishing the reliability of the most sensible components (compressors, blowers, and safety valves), checking the refrigerant/oil blend's behavior, and optimizing the critical control sequences of the system. The experimental dehumidifier dryer operated in extreme temperature, humidity and corrosion conditions, and demonstrated specific feedback phenomena that do not occur in traditional air-forced/heated lumber dryers. Because some of the by-products that could be produced by chemical interactions between the lubricant and working fluid may be acidic and lead to accelerating the corrosion of the system components, periodical controls aimed to determine the chemical behavior of the mixture. After more than 4,000 hours of operation, the refrigerant proved to be thermally stable and chemically inert at the highest temperatures occurring in the system and a first oil chemical analysis proved that there were no problems with the oil breaking down or failing. The oil still showed adequate viscosity and chemical stability as well as a good miscibility with the refrigerant. The initial designed capacity of the heat pumps proved to be too high and consequently, both compressors were slowed down by about 25% which finally resulted in a more adequate capacity, reduced head pressures, and improved efficiency. Because the original employed expansion valves poorly controlled the refrigerant flowing and did not open fast enough, and manufacturer leakage faults were detected, they were later replaced by new generation devices. A crack in the casting of an original pressure relief valve also indicated a manufacturer's defect, and finally both these components have been replaced to 20% higher pressure-limit valves. Other issues included the fact that the kiln was initially poorly insulated and leaky, and some system's components were prematurely corroding.

REFERENCES

Cech, M.J., F. Pfaff. 2000. *Operator Wood Drier Handbook for East of Canada*, edited by Forintek Corp., Canada's Eastern Forester Products Laboratory, Québec city, Québec, Canada.

Fortin, Y. 2009. High-temperature drying strategies for value-added products. *Research and Development Summary: Value to Wood*. Natural Resources Canada (http://www.valuetowood.ca, accessed July 19, 2016).

Girard, B., Y. Fortin, J. Beaulieu. 2005. Conventional and high temperature drying of plantation-grown white spruce for value-added products. In *Proceedings of the 9th International IUFRO Wood Drying Conference,* Nanjing, China, August 21–26, pp. 363–368.

Keey, R.B., J.J. Nijdam. 2002. Moisture movement on drying softwood boards and kiln design. *Drying Technology* 20(10):1955–1974.

Minea, V. 2004. Heat pumps for wood drying—new developments and preliminary results. In *Proceedings of the 14th International Drying Symposium* (IDS 2004), Sao Paulo, Brazil, August 22–25, 2004, Volume B, pp. 892–899.

Minea, V. 2005. Heat pump-assisted high-temperature softwood drying—Part I. Hydro-Québec Research Institute, *Report LTE-RT-2005–0149.*

Minea, V., M. Savard, M. Déry, Y. Lavoie. 2004. High-temperature heat pump for wood drying. ELECTRO-WOOD Research Project, Phase 1. *Report LTE-RT-0490*, Hydro-Québec Research Institute.

17 Present and Future R&D Challenges

One of the present and future key strategic policies concern energy efficient technologies. Among them there is the heat pump technology applied in residential/commercial and industrial sectors for space heating and domestic/industrial hot water heating, and air conditioning, as well as in the field of drying, in general, and of wood drying, in particular.

17.1 HEAT PUMPS

Heat pumps are among the most efficient technologies allowing substantial energy savings. Air-, ground-, and waste-source heat pumps recover pollution-free energy from surrounding air, earth and low-grade industrial or institutional waste heat. They save primary energy and reduce power demand for space heating and cooling, and for domestic/industrial hot water heating, with high utilization factors and coefficients of performance.

In the context of a potential future global energy crisis, R&D and technological activities aiming at developing the next generation of heat pumps are numerous and very challenging.

17.1.1 Recent Achievements

Heat pumps are active heat recovery systems allowing recovering heat at relatively low temperatures and, simultaneously, supplying heat at much higher temperatures for domestic and/or industrial usage. Based on the thermodynamic process known as the reverse Carnot cycle, heat pumping is achieved by consuming external electrical (or thermal, chemical) energy sources, to recover industrial waste heat and, ultimately, reduce the world's use of fossil fuels. Their economic efficiency depends, however, on the relative prices of the driven energy purchased and/or saved.

Even over the past decades, the heat pump technology achieved important advancements that are not applied as widely as they should or could be. Relatively initial high costs, system correct design, and efficient integration remain challenging issues for the heat pump technology. Recent progresses in heat pump systems have focused on the development of advanced cycle, improvement of integrated (hybrid) systems and components, including the choice of working fluids, and extension of practical ranges of application (Chua et al. 2010).

Today, a limited number of manufacturers offer large-scale air-to-air heat pumps for critical applications as for industrial wood drying. With rising costs of fuel and global warming concerns, the interest in heat pumps as a mean of energy recovery will increase throughout the world. The ambient air is the most common heat source for reversible heat pumps. Ground-source (geothermal) heat pumps, which use renewable energy stored in the soil or bedrock, or surface waters and groundwater as heat and sink sources for building heating and cooling are today (and will remain) among the most energy- and cost-efficient and clean heating and cooling systems available.

Several national and international projects aim at actively contributing to the reduction of energy consumption and emissions of greenhouse gases by the increased implementation of industrial heat pumps able to recover large quantities of process waste heat (IEA 2014).

In Canada, for example, during the last three decades, small-, medium- and industrial-scale manufacturing facilities implement heat pumps in order to supply heat for building domestic hot water consumption and/or industrial heating purposes. The objective was to properly design, integrate, and operate industrial heat pumps in various energy-intensive industrial processes able to provide sufficient amount of waste heat at appropriate quality levels, flow rates, and temperatures. In Denmark, several demonstration projects involving supercritical CO_2 and high-pressure, large-scale ammonia heat pumps have been implemented, and other ammonia/water hybrid heat pumps are underway. In France, recent developments have been made to develop industrial-scale (>100 kW of thermal output) high-temperature (>80°C) and very high-temperature heat pumps (>100°C).

In Japan, about 90% of the industrial energy demand is used for processes and building heating, the heat pumps recover most of process waste heat. In this country, technical challenges include the usage of heat pumps working with lower heat-source temperature, higher output temperature, higher system capacity and efficiency, and lower refrigerant charges. Social and economic challenges also include lower system price, standardization, diversification of heat sources, public communication, and development/relaxation of laws and regulations. To achieve the prime target of enhancing the heat pump system efficiency to 1.5 times the current level in 2030 and double in 2050, a Japanese national project aims at developing a super-high-efficiency heat pump. Furthermore, a high-temperature heat pump that uses exhaust heat or simultaneously produces high-temperature steam and low-temperature water as a substitute for existing boilers in factories was selected as a specific technical development theme. According to the *Japanese Technology Development Road Map of Cool Earth Innovative Energy Technology Program*, industrial heat pumps producing 120°C heat should be targeted to achieve higher efficiency by 1.3 times the current level, and those producing 180°C high-temperature heat targeted to exceed the efficiency of existing heating systems (such as boiler systems). Other Japanese programs aiming at promoting heat pumps (*urgent project for promoting the introduction of next-generation heat utilization systems*) support industrial facilities to introduce next-generation heat utilization systems to recover and reuse low-temperature waste heat at 300°C or less. For example, low-temperature exhaust heat-driven absorption refrigerating machines and steam generating heat pumps are covered. Subsidies up to a half of the expenses required for waste heat recovery from gas/steam at less than 140°C, or up to one-third for waste heat recovery at temperatures between 140°C and 300°C were allowed in Japan. In Korea, several activities are underway to improve the efficiency of industrial processes and recover waste heat with heat pumps in order to produce process high temperature hot water. In the Netherlands, new developments such as thermo-acoustic, compression–resorption, and adsorption heat pumps have been reported during the last decade (IEA 2014).

17.1.2 Future Trends and R&D Challenges

In spite of recent advancements, continuous efforts will be required in the future in order to optimize energy use and reduce carbon footprint of many energy-intensive industrial operations by using heat pumps.

Because in the future, most countries will continue to be net energy importers exposed to supply security risks, the heat pumps may help reduce these risks by using electricity as a multi-fuel-based energy. End uses will consequently be less dependent on any particular fuel source, since electricity can be generated from a wide range of fossil and renewable sources.

The next-generation heat pumps thus have to use electrical energy effectively to reduce fossil energy demand for comfort heating as well as for drying and other industrial processes. This will put heat pumps in a good position compared to wind, biomass, and solar energy installations. For example, by 2030, almost all new institutional and commercial buildings and 75% of retrofit buildings will be equipped with heat pump systems in order to use 25% less energy, while 25% of industrial waste heat will be recovered (IEA 2014).

Among the future R&D trends and challenges for heat pumps, the following can be mentioned (Minea 2009; Mujumdar 2012; IEA 2014):

a. Air-source residential/institutional heat pumps have to improve, among others, the defrost methods in order to reduce the demand for auxiliary heating, and thus increase the seasonal energy performances, especially in moderate and cold climates.
b. Ground-source (geothermal) heat pumps will continue to efficiently reduce energy consumption in district (community) and large-scale building applications (as well as associated greenhouse gas emissions) by up to 44% compared to air-source heat pumps and up to 72% compared to standard electric resistance heaters; the design optimization and the reduction of the capital cost of ground heat exchangers will be among the most challenging tasks.
c. Increase the annual performance factors of air- and ground-source heat pumps versus present levels by 50% in 2030 (Europe's target) and by 200% in 2050 (Japan's target)
d. Compared to 2000, reduce the total cost of heat pump systems by 15% in 2020, by 25% in 2030, and by 50% in 2050 (IEA 2014), mainly by optimizing the design and developing new concepts, cost-effective packaged units, materials, and working fluids.
e. Efficient integration of air-, ground-, and waste heat-source heat pumps in low/net-zero energy houses (e.g., with photovoltaic and solar thermal heat recovery devices and passive thermal storage and cogeneration units), industrial processes (e.g., wood, food, pulp, and paper drying), and large district heating and cooling systems.

f. Reduce energy consumption of heat pump systems by as much as 80%.
g. Develop new, combined, high-efficiency heat pumping cycles, as well as extensively using low GWP (such as ammonia, hydrocarbons, CO_2, HFO fluids, and refrigerant mixtures) and nanorefrigerants.
h. Develop new types of (reciprocating, scroll, screw, rotary) reliable compressors working with environment-friendly refrigerants and equipped with inverters, and with or without refrigerant injection.
i. Develop new compact heat exchangers (e.g., cost-effective and efficient micro- and mini-channel coils, and gas coolers for supercritical CO_2 heat pumps) to accommodate the nonlinear behavior of new refrigerant mixtures during phase change and reduce the refrigerant charges.
j. Recover expansion work by using expanders.
k. Improve the (compression, absorption) heat pump cycles and coefficients of performance by integration of simple or cascade ejectors.
l. Improve the efficiency and reliability of heat pump systems with distributed liquid refrigerant and associated charge control.
m. Develop new user-friendly lubricants to replace the existing synthetic products.
n. Improve the market penetration of future electric and gas-fired heat pumps via closer collaboration between utilities, manufacturers, and governments.

17.1.3 Environmental Impacts

Finding new ways to produce and use energy to minimize its environmental impact is one of the key challenges the world faces in the 21st century.

Heat pumps offer one of the most efficient and environmentally friendly solutions to the target of global reduction of greenhouse negative effects (in particular of CO_2 emissions) on environment. Heat pumps can efficiently help to reduce the carbon footprint of buildings and industries because of the synergy between heat pumping and lower-carbon electricity supply.

The wider use of heat pumps will reduce CO_2 emissions because they are more carbon efficient than the direct use of fossil fuel for the same purposes. For example, a CO_2 emission factor for electricity generation of $0.55\frac{kg_{CO_2}}{kWh_{el}}$, an electric-driven heat pump reduces the CO_2 emissions by 45% compared to oil boilers, or by 33% compared to natural gas-fired boilers.

Heat pumps could reduce by up to 50% of new and retrofit building sector and 5% of industrial sector CO_2 emissions, which means that 1.8 billion tonnes of CO_2 per year could be saved in 2020, corresponding to nearly 8% of total global CO_2 emissions (IEA 2014).

It is expected that new heat pump technology-related business opportunities over the next few decades will represent several tens of trillions of dollars. Heat pumping technologies certainly have a promising, brilliant future in the context of the world's next eventual energy crisis.

17.2 HEAT PUMP DRYING

Drying, a key operation in wood processing, is a complex, energy intensive, and highly nonlinear coupled heat and mass transfer process involving material, heat, and mass transfer sciences as well as physical, chemical, and biochemical phenomena. It occurs by supplying heat to wet materials and by vaporizing the liquid contained inside. About 85% of industrial driers are of convective type with hot air as drying medium and 99% of them involve removal of water.

Most heat losses in industrial wood dryers are due to the discharge of moist air and the poor thermal insulation of the drying enclosures. Because of energy losses, attention has focused on reducing as well as recovering wasted heat. Among other heat recovery devices, heat pumps used in convective drying applications have the potential to save part of the primary energy used for drying and thus improve the energy efficiency of drying systems.

Thermal drying has a direct impact on the quality of most commercially dried products. Escalating energy costs, demand for eco-friendly and sustainable technologies and the rising consumer demand for higher quality products have given new incentives to industry and academia to provide increasing efforts to drying R&D.

Generally, the innovation in drying technologies has been nearly exponential growth in drying R&D activities throughout the world especially after the energy crisis of the early 1970s, although it has not accelerated because of the long life cycles of dryers and relatively unchanged fuel costs over the past decade. Many innovations in heat pump-assisted drying, aiming at reducing the capital and running costs of heat pumps and dryers, were based on existing knowledge.

17.2.1 Outlook on Past and Present Achievements

Over recent years, product quality improvement as well as reduction of environmental effects remain the main objectives along with the improvement of drying operation to save more energy. New drying technologies, better operational strategies and control of industrial dryers, as well as improved and more reliable scale-up methodologies have contributed to better cost-effectiveness and better quality of dried products.

Past and present R&D and development works, both fundamental and applicative in drying areas, can be summarized as follows (Mujumdar 2007; Mujumdar and Huang 2007; Minea 2013) (note: listed not in order of importance or relevance):

a. Many materials of various forms are dried commercially over wide ranges of drying times and under specific quality constraints, with thermal efficiencies varying from a low of 20% to a high of 80%.
b. Although the innovation has not accelerated because of the long technical life of dryers (typically, 20–40 years) and relatively unchanged fuel costs over the past decade, no disruptive radical drying technologies have emerged and, hence, the need for replacement with new equipment is still limited.

c. Fundamental researches on theory of drying and dryers' modeling approaches, based on simple materials and equipment, have progressed during the last two decades.
d. Optimized controls of dryers based on models, as well as the development of new dryers and drying strategies, have incrementally advanced, but at a less satisfactory level; some novel proposed drying technologies are not readily accepted by industry and this phenomenon hinders the introduction of new technologies.
e. Although energy costs of drying continue to be an important factor, past and current R&D works are driven by the need for new products and processes basically promoted by market pull (e.g., heat pump-assisted drying).
f. Because most models are applicable for specific product–equipment combinations, a huge problem is still associated with modeling and scaling-up quality criteria that depend not only on the transport phenomena but also on the wet materials being dried.
g. Some R&D studies aim at improving the existing knowledge about chemical heat pump-assisted drying for batch and/or semicontinuous drying (e.g., Ogura et al. 2003).
h. A relatively low number of commercial heat pump-assisted dryer suppliers around the world provide dryers for drying wood.

17.2.2 Future R&D Needs and Challenges

Empirical innovations are incremental improvements of prior technologies, while innovations arising from fundamentals can be incremental or radical.

During the next few decades, R&D needs and challenges for drying industry, in general, and for wood drying, in particular, could be summarized as follows (Mujumdar 2007; Mujumdar and Huang 2007; Genskow 2007; Mujumdar 2012; Minea 2013):

a. Continue to reduce energy consumption in industrial dryers in the context of rising energy prices and growing concerns about climate change; efforts will be required to market any new or improved technical solution.
b. Both evolutionary and revolutionary technological innovations in drying must (i) better understand the actual moisture transport processes occurring at the microscale level; (ii) better understand moisture binding and its consequences on product quality; (iii) develop advanced analytical models and tools for the design, analysis, optimization, and control of industrial dryers; (iv) extend the R&D learning from laboratory scale to industrial scale; (v) develop advanced simulation models covering the entire drying process; (vi) develop better hardware (vii) develop increasing computing power at decreasing costs (viii) develop higher drying capacities; (ix) develop methods for better quality control; (x) reduce environmental impacts; (xi) provide safer operation of dryers; (xii) achieve better drying efficiency resulting in lower investment and running costs; (xiii) improve automatic control of dryers by using new measurement technique; and

(xiv) develop and operate smart dryers that will be able to adjust the local drying environment to obtain the desired specifications.

c. The requirements of the drying industry will continue to drive the future R&D and demonstration works.
d. Some emerging drying technologies will replace traditional technologies, but most of them still will require significant R&D efforts to become marketable.
e. Develop new generations of highly qualified research and technical personnel in drying.
f. Continue the promotion of a close, win-win, and cost-effective industry–academia, cross-disciplinary collaborations via fundamental studies and simulations.
g. Develop advanced strategies for optimum integration of dryers with heat pumps and other technologies and/or energy sources (e.g., solar).
h. Disseminate the existing and coming knowledge in order to avoid future failed in-field industrial applications.
i. For any kind of materials, the air-drying curves, indicating the product drying behavior, must be theoretically or experimentally determined.
j. Future research studies have to carefully consider that air velocity on the product surface must be optimized to produce rapid air changes, avoid the formation of death zones, and provide uniform drying.
k. In regard to selection of dryers and heat pumps, future R&D works, after having investigating the drying mechanism of each specific material, must decide whether conventional convective dryers can be integrated with heat pumps.

17.3 HEAT PUMP-ASSISTED WOOD DRYING

Wood drying is a key operation in sawmill production, affecting product quality, value yield, and wood material suitability to different end-use products.

The heat pump-assisted (dehumidification) systems are similar to heat-and-vent kilns, in that they operate with hot air at normal atmospheric pressures, but they condense the moisture (water vapor) evaporated from the lumber to recover a part of the heat of evaporation from the moist air and recycle it into the kiln at higher temperature in drying process. This is a method of improving the energy efficiency and reducing primary energy consumption of industrial drying processes.

In other words, the principle of heat pump dryers is based on energy conservation, which is more suitable for drying solid products. The systems are energetically closed and, consequently, as opposed to conventional drying systems, drying heat pumps discharge little quantities of warm, humid air into the environment. Unsaturated warm, dry air is led over the surface of the product to be dried, and it takes up the water adhering to the surface. The water vapor removed condensates by passing over the heat pump cold coil (evaporator) and is then warmed again by passing through the heat pump hot heat exchanger (condenser).

Heat pump-assisted drying based on the mechanical vapor compression cycle are largely the most used. Dryers coupled with heat pumps are complex systems due to the interdependency of many components and thermodynamic parameters. Any change occurring in one component or parameter will inevitably influence the others. The principal advantage of drying heat pumps emerges from the hot and humid air as well as from their ability to control the drying air temperature and relative humidity. However, it is necessary to optimize the system components and design to increase the energy efficiency of heat pump dryers.

The future R&D activities in wood drying will focus on objective such as (i) use of piezoelectric sensors on small samples to determine the dynamic sorption behavior and diffusion coefficients; (ii) use of nondestructive, high-resolution imaging monitoring methods of drying process, including neutron imaging and combinations of impedance and acoustic emissions from boards, to determine diffusions coefficients and moisture contents; (iii) model the free water movement in the wood capillary network as well as inclusion of sorption hysteresis and mechanical behavior; (iv) develop homogenization models for material properties for determining diffusion properties of wood at steady state conditions; (v) optimize the lumber sorting strategies based on empirical relationship between lumber moisture content and drying degrade in order to increase sales depending on kiln capacity and market conditions; (vi) optimize the wood drying schedules in order to reduce the overall drying costs; (vii) develop new, improved industrial drying operations; and (viii) develop advanced drying methods such as superheated steam, vacuum drying, microwave and radio-frequency heating for redistribution of pre-dried softwood.

Among future R&D needs for heat pump-assisted drying operations in general, and for wood drying, in particular, can be mentioned (Mujumdar 2012; Minea 2013; Vikberg 2015):

a. To make relevant decisions about using or not heat pumps as dehumidification devices, the drying mechanism of each product has to be relatively well known; data on the physical properties of the dried material is also required to develop new drying heat pump concepts; such information is needed not only for improving the design and control of the heat pump drying systems, but also for setting standards for different categories of dried products.
b. To make heat pump dryers more reliable and efficient, relevant drying data must be developed for each material or group of similar materials.
c. To achieve high drying performance levels and be able to validate them, future published R&D studies on heat pump dryers have to provide minimum information such as (i) dryer's main characteristics, such as the location (hot or cold weather, indoor, and outdoor), volume (or mass) capacity, configuration of air circulation through both dryer and heat pump (fans, blowers, and air vents), and percentage of dryer heat leakage rates (conduction, air ex- and infiltration); (ii) drying schedules for each type of dried material, including the product initial and required final moisture contents, as well as input and output masses (or volumes), preheating requirements and type of supplementary (back-up) heating (if applicable), and for each drying step, air setting dry- and wet-bulb temperatures, or air

temperature depression, and air relative humidity; (iii) heat pump's critical operating parameters, for example, type of refrigerant used, maximum and minimum allowed discharge or suction pressures and temperatures, drying airflow rate and pattern through both the dryer and heat pump heat exchangers, and heat transfer rates of the heat pump evaporator (dehumidification capacity), condenser (heating capacity), and heat rejection device (when required); (iv) schematically but correctly represent or, at least, explain how the drying rate and drying air main parameters are controlled based on both dryer and heat pump physical integration; to do so, it must carefully indicate how the drying medium flows through both the drying chamber and heat pump heat exchangers, and how outside heat rejection is minimized or avoided; this is because operation at low drying rates requires careful design of the heat pump dehumidification capacity; if no correct design is achievable, heat pump drying may fail and, consequently, such devices may not be applied in the industry.

d. Because the single-stage heat pump-assisted drying systems are complex enough, develop new heat pump drying systems as simple as possible.
e. Because the first scope of R&D works is to provide industry with workable solutions, the future research studies have to make heat pump drying systems reliable for each specific drying process; the real challenge is to make these systems feasible by providing relevant information about integration, schedules, control strategies, operating parameters, and energy performances; obviously, without improving the integration of dryers and heat pumps and their coupled operation throughout the whole drying process, no real technology advancement can be achieved; the lack of reliable integrated systems may compromise the future applicability of standard and/or innovative heat pump drying concepts; in other words, scientific studies must not unnecessarily complicate the technology with unworkable solutions.
f. There is a need for a specific type of heat pump dryer per product or group of similar products, that is, agro-food, wood, heat sensitive, biological, solid, liquid, etc., and for each specific range of drying temperatures.
g. Most of the future improvements have to focus on the simplification of the concepts in order to increase the reliability and dehumidification efficiency of industrial drying applications; inadequate integration and wrong operation parameters may provide troubles as compressor high discharge pressures, low dehumidification efficiency, and even compressor mechanical damages.
h. After the knowledge of drying mechanisms and dryer characteristics, and after achieving improvements at the level of the dryer–heat pump integration, the control of the dehumidification processes using drying heat pumps is the following major challenge; both system integration and control strategy are strongly interconnected.
i. For a given dehumidification capacity of the heat pump, it is mandatory to determine the required quantity of the dried material to provide high energy performances.

j. Develop advanced methods to determine with greater accuracy the average moisture content of timber prior and after the drying process (e.g., by using X-ray and microwave).
k. Future studies must explain how better control of temperature and relative humidity of the drying process has been achieved when relatively high dehumidifying performances have been reported; the complexity of controlling integrated drying heat pumps comes from the requirement to match the material thermal dewatering rate with the surface moisture vaporization rate and the heat pump dehumidification capacity throughout the entire drying cycle; it is one of the most critical issues for drying processes using heat pumps.
l. Carefully analyze whether conventional low-temperature drying methods could provide the same results as less cost-effective drying heat pumps.
m. Heat pump drying technology is an interdisciplinary technology involving both drying, heat pump, and control experts; as a consequence, future R&D studies provide numerous and fertile R&D challenges in terms of inter- and multidisciplinary interference of various scientific and engineering fields; in practice, drying specialists cannot replace heat pump and control specialists and vice versa; in the future, they have to cooperate more closely in order to provide viable industrial application solutions; based on such cooperation, theory should be further combined with laboratory and field industrial researches and demonstrations.
n. Academia must develop modeling and simulation techniques for optimizing dryer operation, while industry has to consider more complex approaches for the selection and implementation of drying technologies.
o. Academic R&D studies must provide industry with relevant and applicable heat pump drying systems; the academic community has thus a responsibility in deploying more R&D effort to develop general theories concerning drying, as well as reliable mathematical models to simplify the scale-up of highly nonlinear drying processes.

REFERENCES

Chua, K.J., S.K. Chou, W.M. Wang. 2010. Advances in heat pump systems: A review. *Applied Energy* 87(12):3611–3624.

Genskow, L.R. 2007. Guest Editorial: Drying Technology—The next 25 years. *Drying Technology* 25:13–14.

IEA. 2014. HPP-IETS Annex 35/13. 2014. Application of Industrial Heat Pumps. Final report.

Minea, V. 2009. Challenging future of heat pumps. *IEA Heat Pump Centre Newsletter* 27(4):8–12.

Minea, V. 2013. Heat pump-assisted drying: recent technological advances and R&D needs. *Drying Technology* 10:1177–1189.

Mujumdar, A.S. 2007. An overview of innovation in industrial drying: current status and R&D needs, in *Drying of Porous Materials*, Edited by S.J. Kowalski, ISBN-978-1-4020–5479-2 (HB) ISBN-978-1-4020–5480–8 (e-book), Reprinted from *Transport in Porous Media*, Volume 66(1–2), Published by Springer.

Mujumdar, A.S. 2012. Editorial: Some challenging ideas for future drying R&D. *Drying Technology* 30:227–228.

Mujumdar, A.S., L.X. Huang. 2007. Global R&D Needs in Drying. *Drying Technology* 25:647–658.

Ogura, H., T. Yamamoto, H. Kage. 2003. Efficiencies of $CaO/H_2O/Ca(OH)_2$ chemical heat pump for heat storing and heating/cooling. *Energy* 28(14):1479–1493.

Vikberg, T. 2015. Industrial wood drying—airflow distribution, internal heat exchange and moisture content as input and feedback to the process. *PhD Thesis*, Division of Wood Science and Engineering Department of Engineering Sciences and Mathematics, Luleå University of Technology, Sweden.

Index

Note: Page numbers followed by '*f*' and '*t*' indicate figures and tables respectively.

D

E

F

I

J

K

L

M

N

S

T

X

Y

Z

Milton Keynes UK
Ingram Content Group UK Ltd.
UKHW021935071024
449327UK00022B/1821

9 780367 780982